Communications
in Computer and Information Science 40

T0189870

Sanjay Ranka   Srinivas Aluru
Rajkumar Buyya   Yeh-Ching Chung
Sandeep K. S. Gupta   Ananth Grama
Rajeev Kumar   Vir V. Phoha
Sumeet Dua (Eds.)

Contemporary Computing

Second International Conference, IC3 2009
Noida, India, August 17-19, 2009
Proceedings

Volume Editors

Sanjay Ranka
University of Florida, Gainesville, FL, USA
E-mail: ranka@cise.ufl.edu

Srinivas Aluru
Iowa State University, Ames, IA, USA
E-mail: aluru@iastate.edu

Rajkumar Buyya
The University of Melbourne, Australia
E-mail: raj@csse.unimelb.edu.au

Yeh-Ching Chung
National Tsing Hua University, Taiwan
E-mail: ychung@cs.nthu.edu.tw

Sandeep K. S. Gupta
Arizona State University, Tempe, AZ, USA
E-mail: sandeep.gupta@asu.edu

Ananth Grama
Purdue University, W. Lafayette, IN, USA
E-mail: ayg@cs.purdue.edu

Rajeev Kumar
Indian Institute of Technology Kharagpur, India
E-mail: rkumar@cse.iitkgp.ernet.in

Vir V. Phoha
Louisiana Tech University, Ruston, LA, USA
E-mail: phoha@coes.latech.edu

Sumeet Dua
Louisiana Tech University, Ruston, LA, USA
E-mail: sdua@coes.latech.edu

Library of Congress Control Number: Applied for

CR Subject Classification (1998): F.2, G.1, H.4, J.3, I.5

ISSN 1865-0929
ISBN-10 3-642-03546-9 Springer Berlin Heidelberg New York
ISBN-13 978-3-642-03546-3 Springer Berlin Heidelberg New York

This work is subject to copyright. All rights are reserved, whether the whole or part of the material is concerned, specifically the rights of translation, reprinting, re-use of illustrations, recitation, broadcasting, reproduction on microfilms or in any other way, and storage in data banks. Duplication of this publication or parts thereof is permitted only under the provisions of the German Copyright Law of September 9, 1965, in its current version, and permission for use must always be obtained from Springer. Violations are liable to prosecution under the German Copyright Law.

springer.com

© Springer-Verlag Berlin Heidelberg 2009
Printed in Germany

Typesetting: Camera-ready by author, data conversion by Scientific Publishing Services, Chennai, India
Printed on acid-free paper SPIN: 12725717 06/3180 5 4 3 2 1 0

Preface

Welcome to the Second International Conference on Contemporary Computing, which was held in Noida (outskirts of New Delhi), India. Computing is an exciting and evolving area. This conference, which was jointly organized by the Jaypee Institute of Information Technology University, Noida, India and the University of Florida, Gainesville, USA, focused on topics that are of contemporary interest to computer and computational scientists and engineers.

The conference had an exciting technical program of 61 papers submitted by researchers and practitioners from academia, industry and government to advance the algorithmic, systems, applications, and educational aspects of contemporary computing. These papers were selected from 213 submissions (with an overall acceptance rate of around 29%). The technical program was put together by a distinguished international Program Committee. The Program Committee was led by the following Track Chairs and Special Session Chairs: Srinivas Aluru, Rajkumar Buyya, Yeh-Ching Chung, Sumeet Dua, Ananth Grama, Sandeep Gupta, Rajeev Kumar and Vir Phoha. I would like to thank the Program Committee, the Track Chairs and Special Session Chairs for their tremendous effort.

I would like to thank the General Chairs, Sartaj Sahni and Sanjay Goel for giving me the opportunity to lead the technical program.

Sanjay Ranka

Organization

Chief Patron

Shri Jaiprakash Gaur

Patron

Shri Manoj Gaur

Advisory Committee

S.K. Khanna	Jaypee Institute of Information Technology University, India
C.S. Jha	Jaypee Institute of Information Technology University, India
M.N. Farruqui	Jaypee Institute of Information Technology University, India
Y. Medury	Jaypee Institute of Information Technology University, India
J.P. Gupta	Jaypee Institute of Information Technology University, India
T.R. Kakkar	Jaypee Institute of Information Technology University, India
S.L. Maskara	Jaypee Institute of Information Technology University, India

General Co-chairs

Sartaj Sahni	University of Florida, USA
Sanjay Goel	Jaypee Institute of Information Technology University, India

Technical Program Committee

Sanjay Ranka	University of Florida, USA, Program Chair
Ananth Grama	Purdue University, Indiana, Track Co-chair-Algorithms
Rajeev Kumar	IIT Kharagpur, India, Track Co-chair-Algorithms
Sandeep Gupta	Arizona State University, USA, Track Co-chair-Applications
Srinivas Aluru	Iowa State University, USA, Track Co-chair-Applications
Raj Kumar Buyya	University of Melbourne, Australia, Track Co-chair-Systems
Yeh-Ching Chung	National Tsinghua University, Taiwan, Track Co-chair-Systems
Vir V. Phoha	Louisiana Tech University, USA, Track Co-chair-Analytics for Online Social Networks
Sumeet Dua	Louisiana Tech University, USA, Track Co-chair- Bioinformatics

Devesh Kumar Bhatnagar	Landis&Gyr, Noida, Track Co-chair - Industry Experience Reports
Veena Mendiratta	Alactel-Lucent,USA, Tutorial Chair
Ishfaq Ahmad	University of Texas at Arlington, USA
Scott Emrich	University of Notre Dame, USA
Kanav Kahol	Arizona State University, USA
Randal Koene	Boston University, USA
Manimaran Govindarasu	Iowa State University, USA
Bertil Schmidt	Nanyang Technological University, Singapore
Sudip Seal	Oak Ridge National Laboratory, USA
George Varsamopoulos	School of Computing and Informatics, USA
Cheng-Zhong Xu	Wayne State University, USA
Philippe O. A. Navaux	Universidade Federal do Rio Grande do Sul, Brazil
Beniamino DiMartino	Seconda Universita' di Napoli, Italy
Francisco Massetto	University of São Paulo, Brazil
Rodrigo Mello	University of São Paulo, Brazil
Howie Huang	George Washington University, USA
Robert Hsu	Chung Hua University, Taiwan
Hung-Chang Hsiao	National Cheng Kung University, Taiwan
Adnan Gutub	King Fahd University of Petroleum & Minerals, Saudi Arabia
Tzung-Shi Chen	National University of Tainan, Taiwan
Wenguang Chen	Tsinghua University, China
Jiannong Cao	Hong Kong Polytechnic University, SAR China
Ivona Brandic	Vienna University of Technology, Austria
Jemal Abawajy	Deakin University, Australia
Song Fu	New Mexico Tech, USA
Kuan-Chou Lai	National Taichung University, Taiwan
Cho-Li Wang	The University of Hong Kong, SAR China
Pradeep Chowriappa	Louisiana Tech University, USA
Manas Somaiya	University of Florida, USA
Rama Sangireddy	University of Texas at Dallas, USA
Nirmalya Bandyopadhyay	University of Florida, USA
Vinod Vokkarane	University of Massachusetts Dartmouth, USA
Rudra Dutta	North Carolina State University, USA
Tridib Mukherjee	Arizona State University, USA
Krishna Kumar Venkatasubramanian	Arizona State University, USA
Su Jin Kim	Arizona State University, USA
Gianni Giorgetti	Universita' di Firenze and Arizona State University, Italy/USA

Qinghui Tang	Texas Instruments, USA
Ayan Banerjee	Arizona State University, USA
Costas Bekas	IBM Zurich Research Laboratory, Switzerland
Guofeng Deng	Google, USA
Yu Du	Motorola, USA
Chris Gentle	Avaya, Australia
Rajendra Acharya	Ngee Ann Polytechnic, Singapore
Roberto Rojas-Cessa	New Jersey Institute of Technology, USA
Ananth Kalyanaraman	Washington State University, USA
Aaron Striegel	University of Notre Dame, USA
Nelima Gupta	Delhi University, India
Ashok Srinivasan	Florida State University, USA
Rajiv Ranjan	University of Melbourne, Australia
Kuan-Ching Li	Providence University, USA
Yong-Kee Jun	Gyeongsang National University, Korea
Bharat Madan	Applied Research Laboratory - Penn State University, USA
Yenumula Reddy	Grambling State University, USA
Christian Duncan	Louisiana Tech University, USA
M.D. Karim	Louisiana Tech University, USA
Jean Gourd	Louisiana Tech University, USA
Krishna Karuturi	Genome Institute of Singapore
Hilary Thompson	LSU Health Science, USA
Seetharama Satyanarayana-Jois	University of Louisiana, USA
Nikola Stojanovic	University of Texas in Arlington, USA
Xiaofeng Song	Nanjing University of Aeronautics and Astronautics, China
Pramod Singh	ABV-IIITM, Gwalior, India
Ratan Ghosh	Indian Institute of Technology, Kanpur, India
Manoj Gaur	Malaviya National Institute of Technology, India
Peter Rockett	University of Sheffield, UK
Shyam Gupta	IIT Delhi, India
Sanjay Chaudhary	Dhirubhai Ambani Institute of Information and Communication Technology, India
Vasant Patil	Indian Institute of Technology, Kharagpur, India
Anil Tiwari	The LNM IIT, India
Shankar Lall Maskara	JIIT University, India
R.C. Jain	JIIT University, India
Bani Singh	JIIT University, India
D.P. Mohapatra	NIT RKL, India
G. Sanyal	NIT DGP, India
Sandip Aine	Mentor Graphics, India
N.N. Jha	Alcatel-Lucent India Limited, India
Deqing Zou	Huazhong University of Science and Technology
Sathish Vadhiyar	Indian Institute of Science, India

S. Selvi	Anna University, India
China Vudutala	Centre for Development of Advanced Computing, India
Xiangjian He	University of Technology, Sydney, Australia
Frode Eika Sandnes	Oslo University College, Norway
Raja Logantharaj	University of Louisiana at Lafayette, USA
Atal Chaudhuri	Jadavpur University
Bhawani Panda	IIT Delhi, India
N.P. Gopalan	National Institute of Technology, India
Pabitra Mitra	Indian Institute of Technology, Kharagpur, India
Banshidhar Majhi	National Institute of Technology Rourkela, India
Chandan Mazumdar	Jadavpur University, India
A. Turuk	NIT Rourkela, India

Publicity Co-chairs

Rajkumar Buyya	University of Melbourne, Australia
Mario Dantas	Federal University of Santa Catarina, Brazil
Suthep Madarasmi	King Mongkut's University of Technology, Thailand
Koji Nakano	Hiroshima University, Japan
Masoud Sadjadi	Florida International University
Divakar Yadav	JIIT University, India

Website

Sandeep K. Singh	JIIT University, India
Sangeeta Malik	JIIT University, India
Shikha Mehta	JIIT University, India

Publications Co-chairs

Sushil Prasad	Georgia State University, USA
Dr. Vikas Saxena	JIIT University, India

Publication Committee

Alok Aggarwal	JIIT University, India
Chetna Dabas	JIIT University, India
Muneendar Ojha	JIIT University, India

Registration Committee

Dr. Krishna Asawa (Coordinator)	JIIT University (Chair), India
Manisha Rathi	JIIT University, India

Archana Purwar	JIIT University, India
Purti Kohli	JIIT University, India
Anshul Gakhar	JIIT University, India

Poster Session Committee

Prakash Kumar	JIIT University (Chair), India
Hima Bindu	JIIT University, India
Sangeeta Mittal	JIIT University, India
Jolly Shah	JIIT University, India
Rakhi Himani	JIIT University, India
Priyank Singh	Firmware Developer at Marvell Semiconductor, India
Siddarth Batra	Co-Founder & CEO at Zunavision, USA
Nikhil Wason	Orangut, India
Kumar Lomash	Adode Systems, India
Antariksh De	Xerox, USA

Student Volunteers Chair

Manish Thakur	JIIT University, India

Local Arrangements Committee

Manoj Bharadwaj	JIIT University, India
O.N. Singh	JIIT University, India
S.J.S. Soni	JIIT University, India
Pavan Kumar Upadhyay	JIIT University, India
Adarsh Kumar	JIIT University, India
Tribhuvan K. Tiwari	JIIT University, India
Yamuna P. Shukla	JIIT University, India
Hema N.	JIIT University, India
K. Raj Lakshmi	JIIT University, India
Mukta Goel	JIIT University, India
Meenakshi Gujral	JIIT University, India
Suma Dawn	JIIT University, India
Kavitha Pandey	JIIT University, India
Indu Chawla	JIIT University, India
Shoma Chattergey	JIIT University, India
Anuja Arora	JIIT University, India
Arti Gupta	JIIT University, India
Parmeet Kaur	JIIT University, India
Prashant Kaushik	JIIT University, India
Akhilesh Sachan	JIIT University, India
Sanjay Kataria	JIIT University, India
S. Bhaseen	JIIT University, India

Table of Contents

Technical Session-1: Algorithm-1 (AL-1)

Technical Session-2: Algorithm-2 (AL-2)

Technical Session-3: Algorithm-3 (AL-3)

Technical Session-4: Algorithm-4 (AL-4)

Technical Session-5: Application-1 (AP-1)

Technical Session-6: Application-2 (AP-2)

Technical Session-7: Application-3 (AP-3)

Technical Session-8: Application-4 (AP-4)

Technical Session-9: Bioinformatics-1 (B-1)

Technical Session-10: Bioinformatics-2 (B-2)

Technical Session-11: System-1 (S-1)

Technical Session-12: System-2 (S-2)

A Hybrid Grouping Genetic Algorithm for Multiprocessor Scheduling

Alok Singh[1], Marc Sevaux[2], and André Rossi[2]

[1] Department of Computer and Information Sciences,
School of Mathematics and Computer/ Information Sciences, University of Hyderabad,
Hyderabad – 500046, Andhra Pradesh, India
alokcs@uohyd.ernet.in

[2] Lab-STICC, Université de Bretagne-Sud, UEB, Centre de Recherche, BP 92116,
F 56321 Lorient Cedex, France
{marc.sevaux,andre.rossi}@univ-ubs.fr

Abstract. This paper describes a hybrid grouping genetic algorithm for a multiprocessor scheduling problem, where a list of tasks has to be scheduled on identical parallel processors. Each task in the list is defined by a release date, a due date and a processing time. The objective is to minimize the number of processors used while respecting the constraints imposed by release dates and due dates. We have compared our hybrid approach with two heuristic methods reported in the literature. Computational results show the superiority of our hybrid approach over these two approaches. Our hybrid approach obtained better quality solutions in shorter time.

Keywords: Combinatorial optimization, grouping genetic algorithm, heuristic, multiprocessor scheduling.

1 Introduction

Given a list of n tasks, that has to be scheduled on identical parallel processors. For each task i, a release date r_i, a due date d_i, and a processing time p_i are given. Let s_i be the start time of task i. A task i can not start before its release date ($s_i \geq r_i$) and has to be finished before its due date ($s_i + p_i \leq d_i$). The objective is to minimize the number of processors used while respecting the constraints imposed by release dates and due dates. Clearly, this objective is directly linked with minimizing the hardware cost. Such kind of scheduling problems find application in the field of High-Level Synthesis (HLS) [1, 2], where tasks must be executed on the cheapest possible architecture.

This problem is NP-Hard, because, if we set all release dates to zero and all due dates to a unique date greater than or equal to the processing time of the longest task, then this problem reduces to one-dimensional bin packing problem, which is a well-known NP-Hard problem. Moreover, to the best of our knowledge, no approximation scheme or performance guaranteed algorithm exists for this problem.

Sevaux and Sörensen [3] proposed a tabu search method for the problem. The origin of this method comes from [4]. This method begins by computing a lower bound m on the number of processors and then it solves the classical $Pm| \, r_i \, | \, \Sigma \, U_i$ scheduling

S. Ranka et al. (Eds.): IC3 2009, CCIS 40, pp. 1–7, 2009.
© Springer-Verlag Berlin Heidelberg 2009

problem using tabu search. The $Pm| r_i | \Sigma U_i$ problem seeks an allocation of tasks on m processors so as to minimize the number of tasks that can not be finished before their due dates. Such tasks are called late tasks and $U_i = 1$ if task i is late, 0 otherwise. If a task is not late then that task is called an early task. Clearly, if $\Sigma U_i = 0$ then a solution to the original scheduling problem is found and the method stops, otherwise the value of m is increased by 1 and tabu search procedure is repeated again. This process continues until a solution with $\Sigma U_i = 0$ is found. Tabu search procedure consists of either inserting a late task in between two early tasks or replacing a late task with some consecutive early tasks. This method uses sum of processing times of late tasks as tabu criterion. This method has an inbuilt cycle detection mechanism and adjusts the tabu tenure dynamically. Hereafter, this method will be referred to as TS-I.

Sevaux et al. [5] also describe a tabu search based method. However, this method follows a reverse approach in comparison to [3]. Starting from an initial feasible solution to the problem with m processors, it randomly removes one or two processors from the solution thereby creating a set of unallocated tasks. These unallocated tasks are tried in some random order for allocation on processors by tabu search. If a solution with m or fewer than m processors is obtained then it becomes the new initial solution and the whole process is repeated again, otherwise the new solution is discarded and the original solution is perturbed again. This process is repeated for a fixed number of iterations. Similar to [3], here also tabu search procedure consists of inserting an unallocated task between two allocated tasks or exchanging an unallocated task with some consecutively allocated tasks. However, in this approach an exchange operation is performed only when processing time of each consecutively allocated task is smaller than the processing time of the unallocated task to be exchanged. Moreover, once an unallocated task becomes allocated, it cannot be removed for the duration determined by the tabu tenure. So this approach uses a completely different tabu criterion in comparison to [3]. There is no need for cycle detection in this approach as an unallocated task is exchanged with smaller tasks only. Hereafter, this approach will be referred to as TS-II. In fact, TS-II is a more general approach in the sense that it can handle precedence constraints among tasks also.

In this paper we have proposed a hybrid approach for the problem combining a grouping genetic algorithm [6, 7] with a heuristic. We shall call our hybrid approach the HGGA-SSR. We have compared HGGA-SSR with TS-I and TS-II. On the test instances considered, HGGA-SSR obtained better quality solutions than TS-I and TS-II in shorter time.

The rest of this paper is organized as follows: Section 2 describes our hybrid grouping genetic algorithm. Computational results are presented in section 3, whereas section 4 outlines some conclusions.

2 The Hybrid Approach (HGGA-SSR)

Our hybrid approach is a combination of a steady-state grouping genetic algorithm and a heuristic. After the application of genetic operators some tasks are left unallocated, which are then allocated to processors by the heuristic.

2.1 The Grouping Genetic Algorithm

We have developed a steady-state grouping genetic algorithm for the problem. A steady-state genetic algorithm uses steady-state population replacement method [8]. In this method genetic algorithm repeatedly selects two parents, performs crossover and mutation to generate a single child that replaces a less fit member of the population. This is different from generational replacement, where an entirely new population of children is created and the whole parent population is replaced. The steady-state population replacement method has an advantage over the generational method due to the fact that the best solutions are always retained in the population and the child is immediately available for selection and reproduction. Hence, better solutions can be found faster. Moreover, with steady-state population replacement method we can easily prevent the multiple copies of the same individual from coexisting in the population. In the generational approach multiple copies of the same individual may exist in the population. Though, these individuals are usually the best individuals, they can rapidly dominate the whole population. In this situation, solution quality can not be improved without mutation, and often, a much higher mutation rate is required for any further improvement. In the steady-state approach the child can easily be compared to the existing population members, and, if it is identical to any existing individual in the population then it is discarded. In this way the problem of premature convergence is obviated by forbidding the multiple copies of the same individual to coexist in the population. The other main features of our grouping genetic algorithm are described subsequently.

Chromosome representation. We have represented a chromosome by a set of single-processor schedules, i.e., there is no ordering among the single-processor schedules. Any schedule can be executed on any processor as the processors are identical. This eliminates redundancy in chromosome representation. We have used single-processor schedule and processor interchangeably in the rest of the paper.

Fitness. The fitness of a chromosome is determined using a combination of two fitness functions. The primary fitness function F is the objective function, i.e., the number of processors used in the solution. The secondary fitness function f is computed by summing up the square of the relative idle time (idle time in a schedule divided by the duration of the schedule) over all the single-processor schedules. A solution C is more fit than another solution C' if $(F(C)<F(C'))$ or $((F(C)=F(C'))$ and $(f(C)>f(C'))$. The secondary fitness function is needed because there can be many solutions with the same number of processors. Among these solutions, the secondary fitness function prefers solutions which have some schedules with little or no relative idle time and some schedules with large relative idle time over solutions which have all average relative idle time schedules. The reason for this preference is that our genetic operators, specially crossover, are designed in such a way that they can transmit schedules with little or no relative idle time to the child unaltered while avoiding, as far as possible, transferring schedules with large relative idle time.

Selection. We have used k-ary tournament selection for selecting chromosomes for crossover and mutation. For selecting the two parents for crossover, we have used $k = 2$ and selected the more fit candidate with probability p_{better}. To select a chromosome for mutation we have used $k = 3$ and always selected the candidate with best fitness among the 3 candidates. The reason for using an aggressive selection strategy for mutation is that our mutation operator is designed in such a way that more fit chromosome has greater chance of generating a better chromosome after mutation.

Crossover. Our crossover operator is derived from the crossover operator proposed in [9]. It constructs the child iteratively. During each iteration, it selects one of the two parents uniformly at random and copies the schedule with smallest relative idle time from this parent to the child. Next, it deletes all the tasks belonging to this schedule from the schedules of both the parents and relative idle times of schedules altered are updated accordingly. Let m_1 and m_2 be the number of processors in the two parent solutions then crossover iterates for $\min(m_1, m_2) - 2$ iterations. Clearly, with this crossover scheme some tasks will be left unallocated which will be reallocated using the heuristic. For the purpose of this crossover operator any empty schedule in parents is assumed to have infinite relative idle time.

Mutation. The mutation operator randomly removes some schedules from the solution thereby creating a number of unallocated tasks. These tasks are reallocated using the heuristic. The schedule with largest relative idle time is always removed. Other schedules are selected uniformly at random for removal.

Similar to [9], here also crossover and mutation are used in a mutually exclusive manner, i.e., at each generation either the crossover is used or the mutation is used but not both of them. Crossover is applied with probability p_c, otherwise mutation is used.

Replacement policy. The generated child is first tested for uniqueness in the existing population members. If it is unique then it always replaces the least fit member of the population, otherwise it is discarded.

Initial population generation. Each member of the initial population is generated by following an iterative procedure. During each iteration an unallocated task is selected uniformly at random for allocation. Among all possible positions, a task is inserted at a place on a processor where idle time left after insertion is as small as possible. A task is considered for insertion at the end of processors only when it is not possible to insert the task somewhere in the middle on available processors. If it is neither possible to allocate a task in the middle nor at the end on available processors then a new processor is added to the solution and the task is allocated to it. A task once allocated on a processor at a particular place can not be shifted from that place. This process is repeated until all tasks are allocated.

Each newly generated chromosome is checked for uniqueness in the population members generated so far, and, if it is unique then it is included in the initial population, otherwise it is discarded.

2.2 The Heuristic

After the application of genetic operators some tasks are left unallocated. The heuristic allocates these left out tasks to processors. The heuristic used here is a modified version of the heuristic used in tabu search procedure of [5]. The unallocated tasks are reallocated one by one in an iterative fashion by the heuristic. During each iteration an unallocated task is selected for allocation in one of the two equally likely ways - either the least flexible unallocated task is selected for allocation or a task is selected uniformly at random from the set of all unallocated tasks for allocation. The flexibility of a task i is defined as $(d_i - r_i - p_i)/p_i$.

The reallocation of the selected task can be done in two ways. First, reallocation of a task is tried by inserting it on some processors by pushing other tasks. A task can be inserted at a place on a processor only when there is enough free time span available after pushing other tasks and none of the constraints are violated. Among many possible insertion positions, a task is inserted at a place where idle time left after insertion is smallest. While considering the insertion of task k between tasks i and j on processor p, we require earliest possible completion time (ECT) of task i and latest possible start time (LST) of task j. The ECT of a task allocated to a processor is the earliest time at which a task can be completed considering the present allocation of tasks on that particular processor. LST can be defined analogously. Obviously, computing the ECT of a task requires computing ECTs of all tasks preceding it and computing the LST of a task requires computing LSTs of all tasks that are succeeding it. After inserting the task, all tasks that are preceding that task are rescheduled (pushed left) so that they can be completed by their ECT and all tasks that are succeeding are delayed only by the required minimum amount of time. This is done following the policy of scheduling the tasks as early as possible.

If it is not possible to insert the task on available processors then we try to reallocate the task by exchanging this task with some consecutively allocated tasks. However, an exchange is considered feasible only when flexibility of each consecutively allocated task is greater than the flexibility of the unallocated task in consideration. The motivation behind this condition is that more flexible tasks are easier to schedule than the less flexible ones. Among all possible exchanges, the exchange that leaves the smallest idle time is selected.

If it is neither possible to insert the task somewhere on available processors nor exchange the task with some allocated tasks then a new processor is added to the solution and the task is allocated to it. This process is repeated until all the tasks are allocated.

Table 1. Rules for the instance generator

Parameter	Value
Problem size n	{20, 40, 60, 80, 100}
Horizon T	80
Release date r_i	$\Gamma(0.2, 4)$
Due date d_i	$T - \Gamma(0.2, 4)$
Processing Time p_i	$U(1, d_i - r_i)$

3 Computational Results

The HGGA-SSR has been coded in C and executed on a Linux based 3.0 GHz Pentium 4 system with 512 MB of RAM. In all our computational experiment with HGGA-SSR, we have used a population of 100 chromosomes, $p_c = 0.8$, $p_{better} = 0.9$. We have allowed HGGA-SSR to execute for $100n$ generations on a problem instance with n tasks. For problem instances up to 60 tasks, mutation removes three schedules from the selected chromosome, otherwise it removes four schedules. HGGA-SSR is executed once on each instance.

We have used the same 100 test instances as used in [3] and [5] to test HGGA-SSR. These instances were first used in Sevaux and Sörensen [3]. These instances were generated according to the following rules. A total horizon of 80 units of time was considered. The probability of release dates arising at the beginning of the horizon was greater than the same at the end (the reverse is true for due dates). Therefore, release dates were generated using the Gamma distribution ($\Gamma(0.2, 4)$ that gives a mean of 20 and a standard deviation of 10) and due dates were also generated using the same distribution, but considering the horizon from the end. If $r_i \geq d_i$ then the associated task was discarded and a new task was generated in its place. To generate only feasible tasks, processing time was uniformly generated in the interval $[1, d_i - r_i]$. 20 different instances were generated for each value of n in {20, 40, 60, 80, 100}. Therefore, altogether 100 instances were generated. The rules of the problem instance generator are summarized in table 1, where $U(l, u)$ denotes a uniform distribution with lower bound l and upper bound u, and, $\Gamma(a,b)$ denotes a gamma distribution.

Table 2 compares the performance of HGGA-SSR with TS-I and TS-II in terms of solution quality, where *Opt Av.* column shows for each task size the average value of the optimal number of processors required for all 20 instances of that size. The optimum solutions of all these instances are obtained through a mixed integer linear programming (MILP) based solver [10]. For each method the *Av.* column reports the average number of processors required by that method, the *Dev.* column gives the relative deviation of *Av.* with *Opt Av.* in percentage, *#Opt* column reports the number of instances solved to optimality. Table 3 reports the average execution time of the three methods on instances of each size in seconds. Data for TS-I and TS-II are taken from [3] and [5] respectively.

Table 2. Comparison of HGGA-SSR with TS-I and TS-II in terms of solution quality

n	Opt. Av.	TS-I			TS-II			HGGA-SSR		
		Av.	Dev.	#Opt	Av.	Dev.	#Opt	Av.	Dev.	#Opt
20	9.30	9.3	0.0	20	9.30	0.00	20	9.30	0.00	20
40	17.65	17.8	0.6	16	17.65	0.00	20	17.65	0.00	20
60	24.90	25.8	3.6	7	24.95	0.20	19	24.90	0.00	20
80	32.85	34.4	4.6	0	33.00	0.46	17	32.90	0.15	19
100	40.40	42.6	5.5	2	40.80	0.99	12	40.50	0.25	18

Table 3. Average execution times of HGGA-SSR, TS-I and TS-II in seconds

n	TS-I Average Time	TS-II Average Time	HGGA-SSR Average Time
20	2.2	0.45	0.05
40	8.8	1.73	0.17
60	15.5	2.86	0.40
80	37.6	3.94	0.88
100	49.9	5.24	1.32

Tables 2 and 3 clearly show the superiority of HGGA-SSR over TS-I and TS-II. In comparison to these two methods, HGGA-SSR returned better quality solutions in shorter time. Out of 100 instances it solves 97 instances optimally, whereas TS-I and TS-II respectively solve 45 and 88 instances optimally. For the remaining three instances HGGA-SSR requires only one processor more than the optimal number of processors. It is also much faster than TS-I and TS-II.

4 Conclusions

In this paper we have developed a hybrid approach, combining a steady-state grouping genetic algorithm with a heuristic, for a multiprocessor scheduling problem where the objective is to minimize the total number of processors used. We have compared our hybrid approach with two tabu search based approaches. Our hybrid approach outperformed both the approaches in terms of solution quality as well as running time. As a future work we plan to extend our hybrid approach to the version of the problem where there are precedence constraints among tasks.

References

1. Gajski, D.D., Dutt, N.D., Wu, A.C.-H., Lin, S.Y.-L.: High Level Synthesis: Introduction to Chip and System Design. Kluwer Academic Publishers, Norwell (1992)
2. Lee, J.-H., Hsu, Y.-C., Lin, Y.-L.: A New Integer Linear Programming Formulation for the Scheduling Problem in Data Path Synthesis. In: Proceedings of the 1989 IEEE International Conference on Computer-Aided Design, pp. 20–23. IEEE Computer Society Press, New York (1989)
3. Sevaux, M., Sörensen, K.: A Tabu Search Method for High Level Synthesis. In: Proceedings of the Francoro V / Roadef 2007, Grenoble, France, pp. 395–396 (2007)
4. Sevaux, M., Thomin, P.: Heuristics and Metaheuristics for Parallel Machine Scheduling: A Computational Evaluation. In: Proceedings of the 4th Metaheuristics International Conference (MIC 2001), Porto, Portugal, pp. 16–20 (2001)
5. Sevaux, M., Singh, A., Rossi, A.: Tabu Search for Multiprocessor Scheduling: Application to High Level Synthesis. Communicated to Asia Pacific Journal of Operational Research (2009)
6. Falkenauer, E.: New Representations and Operators for GAs Applied to Grouping Problems. Evolutionary Computation 2, 123–144 (1992)
7. Falkenauer, E.: Genetic Algorithms and Grouping Problems. John Wiley & Sons, Chichester (1998)
8. Davis, L.: Handbook of Genetic Algorithms. Van Nostrand Reinhold, New York (1991)
9. Singh, A., Gupta, A.K.: Two Heuristics for the One-Dimensional Bin-Packing Problem. OR Spectrum 29, 765–781 (2007)
10. Rossi, A., Sevaux, M.: Mixed-Integer Linear Programming Formulation for High Level Synthesis. In: Proceedings of the Eleventh International Workshop on Project Management and Scheduling, Istanbul, Turkey, pp. 222–226 (2008)

PDE Based Unsharp Masking, Crispening and High Boost Filtering of Digital Images

Rajeev Srivastava[1], J.R.P. Gupta[2], Harish Parthasarthy[2], and Subodh Srivastava[3]

[1] Department of Computer Engineering, Institute of Technology, BHU, Varanasi, India
rajeev.cse@itbhu.ac.in, rajeev_sri@yahoo.com
[2] Netaji Subhas Institute of Technology, Sector-3, Dwarka, New Delhi, India
[3] UNSIET, India

Abstract. A partial differential equation (PDE) based technique is proposed and implemented to perform unsharp masking, crispening and high boost filtering of digital images. The traditional concept of unsharp masking and crispening of edges which uses Laplacian as intermediate step for smoothening the image has been extended and modified using the idea of Perona and Malik [1] which overcomes the disadvantages of Laplacian method. For descretization, finite differences scheme has been used. The scheme has been implemented using MATLAB 7.0 and performance is tested for various gray images of different resolutions and the obtained results justify the applicability of proposed scheme.

Keywords: Unsharp filter, Crispenening, Sharpening of images, Laplacian, PDE, High Boost Filtering.

1 Introduction

In many digital imaging related applications sometimes it is desired to enhance or crispen the edges and highlight other high frequency components of an image in addition to sharpening the same. The unsharp filtering technique is commonly used in the photographic and printing industries for crispening of the edges. Other prominent areas of application are medical imaging and biometrics where it is required to sharpen the image along with enhancement of other high variations details such as edges and other features of interest. Since the high variations or high frequency components of an image also belong to noise in an image, therefore it is desired to suppress these noisy components and enhance only the actual features of interest within the image. The unsharp filter is a simple sharpening operator that enhances edges and other frequency components in an image through a procedure which subtracts an unsharp or blurred version of an image from the original one, [2] and [3].

$$I_{sharp}(x, y) = I(x, y) - I_{blurred}. \tag{1}$$

The basic procedure, [2], [3] and [4], for unsharp masking and crispening the image is as follows:

At first the original image is available as input. In second step, a low pass filter is applied on the original image for smoothening the same. In third step, an edge description and other desired high frequency components of an image are calculated by subtracting the smoothened image obtained in second step from the original image. At

S. Ranka et al. (Eds.): IC3 2009, CCIS 40, pp. 8–13, 2009.
© Springer-Verlag Berlin Heidelberg 2009

last, the edge image obtained in step three is used for sharpening the edges and other high variation components of original image by adding back it to the original signal.

The unsharp masking produces an edge image $I_e(x, y)$ from an input image $I(x, y)$ via

$$I_e(x, y) = I(x, y) - I_{smooth}(x, y) \tag{2}$$

where $I_{smooth}(x, y)$ is the smoothened version of $I(x, y)$.

The complete unsharp masking operator reads

$$I_{sharp}(x, y) = I(x, y) + k^* I_e(x, y). \tag{3}$$

Where k is a scaling constant, k > 0.The reasonable values for k varies between 0.2 to 0.8, with the larger values providing increasing amount of sharpening.

Equation (3) may also be interpreted as; this procedure is equivalent to adding the gradients, or a high pass signal to an image.

High boost filtering [2] is slight further generalization of unsharp masking. A high boost filtered image, I_{hb}, is defined at any point (x, y) as

$$I_{hb}(x, y) = \lambda * I(x, y) - I_{blurred}(x, y). \tag{4}$$

Where λ greater or equal to 1.

From equation (1) and equation (4) the high boost filtering process can be expressed [2] as

$$I_{hb}(x, y) = (\lambda\text{-}1) * I(x, y) + I_{sharp}(x, y). \tag{5}$$

A commonly used gradient function for smoothening the image i.e. $I_{smooth}(x, y)$ and the unsharp masks for producing an edge image is negative discrete Laplacian filter which is a second order derivative of an image taken in both x and y directions.

$$I_{smooth}(x, y) = \nabla^2 I(x, y) = [I(x-1, y) + I(x, y-1) + I(x+1, y) + I(x, y-1) - 4I(x, y)]. \tag{6}$$

Therefore, equation (2) reads

$$\begin{aligned} I_e(x, y) &= I(x, y) - I_{smooth}(x, y) \\ &= I(x, y) - \nabla^2 I(x, y). \end{aligned}$$

And equation (3) for unsharp masking and crispening reads

$$I_{sharp}(x, y) = (k+1)^*I(x, y) - k^* \nabla^2 I(x, y). \tag{7}$$

Another method used in place of discrete Laplacian is Laplacian of Gaussian (LoG). In this case since the kernel peak is positive, the edge image is subtracted, rather than added back to the original image.

Disadvantages of these schemes:

Gradient images produced by both filters, Laplacian and Laplacian of Gaussian (LoG), produces the side effects of ringing or introduction of additional intensity image structure and this ringing occurs at high contrast edges. In comparing unsharp mask defined using the Laplacian with LoG, the LoG is more robust to noise, as it has been designed explicitly to remove noise before enhancing edges. Therefore, one can obtain a slightly less noisy, but also less sharp image using a smaller Laplacian kernel.

Hence, the unsharp filter is a powerful sharpening operator, but it also produces a poor result in the presence of noise.

In this paper, a PDE based approach is proposed to overcome these problems.

2 Proposed PDE Based Model

The equation (6) that is Laplacian which is used as unsharp mask to produce the edge image is defined as,

$I_{smooth}(x, y) = \nabla^2 I(x, y)$ and this is a Heat equation [7] which performs the isotropic diffusion to de-noise the image.

The equation (6), can be regarded as an evolution process governed by a PDE that performs regularization of the image, [5] and [6].

$$\frac{\partial I}{\partial t} = \nabla^2 I(x, y) \tag{8}$$

Regularization, [1], [5] and [6], means simplifying data in a way that only interesting features are preserved and it can be used for removing the noisy within the image. The regularization term R in PDE formulations like $\frac{\partial I}{\partial t} = R$, introduces additional notion of scale space i.e. the data are iteratively regularized and a continuous sequence of smoother images is generated as time *t* goes by. Regularization PDEs may be seen as nonlinear filters that simplify the image little by little and minimize the image variations. The Laplacian represents a regularization process that smoothen the image with the same weight in all the spatial directions i.e. it performs isotropic diffusion, [1] and [6].

Therefore, to effectively remove the noise from the image and preserving as well as enhancing the edge structure of an image, the equation (6) can be modified according to Perona and Malik [1] who proposed a nonlinear diffusion method to avoid blurring and localization problem of linear diffusion filtering. This achieves both noise removal and edge enhancement through the use of a non-uniform diffusion which acts as unstable inverse diffusion near edges and as linear heat equation like diffusion in homogeneous regions without edges. The basic idea is that heat equation (8) for linear diffusion can be written in divergence form:

$$\frac{\partial I}{\partial t} = \nabla^2 I = div(gradI) = \vec{\nabla}.\vec{\nabla} I \tag{9}$$

The introduction of a conductivity coefficient c in the above diffusion equation makes it possible to make the diffusion adaptive to local image structure [1]:

$$\frac{\partial I}{\partial t} = \vec{\nabla}.c\vec{\nabla} I \tag{10}$$

Where the function $c = c(I, I_{x,} I_{xx}, ...)$ is a function of local image differential structure that depends on local partial derivatives.

Equation (10) can be written as

$$\frac{\partial I}{\partial t} = \frac{I_{smooth}(x,y) - I(x,y)}{\Delta t} = \vec{\nabla}.c\vec{\nabla}I$$

$$I_{smooth}(x,y) = I(x,y) + \lambda i(\vec{\nabla}.c\vec{\nabla}I) \tag{11}$$

Where $\lambda i = \Delta t = 0 - 0.25$ for stability purposes.

The R.H.S. of equation (11) can be discretized using centred difference scheme as proposed in [1].

Therefore the algorithm for Image Unsharp masking and crispening is as follows:

Algorithm - Image Unsharp masking and crispening

1. Take the input image $I(x, y)$, it may be noisy.
2. De-noise or smoothen the image using equation (11).
   For n=1 to niterations
   $$I_{smooth}(x,y) = I(x,y) + \lambda i(\vec{\nabla}.c\vec{\nabla}I)$$
   End
3. Obtain the edge description of the image according to equation (2)
   $I_e(x, y) = I(x, y) - I_{smooth}(x, y)$
4. The complete unsharp masking and crispening operator reads
   $I_{sharp}(x, y) = I(x, y) + k * I_e(x, y)$.

   Which is used to obtain the sharpened with crisped edges and k is a scaling constant, k > 0.The reasonable values for k varies between 0.2 to 0.8, with the larger values providing increasing amount of sharpening.

In a similar way equation (5) can also be modified to produce high boost filter.

3 Results

The algorithm for unsharp masking and crispening was implemented in MATLAB 7.0. For digital implementations, the resulting PDE was discretized using finite differences scheme as proposed in [1]. The value of k was set to 0.7; λ was set to 0.25 for stability purposes .Conductivity coefficient, c(x, y, t) that controls the diffusion in various regions of the image is used as proposed in [1]. The scheme was tested for various gray images of various resolutions and the results obtained justify the applicability of the scheme. Results for two different gray images moon.tif (512x512) and cameraman.tif (512x512) are shown in this paper.

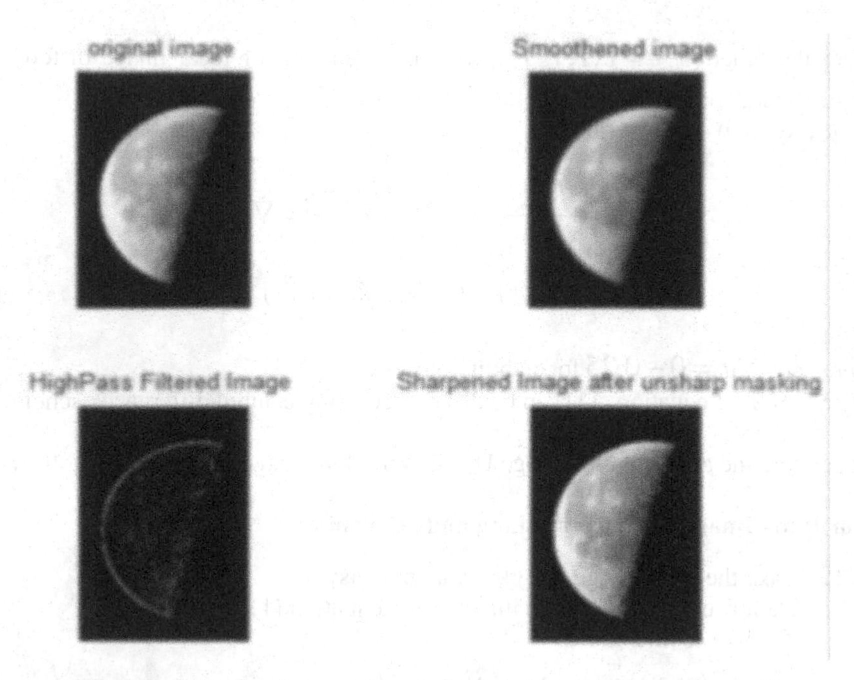

Fig. 1. Results of unsharp masking and crispening of the image, moon.tif

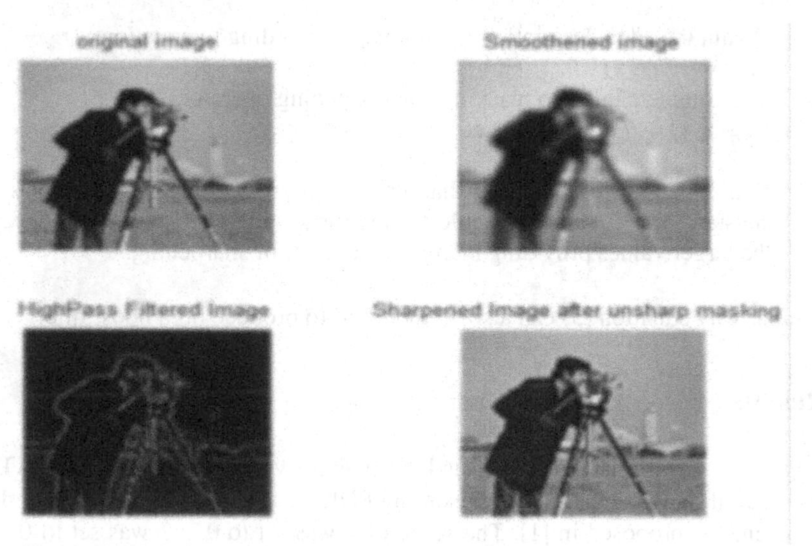

Fig. 2. Results of unsharp masking and crispening of the image, cameraman.tif

4 Conclusion

A partial differential equation (PDE) based technique was proposed to perform Unsharp Masking, Crispening and High Boost Filtering of Digital Images. For discretization, of the resulting PDE finite differences techniques were used. Finally, the proposed scheme was implemented in MATLAB and tested for various gray images of different resolutions and the results obtained justifies the applicability of the scheme. The basic advantage of this scheme is that image processing can be done in continuous domain itself and it is capable of eliminating noises from the image in an effective manner which was not completely possible by applying only Laplacian for smoothening purposes. In addition, this technique also eliminates other problems related with Laplacian as discussed in this paper.

References

1. Perona, P., Malik, J.: Scale space and edge detection using anisotropic diffusion. IEEE transactions on Pattern Analysis and Machine Intelligence 12, 629–639 (1990)
2. Gonzalez, R.C., Wintz, P.: Digital Image Processing, 2nd edn. Academic Press, New York (1987)
3. Jain, A.K.: Fundamentals of Digital Image Processing. PHI (2005)
4. Sonka, M., et al.: Image Processing, Analysis and Machine Vision, 2nd edn. PWS Publishing (2007)
5. Caselles, V., Morel, J., Sapiro, et al: Introduction to the special issue on partial differential equations and geometry driven diffusions in image processing. IEEE transactions on image processing 7(3) (March 1998)
6. Witkin, A.P.: Scale space filtering. In: Proc. Int. Joint Conf. on Artificial Intelligence, Germany, pp. 1019–1023 (1983)
7. McOwen, R.C.: Partial Differential Equations-Methods and Applications, 2nd edn. Pearson Education, London (2005)
8. Numerical Recipes in C: The Art of Scientific Computing. Cambridge University Press (1992) ISBN 0-521-43108-5

A New Position-Based Fast Radix-2 Algorithm for Computing the DHT

Gautam A. Shah[1] and Tejmal S. Rathore[2]

[1] Graduate Student Member, IEEE, Department of E&TC
MPSTME, NMIMS University, Mumbai, 400 056, India
gautamshah@ieee.org

[2] Senior Member, IEEE, Department of E&TC
St. Francis Institute of Technology, Mumbai, 400 103, India
tsrathor@ee.iitb.ac.in

Abstract. The radix-2 decimation-in-time fast Hartley transform algorithm for computing the DHT has been introduced by Bracewell. A set of fast algorithms were further developed by Sorenson et al. A new position-based fast radix-2 decimation-in-time algorithm that requires less number of multiplications than that of Sorenson is proposed. It exploits the characteristics of the DHT matrix and introduces multiplying structures in the signal flow-diagram (SFD). It exhibits an SFD with butterflies similar for each stage. The operation count for the proposed algorithm is determined. It is verified by implementing the program in C.

Keywords: Algorithm, decimation-in-time, discrete Hartley transform, matrix approach, radix-2.

1 Introduction

Over the years, the DHT has been established as a potential tool for signal processing applications [1]-[3]. It is popular due to its real-valued and symmetric transform kernel that is identical to its inverse [4]. Several algorithms have been reported in the literature for its fast computation [5]-[10]. An algorithm using a mixture of radix-2 and radix-8 index maps [11] is used in the computation of the DHT of an arbitrary length. The algorithm is expressed in simple matrix form and it facilitates easy implementation of the algorithm and allows for an extension to multidimensional cases. A systolic algorithm that uses the advantages of cyclic convolution structure has been presented in [12] for the VLSI implementation of a prime length DHT. Recently a new formulation using cyclic convolutions has been presented in [13], which leads to modular structures consisting of simple and regular systolic arrays for concurrent memory-based realization of the DHT and increases the throughput. A variety of vector-radix and split vector-radix algorithms have been extended for the fast computation of two-dimensional [14], three-dimensional [15]-[16] and multidimensional [17]-[18] DHTs, each possessing their own properties in terms of operational complexity, ease of implementation, regularity of butterfly structure and in place computation that are highly desirable for multidimensional applications. Nevertheless, there is a strong need to compute the transform at a high speed to meet the requirements of

S. Ranka et al. (Eds.): IC3 2009, CCIS 40, pp. 14–25, 2009.
© Springer-Verlag Berlin Heidelberg 2009

real-time signal processing. The fast Hartley transform (FHT) algorithm [5] by Bracewell performs the DHT in a time proportional to $N\log_2 N$. Sorenson et al [6] further analyzed it using the index mapping approach and implemented the algorithm with the same decomposition. However, both the approaches require stage dependent cosine and sine coefficients. Bracewell's SFD requires cross-over boxes for the sine channels.

The proposed algorithm utilizes either the cosine or sine coefficients which are stage independent and eliminates the crossover box. It introduces multiplying structures (MSs), and results in an SFD with butterflies similar in each stage structure (SS). In this endeavor, the position-based method [19] is extended to the existing FHT algorithm, and leads to simplification of the stage computations and reduces the operational complexity.

2 Discrete Hartley Transform

An N-point DHT X_H of a sequence $x(n)$ is defined as

$$X_H(k) = \sum_{n=0}^{N-1} x(n)\,\mathrm{cas}\left(\frac{2\pi kn}{N}\right),\ k = 0, 1, \ldots, N-1$$

where cas (.) = cos (.) + sin (.).

Using the matrix approach [20], the DHT can be expressed as

$$\begin{bmatrix} X_H(0) \\ X_H(1) \\ \vdots \\ X_H(N-1) \end{bmatrix} = \begin{bmatrix} h_{0,0} & h_{0,1} & \cdots & h_{0,N-1} \\ h_{1,0} & h_{1,1} & \cdots & \vdots \\ \vdots & \cdots & \cdots & \vdots \\ h_{N-1,0} & \cdots & \cdots & h_{N-1,N-1} \end{bmatrix} \cdot \begin{bmatrix} x(0) \\ \vdots \\ \vdots \\ x(N-1) \end{bmatrix}.$$

In the expression $[X_H] = [H_N] \cdot [x]$, H_N is the $N \times N$ Hartley matrix and its elements are given by

$$h_{i,j} = \mathrm{cas}\left(\frac{2\pi ij}{N}\right) \tag{1}$$

where indices i and j are integers from 0 to $N-1$ [21].

The FHT algorithm [5] by Bracewell performs the DHT of a data sequence of N elements in a time proportional to $N\log_2 N$, where $N = 2^P$. A condensed view of the FHT algorithm operation may be gained as a sequence of matrix operations on the data. The N-point DHT $X_H = N^{-1}L_P L_{P-1} \cdots L_1 P_r x(n)$, where P_r is the permutation matrix, that rearranges the sequence of data $x(n)$ in a bit reversed pattern and L_S (S = 1 to P) are the stage matrices. For every stage matrix L_S, there is a matrix L_Y having dimensions $Y \times Y$, where $Y = 2^S$ which repeats itself along the forward diagonal N/Y times. Splitting L_Y into 4 quadrants as $\begin{array}{c|c} Q_1 & Q_3 \\ \hline Q_2 & Q_4 \end{array}$, Q_1 and Q_2 are forward unity diagonal matrices, whereas Q_3 and Q_4 have the cosine, sine and cas coefficients placed in a fixed pattern. The cosine coefficients (CCs) are distributed along their forward diagonal and the sine coefficients (SCs) run the other way, the element of intersection of the two coefficients is the cas coefficient (KC). Figures 1(a) - 1(d) show the permutation and stage matrices involved in the computation for $N = 8$, where $C_n = \cos\left(\frac{2\pi n}{2^S}\right)$, $S_n = \sin\left(\frac{2\pi n}{2^S}\right)$ and $K_n = C_n + S_n$ are stage dependent.

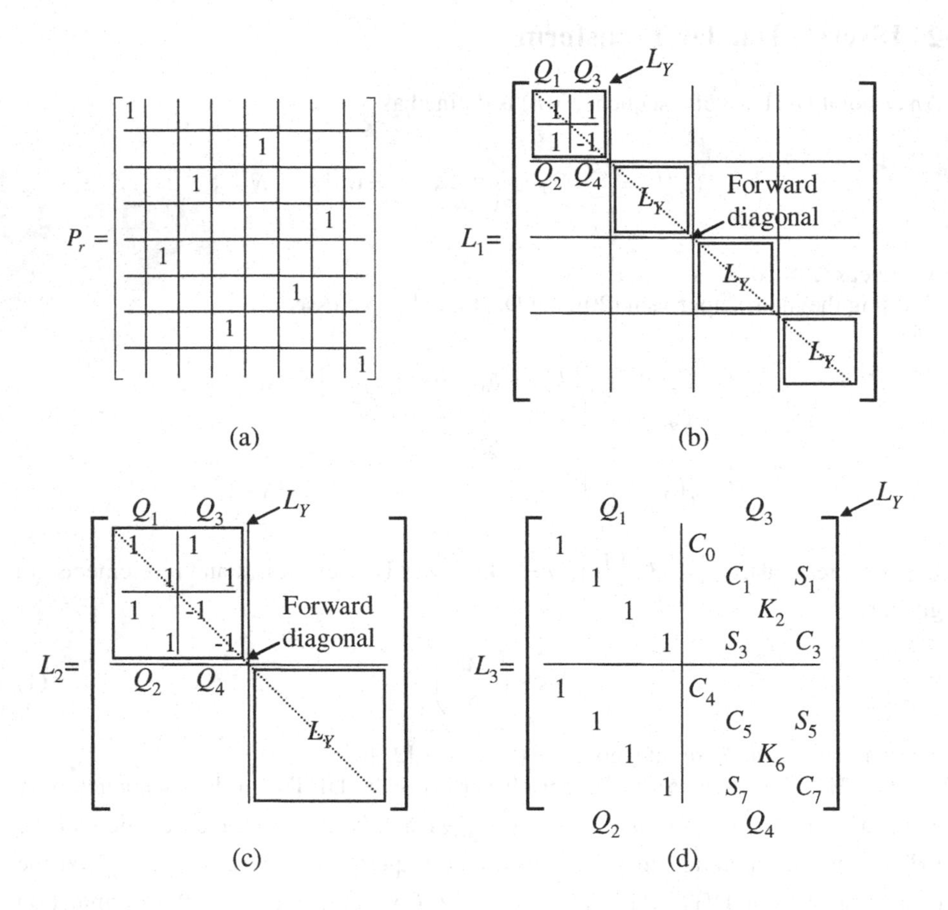

Fig. 1. (a) Permutation Matrix for $N = 8$. (b) Matrices L_S and L_Y for $N = 8$ and $S = 1$. (c) Matrices L_S and L_Y for $N = 8$ and $S = 2$. (d) Matrices L_S and L_Y for $N = 8$ and $S = 3$.

Another way of proceeding to evaluate the DHT is shown in the SFD of Fig. 2 for the case $N = 8, P = 3$. Dashed lines represent transfer factors –1 while full lines represent unity factors. C_n and S_n represent the CCs and SCs and $x_s(n)$ represents the stage output.

The PERMUTE operation rearranges the sequence of data in a bit reversed pattern $x_{pe}(n)$. The first two stages involve only addition/subtraction of data as the CCs and SCs are 0, +1 or –1. In the final stage, cosine and sine channels with the cross over box take care of multiplication with the CCs and SCs. The final transform is obtained by summing the three sets of eight inputs each in the COMBINE operation. Following permutation on the data by P_r which permutes $x(n)$ into $x_{pe}(n)$ come a succession of P operations and outputs through $x_1(n)$ to $x_P(n)$ leading stage by stage to the final transform output X_H.

Sorenson et al [6] analyzed Bracewell's [5] FHT algorithm using the index mapping approach. Their Hartley butterfly is shown in Fig. 3. Their program implements the FHT algorithm with operation counts of N_A additions and N_M multiplications given by

$$N_A = \frac{(3N\log_2 N - 3N + 4)}{2} \text{ and } N_M = N\log_2 N - 3N + 4 \ . \tag{2}$$

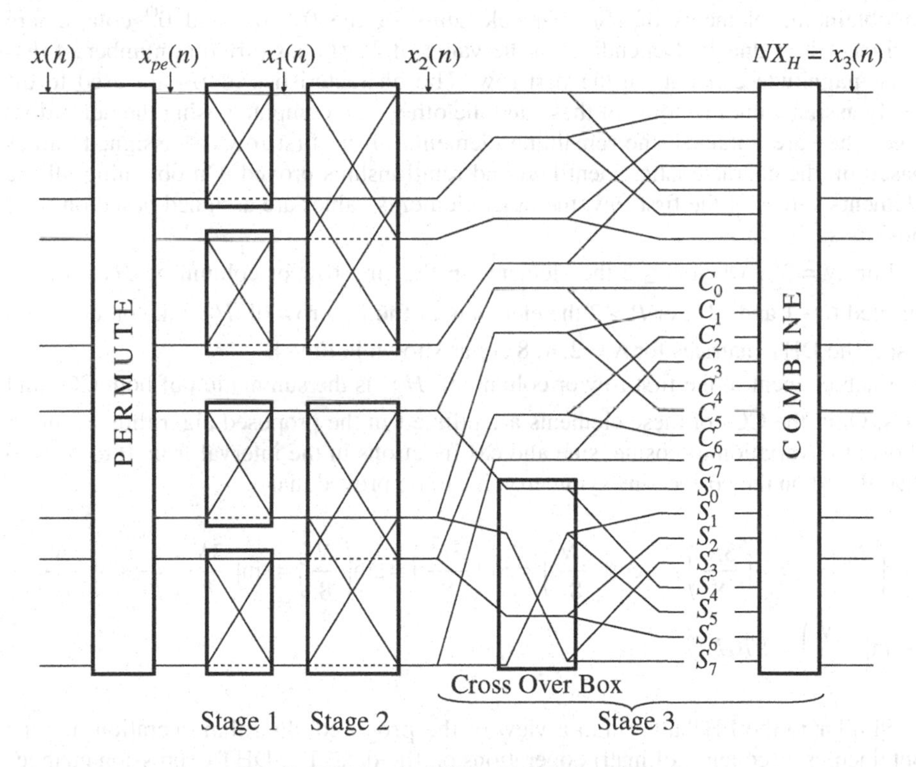

Fig. 2. SFD for the FHT with $N = 8$ and $P = 3$

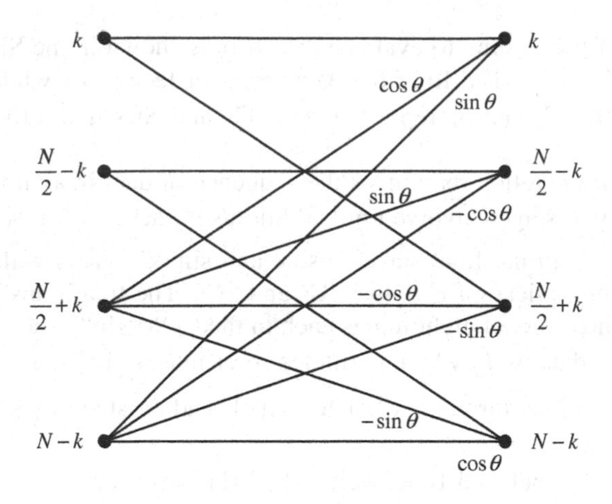

Fig. 3. Hartley butterfly

3 Proposed Algorithm

The position-based method (PBM) [19] makes use of the matrix characteristics to obtain the elements of H_N. The elements of the 0th row and 0th column are assigned the value 1. Depending on the value of *N*, H_N has different number of distinct magnitude elements in the first row. The characteristics of H_N are used to directly assign values to some of these and the others are computed using the definition. Once they are obtained, the remaining elements of the first row are assigned values based on the characteristics identified and relationships proved. On obtaining all the elements values of the first row, the other elements values are assigned based on their positions.

For $N = 2^P$, when $P \leq 2$ the elements in the first row or column of H_N are restricted to +1 and –1. For $P > 2$ the elements in the first row of H_N take other values also. The DHT matrices for $N = 2, 4, 8$ are as shown in Fig. 4.

Each element in the first row or column of H_N is the summation of both CCs and SCs. Only the CCs of these elements are utilized in the proposed algorithm. Figure 5 shows the variation of cosine, sine and cas functions in the interval from 0 to *N* (0 to 2π). Based on the cosine-sine symmetry, it can be proved that

$$\cos\left(\frac{N}{8}\right) = -\cos\left(\frac{3N}{8}\right) = -\cos\left(\frac{5N}{8}\right) = \cos\left(\frac{7N}{8}\right) = \sin\left(\frac{N}{8}\right) = \sin\left(\frac{3N}{8}\right) = -\sin\left(\frac{5N}{8}\right) = -\sin\left(\frac{7N}{8}\right) = 0.707\,.$$

Similar to the FHT algorithm a view of the proposed algorithm operation may be obtained as a sequence of matrix operations on the data. The DHT expression changes

$$H_2 = \begin{array}{c|cc|} \overset{i}{\downarrow}\; j\rightarrow & 0 & 1 \\ \hline 0 & 1.000 & 1.000 \\ 1 & 1.000 & -1.000 \end{array}
\qquad
H_4 = \begin{array}{c|cccc|} \overset{i}{\downarrow}\; j\rightarrow & 0 & 1 & 2 & 3 \\ \hline 0 & 1.000 & 1.000 & 1.000 & 1.000 \\ 1 & 1.000 & 1.000 & -1.000 & -1.000 \\ 2 & 1.000 & -1.000 & 1.000 & -1.000 \\ 3 & 1.000 & -1.000 & -1.000 & -1.000 \end{array}$$

$$H_8 = \begin{array}{c|cccccccc|} \overset{i}{\downarrow}\; j\rightarrow & 0 & 1 & 2 & 3 & 4 & 5 & 6 & 7 \\ \hline 0 & 1.000 & 1.000 & 1.000 & 1.000 & 1.000 & 1.000 & 1.000 & 1.000 \\ 1 & 1.000 & 1.414 & 1.000 & 0.000 & -1.000 & -1.414 & -1.000 & 0.000 \\ 2 & 1.000 & 1.000 & -1.000 & -1.000 & 1.000 & 1.000 & -1.000 & -1.000 \\ 3 & 1.000 & 0.000 & -1.000 & 1.414 & -1.000 & 0.000 & 1.000 & -1.414 \\ 4 & 1.000 & -1.000 & 1.000 & -1.000 & 1.000 & -1.000 & 1.000 & -1.000 \\ 5 & 1.000 & -1.414 & 1.000 & 0.000 & -1.000 & 1.414 & -1.000 & 0.000 \\ 6 & 1.000 & -1.000 & -1.000 & 1.000 & 1.000 & -1.000 & -1.000 & 1.000 \\ 7 & 1.000 & 0.000 & -1.000 & -1.414 & -1.000 & 0.000 & 1.000 & 1.414 \end{array}$$

Fig. 4. DHT matrices for $N = 2$, 4 and 8

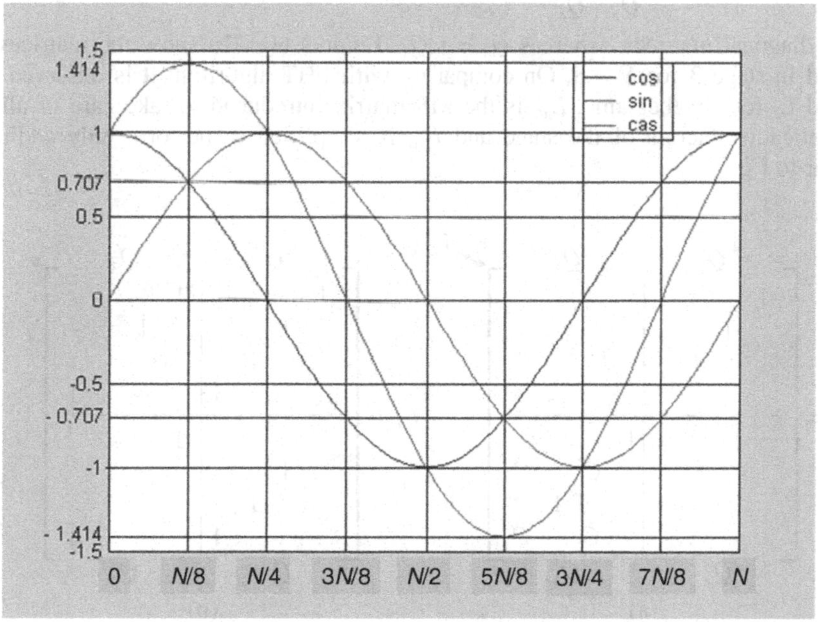

Fig. 5. Plots of the cosine, sine and cas functions

to $X_H = N^{-1} L_{PB} L_{PA} L_{(P-1)B} L_{(P-1)A} \cdots L_{3B} L_{3A} L_2 L_1 P_r x(n)$, where SS matrices (SSMs) are L_S for $S \leq 2$ and L_{SB} for $2 < S < P$. Note that an MS matrix (MSM) L_{SA} is introduced for each L_{SB} .The MSMs L_{SA} are introduced for all stages from 3 to P. For each L_{SA}, matrix L_Z having dimensions $Z \times Z$, where $Z = 2^S$, repeats itself along the forward diagonal N/Z times. Splitting L_Z into 4 quadrants Q_1 to Q_4 as $\begin{array}{c|c} Q_1 & Q_3 \\ \hline Q_2 & Q_4 \end{array}$, only Q_1 and Q_4 have some non-zero elements. Q_1 is a forward unity diagonal matrix, whereas elements which appear in Q_4 are related to the CCs of the elements in the first row or column of H_N . The CCs appear along the forward diagonal of Q_4. The SCs which appear in the corresponding matrix of the FHT algorithm are replaced by other CCs, along the inverse diagonal of the sub-matrix formed within Q_4 after deleting the 0th row and 0th column of Q_4. The intersection element of both these elements which is a cas (K) term in the corresponding matrix of the FHT algorithm, is replaced by a unity factor. Due to the cosine-sine symmetry, there is a reduction in the number of these elements to only a few CCs and a unity factor. While computing L_Z it is sufficient to compute Q_4. The permutation and stage matrices involved in the computation remain the same as in FHT algorithm till stage 2, after which the MSMs are introduced and the SSMs are modified.

For each SSM L_S and/or L_{SB}, there is a matrix L_Y having dimensions $Y \times Y$, where $Y = 2^S$ which repeats itself along the forward diagonal N/Y times. If L_Y is split into 4 quadrants Q_1 to Q_4 as $\begin{array}{c|c} Q_1 & Q_3 \\ \hline Q_2 & Q_4 \end{array}$, Q_1 Q_2 and Q_3 have the same elements and are forward unity diagonal matrices, whereas Q_4 = - Q_1. Figures 6(a)-(b) show the matrices involved in stage 3 for N = 8. On comparing with FHT algorithm it is observed that L_1 and L_2 remain the same. L_{3A} is the MS matrix introduced to take care of all the multiplication factors of the stage and L_{3B} is simplified to perform only additions similar to L_2.

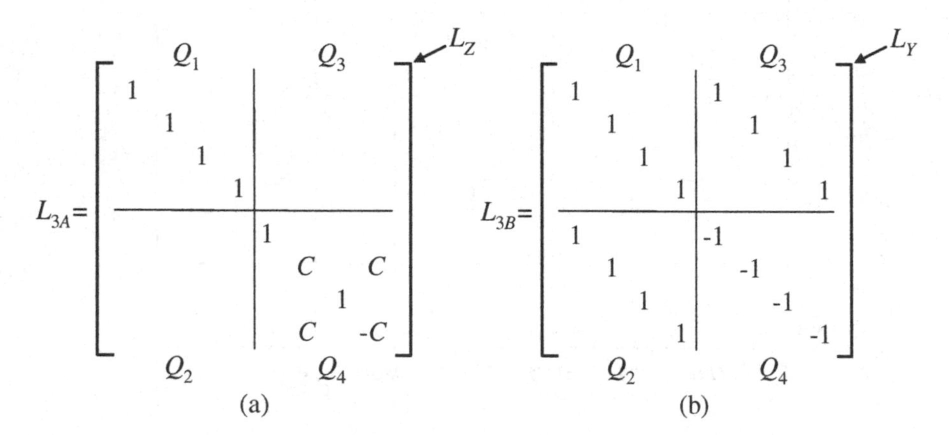

Fig. 6. (a) Matrices L_{SA} and L_Z for $N = 8$ and $S = 3$. (b) Matrices L_{SB} and L_Y for $N = 8$ and $S = 3$.

The introduction of L_{SA} in the expression for the DHT reflects in the form of an MS in the SFD. Fig. 7 depicts a partial SFD showing a generalized MS and SS. For each MS, there are multipliers for different elements as follows:

Elements from 0 to $m/2$ have no multipliers.

Each element $(m/2) + i$ has no multiplier for $i = m/4$, a multiplier $C = 0.707$ for $i = m/8$ and multipliers $C\left[i\frac{N}{m}\right]$, $C\left[\left(\frac{m}{4}-i\right)\frac{N}{m}\right]$ for $i = 1$ to $(m/8) - 1$ and $(m/8) + 1$ to $(m/4) - 1$.

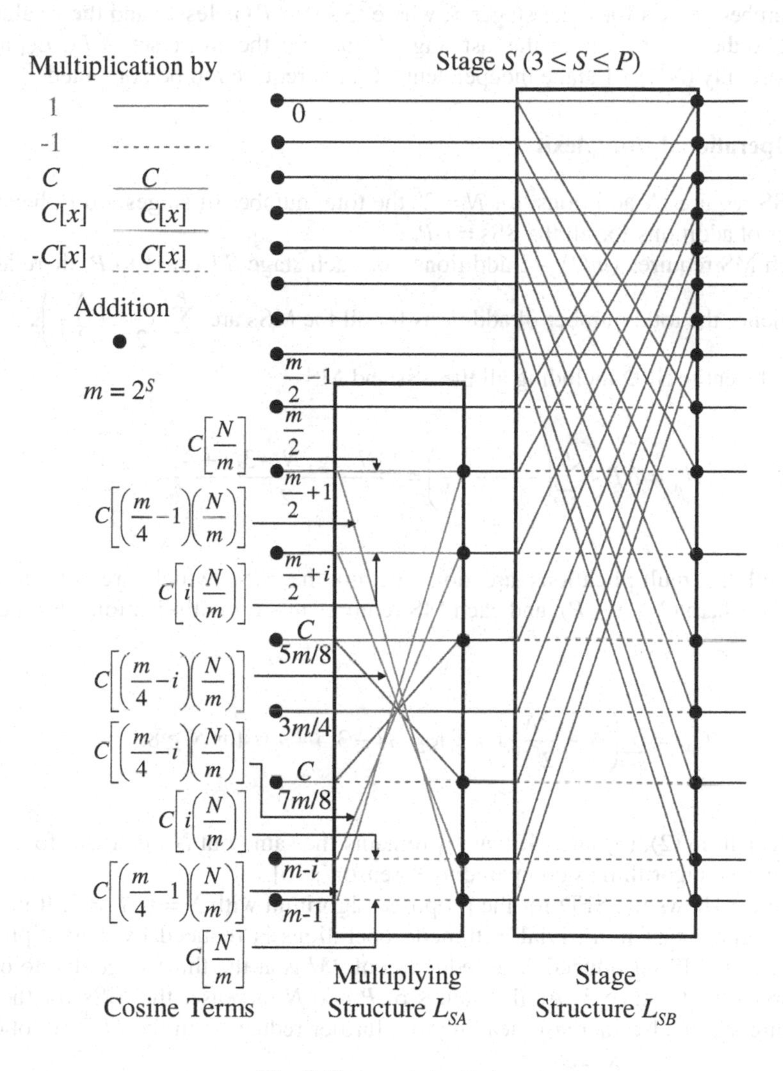

Fig. 7. Proposed structure

Each element $m - i$ has a multiplier $C = 0.707$ for $i = m/8$ and multipliers $-C\left[i\frac{N}{m}\right]$ and $C\left[\left(\frac{m}{4}-i\right)\frac{N}{m}\right]$ for $i = 1$ to $(m/8) - 1$ and $(m/8) + 1$ to $(m/4) - 1$, where $C[x] = \cos\frac{2\pi x}{N}$ represents the CCs.

For each stage S from 3 to P, the CCs required in the corresponding MS are given by $C[x]$ where $x = 2^{(P-S)}n$ and $n = 1$ to $[2^{(S-2)} - 1]$. The maximum number of CCs are required for the last stage when $S = P$. Here $m = N$ and x ranges from 1 to $(N/4) - 1$. The number of CCs for other stages S, where $(3 < S < P)$ is lesser and their values will belong to the set of CCs for the last stage. Hence for the entire set of L_S, L_{SA} and L_{SB} matrices, only $(N/4)$ - 1 stage independent CCs are required to be computed.

3.1 Operational Complexity

Each SS requires N additions, for $N = 2^P$ the total number of stages are P, hence total number of additions for all the SSs = NP.

Each MS requires $(m/2) - 2$ additions, for each stage S from 3 to P, there are N/m MSs, hence the total number of additions for all the MSs are $\sum_{S=3}^{P}\left(\frac{N}{2}-\frac{N}{2^{S-1}}\right)$.

For the entire SFD including all the SSs and MSs

$$N_A = NP + \sum_{S=3}^{P}\left(\frac{N}{2}-\frac{N}{2^{S-1}}\right) = \frac{(3N\log_2 N - 3N + 4)}{2}. \tag{3}$$

Since all the multiplications are done within the MSs which are introduced for stages S, where $(3 \leq S \leq P)$, and each MS requires $m - 6$ multiplications, for the entire SFD

$$N_M = \sum_{S=3}^{P}\left(N - \frac{3N}{2^{S-1}}\right) = N\log_2 N - 3.5N + 6 \text{ for } N \geq 8. \tag{4}$$

It is clear from (2), (3) and (4) that N_A remains the same but N_M is lesser for $N \geq 8$ in the proposed algorithm as compared to Sorenson's [6].

Figure 8 shows the SFD for the proposed algorithm with $N = 8$, $P = 3$. It is evident that the number of non-trivial arithmetic operations is reduced by 2 multiplications (M) for each MS introduced. The reduction of $2M$ is at the third stage due to one MS corresponding to stage 3. As the values of P and N increase, the MSs for the corresponding stages also increase, leading to a further reduction in the M. The total number of M reduces by $\frac{N-4}{2}$ for $N \geq 8$.

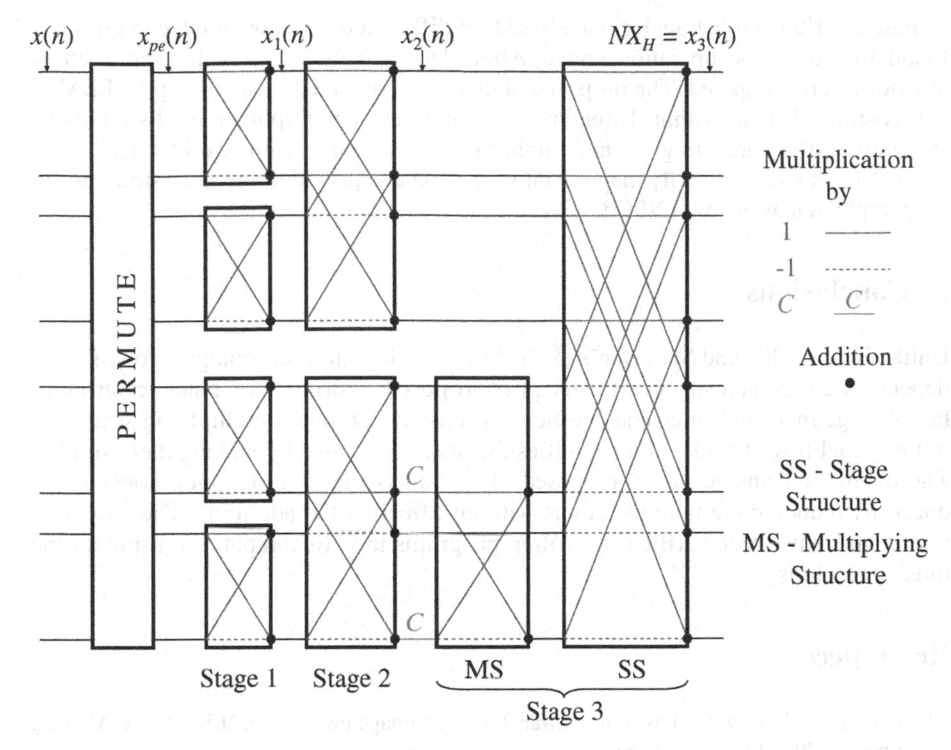

Fig. 8. SFD for proposed algorithm with $N = 8$, $P = 3$

4 Results

The number of operations for different sequences of length N is shown in Table 1.

Table 1. Comparison of Operation Complexity

Length N	Radix-2 FHT algorithm (Sorenson et al)			Position-Based Fast Radix-2 Algorithm		
	N_M	N_A	Total	N_M	N_A	Total
8	4	26	30	2	26	28
16	20	74	94	14	74	88
32	68	194	262	54	194	248
64	196	482	678	166	482	648
128	516	1154	1670	454	1154	1608
256	1284	2690	3974	1158	2690	3848
512	3076	6146	9222	2822	6146	8968
1024	7172	13826	20998	6662	13826	20488
2048	16388	30722	47110	15366	30722	46088
4096	36868	67586	104454	34822	67586	102408

Fast algorithms of a transform are based on different decomposition techniques and found to reduce the operation count. After decomposition eventually short length transforms are computed. The proposed algorithm is obtained by applying the PBM to the existing FHT algorithm. It requires less number of multiplications. Its recursive structure allows generating the next higher order transform from two identical lower order ones. It can be directly mapped into the SFD and provides a regular structure for easy implementation in VLSI [7].

5 Conclusions

Unlike Bracewell's and Sorenson's algorithms which require the computation of stage dependent cosine and sine coefficients, the proposed algorithm computes lesser number of stage independent cosine coefficients only. It introduces multiplying structures in the signal flow diagram and simplifies the stage structures by making them similar. The distinct advantage of the proposed algorithm is that the operation count is reduced by reducing the multiplications without affecting the additions. The computation counts have been verified by writing programs in C to compute X_H using all the three algorithms.

References

1. Paik, C.H., Fox, M.D.: Fast Hartley transform for image processing. IEEE Trans. Medical Imaging 7(6), 149–153 (1988)
2. Wu, J.L., Shiu, J.: Discrete Hartley transform in error control coding. IEEE Trans. Signal Processing 39(10), 2356–2359 (1991)
3. Meher, P.K., Panda, G.: Unconstrained Hartley-domain least mean square adaptive filter. IEEE Trans. Circuits and Systems-II: Analog and Digital Signal Processing 40(9), 582–585 (1993)
4. Meher, P.K., Srikanthan, T., Patra, J.C.: Scalable and modular memory-based systolic architectures for discrete Hartley transform. IEEE Trans. Circuits and Systems-I: Fundamental Theory & Applications 53(5), 1065–1077 (2006)
5. Bracewell, R.N.: The fast Hartley transform. Proc. IEEE 72(8), 1010–1018 (1984)
6. Sorensen, H.V., Jones, D.L., Burrus, C.S., Heideman, M.T.: On computing the discrete Hartley transform. IEEE Trans. Acoustics, Speech, and Signal Processing 33(4), 1231–1238 (1985)
7. Hou, H.S.: The Fast Hartley Transform Algorithm. IEEE Trans. Computers 36(2), 147–156 (1987)
8. Malvar, H.S.: Fast computation of the discrete cosine transform and the discrete Hartley transform. IEEE Trans. Acoustics, Speech, and Signal Processing 35(10), 1484–1485 (1987)
9. Prabhu, K.M.M., Nagesh, A.: New radix-3 and -6 decimation in frequency fast Hartley transform algorithms. Can. J. Elect. and Comp. Eng. 18(2), 65–69 (1993)
10. Bi, G., Chen, Y.Q.: Fast DHT algorithms for length $N = q*2^m$. IEEE Trans. Signal Processing 47(3), 900–903 (1999)
11. Bouguezel, S., Ahmad, M.O., Swamy, M.N.S.: A new split-radix FHT algorithm for length-$q*2^m$ DHTs. IEEE Trans. Circuits Syst. I, Reg. Papers 51(10), 2031–2043 (2004)

12. Chiper, D.F., Swamy, M.N.S., Ahmad, M.O.: An efficient systolic array algorithm for the VLSI implementation of a prime-length DHT. In: Proc. Int. Symp. Signals, Circuits Syst (ISSCS 2005), July 2005, pp. 167–169 (2005)
13. Meher, P.K., Patra, J.C., Swamy, M.N.S.: High-throughput memory-based architecture for DHT using a new convolutional formulation. IEEE Trans. Circuits Syst. II, Exp. Briefs 54(7), 606–610 (2007)
14. Wu, J.S., Shu, H.Z., Sehnadji, L., Luo, L.M.: Radix – 3 × 3 algorithm for the 2 –D discrete Hartley transform. IEEE Trans. Circuits Syst. II, Exp. Briefs 55(6), 566–570 (2008)
15. Bouguezel, S., Ahmad, M.O., Swamy, M.N.S.: An efficient three-dimensional decimation-in-time FHT algorithm based on the radix-2/4 approach. In: Proc. IEEE (ISSPIT), December 2004, pp. 52–55 (2004)
16. Bouguezel, S., Ahmad, M.O., Swamy, M.N.S.: A split vector-radix algorithm for the 3-D discrete Hartley transform. IEEE Trans. Circuits Syst. I, Reg. Papers 53(9), 1966–1976 (2006)
17. Bouguezel, S., Ahmad, M.O., Swamy, M.N.S.: An efficient multidimensional decimation-in-frequency FHT algorithm based on the radix-2/4 approach. In: Proc. IEEE (ISCAS), May 2005, pp. 2405–2408 (2005)
18. Bouguezel, S., Ahmad, M.O., Swamy, M.N.S.: Multidimensional vector radix FHT algorithms. IEEE Trans. Circuits Syst. I, Reg. Papers 53(4), 905–917 (2006)
19. Shah, G.A., Rathore, T.S.: Position-based method for computing the elements of the discrete Hartley transform matrix. In: Proc. Internat. Conf. IEEE Region 10, TENCON-2008, Hyderabad, India (2008)
20. Rathore, T.S.: Hartley transform – Properties and algorithms. In: Proc. National Conf. Real Time Systems, Indore, India, pp. 21–30 (1990)
21. Culhane, A.D., Peckerar, M.C., Marrian, C.R.K.: A neural net approach to discrete Hartley and Fourier transforms. IEEE Trans. Circuits and Systems 36(5), 695–703 (1989)

Study of Bit-Parallel Approximate Parameterized String Matching Algorithms

Rajesh Prasad and Suneeta Agarwal

Department of Computer Science & Engineering,
Motilal Nehru National Institute of Technology,
Allahabad-211004, India
rajesh_ucer@yahoo.com, suneeta@mnnit.ac.in

Abstract. In the parameterized string matching, a given pattern P is said to match with a substring t of the text T, if there exist a bijection from the symbols of P to the symbols of t. This problem has an important application in software maintenance, where we wish to find the equivalency between two sections of codes. Two sections of codes are said to be equivalent, if one can be transformed into the other by renaming identifiers and variables. In the approximate parameterized matching, a given pattern P matches the given substring t of the text T with $k \geq 0$ errors, if P can be transformed into t with at most k modifications (insertion, deletion, replacement). In this paper, we extend Myers Bit-Parallel algorithm and Approximate String Matching by using Bit-Parallel NFA (both for approximate matching), for parameterized string matching problem. These extended algorithms are known as PAMA and PABPA respectively. Theoretically, PAMA algorithm is faster than PABPA algorithm. The above algorithms are applicable only when pattern length (m) is less than word length (w) of computer used (i.e. $m \leq w$).

Keywords: Algorithm, finite automata, bit-parallelism, approximate matching, Non-deterministic finite automata, prev-encoding and parameterized matching.

1 Introduction

In the traditional string matching problem, all the occurrences of a pattern P [0…m-1] in the text T [0…n-1] are to be reported. Many algorithms for solving this problem exists [1] [2]. In [1], a bit-parallel string matching (shift-or) was developed. In the approximate string matching [3] [4] [5] problem, the goal is to find all the positions of a text where a given pattern occurs, allowing a limited number of "errors" in the matches. One of the studied particular case of the error model is the edit distance, which allows to delete, insert and replace simple characters (by a different one) in both strings. If different operations have different costs, we speak of general edit distance. Otherwise, if all the operations cost 1, we speak of simple edit-distance or edit distance (ed). In this case we simply seek for the minimum number of insertions, deletions and replacement to make both the strings equal. For example ed ("survey", "surgery") = 2.

S. Ranka et al. (Eds.): IC3 2009, CCIS 40, pp. 26–36, 2009.
© Springer-Verlag Berlin Heidelberg 2009

In the parameterized string matching [6], there are two disjoint alphabets Σ: for fixed symbols and Π: for parameterized symbols. The symbols of *pattern* P and *text* T are taken from Σ∪Π. In this type of string matching, while looking for occurrences of P in the substring t of the text T, the symbols of Σ must match exactly whereas the symbols of Π can be renamed. A given pattern P is said to match with a substring t of the text T, if ∃ a bijection from the symbols of P to the symbols of t. This problem has an important application in software maintenance, where it is required to find the equivalency between two sections of codes. Two sections of codes are said to be equivalent if one can be transformed into the other via one-to-one correspondence. In [5], parameterized on-line matching algorithm for a single pattern was developed. In [7], a bit-parallel parameterized string matching algorithm (shift-or) was developed. In [8], a fast parameterized matching with q-gram was developed. Approximate parameterized string matching (APSM) [9] [10] is the problem of finding at each location of the text T, a bijection that maximizes the number of characters that are mapped from P to the approximate |P|-length substring of text T.

In this paper, we extend: Approximate Myers algorithm (AMA) [11] [12] and Approximate String matching by using Bit-Parallel Non-deterministic finite automata (ABPA) [13] for approximate parameterized string matching. We call these algorithms PAMA and PABPA respectively. We analyze these algorithms theoretically and found that PAMA algorithm is better than PABPA algorithm.

2 Related Concepts

In this section, we present the related concepts and algorithms used in development of our proposed algorithms.

2.1 Parameterized String Matching Problem

We assume that *pattern* is P[0…m-1] and *text* is T[0…n-1]. All the symbols of P and T are taken from Σ∪Π, where Σ is fixed symbol alphabet of size σ and Π is parameter symbol alphabet of size π. The pattern P matches the text substring T[j…j+m-1] if and only if ∀ i ∈ {0, 1, 2…m-1}, f_j (P[i] = T[j+i]), where $f_j(.)$ is a bijective mapping on Σ∪Π. It must be identity on Σ but need not be identity on Π. For example, let Σ = {A, B} and Π = {X, Y, Z, W} and P = XYABX. P matches the text substring ZWABZ with bijective mapping X → Z and Y → W. This mapping can be simplified by *prev* encoding [6]. For any string S, *prev*(S) maps its each parameter symbols to a non-negative integer p, where p is the number of symbols since the last occurrences of s in S. The first occurrence of any parameter symbol in *prev* encoding is encoded as 0 and if s∈Σ it is mapped to itself (i.e. to s). For example, *prev* (P) = 00AB4, *prev*(ZWABZ) = 00AB4. With this scheme of *prev* encoding, the problem of parameterized string matching can be transformed to traditional string matching problem, where *prev*(P) is matched against *prev*(T [j…j+m-1]) for all j = 0, 1, 2...n-m. The *prev* (P) and *prev*(T [j…j+m-1]) can be recursively updated as j increases with the help of lemma given in [6].

Lemma 1. Let S′ = prev(S). Then for S″ = prev(S[j…j+m-1]), ∀ i such that S[i] ∈ Π it holds that S″[i] = S′[i] if and only if S″[i] < m, otherwise S″[i] = 0.

2.2 Approximate String Matching Problem

The general goal is to perform string matching of a pattern in a text where one or both of them have suffered some kind of (undesirable) corruption [5]. Some examples are: recovering the original signals after their transmission over noisy channels, Finding DNA subsequences after possible mutations, and text searching under the presence of typing or spelling errors. The *approximate string matching* (*string matching allowing errors*) is to find the positions of a text where a given pattern occurs, allowing a limited number of "errors" in the matches.

Distance Function

The distance d (x, y) between two strings x and y is the minimal cost of a sequence of operations that transform x into y (and ∞ if no such operation exist). The cost of sequence of operations is the sum of the costs of the individual operations. The operations are a finite set of rules of the form δ (z, w) = t, where z and w are different strings and t is a non-negative real number.

In most applications, the set of possible operations are restricted to:

– Insertion: δ (∈ , a), i.e. inserting the letter a
– Deletion: δ (a, ∈), i.e. deleting the letter a
– Replacement or Substitution: δ(a, b) for a ≠ b
– Transposition: δ(ab, ba) for a ≠ b

Edit Distance Functions

Levenshtein or Edit distance: Allows insertion, deletion and replacements. In the simplified definition, all the operations cost 1. In the literature, the search problem in many cases called "string matching with k differences".

Hamming distance: allows only replacements, which cost 1 in the simplified definition. In the literature, the search problem in many cases called "string matching with k mismatches"

Episode distance: allows only insertions, which cost 1. In the literature, it is called "episode matching".

Longest Common Subsequence distance: allows only insertions and deletions, all casting 1.

In the present discussion, we consider only Edit distance as a measure of error.

2.3 Dynamic Programming Algorithms

In this section, we discuss the dynamic programming based [14] algorithm to solve the approximate string matching problem. This algorithm computed the edit distance, and it was converted into a search algorithm by Sellers [14].

We first show, how to compute the edit distance between two strings x and y.

Computing Edit Distance

In order to compute ed (x, y), algorithm form a matrix $C_{0...|x|,\ 0...|y|}$ where $C_{i,j}$ represents the minimum number of operations needed to match $x_{1...i}$ to $y_{1...j}$. This is computed as follows:

$C_{i,\,0} = i$
$C_{0,\,j} = j$
$C_{i,\,j} = \text{if}(x_i = y_j) \text{ then } C_{i\text{-}1,\,j\text{-}1}$
$\qquad\qquad \text{else } 1+ \min\,(C_{i\text{-}1,\,j}, C_{i,j\text{-}1}, C_{i\text{-}1,\,j\text{-}1})$

Where at the end $C_{|x|,\,|y|} = ed(x, y)$

Example 1. Let **x** = *survey* and **y** = *surgery*. Fig. 1 shows the ed (x, y).

		0	1	2	3	4	5	6	7
			s	u	r	g	e	r	y
0		**0**	1	2	3	4	5	6	7
1	s	1	**0**	1	2	3	4	5	6
2	u	2	1	**0**	1	2	3	4	5
3	r	3	2	1	**0**	1	2	3	4
4	v	4	3	2	1	**1**	2	3	4
5	e	5	4	3	2	2	**1**	**2**	3
6	y	6	5	4	3	3	2	2	**2**

Fig. 1. Edit Distance between "survey" and "surgery". The bold entries show the path to the final result.

Text Searching

The algorithm is basically the same, with x = P and y = T. The only difference is that we must allow that any text position is the potential start of a match. This is achieved by setting $C_{0,\,j} = 0$ for all $j \in 0\ldots n$. That is the empty pattern matches with zero errors at any text position.

The algorithm then initializes its column $C_{0..m}$ with the values $C_i = i$, and processes the text character by character. At each new text character T_j, its column vector is updated to $C'_{0..m}$. The update formula is

$$C'_i = \text{if } (P_i = T_j) \text{ then } C_{i\text{-}1}$$
$$\qquad \text{else } 1+\min(C'_{i\text{-}1}, C_i, C_{i\text{-}1})$$

and the text position where $C_m \le k$ are reported. Fig. 2 shows the table.

		0	1	2	3	4	5	6	7
			s	u	r	g	e	r	y
0		0	0	0	0	0	0	0	0
1	s	1	0	1	1	1	1	1	1
2	u	2	1	0	1	2	2	2	2
3	r	3	2	1	0	1	2	2	3
4	v	4	3	2	1	1	2	3	3
5	e	5	4	3	2	2	1	2	3
6	y	6	5	4	3	3	**2**	**2**	**2**

Fig. 2. Searching "survey" in "surgery". Bold entries indicate matching text positions.

2.4 Algorithms Based on Finite Automata

In this section, we discuss Bit-Parallel Automata (BPA) [13] approach to approximate string matching. An alternative and very useful way to consider the problem is to model the search with a non-deterministic automaton (NFA) [13]. NFA for k = 2 is shown in the Fig. 3.

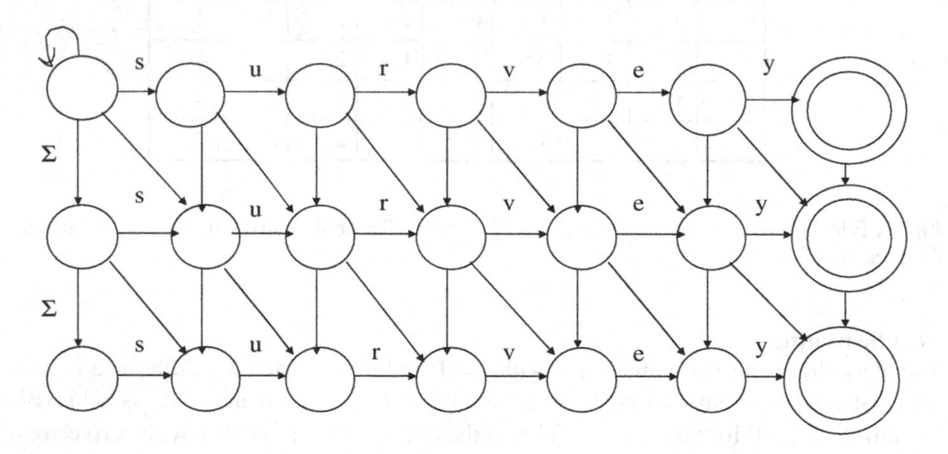

Fig. 3. An NFA for approximate string matching for the pattern "survey". Diagonal arrow = $\Sigma \cup \in$, vertical = Σ , row 1 = 0 error, row 2 = 1 error, row 3= 2 error.

We can use the (k+1)- row NFA for searching purposes. Every time the final state of the bottom row is active, we declare the final position of a pattern occurrence within distance threshold k. A bit-parallel approach due to Wu and Manber [13] is used to simulate the above (k+1) state NFA as follows: It compute table B as in shift-and [1] algorithm. It sets up k+1 bit masks, $D_0 \ldots D_k$, each representing the active bits of each NFA row. Initialize them as: $D_i = (1<<i)-1$ for $0 \leq i \leq k$.

For each new text character c, compute the new value for D0 as $D'_0 \leftarrow ((D_0 << 1) \mid 1)$ & B[c] and then for $i \in 1\ldots k$ in increasing order, compute the new value of Di as:

$$D'_i \leftarrow ((D_i << 1) \,\&\, B[c]) \mid D_{i-1} \mid (D_{i-1} << 1) \mid (D'_{i-1} << 1) \mid 1$$

Finally, we declare the match whenever $D_k \,\&\, (1<<(m-1)) \neq 0$.

2.5 Myers' Bit-Parallel Algorithm

In this section, we discuss Myers algorithm [11] for approximate string matching problem. The algorithm is based on representing the dynamic programming table D with vertical, horizontal and diagonal differences and pre-computing the matching positions of the pattern into an array of size σ. The algorithm has been implemented by using the following length-m bit vectors and taking the advantages of the dynamic programming matrix D as follows:

The Diagonal Property: D [i, j] – D[i-1, j-1] = 0 or 1
The Adjacency Property: D [i, j] – D[i, j-1] = -1, 0, or 1
D [i, j] – D[i-1, j] = -1, 0, or 1

This is done by using the following length-m bit vectors:
Vertical positive delta: VP[i]=1 at text position j iff D [i, j]-D[i-, j]=1
Vertical negative delta: VN[i]=1 at text position j iff D [i, j]-D[i-1, j]=-1
Horizontal positive delta HP[i]=1 at text position j iff D[i, j]-D[i, j-1]=1
Horizontal negative delta HN[i]=1 at text position j iff D[i, j]-D[i, j-1]=-1
Diagonal zero delta: $D_0[i]=1$ at text position j iff D[i, j] = D[i-1, j-1]
Pattern match vector PM_λ for each $\lambda \in \Sigma$: $PM_\lambda[i]=1$ iff $P_i=\lambda$

Searching Algorithm

currentdist←m
For j=1 to n
if HP[j] & $10^{m-1} = 10^{m-1}$
  then currentdist←currentdist+1
else if HN[j] & $10^{m-1} = 10^{m-1}$
    then currentdist←currentdist-1
if currentdist ≤ k
  then report occurrence at j

2.6 Parameterized Bit-Parallel String Matching Algorithm

In this section, we discuss parameterized shift-and [15] string matching algorithm. The algorithm is a variant of the algorithm explained in [6]. In this algorithm:

(i) The pattern P is encoded by *prev*-encoding and stored as *prev*(P). To compute *prev*(P), an array *prev*[c] is formed, which for each symbol $c \in \Sigma$, stores the position of its last occurrence in P. For example, let *pattern* P = XAYBX on fixed alphabet Σ = {A, B} and parameter alphabet Π = {X, Y}. Here prev(P) = 0A0B4.

(ii) For all j = 0, 1, 2...n-m, *prev*(T[j…j+m-1]) can be efficiently *prev*-encoded by lemma 1.
(iii) The table B is built such that all parameterized pattern prefixes can be searched in parallel.

To simplify indexing into B array, it is assumed that $\Sigma = \{0, 1\ldots\sigma\text{-}1\}$, and *prev* encoded parameter offsets are mapped into the range $\{\sigma\ldots\sigma\text{+m-1}\}$. For this purpose, an array $A[0\ldots\sigma\text{+m-1}]$ is formed, in which the positions $0\ldots\sigma\text{-1}$ are occupied by element of Σ and the rest positions are occupied by *prev* encoded offsets. For example, if we take pattern P = XAYBX on $\Sigma = \{A, B\}$ and $\Pi = \{X, Y\}$ with $\sigma = 2$, m = 5, then *prev*(P) = 0A0B4. Array *A* looks like as:

A	B	0	1	2	3	4
0	1	2	3	4	5	6

Let P′ is *prev*(P) and T′ is *prev*(T). Searching for P′ in T′ can't be done directly as explained below. Let P = XAXX and T = ZZAZZAZZ then P′ = 0A21 and T′ = 01A21A21. Clearly, P has two overlapping parameterized occurrences in T (one with shift = 1 and other with shift = 4), but P′ does not have any occurrences in T′ at all. The problem occurs because of when searching for all the length m prefixes of the text in parallel, then some non-zero encoded offset p in T′ should be interpreted as zero in some case. For example, when searching for P′ in T′[1…4] = **1**A21, **1**(from left) should be zero. The solution to this problem is that, the lemma 1 should be applied in parallel to all m-length sub strings of T′. This is achieved as follows:

The bit vector $B[A[\sigma+i]]$ is the match vector for *A*[i], for $0 \le i \le \sigma+m-1$. If the jth bit of this vector is 1, it means that P′[j] = *A*[i]. If any of the ith least significant bit of $B[A[\sigma]]$ is 1, corresponding bit of $B[A[\sigma+i]]$ is also set to 1. This is achieved as follows:

$$B[A[\sigma+i]] \leftarrow B[A[\sigma+i]] \mid (B[A[\sigma]] \ \& \ \sim(\sim 0 \ll i))$$

which, signifies that, for $\sigma \le i \le \sigma+m-1$, *A*[i] is treated as *A*[i] for prefixes whose length is greater than A[i] and as zero for shorter prefixes thus satisfies lemma 1.

Example 2
Let P = XAXX and T = ZZAZAZZAZZ on $\Sigma = \{A\}$ and $\Pi = \{X, Z\}$. Here $\sigma = 1$ and $\pi = 2$. *prev*(P) is equal to P′ = 0A21 and *prev*(T) = 01A2A21A21. On processing P′, we get B[0] = 0001, B[1] = 1000, B[A] = 0010, B[2] = 0100, and B[3] = 0000. Initially, D = 0000. From the preceding discussions, the placing of element of Σ and prev-encoded symbols are shown in the following array *A*

A	0	1	2	3
0	1	2	3	4

On processing the algorithm, we get,
$B[A[2]] = B[A[2]] \mid (B[A[1]] \ \& \ \sim(\sim 0 \ll 1))$
$B[1] = 1000 \mid (0001 \ \& \ 0001)$
$\quad = 1000 \mid 0001 = 1001$

Similarly,
B[2] = 0100 | (0001 & 0011) = 0101
B[3] = 0000 | (0001 & 0111) = 0001
// Begin the algorithm
Step 1: D←((D<<1) | 0^{m-1} 1) & B[prev[T[0]]]
D = (0000 | 0001) & B[0]
D = (0000 | 0001) & 0001= 0001
Similarly,
Step 2: D = (0010 | 0001) & 1001
= 0011 & 1001 = 0001
Step 3: D = (0010 | 0001) & 0010 = 0010
Step 4: D = (0100 | 0001) & 0101 = 0101
Step 5: D = (1010 | 0001) & 0010 = 0010
Step 6: D = (0100 | 0001) & 0101 = 0101
Step 7: D = (1010 | 0001) & 1001 = 1001
3rd bit is set to 1; therefore pattern occurs with shift 7-4 = 3.
Step 8: D = (0010 | 0001) & 0010 = 0010
Step 9: D = (0100 | 0001) & 0101 = 0101
Step 10: D = (1010 | 0001) & 1000 = 1000
3rd bit is set to 1; therefore pattern occurs with shift 10-4 = 6.

2.7 Approximate Parameterized String Matching (APSM) Problem

In the Approximate parameterized string matching (APSM) [9, 10] problem, our task is to find at each location of the text T, a bijection that maximizes the number of characters that are mapped from P to the approximate |P|-length substring of text T.

3 Proposed Algorithms

In this section, we present the proposed algorithms: APSM algorithm based on finite automata and Myers's algorithm for approximate parameterized string matching problem.

3.1 Algorithm for APSM Problem Based on Finite Automata

In this section we extend approximate BPA algorithm [13] discussed in section 2.4 for parameterized string matching. We call this algorithm as PBPA. The problem of approximate parameterized string matching (APSM) can be solved with the help of bit-parallel automata (discussed in section 2.4) and prev-encoding both pattern P and text T (discussed in section 2.6). Concept of section 2.6 is used to prev-encode both pattern and text. Pattern P is prev-encoded offline and text T is prev-encoded on line. We use same indexing mechanism as discussed in section 2.6 for prev-encoding the text. This algorithm construct the table B in the same way as the shift-and (discussed in section 2.6) algorithm. Similar to BPA algorithm discussed in section 2.4, this algorithm also sets up k+1 bit masks, $D_0 \ldots D_k$, each representing the active bits of each NFA row and initializes them as: $D_i = (1<<i)-1$ for $0 \le i \le k$.

For each new text character c, compute the new value for D_0 as $D'_0 \leftarrow ((D_0 << 1) \mid 1)$ & $B[prev[c]]$ and then for $i \in 1 \ldots k$ in increasing order, compute the new value of D_i as:

$$D'_i \leftarrow ((D_i << 1) \,\&\, B[prev[c]]\,) \mid D_{i-1} \mid (D_{i-1} << 1) \mid (\,D'_{i-1} << 1) \mid 1$$

Finally, a match is declared with error k at position i, whenever $D_k \,\&\, (1 << (m-1)) \neq 0$.

Example 3
Let P=x a x and T = z z a z, $\Sigma = \{a\}$, $\Pi = \{x., z\}$ Error k = 2, prev (P) = 0 a 2, prev (T) = 0 1 a 2, B[0]=001, B[1]=000, B[a]=010, B[2]=100

Updating of B table (in the same manner as in section 2.6)

B[1] = 000 | (001 & ~(111<<1)) = 001, B[2] = 100 | (001 & ~ (111<<2)) = 101

Initial values:

$D_0 = 000$, $D_1=001$, $D_2=011$

Step 0: on input 0
D'_0 = 001 & 001 = 001, D'_1 = (001 & 001) | 000 | 000 | 010 | 1 = 011
D'_2 = (110 & 001) | 001 | 010 | 110 | 1
= **111** ⇒ *Pattern occurs with* ***error 2*** *at text position 0*

Step 1: on input 1
D'_0 = (010 | 1) & 001 = 001, D'_1 = (110 & 001) | 001 | 010 | 010 | 1 = 011
D'_2 = (110 & 001) | 011 | 110 | 110 | 1
= **111** ⇒ *Pattern occurs with* ***error 2*** *at text position 1*

Step 2: on input a
D'_0 = (011)& 010 = 010, D'_1 = (110 & 010) | 001 | 010 | 100 | 1
= **111** ⇒ *Pattern occurs with* ***error 1*** *at text position 2*
D'_2 = (110 & 010) | 011 | 110 | 110 | 1
= **111** ⇒ *Pattern occurs with* ***error 2*** *at text position 2*

Step 3: on input 2
D'_0 = (100 | 1)& 101 = 101 ⇒ *Pattern occurs with* ***error 0*** *at text position 3*
D'_1 = (110 & 101) | 010 | 100 | 110 | 1
= **111** ⇒ *Pattern occurs with* ***error 1*** *at text position 3*
D'_2 = (110 & 101) | 111 | 110 | 110 | 1
= **111** ⇒ *Pattern occurs with* ***error 2*** *at text position 3*

Analysis
Since for each k row, we have to process the text of length n, therefore running time of the algorithm is $O(k \lceil m / w \rceil n)$ in the worst and average case which is O (k n) when $m \leq w$.

3.2 Algorithm for APSM Problem Based on Myers' Algorithm

In this section we extend Myers algorithm [11] for approximate parameterized matching. We call this algorithm as PAMA. The problem of approximate parameterized matching can be solved with the help of Myers algorithm [11] (discussed in section 2.5) and prev-encoding both pattern P and text T (discussed in section 2.6). The algorithm constructs the table B in the same way as in section 2.6. The algorithm is modified as follows:

In order to compute ed(x, y), we form a matrix $C_{0...|x|,\ 0...|y|}$ where $C_{i,,j}$ represents the minimum number of operations needed to match $x_{1...i}$ to $y_{1...j}$, where x and y are prev-encoded strings .This is computed as follows:

$C_{i,0} = i,\ C_{0,j} = j,$
$C_{i,j} =$ if($\mathbf{B[x_i] \wedge B[y_j] \neq 0}$) then $C_{i-1,\ j-1}$
else $1+ \min(C_{i-1,\ j},\ C_{i,j-1},C_{i-1,\ j-1})$

Where at the end $C|x|, |y| = ed(x, y)$

The searching algorithm is basically the same, with x = prev(P) and y = prev(T). The only difference is that we must allow that any text position is the potential start of a match. This is achieved by setting $C_{0,j} = 0$ for all $j \in 0...n$. Here is an example illustrating the algorithm.

	D		0	1	a	2
			0	1	2	3	4
0	0	0	0	0	0	0
1	0	1	0	0	1	0
2	a	2	1	1	0	1
3	2	3	2	1	1	0

	HP		0	1	a	2
		0	1	2	3	4
0						
1	0		0	0	1	0
2	a		0	0	0	1
3	2		0	0	0	0

HN		0	1	a	2
0		1	0	0	0
a		1	0	1	0
2		1	1	1	1

Fig. 4. The Dynamic Programming Matrix to search parameterized pattern P=x a x into parameterized text T= zzaz. Bold entries indicates that pattern can match in the text within maximum error=2.

Example 4

Let P=x a x and T= z z a z, prev(P) =0a2 and prev (T) = 0 1 a 2
B[0]=001, B[1] = 001, B[2]=101, B[a]=010. Fig. 4. shows the table.

Analysis

Running time of the algorithm is $O(\lceil m / w \rceil n)$ in the worst and average case which is O(n) when $m \leq w$. Clearly this algorithm is an improved version of the BPA algorithm.

4 Conclusions and Future Work

The Algorithm PAMA is better than PBPA when running on the same set of $\Sigma \cup \Pi$. These algorithms can be implemented to verify their running time. Furthermore, we can analyze the performance of these algorithms with respect to duplicity present in the code.

References

1. Baeza-Yates, R.A., Gonnet, G.H.: A new approach to text searching. Communication of ACM 35(10), 74–82 (1992)
2. Boyer, R.S., Moore, J.S.: A fast string-searching algorithm. Communication of ACM 20(10), 762–772 (1977)
3. Baeza-Yates, R., Navarro, G.: Faster approximate string matching. Algorithmica 23(2), 127–158 (1999)
4. Hyyro, H., Navarro, G.: Faster bit-parallel approximate string matching. In: Apostolico, A., Takeda, M. (eds.) CPM 2002. LNCS, vol. 2373, pp. 203–224. Springer, Heidelberg (2002)
5. Navarro, G.: A guided tour to approximate string matching. ACM Computing Survey 33(1), 31–88 (2001)
6. Baker, B.S.: Parameterized duplication in string algorithm and application in software maintenance. SIAM J. Computing 26(5), 1343–1362 (1997)
7. Fredriksson, K., Mozgovoy, M.: Efficient parametrized string matching. Information Processing Letters (IPL) 100(3), 91–96 (2006)
8. Salmela, L., Tarhio, J.: Fast Parameterized Matching with q-grams. In: Proc. of 7th Combinatorial Pattern Matching (CPM), pp. 354–364 (2006)
9. Apostolico, A., Erdos, P., Lewenstein, M.: Parameterized matching with mismatches. Journal of discrete algorithms (to appear)
10. Baker, B.S.: Parameterized diff. In: Proc. 10th Symposium on Discrete Algorithm (SODA), pp. 854–855 (1999)
11. Myers, G.: A fast bit-vector algorithm for approximate string matching based on dynamic programming. Journal of the ACM 46(3), 395–415 (1999)
12. Hyyro, H.: Explaining and extending the bit-parallel approximate string matching algorithm of Myers. Tech. Rep. A-2001-10, Department of Computer Science and Information Sciences, University of Tampere, Tampere, Finaland (2001)
13. Wu, S., Manber, U.: Fast text searching allowing errors. Communication of the ACM 35(10), 83–91 (1992)
14. Sellers, P.: The theory and computation of evolutionary distances: pattern recognition. Journal of Algorithms 1, 359–373 (1980)
15. Prasad, R., Agarwal, S.: Parameterized shift-and string matching algorithm using super alphabet. In: Proc of the International Conference on Computer and Communication Engineering (available on IEEE Xplore), pp. 937–942 (2008)

Optimization of Finite Difference Method with Multiwavelet Bases

Eratt P. Sumesh[1] and Elizabeth Elias[2]

[1] Department of Electronics and Communication Engineering, Amrita School of Engineering, Ettimadai, Coimbatore, India
[2] Department of Electronics and Communication Engineering, National Institute of Technology, Calicut, Kerala, India
ep_sumesh@ettimadai.amrita.edu, elizabeth@nitc.ac.in

Abstract. Multiwavelet based methods are among the latest techniques to solve partial differential equations (PDEs) numerically. Finite Difference Method (FDM) – powered by its simplicity – is one of the most widely used techniques to solve PDEs numerically. But it fails to produce better result in problems where the solution is having both sharp and smooth variations at different regions of interest. In such cases, to achieve a given accuracy an adaptive scheme for proper grid placement is needed. In this paper we propose a method, 'Multiwavelet Optimized Finite Difference Method' (MWOFD) to overcome the drawback of FDM. In the proposed method, multiwavelet coefficients are used to place non-uniform grids adaptively. This method is highly converging and requires only less number of grids to achieve a given accuracy when contrasted with FDM. The method is demonstrated for nonlinear Schrödinger equation and Burgers' equation.

Keywords: Finite Difference Method, multiwavelets, partial differential equation discretization, grid size, iteration.

1 Introduction

Numerical methods are widely accepted tools for the solution of physical problems modeled as partial differential equations (PDEs). In numerical methods, the integral or differential equations are converted into a system of linear algebraic equations which are mostly sparse block diagonal matrices. This process is known as the discretization of PDEs, which is done by taking only finite number of grid points [1, 2] in the region of interest. When the geometry of the problem becomes more complicated, more grid points are required and the resulting matrix size increases drastically. In these situations, the schemes with sparse diagonal matrix will become more profound. Finite difference method (FDM) [3–5] is a sub-sectional basis method universally applied in engineering. This method can lead to sparse representation of differential operator but do not necessarily convert the integral operator or integro-differential operators into sparse systems of linear algebraic equations [6–8]. Numerical solution of differential and integral equations using wavelets is described in [4–6, 9–12]. The

S. Ranka et al. (Eds.): IC3 2009, CCIS 40, pp. 37–48, 2009.
© Springer-Verlag Berlin Heidelberg 2009

use of multiwavelets in partial differential equation is demonstrated by Alpert [7, 13] using Legendre multiwavelet bases. Leland Jameson [14] proposed a scheme known as 'Wavelet Optimized Finite Difference method' (WOFD) which uses wavelets in FDM. Ole. M. Nielson [15] proved that WOFD scheme improves the accuracy of the FDM for a given number of grid points. The proposed 'Multiwavelet Optimized Finite Difference Method' (MWOFD) provides further improvement in the performance of the FDM applied to PDEs.

The use of multiwavelets offers several additional advantages over scalar wavelets. Multiwavelets with m vanishing moments maintain convergence of order $m-1$ up to the boundary [13, 16]. Due to the higher convergence order, these bases are suitable in problems where the boundary error has to be kept a minimum. The interpolating property of multiwavelet scaling functions make the coefficient values same as the solution while solving PDEs, a unique property not satisfied by the scalar wavelets. An advantage of the wavelet and multiwavelet bases is their localization property. By this property it is possible that each expansion coefficient affects the approximation of the unknown function only over a sub-domain of the region of interest. When the region of interest is divided into sub-regions based on the resolution needed, this property simplifies the computation and increases the convergence of the solution [4,5,9]. In wavelet based solvers, orthogonality of wavelet is attained only when there is no overlap between sub-domains of interest. Higher continuity order of bases is required when the solution is a smooth function. Higher the continuity order, larger the supporting region required for the bases. This is a bottleneck in wavelet based techniques which can be largely overcome by the use of multiwavelets. The trade-off between orthogonality and continuity is well balanced in the case of multiwavelets. Multiwavelet bases retain some properties of wavelet bases, such as vanishing moments, orthogonality and compact support. Multiwavelets posses extra advantages as they consist of bases that do not overlap on a given scale and are organized in small groups of several filter functions sharing a given support. The vanishing moment property of the bases makes a wide class of integro-differential operators having sparse representation in these bases. This property makes them more suitable for adaptive solvers of PDEs subject to boundary conditions [5,7].

This paper is organized as follows. PDE discretization methodology is described in section 2. Grid generation method in the proposed MWOFD is described in section 3. The methodology and numerical results for the nonlinear Schrödinger equation and Burgers' equation are presented in section 4 and section 5 gives the conclusion.

2 Discretization of PDEs

Multiwavelet bases provide an efficient mechanism for adaptive placement of the grid-points. Ideal choice of grid-points in a finite difference scheme is to place sparse grids in regions of the domain where the solution is smooth and finer grids are to be used in regions of the domain where the solution is turbulent. The FDM is illustrated in detail in [9]. In WOFD method, wavelet coefficients are used to generate an irregular grid and these grid points are used to solve the PDE using FDM [4]. A new method for adaptively arranging the grid size of FDM using multiwavelet coefficients is proposed here. We name the proposed scheme as 'Multiwavelet Optimized Finite Difference method' (MWOFD) and the methodology is described in this section.

Discretization scheme for the periodic initial-value problem – the nonlinear Schrödinger equation describing the propagation of a pulse envelope on an optical fiber is presented here.

$$\left.\begin{aligned} &u_t = \zeta u + i\gamma|u|^2 && u,t>0 \\ &u(x,0)=h(x) && \\ &u(x,t)=u(x+L,t), && t\geq 0 \end{aligned}\right\} x\in\mathbb{R} \tag{1}$$

where $h(x)=h(x+L)$, γ the magnitude of the nonlinear term which counteracts dispersion and $\zeta = -\frac{i}{2}\beta_2\frac{\partial^2}{\partial x^2}-\frac{\alpha}{2}$ where β_2 is the dispersion parameter and α the attenuation. Assuming u is a periodic function with period L, it is sufficient to consider the function in the range $[-L/2, L/2]$. Divide the range into N points, where $N=2^j$, j is an integer. Defining a grid consisting of points $x_l = \left(\frac{l}{N}-\frac{1}{2}\right)L,\ \ l=0,1,2,\ldots,N-1$, and the vector $u(t)$ such that

$$u_l(t) = u_J(x_l,t),\ \ l=0,1,2,\ldots,N-1 \tag{2}$$

where $u_J(x_l,t)$ is an approximate solution of Eqn. (1). Equation (1) can be written as the following initial problem[15] for u formulated with respect to the physical space;

$$\frac{d}{dt}u(t) = \zeta u(t) + N(u(t))u(t),\ \ t\geq 0 \tag{3}$$

and

$$u(0) = h \equiv [h(x_0, h(x_1),\ldots,h(x_{N-1})]^T, \tag{4}$$

where

$$\zeta = -\frac{i}{2}\beta_2\frac{D_p^{(2)}}{L^2}-\frac{\alpha}{2}I, \tag{5}$$

$$N(u(t)) = i\gamma diag(|u_l(t)|)^2,\ \ l=0,1,2,\ldots,N-1. \tag{6}$$

$D_p^{(2)}$ is the differentiation matrix [17] and I is the unit matrix of appropriate size. Equation (3) is approximated by Crank-Nicolson [2] method and is given by

$$\frac{u(t+\Delta t)-u(t)}{\Delta t} = \zeta\frac{u(t+\Delta t)+u(t)}{2} + N\left(\frac{u(t+\Delta t)+u(t)}{2}\right)\frac{u(t+\Delta t)+u(t)}{2}. \tag{7}$$

The time stepping procedure described in [18,19] gives

$$Au_{n+1} = Bu_n + \Delta t.N\left(\frac{u_{n+1}+u_n}{2}\right)\frac{u_{n+1}+u_n}{2}, \quad n = 0,1,2,\dots,n_1 - 1, \tag{8}$$

$$u_0 = h, \tag{9}$$

where

$A = I - \frac{\Delta t}{2}\zeta, B = I + \frac{\Delta t}{2}\zeta, u_n = u(n\Delta t)$ and $n_1 = \max\{t\}/\Delta t$. An iterative procedure improves the accuracy of the result obtained [15]. The iterative method is implemented by Algorithm 1 below.

Algorithm 1. Iteration of solution obtained by time stepping procedure

Step1: $u_{n+1}^{(0)} = u_n$

Step2: $Au_{n+1}^{(q+1)} = Bu_n + \Delta tN\left(\frac{u_{n+1}^{(q)}+u_n}{2}\right)\frac{u_{n+1}^{(q)}+u_n}{2}, \quad q = 0,1,2,\dots$

Step3: Step 2 is repeated until u_{n+1} converges.

3 Grid Generation Using Multiwavelets

Let the multiwavelet functions given by $\Psi(x) := [\psi_0(x), \psi_1(x), \psi_2(x), \dots, \psi_{r-1}(x)]^T$ have vanishing moments $[P_0(x), P_1(x), P_2(x), \cdots, P_{r-1}(x)]^T$. The multiwavelet coefficients $d_{i,j,k}$ – corresponding to the ith multiwavelet function $\psi_i(x)$ with a scale j and a shift k –decrease rapidly for a smooth function. If a function has a discontinuity in one of its derivative, then the wavelet coefficients will decrease slowly in the vicinity of that discontinuity and it decays rapidly in those regions where the function is smooth. The following Theorem 1 gives the decay of the multiwavelet coefficients.

Theorem 1. Let $P_i = D_i/2$ be the number of vanishing moments for the ith multiwavelet function in the jth scale [20] and let $f \in C^{P_i}(\mathbb{R})$. The multiwavelet coefficients of f decay as follows:

$$\left|d_{i,j,k}\right| \le \frac{1}{L^{P_i}} C_{P_i} \, r^{-j(P_i+\frac{1}{r})} \max_{\xi \in I_{i,j,k}} \left|u^{(P_i)}(\xi)\right|,$$

where r is the multiplicity of the multiwavelets, C_{P_i} is a constant independent of j, k and f, $I_{i,j,k}$ is the support of $\psi_{i,j,k}$ given by $I_{i,j,k} = \left[\frac{k}{r^j}, \frac{k+D_i-1}{r^j}\right]$

Proof. We modify the Theorem describing the decay of wavelet coefficients given in [15] to prove Theorem1. The modified proof is given below.

Expressing the function f around $x = \frac{k}{r^j}$ for $x \in I_{i,j,k}$;

$$f(x) = \left(\sum_{p=0}^{P_i-1} f^{(p)}(k/r^j) \frac{(x-k/r^j)^p}{p!} \right) + f^{(P_i)}(\xi) \frac{(x-k/r^j)P_i}{P_i!} \tag{10}$$

where $\xi \in \left[k/r^j, x \right]$. The multiwavelet coefficients $d_{i,j,k}$ can be expressed as

$$d_{i,j,k} = \int_{I_{i,j,k}} f(x) \psi_{i,j,k}(x) dx \tag{11}$$

$$= \left(\sum_{p=0}^{P_i-1} f^{(p)}(k/r^j) \frac{1}{p!} \int_{I_{i,j,k}} (x-k/r^j)^p \psi_{i,j,k}(x) dx \right) + \frac{1}{P_i!} \int_{I_{i,j,k}} f^{(P_i)}(\xi)(x-k/r^j) P_i \psi_{i,j,k}(x) dx. \tag{12}$$

We use $I_{i,j,k} = \left[\frac{k}{r^j}, \frac{k+D_i-1}{r^j} \right]$, $p = 0,1,\ldots,P_i-1$ and letting $y = r^j x - k$

$$\int_{k/r^j}^{(k+D_i-1)/r^j} \left(x - \frac{k}{r^j} \right)^p r^{j/r} \psi_i(r^j x - k) dx = 0,$$

by the property of vanishing moments [20], $p = 0,1,\ldots,P_i-1$.

Equation (12) reduces to

$$\left| d_{i,j,k} \right| = \frac{1}{P_i!} \left| \int_{I_{i,j,k}} f^{(P_i)}(\xi)(x - \frac{k}{r^j})^{P_i} r^{j/r} \psi_i(r^j x - k) dx \right|$$

$$\leq \frac{1}{P_i!} \max_{\xi \in I_{i,j,k}} \left| f^{(P_i)}(\xi) \right| \int_{I_{i,j,k}} \left| (x - \frac{k}{r^j})^{P_i} r^{j/r} \psi_i(r^j x - k) \right| dx$$

$$\left| d_{i,j,k} \right| \leq r^{-j(P_i + 1/r)} \frac{1}{P_i!} \max_{\xi \in I_{i,j,k}} \left| f^{(P_i)}(\xi) \right| \int_0^{D_i-1} \left| y^{P_i} \psi_i(y) \right| dy \tag{13}$$

$$\text{defining } C_{P_i} = \frac{1}{P_i!} \int_0^{D_i-1} \left| y^{P_i} \psi_i(y) \right| dy \text{ and } L^{P_i} = \frac{1}{r^{-j(P_i+1/r)}} \tag{14}$$

completes the proof of Theorem 1.

Theorem 1 concludes that if f can be expressed as a polynomial of degree less than P_i in the interval $I_{i,j,k}$, then $f^{(P_i)} \equiv 0$ and the corresponding wavelet coefficient $d_{i,j,k}$ is zero. If $f^{(P_i)} \neq 0$, then $d_{i,j,k}$ decay exponentially with respect to the scale j. If f has a discontinuity in a derivative of order less than or equal to P_i, then Theorem 1 does not hold for $D_i - 1$ values of $d_{i,j,k}$ near the discontinuity. Owing to the localization property of multiwavelets the coefficients away from the discontinuity are not affected. Smaller $d_{i,j,k}$ values imply that larger grid size can be adapted to those regions and a higher $d_{i,j,k}$ means finer grid size is required to represent the solution. We make use of this property to adaptively change the grid size in the proposed multiwavelet optimized finite difference method.

The elements of $u(t)$ approximate the function values $u(x_k,t)$, $k=0,1,2,...,N-1$. The error in the approximation reduces if a grid which is dense where $u(t)$ is highly varying and sparse where it is smooth is employed [15, 17]. The error at k/r^j depends on the size of the neighboring intervals. A large value of $|d_{i,j,k}|$ mean the grid spacing $1/r^j$ is too large to represent $u(t)$ in the interval $I_{i,j,k}$. Points with spacing $1/r^{j+1}$ is evenly added about position k/r^j over the entire interval $I_{i,j,k}$ when the value of $|d_{i,j,k}|$ is large. This reduces the error locally. The addition of new grid point over the entire interval $I_{i,j,k}$ is done due to the fact that large variation of $u(t)$ may happen anywhere within the interval $I_{i,j,k}$. The algorithm is realized by introducing an equidistant grid at an arbitrary level j. If the solution vector u is defined for a coarser grid, interpolation is used to compute the values on the fine grid. MWOFD grid is generated by choosing those grid points which correspond to multiwavelet coefficients above a threshold. Once the grid refinement is done, the finite difference equations are constructed as given in Eqn. (8) and a number of time steps are computed. We propose a method for the grid refinement as Algorithm 2.

Algorithm 2. Multiwavelet based grid refinement

Step1: Interpolate u from current coarse grid to fine equidistant grid at scale j with grid size $1/r^j$ with u_0 is defined in Eqn. (4).

Step2: Modify grids to equidistant grids of size $1/r^j$.

Step3: Do the multiwavelet transform of u to obtain the multiwavelet expansion coefficients $d_{i,j,k}$.

Step 4: Insert new grid points evenly in the interval $I_{i,j,k}$ for values of k where $|d_{i,j,k}| > \varepsilon$, where ε is the value of threshold.

Step 5: Construct A and B as given in Eqn. (8) to obtain u_{n+1}.

Step 6: Perform time stepping procedure in Algorithm 1 till the solution converges.

4 Results

In this section we present the results obtained when the proposed MWOFD scheme is applied in nonlinear Schrödinger equation and Burgers' equation. These equations are well known for their mixed elliptic, parabolic, and hyperbolic behavior [21]; they are widely used as test problems for numerical computation.

Nonlinear Schrödinger Equation

MWOFD method described in sections 2 and 3 is used to solve Eqn. (1) at $t = 0.5$ Sec. The parameters employed are

$$\beta_2 = -2,\ \gamma = 2,\ \alpha = 0,\ L = 64,\ h(x) = 2sech(x),\ N = 1024,$$
$$\lambda = 10, \varepsilon = 10^{-4}, p = 5,\ \Delta t = 1/2^{10} \text{ and } n_1 = 512,$$

where N = total number of grid points employed, λ the transform depth, *tol* = tolerance of multiwavelet coefficients, ε = tolerance in the nonlinear term. The MWOFD is applied with Daubechies balanced multiwavelets of order 4 and multiplicity 2.

The method starts with equal grid points in the range of interest of x and proceeds as follows:

1. Obtain the FDM solution u with equally spaced grid points.
2. Iterate with the solution u as specified in Algorithm 1 and obtain the converged solution.
3. Compute the multiwavelet coefficients for the selected number of levels – λ.
4. Prepare a histogram of the multiwavelet coefficients.
5. In the histogram, identify the number – υ – and locations of the coefficients whose magnitudes exceed the threshold. Reduce the grid size at these locations by a factor of 2.
6. In the histogram, again identify the locations of the υ coefficients having the lowest magnitudes. Increase the grid size at these locations by a factor of 2. This along with step 5 above ensures that the total number of grid points remain unaltered.
7. Compute the solution u with the new set of grid points.
8. Repeat steps 2 to 7 iteratively until convergence – say at the nth iteration. Here convergence can be any one of two ways.
   - The grid does not get readjusted as one proceeds to the $(n + 1)^{th}$ iteration. In turn the solution in the $(n + 1)^{th}$ iteration remains identical to that at the nth iteration.
   - The maximum difference between the solution in the nth iteration and that in the $(n + 1)^{th}$ iteration is less than the error threshold.

The solution of Eqn. (1) is shown in Fig. 1, the benefit of using closer grid points in the region of turbulence is clearly brought out here – with the help of an enlarged scale. The solution for the system – with the same parameters and Db 6 wavelets – has been obtained using WOFD method described in [15]; Fig. 2 shows the error introduced by MWOFD and WOFD methods and their contrast. A close examination of the results brings out the following:

1. The error with WOFD is of the order of 10^{-4}. In contrast, MWOFD gives a solution whose error is of the order of 10^{-9} or better in the whole region. The computational efforts in both the cases are almost the same (as brought in Table 1). To bring out the contrast between the two methods, their errors are shown together in Fig. 2. The improvement brought about by MWOFD can be seen to be remarkable.

2. Figure 2 shows the error in the turbulence region in an expanded scale for the solution with MWOFD. The error can be seen to be less than 10^{-11}; In contrast the error with WOFD remains above 10^{-5} throughout the turbulence region. The fact that the solution in the turbulence region is of more interest, makes the MWOFD method all the more attractive.

In the proposed MWOFD scheme the boundaries are treated as a discontinuity. The method readjusts the grid size near the boundaries adaptively. This avoids the need for further processing at the boundaries. In case of the nonlinear Schrödinger equation, there is no discontinuity near the boundary. It is clear from the Fig. 2 that the error introduced at the boundaries are negligible compared to that at the interior points. In summary, MWOFD method takes care of boundary error during the grid adaption stages.

Table 1. Computation time† and maximum error in the solution of nonlinear Schrödinger equation in MWOFD and WOFD methods

method	tolerance	epsilon	Db 4/ Daub bal-4		Db 6/ Daub bal-6	
			max.error	comp.time	max.error	comp.time
MWOFD	0.1	1.00E-15	1.06E-05	21.05	0.03406	5.75
WOFD	0.1	1.00E-15	0.1543	16.55	4.63E-01	19.39
MWOFD	1.00E-03	1.00E-10	9.31E-08	17.31	9.14E-04	6.203125
WOFD	1.00E-03	1.00E-10	4.61E-04	19.05	2.10E-03	15.30
MWOFD	1.00E-05	1.00E-15	7.28E-10	24.20	1.43E-06	7.84375
WOFD	1.00E-05	1.00E-15	4.61E-08	26.58	4.78E-06	23.06
MWOFD	1.00E-06	1.00E-15	9.43E-11	25.25	2.47E-08	9.296875
WOFD	1.00E-06	1.00E-15	No convergence		3.46E-07	22.00
MWOFD	1.00E-12	1.00E-15	8.82E-16	30.00	2.77E-14	1.48E+01
WOFD	1.00E-12	1.00E-15	2.65E-13	34.78	4.19E-13	26.28

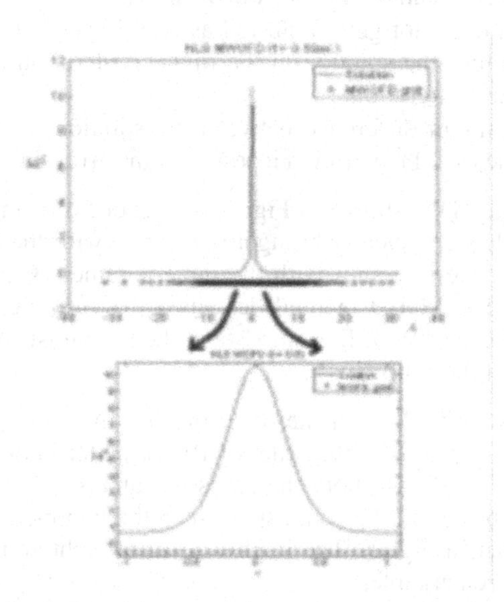

Fig. 1. Solution of nonlinear Schrödinger equation at time $t = 0.5$ *Sec.* using MWOFD – showing the turbulence region in an enlarged scale

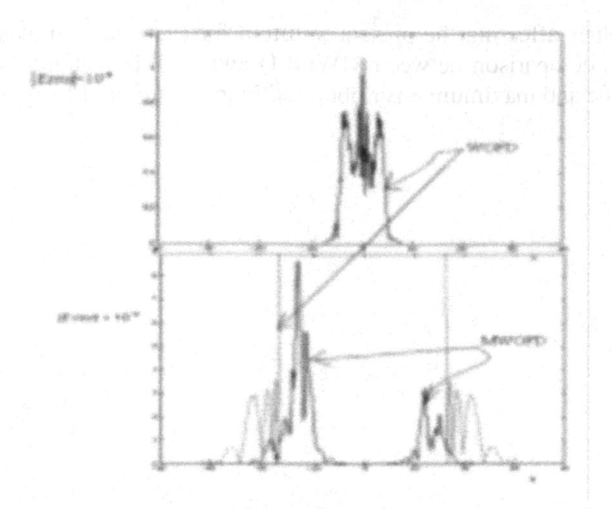

Fig. 2. Error introduced by MWOFD and WOFD methods and their contrast

The Burgers' equation is given by

$$\left.\begin{aligned} &u_t = \nu u_{xx} - (u+\rho)u_x, \quad t>0 \\ &u(x,0) = h(x) \\ &u(x,t) = u(x+1,t), \quad t \geq 0 \end{aligned}\right\} x \in \mathbb{R} \tag{15}$$

where ν is a positive constant, $\rho \in \mathbb{R}$, and $h(x) = h(x+1)$. Burgers' equation with $\rho = 0$ describes the evolution of u under linear dissipation and nonlinear advection, $\rho \neq 0$ add linear advection to the system. Discretizing Eqn. (15) in the same manner as in Section 2, we obtain the system

$$\frac{d}{dt}u(t) = \zeta u(t) + N(u(t))u(t), \quad \mathrm{t} \geq 0 \tag{16}$$

and

$$u(0) = h \equiv [h(x_0, h(x_1), ..., h(x_{N-1})]^T,$$

where

$$\zeta = \nu D^{(2)} - \rho I \text{ and } N(u(t)) = -diag(D(u(t)).$$

The operator D=$D^{(1)}$ and $D^{(2)}$ have the same meaning as given in Eqn. (6). LU factorization method [1] available in MATLAB® is used to solve $N(u(t))$. Now the problem can be solved in a similar manner as Schrödinger equation described in section 2. The solution obtained is shown in Fig. 3. The parameters are ν=0.002, ρ=0, $h(x) = \sin(2\pi x)$, $N=2^{11}$, J=11, λ=J, $\varepsilon=10^{-6}$, p=3, $\Delta t=1/2^9$ and n_1=256. λ is the transform depth (number of level of wavelet or multiwavelet transform), p=number of adjacent values affecting the present solution for FDM known as stencil size[1]. Table 2 gives a comparison between MWOFD and WOFD methods in terms of the computation time and maximum error obtained in the solution of Burgers' equation.

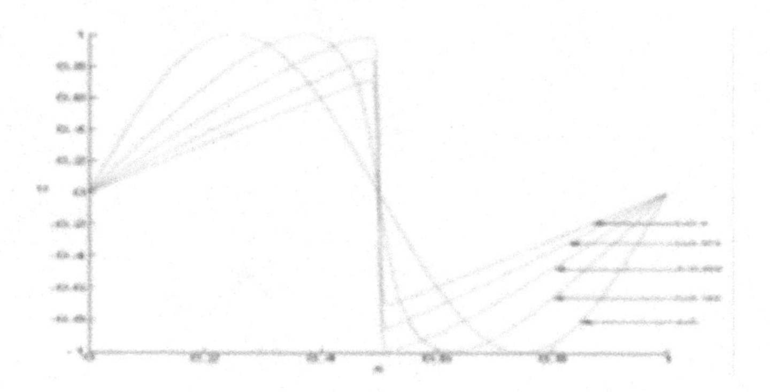

Fig. 3. Solution of Burgers' equation using MWOFD method at various time steps

Table 2. Computation time and maximum error obtained in the solution of Burgers' equation in MWOFD and WOFD methods

Method	epsilon	Eval.time	Soln.time	grd.time	Ref.time	max.error	comp.time
MWOFD	0.0010	5.17	0.61	0.1875	0.66	0	6.44
WOFD	0.0010	4.97	0.15	0.0469	0.77	0.02	5.86
MWOFD	0.0100	6.73	0.84	0.1875	0.64	5.60E-19	8.80
WOFD	0.0100	3.06	0.05	0.0313	0.78	4.99E-02	3.91
MWOFD	0.06	8.09	0.52	0.1719	0.75	4.62E-09	9.36
WOFD	0.06	3.97	0.05	0.0313	0.72	0.7843	4.73
MWOFD	0.1000	8.05	0.66	0.1563	0.70	3.24E-06	9.41
WOFD	0.1000	Converging up to 0.3125sec. only, error at 0.3125sec. = 0.7884					

All the parameters have the same meanings as given in Table1.

Table 3. Computation time and maximum error in the solution of nonlinear Schrödinger equation with MWOFD method using different Multiwavelets

Multiwavelets	max.error	comp.time
Daub bal-4	7.283E-10	24.20
Chui-Lian2	6.244E-10	19.59
Chui-Lian3	6.144E-10	12.81
Hermite-Cubic	3.721E-9	15.50
DGHM	7.782E-10	17.43
BAT 01	2.798E-7	14.97

Tables 1 and 2 summarize the results of computations carried out by both the methods – MWOFD and WOFD – for Nonlinear Schrödinger equation and Burgers' equation respectively. By a comparison of the methods, the following observations are in order here:

1. In the case of nonlinear Schrödinger equation for a given tolerance in the nonlinear part and wavelet coefficients, MWOFD method provides considerable improvement in terms of accuracy of results as compared with WOFD method.
2. For a given error threshold, computational time required for MWOFD method is marginally higher than that for the WOFD method. Note that the increase in computational time is essentially due to the computation of multiwavelet transform.
3. Consider first two rows of Table 2 which show the results of both the methods for the same value of threshold (threshold for transform coefficients) namely 0.001. The error with MWOFD method is substantially lower than that with WOFD, the computation time being practically the same for both. A similar and consistent difference is observed in all the cases in Table 1 as well as Table 2.
4. As the value of threshold increases, the computation time being comparable, the error values increase in both cases. However in all the cases, the error with MWOFD is conspicuously lower than that with WOFD (in fact in one of the cases the solution with WOFD does not converge at all).

Table 3 presents the solution of nonlinear Schrödinger equation with commonly used multiwavelet bases. Table 3 clearly brings out the consistency of the MWOFD method in terms of accuracy and computation time.

5 Conclusion

The results obtained imply that for a given pair of thresholds for transform coefficients and for acceptable error, one need to use only less number of grid points with MWOFD than with WOFD. In turn, the complexity of the computation scheme – computation time, number of iterations, FDM time *etc.* – reduces drastically with MWOFD. As the acceptable error in the solution is reduced, MWOFD converges faster compared to WOFD in all cases. The benefits with the use of MWOFD are conspicuous in terms of accuracy and computational effort. Improvement in accuracy in the turbulent regions of the solution is substantial. All these go to show that MWOFD is a powerful method of solving different types of PDEs.

References

1. Booton, R.C.: Computational methods for Electromagnetics and Microwaves. Wiley series in Microwave and Optical Engineering. Wiley-IEEE Press (1992)
2. Smith, G.D.: Numerical Solution of Partial Differential equations: Finite Difference Methods. Oxford Univ. Press, Oxford (1985)
3. Ruppa, K.T., Parimal, T., Mohammad, R.R.: Wavelet optimized finite- difference Approach to solve jump-diffusion type partial differential equation for option pricing. Computing in Economics and Finance 2005 series, Soc. for Computational Economics, vol. 471 (2005)

4. Kumar, V., Mehra, M.: Wavelet optimized finite difference method using interpolating wavelets for solving singularly perturbed problems. Journal of Wavelet Theory and Applications 1(1), 83–96 (2007)
5. Karami, A., Karimi, H.R., Moshiri, B., Maralani, P.J.: Investigation of Interpolating Wavelets Theory for the Resolution of PDEs. Int. J. Contemp. Math. Sci. 3(21), 1017–1029 (2008)
6. Beylkin, G., Keiser, J.M.: On the adaptive numerical solution of nonlinear partial differential equations in the wavelet bases. J. Comp. Physics 132, 233–259 (1997)
7. Alpert, B., Beylkin, G., Gines, D., Vozovoi, L.: Adaptive solution of partial differential equations in multiwavelet bases. J. Comput. Phys. 182 (2002)
8. Fann, G., Beylkin, G., Harrison, R.J., Jordan, K.E.: Singular operators in multiwavelet bases. IBM J. Res. and Dev. 48(2) (2004)
9. George, W.P.: Wavelets in Electromagnetics and Device Modeling. Wiley series in Microwave and Optical Engineering. Wiley-IEEE Press (2003)
10. Beylkin, G., Coifman, R.R., Rokhlin, V.: Fast wavelet transform and numerical algorithms. J. Comm. Pure Appl. Math. 44, 141–183 (1991)
11. Wang, G., Pan, G.: Full wave analysis of microstrip floating line structure by wavelet expansion method. IEEE Trans. Microwave Theory Tech. 43, 131–142 (1995)
12. Steinberg, B., Leviatan, Y.: On the use of wavelet expansions in the method of moments. IEEE Trans. Ant. Propg. 41(5), 610–619 (1993)
13. Alpert, B.: A class of bases in L2 for the sparse representation of integral operators. SIAM J. Appl. Math. Anal. 24(1), 246–262 (1993)
14. Jameson, L.: On the wavelet-optimized finite difference method. Technical report, NASA CR-191601, ICASE Report no. 94-9 (1994)
15. Ole, M., Nielson: Wavelets in scientific computing. Ph.D. Thesis, Department of Mathematical modeling, Technical University of Denmark (1998)
16. Capdeboscq, Y., Vogelius, M.S.: Wavelet based homogenization of a 2 dimensional elliptic problem. Hal-00021467, Version 1, Mardi 21 Mars (2006)
17. Jameson, L.: On the Daubechies-based wavelet differentiation matrix. J. Sci. Comp. 8(3), 267–305 (1993)
18. Charton, P., Perrier, V.: A pseudo-wavelet scheme for the two dimensional Navier-Stokes equations. Mathematica Aplicada e Computational 15(2), 139–160 (1996)
19. Dorobantu, M.: Wavelet-based algorithms for fast PDE solvers. Ph.D. Thesis, Royal Institute of Technology, Stockholm (1995)
20. Keinert, F.: Wavelets and Multiwavelets. Chapman & Hall/CRC Press, Boca Raton (2003)
21. Chung, T.J.: Computational fluid dynamics. Cambridge University press, Cambridge (2002)

A Novel Genetic Algorithm Approach to Mobility Prediction in Wireless Networks

C. Mala[1], Mohanraj Loganathan[2], N.P. Gopalan[3], and B. SivaSelvan[4]

[1] National Institute of Technology, Tiruchirappalli, TN-620015
mala@nitt.edu
[2] Department of Computer Science & Engineering
National Institute of Technology, Tiruchirappalli, TN-620015
raj_mohan30@yahoo.com
[3] Department of Computer Applications
National Institute of Technology, Tiruchirapalli, TN-620015
gopalan@nitt.edu
[4] IIITDM Kancheepuram, IIT Madras Campus, Chennai-600036
sivaselvanb@iiitdm.ac.in

Abstract. Wireless networks are required to support nomadic computing, providing seamless mobility without call drops. The number of mobile users must be known apriori for allocating the channel bandwidth, which is a function of the number of mobile users already admitted into the system and direction of their movement. If the user's next movement can be predicted, it may be used to allocate a channel in the neighboring cell, thereby reducing the call dropping rate and leading to seamless mobility. Most of the mobility prediction algorithms offer reduced performance as a result of being movement history based and also do not give accurate results when users move towards the corner of the hexagonal cell structure but for the sectorized ones. Substantial improvement is demonstrated with the use of the proposed novel genetic algorithm based approach presented in the paper. Simulation results show that the proposed algorithm gives accurate results independent of the user's movement pattern.

Keywords: Seamless mobility, Mobility prediction, Channel allocation, Genetic algorithm, Call dropping rate, HHO region.

1 Introduction

In this modern age of global connectivity, it is required to support the increasingly nomadic lifestyle of people. Wireless networks allow a more flexible communication model than traditional networks as the users are not limited to a fixed physical location. They are more sensitive to call dropping when they move from one base station's coverage area to another. Traditional channel allocation schemes are focused on the tradeoff between new call blocking and handoff call blocking probabilities[1,2]. In order to efficiently utilize the network resources and to provide seamless mobility, the base station should be aware of the incoming call arrival rate. Usually, the hand off calls are given higher priority than

S. Ranka et al. (Eds.): IC3 2009, CCIS 40, pp. 49–57, 2009.
© Springer-Verlag Berlin Heidelberg 2009

new calls to support continuity of service. Resources may be reserved if the base station knows in advance the number of users arriving. If the prediction is incorrect the reserved resources will be underutilized. An approach using Genetic Algorithm (GA) optimization technique to acquire the knowledge of the next hand off movement of the user is presented in this paper. It has the following characteristics:

1. Accuracy in prediction avoiding the overheads incurred by traditional prediction algorithms.
2. Elimination of a history base.

The rest of the paper is organized as follows: Section 2 gives an overview of existing schemes. The proposed genetic algorithm based mobility prediction scheme is discussed in Section 3. Simulation results and performance analysis of the proposed algorithm is presented in Section 4. Finally Section 5 concludes with a summary of the work done and results achieved.

2 Mobility Prediction Techniques

A number of user mobility prediction algorithms are proposed in [1,3,4,5,6]. Most of these algorithms make use of a history base that has a record of the previous movements of users. These algorithms make use of the probability of movements along with the direction of motion and velocity and produce inaccurate results. In the sectorized mobility prediction algorithm [3], for a user moving in a regular path, mobility history base details are used for prediction. Even a slight change in user movement is not tolerated in such schemes. For random user movements, a cell sector scheme is employed and both these approaches have the overhead of maintaining the history bases.

The mobility prediction problem addressed to the sectorized cell structure is shown in Figure 1. The base station area is divided into six equal sectors. A mobile node is represented as a black dot and is in the Cell0 Sec6. As per

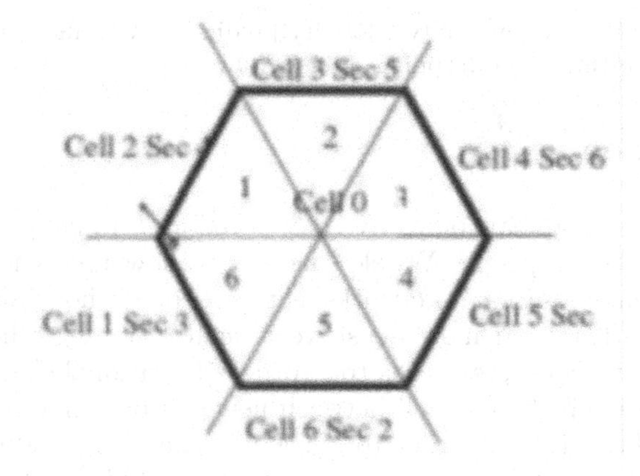

Fig. 1. Sectorized Cell Structure

the sectorized mobility prediction algorithm the next hand off will be to Cell2 Sec4. But there is a possibility that the user might move towards Cell1 Sec3 or Cell0 Sec1. So the sectorized mobility prediction algorithm may not give accurate results for all the movements of the user, especially when the user moves towards the corners. The faster convergence of genetic algorithm has been advantageously utilized in this paper for relatively accurate prediction.

3 Proposed Approach

The algorithm in [3] considers the sector as a single entity, which leads to inaccurate prediction of next hand-off movement of the user. The scheme of splitting a sector into HHO (HHO) and Low handoff (LHO) regions is proposed in this paper to overcome this drawback to a certain extent. The HHO region is defined as the 100 angled areas on either side of the sector boundary in the same cell and the area covered by 0.1r, where r is the radius of the cell, on either side of the sector boundary with the adjoining cell. The Received Signal Strength(RSS) of the mobile unit will be low in this region. The remaining area in the sector is defined as the LHO and is shown in Figure 2. RSS will be high in this region. Splitting the sector into two regions allows the use of two different prediction algorithms effectively. The conventional sectorized mobility prediction algorithm [3] is employed in LHO while a new GA based approach is used in the HHO regions.

3.1 GA Approach to Predict the Next Movement of the User

Genetic Algorithm's for mobility prediction make use of an objective function of three variables to compute the possibility of a user's handoff to any neighboring sector as given by the following expression

$$g = \alpha \times NeighborSector + \beta \times Speed + \gamma \times Traffic \quad (1)$$

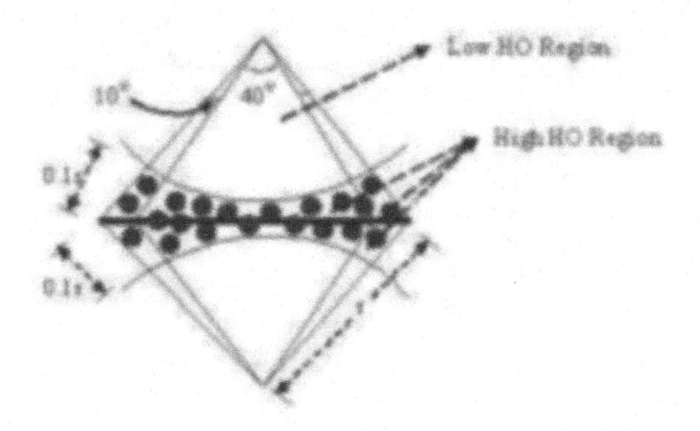

Fig. 2. Cell Handoff Regions

where,
α is the probability of a user being in a sector,
NeighborSector is one among the six sectors to which the user may handoff
β is the probability of a user getting into a sector with a specific speed, and
γ is the probability of a user moving into a nearby sector during a specific traffic.

The fitness function used in this GA approach is a minimization function, which is the inverse of the objective function given in (1).The data flow diagram for the GA approach is given in Figure 3.

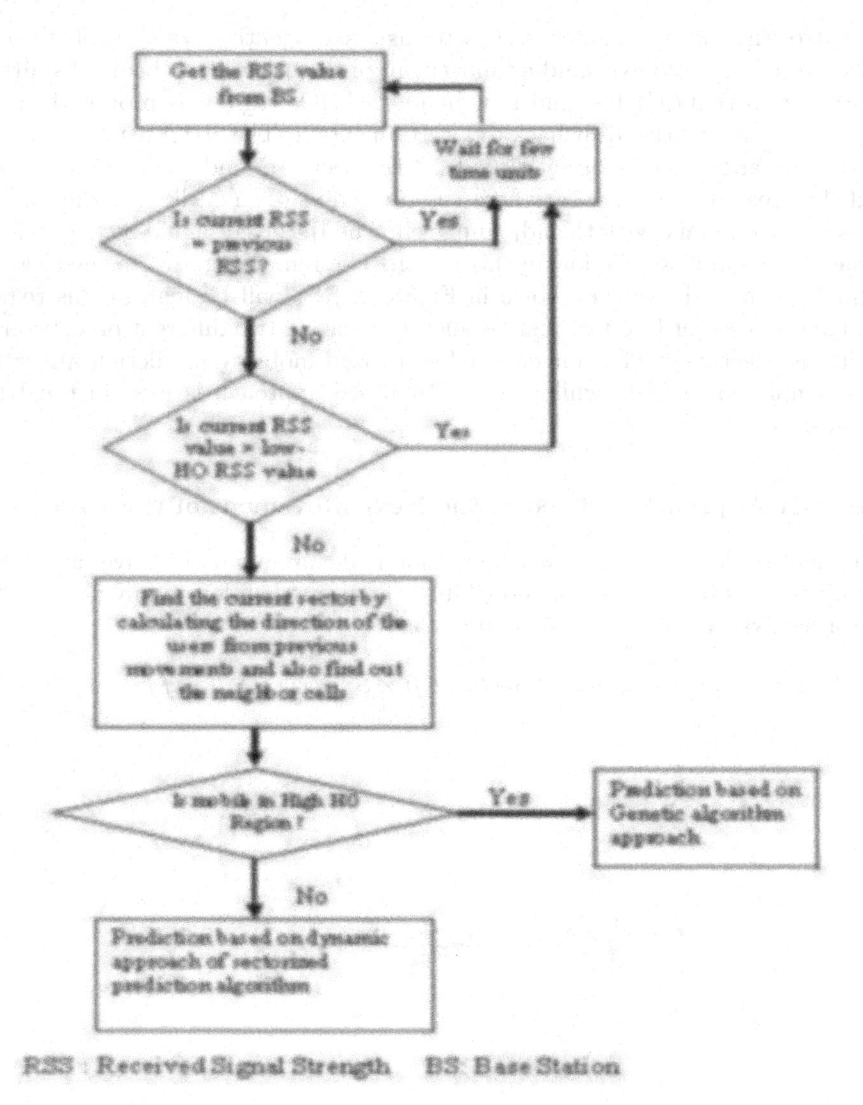

Fig. 3. Data Flow Diagram for the Genetic Algorithm Approach

3.2 GA Methodology

Genetic algorithm [7] is a computerized search and optimization algorithm based on the mechanics of natural genetics and selection. The steps involved in it are:

1. Choose a coding scheme to represent problem parameters, selection operators, a cross over operator; and a mutation operator. Choose population size n, crossover probability p_c and mutation probability p_m. Initialize a random population of strings of size l. Choose a maximum allowable generation number t_{max}. Set t=0.
2. Evaluate each string in the population.
3. If $t > t_{max}$ or if other termination criteria are satisfied, terminate.
4. Perform reproduction on the population.
5. Perform crossover on a pair of strings with probability p_c.
6. Perform mutation on strings with a probability p_m.
7. Evaluate the strings in the new population. Set t = t + 1 and go to step 3.

There are two kinds of optimization techniques, viz., maximization and minimization. In minimization[8] the objective function, is defined as one whose maximum corresponds to the optimal solution. Maximization is inverse of the minimization. In this paper, the objective function selected is a minimization function.

$$GA \quad Minimization \quad Fitness \quad Function \quad = \frac{1}{ObjectiveFunction} \tag{2}$$

The Fitness function used is given in Equation 3.

$$f(x) = \frac{1}{g(x)} \tag{3}$$

The parameters used by the GA are as shown in Table 1

Table 1. GA Parameters Used

Parameter	Value
Population size	30
Length of chromosome	20
Selection operator	Rank order
Crossover Operator	Single point operator
Crossover probability	0.8
Mutation Probability	0.01
Fitness Parameter	probability of handoff

3.3 GA Based Mobility Prediction Algorithm

1. Choose a coding scheme to represent problem parameters, selection operator, and a crossover operator. Each problem parameter is represented in a binary string format. The total length of the string is 20 in which first 10 bits are

used for speed representation and next 10 bits represent the traffic variable. The speed and traffic are represented as sub strings in the chromosome. The strings (0000000000) and (1111111111) represent the lower and upper limits of speed and traffic. Rank order selection operator is used for selection and single point crossover is selected as crossover operator.
2. Choose population size n=30, crossover probability $p_c = 0.8$ and mutation probability p_m =0.01. The mutation probability should be very low when compared to crossover probability.
3. Initialize a random population of strings of size l=32.
4. Choose a maximum allowable generation number $t_{max} = 8$. Set t=0.
5. Evaluate each string in the population ie, a fitness value is computed for each string in the population by equation (3).
6. If $t > t_{max}$ terminate.
7. Perform reproduction on the population.
8. Perform crossover on a pair of strings with the crossover probability p_c.
9. Perform mutation on each string with the mutation probability p_m.
10. Evaluate each string in the new population using Equation 3. The new population obtained is considered to be the parents for the next generation. Set t = t + 1 and go to step 6.

The maximum probability of handoff is selected from the last iteration from the above steps. Similarly, the maximum probability of handoff in the other two cells is also selected. Out of these three probability values, the cell with the maximum probability value is selected as the predicted cell completing the procedure.

4 Simulation Results and Performance Analysis

To analyze the performance of the proposed algorithm a simulation tool using Java AWT has been developed with the following constraints and parameters:

1. Each cell is assigned 10 channels.
2. Each call requires only one channel for service.
3. For both new calls and handoff calls, the call duration times are exponentially distributed with mean 240 seconds and the cell dwell times are also exponentially distributed with mean 120 seconds.
4. Mobile movement patterns are selected accordingly to check the prediction accuracy with respect to, with and without using GA mobility prediction approach.
5. The accuracy is calculated using the formula given in Equation 4.

$$Accuracy = \frac{Number \quad of \quad Handoffs}{Number \quad of \quad Predictions + Misses} \tag{4}$$

4.1 Effect of GA Based Mobility Prediction on Call Dropping

Figure 4 shows the performance of Mobility Prediction (MP) using sectorized mobility prediction algorithm, with Mobility prediction algorithm using GA approach and also without Mobility Prediction. In Figure 4, it is seen that the call dropping probability increases with increase in traffic. Also, it can be seen that for a call arrival rate of 60 calls per unit time, the call dropping percentage is 12, 24, and 38 in the algorithms without MP, MP without GA, and MP with GA respectively. So, it is inferred from the graph that the call dropping is more when there is no mobility prediction. The sectorized mobility prediction reduces the call-dropping rate by splitting the sector into three regions based on the closeness to the base station. GA based mobility prediction further reduces the call-dropping rate as it splits the sector into two regions, viz., High HO region and Low HO region.

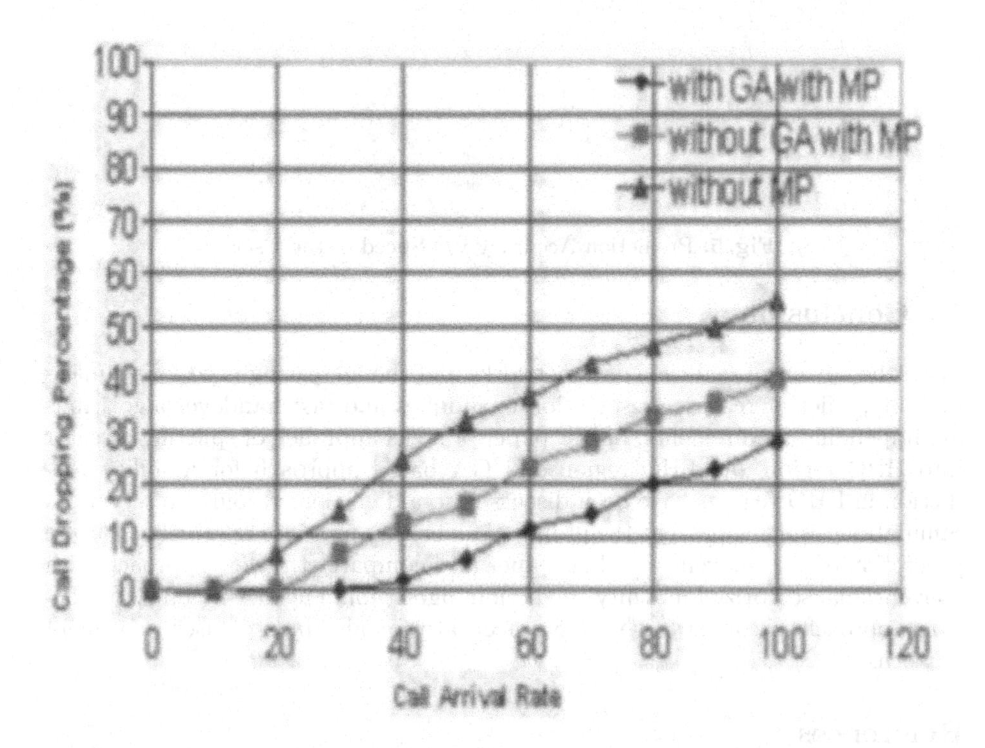

Fig. 4. Call Arrival Rate v/s Call Dropping Percentage

4.2 Accuracy of Prediction

Prediction Accuracy depends on the speed of the user and the tracking area which is the HHO region of the Base Station. From Figure 5 it can be seen that as the speed of the user increases, the mobility prediction accuracy decreases. For the speed of the user beyond 120 kilometers, the accuracy of sectorized mobility prediction algorithm approaches zero as a search in the history base needs to be done and this may not get completed within a short time interval.

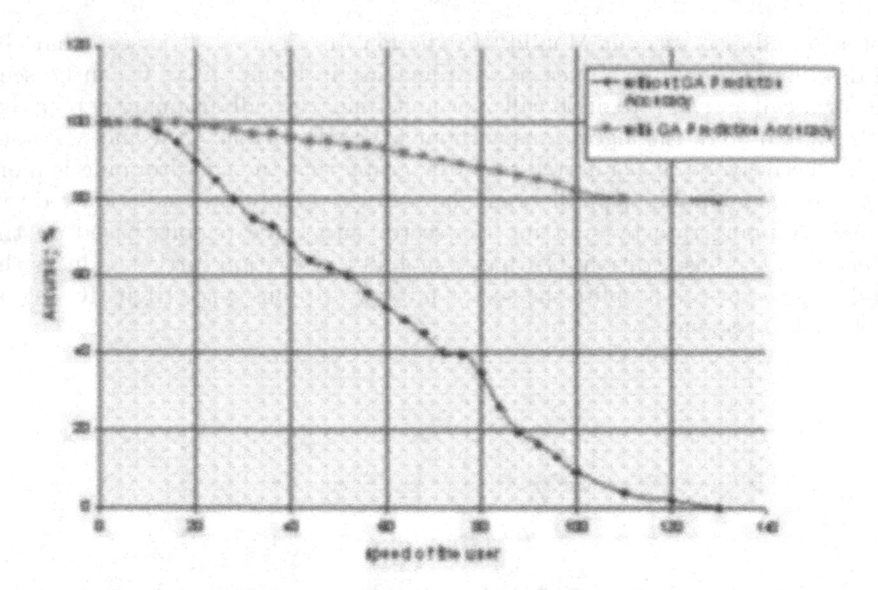

Fig. 5. Prediction Accuracy v/s Speed of the User

5 Conclusion

With the advent of real-time data networks and the increasing need for seamless mobility, efficient resource reservation techniques and fast handover algorithms are impending requirements. In this paper, a novel approach of splitting a sector into HHO region and LHO region and GA based approach for mobility prediction in HHO regions have been discussed to aid efficient resource reservation. Simulation results show that the proposed algorithm provides relatively accurate prediction of user movements. This paper also compares the new approach with conventional sectorized mobility prediction algorithm. The GA mobility prediction approach is comparatively more accurate for all kinds of user movement patterns.

References

1. Chan, J., Seneviratne, A.: A practical user mobility prediction algorithm for supporting adaptive qos in wireless networks. In: IEEE International Conference ICON 1999, September 1999, pp. 104–111 (1999)
2. Yang, J., Jiang, Q., Manivannan, D.: A fault-tolerant distributed channel allocation scheme for cellular networks. IEEE Transactions on Computers 54(5), 616–629 (2005)
3. Chellappa, R., Jennings, A., Shenoy, N.: The sectorized mobility prediction algorithm for wireless networks. In: International Conference on Information and Communication Technologies (ICT 2003) (April 2003)
4. Liu, T., Bahl, P., Chlamtac, I.: A hierarchical position prediction algorithm for efficient management of resources in cellular networks. In: Global Telecommunication conference, GLOBECOM 1997, November 1997, pp. 982–986 (1997)

5. Liu, G., Maguire Jr., G.: A class of mobile motion prediction algorithms for wireless mobile computing and communications. Mobile Networks and Applications 1(2), 113–121 (1996)
6. Pathirana, P., Savkin, A., Jha, S.: Mobility modeling and trajectory prediction for cellular networks with mobile base station. In: Mobihoc 2003, Maryland, USA (June 2003)
7. Goldberg, D.E.: In: Genetic Algorithms in Search, Optimization, and Machine Learning. Prentice-Hall, Englewood Cliffs (2006)
8. http://rkb.home.cern.ch/rkb/AN16pp/node1w73.html#172
9. Dasilva, L.A.: Pricing for qos-enabled networks: a survey. IEEE Communication Surveys and Tutorials 3(2), 2–8 (2000)
10. Das, S.K., Sen, S.K.: New location update strategy for cellular networks and its implementation using a genetic algorithm. In: MOBICOM 1997, Budapest (1997)

A Beaconless Minimum Interference Based Routing Protocol for Mobile Ad Hoc Networks

Natarajan Meghanathan and Meena Sugumar

Jackson State University
Jackson, MS 39217, USA
{nmeghanathan,meena.sugumar}@jsums.edu

Abstract. We propose a novel on-demand minimum interference based routing protocol (MIF) that minimizes the end-to-end delay per data packet for mobile ad hoc networks. MIF does not require periodic exchange of beacons in the neighborhood. During the broadcast of the Route Request (RREQ) messages, each node inserts its identification and location information. The interference index of a link is the number of interfering links surrounding it. Two links are said to interfere with each other, if the distance between the mid points of the two links is within the interference range. The interference index of a path is the sum of interference index values of the constituent links. The destination uses the RREQs to locally construct a weighted graph of the network topology and selects the path with the minimum interference index value. In addition to end-to-end delay, MIF provides higher route stability and packet delivery and lower energy consumption.

Keywords: Interference, Ad hoc Networks, Routing Protocols, Simulation.

1 Introduction

A mobile ad hoc network (MANET) is a dynamic distributed system of wireless nodes that move independent of each other. The nodes operate with a limited battery charge and as a result have a limited transmission range. Routes in MANETs are often multi-hop in nature; thus each node is capable of serving both as a forwarding node as well as a source/destination of a data communication session. MANET routing protocols are of two types: proactive vs. reactive. The proactive routing protocols predetermine routes for every possible source-destination pair irrespective of the requirement. The reactive routing protocols determine a route from a source to destination only when required. In dynamically changing mobile environments, reactive on-demand routing incurs significantly less overhead than proactive routing. Hence, we restrict ourselves to reactive routing protocols for the rest of this paper.

Interference between the radio signals significantly influences the throughput of wireless ad hoc networks. With multi-hop routing so common in MANETs, interference-aware routing is essential for these networks. The strength of the signal received at a node is the sum of the strength of the attenuated signals transmitted from nodes that are within the neighborhood of the receiver node. Two signals are said to

S. Ranka et al. (Eds.): IC3 2009, CCIS 40, pp. 58–69, 2009.
© Springer-Verlag Berlin Heidelberg 2009

interfere with each other, if the sum of their Signal-to-Noise Interference Ratios, *SINRs*, is appreciably different from their individual *SINRs*. Thus, signals of two nodes that are not within the transmission range of each other can still interfere with each other. In [1], interference between neighboring nodes has been modeled using a conflict graph which indicates the group of links that mutually interfere and cannot be active simultaneously. Performance studies suggest that network throughput can be significantly improved by modeling the network as a conflict graph and by employing an interference-aware routing protocol that uses conflict graphs.

In this paper, we propose an on-demand routing protocol called Minimum Interference (MIF) based routing protocol that makes use of the conflict-graph modeling for ad hoc networks. According to this model, two links are said to interfere (or conflict) with each other, if the distance between the mid-points of these links is less than or equal to a parameter called the *Interference Range per Link* (IRL). MIF works as follows: When the source does not have a path to the destination node, the source broadcasts a Route Request (RREQ) message throughout the network. Each intermediate node, upon receiving the RREQ message for the first time, will include its location and identification information in the RREQ message before broadcasting the message in its neighborhood. The destination receives several RREQ messages, each of them having traversed over a different path. The destination constructs a weighted graph, called the interference graph, in which the vertices of the graph represents the nodes in the network and there is an edge (i.e., link) between any two vertices, if the Euclidean distance between the two vertices is less than or equal to the transmission range. The weight of each link (called the interference index) in the graph is the number of interfering links within the interference range of the link. The interference index of a path is the sum of the interference index values of the constituent links of the path. The destination runs the minimum weight path Dijkstra algorithm [2] on the interference graph and determines the minimum weight path. A Route Reply (RREP) packet is sent to the source along the minimum weight path determined. The source begins to transmit the data packets on the path learnt. MIF does not require the periodic exchange of beacons in a local one-hop neighborhood.

We compare the performance of MIF with that of the load balancing routing (LBAR) protocol [3] that balances the load in the network using the activity level at each node in the network. In LBAR, a node learns the number of source-destination (*s-d*) sessions each of its neighbors is part of through periodic beacons received from the neighbors. The traffic interference metric for a node in LBAR is recorded in the RREQ messages and it is the sum of the number of active *s-d* sessions in which the node and its neighbors are part of. The traffic interference metric for a path is the sum of the traffic interference metric values of the constituent nodes, except the source and destination, of the path. The destination chooses the path that has the smallest traffic interference metric value. Simulation results in [3] show that LBAR incurs significantly lower delay per packet when compared to the well-known minimum-hop based Dynamic Source Routing (DSR) [4] and the Ad hoc On-demand Distance Vector (AODV) [5] routing protocols.

The rest of the paper is organized as follows: Section 2 presents our design of the MIF protocol along with details of the packet headers and the different phases of the protocol. Section 3 presents the simulation environment and the simulation results comparing MIF with LBAR. Section 4 concludes the paper.

2 Minimum Interference Based Routing Protocol

The objective of the MIF protocol is to choose routes whose constituent links suffer the minimum interference. We assume each node is aware of its current location in the network using techniques like Global Positioning System (GPS) and etc. The routing protocol does not require periodic beacon exchange among the nodes in a neighborhood, which has been an essential requirement for many of the other interference-aware routing protocols, including LBAR. The following sections describe in detail, the working of the MIF routing protocol.

2.1 Route Request-Reply Cycle to Collect Location Update Vectors

When a source node needs to send data packets to a destination and is not aware of any route to that node, it initiates a flooding-based route discovery by broadcasting a Route-Request (RREQ) packet to its neighbors. The source maintains a monotonically increasing sequence number for the flooding-based route discoveries it initiates to reach the destination. Each node on receiving the first RREQ of the current flooding process from the source to the destination (i.e., an RREQ packet with a sequence number greater than those seen before), includes its location update vector, LUV (includes the node ID and location: the X and Y co-ordinates) in the RREQ packet before further rebroadcasting the packet in its neighborhood. The structure of the LUV and RREQ packet is shown in Figures 1 and 2 respectively.

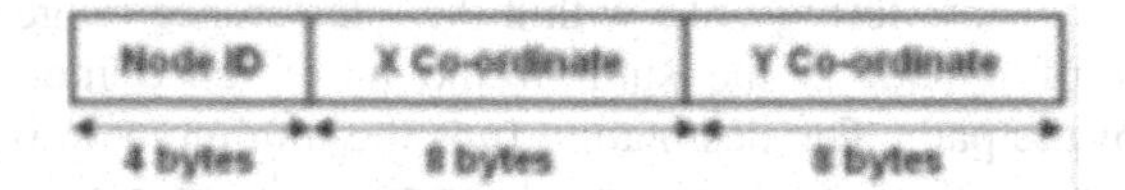

Fig. 1. Location Update Vector (LUV) per Node

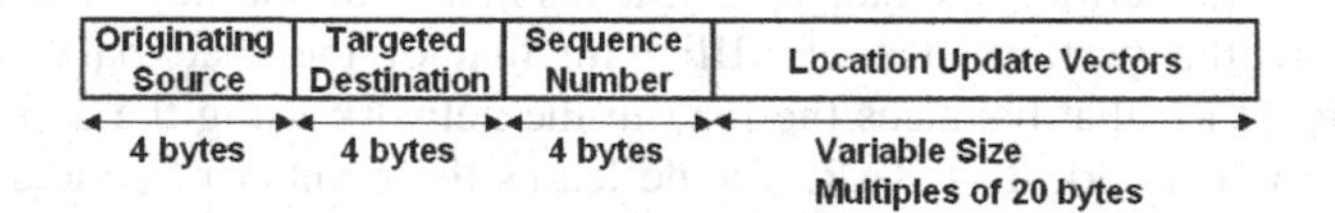

Fig. 2. Route Request (RREQ) Packet with LUVs

2.2 Route Selection and Route Reply

The RREQ packets propagate through several paths. After receiving the first RREQ packet, the destination waits for a certain time to receive the RREQ packets from all the paths. The destination then employs the following procedure to choose the best route to notify the source: Using the LUVs collected from each node, the destination locally constructs a weighted interference graph $G = (V, E)$ of the whole network. The vertices (set V) in the graph are the nodes that identified themselves in the RREQ packets. There exists an edge (in the set E) between any two vertices in the graph if and only if the distance between the corresponding nodes is less than or equal to the transmission range for a node. The weight of each edge is quantified using the *interference index*, which represents the number of interfering edges. To determine the interference index for an edge (v_i, v_j) between vertices v_i and v_j, we adopt the following algorithm (illustrated in Figure 3): We consider the set of all edges in E and determine the distance between the mid-points of the edge (v_i, v_j) and that of each other edge in E. If the distance between the mid-points of (v_i, v_j) and an edge (v_a, v_b) is less than or equal to the interference range of the edges, then we say the two edges (v_i, v_j) and (v_a, v_b) interfere with each other. The interference index of an edge is the number of surrounding edges interfering with its transmission.

Input: Edge (v_i, v_j)
Interference Graph $G = (V, E)$ // stores the (X, Y) co-ordinates of each node
Interference Range per Link *IRL*
Output: Interference Index II_{ij}

Begin Algorithm *Interference_Index*
$II_{ij} \leftarrow 0$
$X_{ij} \leftarrow (X_i + X_j)/2$
$Y_{ij} \leftarrow (Y_i + Y_j)/2$
for every edge $(v_a, v_b) \in E$
$X_{ab} \leftarrow (X_a + X_b)/2$
$Y_{ab} \leftarrow (Y_a + Y_b)/2$
Mid-pt-distance[(v_i, v_j), (v_a, v_b)] =
$$\sqrt{(X_{ab} - X_{ij})^2 + (Y_{ab} - Y_{ij})^2}$$
if (Mid-pt-distance[(v_i, v_j), (v_a, v_b)] $\leq IRL$)
$II_{ij} \leftarrow II_{ij} + 1$
end if
end for
return II_{ij}
End Algorithm *Interference_Index*

Fig. 3. Algorithm to Find Interference Index of an Edge in the Interference Graph

The interference index of a path is the sum of the interference index values of its constituent edges. The destination selects the path with the smallest interference index by running the Dijkstra minimum-weight path algorithm [2] on the interference graph where the weight of an edge is its interference index. The destination sends a Route Reply (RREP) packet along the chosen minimum interference index path towards the source. The RREP packet structure is shown in Figure 4.

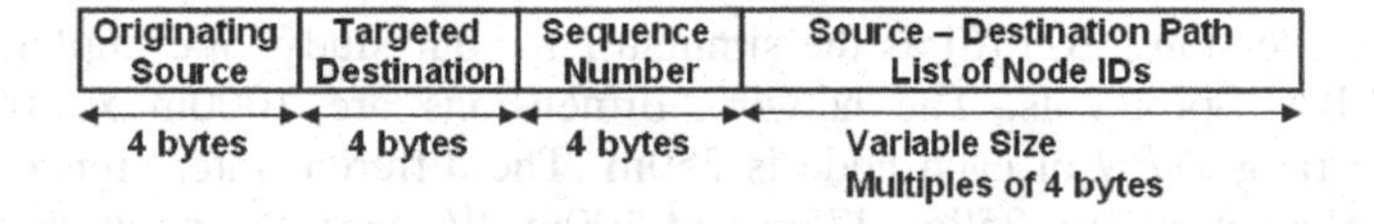

Fig. 4. Route Reply (RREP) Packet

All the nodes receiving the RREP packet update their routing tables to forward the incoming data packets (coming from the source to the destination) to the node that sent the RREP packet. Note that upon receiving a RREQ packet, we do not let an intermediate node to immediately generate a RREP packet to the source, even though the intermediate node might know of a valid route to the destination. We intentionally do this so that we could collect the latest location information from each node in the network through the LUVs as part of the RREQ packets and let the destination choose the best route according to the above route selection procedure. The overhead associated with including the node location and identification information is only 20 bytes per node. All the other information included in the RREQ and RREP packets are similar to those included in the well-known MANET routing protocols.

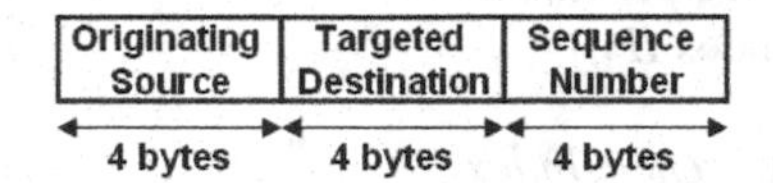

Fig. 5. Structure of the Header of the Data Packet

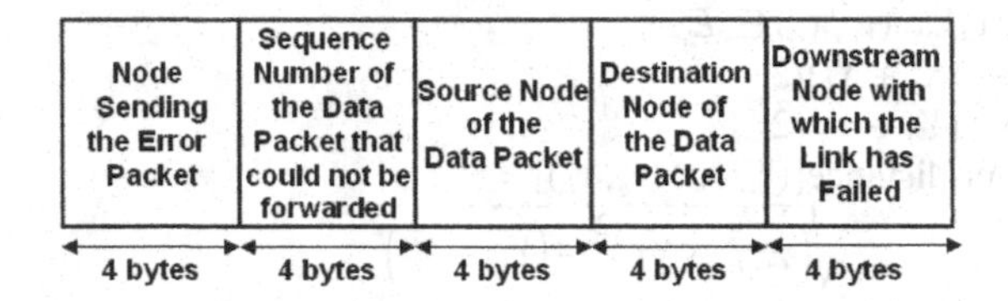

Fig. 6. Structure of the Route Error (RERR) Packet

2.3 Data Packet Transmission and Route Maintenance

The source starts sending the data packets to the destination on the route learnt through the RREP packet. The header of the data packet (structure shown in Figure 5) contains the usual fields of the source and destination IDs and the monotonically increasing sequence number for the data packets. If a link failure occurs due to the two nodes constituting the link drifting away, the upstream node of the broken link informs about the broken route to the source node through a Route Error (RERR) packet whose structure is shown in Figure 6. The source node on learning the route failure will stop sending the data packets and will initiate another global-flooding based route discovery.

3 Simulations

We use ns-2 (version 2.32) [6] as the simulator for our study. We implemented the MIF and LBAR protocols. The network dimensions are 1000m x 1000m. The transmission range *TRN* of each node is 250m. The different interference range per link, *IRL*, values used are: 250m, 375m and 500m. We vary the network density by conducting simulations with 25 nodes (low network density, with an average of 5 neighbors per node) and 50 nodes (high network density, an average of 10 neighbors per node). The energy level at each node is initially 1500 Joules. In the case of LBAR, each node periodically broadcasts to its one-hop neighbors information about its activity (i.e., the number of source-destination sessions that the node is part of) through beacons and each node keeps track of the activity of each of its neighboring nodes. The medium access control (MAC) layer uses the distributed co-ordination function (DCF) of the IEEE 802.11 standard [7] for wireless LANs.

Traffic sources are continuous bit rate (CBR). Number of source-destination (*s-d*) sessions used is 15 (low traffic load) and 30 (high traffic load). Data packets are 512 bytes in size; the packet sending rate is 4 data packets per second. In [8], it has been shown that no real optimization in energy consumption can be achieved when the energy lost in idle state is considered. In this paper, we do not consider the energy lost in the idle state and focus only on the energy consumed during the transmission and reception of messages (the data packets, the MAC layer RTS-CTS-ACK packets and the periodic beacons), and the energy consumed due to route discoveries. There is no transmission power control. For all situations, the transmission power per hop is 1.4 W and the reception power per hop is 0.967 W [9]. The node mobility model used is the Random Waypoint model [10]: each node starts moving from an arbitrary location to a randomly selected destination location at a speed uniformly distributed in the range $[0,\ldots,v_{max}]$. The v_{max} values used are 5 m/s, 25 m/s and 50 m/s representing low, moderate and high mobility scenarios respectively. Pause time is 0 seconds.

Each data point in Figures 7 through 13 is an average of data collected using 5 mobility trace files and 5 sets of randomly selected 15 and 30 *s-d* sessions. MIF results reported in Figures 7 – 11 are obtained when $IRL = TRN =$ 250m. Figures 12 and 13 illustrate that there is no appreciable difference in the performance of MIF for values of $IRL \geq TRN$. We study the following performance metrics for both MIF and LBAR:

(i) *End-to-End Delay per Data Packet* – the average of the delay incurred by the data packets that originate at the source and delivered at the destination. The delay incurred by a data packet includes: buffering delay due to the route acquisition latency, queuing delay at the interface queue to access the medium, transmission delay, propagation delay, and retransmission delays due to MAC layer collisions.
(ii) *Average Route Lifetime* – the average of the lifetime of the routes per *s-d* session, averaged over all the *s-d* sessions of a simulation run.
(iii) *Hop Count per Route* – the time-averaged hop count of the routes in all the *s-d* sessions. For example, if a routing protocol uses paths P1 and P2 of hop counts 3 and 5 for time 10 seconds and 5 seconds respectively, the time-averaged hop count for the total time of 15 seconds is (3*10+5*5)/15 = 3.67 seconds and not simply 4 seconds.
(iv) *Packet Delivery Ratio* – the ratio of the data packets delivered to the destination to the data packets originated at the source, computed over all the *s-d* sessions.
(v) *Energy Consumed per Node* – the average of the energy consumed at all the nodes in the network. A node loses energy due to transmission and reception of data packets, MAC layer packets, beacon exchange and propagation of RREQ and RREP packets during route discoveries.

3.1 End-to-End Delay per Data Packet

MIF incurs a significantly lower end-to-end delay per data packet compared to that incurred using LBAR. This illustrates the effectiveness of the interference index mechanism and the beaconless nature of the MIF protocol. Figures 7(a) and 7(b) illustrate the end-to-end delay per data packet incurred with the two routing protocols under different conditions of node density, node mobility and offered traffic load.

When the offered data traffic load is low, for a given node mobility, the end-to-end delay per data packet incurred by LBAR is about 30% more than that incurred by MIF in low-density networks and is about 60-90% more than that incurred by MIF in high-density networks. When the offered data traffic load is high, for a given node mobility, the end-to-end delay per data packet incurred with LBAR is about 15-20% more than that incurred by MIF in networks of low-density and is about 30-60% more than that incurred by MIF in high-density networks. Thus, MIF is effective in both low and high traffic scenarios.

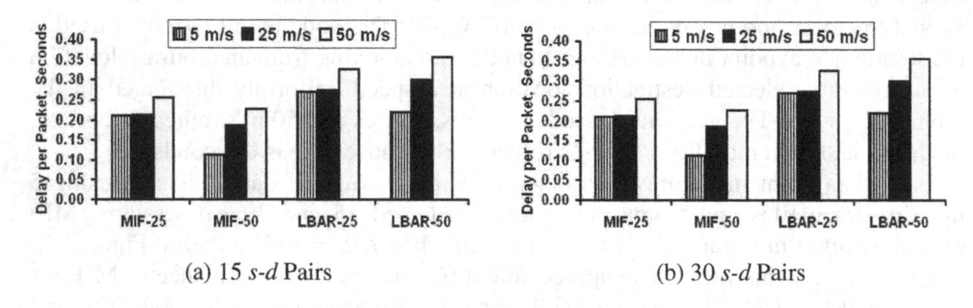

(a) 15 *s-d* Pairs (b) 30 *s-d* Pairs

Fig. 7. Average End-to-end Delay per Data Packet for MIF and LBAR

For a given network density, the end-to-end delay per data packet for both protocols increases with increase in the offered data traffic load. For a given network density, the increase in the delay per data packet for MIF and LBAR is by factor of 2 to 2.5 and 1.75 to 2.2 respectively. With a lower offered data traffic load, as node mobility is increased from v_{max} values of 5 m/s to 50 m/s, the delay per data packet for MIF and LBAR increases by a factor of 1.2 each in low-density networks and by a factor of 2.0 and 1.6 respectively in high-density networks. With a higher offered data traffic load, as node mobility is increased from v_{max} values of 5 m/s to 50 m/s, the delay per data packet for MIF and LBAR increases by a factor of 1.7 each in low-density networks and by a factor of 2.3 and 1.9 respectively in high-density networks. The increase is attributed to frequent route discoveries at high node mobility.

3.2 Route Lifetime

The lifetime of MIF routes is about 2-7% more than that of LBAR under most of the simulation conditions (Figures 8(a) and 8(b)). At the worst case, the lifetime of MIF routes is equal to that of LBAR routes. Hence, MIF routes are as stable as LBAR routes, if not better. For a given offered data traffic load, as the maximum node velocity is increased from 5 m/s to 50 m/s, the lifetime of MIF and LBAR routes decreases by a factor of 5.4-6.5. Both protocols incur relatively more stable routes in high-density networks. Since the decrease in the route lifetime is sub-linear with increase in the maximum node velocity, both the protocols are scalable with increase in node mobility.

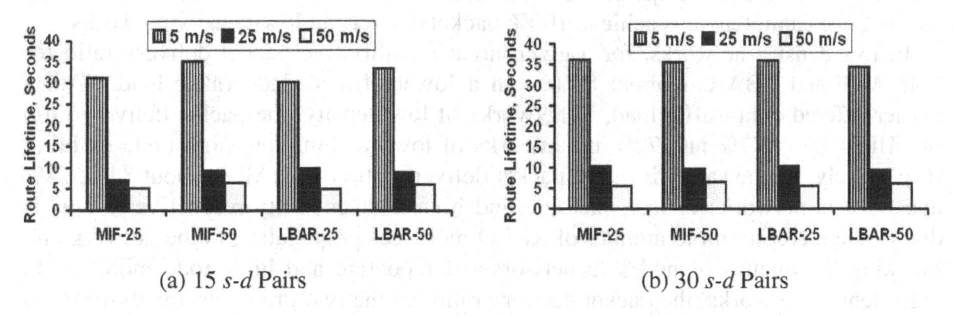

(a) 15 *s-d* Pairs (b) 30 *s-d* Pairs

Fig. 8. Average Route Lifetime for MIF and LBAR

3.3 Average Hop Count per Path

The average hop count per path incurred by both MIF and LBAR is almost the same for all the simulation conditions tested (Figures 9(a) and 9(b)). Both the protocols incur a reduction in the hop count per path with increase in network density. This is attributed to the presence of several alternate paths in networks of high density and there is relatively a higher chance of finding paths with a lower interference index value (for MIF) or lower traffic interference value (for LBAR) in high-density networks. Since both MIF and LBAR use a cost function that is additive in nature, both the protocols tend to find paths with a lower hop count. Note that for a given network density, the hop count of the routing protocols does not change much with node mobility and offered data traffic load.

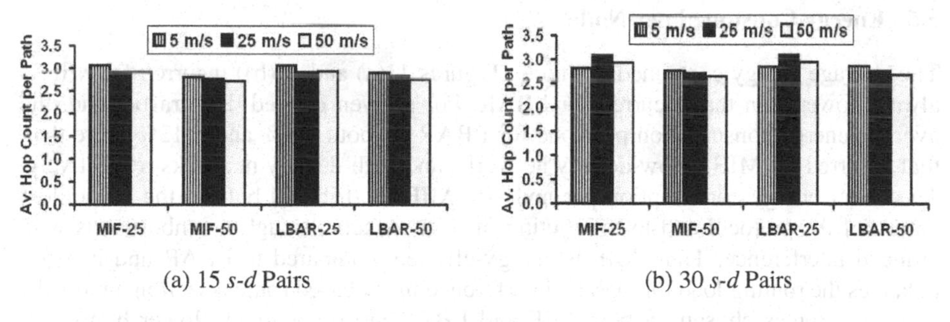

(a) 15 *s-d* Pairs (b) 30 *s-d* Pairs

Fig. 9. Average Hop Count per Route for MIF and LBAR

3.4 Packet Delivery Ratio

In all of our simulations, MIF incurs a higher packet delivery ratio compared to that of LBAR (Figures 10(a) and 10(b)). As MIF also incurs a significantly lower end-to-end delay per data packet than LBAR, we can conclude that MIF incurs a relatively higher throughput than that incurred with LBAR. The packet delivery ratio of the routing protocols depends on two factors: (i) network density and (ii) offered data traffic load. Note that the size of the queue at each node is 200 and is fixed for all the simulations. Low density networks may not remain connected all of the time. As a result, even if the node queue size and offered data traffic load are not limiting factors, one cannot always achieve 100% packet delivery in low-density networks.

In low-density networks, for a given node mobility, the packet delivery ratio for both MIF and LBAR is about 88% with a lower offered data traffic load. With a higher offered data traffic load, in networks of low-density, the packet delivery ratio of MIF is 86%, 77% and 72% in networks of low, medium and high node mobility respectively. At the same time, the packet delivery ratio of LBAR is about 84%, 75% and 68% in networks of low, medium and high node mobility respectively. This is due to the increase in the number of RREQ messages propagating in the network and blocking the queues of nodes in networks of moderate and high node mobility. In high-density networks, the packet delivery ratios of the two protocols are above 95%, with MIF having almost unit packet delivery ratio.

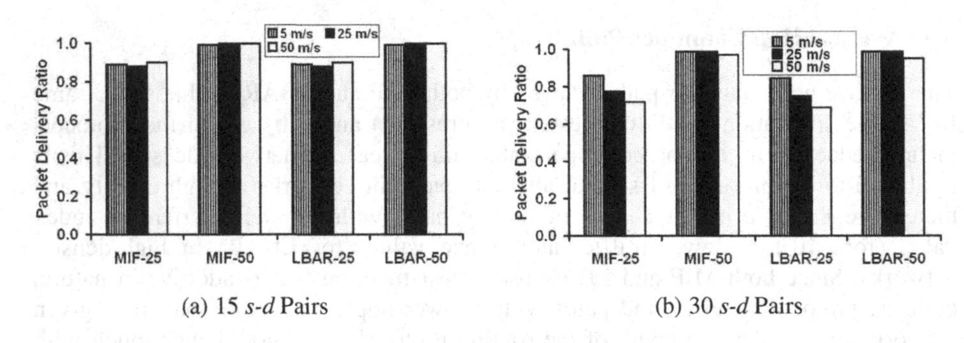

(a) 15 *s-d* Pairs (b) 30 *s-d* Pairs

Fig. 10. Average Packet Delivery Ratio for MIF and LBAR

3.5 Energy Consumed per Node

The average energy consumed per node (Figures 11(a) and 11(b)) incurred for MIF is always lower than that incurred for LBAR. For a given offered data traffic load, the average energy consumption per node for LBAR is about 3-4% and 9-12% more than that incurred for MIF in low-density networks and high-density networks respectively. The lower energy consumption per node for MIF is attributed both to the beaconless nature of the protocol and to the routing of data packets through neighborhoods with reduced interference. Thus, MIF is energy-efficient compared to LBAR and it better balances the routing load through its interference index based route selection approach.

As the routes chosen by both MIF and LBAR are of relatively lower hop count (with a larger physical distance between the constituent nodes of a hop) in networks of high density (compared to those chosen in low-density networks), the routes are relatively less stable in high-density networks. Thus, both MIF and LBAR spend relatively more energy for route discoveries in high-density networks. In high-density networks, as the maximum node velocity is increased from 5 to 50 m/s, the average energy consumed per node for both the protocols increases by about 13-17%.

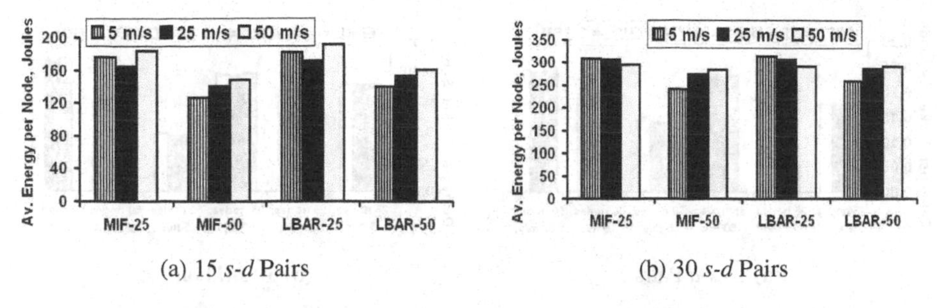

(a) 15 *s-d* Pairs (b) 30 *s-d* Pairs

Fig. 11. Average Energy Consumed per Node for MIF and LBAR

Another significant observation is that for a given routing protocol and an offered data traffic load, the average energy consumption per node is relatively more for low-density networks. This is attributed to the design of the two routing protocols to share the offered data traffic load equally across all the nodes in the network. In the case of MIF, the average energy consumption per node in low-density networks is about 20-40% and 10-30% more than that incurred in high-density networks for low and high offered data traffic load respectively. In the case of LBAR, the average energy consumption per node in low-density networks is about 10-30% and 7-20% more than that incurred in high-density networks for low and high offered data traffic load respectively.

3.6 Impact of Interference Range per Link

We conducted extensive simulations for MIF with three different values of interference range per link (250m, 375m, and 500m) and the transmission range of a node is 250m for all the three cases. The simulation results for packet delivery ratio and end-to-end delay per data packet presented in Figures 12 and 13 respectively indicate that the different interference range for the links does not significantly influence the performance of MIF. In practical terms, as long as the interference range value considered for the links during route selection is at least the transmission range per node, there is no significant change in the *SINR* of the transmissions.

The interference index values for every link increases significantly, but almost in the same rate, as we increase the interference range per link, *IRL*. Also, note that we

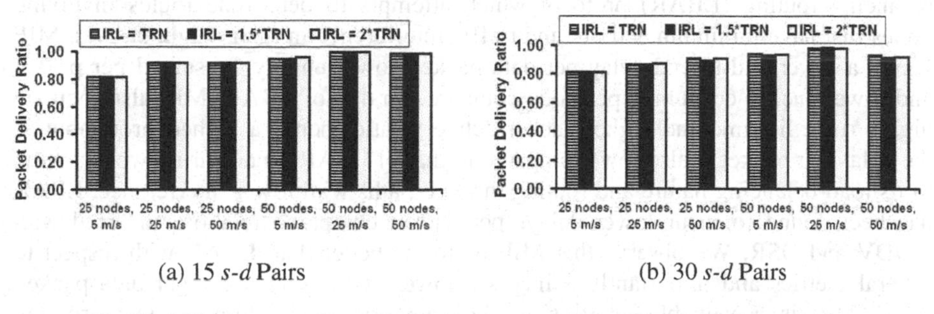

(a) 15 *s-d* Pairs (b) 30 *s-d* Pairs

Fig. 12. MIF: Impact of Interference Range per Link on Packet Delivery Ratio

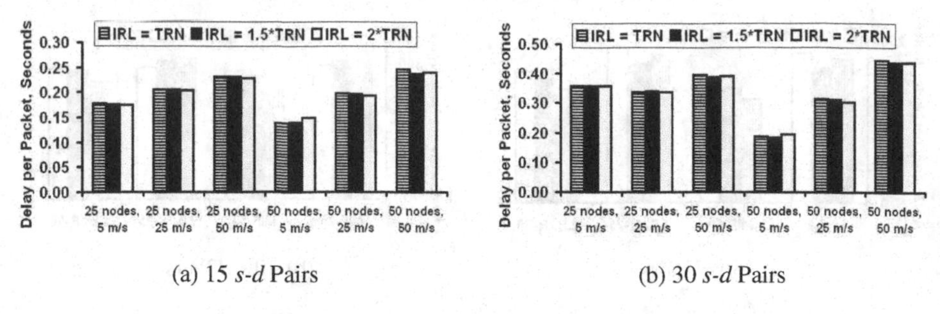

(a) 15 *s-d* Pairs (b) 30 *s-d* Pairs

Fig. 13. MIF: Impact of Interference Range per Link on End-to-End Delay

consider a square network for all our simulations and the impact of border effect is very minimal in such networks. MIF computes the sum of the interference index values for the constituent links of a path. The additive cost function by MIF could also be attributed to lack of much difference in the quality of the routes chosen for different *IRL* values.

4 Conclusions

The high-level contribution of this paper is the design and development of a beaconless, minimum interference based routing protocol (called MIF) for MANETs. Two links are said to interfere with each other if the distance between the midpoints of the two links is less than or equal to the interference range per link (*IRL*). Through extensive simulations, we find that MIF performs very well as long as the *IRL* value is at least the transmission range for the nodes in the network. MIF uses the notion of location update vectors (LUVs) to collect the location information of every node in the network during the propagation of RREQ messages to discover a source-destination route. The destination receives the RREQ messages along several paths and locally constructs a global topology (called the interference graph) using the location information of the nodes learnt through the LUVs. The weight of a link, referred to as the interference index of a link, in the interference graph is the number of interfering links surrounding it. The interference index of a path is the sum of the interference index values of its constituent links.

The performance of MIF is compared with that of the interference-aware load-balancing routing (LBAR) protocol which attempts to determine routes involving nodes that have minimum activity and traffic interference in their neighborhood. MIF incurs a lower end-to-end delay per data packet, lower energy consumed per packet and lower energy-consumed per node compared to that of LBAR. MIF also incurs a higher route lifetime and higher packet delivery ratio (hence a higher throughput as the delay per packet is also low) compared to that of LBAR. In an earlier work, owing to its load-balancing nature and routing through paths with lower interference, LBAR has been found to incur lower delays per packet compared to those incurred with AODV and DSR. We observe that MIF performs better than LBAR with respect to several metrics and importantly it incurs a lower end-to-end delay per data packet. Thus, MIF is a valuable addition to the category of interference-aware routing protocols for mobile ad hoc networks.

References

1. Jain, K., Padhye, J., Padmanabhan, V., Qiu, L.: Impact of Interference on Multi-hop Wireless Network Performance. In: 9th International Conference on Mobile Computing and Networking, pp. 66–80. ACM, San Diego (2003)
2. Cormen, T.H., Leiserson, C.E., Rivest, R.L., Stein, C.: Introduction to Algorithms, 2nd edn. MIT Press/ McGraw Hill, New York (2001)
3. Hassanein, H., Zhou, A.: Routing with Load Balancing in Wireless Ad hoc Networks. In: The 4th International Workshop on Modeling, Analysis and Simulation of Wireless and Mobile Systems, pp. 89–96. ACM, Rome (2001)
4. Johnson, D.B., Maltz, D.A., Broch, J.: DSR: The Dynamic Source Routing Protocol for Multi-hop Wireless Ad hoc Networks. In: Perkins, C.E. (ed.) Ad hoc Networking, pp. 139–172. Addison-Wesley, New York (2001)
5. Perkins, C.E., Royer, E.M.: The Ad hoc On-demand Distance Vector Protocol. In: Perkins, C.E. (ed.) Ad hoc Networking, pp. 173–219. Addison-Wesley, New York (2000)
6. Fall, K., Varadhan, K.: NS-2 Notes and Documentation, The VINT Project at LBL, Xerox PARC, UCB, and USC/ISI (2001)
7. Bianchi, G.: Performance Analysis of the IEEE 802.11 Distributed Coordination Function. IEEE Journal of Selected Areas in Communication 18(3), 535–547 (2000)
8. Kim, D., Garcia-Luna-Aceves, J.J., Obraczka, K., Cano, J.-C., Manzoni, P.: Routing Mechanisms for Mobile Ad hoc Networks based on the Energy Drain Rate. IEEE Transactions on Mobile Computing 2(2), 161–173 (2003)
9. Feeney, L.M.: An Energy Consumption Model for Performance Analysis of Routing Protocols for Mobile Ad hoc Networks. Journal of Mobile Networks and Applications 3(6), 239–249 (2001)
10. Bettstetter, C., Hartenstein, H., Perez-Costa, X.: Stochastic Properties of the Random-Way Point Mobility Model. Wireless Networks 10(5), 555–567 (2004)

An Optimal, Distributed Deadlock Detection and Resolution Algorithm for Generalized Model in Distributed Systems

S. Srinivasan[1], Rajan Vidya[1], and Ramasamy Rajaram[2]

[1] Department of Information Technology, Thiagarajar College of Engineering, Madurai, India
ssnit@tce.edu, rajan.vidhu@gmail.com
[2] Department of Computer Science and Engg, Thiagarajar College of Engineering, Madurai, India
rrajaram@tce.edu

Abstract. We propose a new distributed algorithm for detecting generalized deadlocks in distributed systems. The algorithm records the consistent distributed snapshot of global wait-for-graph (WFG) through propagating the probe messages in the forward phase and reducing the WFG to determine the entire set of deadlocked processes in the backward phase. The reducibility of each process is decided based on the information in replies that have received from its immediate wait-for processes during the reduction. We also formally prove the correctness of our algorithm. It has a worst case time complexity of 2d time units and the message complexity of 2e, where d is the diameter and e is the number of edges of the distributed spanning tree induced by the algorithm. The significant improvement of proposed algorithm over earlier algorithms is that it achieves optimum results using fixed sized messages and minimizes the messages to resolve deadlocks.

Keywords: Distributed Systems, Deadlock Detection, Deadlock Resolution, Distributed algorithms, Generalized Deadlocks, Wait-For Graph.

1 Introduction

Deadlock is an important resource management problem in distributed systems since it reduces the throughput by minimizing the available resources. In general, deadlock is defined as a system state, in which every process in a set is waiting indefinitely for other processes in the subset of processes. The interdependency among the distributed processes is commonly represented by a directed graph, known as wait-for- graph (WFG)[2]. In the WFG, each node represents a process and an arc represents dependency relation by starting from a process which waits for a resource to a process which is presently holding it. The deadlock detection algorithms in distributed systems is classified based on the underlying resource request models such as single

S. Ranka et al. (Eds.): IC3 2009, CCIS 40, pp. 70–80, 2009.
© Springer-Verlag Berlin Heidelberg 2009

Resource model, OR model, AND model and P-out-of- Q model [1]. In the AND model, processes requires all the requested resources to proceed the execution while in the OR model, processes requires few resources from the requested resources. In the P-out-of-Q model, every process requires "P" number of resources among "Q" resources requested for processing. Since AND and OR models are a special case of P-out-of-Q model, they are also referred as a generalized model. A deadlock based on the P-out-of-Q model is called as generalized deadlock. The generalized deadlock is occurred in many areas such as resource allocation in distributed operating systems, communicating processes and quorum consensus algorithm for distributed databases [11]. The AND and OR deadlocks are detected by existence of cycles and knot in the global WFG. For the generalized model, the cycle is necessary but not sufficient and knot is sufficient and not necessary to detect the deadlocks. Hence it involves the detection of most complex topology like generalized tie.

Very few algorithms have been proposed in the literature to detect and resolve generalized deadlocks. And most of them used diffusion computation technique to detect deadlocks. In diffusion computing [7], a special process "initiator" sends one or more messages. All the dependent processes will send the message only after receiving a message form the initiator. Thus the computation is terminated when every process is idle and waiting for every other process. The WFG is built by using the replies received by the initiator. Generalized deadlock detection algorithms are grouped into two categories namely centralized and distributed algorithms based on the existence of WFG. In the centralized algorithm [9,13], the probe messages carry the information required to built the WFG at the initiator before search for a deadlock. But the WFG constitute by multiple sites in distributed algorithms [4,5,,6,11,12]. The distributed algorithms follows single or two phases to record and reduce the global WFG. In a two phase algorithm [4,5,6], the WFG is recorded at the first phase through propagating the probe messages to all blocked processes and remove the unblocking processes during the second phase. But the overlapped outward and inward sweeps record as well as reduce the WFG in a single phase algorithms [11,12].

1.1 Previous Work on Distributed Deadlock Detection Algorithms

In Bracha and Toueg [4], the initiator propagates the probe messages to record the WFG in the first phase and collects the replies form all unblocked processes in the second phase. The insufficient arrival of replies to become unblocked indicates the existence of generalized deadlock. Here, the first phase terminates only after the second phase has terminated. It exchanges 4e messages to know whether the initiator is in deadlock with the delay of 4d. The Wang et al[5] uses an effective termination technique to detect the end of the first phase after recording all the processes in initiator's reachable set. The second phase reduces the graph that has recorded earlier to verify the existence of deadlock. It uses 6e messages and detects deadlocks within 3d+1 time units. Brzezinski et al[6] arranges all the processes in distributed snapshot into logical ring and circulates the probe messages. The algorithm declares deadlock when the processes does not change their states in two consecutive rounds. It uses weak termination technique compared to earlier algorithms. However it detects deadlocks using ½ n^2 messages in 4n time units. Unlike in [4,5], Kshemkalyani et al [11] algorithm recorded as well as reduced the WFG simultaneously and detects all deadlocked processes. It records the snapshot in outward sweep and reduce the processes which grants the resources in inwards sweep. Even though it detects the generalized deadlock through 4e-2n+4l messages within 2d time units, it might consider the situation in which the reduction of a unblocked process has begun before recording all the edges inclined in it. The Kshemkalyani et al [12] algorithm uses the lazy evaluation technique to allow the reduction of all unblocked processes until the initiator receives the replies from all its request messages. It detects deadlock using within 2d+2 time units and 2e messages only. In addition, the initiator retains additional information to resolve the deadlock to avoid additional e messages in [14].

1.2 Paper Objectives

We present a new decentralized algorithm for detecting generalized deadlock along with the correctness proof. This single phase algorithm records the consistent distributed snapshot comprising all processes reachable from the initiator through propagating the CALL message in the outward sweep. In the inward sweep, a blocked process sends the REPORT message to its wait by processes only after collecting all REPORT messages from wait for processes. The reduction of WFG is initiated by the active processes by responding the CALL message immediately during the inwards sweep. Here, the inward sweep begins only after an initiator records the distributed spanning tree in the outward sweep. A process is reduced if and only if it satisfies its p-out-of q resource requests. After the reduction process, all the blocked processes in the distributed snapshot are declared as deadlocked processes.

1.3 Differences between the Proposed Algorithm and Earlier Algorithm

Since the reducibility of each node is decided before it responds to its own CALL message unlike in [12] and it does not require to construct the complete WFG in the worst case or partial WFG at the initiator. Unlike in [11], the termination of the algorithm is detected based on Dijistra Scholten algorithm [3] instead of weight throwing and the outward and inward sweeps are not overlapped. In addition, the initiator knows the identity of all blocked processes at the end of the termination and hence simplifies the deadlock resolution by choosing the appropriate victim.

This paper is organized as follow. Section 2 describes the underlying computation model and section 3 discusses the proposed algorithm in detail. Section 4 analyses the performance and compare it with earlier algorithms. Finally, we conclude the paper in Section 5.

2 Preliminaries

The system consists of *N* processes and each has unique system wide identifier. The processes are communicating through a logical communication channel only by message passing. There is no shared memory in the system. The messages are delivered at the destination as per the order they sent in the sender within arbitrary but finite delay. The messages are neither lost nor duplicated and the entire system is fault-free. The system is subject to both internal and external event during the computation such as sending or receiving the message. The events are time stamped using Lamport's logical clock[8] and they are classified into computation events and control events. The computation event triggers the computational messages such as REQUEST, REPLY, CANCEL and ACK due to the execution of applications. And the control event generates the control messages including CALL and REPORT as a result of deadlock detection algorithm.

In a generalized model, the process resource request is expressed as a predicate using logical AND and OR operators. For example, a process resource requirement $a \wedge (b \vee c)$ specifies that it requires a resource from a and a resource from either b or c. Here, the generalized resource requirement of a process 'i' is represented as a function Fi. The function is evaluated only after it receives replies from all processes $j \in$ OUTi, The process *i* substitute true if a process grants a resource and false if it denies it.

Each process i maintain its local state using the following data structure

t_block_i : the logical time at which i was last blocked(0)
IN_i : the set of tuples $<k, t\_block_k>$ where k is a process waiting for i and
t_block_k : is the logical time at which k sent its request to i.(ϕ)
OUT_i : set of processes for which i is waiting since the last $t\_block_i(\phi)$

When a process i blocks on pi out of q_i requests, it sends a REQUEST message to qi processes. If a node receives pi REPLY messages, it immediately sends qi-pi CANCEL messages to withdraw its request. As a result, a process may be active or block state at any instant. An active process can send both communication and control messages while the blocked process can send either control messages or ACK. The blocked process cannot require additional resources and unblock abnormally. It is essential to ensure the record of consistent snapshot and the detection of false deadlock by the algorithm.

Whenever a process 'i' receives the REQUEST message from any blocked process $j \in$ OUTi, it records the request along with t_blocki in IN_i . A blocked process j sends ACK message immediately as a receipt of REQUEST. The communication messages are properly time stamped according to [18] to achieve synchronization. The distributed snapshot contains the local states of all the processes in initiator's reachable set. Each processes i in a snapshot records the sending or receiving of REQUEST, REPLY, and CANCEL messages consistently in its data structure.

Definition 1: A snapshot of process i is a collection of local states of all processes reachable from it. A consistent snapshot records both the sending and receiving of REQUEST, REPLY and CANCEL message. If the receipt of the message is exist in the snapshot, the sending of corresponding event must be recorded.

Definition 2: A generalized deadlock is a sub graph (D,K) of WFG (V,E) in which

i) $\forall i \in D$, i blocks on p_i out of q_i request $\wedge$ $D \neq \phi$
ii) $\exists C_i$, $C_i \subseteq OUT_i \wedge |C_i| \geq q_i - p_i + 1$
iii) $\forall i \in D$, $\forall j \in OUT_i$, no REPLY message is in underlying communication channel from j to i.

The correctness of any deadlock detection algorithm depends on the following two criteria.

Liveness: The algorithm detects the deadlock within a finite time after its formation in the system.
Safety: The algorithm reports the deadlock iff it actually exists in the system.

3 Distributed Deadlock Detection Algorithm

When a node i blocks on a P-out-of-Q request, it initiates the deadlock detection algorithm and became an initiator of that instance. The initiator i records a distributed snapshot comprising all the nodes along with their the wait-for relations in its reachable set. Although our algorithm supports the execution of multiple instances simultaneously, each instance is isolated using unique time stamps. Hence, each blocked node maintains a local snapshot to detect deadlocks. Here, we focus on the execution of single instance of our algorithm for simplicity.

3.1 The Description

This two phase algorithm records all the nodes reachable from the initiator at the first phase and reduces it in the second phase to find out all deadlocked processes. The initiator records the distributed snapshot of the WFG through diffusing the CALL messages in the first phase. It sends the CALL messages to its immediate successors which in turn propagate the message to their own successors. The edges of the WFG on which the first CALL message is received by each node induce a directed spanning tree of the WFG. When a blocked node i receives the first CALL message it keeps the sending node as its father before forwarding the probe. Deadlock detection as well as detecting the termination of the algorithm are performed simultaneously based on the REPORT message received from the active nodes.

The reduction phase is initiated by sending of replies to the CALL messages, i.e. REPORT messages from the leaf nodes. The REPORT message sent by an active node and a blocked node is differentiated by the field in the REPORT message, p which is equal to zero in the case of an active node and in the case of a blocked node its value is equal to the number of resources for which it is waiting. An active node immediately replies to all its CALL messages sent whereas a blocked node sends replies to all its wait by edges except the father until it receives the REPORT messages from all of its successors. A node gets reduced when it gets reply from active nodes which reduces its p value to zero. The REPORT message received from a blocked node do not reduces a node but it decreases the n value used for termination of the algorithm. The algorithm gets terminated when all the nodes receive the REPORT messages and thus the value of n becomes zero for all the nodes. After the algorithm terminates there exist some nodes which do not get reduced. These resulting blocked processes which do not get reduced on the receipt of the REPORT messages are declared to be deadlocked.

3.2 Formal Specification

Additional Datastructure

Ls_i : array [1..N] of record /*Snapshot Datastucture*/

Node data structure

out_i : set of integer $\leftarrow$ out_i /*nodes for which node i is waiting*/

in_i :set of integer $\leftarrow$ ϕ /*nodes waiting for i*/

t_i : integer $\leftarrow$ 0 /*time when init initiated snapshot*/

p : integer $\leftarrow$0 /*value of p as seen in snapshot*/

q_i :integer $\leftarrow$0 /*value of q as seen in snapshot*/

$father_i$: integer$\leftarrow$0 /*parent of node i*/

n_i :integer$\leftarrow$| out_i | /*no of successors of node i*/

Algorithm

Initiate a snapshot

/*executed by process i to detect deadlock*/

$init \leftarrow i$

$Ls_i\ t . \leftarrow t_i$

$Ls_i . out \leftarrow out_i$

$Ls_i . in \leftarrow \phi$

$Ls_i . p \leftarrow p_i$

$Ls_i . q \leftarrow q_i$

$Ls_i . father_i \leftarrow$ i

send call (i , i , t_i) to each in j in out_i

on sending call($i, init, t_i$)

/*executed by process i when it blocks on a p out of q request model*/.

For each node j of out_i nodes on which i blocks

send call (i , i , t_i) to j

on receiving call($j, init, t_i$)
/* executed by process i on receiving call from j*/
$Ls_i\ out \leftarrow out_i$
$Ls_i .in \leftarrow \{ j \}$
$Ls_i .t \leftarrow t_i$
$Ls_i .p \leftarrow p_i$
$Ls_i .q \leftarrow q_i$
$Ls_i .n \leftarrow n_i$

case(i)
if ($father_i = udef \wedge p_i$) $\rightarrow$
/*node i is a unvisited intermediate node*/
$father_i \leftarrow j$
send call (i , $init$, t_i) to each j in out_i /* i is blocked*/

case(ii) /*node i is a blocked visited intermediate node*/
if($father_i = def$) ^ (p_i)$\rightarrow$
send report ($father_i, i, t_i, p_i, |in_i|, victim$) to j
/* i is blocked visited intermediate node*/

case(iii) /*node i is a reduced visited intermediate node*/
if($father_i = def$) ^($\neg p_i$) $\rightarrow$
send report ($father_i, i, t_i$,0,0, ϕ) to j

case(iv) /*node i is a leaf node hence no first edge*/
if($\neg p_i \wedge \neg q_i$)
send report ($father_i, i, t_i$,0,0, ϕ) to j /* i is unblocked leaf node*/

on receiving report($father_i, i, t_i, p_j, |in_j|, victim$)

/*process i on receiving a report from unblocked node j*/
if($\neg p_j$) then $\rightarrow$ /*report from a unblocked node*/
$p_i \leftarrow p_i$ -1
$n_i \leftarrow n_i$ -1 /*process i on receiving a report from blocked node j*/
else
$n_i \leftarrow n_i$ -1
if ($|in_j| >= |in_i|$)$\rightarrow$
$victim \leftarrow victim \cup \{ j \}$
if (n=0) then $\rightarrow$

send report($father_i, i, t_i, p_i, |in_j|, victim$) to $father_i$
/*node with highest in array degree is chosen as victim*/
else
send report($father_i, i, t_i, p_i, |in_j|, victim$) to other in array nodes
else
$victim \leftarrow victim \cup \{i\}$
if (n=0) then →
send report($father_i, i, t_i, p_i, |in_i|, victim$) to $father_i$
else
send report($father_i, i, t_i, p_i, |in_i|, victim$ to other in array nodes

3.3 Example Execution

We now illustrate the algorithm with the help of an example shown in fig.1 shows a distributed WFG that spans ten nodes labeled a through j.

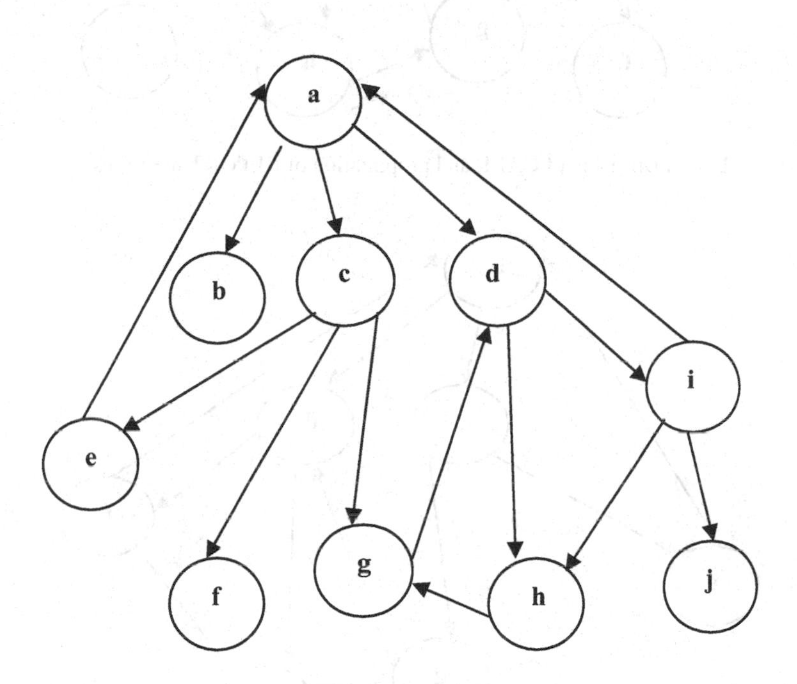

Fig. 1. The WFG

Node a initiates deadlock detection algorithm and sends CALL messages to nodes b, c and d. Fig.2 shows the diffusion of CALL messages through the WFG and the propagation of REPORT messages by the leaf nodes. Fig.3 shows the reduced WFG which contains all the unreduced nodes that are declared to be deadlocked.

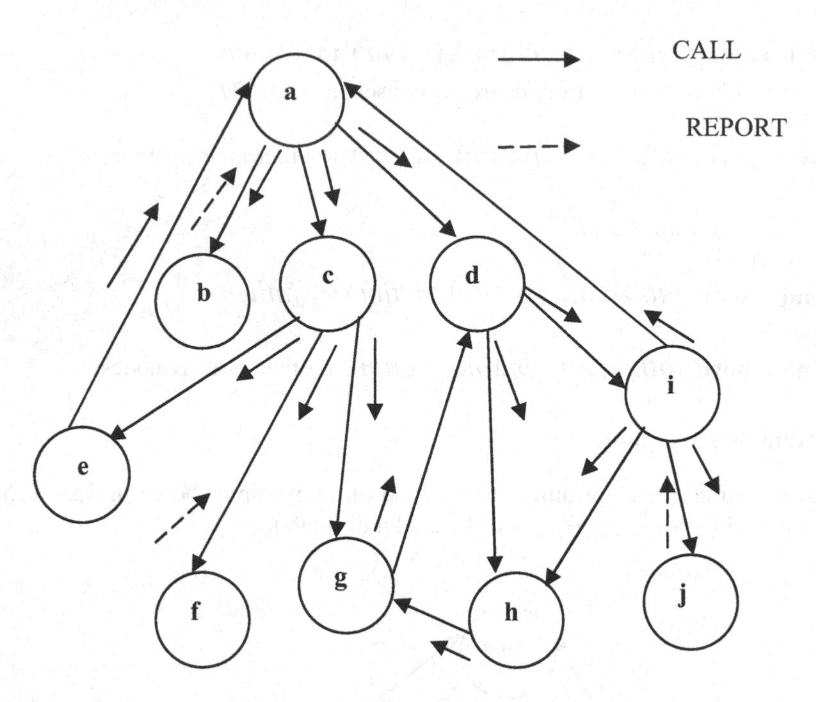

Fig. 2. Diffusion of CALL and propagation of REPORT messages

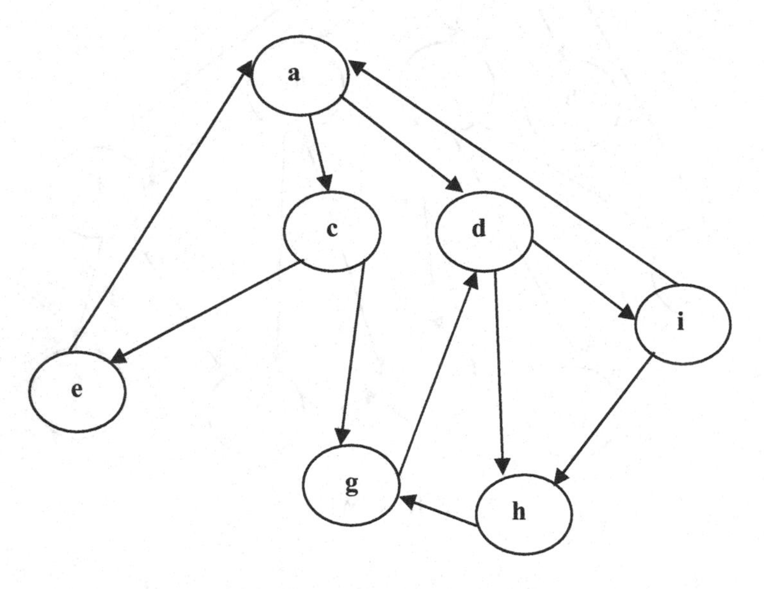

Fig. 3. The Snapshot

Node a initiates the algorithm and propagates CALL messages to all its successors. The nodes b, c, d marks a as the father node since a sends the first CALL messages to them. The nodes b, c, d in turn diffuses the CALL messages to its successors. The process repeats until all the nodes reachable from a have received the CALL messages. The leaf nodes e, f, j sends the REPORT messages to its IN array nodes and gets reduced from the WFG. The blocked nodes like e, c, d, g, h, i sends REPORT messages to its entire wait by edges except the father until it gets REPORT from all its successors. These nodes remain without getting reduced because their p value is not decreased to zero on the receipt of REPORT messages. Thus these blocked nodes are declared to be deadlocked processes.

3.4 Deadlock Resolution

The victims are selected based on the number of edges inclined in a node during the propagation of replies back to the root/initiator. The node which makes many processes to wait for it is chosen as the candidate to be removed. The number of processes waiting for a node is compared with other nodes and the node with greater value is chosen and its IN array degree is propagated in the REPORT messages to its ancestor nodes. A comparison is again made with ancestor nodes and neighboring nodes and the nodes with highest degree for the IN array are chosen and sent in victim array to the initiator. The initiator and the first element of the victim array are killed in order to remove deadlock.

4 Performance Analysis

We analyze the performance of our algorithm based on the following assumption. Each message transmission takes one time unit. In our algorithm, the initiator propagates the CALL messages along the edges of WFG and in response all the nodes propagate the REPORT messages. Hence, the message complexity is 2e where e is equal to the number of edges .Since the entire nodes are visited twice. The time complexity is 2d .The message length is O(1) unlike in [14,15] since it uses fixed size messages. The following table Table.1 compares the performance of our algorithm with other distributed algorithms.

Table 1. Performance Comparison of Distributed Deadlock Detection algorithm in the worst case

Algorithm	Delay	Messages	Message Length	Resolution
Bracha and Toueg[4]	4d	4e	O(1)	No scheme
Wang.et.al[5]	3d+1	6e	O(1)	No scheme
Brzezinski[6]	S^2	S^2	O(n)	No scheme
Kshemkalyani.et.al[11]	2d+2	4e-2n+2l	O(1)	e message
Kshemkalyani.et.al[12]	2d+2	2e	O(e)	No Scheme
Proposed Algorithm	**2d**	**2e**	**O(1)**	**1 message**

Here, S represents the total number of processes in the System, n represents the number of nodes, e represents the number of edges and d represents the diameter of WFG (V, E).

5 Conclusion

We have presented a new distributed algorithm to detect and resolve generalized deadlocks. The algorithm detects all deadlocked processes in the system by diffusing CALL messages to all processes in its reachable set. And each process propagates the REPORT towards the initiator. The algorithm determines the reducibility of each once it collects replies from all its successors. We also proved its correctness using an example. This algorithm efficiently resolves a deadlock unlike earlier distributed algorithms. In addition to minimize the time and message complexities, it drastically reduces the data traffic complexity into optimal compare to any existing algorithms.

References

1. Knapp, E.: Deadlock Detection in Distributed Database System. ACM Computing Surveys 19(4), 303–327 (1987)
2. Singhal, M.: Deadlock detection in distributed systems. IEEE Computer 22, 37–48 (1989)
3. Dijkstra, E.W., Scholten, C.S.: Termination Detection for Diffusing Computations. InformationProcessing Letters 11(1), 104 (1980)
4. Bracha, G., Toueg, S.: A Distributed algorithm for Generalized Deadlock Detection. Distributed Computing 2, 127–138 (1987)
5. Wang, J., Huang, S., Chen, N.: A distributed algorithm for detecting generalized deadlocks, Tech.Rep., Dept. of Compter Science, National Tsing-Hua Univ. (1990)
6. Brzezinski, J., Helary, J.M., Raynal, M., Singhal, M.: Deadlock Models and a General Algorithm for Distributed Deadlock Detection. J. Parallel and Distributed Computing 31(2), 112–125 (1995)
7. Chandy, K.M., Misra, J., Hass, L.: Distributed Deadlock Detection. ACM Trans. On Computing Surveys 1(2) (May 1983)
8. Chandy, K.M., Lamport, L.: Distributed Snapshots: Determing global states of istributed systems. ACM Trans. Computer Systems 3(1), 63–75 (1985)
9. Chen, S., Deng, Y., Attie, P.: Deadlock detection and resolution in distributed systems based on locally constructed wait-for graphs, Tech. Report, School of Computer Science, Florida International University (August 1995)
10. Kshemkalyani, A.D.: Characterization and correctness of distributed deadlock detection and resolution, Ph.D dissertation, Ohio state Univ. (August 1991)
11. Kshemkalyani, A.D., Singhal, M.: Efficient detection and Resolution of generalized distributed deadlocks. IEEE Trans. Soft. Eng. 20(1), 43–54 (1994)
12. Kshemkalyani, A.D., Singhal, M.: A One-Phase Algorithm to detect distributed deadlocks in Replicated Databases. IEEE Trans. Knowledge and Data Eng. 11(6), 880–895 (1999)
13. Lee, S.: Fast, Centralized detection and resolution of distributed deadlocks in the Generalised model. IEEE Trans. Soft. Eng. 30(9), 561–573 (2004)

Throughput Considerations of Fault-Tolerant Routing in Network-on-Chip

Arshin Rezazadeh and Mahmood Fathy

Department of Computer Engineering
Iran University of Science and Technology
University Road, Narmak, Tehran, Iran 16846-13114
rezazadeh@comp.iust.ac.ir, mahfathy@iust.ac.ir

Abstract. Fault-tolerant routing is the ability to survive beyond the failure of individual components and usually uses several virtual channels (VCs) to pass faulty nodes or links. A well-known wormhole-switched routing algorithm for 2-D mesh interconnection network, f-cube3, uses three virtual channels to pass faulty blocks such as f-ring and f-chain, while only one virtual channel is used when a message does not encounter any fault. One of the key issues in the design of NoCs is the development of an efficient communication system to provide high throughput and low latency interconnection networks. We have evaluated a new fault-tolerant routing algorithm based on f-cube3 as a solution to increase the throughput of physical links more than f-cube3 with lower message delay which uses less number of VCs compared to f-cube3 by reducing required virtual channels to two. Simulation of both f-cube3 and improved algorithm, if-cube2 (improved f-cube2) for the same conditions presented. As the simulation results show, if-cube2 has a higher performance than f-cube3 algorithm even with one less VC. The results also show that our algorithm has less exist packets in network and better performance with 100% traffic load in Network-on-Chip.

Keywords: Network-on-Chip, throughput, performance, interconnection, fault-tolerant routing, wormhole switching, virtual channel, deterministic routing.

1 Introduction

Interconnection networks have become integral issues to interconnect components of parallel computers. In these networks, nodes are connected to only a few nodes, its neighbors, according to the topology of the network and communicate with each other by passing messages. The 2-dimensional (2-D) mesh network is currently one of the most popular topologies for interconnection systems. Low-dimensional mesh networks, due to their low node degree, are more popular in comparison with the high dimensional mesh networks [18].

A possible approach for getting over the limiting factor in future system-on-a-chip designs is to use an on-chip interconnection network instead of a global wiring [8].

S. Ranka et al. (Eds.): IC3 2009, CCIS 40, pp. 81–92, 2009.
© Springer-Verlag Berlin Heidelberg 2009

On-chip networks relate closely to interconnection networks for parallel computers, in which each processor is an individual chip [19]. The tile-based network-on-chip architecture is known as a suitable solution for overcoming communication problems in future VLSI circuits [8] [12] [14]. Such chips are composed of many tiles regularly positioned in a grid where each tile can be, for example, an embedded memory, or processor, connected to its adjacent tiles through routers [8][16]. Each tile has two segments to operate in communication and computation modes separately [11].

This enables us to use packets for transferring information between tiles without requiring dedicated wirings. A NoC is a regular or irregular set of routers that are connected to each other on a point to point link in order to provide a communication backbone to the cores of a SoC. The most common template for NoC is a 2-D mesh where each resource or set of resources is connected with a router [1]. In brief, NoCs present the scalable performance needed by systems which grow with each new generation [3]. They allow the wiring energy consumption to be reduced by avoiding the use of long global wires. Furthermore, NoCs are reusable templates and aid to reduce design productivity gap. Finally, bus architecture will not meet all the requirements of future SoCs, as NoCs assure to do. In on-chip networks, information of routing is distributed, and the determination of the network status is distributed among the nodes which exchange information with each other. This type of algorithm is used in the large-scale networks [18].

The NoC system is composed of a large number of interconnected components (such as processors, embedded memories, and DSP cores) where communication of nodes is achieved by sending messages over a scalable interconnect network. In these networks, source nodes (a term for IP-Cores), generate packets that include headers as well as data, then routers transfer them through connected links, and destination nodes decompose them [16].

The wormhole (WH) switching technique proposed by Dally and Seitz [7] has been widely used in the interconnections such as [13] [18] [17]. In the WH technique, a packet is divided into a sequence of fixed-size units of data, called flits. If a communication channel transmits the header flit of a message, it must transmit all the remaining flits of the same message before transmitting flits of another message. As the header flit containing routing and control information advances along a specific route; the sequential flits follow it in a pipelined fashion [5]. Each flit of a packet is transferred and stored in the buffers of each router [16]. When the header flit is blocked due to lack of output channels, all of the flits wait at their current nodes for available channels.

Routing is the process of transmitting data from one node to another in a given system. Most past interconnection networks have implemented deterministic routing where messages with the same source and destination addresses always take the same network path. This form of routing has been popular because it requires a simple deadlock-avoidance algorithm, resulting in a simple router implementation [8]. Wormhole routing requires the least buffering (flits instead of packets) and also allows low-latency communication. To avoid deadlocks among messages, multiple virtual channels (VC) are simulated on each physical channel [4].

All routers have five ports, where one is used for its core and the other four ports are used for communication channel between neighbor switches. As a consequence of insensitivity to distance, pipelined flow of messages, and small buffer requirements, we have used wormhole technique for the switching [4]. After receiving the packet header, first the routing unit determines which output should be used for routing this packet according to its destination and then the arbiter requests for a grant to inject the packet to a proper output using the crossbar switch.

Currently, most of the proposed algorithms for routing in NoCs are based upon deterministic routing algorithms which in the case of link failures, cannot route packets. Since adaptive algorithms are very complex for NoCs, a flexible deterministic algorithm is a suitable one [9]. Deterministic routing algorithms establish the path as a function of the destination address, always applying the same path between every pair of nodes. The routing decision is independent of the state of the network. This routing algorithm is known as dimension-order routing. This routing algorithm routes packets by crossing dimensions in strictly increasing (or decreasing) order, reducing to zero the offset in one dimension before routing in the next one [10]. This paper evaluates a fault-tolerant routing algorithm based on Sui-Wang's [20] deterministic algorithm.

In this paper we show enhancements of the use of virtual channels in a physical link when a message is not blocked by fault. In our method, a message uses only two VCs, in non-faulty conditions. Our simulation results show that performance of *f*-cube3 algorithm is worse than the improved one, *if*-cube2. We simulate the algorithms for 5% and 10% of all links faulty. Results for all of situations show that our algorithm has better (1) throughput, (2) less message delay, (3) less existing messages rate, and (4) throughput and average message delay per faults than based algorithm. Also, it can work in higher message injection rates, with higher saturation point.

The rest of paper is structured as follows. In section 2 the fault-tolerant routing algorithms used in NoC with a comparison of *f*-cube3 fault-tolerant routing algorithm and our enhanced fault-tolerant routing algorithm are described followed by section 3 in which presents how we have simulated these two algorithms and defines performance parameters that are considered. Finally, section 4 summarizes and concludes the results given in this paper.

2 Fault-Tolerant Routing Enhancement

Fault tolerance is the ability of the network to function in the presence of component failures. The presence of faults renders existing routing algorithms to deadlock- and livelock-free routing ineffective. The fault-tolerant routing has been proposed as an efficient routing algorithm to preserve both performance and fault-tolerant demands in Network-on-Chips. Many algorithms have been suggested to operate in faulty conditions without deadlock and livelock. Some of these algorithms like [4] [17] [20] are based on deterministic algorithms. In [20], Sui and Wang have improved the Boppana and Chalasani's [4] deterministic algorithm using less VCs. This improved algorithm uses three VCs to pass the both single and block faults. Since our evaluated fault-tolerant routing algorithm is based on this algorithm improving wire delays and throughput of VCs. In the next subsection we are going to describe how Sui and Wang's deterministic algorithm works by three VCs with our modification that reduces the number of required VCs to two.

Since adaptive routing requires many resources, if cyclic dependencies between channels are to be avoided, deterministic based fault-tolerant routing algorithms are more useful for NoCs because of simple implementation, simple structure of routers and more speed, low power consumption and lower space used by routers. The deterministic e-cube algorithm is the most commonly implemented wormhole routing algorithm. Boppana and Chalasani's deterministic algorithm [4] has used e-cube technique.

Since some buffers are required to store fragments of each packet, or even the whole packet, at each intermediate node or switch, each packet whose header has not already arrived at its destination requests some buffers while keeping the buffers currently storing the packet, a deadlock may arise. A deadlock occurs when some packets cannot advance toward their destination because the buffers requested by them are full. All the packets involved in a deadlocked configuration are blocked forever. A different situation arises when some packets are not able to reach their destination, even if they never block permanently. A packet may be traveling around its destination node, never reaching it because the channels required to do so are occupied by other packets. This situation is known as livelock. It can only occur when packets are allowed to follow non-minimal paths [10]. However, our improved fault-tolerant routing algorithm does not use these paths.

2.1 *f*-cube3 Algorithm

The algorithm presented by Sui and Wang [20], *f*-cube3, uses one less VC in comparison with the algorithm is discussed in [4]. Since each node in this algorithm needs fewer buffers, the area used by buffers of a chip would be reduced. Such an algorithm is able to pass *f*-ring and *f*-chain even with overlapped faulty regions (f-regions) [20]. Each message is injected into the network as a row message and its direction is set to null. Messages are routed along their e-cube hop if they are blocked by faults. When faults are encountered, depending on the message type and the relative position of the destination nodes to the source nodes, the direction of messages are set to clockwise or counter-clockwise. To do this, *f*-cube3 uses Set-Direction(M) procedure as given in [20]. Messages are routed on *f*-rings or *f*-chains according to the specified directions. The direction of a message which is passed the f-region would be set to null again. When an end point of *f*-chain is reached, messages take a u-turn and their directions are reversed.

Sui and Wang used four types of messages; WE, EW, NS, and SN. In their method, all messages use virtual channel number 0, if not blocked by faults; otherwise number 1 and 2. With this channel assignment, three virtual channels per physical channel are needed. It is noted that WE, EW, NS, and SN messages use disjoint sets of virtual channels. An entire column fault disconnects meshes and is not considered in this algorithm. Finally, they proved deadlock and livelock freeness of their algorithm. In the next section we will explain our improved fault-tolerant routing algorithm, *if*-cube2.

2.2 *If*-cube2 Algorithm

First, we show how to enhance the well-known *f*-cube3 routing algorithm. This algorithm is based on *f*-cube3, and able to pass *f*-ring and *f*-chain. Like [20] each message is injected into the network as a row message and its direction is set to null and e-cube routes a message in a row until the message reaches a node that is in the same column as its destination, and then routes it in the column. Then it would be changed as a column message to reach the destination. At each point during the routing of a message, the e-cube specifies the next hop to be taken by the message. A column message could not change its path as a row message, unless it encounters with fault. The message is said to be blocked by a fault, if its e-cube hop is on a faulty link. In such a faulty situation, a column message could change its direction into clockwise or counter-clockwise.

In *if*-cube2, at first, each message should be checked if it has reached to destination node. Else, if this message is a row message and has just reached to the column of destination node, it would be changed as a column message. Next, if a message encountered with fault, the Set-Direction(M) procedure would be called to set the direction of the message. The role of this procedure is to pass f-region by setting the direction of message to clockwise or counter-clockwise. Again, the direction of the message will be set to null when the message passed f-region. While the direction of a message is null, we use only two VCs for routing process by using every two VCs and no need to any extra virtual channels for routing. A message encountered with f-region, uses predefined virtual channels as given in [20].

In [20], it is explained that how virtual channels would be used when a message is passing faulty region. With this channel assignment only two virtual channels per physical channel are needed in our method. It is noted that WE, EW, NS, and SN messages use disjoint sets of virtual channels. Note that in this algorithm a message which passed a faulty region or not encountered with fault can use all VCs while only one extra VC is used in primitive algorithm except that two VCs to pass the fault. Since this algorithm is based on [20], deadlock freeness and livelock freeness features of this fault-tolerant routing algorithm is proven in [20].

The technique evaluated in this paper has one primary advantage over the one presented in the previous work. According to [20], a flit always uses one of three virtual channels. However, in the current paper, a flit is allowed to use all virtual channels instead of just one fixed virtual channel when does not encounter any fault. Using this modification, simulations are performed to evaluate the performance of the enhanced algorithms in compared to the algorithm proposed in prior work. Simulation results indicate an improvement in the average message delay and average exist messages and throughput for different fault rates and different message lengths. Furthermore, the enhanced approach can handle higher message injection rates (i.e., it has a higher saturation rate). This modification allows us to use only two virtual channels instead of three because when a message routes on a faulty condition, it uses one of predefined virtual channels mentioned in [20], and while routed in non-faulty hops, it uses that two virtual channels.

If-cube2 algorithm has been showed in the reminder of this section.

```
if-cube2 algorithm
begin
/* the current host of message M is (sx, sy) and its
destination is (dx, dy). */
set message type (EW, WE, NS, or SN) based on the
address of the destination
route message M as follows:
        if sx = dx and sy = dy, deliver the message
        and return
        if the message M has reached the destination
        column then set the message type to NS or SN
        according to destination node
        if exist fault-free path then the message M is
        forwarded along the e-cube path and use every
        2 virtual channels
        if the message has encountered a fault region
        then the message M picks a direction to follow
        along the fault ring according to [20] and the
        relative location of the destination
end.
```

3 Performance Evaluation and Simulation Results

In this section, we describe how we perform the simulation and acquire results from simulator. On-chip network is by far resources limited, e.g. limited by wiring, buffer and area. In the standard computer networks, performance is drastically impacted by the amount of buffering resources available, especially when the traffic is congested. In contrast, on-chip networks should use the least amount of buffer space due to the limited energy and area budget [15]. The more existing packets in the network during one period of time, will be consumed more buffers. One of the parameters we have considered as a performance metric of our algorithm, related to free buffer space, is the number of existing packets in the network. The less existing packets, the more free buffers will be available. Network throughput can be increased by organizing the flit buffers associated with each physical channel into several virtual channels. A virtual channel consists of a buffer, together with associated state information, capable of holding one or more flits of a message [6].

Some parameters we have considered are average of exist messages (AEM) in each period of time and average delay of messages per faults. Another examined parameter in this paper is the throughput of the network which is using our routing algorithm, *if*-cube2. Throughput illustrates the number of packets in each cycle which passed from a node [21]. We have examined throughput over message injection rate (MIR), average message delay (AMD) over throughput and throughput over percentage of faults at 100% traffic load for all sets of faulty links. We have also show the AMD per faults to illustrate the influence of modification for primitive routing algorithm.

3.1 Simulation Methodology

We have simulated a flit-level 16 × 16 mesh with 32 and 48 flit packets for the uniform traffic pattern – the source node sends messages to any other node with equal probability. This simulator can be used for wormhole routing in meshes. The virtual channels on a physical channel are demand time-multiplexed, and it takes one cycle to transfer a flit on a physical channel. We record statistics measured in the network with the time unit equal to the transmission time of a single flit, i.e. a clock cycle. Our generation interval has exponential distribution which leads to Poisson distribution of number of generated messages per a specified interval.

In this simulation we have considered 5% and 10% of the total network links faulty. The network includes 480 links. Specifically for the 10% case, we have set, randomly, 12 nodes faulty; since four links are incident on a node. For the 5% fault case, we have set six nodes - 24 links in total - faulty. In each case, we have randomly generated the required number of faulty nodes and links. A node, in which its all links are faulty, never sets as a destination of messages.

The number of messages generated for each simulation result is 1,000,000 messages. The simulator has two phases: start-up, steady-state. The start-up phase is used to ensure the network is in steady-state before measuring parameters. For this reason we do not gather the statistics for the first 10% of generated messages. All measures are obtained from the remaining of messages generated in steady-state phase.

Finally, in the remaining of this section, we study the effect of using two VCs on the performance of *if*-cube2. We perform this analysis under a different traffic distribution pattern, i.e. hotspot and matrix transpose. It is noted that due to lack of space, only parts of simulation results are presented in this paper.

3.2 Comparison of *f*-cube3 and *if*-cube2

We defined throughput as the major performance metric. For an interconnect network, the system designer will specify a throughput requirement. Figures 1 to 5 show the simulation results for two different fault cases, 5 and 10 percent with 32 and 48 flits on 16 × 16 mesh. Figure 1 shows the throughput over the message injection rate for two cases of faults. As we can see, the throughput of the network which uses *f*-cube3 algorithm is lower while the *if*-cube2 algorithm has higher network throughput. As an example in the 5% case of *f*-cube3 with 32 flit packets, the throughput for 0.002 MIR is 0.0013 at 100% traffic load; however, the other algorithm, *if*-cube2, could achieve 0.0025 throughput in 0.00325 MIR at high traffic load – more than 90% improvement in this case. In fact our fault-tolerant routing algorithm has higher throughput for higher MIRs.

The most valuable comparison we have done between these two algorithms is the rate of average message delay over throughput. Comparative performance across different fault cases in figure 2 is specific to the several fault sets used. For each case, we have simulated previous fault sets for 100% traffic load. The injection control helps us here; otherwise, we would have to perform the tedious task of determining the saturation point for each fault set.

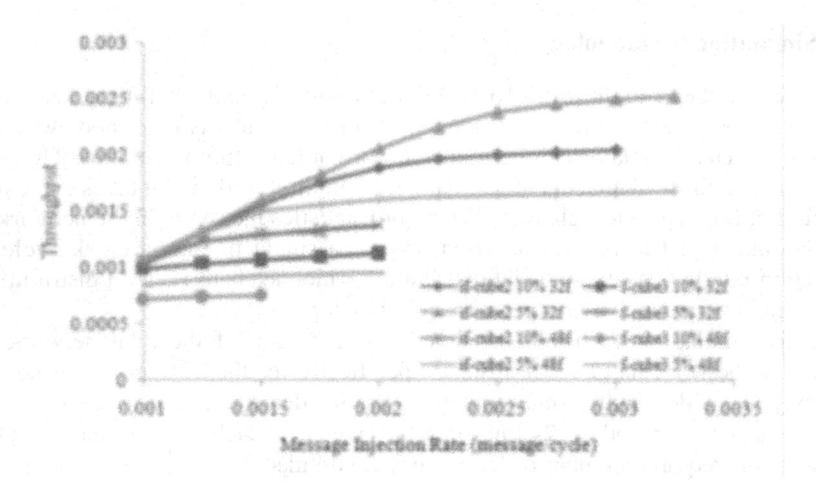

Fig. 1. Throughput of f-cube3 and if-cube2 in uniform traffic for 5% and 10% of total network links faulty by 32 and 48 flit packets

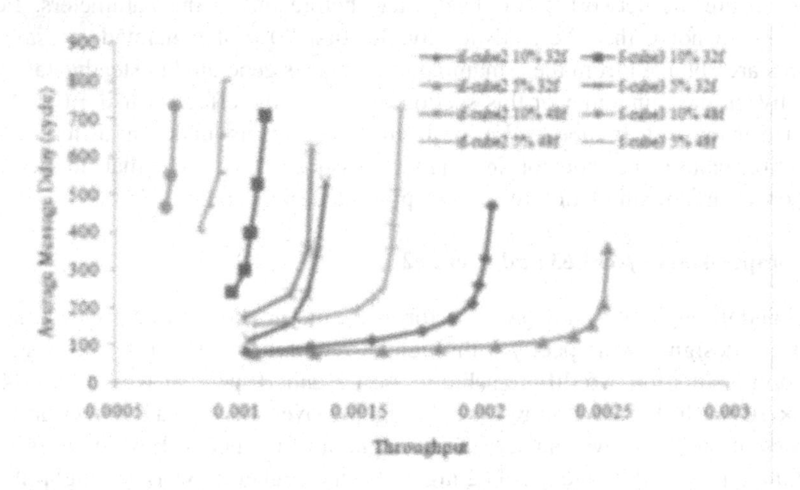

Fig. 2. Performance of f-cube3 and if-cube2 in uniform traffic for 5% and 10% of total network links faulty by 32 and 48 flit packets

As an example in this figure, we can look at the amount of average message delay for both algorithms with 0.0013 throughputs for 5% faults and 32 flit packets. In this point of throughput, the network which is using *f*-cube3 has more than 530 AMD at 100% traffic load while the other network, using *if*-cube2, has less than 80 AMD, and it has not been saturated. Comparing the throughput of these algorithms for 100% traffic load, it is obvious the network using *if*-cube2 has 0.00251 throughputs whereas the other one has just 0.00136 throughputs. We have improved throughput of network more than 84% by our proposed algorithm at 100% traffic load.

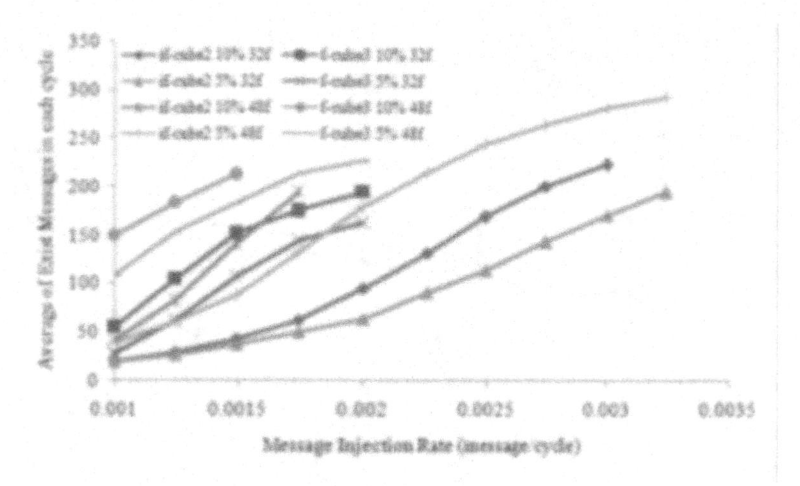

Fig. 3. Average of exist messages of f-cube3 and if-cube2 in uniform traffic for 5% and 10% of total network links faulty by 32 and 48 flit packets

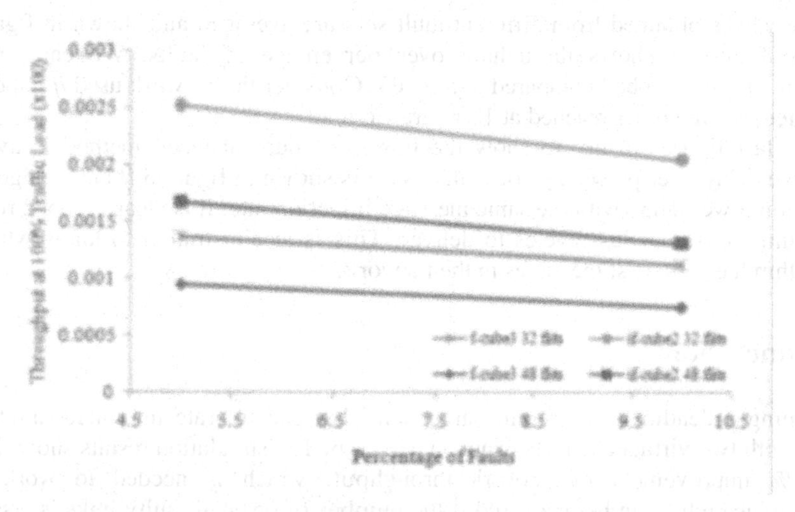

Fig. 4. Throughput-Faults of f-cube3 and if-cube2 in uniform traffic for 5% and 10% of total network links faulty by 32 and 48 flit packets

We have also shown that the average of exist messages in each cycle has been decreased by our algorithm. This parameter shows the busy input and output buffers on the network. If nodes of communication system have more free buffers, messages may deliver simply across the interconnection network. Figure 3 shows the average number of existing messages per each period of time. As an illustration, we can see that for 0.002 MIR, the AEM of our algorithm is 62.6 with 5% faults and 32 flit packets, where this value is 161.3 for *f*-cube3 – a great enhancement.

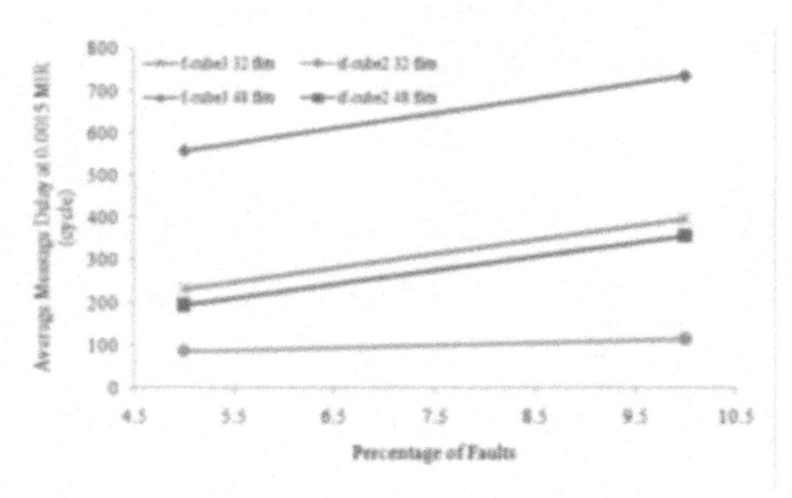

Fig. 5. AMD-Faults of f-cube3 and if-cube2 in uniform traffic for 5% and 10% of total network links faulty by 32 and 48 flit packets

The values obtained from different fault sets are averaged and shown in figures 4 and 5. Figure 4 shows throughput over percentage of faults. We can see the performance of *if*-cube2 compared to *f*-cube3. Consider the network used *if*-cube2 not saturated, but the other reached at 100% traffic load.

The last figure we use to show the power of our enhanced method is average message delay over percentage of faults. As it is shown in figure 5 *if*-cube2 algorithm works on lower delay with the same message injection rate. It is clear *if*-cube2 routing algorithm consumes less cycles to deliver. This is an illustration to know why this algorithm has less exist messages in the network.

4 Conclusion

Designing a deadlock-free routing algorithm that can tolerate unlimited number of faults with two virtual channels is not an easy job. The simulation results show that up to 90% improvement of network throughput, which is needed to work with rectangular faults, can be recovered if the number of original faulty links is less than 10% of the total network links.

In this paper, for the purpose of reducing the number of virtual channels, we evaluated a method to shrink, by using two virtual channels, these block faults.

We also showed that in different message lengths these block faults can be handled. The deterministic algorithm is enhanced from the non-adaptive counterpart by utilizing the virtual channels that are not used in the non-faulty conditions. The method we used for enhancing the *if*-cube2 algorithm is simple, easy and its principle is similar to the previous algorithm, *f*-cube3. There is no restriction on the number of faults tolerated and only two virtual channels per physical channel are needed in the proposed algorithm.

We presented that the studied parameters have acceptable results using *if*-cube2; our algorithm, called *if*-cube2, however, works better by using only two virtual channels – one less VC. We have been simulated both *f*-cube3 and *if*-cube2 algorithms for the same message injection rates, fault situations, message lengths, network size, and the percentage of faulty links. All of parameters we have examined show better results for *if*-cube2 in comparison with *f*-cube3 algorithm.

We could achieve higher performance by modifying fault-tolerant algorithms, especially with focusing on how messages passed the faulty regions. Both delay of mesh network by different message injection rates, utilization of interconnection network and power consumption, are important parameters to consider in fault-tolerant routing algorithms for Network-on-Chips.

References

1. Ali, M., Welzl, M., Zwicknagl, M., Hellebrand, S.: Considerations for fault-tolerant network on chips. In: The 17th International Conference on Microelectronics, December 13-15, pp. 178–182 (2005)
2. Banerjee, N., Vellanki, P., Chatha, K.S.: A Power and Performance Model for Network-on-Chip Architectures. In: Proceedings of the Design, Automation and Test in Europe Conference and Exhibition (DATE 2004), February 16-20, vol. 2, pp. 1250–1255 (2004)
3. Benini, L., De Micheli, G.: Networks on chips: A new SoC paradigm. IEEE Computer, 70–78 (January 2002)
4. Boppana, R.V., Chalasani, S.: Fault-tolerant wormhole routing algorithms for mesh networks. IEEE Trans. Computers 44(7), 848–864 (1995)
5. Dao, B.V., Duato, J., Yalamanchili, S.: Dynamically configurable message flow control for fault-tolerant routing. IEEE Transactions on Parallel and Distributed Systems 10(1), 7–22 (1999)
6. Dally, W.J.: Virtual channel flow control. IEEE TPDS 3(2), 194–205 (1992)
7. Dally, W.J., Seitz, C.L.: Deadlock-free message routing in multiprocessor interconnection networks. IEEE Trans. Computers 36(5), 547–553 (1987)
8. Dally, W.J., Towles, B.: Principles and practices of interconnection networks. Morgan Kaufman Publishers, San Francisco (2004)
9. Dally, W.J., Towles, B.: Route packets, not wires: On-chip interconnection networks. In: Proceedings. Design Automation Conference, Las Vegas, NV, USA, June 18-21, pp. 684–689 (2001)
10. Duato, J., Yalamanchili, S., Ni, L.: Interconnection networks: An engineering approach. Morgan Kaufmann, San Francisco (2003)
11. Guerrier, P., Greiner, A.: A generic architecture for on-chip packet-switched interconnections. In: Proceedings. Design Automation and Test in Europe Conference and Exhibition, Paris, France, March 27-30, pp. 250–256 (2000)
12. Hemani, A., Jantsch, A., Kumar, S., Postula, A., Oberg, J., Millberg, M., Lindqvist, D.: Network on chip: an architecture for billion transistor era. In: IEEE NorChip Conf., November 2000, pp. 120–124 (2000)
13. Kiasari, A.E., Sarbazi-Azad, H.: Analytic performance comparison of hypercubes and star graphs with implementation constraints. Journal of Computer and System Sciences 74(6), 1000–1012 (2008)
14. Kumar, S., Jantsch, A., Millberg, M., Oberg, J., Soininen, J., Forsell, M., Tiensyrj, K., Hemani, A.: A network on chip architecture and design methodology. In: Symposium on VLSI, April 2002, pp. 117–124 (2002)

15. Lap-Fai, L., Chi-Ying, T.: Optimal link scheduling on improving best-effort and guaranteed services performance in network-on-chip systems. In: DAC 2006, San Francisco, California, USA (July 2006)
16. Matsutani, H., Koibuchi, M., Yamada, Y., Jouraku, A., Amano, H.: Non-minimal routing strategy for application-specific networks-on-chips. In: ICPP 2005, International Conference Workshops on Parallel Processing, June 14-17, pp. 273–280 (2005)
17. Rezazadeh, A., Fathy, M., Hassanzadeh, A.: If-cube3: an improved fault-tolerant routing algorithm to achieve less latency in NoCs. In: IACC 2009, IEEE International Advanced Computing Conference, March 6-7, pp. 278–283 (2009)
18. Rezazadeh, A., Fathy, M., Rahnavard, G.A.: An enhanced fault-tolerant routing algorithm for mesh network-on-chip. In: ICESS 2009, 6th International Conference on Embedded Software and Systems, May 25-27, pp. 505–510 (2009)
19. Srinivasan, K., Chatha, K.S.: A technique for low energy mapping and routing in network-on-chip architectures. In: ISLPED 2005, San Diego, California, USA, August 8-10, pp. 387–392 (2005)
20. Sui, P.H., Wang, S.D.: An improved algorithm for fault-tolerant wormhole routing in meshes. IEEE Trans. on Computers 46(9), 1040–1042 (1997)
21. Yang, S.G., Li, L., Xu, Y., Zhang, Y.A., Zhang, B.: A power-aware adaptive routing scheme for network on a chip. In: ASICON 2007, 7th International Conference on ASIC, October 22-25, pp. 1301–1304 (2007)

A New Approach towards Bibliographic Reference Identification, Parsing and Inline Citation Matching

Deepank Gupta[1], Bob Morris[2], Terry Catapano[3], and Guido Sautter[4]

[1] Netaji Subhas Institute of Technology, Plazi
[2] University of Massachusetts, Boston, Plazi
[3] Columbia University, Plazi
[4] University of Karlsruhe, Plazi

Abstract. A number of algorithms and approaches have been proposed towards the problem of scanning and digitizing research papers. We can classify work done in the past into three major approaches: regular expression based heuristics, learning based algorithm and knowledge based systems. Our findings point to the inadequacy of existing open-source solutions such as Paracite for papers with "micro-citations" in various European Languages. This paper describes the work done as part of the Google Summer of Code 2008 using a combination of regular-expression based heuristics and knowledge-based systems to develop a system which matches inline citations to their corresponding bibliographic references and identifies and extracts metadata from references. The description, implementation and results of our approach have been presented here. Our approach enhances the accuracy and provides better recognition rates.

Keywords: Bibliographic Reference Parsing, Inline Citation Matching, Regular Expression, Metadata Extraction, Knowledge-based Systems, Micro-citations.

1 Introduction

Scientific research never happens in a vacuum and always builds upon previous work. Thus, we say that a scientist always stands on the shoulders of Giants. The previous work upon which research is done is cited through references in scientific papers and journals.

This paper describes automatic recognition, parsing and normalization of bibliographic references to enable easy search and retrieval of related information content. Since the fields of scholarship are spread out with work taken up in different countries in different times there has never been a single rigid method of referencing. So, the study of legacy methods of referencing must be carried out and tools must be provided which will be able to extract the information from citations so that it may be utilized irrespective of the format of referencing followed by various journals.

As in most scholarly publications, papers in the field of biological taxonomy contain many inline citations that are severely abbreviated. From an information processing point of view, these are actually references to a complete entry in a

S. Ranka et al. (Eds.): IC3 2009, CCIS 40, pp. 93–102, 2009.
© Springer-Verlag Berlin Heidelberg 2009

bibliography elsewhere in the paper. The nature of the abbreviation, and the possible requirement to locate and parse the full reference, can make these citations difficult to identify, parse, and extract information from. Taxonomists sometimes call these "micro-citations", although they go by many names in published style guides. Since, we are doing computerized reading of papers, we often do not get reliable information from formatting and thus the formatting cues have to be largely ignored. Apart from this, the separators are semantically overloaded with information and often serve more than one purpose.

In this paper, we will look at a unique combination of information obtained from separators, domain-specific knowledge and localized knowledge of a paper; to obtain good accuracies in the field of parsing and normalizing references.

2 Previous Works

The problem of digitizing and categorizing the world's information is not new. It has been studied by many scientists over the last decade or so. A number of suitable algorithms and approaches have been proposed towards the problem of scanning and digitizing research papers, extracting references, extracting metadata from references and citation matching. Some of the notable projects which address the problems presented above are ParaCite, CiteSeer and Google Scholar.

We can classify the work done in the past into three major approaches. Firstly, the regular-expression based heuristics as discussed in [1], [9] have been applied to extract metadata from references. ParaCite Toolkit [1] which is a collection of perl modules often termed as ParaTools uses a standard set of templates to extract metadata from the references. The technique was further perfected by the application of a protein sequence analyzer namely BLAST [9] to generate a more comprehensive set of around 2500 templates against which the citations were matched and indexed. Another novel application of a regular expression based parser has also been developed in [10] to find references between Dutch Laws.

The second approach is to apply learning based algorithms such as Hidden Markov Model, Conditional Random Fields and Support Vector Machines which utilize machine learning and get better with more training data. Hidden-Markov Model [7, 8] is a probabilistic model in which there is transition between a finite set of states accompanied by a corresponding output usually as a symbol from the character set. They are known as hidden as the output is known and the task it to determine the sequence of states through which the model goes to result in the emission sequence.

Lastly, knowledge based systems proposed by Giufridda [2] reports a good accuracy from a narrowed down corpus of computer science based research papers. He uses the spacial/visual knowledge principle for extracting metadata from scientific papers stored as PostScript files. Another interesting work on reference and citation extraction was done by Brett Powley and Robert Dale [3] in which localized information was used from the research paper to identify and match citations inside it.

The three different approaches have all given different results. This paper describes a combination of the conventional approaches of regular-expression based heuristics and knowledge-based systems to develop a system which converts the input files into a computer-readable schema; identification and extraction of metadata from references and citation matching. This paper has been divided into sections as described here. The General Approach and algorithms have been described in the Section 3. This section will describe the approaches followed towards various problems in the various stages. This will be followed by the results obtained with the test corpus in Section 4 and related observations. Section 5 focuses on the conclusions and the scope of future work regarding the same.

3 Terms and Notations

The following terms will be frequently used through-out the paper:

- **Reference:** A Reference appears in the list of works containing the full bibliographic information such as 'Author Name(s)', 'Year of Publication', 'Title of Work' and 'Journal Name' about a cited work.
- **Citation:** Present inside the text of the document and contains enough information to identify reference uniquely from the list of references in the document.
- **Inline-textual Citation:** This type of citation is usually a part of sentence and contains an Author, Year Pair to uniquely identify the reference from the list of references in the document.
- **Separator:** It refers to the special character used to separate two different fields of a reference
- **Reference Block:** The section of a document which contains the list of bibliographic references referred to in the document is known as a reference block.
- **Keywords:** These are the special words which mark the start of a reference block e.g. "References" marks the start of the reference block of this document.

4 Our Work

The approach and system employed has been described in this section of the paper. The process can be divided into 4 stages as described in diagram 1: Obtaining Input Files in the TaxonX Schema [5] from the scanned documents; obtaining a reference block in the document and identifying references; parsing the references into Author Name, Year, Title, Publication and other metadata; and, identification and matching of the corresponding citations with the references in the document.

4.1 Input Files

There are 3 distinct features of our test corpus. Firstly, our test corpus consists largely of papers from the field of Zoology. Secondly, the test-corpus is not limited to English Language, containing papers written in many languages. And lastly, there are papers ranging from 18th century to the present in our corpus which papers in different formats and following different conventions. Although, our test corpus is mainly

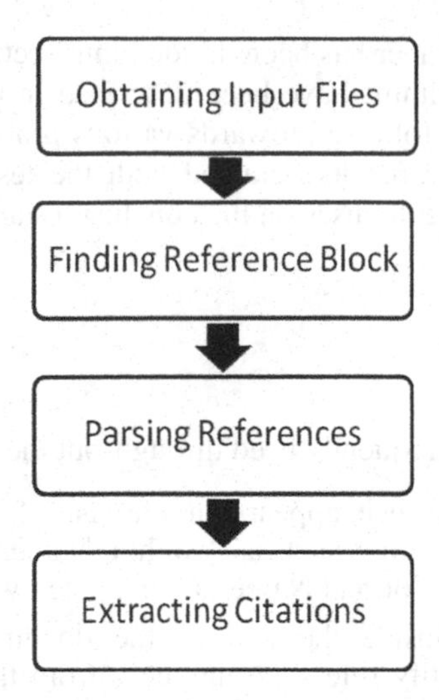

Fig. 1. Process Outline

restricted to papers related to Zoology and various ant species the algorithm described in this paper can handle papers presented on other subjects also.

The basic strategy followed in converting the input files are as follows: At first, the documents are hand scanned using Optical Character Recognition (OCR) and converted into a PDF Format. The PDF documents are then converted into html using Abby PDF Reader which is propriety software for converting PDFs. Since the text of the PDFs are produced by OCR, there can be errors due to scanning in the test corpus. The formatting of the text is usually not consistent. Also, the capitalization of the text cannot be relied upon as useful indicator for parsing. Certain spelling mistakes are also encountered whenever the print is unclear. After this, the GoldenGate Editor [11] is used to add annotations and then export to an internal format which is converted to TaxonX [5][6] by a purpose built XSLT transformation which is being used internally in Plazi. The TaxonX Schema based xml files are then used by the Reference Block Identifier as an input.

4.2 Reference Block Identification

The document is then processed to identify reference block. A reference block can be defined as a set of paragraphs in the document contains all the references to other papers. Usually the reference block is present in the end of a research paper and is often preceded by a keyword: "References". This is often followed by references preceded by index such as 1,2,3.... ; [1], [2], ...;A, B, ... and so on.

It has been observed that every reference block starts with a verbal cue which we will refer to as a keyword. The keywords differ with various languages and publications such as “References”, “Works Cited” etc. Thus, a very simple approach for the reference extraction, (also employed by the ParaCite Project) is to look for Keywords such as “References” in the research paper and then taking the reference block to be as the block from the start to the end of the paper. This approach although correct in essentials often misses out on certain Reference blocks which are preceded by keywords not present in the database like: "Bibliography", "Works Cited", "Bibliographia" etc. Thus, a minor improvement to the approach will be to scan for all such keywords to identify the start of a reference block. Thus, an editable list of keywords should be maintained which helps to identify the start of a reference block. But with reliance on keyword alone will mean that there will always be a chance that the program might encounter a keyword not present in our database.

Also, a document may only have different keyword to denote the start of a reference block; it will not necessarily contain only references until the end of the paper. For instance, a document can contain a list of tables, a list of figures or appendix information after the reference block. Also, a document such as a journal or a conference proceeding often contains more than one paper and hence, more than one reference blocks in it. Thus, our second assumption that a reference block once started continues till the end of the document is also not correct every-time. A minor improvement to the above mentioned algorithm will be to scan for words such as "Tables", "Figures", and “Appendix" and stop the reference block when they are encountered. Since, we cannot rely on formatting information we cannot provide the start of another section by checking the difference in the format of the words. This modification too, will not give us very satisfactory results and will lead to errors.

Thus, a probability model based approach in which we look at each paragraph as a potential candidate for being a reference may be preferable. Every paragraph which is a reference will contain a year to denote the year of publication. By manual inspection, it has been found, that every reference contains a year and thus the absence of a year from a paragraph positively means that the paragraph is not a reference. But the inverse is not always true, since a general paragraph might also contain a reference to a year, or might even cite a reference in the format such as: (Allen, 1987). Thus, the presence of the year in the normal paragraph (i.e. a paragraph which is not a reference) will either be in a sentence or in the form of a citation. Thus, combining this information along with the locality of a reference block i.e. the references being present one after another, and being preceded by a keyword, we can make a good probability model. The other criteria we put in into the model are the conformance of a paragraph to a reference structure. Usually, by reference structure, it is meant that a paragraph denoting a reference is small in length and it follows a certain template.

If we classify the parameters as:

P1: Absence of year
P2: Presence of keyword
P3: Previous words of a year conforming to Reference Structure
P4: Correlation in the length of paragraph with Average Reference Length
P5: Number of reference paragraphs directly above the candidate paragraph
P6: Number of reference paragraphs directly below the candidate paragraph
P7: Correlation in Reference Template with the candidate paragraph

And we classify the weights of the parameters as: W1, W2, W3, W4, W5, W6, W7 correspondingly.

The probability of a candidate paragraph to be a reference can then be defined as:

$P(\text{Candidate}) = \sum W_i P_i / \sum W_i$
If P(Candidate) >= Threshold Value, then Candidate paragraph is a reference.

Thus, the first step is to search for a year which is a four digit number from 1800 to the current year. Apart from this, we make use of other parameters such as presence of a keyword before the paragraph, references immediately before and after the paragraph being considered, length of the paragraph and the perceived author name from the paragraph. All this information has been put in a probability model which is customized to get the overall probability of a particular paragraph to be a potential reference. If the probability is above a certain threshold(0.7 in our case), we assume it to be a reference. All the references thus extracted are cleaned up to remove the extraneous formatting and redundant information. The cleaned up references are put into a temp file for internal use by the next module.

4.3 Reference Parsing

As discussed earlier, the presence of abbreviations, inconsistent formatting and semantically overloaded punctuation and separators have presented difficulties. Thus, we have followed an integrated approach towards this problem. The approach consists of:

1. Template matching i.e. use of regular expressions to classify various portions of the text as particular fields.
2. Use of domain based information like a list of publications to classify a portion as a Publication in case of failure by the first approach.

The parsing for a reference starts simultaneously from left and right. The parsing is started from the left to obtain the Author Name and Year information. It has been observed from the test corpus that every reference has author name(s) being immediately followed by the year of publication. Often, there are more than one Authors followed by a year. A year is a four digit number which will have the value within the range of normal year values. A year can also be followed by a special character as 1996a or 1996b. A reference might also contain more than one year in a single line such as 1995, 1996 to denote two papers published by the same authors. Keeping all this in mind, the year and author names are extracted from the reference. Simultaneously, parsing is started from the right to look for page-numbers and related cues. For instance, a page number is represented as pp. p. or without any prefix, simply as two numbers separated by a hyphen. Regular expressions are used thus to find out Author Names, Year of Publication, Volume and Page numbers with the reference being parsed both from the front and the back. The parsed information is then removed from the reference paragraph.

For the other parts of the reference, the approach is not so simple. The use of separators often cannot be used as a reliable measure for distinguishing between Title and Publication. This is because no single separator is used consistently to distinguish a title from a publication. In many cases, a dot (.) is used both as a separator and also as an abbreviation marker. Similarly, common separators like comma (,), hyphen (-) and colon (:) are all semantically overloaded serving both as a field separator as well as working as punctuations. It has also been observed that many references have all the words of a title capitalized, while some have every character capitalized and the rest have every character in small-case letters. Since, capitalization is neither preserved during scanning and nor is a uniform means of distinguishing, it cannot be used for the parsing. Thus, capitalization, formatting or punctuations cannot be used as a reliable means for parsing.

A combination of domain specific knowledge and a prediction model based on the length of the typical fields has been followed with automatic detection of certain false results. For this, a huge database of existing Zoological Publications has been formed largely from Zootaxa and the Natural History Museum Database to find out the publications present in a reference. We use a word-to-word based matching algorithm which matches the publication with the most eligible candidate i.e. the publication with most matched words. It has been noted that some of the words are redundant such as various prepositions and can generate false matches. Thus, those words are not taken into account. When we have a single word match, we largely ignore it, but we take it into account only when the publication has a single word in it. Note that a match is only considered for consecutive words, i.e. there can be no gap between two matched words except that of whitespace or punctuations.

If the publication finds no match, then we take the prediction model into account. The prediction model is based on the average publication length, their distance from the start of the reference and the major separators used. The statistics of the correctly parsed references are logged in while parsing. These are then used for the references that fail. Sometimes, a reference might not have any publication listed; in such cases also, we need to correctly parse the titles.

Thus, with the help of domain knowledge, regular expressions and prediction models, we achieve good accuracy in parsing.

4.4 Citation Parsing

It has been observed that a citation always consists of an author name followed by a year. The indexed citations are easy to associate with a corresponding reference. A textual citation, however requires more work. It can be found in a paragraph by looking for years in the paragraph. After finding a year, we need to find out whether it is preceded by an author name or not. It has been observed that there are certain temporal prepositions which commonly appear before years such as "in, since, during, until and before". If these temporal prepositions come in front of a year, it automatically means that this phrase is not a citation.

Identification of author names is done by matching of Authors field from parsed references. Every citation corresponds to a certain reference and thus the localized information obtained by parsing of the Reference section can be used to identify the Author. If the Author match is found, the corresponding match is taken as a citation, otherwise, the phrase and the year succeeding it is not considered as a citation. The punctuations are not at all considered in this process. Thus, this algorithm is able to identify both the textual as well as parenthetical citations in the paper. At the end of this process, an output file is generated which contains all the parsed information consisting of all the references and the corresponding citations in the research paper.

5 Experiment and Results

Our test corpus consists of papers for which a TaxonX schema based document has already been generated using the above-mentioned approach. The corpus consists of papers written in various different languages and are encoded in UTF-8 encoding. The corpus has been classified according to the languages as in Figure 2. The algorithm provides an accuracy which is considerably better than the accuracy provided by the various approaches used individually.

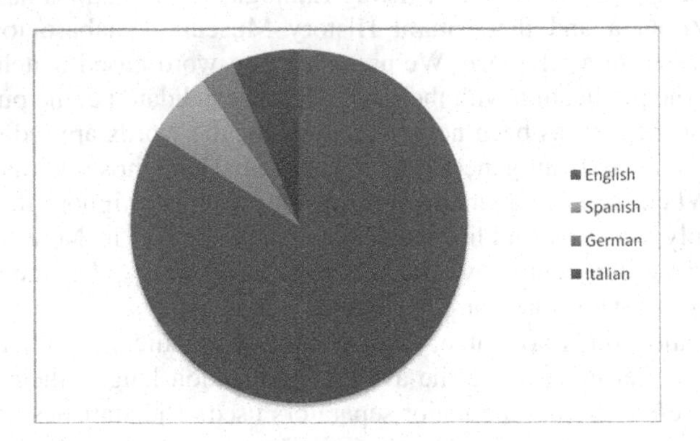

Fig. 2. Languages in Corpus

Table 1. Reference Block Identification

Title	ParaCite Unmodified	Modified Code I	Modified Code II	Our Approach
Number of References(Hand Counted)	664	664	664	664
True Positives Identified	330	660	660	660
False Positives Identified	310	550	86	3
Percentage of False Positives	48.4%	45.5%	13.0%	0.5%
Percentage of False Negatives	50.4%	0.6%	0.6%	0.6%

Table 1 presents the number of references detected by various programs from a research paper in our test corpus. The unmodified ParaCite code only identified 330 of 660 references of the total references present in the document. Thus, the original ParaCite code has low precision and low recall. Our first modification improved the recall, but not the precision. Our second modification improved the recall to 100%, the same as our own, but with 10 false positives compared to our 3. ParaCite suffered its low recall because it identifies references by looking only specific keywords like "References" or "Works Cited". Our first modification enhanced knowledge base of reference block candidates to include all possible keywords. Our second set of modifications reduces false positives by excluding several common types of document segments that commonly follow a reference block, but which do not contain references. Such sections include a list of figures and tables or appendices. Our own method adopts some of these enhancements, along with altogether different probability based-model, which we feel is more widely applicable. See section 3.2.

Table 2. Reference Parsing

Fields		**Number Correctly Identified**	**Percentage Correctly Identified**
Authors		660	99.5%
Year of Publication		660	99.5%
Title		504	76.0%
Publication	**Combined Approach**	490	73.9%
	Regular Expression Heuristic	445	91.8%
	Knowledge Based System	45	9.2%

The Table 2 represents the number of fields correctly parsed in the references parsed by the algorithm. The various approaches followed while parsing Publications are shown in the columns Domain Based and Prediction Based. The domain-based knowledge obtained from the database of reputed publications for taxonomic papers was used to identify the Publisher/journal effectively.

Apart from this, in case no such match was found, prediction was made using the localized knowledge such as the average length, starting point of publications, separator symbols used etc. and the algorithm tried to predict the publication using it. It was often found that the prediction based knowledge supplemented the first technique to give us better results. It should be noted however, that the existing system will need the domain-based knowledge for this technique to work. Thus, a corpus must always be supplemented with this knowledge, otherwise, the results might not be as good as shown above.

While the identification of "Title" and "Publication" has somewhat lower accuracy, "Author" and "Year of Publication" show high accuracy results. This is because many of the references in our corpus are micro-citations and have semantic overloading of information in the punctuation marks like dot(.). An example of the same will be:

Simon, E. 1891. On the spiders of the island of St. Vincent. Pt. 1 Proc. Zool. Soc. of London, Nov. 17, 1891: 549 - 575

In this, the title ends at St. Vincent. But the title itself contains dot as an abbreviation specifier and also as a separator between Title and Publication. We encounter many such cases of micro-citations in our test-corpus.

6 Conclusion and Future Scope

This paper studies the limitations of existing open-source solutions such as Paracite in bibliographic reference parsing along with the description and implementation of a new approach. This approach combines multiple techniques to enhance the accuracy and provide better recognition rates. More work needs to be done to achieve 100% accuracy results. The future scope includes the integration of a more efficient self-learning model into this approach to make the program learn from its mistakes. The refinement of the project can further be done by using a bigger test corpus. Also, the requirement of an existing knowledge domain can be lessened in the future by training it with multiple domains of human knowledge.

References

1. Jewel, M.: Paracite (2003), http://paracite.eprints.org/developers
2. Giuffrida, G., Shek, E.C., Yang, J.: Knowledge-based metadata extraction from PostScript files. In: DL 2000: Proceedings of the fifth ACM conference on Digital libraries, pp. 77–84. ACM Press, New York (2000)
3. Powley, B., Dale, R.: Evidence-based information extraction for high accuracy citation and author name identification. In: Proceedings of RIAO 2007: The 8th Conference on Large-Scale Semantic Access to Content, Pittsburgh, Pa., USA (2007)
4. Sautter, G., Böhm, K., Agosti, D.: A combining approach to find all taxon names (FAT). Biodiv. Inf. 3, 46–58 (2006)
5. Sautter, G., Böhm, K., Agosti, D.: A Quantitative Comparison of XML Schemas for Taxonomic. Biodiversity Informatics (2007)
6. McCallum, A., Nigam, K., Ungar, L.H.: Efficient clustering of high-dimensional data sets with application to reference matching. In: Knowledge Discovery and Data Mining, pp. 169–178 (2000)
7. Hetzner, E.: A simple method for citation metadata extraction using hidden markov models. In: Proceedings of the 8th ACM/IEEE-CS joint conference on Digital Libraries (2008)
8. Takasu: Bibliographic Attribute Extraction from Erroneous References Based on a Statistical Model. In: Proceedings of Joint Conference on Digital Libraries (2003)
9. Huang, I.A., Jan-Ming, H., Kao, H.Y., Lin, S.: Extracting citation metadata from online publication lists using BLAST. In: Dai, H., Srikant, R., Zhang, C. (eds.) PAKDD 2004. LNCS, vol. 3056, pp. 539–548. Springer, Heidelberg (2004)
10. Matt, E.D., Winkels, R., Van Engers, T.: Automated Detection of Reference Structures in Law. In: Proceedings of the Conference at University Pantheon, Assas, Paris II France, pp. 41–50 (2006)
11. Sautter, G., Agosti, D., Böhm, K.: Semi-Automated XML Markup of Biosystematics Legacy Literature with the GoldenGATE Editor. In: Proceedings of PSB, Wailea, HI USA (2007)

Optimized Graph Search Using Multi-Level Graph Clustering

Rahul Kala, Anupam Shukla, and Ritu Tiwari

Department of Information Technology, Indian Institute of Information Technology and Management Gwalior, Gwalior, MP, India
rahulkalaiiitm@yahoo.co.in, dranupamshukla@gmail.com, rt_twr@yahoo.co.in

Abstract. Graphs find a variety of use in numerous domains especially because of their capability to model common problems. The social networking graphs that are used for social networking analysis, a feature given by various social networking sites are an example of this. Graphs can also be visualized in the search engines to carry search operations and provide results. Various searching algorithms have been developed for searching in graphs. In this paper we propose that the entire network graph be clustered. The larger graphs are clustered to make smaller graphs. These smaller graphs can again be clustered to further reduce the size of graph. The search is performed on the smallest graph to identify the general path, which may be further build up to actual nodes by working on the individual clusters involved. Since many searches are carried out on the same graph, clustering may be done once and the data may be used for multiple searches over the time. If the graph changes considerably, only then we may re-cluster the graph.

Keywords: Clustering, Searching, Graph theory, Social Networking Analysis, Web Search Results.

1 Introduction

Graphs are defined as a collection of vertices and edges or G(V,E). In the past years a lot of research in this field has led to many algorithms that are highly efficient. Searching in a graph refers to the algorithm of finding out a goal node, starting from a source node. A key emphasis is given on the length of the path from source to goal and the time taken to reach the goal node. For a good algorithm, the length of the path traversed as well as the time taken should be minimal. Searching is one of the most trivial operations carried out in graphs. Various search algorithms exist like Breadth First Search, Depth First Search, Best First Search, A* Algorithm etc.

Clustering in a graph refers to the grouping of closer nodes to form one cluster. Likewise by constant grouping, various clusters of considerable sizes may be made. Clustering is of ample of use in data analysis in various domains like biomedical, pattern recognition, etc. Various algorithms exist for the clustering of the graph. Some of the prominent ones are C-means fuzzy clustering, k-means clustering, hierarchical clustering.etc.

S. Ranka et al. (Eds.): IC3 2009, CCIS 40, pp. 103–114, 2009.
© Springer-Verlag Berlin Heidelberg 2009

Graphs are being used to model various problems. They have been extensively used to solve many day-to-day problems and find optimal solutions by various algorithms. One of the major applications is in the formation of social network graphs. This is especially motivated by their ample use at the social networking sites. These sites offer each individual to open their account. He may link to various other people by adding them as friends and build his social network. We consider each such user as a vertex of graph. If a person x adds a person y as a friend, we represent it as an edge between x and y. Hence this forms a massive graph of the order of millions of edges. Here we assume that a person links to only those people whom he knows reasonably well. In the absence of this assumption, the number of edges would become very high and the performance of Breadth First Search may exceed all known algorithms.

Another application of the algorithm is the search performed by the search engines. The search engines crawls the web to find results that match the given keyword. A typical search engine ranks the web pages on the basis of their closeness to the entered search query. This may give various kinds of pages. We may also cluster the search results. Very similar web pages may be put in one cluster. E.g. the search 'sun' might refer to sun Microsystems, solar system, sun java etc. These are the various classes the search result may be clustered in. We may hence represent the entire search result as a graph with the edges between any two pages depending upon their closeness to each other. Hence we would be able to form clusters. Initial results may show pages from cluster that is closer to the search query. As the user clicks some of the pages, we would get to know his interests. Then we may show more results from his preferred cluster. Some results may also be shown from the nearby clusters.

Whichever problem we take, we find that many times a series of search operations are called quite frequently, without considerable change in the graph. Every time the search algorithm tries to reach the goal node starting from the source node. In this paper we propose the graph to be clustered once at the starting. The entire graph is clustered to form clusters of some size. These clusters are replaced by new vertices. The edges are placed from this new vertex to any other vertex, depending whether it was connected in any manner to that vertex.

We cluster the graph in a multi layer method. Once the graph is clustered, to give us a smaller graph, we apply the clustering algorithm again to give us a still smaller graph. This process is repeated multiple times unless it is not possible to further cluster the graph (the individual cluster cannot be larger than the desired threshold size).

Instead of applying the searching to the entire graph, we apply it onto the clustered graph, to find the appropriate path. Once we know this, we iterate down level-by-level to every cluster. In these steps, we further come to the actual nodes rather than the ones generated while clustering. In the end we reach the first level. At this level the entire solution is a collection of points in the actual graph given as input.

The clustering may be done once and may not be repeated every time. Even if we wish to add nodes or modify them, we may easily do the same in the various levels of clusters formed. This would ensure that we do not have to re-cluster the graph for a long time ahead.

2 Related Work

With the ever increasing applications of graph and graph theory, it is evident that a lot of research is being carried out in this field. We find various algorithms being developed and various optimizations being carried out in order to improve the searching technique [2, 5, 10, 15]. The traditional searching involves algorithms like Breadth First Search, Depth First search, Heuristic Search etc.

Clustering is another rapidly developing area. In the past few years a lot of work has been done to improve the clustering algorithms and applying them to the graphs [1, 3, 8]. The metrics for the performance evaluation of graph clustering is another major area of research [6, 11]. We find various clustering algorithms like k-Means Clustering, C-Means Fuzzy Clustering, etc. that are being widely used and improved.

Social Networking has got a huge importance because of the ever increase in the number of users [13]. Various kinds of work are being done for their analysis and developing searching techniques. Similarly a lot of study is going on in the field of web searches and website navigation [4, 7, 9]. People are trying to cluster the results using various forms of optimizations [14]. There are works going on to build optimized graphs for the searching of results through keywords. Though this paper is not only limited to the these networks, but we will make a special mention as these networks are the motivation and provide the basic design of the search algorithm.

It may be noted that we assume here that the vertices of the graph are connected to many, but limited number of vertices. If we get vertices that are connected to most of the vertices, the total length of any path from source to destination would be very small. In such a situation, the Breadth First Search would perform very well. In the case of social networking sites, we assume that a person is linked to the other person, only in case he knows the other person relatively well. The absence of this case has lead researchers to conclude social networking graphs as low width graphs.

In this paper, rather than using an already built clustering algorithm, we build a separate algorithm. The reason for this is that we do not need precise clustering, where clusters need to be neatly well apart. We know the kind of application areas in which these algorithms would be put into. Hence we can get a lot of facts to develop faster clustering algorithms that take a lot less time.

The search algorithm is a simple breadth first search, but it is applied on the clustered graphs. This reduces the size of the problem and hence gives faster results. We need to apply the breadth first search at every level of clustering. When we finish, we get the final path that is almost the most optimal path.

3 Algorithm

The whole process of this algorithm is divided into 2 parts. First we cluster the graph. In the second part, we use the multi level clustered graph and search for the goal node starting from the source vertex. We discuss both of these in the next sections.

3.1 Graph Clustering

The first step is to form clusters of the given graph. While forming clusters, we do not lay stress on the quality of the clusters produced, as is the case with many other clustering algorithms. Rather stress is given on the speed of the algorithm. We also have no initial idea as to how many clusters would be formed. This further urges the need of a new clustering algorithm. We discuss clustering in next sections. First we have a look at the multi layer concept and then we present the way we form clusters.

Multi Layer Clustering: Here we form multi layered clusters. The given graph is clustered to bring together vertices that lie close to each other. Each of the clusters is replaced by a single vertex. Hence after clustering we get a graph that is smaller in size as compared to the initial graph. Many vertices are clustered and replaced by one vertex. In the graph that comes out, we again apply the same algorithm. This process of applying algorithm goes on until it is not possible to cluster the graphs up any further, subjected to a maximum of A times. The general procedure of the multi layered clustering is given below. The algorithm is also summarized in figure 1.

MultiLayeredClustering
Step1: g ← original graph
Step2: for i = 1 to A
Step3: g ← makeCluster(g)
Step4: if there is no change in g
Step 5: break
Step 6: add g to graphs

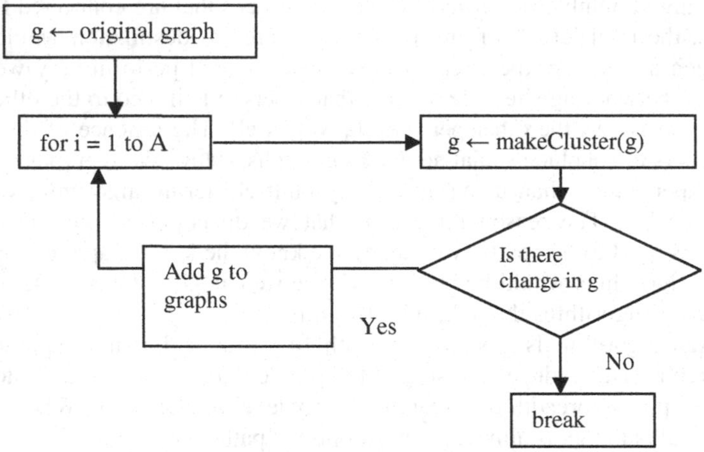

g ← original graph
for i = 1 to A
g ← makeCluster(g)
Is there change in g
Add g to graphs
Yes
No
break

Fig. 1. Algorithm of Multi Layer Clustering

The entire concept of multi-layer clustering can be easily understood with the help of the special case of social network graphs. At the lowest level each user is a vertex. We know that all users who belong to same department in an institute would be connected to each other very well. Hence this may be taken as a cluster separately. Like this we would be having many separate clusters each belonging to some particular department.

We also know that in the resultant graph, all departments of the institute would be strongly connected to each other. Hence we form clusters of every institute. Hence at the second level, graph represents various institutes as vertices. Likewise at the third level, clustering may be in form of state, in fourth level by country. In any general graph, clustering in this manner will reduce the graph size.

Making Clusters: Graph is given as an input to the clustering algorithm. The algorithm divides it into clusters of sufficient size. After we have finished forming clusters one after the other, some vertices are left. These vertices do not belong to any of the clusters. Each cluster is deleted from the graph. Here we remove all the edges and vertices that lie within this cluster. This is then replaced by a single new vertex that represents the entire cluster. This new vertex is connected by edges to all the vertices in the new graph that were connected to any of the vertex in the cluster in the original graph. E.g. figure 2 shows a clustered graph. The graph formed after the vertices and the edges of the new graph have been found out is given in figure 3.

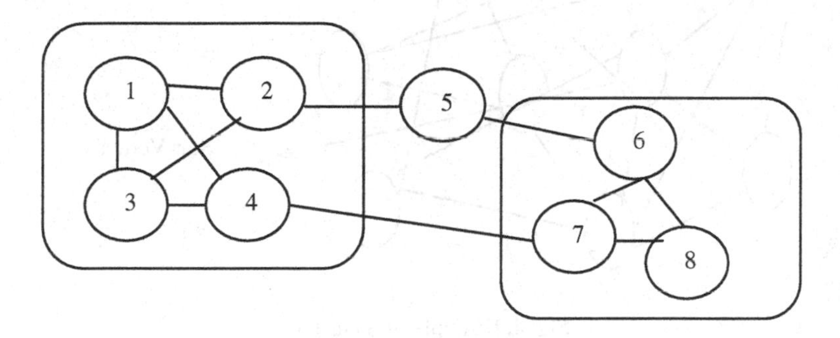

Fig. 2. Original graph

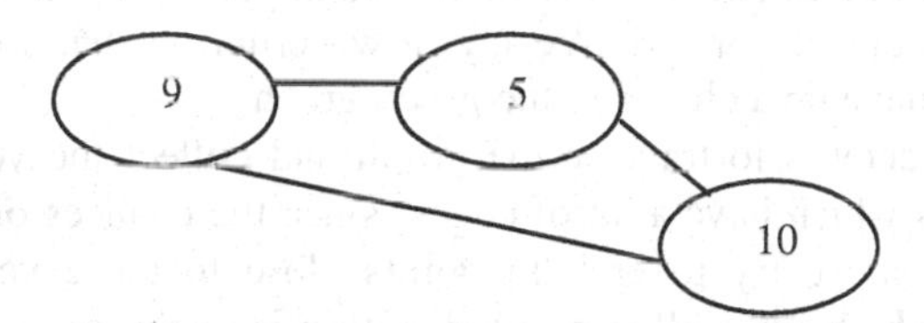

Fig. 3. Graph after clustering

The algorithm for making clusters is as given below.

Make Clusters()
Step1: While more clusters are possible
Step2: c ← getNextCluster()
Step3: for each vertex v in c
Step4: Delete v from graph and delete all edges from/to it
Step5: Add a new unique vertex v_2
Step6: Add edges from/to v_2 that were deleted from the graph
Step7: Add information of cluster v_2 to set of clusters in the particular level

We need to keep a record of all the vertices a cluster contains. Every cluster has a unique identity that is the identity of the vertex that substitutes the cluster. We register the cluster by mentioning all the vertices it contains. This information would be needed later in searching process.

For each cluster, we also register another parameter. This is the star vertex of the cluster. Every cluster has a star vertex. This is the vertex that is centrally located in the cluster. Hence all the vertices of the cluster are very easily reachable from the star vertex. This vertex is usually the vertex with the maximum edges. It is possible for the graph to have many star vertices in it, but we select only one of them. This is explained in figure 4.

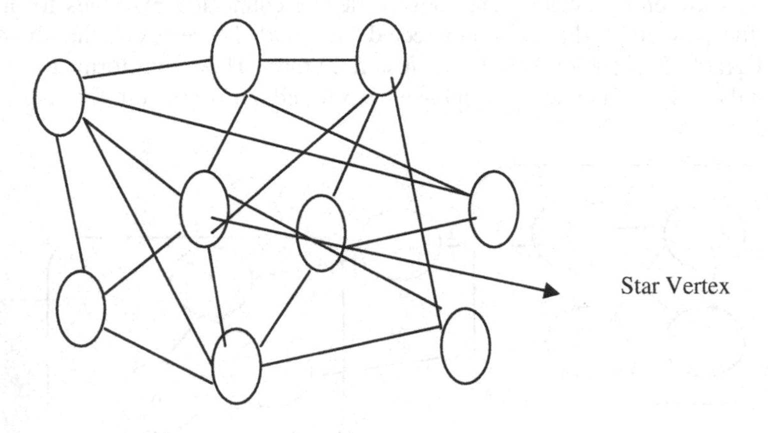

Fig. 4. Example of a cluster

Choosing vertices to form cluster: This is the main part of the algorithm that decide how the clusters would be formed. We lay down criterion that would enable us to select vertices that make up a cluster in the given graph

The general tendency is to traverse the graph and collect the vertices. We try to locate those vertices which have a lot of edges, since the chances of clusters are high at those points. Then we try to grab in points close to these vertices. We try to separate the areas which are well connected within it to the areas which are not so well connected.

In this problem, we define a cluster to be a set of vertices such that almost all the vertices are connected to each other. These vertices form a kind of well-connected architecture. Along with the edges within the members of the cluster, a vertex may have edges pointing to vertices outside the cluster. It is very possible for a vertex to be members of 2 clusters, but while implementing this algorithm, we do not allow any 2 clusters to have a common vertex.

The algorithm to get a cluster from a graph is as given below. In any step we do not regard the vertices representing clusters (of same level) as normal vertices. They are treated as completely different entities.

getNextCluster()
Step1: Find the vertex v in graph with maximum edges
Step2: If the maximum edges are less than α then return null
Step3: c $\leftarrow$ all vertices that are at a maximum distance of 2 units away from v
Step4: Sort c in order of decreasing number of edges of vertices
Step5: Select any 3 vertices v_1, v_2, v_3 in c such that all 3 vertices are connected to β common vertices
Step6: $c_2 \leftarrow$ all vertices in c that are connected to v_1 and v_2 and v_3
Step7: Add all vertices in c to c_2 that are connected to at least 4 vertices already present in c_2
Step8: Return c_2

Here we have tried to get appropriate clusters that are distinctly visible from the surroundings, without making complex steps that would have consumed time. The inspiration of such an algorithm lies from the study of the traits of the data on which it is going to be applied.

As explained above, a cluster is a well-connected network of vertices. We first try to find a vertex that may be fit to serve as a star vertex. The best option is to use the vertex with the maximum edges (Step 1). Then we try to collect all possible points that are near to the star vertex. Since we need to keep all the vertices very close to the star vertex, we travel a maximum 2 unit distance (Step 3).

Now we know that our solution is a subset of this collection. It is very possible that the star vertex collected may be part of more than one cluster. If we do not shorten our collection, we are likely to get the union of all clusters as answer.

In order to restrict ourselves to only one cluster, we select any 3 vertices with the hope that they all belong to same cluster (Step 5). If they belong to the same cluster, they would be having many common vertices to which they are connected (Step 5). Also the common vertices to which they are connected would all be members of the cluster (Step 6). Here we assume that the 3 vertices selected all had high connectivity within the graph. The three vertices selected were also probable candidates for the star vertex.

In this manner we may select members of the cluster. In this method, it may be possible that a vertex is having less number of edges. It may be connected to only a few members of the cluster. Hence we again iterate through all the vertices in collection. If any of them is connected to sufficient number of vertices in the cluster, then we accept this vertex also to be a member of the cluster (Step 7). In this manner we are able to distinguish a cluster from its surroundings.

3.2 Searching

After the clustering is over, we should be able to perform the search operation. Just like the clustering was in multi levels, the searching also would be conducted separately in the entire levels one after the other. We would start with the highest level and reach the lowest level.

The first job in this regard is to find out the source and the destination at the various levels. We know that a source might have been replaced by a cluster at the first level. This new cluster might have been replaced by another cluster at the second level and so on. Hence we need to start from the basic given graph and trace the source and the destination at each level. At the end of this process, we would get to know at each level the points from where searching needs to start and the point where we end. The algorithm for finding out these points is as given below. It can even be summarized as shown in figure 5.

PointSearch
Step1: Level ← 0
Step2: While there are graphs in current level
Step3: If source is a member of any cluster of this level
Step4: Source ← Cluster number where it was found
Step5: If goal is a member of any cluster of this level
Step6: Goal ← Cluster number where it was found
Step7: Add source, destination to point set
Step8: Level ← Level + 1

Now we know the source and the goal node at each and every level. This means that our task is now to start the search from the source and try to reach the goal. We take a solution vector that would be storing the solutions of any level. When we find out the solution at any level, this is stored in this solution vector. When we move from a higher level to a lower level, then we pass this solution vector between levels. The lower level works on the solution vector generated from the higher level graph. It tries to put the points found in its graph, removing the ones found in the higher level clusters.

The algorithm for the search is given by the following algorithm

Search()
Step1: Solution ← null
Step2: For each (source, destination) in point set
Step3: $Solution_2$ ← start + all vertices in solution + destination
Step4: If any vertex in $solution_2$ is a cluster of the higher level
Step5: Replace that vertex with the star vertex of that cluster
Step6: Solution ← null
Step7: For all adjacent vertices (v_1,v_2) in $Solution_2$, taken in order
Step8: Solution ← Solution + bfs(current level graph,v_1,v_2) – v_2
Step9: Solution ← Solution + destination

Here function "bfs(current level graph,v_1,v_2)" refers to the standard breadth first search algorithm that acts on the current graph, takes the source as v_1 and destination as v_2. This algorithm returns the collection of vertices traversed from v_1 to v_2. In case it returns null, this means that the path is not possible. In our system if we get a null in the BFS algorithm, we break out of the complete algorithm, stating that the path was not possible.

In the algorithm we have purposefully deleted all destinations of the BFS algorithm. This is because the destination of the current step would become the source of the next step and would be repeated.

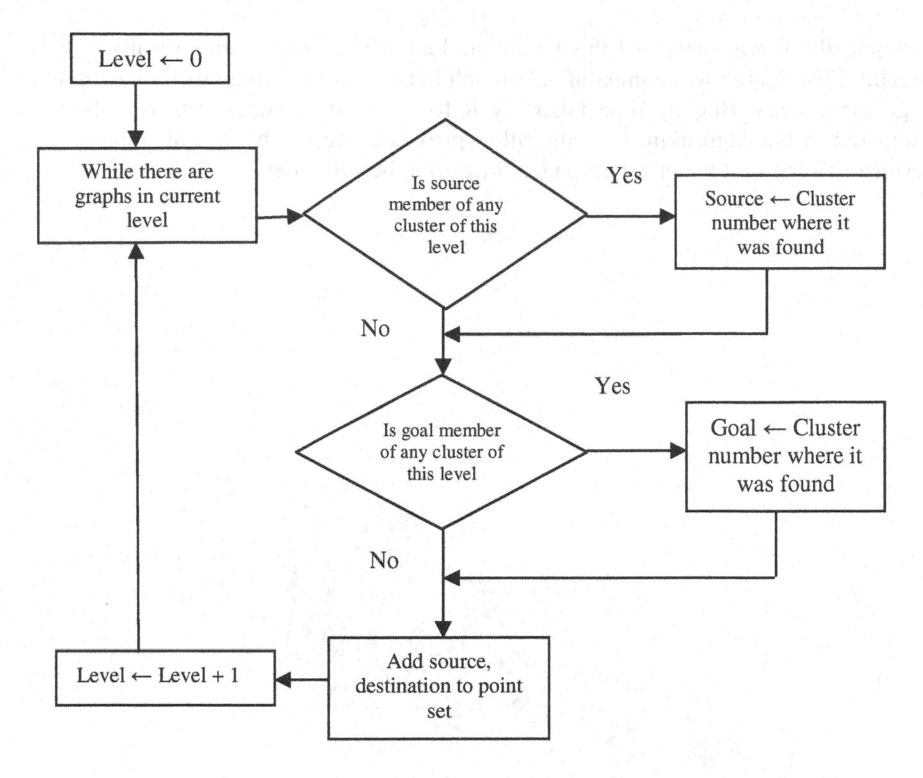

Fig. 5. The point searching algorithm

4 Results

In order to test the working of the algorithm, we coded the algorithm using JAVA and made the program to run in various test cases. Synthetic data was generated to test the algorithm. This data was generated keeping in mind the conditions over which the algorithm would be applied. Multiple search queries were fired to the algorithm. For all of these we found that the algorithm was able to solve the problem and come up with a path from the source to the destination. In order to validate the results, we even coded the standard breadth first search and depth first search and compared the solutions generated by these three algorithms. It was observed that the breadth first search came up with the shortest paths. The depth first search used a lot of nodes and came up with complicated paths. Our algorithm came up with the shortest paths as well, but the algorithm does not guarantee the paths being the shortest.

When we increased the input size, the difference between the breadth first search and the depth first search and our algorithm was clear. The depth first search kept complicating. Our algorithm kept following the paths of optimal solutions and generated good results.

Time saving was the key point of the algorithm. We also studied the time taken by the three search algorithms. In the starting there was not much difference in the running times. As we increased the input size, the depth first search time increased dramatically. It was clear that this technique badly failed to generate results in time. Breadth First Search was comparatively much better and performed well in even quite large input sizes. But the time taken by it for large input sizes was very high as compared to our algorithm. Our algorithm proved to be the best of all three which performed very well when the graph had huge number of nodes.

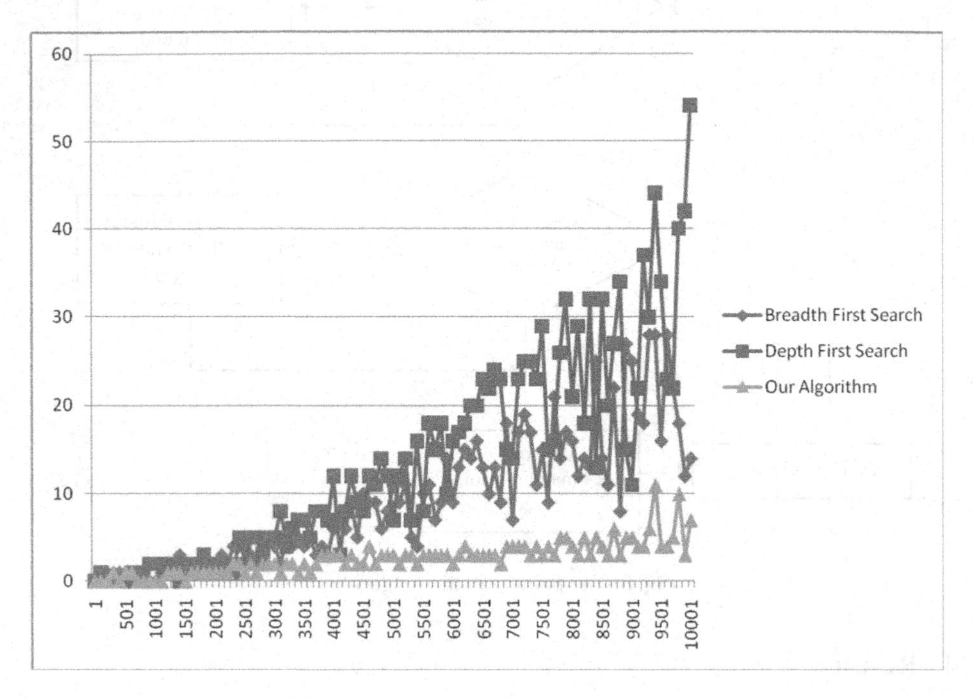

Fig. 6. Time v/s input size graph for the 3 algorithms

We plotted the time taken by the three algorithms with respect to input size. This is given in figure 6. It may be noted here that the search nodes were opted randomly rather than those being the most distant or normally located. In the graph we have omitted the results which we got from the points that were found to be too close to the goal.

5 Conclusion

In this paper we have seen that we were able to search the graphs in almost the most optimal way. We used the multi layered clustering of graph that saved a lot of time that we could have wasted in searching the wrong nodes. Clustering shortened the graph to a great extent. Only the good areas, or the areas required were searched rather than the whole graph.

The experimental results clearly prove the efficiency of the algorithm. When we coded the algorithm and matched its efficiency to the existing algorithms, we clearly saw that our algorithm was very efficient and solved the problem very efficiently.

The basic motive of the algorithm is to apply it to huge amount of data in which searching may be a common activity. The algorithm would prove to be very useful in such situations. The algorithm of clustering is quite flexible and may be customized while implementing to match the needs of the particular database.

The basic construction of the algorithm is made keeping in mind the situation in which it would be applied. It may be noted that the algorithm would work for any general graph as well with better efficiencies. The application of this algorithm in other area needs to be studied. As well as clustering algorithms may be developed to adapt to the specific requirements.

A better study of the individual problems and the algorithm adaptation to these problems may be done in the future. Further we have not modeled the insert and delete operation that may affect the search operation. These operations need to be incorporated in the model and may be done in the future.

References

1. Karl-Heinrich, A.: A Hierarchical Graph-Clustering Approach to find Groups of Objects. In: ICA Commission on Map Generalization, 5th Workshop on Progress in Automated Map Generalization (2003)
2. Arcaute, E., Chen, N., Kumar, R., Liben-Nowell, D., Mahdian, M., Nazerzadeh, H., Xu, Y.: Deterministic Decentralized Search in Random Graphs. In: Bonato, A., Chung, F.R.K. (eds.) WAW 2007. LNCS, vol. 4863, pp. 187–194. Springer, Heidelberg (2007)
3. Ulrik, B., Marco, G., Dorothea, W.: Experiments on graph clustering algorithms. In: Di Battista, G., Zwick, U. (eds.) ESA 2003. LNCS, vol. 2832, pp. 568–579. Springer, Heidelberg (2003)
4. Nick, C., Martin, S.: Random Walks on the Click Graph. iN: SIGIR Conf Research and Development in Information Retrieval 239 (2007)
5. Goldberg Andrew, V., Chris, H.: Computing the Shortest Path: A* Search Meets Graph Theory. In: Proceedings of SODA, pp. 156–165 (2005)
6. Simon, G., Horst, B.: Validation indices for graph clustering. Pattern Recognition Letters 24, 1107–1113 (2003)
7. Hao, H., Haixun, W., Jun, Y., Yu Philip, S.: BLINKS: Ranked Keyword Searches on Graphs. In: The ACM International Conference on Management of Data (SIGMOD), Beijing, China (2007)
8. Adel, H., Shengrui, W.: A Graph Clustering Algorithm with Applications to Content-Based Image Retrieval. In: Proceedings of the Second International Conference on Machine Learning and Cybernetics, Xi'an (2003)
9. Varun, K., Shashank, P., Soumen, C., Sudarshan, S., Rushi, D., Hrishikesh, K.: Bidirectional Expansion For Keyword Search on Graph Databases. In: ACM Proceedings of the 31st international conference on Very large data bases Trondheim, Norway (2005)
10. Marc, N., Wiener Janet, L.: Breadth-First Search Crawling Yields High-Quality Pages. In: ACM Proceedings of the 10th international conference on World Wide Web, Hong Kong (2001)
11. Rattigan Matthew, J., Marc, M., David, J.: Graph Clustering with Network Structure Indices. In: ACM Proceedings of the 24th international conference on Machine learning Corvalis, Oregon (2007)
12. Tadikonda Satish, K., Milan, S., Collins Steve, M.: Efficient Coronary Border Detection Using Heuristic Graph Searching. IEEExplore

13. Yonggu, W., Xiaojuan, L.: Social Network Analysis of Interaction in Online Learning Communities. In: ICALT 2007 Seventh IEEE International Conference on Advanced Learning Technologies (2007)
14. Jing, Y., Baluja, S.: PageRank for Product Image Search. In: WWW 2008. Refereed Track: Rich Media (2008)
15. Rong, Z., Hansen Eric, A.: Sparse-Memory Graph Search. In: 18th International Joint Conference on Artificial Intelligence, Acapulco, Mexico (2003)

An Integrated Framework for Relational and Hierarchical Mining of Frequent Closed Patterns

B. Pravin Kumar, V. Divakar, E. Vinoth, and Radha SenthilKumar

Department of Information Technology, MIT Campus, Anna University, Chennai, India
{piravinjb,vdiva.mit,vinoth69.mit,radhasenthilkumar}@gmail.com

Abstract. This paper addresses an Integrated Framework for relational and hierarchical mining of Frequent Closed Pattern. Large data banks have created the necessity to formulate a system for effective retrieval of data patterns. The major issues that have to be dealt here are granularity of patterns, effectiveness of patterns and time taken for retrieval. Here we discuss Inter-related generalized self-organizing map (IGSOM) and relational attribute-oriented induction (RAOI), which are focused on pattern extraction along with CC-MINER, a hierarchical mining technique for exploring Frequent Closed Pattern from very dense data sets. We further provide implementation results for education data set and prostrate cancer data set.

Keywords: IGSOM, RAOI, CC-Miner, GSOM, EAOI, Cluster, Pattern, Frequent Closed Pattern.

1 Introduction

Data mining uncovers hidden, previously unknown, and potentially useful information from large amounts of data [1]. Compared to the traditional statistical and machine learning data analysis techniques, data mining emphasizes on providing a convenient and complete environment for the data analysis [9]. The proposed framework involves three major phases. They are Inter-related generalized self organizing map (IGSOM), relational attribute-oriented induction (RAOI) and Controlled compact matrix division (CC-Miner). IGSOM and RAOI are mainly concerned with providing relational characteristics (proposed by E.F CODD for relational data bases [4]) to the existing GSOM and EAOI system [2]. CC-Miner performs a controlled mining of the subtasks to generate FCP's.

The important task in the proposed system is to identify the relationship of each and every attribute with the nodes not only in same cluster but also from other distinct clusters. This relationship helps in making the pattern extraction process more diversified with discovery of hidden patterns thus making the system effective. The patterns thus generated needs to be treated for extracting frequent closed patterns over these large data sets. The weight and confidence metrics that are discovered in the RAOI process helps to determine the characteristics of the clusters and perform a controlled mining of the subtasks. Thus we can achieve better performance time.

S. Ranka et al. (Eds.): IC3 2009, CCIS 40, pp. 115–126, 2009.
© Springer-Verlag Berlin Heidelberg 2009

2 Related Works

The self-organizing map (SOM) [5] is an unsupervised neural network which projects high-dimensional data onto a low dimensional grid. To directly handle categorical data such that the system not only faithfully reflects the topological structure of mixed data, but also becomes more user-friendly Generalized Self Organized Map (GSOM) was proposed in [2]. Attribute-oriented induction [6] extracts data patterns in a large amount of data and produces a set of concise rules, which represent the general patterns hidden in the data. The traditional AOI is incapable of revealing major values and suffers from discrediting numeric attributes. To overcome the problems of major values and numeric attributes, an extension to the conventional AOI is proposed in [2] with the name Extended Attribute Oriented Induction (EAOI). The basic idea of the hierarchical FCP [7] mining framework comprises three phases data compaction phase, subtask generation phase and subtask mining phase. The whole mining task is split into independent subtasks and mined individually. Some FCP mining's include Closet [8], CHARM, and CARPENTER. Although they have showed worthy they failed to handle large data. This made the authors review them and create C-miner [3].

3 Contributions of This Paper

This paper provides a comprehensive framework for pattern extraction. The enhancements required for the previous works like GSOM, EAOI and C-miner are addressed here.

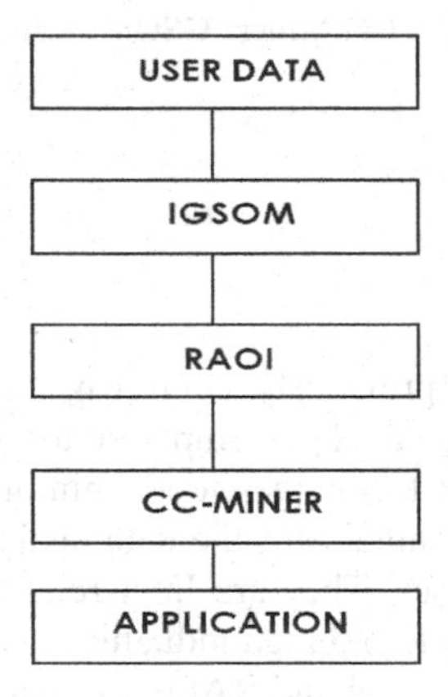

Fig. 1. CPF Framework

Relational structures are identified and applied to clusters in IGSOM phase, while the major enhancement to RAOI is identifying the relational patterns along with characteristics and confidence. CC-miner's purpose is to generate subtask from the relational pattern and perform controlled mining of subtasks.

3.1 CPF Framework

A layered architecture has been devised for the Cluster, Pattern extraction and Frequent closed pattern (CPF) Frameworks. Fig. 1 shows the proposed architecture.

3.2 User Data

This is the raw data that is entered into the system. Input can be numerical data, categorical data or design structures that are used in image recognition process.

3.3 Inter-related Generalized Self Organized Mapping (IGSOM)

The addition of relational structures to GSOM is as follows. Each node in the cluster is provided with a relational structure as shown in the Fig. 2 This structure contains data regarding the relationship with nodes that are present in other clusters. The relation with other cluster can be in any level of the hierarchical tree.

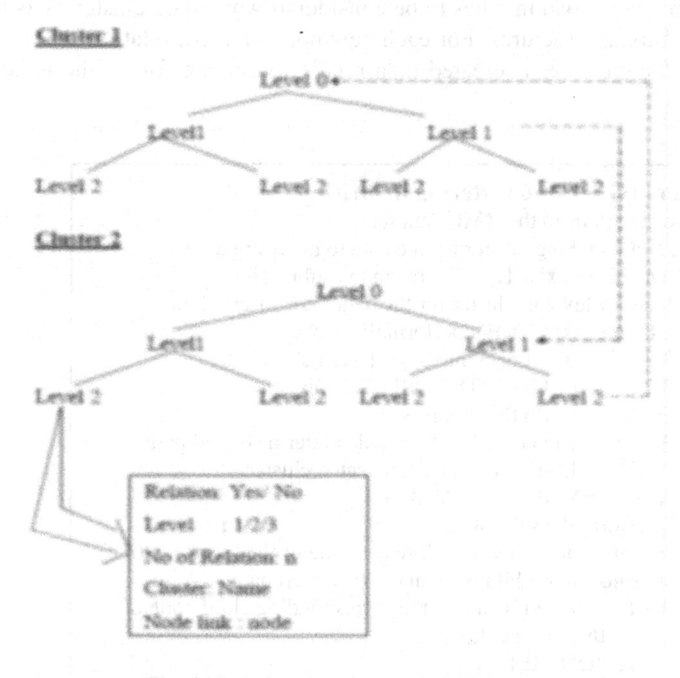

Fig. 2. Relational structures in IGSOM

The Fig. 2 shows that a node in level 1 of cluster 1 contains a relation with a node in level 1 of cluster 2. Again a node in level 2 of cluster 2 contains a relation map to a node at level 0 of cluster 1 this would mean that all the child nodes under level 0 of cluster 1 will fall into this relation extended by the node at level 2 of cluster 2.

Consider an educational information warehouse. One cluster contains data regarding courses available in different geographical region. Another cluster contains data regarding age group of student and number of students enrolled for particular courses. Using the new relational structures we can arrive at patterns like distribution of students over different courses and various regions. Thus the relation between clusters can be effectively used to unhide many more patterns.The objects of the relation structure denote relation with different levels of different clusters. The relation structure contains following attributes

- **Relation:** Specifies whether relation exist or not.
- **Level:** With which level of the cluster does the relationship exists.
- **Number of Relation:** With how many nodes in the level does the relationship exists.
- **Cluster:** Name of the cluster where the relation is dealt with.
- **Node-link:** Identifiers for the nodes

The main algorithm IGSOM makes use of two more algorithms namely Identify_Relation and Relation_all. The main function of IGSOM is to find a specific cluster for the given piece of data. If such cluster doesn't exist then a new cluster has to be created and further specific hierarchy has to be created to accommodate the data. The additional workload that has to be considered while data clustering is the identification of relational structures. For each new node created, relation structure has to be created. If a new cluster is created then relation structure for all the nodes has to be updated.

Algorithm. IGSOM, Clustering algorithm

1. Retrieve data from the XML data sets
2. Identify a matching cluster for the data to be inserted into.
3. IF such cluster exist THEN perform the subtasks
    3.1. Identify level in cluster for the data to be inserted into
    3.2. IF level exist THEN perform the subtasks
        3.2.1. insert data into the selected cluster
        3.2.2. call IDENTIFY_RELATION
    3.3. ELSE perform the subtasks
        3.3.1. Create a new child node under traversed path
        3.3.2. Insert data into the selected cluster
    3.4. call IDENTIFY_RELATION
4. ELSE perform the subtasks
    4.1. Create a new cluster with select node of the hierarchy tree
    4.2. Create a new child node under traversed path
        4.2.1. Insert data into the corresponding child node of the current cluster
5. call RELATION_ALL

Identify_relation. is used to find the relation that exists between a particular node that is selected and all possible nodes that may exhibit a relation with the selected node. Relation_all. is used to perform identify_relation for all the nodes when a new node is created and it can disturb the entire relation structure of the cluster system.

3.4 Relational Attributed Oriented Induction (RAOI)

This focuses on extracting hidden patterns. The main tasks that will be involved in this stage are analyzing the data, obtaining relationship over results of analysis and discovering new patterns. This stage is the core for output stage and shows the major difference between traditional data bases and data mining systems. The patterns that are obtained in this stage are used by CC-Miner for extracting FCP's [7]. The relationship inducted in the IGSOM stage will be used in this stage

For example consider a city's data warehouse. The system will contain data about all the individuals in the city. This can be used to find hidden data from it like majority of individuals are middle aged men. Major cause for resent death in the city is due to a particular epidemic disease. What is the impact of new policies, etc...

A direct relationship exists between maximum number of patterns that can be extracted from the nodes and total number of relationship that exist is shown in (1)

$$P = 2^R - 1 \tag{1}$$

Where R is the number of relationships that exist from a node and P is the total number of patterns that can be extracted from the given node. The formula is derived from subset theory, where maximum subset that can be derived from a set is 2^n. But for us the maximum pattern will be one less that that because empty set can't be included.Two attributes of patterns are extracted they are Relational characteristics and Relational Confidence index

Relational Characteristics. RAOI can be used in extracting following characteristics from each cluster Ck, these Relation characteristics can describe the influence of one character over other and there extent. (2) Gives the characteristic of cluster.

$$C_k = \{[\{a_{i,j} = u \mid a_{i,j}(\mu,\sigma)\},\{ b_{i,j} = v \mid b_{i,j}(\mu,\sigma)\}];W_{a,b}\} \tag{2}$$

Where $a_{i,j}$ is the jth component of the ith pattern extracted from C_k and $b_{i,j}$ is the jth component of the ith pattern extracted from relation that exist between the two nodes. Here we have given $W_{a,b}$ which denotes the weight of the relation that exists between the nodes. The value of $a_{i,j}$ or $b_{i,j}$ is categorical, represented by a concept, or numeric, represented by a pair of mean deviation. Example, North_TN (C_1) = { [{ div = Chennai } , { dept = Information tech }] ; 0.6 } Represents patterns, which take 60% percent supports, extracted from relation.

Relational Confidence. The confidence is used to determine how far a relation can be used over a particular pattern and its corresponding counterpart. (3) Gives the confidence index derivation of a pattern. It determines the usability of a relation and its discovered pattern

$$C_k = \{[\{a_{i,j} = u \mid a_{i,j}(\mu,\sigma)\},\{ b_{i,j} = v \mid b_{i,j}(\mu,\sigma)\}];W_{a,b,}\}\text{->}\{D_{a,b}\} \tag{3}$$

Where $D_{a,b}$ denotes the confidence level that we are talking about in this section. This can provide a valuable property in deriving patterns. North_TN (C_1) = { [{ div = Chennai } , { dept = Information tech }] ; 0.6 }->{0.8}.

Weight is determined by the ratio of number of data satisfying the jth pattern of the relation applied over C_k to total pattern of the relation applied over C_k. Confidence is a machine learning process. All the confidence values are initialized to 1 and they are adjusted based on their usage.

Algorithm. Major Value

1. For each [attribute Ai to be generalized in W]
   1.1. Determine whether Ai should be removed based on the value of β
   1.2. If removed, Construct its major-value set and rearrange the hierarchy of Ai
2. END FOR

Algorithm. Relational_Induction

1. For each [attribute Ai generalized in W]
   1.1. Determine its relation that exists through relation structure
   1.2. Determine its induction according to relation value R.
2. END FOR
3. Calculate weight Wa,b and confidence Da,b of the relation that exist

3.5 CC-MINER

Frequent Closed patterns are discovered through CC-Miner phase. CC-miner is a three staged process and involves Controlled compaction phase, Subtask generation phase and Subtask mining phase.

Table 1. Relational compaction phase in CC-Miner

RL/C	C0	C1	C2	C3	C4	C5	C6	C7
Rl1(r1,r2,3)	0.2	0.4	0	0.4	0.1	0.2	0	0.1
Rl2(r4)	0.3	0	0.2	0.6	0	0.8	0	0
Rl3(r5,r6)	0.6	0.5	0	0	0	0.1	0.5	0

Controlled Compaction Phase. Table 1 shows compacted rows and the corresponding attribute values obtained using (4). The data value in the matrix can be any value between 0 and 1.Thus instead of describing an attribute by its presence, here we denote that an attribute is present only of it has a sufficient $M_{i.j}$. Using a threshold value we can eliminate less important patens identified by their $M_{i.j}$

$$M_{i.j} = \Sigma_{p=(\text{rows in clus})} \Sigma_{q=(\text{column})} W(r_p, c_q) \times D(r_p) \quad (4)$$

Where $W(r_p, c_q)$ weight of the relation between r and c and $D(r_p)$ is the confidence of the relation. This entry denoted the quality of the compacted clusters.

Subtask Generation Phase. Now subtask generation phase follows compaction phase. This phase is similar to ‘partitioning the mining task’ as discussed in [3]. The subtask s1, s2, s3, s4 are formed from new clusters l1, l2, l3 through splitting process by applying cutters. Here Weight and Confidence are determined from the row set and relations that exist between the patterns along with the relation structure that can show the link that exist between other patterns leading to new patterns that exist.

Subtask Mining Phase. Each of the resulting subtasks can be expanded into their original constituent and the weight and confidence can be used to preserve the extent of decomposition. Determination of decomposition extent is done by using a threshold value α. α is determined by the user or we can even allow the system to determine the value using data from previous values and decomposition effectiveness. This α value is compared with the product of $W(r_p, c_q)$ and $D(r_p)$. The subtasks whose product value lesser than α are eliminated. Thus this system provides a better means of selected subtask mining compared to the previous C-miner.

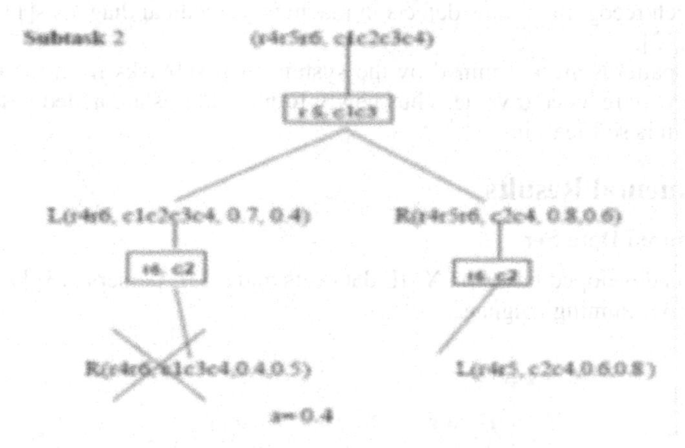

Fig. 3. Subtask mining phase of CC-Miner

Fig. 3 depicts how the controlled mining is carried out. Here we can see that while breaking subtask only the right side portion satisfied the value while the left side doesn't. Hence we don't break left hand side portion further and the current region is sufficient for analysis.

Algorithm. Relational Compaction

1. For each I representing rows in original data set
   - 1.1. IF similar row found in the compaction then perform subtasks
     - 1.1.1. Add it to the previous compaction RC and enter corresponding 1/0 in the matrix
   - 1.2. ELSE perform the following tasks
     - 1.2.1. Add new row cluster, RC and enter corresponding 1/0 in matrix
   - 1.3. Enter original matrix values using formula $M_{i,j} = \Sigma$ p=(rows in clus) Σ q=(column) W(rp ,cq) x D(r p)
2. End For

Algorithm. Subtask Mining

1. For each I a subtask from subtask generation phase
   - 1.1. For J in each row element of subtask I
     - 1.1.1. Extract current element's zero value from matrix
   - 1.2. If confidence greater than or equal to α then perform subtasks
     - 1.2.1. Use the retrieved entry as cutter function
   - 1.3. IF the required FCP's not found and α value can be decremented further THEN
     - 1.3.1. Deferred Subtask analysed by decrementing α value
   - 1.4. End For
2. End For

3.6 Application

The pattern thus extracted can find a wide range of application. Their application includes engineering application like[10] texture recognition, process monitoring, control systems, speech recognition, flaw defects in machinery, medical diagnosis[11] financial management[12].

If desired patterns are not mined by the system then subtasks from the CC-Miner stage are run with reduced α value. Thus newly found patterns are added to the system and the system is self learning.

4 Experimental Results

4.1 Educational Data Set

A Prototype is developed based on XML data sets and XML parsers [13] [14] [15] using java as programming language.

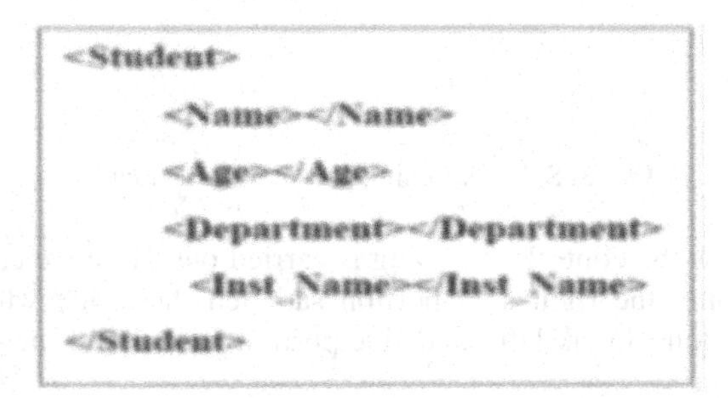

Fig. 4. Student data set schema

A synthetic data set which contains information regarding students, institution, department and strength is used. Fig. 4 shows the sample data set tuple used in the XML data set for synthetic data. In Fig. 5 the cluster created is visualized. Four basic hierarchical clusters are created denoting four parts of the state namely North_TN, South_TN,

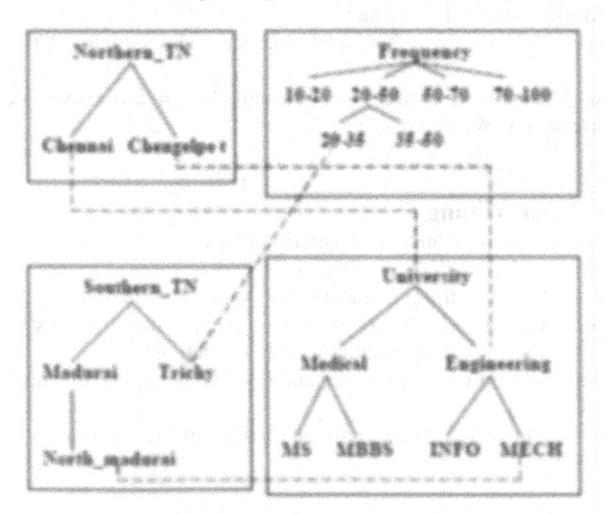

Fig. 5. Clusters created during sample run

East_TN, West_TN under these structures are the cities which fall in these regions. Thus they form a hierarchical structure providing a visual structure of the data sets.

In Fig. 5 we can see that relation exist between Chennai (node) and university cluster's top level (node). This denotes that Chennai will have all the nodes which come under university cluster. Further only medical (node) is available in chengelpet and we can see the frequency relation extended by trichy.

The RAOI helps to increase the number of patterns that are extracted from the clusters formed. The weight and the confidence over relations that are noted over the patterns are tabulated along with the patterns extracted with the help of RAOI in this phase. Table 2 shows the observed pattern and their attributes.

Table 2. Pattern extracted from RAOI along with weight and confidence

Region	City	Dept	Freq	Weight	Confidence
North_TN	Chennai	University	20-30	0.8	0.5
North_TN	Cglpet	Medical	0-10	0.2	0.3
South_TN	Madurai	Info_tech	40_50	0.6	0.4
South_TN	Madurai	Medical	10_20	0.7	0.8
East_TN	Pondy	MBBS	0_10	0.8	0.7
East_TN	Tanavur	MS	35_50	0.6	0.4
West_TN	Kovai	Mech	70_100	0.7	0.9

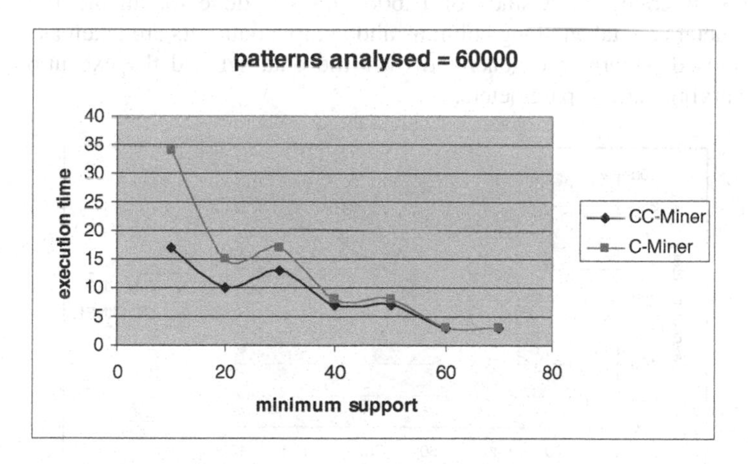

Fig. 6. Execution time vs. minimum support with constant pattern strength of 60000

A comparative analysis of C-Miner and CC-Miner is shown here. In Fig 6 the variation of execution time with minimum support is observed over a constant number of patterns. It is inferred here that the execution time of CC-Miner is less that C-Miner for most of the cases. For higher support values, though it may appear both C-Miner and CC-Miner have same execution time but it is not. Actually the difference is not visible in second's unit. The difference can be visualized only in the milliseconds band.

This performance study shown in Fig. 7 reveals the variation of execution time with number of patterns over constant minimum support. Here we can see that execution time of the mining process increases with the number of patters because more number of patterns implies more time to process them.

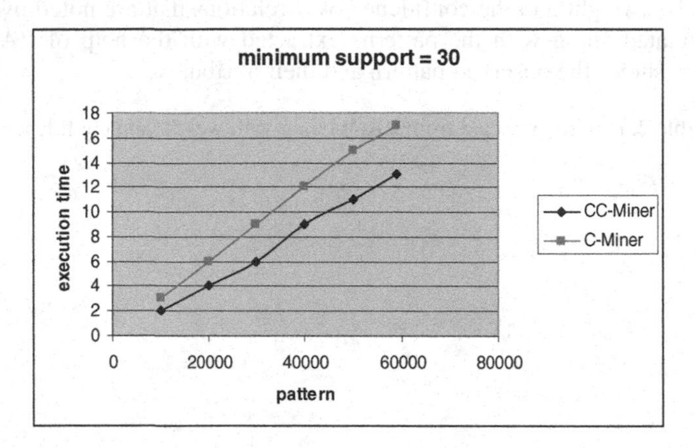

Fig. 7. Pattern strength vs. execution time with a constant minimum support of 30

4.2 Prostrate Cancer Data Set

This data set contains experimental results for testing of presence of prostrate caner in patients. Further the effect of genes over the disease is also studied. A test sample of 102 patients is collected. A study of 12600 genes is done for all the 102 patients. Gene parameters are taken along columns and sample data sets are taken as rows. The Frequent closed patterns are generated over the data set and the execution time is noted by varying various parameters.

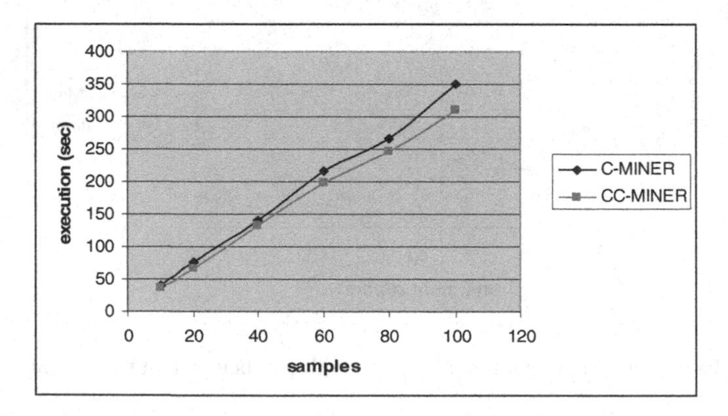

Fig. 8. Samples vs. execution time with minimum support=80 and α=30

Fig. 8 shows the variation of execution time with number of sample data set. We have kept minimum support a constant. The graph is plotted minimum support of 80. The α value is 30. We can see thar the execution time of CC-Miner is better than C-Miner.

5 Conclusion

In this paper we have proposed a novel framework for pattern extraction. The attributes participating in the training of the IGSOM have a significant impact on the result due to the distance metric and relationship structure established in the training algorithm. The outcome of the IGSOM clustering is currently user-dependent, as it is done semi-automatically.The compaction algorithm proposed is designed with the view to reduce the number of traversal while identifying and extracting the patterns. This can be further enhanced to provide data compression in the future. Moreover the sub tasking feature of the CC-Miner can be deployed with parallel computing to achieve optimum output while dealing with enormous amount of data.The number of pattern extracted during RAOI can be limited in auto pattern extraction phase to avoid generation of unnecessary patterns. Whereas the pattern can be generated on demand (when the system is other than real time system) based on our requirements.

References

1. Fayyad, U., Uthurusammy, R.: Data Mining and Knowledge Discovery in Databases. Comm. ACM 39, 24–26 (1996)
2. Hsu, C.-C., Wang, S.-H.: An Integrated Framework for Visualized and Exploratory Pattern Discovery in Mixed Data. IEEE transactions on Knowledge and Data Engineering 18(2) (February 2006)
3. Ji, L., Tan, K.-L., Tung, A.K.H.: Compressed Hierarchical Mining of Frequent Closed Patterns from Dense Data Sets. IEEE transactions on Knowledge and Data Engineering 19(9) (September 2007)
4. Codd, E.F.: Relational Database Model
5. Kohonen, T., Kaski, S., Lagus, K., Salojarvi, J., Honkela, J., Paatero, V., Saarela, A.: Self-Organization of a Massive Document Collection. IEEE Trans. Neural Networks 11(3), 574–585 (2000)
6. Han, J., Cai, Y., Cercone, N.: Data-Driven Discovery of Quantitative Rules in Relational Databases. IEEE Trans. Knowledge and Data Eng. 5, 29–40 (1993)
7. Cong, G., Tan, K.L., Tung, A.K.H., Pan, F.: Mining Frequent Closed Patterns in Microarray Data. In: Proc. Fourth IEEE Int'l Conf. Data Mining (ICDM 2004), pp. 363–366 (2004)
8. Wang, J., Han, J., Pei, J.: CLOSET+: Searching for the Best Strategies for Mining Frequent Closed Itemsets. In: Proc. Ninth ACM SIGKDD Int'l Conf. Knowledge Discovery and Data Mining (KDD 2003), pp. 236–245 (2003)
9. Jain, A.K., Dubes, R.C.: Algorithms for Clustering Data. Prentice Hall, Englewood Cliffs (1988)
10. Kohonen, E., Oja, O., Simula, A.: Engineering Applications of the Self-Organizing Map. Proc. IEEE 84(10), 1358–1384 (1996)

11. Chen, D.R., Chang, R.F., Huang, Y.L.: Breast Cancer Diagnosis Using Self-Organizing Map for Sonography. Ultrasound in Medicine and Biology 1(26), 405–411 (2000)
12. Kasabov, N., Deng, D., Erzegovezi, L., Fedrizzi, M., Beber, A.: On-Line Decision Making and Prediction of Financial and Macroeconomic Parameters on the Case Study of the European Monetary Union. In: Proc. ICSC Symp. Neural Computation (2000)
13. Xpath, w3schools, `http://www.w3schools.com/xpath/default.asp`
14. Xpath traverasals, w3schools, `http://www.w3.org/TR/xpath`
15. XML schemas, `http://www.developer.com/xml/article.php`

A Modified Differential Evolution Algorithm with Cauchy Mutation for Global Optimization

Musrrat Ali, Millie Pant, and Ved Pal Singh

Department of Paper Technology,
Indian Institute of Technology Roorkee, Saharanpur Campus, Saharanpur – 247001, India
musrrat.iitr@gmail.com, millifpt@iitr.ernet.in,
singhvp2@yahoo.co.in

Abstract. Differential Evolution (DE) is a powerful yet simple evolutionary algorithm for optimization of real valued, multi modal functions. DE is generally considered as a reliable, accurate and robust optimization technique. However, the algorithm suffers from premature convergence, slow convergence rate and large computational time for optimizing the computationally expensive objective functions. Therefore, an attempt to speed up DE is considered necessary. This research introduces a modified differential evolution (MDE), a modification to DE that enhances the convergence rate without compromising with the solution quality. In Modified differential evolution (MDE) algorithm, if an individual fails in continuation to improve its performance to a specified number of times then new point is generated using Cauchy mutation. MDE on a test bed of functions is compared with original DE. It is found that MDE requires less computational effort to locate global optimal solution.

Keywords: Differential evolution, Cauchy mutation.

1 Introduction

DE was proposed by Storn and Price [1] in 1995. It soon became a popular tool for solving global optimization problems because of several attractive features like having fewer control parameters, ease in programming, efficiency etc. DE is similar to GAs in the sense that it uses same evolutionary operators like mutation, crossover and selection for guiding the population towards the optimum solution. Nevertheless, it's the application of these operators that makes DE different from GA. The main difference between GAs and DE is that; in GAs, mutation is the result of small perturbations to the genes of an individual while in DE mutation is the result of arithmetic combinations of individuals. DE has been successfully applied to solve a wide range of real life application problems [2–4] and has reportedly outperformed other optimization techniques [5–7].

Despite several positive features, it has been observed that DE sometimes does not perform as good as the expectations. Empirical analysis of DE has shown that it may stop proceeding towards a global optimum even though the population has not converged to a local optimum [8]. The situation when the algorithm does not show any improvement though it accepts new individuals in the population is known as

S. Ranka et al. (Eds.): IC3 2009, CCIS 40, pp. 127–137, 2009.
© Springer-Verlag Berlin Heidelberg 2009

stagnation. Besides this, DE also suffers from the problem of premature convergence. This situation arises when there is a loss of diversity in the population. As a result the entire population converges to a point which may not even be a local optimal solution. It generally takes place when the objective function is multi modal having several local and global optima. Like other EA, the performance of DE deteriorates with the increase in dimensionality of the objective function. Several modifications have been made in the structure of DE to improve its performance [9 – 14]. In the present study we propose two modifications in the basic scheme of DE. The first modification is the concept of *failure_counter* which counts that a particular individual how many times fail to improve its performance and the second modification is to use the Cauchy mutation to generate a new trial vector.

The remaining of the paper is organised as follows; in Section 2, we give a brief description of DE. In Section 3, we describe the proposed DE version. In Section 4, experimental settings and numerical results are given. The paper concludes with section 5. Benchmark problems taken for study are given in the appendix.

2 Basic DE

As mentioned earlier in the previous section, there are several variants of the DE-algorithm [15]. Throughout the present study we shall follow the version *DE/rand/1/bin*, which is apparently the most commonly used version and shall refer to it as basic version. This particular scheme is briefly described as follows:

DE starts with a population of NP candidate solutions which may be represented as $X_{i,G}$, $i = 1, \ldots, NP$, where i index denotes the population and G denotes the generation to which the population belongs. The working of DE depends on the manipulation and efficiency of three main operators; mutation, reproduction and selection which briefly described in this section.

Mutation: The mutation operation of DE applies the vector differentials between the existing population members for determining both the degree and direction of perturbation applied to the individual subject of the mutation operation. The mutation process at each generation begins by randomly selecting three individuals in the population. The ith perturbed individual, $V_{i,G+1}$, is then generated based on the three chosen individuals as follows:

$$V_{i,G+1} = X_{r3,G} + F * (X_{r1,G} - X_{r2,G}) \tag{1}$$

Where, $i = 1, \ldots, NP$, $r_1, r_2, r_3 \in \{1, \ldots, NP\}$ are randomly selected and satisfy: $r_1 \neq r_2 \neq r_3 \neq i$, $F \in [0, 1+]$, F is the control parameter proposed by Storn and Price [1].

Crossover: The perturbed individual, $V_{i,G+1} = (v_{1,i,G+1}, \ldots, v_{n,i,G+1})$, and the current population member, $X_{i,G} = (x_{1,i,G}, \ldots, x_{n,i,G})$, are then subject to the crossover operation, that finally generates the population of candidates, or "trial" vectors, $U_{i,G+1} = (u_{1,i,G+1}, \ldots, u_{n,i,G+1})$, as follows:

$$u_{j,i.G+1} = \begin{cases} v_{j,i.G+1} \; if \; rand_j \leq C_r \vee j = k \\ x_{j,i.G} \qquad otherwise \end{cases} \tag{2}$$

Where, j = 1. . . n, k ∈ {1, . . . , n} is a random parameter's index, chosen once for each *i*, and the crossover rate, Cr ∈ [0, 1], the other control parameter of DE, is set by the user.

Selection: The population for the next generation is selected from the individual in current population and its corresponding trial vector according to the following rule

$$X_{i.G+1} = \begin{cases} U_{i.G+1} \; if \; f(U_{i.G+1}) \leq f(X_{i.G}) \\ X_{i.G} \qquad \qquad otherwise \end{cases} \tag{3}$$

Thus, each individual of the temporary (trial) population is compared with its counterpart in the current population. The one with the lower objective function value will survive from the tournament selection to the population of the next generation. As a result, all the individuals of the next generation are as good as or better than their counterparts in the current generation. In DE trial vector is not compared against all the individuals in the current generation, but only against one individual, its counterpart, in the current generation.

3 Modified Differential Evolution

Differential evolution algorithm described above typically converge relatively rapidly in the initial stage of the search and then slow down as it approach to global minimum and some time a particular individual fails continuously many times to improve its performance. The goal of this research is to allow that individuals of the population escape from the local minima when they are prematurely attracted to local attractor. If there is an improvement in fitness from generation to generation, it means new points generated by DE is better than target and it replace the target vector in next generation, if there is no improvement in fitness of an individual , then it is a sign that for this particle second term of eq (1) is zero. In such a case it is necessary to introduce a jump to a new point in the search space, which may help to escape the local attractor. This can be done by introducing a *failure_counter* which monitors the improvement of the fitness value for each individual for a pre-specified number of generations. If there is no improvement in fitness, then the *failure_counter* is increased by one in each generation till this value achieves a pre-specified value MFC (maximum failure counter), then the individual should jump to a new point. This can be carried out by introducing a mutation operator to change the position of the individual. Here we used Cauchy mutation [16] whose pdf

$$f(x; x_0, \gamma) = \frac{1}{\pi\gamma\left[1+\left(\frac{x-x_0}{\gamma}\right)^2\right]} = \frac{1}{\pi}\left[\frac{\gamma}{(x-x_0)^2+\gamma^2}\right] \tag{4}$$

Is given by equation (4), where x_0 is the location parameter, specifying the location of the peak of the distribution, and γ is the scale parameter which specifies the half-width at half-maximum. The graph of pdf for different values of x_0 and γ is shown in figure 1. A new trial vector $U_{i,G+1} = (u_{1,i,G+1}, \ldots, u_{n,i,G+1})$ by MDE is generated as follows:

$$u_{j,i,G+1} = \begin{cases} x_{j,best,G} + C(\gamma, 0) & if\ rand(0,1) \leq 0.9 \\ x_{j,i,G} & otherwise \end{cases} \tag{5}$$

Where $C(\gamma, 0)$ stands for random number generated by Cauchy probability distribution with scale parameter γ centred at origin. After generation of new point selection criteria is same as for DE.

This modification enables the algorithm to get a better tradeoff between the convergence rate and robustness. Thus it is possible to increase the convergence rate of the differential evolution algorithm and thereby obtain an acceptable solution with a lower number of objective function evaluations. Such an improvement can be advantageous in many real-world problems where the evaluation of a candidate solution is a computationally expensive operation and consequently finding the global optimum or a good suboptimal solution with the original differential evolution algorithm is too time-consuming or even impossible within the time available.

In MDE, we use the concept described above. The basic structure of MDE is same as DE. The major difference between DE and MDE is that MDE uses two mutation operator, one basic De mutation and second Cauchy mutation. The working procedure of the algorithm is outlined below:

Step 1: Initialization: Generate randomly NP vectors, each of n dimensions, $X_{i,j}= X_{min,j} + rand(0, 1)(X_{max,j}-X_{min,j})$, where $X_{min,j}$ and X_{max} are lower and upper bound for j^{th} component respectively, rand(0,1) uniform random number between 0 and 1. Calculate the objective function value $f_i= f(X_i)$ for all X_i , input scaling factor F, crossover rate C_r, *failure_counter[i]=0* and MFC.

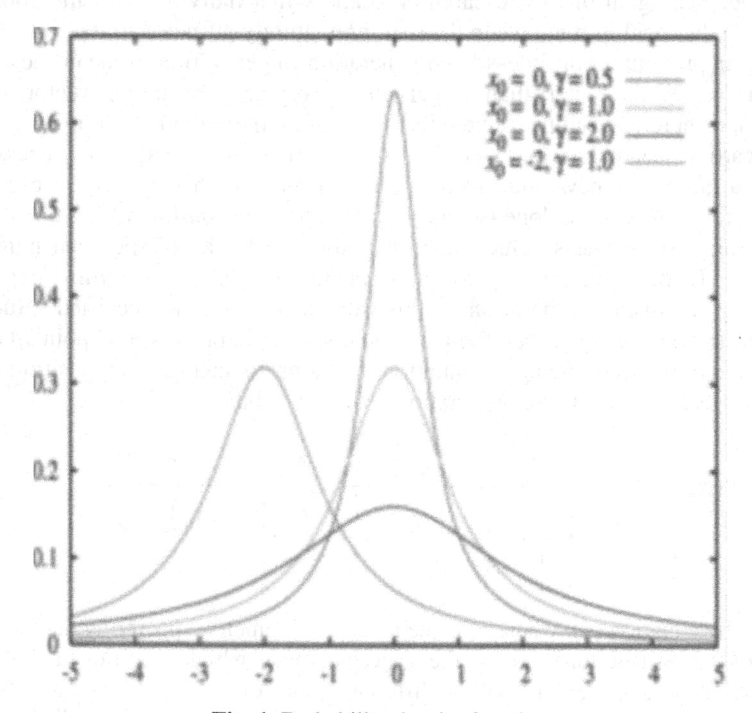

Fig. 1. Probability density function

Step 2: set i=0.
Step 3: i=i+1.
Step 4: If *failure_counter[i]* < MFC then go to step 5 otherwise go to step 7.
Step 5: *Mutation:* Select three points from population and generate perturbed individual V_i using equation (1).
Step 6: *Crossover:* Recombine the each target vector X_i with perturbed individual generated in step 5 to generate a trial vector U_i using equation (2) and go to step 8.
Step 7: Generate a trial vector using equation (5).
Step 8: *Selection:* Calculate the objective function value for vector U_i. Choose better of the two (function value at target and trial point) using equation (3) for next generation if trial vector is selected for next generation then *failure_counter[i]=0* otherwise increase the value of *failure_counter[i]* by one.
Step 9: *Iteration*: If i<NP then go to step 3 otherwise go to step 10.
Step10: Check whether the termination criterion met if yes then stop otherwise go to step 2.

4 Experimental Settings and Numerical Results

With DE, the lower limit for NP is 4 since the mutation process requires at least three other chromosomes for each parent. While testing the algorithms, we began by using the optimized control settings of DE. Population size, NP can always be increased to help maintain population diversity. As a general rule, an effective NP is between 3 * n and 10 * n, but can often be reduced to minimize the NFE. For the present study we performed various experiments with the population size as well as with the crossover rate and mutation probability rate and observed that for problems up to dimension 30 a population size of 10*n is sufficient. Values of F scale outside the range of 0.4 to 1.2 are rarely effective, so F=0.5 is usually a good initial choice. In general C_r should be as large as possible to speed up the convergence so in this study we have taken C_r =0.5. All the algorithms are executed on a PIV PC, using DEV C++, thirty times for each problem. Random numbers are generated using the inbuilt random number generator *rand ( ) function* available in DEVC++. In every case, a run was terminated when the best function value obtained is less than a threshold for the given function or when the maximum number of function evaluation (NFE=10^6) was reached. In order to have a fair comparison, these settings are kept the same for all algorithms over all benchmark functions during the simulations. For MDE the one extra variable MFC is assigned a value five.

4.1 Numerical Results

Table 1 gives the average best fitness function value, standard deviation and t-value obtained after 30 runs. From this Table it can be clearly observed that in terms of average fitness function value the proposed schemes performed at par with the original DE and in some cases even gave better results. The finer performance of the proposed versions is more evident from Table 2 which gives the average number of function evaluations and the time taken by the algorithms to meet the stopping criteria. This Table shows that in almost all the test cases, the proposed versions converged faster than the original DE except for the functions f_{GP} and f_{H3}. Performance curves (convergence graphs) of few selected functions are given in Fig2(a) – Fig2(d). From these illustrations it is evident that the convergence of proposed algorithm is faster than basic DE.

Table 1. Comparison of average fitness function value and standard deviations

Fun	Dim	Fitness Standard deviation		t-value
		DE	MDE	
f_{EP}	2	-0.999999 6.98197e-07	-0.999999 1.25952e-06	0
f_{FX}	2	0.998004 3.62360e-10	0.998004 1.79905e-10	0
f_{CB6}	2	-1.03163 8.17617e-07	-1.03163 8.16449e-07	0
f_{GP}	2	3 1.07046e-06	3 6.80806e-07	0
f_{H3}	3	-3.8623 1.46942e-06	-3.8623 7.19190e-07	0
f_{SP}	30	4.96717e-05 7.77918e-06	4.73394e-05 5.34452e-06	1.33
f_{ACK}	30	0.00014280 1.79218e-05	0.00014619 2.24043e-05	0.63
f_{SWF}	30	0.00072896 3.85744e-06	0.00072826 4.91473e-06	0.60
f_{GW}	30	4.62272e-05 9.03396e-06	4.57326e-05 8.40580e-06	0.21
f_{LM2}	30	8.25033e-05 7.39547e-06	4.95371e-05 7.22451e-06	17.17

Table 2. Comparison of average function evaluation and time in second for all algorithms

Fun	Dim	NFE		% Impr-	Time	
		DE	MDE		DE	MDE
f_{EP}	2	833	670	19.56	0.10	.06
f_{FX}	2	977	482	50.66	0.11	.07
f_{CB6}	2	1020	712	30.19	0.11	.06
f_{GP}	2	970	988	0.00	0.11	.12
f_{H3}	3	1170	1284	0.00	0.10	.11
f_{SP}	30	146400	141600	3.27	10.10	9.3
f_{ACK}	30	259410	232740	10.28	18.90	16.2
f_{SWF}	30	366570	130800	64.31	5.20	1.8
f_{GW}	30	224910	133440	40.66	17.10	10
f_{LM2}	30	182700	133770	26.78	25.2	17.2

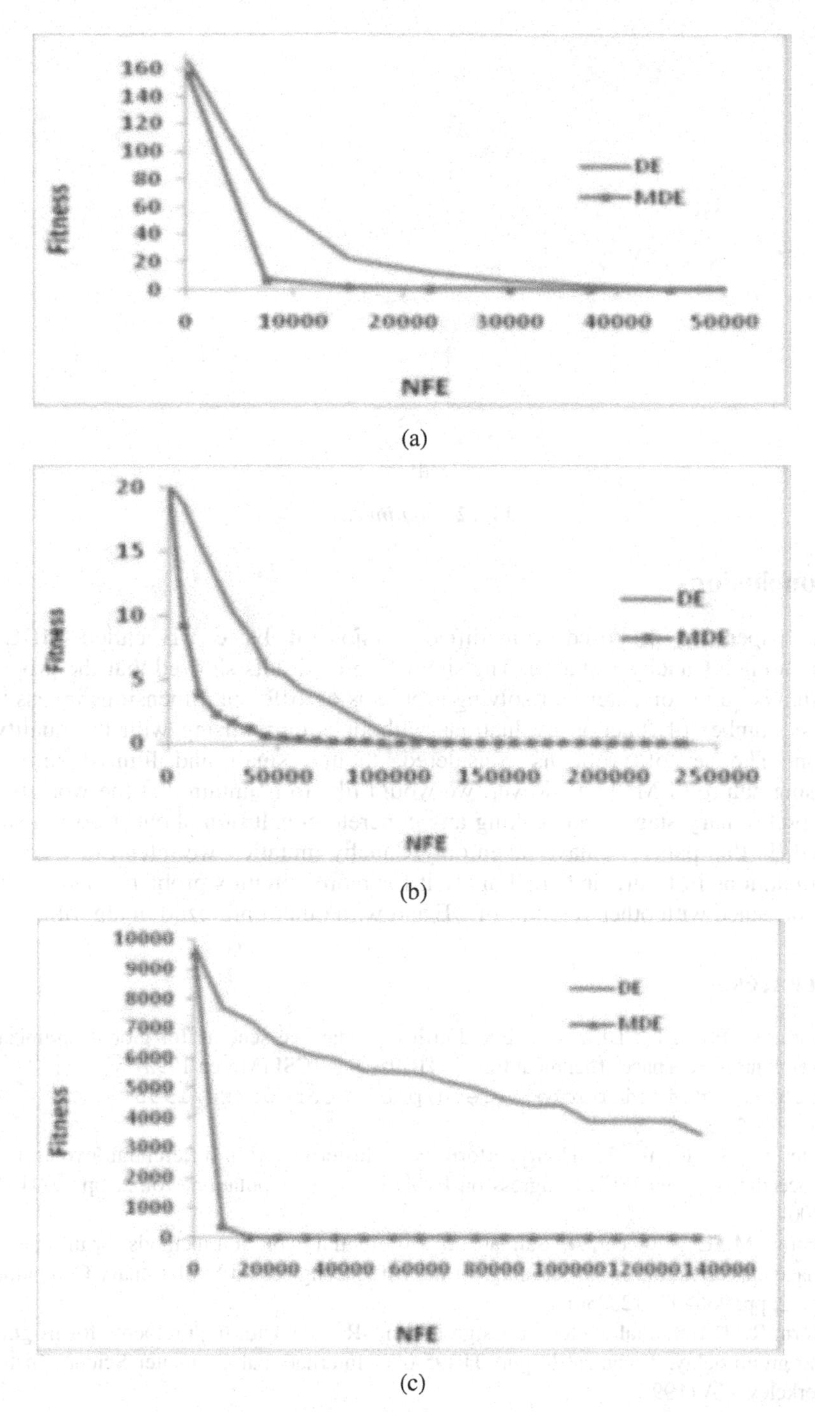

Fig. 2. (a). Performance curves of DE vs. MDE for function f_{SP}. (b). Performance curves of DE vs. MDE for function f_{ACK}. (c). Performance curves of DE vs. MDE for function f_{SWF}. (d). Performance curves of DE vs. MDE for function f_{GW}.

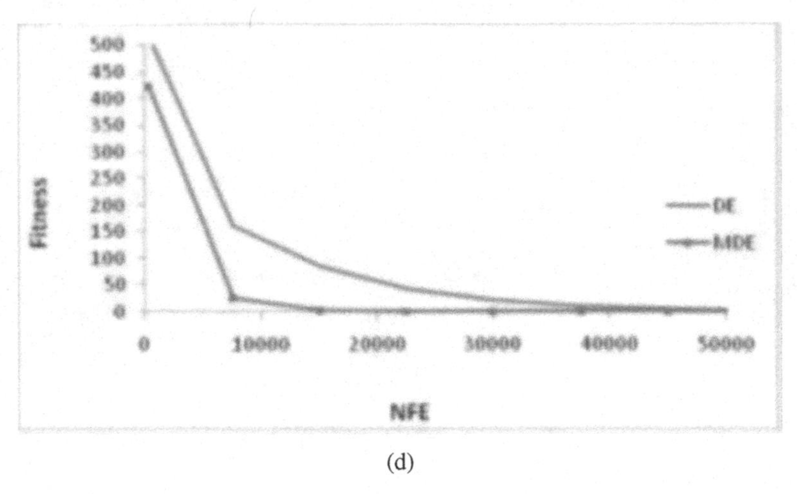

(d)

Fig. 2. *(continued)*

5 Conclusions

In this paper we proposed a modified version of basic DE called MDE by incorporating a Cauchy mutation. The simulation of results showed that the proposed algorithm is quite competent for solving problems of different dimensions in less time and less number of function evaluations without compromising with the quality of solution. The set of problems considered, though small and limited show the promising nature of MDE. However, we would like to maintain that the work is still in the preliminary stages and?making any concrete conclusion about it do not sound justified. In this paper we have taken only Cauchy mutation we intend to work with more mutations in future and shall apply it for more complex problems and compare its performance with other versions of DE and with other optimization algorithms.

References

1. Storn, R., Price, K.: DE-a simple and efficient adaptive scheme for global optimization over continuous space, Technical Report TR-95-012, ICSI (March 1995), `http://ftp.icsi.berkeley.edu/pub/techreports/1995/tr-95-012.ps.Z`
2. Paterlini, S., Krink, T.: High performance clustering with differential evolution. In: Proceedings of the IEEE Congress on Evolutionary Computation, vol. 2, pp. 2004–2011 (2004)
3. Omran, M., Engelbrecht, A., Salman, A.: Differential evolution methods for unsupervised image classification. In: Proceedings of the IEEE Congress on Evolutionary Computation, vol. 2, pp. 966–973 (2005a)
4. Storn, R.: Differential evolution design for an IIR-filter with requirements for magnitude and group delay. Technical Report TR-95-026, International Computer Science Institute, Berkeley, CA (1995)
5. Vesterstroem, J., Thomsen, R.: A comparative study of differential evolution, particle swarm optimization, and evolutionary algorithms on numerical benchmark problems. Proc. Congr. Evol. Comput. 2, 1980–1987 (2004)

6. Andre, J., Siarry, P., Dognon, T.: An improvement of the standard genetic algorithm fighting premature convergence in continuous optimization. Advance in Engineering Software 32, 49–60 (2001)
7. Hrstka, O., Kučerová, A.: Improvement of real coded genetic algorithm based on differential operators preventing premature convergence. Advance in Engineering Software 35, 237–246 (2004)
8. Lampinen, J., Zelinka, I.: On stagnation of the differential evolution algorithm. In: Ošmera, P. (ed.) Proc. of MENDEL 2000, 6th International Mendel Conference on Soft Computing, pp. 76–83 (2000)
9. Zaharie, D.: Control of population diversity and adaptation in differential evolution algorithms. In: Matousek, D., Osmera, P. (eds.) Proc. of MENDEL 2003, 9th International Conference on Soft Computing, Brno, Czech Republic, pp. 41–46 (2003)
10. Abbass, H.: The self-adaptive pareto differential evolution algorithm. In: Proc. of the 2002 Congress on Evolutionary Computation, pp. 831–836 (2002)
11. Omran, M., Salman, A., Engelbrecht, A.P.: Self-adaptive differential evolution. In: Hao, Y., Liu, J., Wang, Y.-P., Cheung, Y.-m., Yin, H., Jiao, L., Ma, J., Jiao, Y.-C. (eds.) CIS 2005. LNCS (LNAI), vol. 3801, pp. 192–199. Springer, Heidelberg (2005)
12. Brest, J., Greiner, S., Boškovic, B., Mernik, M., Žumer, V.: Self-adapting Control parameters in differential evolution: a comparative study on numerical benchmark problems. IEEE Transactions on Evolutionary Computation 10(6), 646–657 (2006)
13. Rahnamayan, S., Tizhoosh, H.R., Salama, M.M.A.: Opposition-Based Differential Evolution. IEEE Transactions on Evolutionary Computation 12(1), 64–79 (2008)
14. Chakraborty, U.K. (ed.): Advances in Differential Evolution. Springer, Heidelberg (2008)
15. Price, K.: An introduction to DE. In: Corne, D., Marco, D., Glover, F. (eds.) New Ideas in Optimization, pp. 78–108. McGraw-Hill, London (1999)
16. Stacey, A., Jancie, M., Grundy, I.: Particle swarm optimization with mutation. In: Proceeding of IEEE congress on evolutionary computation, pp. 1425–1430 (2003)

Appendix

Benchmark Problems:

1. Easom function (EP):

$$f_{EP}(x) = -\cos(x_1)\cos(x_2)\exp\left[-(x_1-\pi)^2-(x_2-\pi)^2\right]$$

With $-10 \le x_i \le 10$, min $f_{EP}(\pi,\pi) = -1$

2. Foxhole function (FX):

$$f_{FX}(x) = \left(\frac{1}{500} + \sum_{i=1}^{25} \frac{1}{i + \sum_{j=1}^{2} (x_j - a_{ij})^6} \right)^{-1}$$

With $-65.536 \le x_i \le 65.536$,

min $f_{FX}(-32,-32) = .998004$

$$a_{ij} = \begin{pmatrix} -32 & -16 & 0 & 16 & 32 & -32 & \dots & -32 & -16 & 0 & 16 & 32 \\ -32 & -32 & -32 & -32 & -32 & -16 & \dots & 32 & 32 & 32 & 32 & 32 \end{pmatrix}$$

It is a multimodal non separable function having several local optima. Many standard optimization algorithms get stuck in the first peak they find.

3. Six hump Camel back function (CB6):

$$f_{CB6}(x) = 4x_1^2 - 2.1x_1^4 + \frac{1}{3}x_1^6 + x_1x_2 - 4x_2^2 + 4x_2^4 \quad \text{With} -5 \le x_i \le 5,$$

$$\min f_{CB6}(0.0898,-0.7126)/(-0.0898,0.7126) = -1.0316285$$

It is a multimodal separable function having two global optima and four local optima.

4. Goldstien problem (GP):

$$f_{GP}(x) = [1+(x_1+x_2+1)^2(19-14x_1+3x_1^2-14x_2+6x_1x_2+3x_2^2)]$$
$$\times[30+(2x_1-3x_2)^2(15-32x_1+12x_1^2+42x_2-36x_1x_2+27x_2^2)]$$

With $-2 \le x_i \le 2$, min $f_{GP}(0,-1) = 3$

It is a multimodal, non separable function having one global minimum and four local minima.

5. Hartman 3 function (H3):

$$f_{H3}(x) = -\sum_{i=1}^{4} c_i exp\left[-\sum_{j=1}^{3} a_{ij}\left(x_j - p_{ij}\right)^2\right] \quad \text{With} \, 0 \le x_i \le 1,$$

$$\min f_{H3}(.114614,.555649,.852547) = -3.862782$$

I	c_i	a_{ij}			p_{ij}		
		$j=1$	2	3	$j=1$	2	3
1	1	3	10	30	0.3689	0.117	0.2673
2	1.2	.1	10	35	0.4699	0.4387	0.747
3	3	3	10	30	0.1091	0.8732	0 .5547
4	3.2	.1	10	35	0.3815	0 .5743	0.8828

It's a multimodal non separable function having four local minima and one global minimum.

6. Sphere function:

$$f_{SP}(x) = \sum_{i=1}^{n} x_i^2 \quad , \text{With} -5.12 \le x_i \le 5.12,$$

$$\min f_{SP}(0,...,0) = 0$$

It is a simple, continuous Unimodal, separable and highly convex function. It serves as test case for validating the convergence speed of an algorithm.

7. Ackley's function (ACK):

$$f_{ACK}(X) = -20 * \exp\left(-.2\sqrt{1/n\sum_{i=1}^{n} x_i^2}\right) - \exp\left(1/n\sum_{i=1}^{n}\cos(2\pi x_i)\right) + 20 + e,$$

With $-30 \le x_i \le 30$, min $f_{ACK}(0,...,0) = 0$

8. Schwefel's problem (SWF):

$$f_{SWF}(x) = 418.9829 \times n - \sum_{i=1}^{n} x_i \sin\left(\sqrt{|x_i|}\right) \quad \text{With} -500 \le x_i \le 500,$$

min $f_{SWF}(s,...,s) = 0$

Where s=420.97

9. Griewenk function (GW):

$$f_{GW}(x) = \frac{1}{4000}\sum_{i=1}^{n} x_i^2 - \prod_{i=1}^{n}\cos\left(\frac{x_i}{\sqrt{i}}\right) + 1 \quad \text{With} -600 \le x_i \le 600,$$

min $f_{GW}(0,...,0) = 0$

10. Levy and Montalvo 2 Problem (LM2):

$$f_{LM2}(x) = \sin^2(3\pi x_1) + \sum_{i=1}^{n-1}(x_i - 1)(1 + \sin^2(3\pi x_{i+1})) + (x_n - 1)(1 + \sin^2(2\pi x_n))$$

With $-5 \le x_i \le 5$, min $f_{LM2}(1,...,1) = 0$.

Zone Based Hybrid Feature Extraction Algorithm for Handwritten Numeral Recognition of South Indian Scripts

S.V. Rajashekararadhya[1] and P. Vanaja Ranjan[2]

[1] Research Scholar
[2] Asst.professor
Department of Electrical and Electronics Engineering, College of Engineering
Anna University Chennai-600 025, India
svr_aradhya@yahoo.co.in

Abstract. India is a multi-lingual multi script country, where eighteen official scripts are accepted and have over hundred regional languages. In this paper we propose a zone based hybrid feature extraction algorithm scheme towards the recognition of off-line handwritten numerals of south Indian scripts. The character centroid is computed and the image (character/numeral) is further divided in to n equal zones. Average distance and Average angle from the character centroid to the pixels present in the zone are computed (two features). Similarly zone centroid is computed (two features). This procedure is repeated sequentially for all the zones/grids/boxes present in the numeral image. There could be some zones that are empty, and then the value of that particular zone image value in the feature vector is zero. Finally 4*n such features are extracted. Nearest neighbor classifier is used for subsequent classification and recognition purpose. We obtained 97.55 %, 94 %, 92.5% and 95.2 % recognition rate for Kannada, Telugu, Tamil and Malayalam numerals respectively.

Keywords: Handwritten Character Recognition, Feature Extraction Algorithm, Nearest Neighbor Classifier, Indian scripts.

1 Introduction

Handwriting recognition has always been a challenging task in pattern recognition. There are five major stages in the handwritten character recognition (HCR) problem: Image preprocessing, segmentation, feature extraction, training & recognition and post processing. Feature extraction method is probably the most important factor in achieving high recognition performance.

India is multi-lingual and multi-script country comprising of eighteen official (Indian constitution accepted) languages. Research in HCR is popular for various practical application potential such as reading aid for the blind, bank cheque, vehicle number plate, automatic pin code reading of postal mail to sort. There is a lot of demand on Indian scripts character recognition and a review of the OCR work done on Indian language is excellently reviewed in [1]. In [2] a survey on feature extraction methods for character recognition is reviewed. Feature extraction method includes

S. Ranka et al. (Eds.): IC3 2009, CCIS 40, pp. 138–148, 2009.
© Springer-Verlag Berlin Heidelberg 2009

Template matching, Deformable templates, Unitary Image transforms, Graph description, Projection Histograms, Contour profiles, Zoning, Geometric moment invariants, Zernike Moments, Spline curve approximation, Fourier descriptors , Gradient feature and Gabor feature.

We will now briefly review the few important works done towards HCR with reference to the Indian language scripts. In [3] Grid based feature extraction method is used to recognize the handwritten Bangla numerals using multifier classifier. The accuracy of 97.23% is reported. Recognition of conjunctive Bangla Characters by Artificial Neural Network is reported in [4]. Recognition of Devanagari characters using gradient features and fuzzy-neural network are reported in [5] and [6] respectively. We found curvature feature for recognizing Oriya characters in [7].

In [8] zone/grid based feature extraction for handwritten Hindi numerals is reported. For extracting the features, each character image is divided in to 24 zones. By considering the bottom left corner of the image as absolute reference, normalized vector distance is then computed. Character recognition for Telugu scripts using multi-resolution analysis and associative memory is reported in [9]. Fuzzy technique and neural network based off-line Tamil character recognition are found in [10].

In [11] author is described a scheme for extracting features from the gray scale images of the handwritten characters on their state-space map with eight directional space variations. In [12] for feature computation, the bounding box of a numeral image is segmented into blocks and the directional features are computed in each of the blocks. These blocks are then down sampled by a Gaussian filter and the features obtained from the down sampled blocks are fed to a modified quadratic classifier for recognition. Off-line HCR for numeral recognition is also found in [13-17].

Recognition of isolated handwritten Kannada numerals based on image fusion method is found in [13]. In [14] Zone and Distance metric based feature extraction is used. The character centroid is computed and the image is further divided in to n equal zones. Average distance from the character centroid to the each pixel present in the zone is computed. Selection of feature extraction method is also a most important factor for achieving efficient character recognition. In this paper we propose a simple and efficient zone based hybrid feature extraction algorithm. Nearest neighbor classifier (NNC) is used for recognition and classification of numeral image. We have tested our method for Kannada, Telugu, Tamil and Malayalam numerals.

The rest of the paper is organized as follows. In Section 2 we discuss about the brief overview of South Indian scripts. In Section 3 we will briefly explain about data set and preprocessing. In Section 4 we will discuss about proposed methodology. Section 5 describes the experimental results and comparative study. Support vector machine (SVM) classifier is also used and the proposed method is tested with standard Bangla numeral data set in Section 5.1. Conclusion is given in Section 6.

2 Brief Overview of South Indian Scripts

In this section, we will explain the properties of four popular South Indian scripts. Most of the Indian scripts are originated from Brahmi script through various transformations. Writing style of Indian scripts considered in this paper is from left to right, and the concept of upper/lower case is not applicable to these scripts.

Kannada is one of the major Dravidian languages of Southern India and one of the earliest languages evidenced epigraphically in India and spoken by about 50 million people in the Indian state of Karnataka, Tamil Nadu, Andra Pradesh and Maharasthra. The script has 49 characters in its alphasyllabary and is phonetic. The characters are classified into three categories: swaras (vowels), vyanjans (consonants) and yogavaahas (part vowel, part consonants). The script also includes 10 different Kannada numerals of the decimal number system.

Tamil is a Dravidian language and is one of the oldest languages in the world. It is the official language of the Indian state of Tamil Nadu, and also has official status in Sri Lanka, Malaysia and Singapore. Tamil script has 10 numerals, 12 vowels, 18 consonants and five grantha letters. The script however is syllabic and not alphabetic. The complete script therefore consists of 31 letters in their independent form, and an additional 216 combining letters representing every possible combination of a vowel and a consonant.

Telugu is the Dravidian language and it is the third most popular script in India. It is the official language of the southern Indian state, Andhra Pradesh and also spoken by neighboring states. Telugu is also spoken in Bahrain, Fiji, Malaysia, Mauritius, Singapore and the UAE. The Telugu script is closely related to the Kannada script. Telugu is a syllabic language. Similar to most languages of India, each symbol in Telugu script represents a complete syllable. Officially, there are 10 numerals, 18 vowels, 36 consonants, and three dual symbols.

Malayalam is a Dravidian language and it is the eighth most popular scripts in India and spoken by about 30 million people in the Indian state of Kerala. Both the language and the writing system are closely related to Tamil. However Malayalam has its own script. The script has 16 vowels, 37 consonants and 10 numerals.

3 Data Set and Preprocessing

At present no standard South Indian numeral scripts database is neither commercially nor freely available for public. Data collection for the experiment has been done from the different individuals. Currently we are developing data set for Kannada, Telugu, Tamil and Malayalam numerals.

We have collected earlier 2000 Kannada numeral samples from 200 different writers [17]. Writers were provided with the plain A4 sheet and each writer has asked to write Kannada numerals from 0 to 9 for one time. Recently we have collected again 2000 Kannada numerals by 40 different writers. In this paper data set size of 4000 Kannada numerals are used. The database is totally unconstrained and has been created for validating the recognition system.

Similarly we have collected 2000 Telugu numeral samples from 200 different writers, 2000 Tamil numeral samples from 200 writers and 2000 Malayalam numeral samples from 200 different writers. The collected documents are scanned using HP-scan jet 5400c at 300dpi which is usually a low noise and good quality image. The digitized images are stored as binary images in BMP format. A sample of Kannada, Telugu, Tamil and Malayalam handwritten numerals from the data set are shown from Fig 1 to Fig 4 respectively.

Fig. 1. Sample handwritten Kannada numerals 0 to 9

Fig. 2. Sample handwritten Telugu numerals 0 to 9

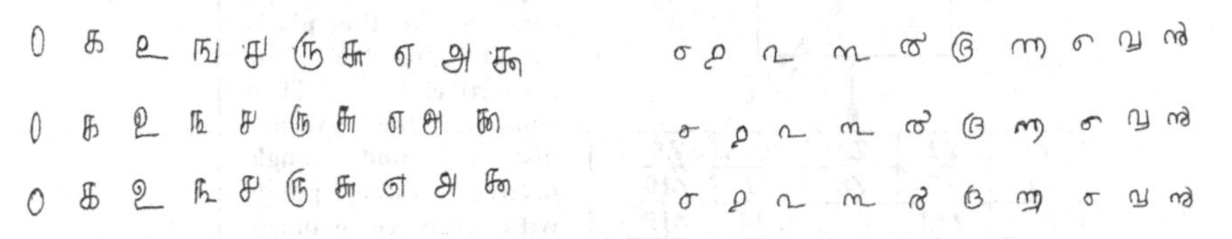

Fig. 3. Sample handwritten Tamil numerals 0 to 9

Fig. 4. Sample handwritten Malayalam numerals 0 to 9

In pre processing step, normalization and thinning are the two important stages. Normalization is required as the size of a numeral varies from person to person and even with the same person from time to time. The input numeral image is normalized to size 50 x 50 after finding the bounding box of the handwritten numeral image. Thinning algorithms have played an important role in the pre processing phase of HCR systems. Thinning is the process of reducing thickness of each line of pattern to just a single pixel. In this work, we have used Morphology based thinning algorithm for better numeral symbol representation. The detail information about the thinning algorithm is available in [18]. Thus the reduced pattern is known as the skeleton and is close to the medial axles, which preserves the topology of the image.

4 Proposed Feature Extraction Methodology

For extracting the feature, the zone based hybrid approach is proposed. The most important aspect of handwriting recognition scheme is the selection of good feature set, which is reasonably invariant with respect to shape variations caused by various writing styles. The major advantage of this approach stems from its robustness to small variation, ease of implementation and provides good recognition rate. Zone based feature extraction method provides good result even when certain preprocessing steps like filtering, smoothing and slant removing are not considered. In this section, we explain the concept of feature extraction method used for extracting features for efficient classification and recognition. The following paragraph explains in detail about the feature extraction methodology.

The character centroid is computed and the image (character/numeral) is further divided in to fifty equal zones as shown in figure 5. Average distance and angle from the character centroid to the pixels present in the zone are computed. Similarly zone centroid is computed. This procedure is sequentially repeated for all the zones/grids/boxes present in the numeral image. There could be some zones that are empty, and then the value of that particular zone image value in the feature vector is zero. Finally 200 such features are used for feature extraction.

Image centroid and zone based distance feature extraction algorithm presented in [14] and zone centroid and angle feature extraction system (ZC-ICZA) presented in [19] are combined and new hybrid feature extraction method is proposed here. The following is the algorithm to show the general working procedure of our proposed method.

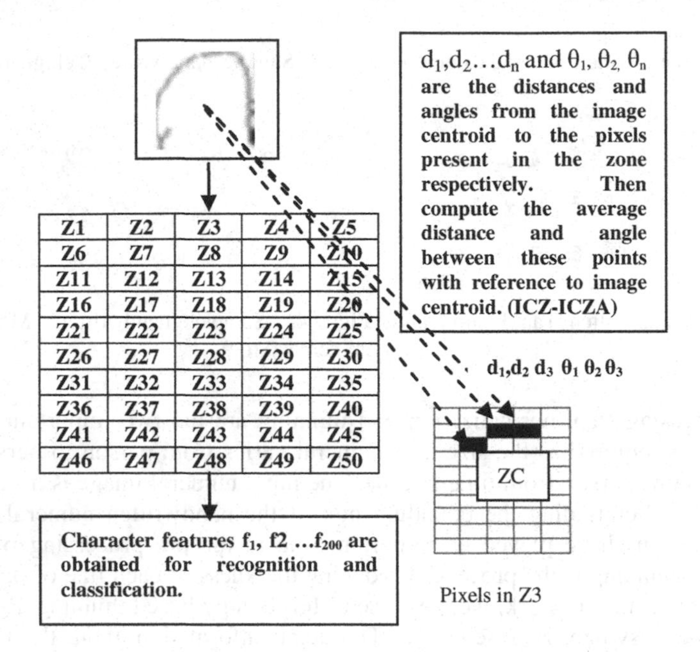

Fig. 5. Procedure for extracting the features from the numeral image

Proposed Algorithm. Zone based feature extraction system
Input. Preprocessed numeral image
Output. Features for Classification and Recognition
Method Begins
Step 1: Compute the input image centroid
Step 2: Divide the input image in to n equal zones
Step 3: Compute the distance between the image centroid to pixel present in the zone.
Step 4: Repeat step 3 for the entire pixel present in the zone.
Step 5: Compute the average distance between these points. (One feature)
Step 6: Compute the angle between the image centroid to pixel present in the zone.
Step 7: Repeat step 6 for the entire pixel present in the zone.
Step 8: Compute the average angle between these points. (One feature)
Step 9: Compute the zone centroid (two features)
Step 10: Repeat the steps 3-9 sequentially for the entire zone.
Step11: Finally, 4*n such features will be obtained for classification and recognition.
Step12: For classification and Recognition NNC is used.
Method Ends

5 Experimental Results and Comparative Study

For large-scale pattern matching, a long-employed approach is the NNC. The training phase of the algorithm consists only of storing the feature vectors of the training samples. In the actual classification phase, the same features as before are computed for the test samples. Distances from the new vector to all stored vectors are computed. Then Classification and recognition is achieved on the basis of similarity measurement. In order to evaluate the performance of the proposed method, we consider handwritten Kannada, Telugu, Tamil and Malayalam Numerals. For recognition and classification purpose NNC classifier is used.

Table 1. Recognition results for Kannada Handwritten Numerals using NNC classifier

Kannada numerals	Recognition rate (in %)
0	99.5
1	100
2	100
3	94
4	99
5	97
6	94.5
7	94.5
8	97
9	100
Average Recognition rate	**97.55**

Table 2. Recognition results for Telugu Handwritten Numerals using NNC classifier

Telugu numerals	Recognition rate (in %)
0	92
1	91
2	95
3	93
4	96
5	86
6	100
7	91
8	98
9	98
Average Recognition rate	**94**

Table 3. Recognition results for Tamil Handwritten Numerals using NNC classifier

Tamil numerals	Recognition rate (in %)
0	97
1	96
2	91
3	91
4	96
5	96
6	86
7	95
8	96
9	81
Average Recognition rate	**92.5**

Table 4. Recognition results for Malayalam Handwritten Numerals using NNC classifier

Malayalam numerals	Recognition rate (in %)
0	97
1	99
2	99
3	97
4	77
5	98
6	89
7	96
8	100
9	98
Average Recognition rate	**95.2**

Table 1 to Table 4 gives Recognition rate for Kannada, Telugu, Tamil and Malayalam Numerals respectively. In Kannada numerals, 3 and 7 are the most confusion pair. In Tamil, numerals 6 and 9 are the most confusion pair. In Malayalam, numerals 4 and 9 are the most confusion pair. This confusion reduces the overall result. This is due to the more similarity between numerals of confusion pair.

Table 5 to Table 8 gives the comparative result of our present work with earlier works for Kannada, Telugu, Tamil and Malayalam Numerals respectively. We have implemented the same feature extraction algorithm [8] and we tested for 4000 Kannada numeral samples. For the same data set we have implemented the proposed hybrid feature extraction algorithm and this procedure is also repeated for Telugu, Tamil and Malayalam numerals.

Table 5. Comparative Results for Kannada Numerals

Kannada handwritten numerals Training samples=2000 & Testing samples= 2000 NNC classifier						
	I fold		II fold		CV	
Ref	RR (In %)	Run time (Seconds)	RR (In %)	Run time (Seconds)	RR (In %)	Run time (Seconds)
[8]	96.4	54.3281	95.3	53.9375	95.85	54.1328
[14]	96.95	53.0113	96.3	57.3750	96.625	55.1932
[19]	97.25	87.5781	96.3	93.0156	96.775	90.2969
Proposed method	**97.55**	114.9219	**96.3**	116.7813	**96.95**	115.8516

Table 6. Comparative Results for Telugu Numerals

Telugu handwritten numerals Training samples=2000 & Testing samples= 2000 NNC classifier						
	I fold		II fold		CV	
Ref	RR (In %)	Run time (Seconds)	RR (In %)	Run time (Seconds)	RR (In %)	Run time (Seconds)
[8]	91.5	11.3281	93.9	11.2813	92.7	11.3047
[14]	93.3	12.5156	95.4	12.2969	94.35	12.4062
[19]	94.4	21.1719	94.8	21.0625	94.6	21.1172
Proposed method	**94**	25.3281	**95**	25.5469	**94.5**	25.4375

Table 7. Comparative Results for Tamil Numerals

Tamil handwritten numerals Training samples=2000 & Testing samples= 2000 NNC classifier						
	I fold		II fold		CV	
Ref	RR (In %)	Run time (Seconds)	RR (In %)	Run time (Seconds)	RR (In %)	Run time (Seconds)
[8]	90.6	11.4063	90.6	11.0625	90.6	11.2344
[14]	91.4	13.4219	90.9	13.4531	91.15	13.4375
[19]	92	21.8125	91	21.0781	91.5	21.4453
Proposed method	**92.5**	25.4063	**91**	25.7188	**91.75**	25.5626

Table 8. Comparative Results for Malayalam Numerals

Malayalam handwritten numerals Training samples=2000 & Testing samples= 2000 NNC classifier						
	I fold		II fold		CV	
Ref	RR (In %)	Run time (Seconds)	RR (In %)	Run time (Seconds)	RR (In %)	Run time (Seconds)
[8]	94.9	11.4219	94.2	11.0625	94.55	11.2422
[14]	95.2	11.5781	94.3	11.6875	94.75	11.6328
[19]	95.3	91.9063	94.4	22.1094	94.85	22.0078
Proposed method	**95.2**	21.8125	**94.4**	25.5469	**94.8**	25.6797

5.1 Experimental Result on Bangla Numeral Database

In this section, we experimentally evaluate the performance of proposed method on well-known ISI Bangla numerals database [20, 21, 22]. The ISI Bangla numeral database has 19,392 training samples and 4000 test samples. Preprocessed samples of ISI Bangla handwritten digits (here size normalization (50x50) and thinning are performed) are shown in Fig 6. We have considered 12000 training samples and 3000 testing samples for our experimental analysis. For the proposed feature extraction method, we achieved **94.33 %** recognition rate using SVM classifier. Table 9 provides the summary of few previous methods and result for handwritten Bangla numerals using proposed method.

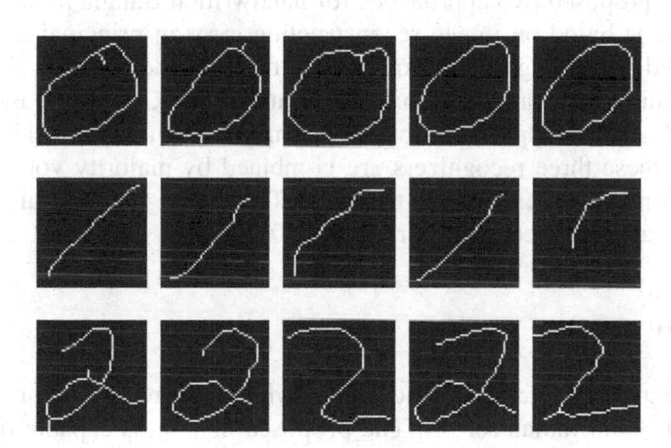

Fig. 6. Preprocessed Sample images of handwritten digits of ISI Bangla database

Pal et al. have reported results of handwritten numeral recognition six popular Indian scripts, including Bangla [12]. For feature computation, the bounding box of a numeral image is segmented into blocks and the directional features are computed in each of the blocks. These blocks are then down sampled by a Gaussian filter and the features obtained from the down sampled blocks are fed to a modified quadratic classifier for recognition. By five fold cross validation (rotation) on 14,650 Bangla numeral samples; they reported a recognition rate of 98.99 %.

Table 9. Summary of existing results and result achieved with proposed method for Bangla script. RR refers recognition rate and CV refers to cross validation

Ref	Feature	Classifier	#Train	#Valid	#Test	RR (in %)	Run time (Seconds)
Proposed method	Zone	SVM	12000	0	3000	**94.33**	691.87
[8]	Zone	SVM	12000	0	3000	93.57	237.69
[12]	Direction	Quadratic Classifier	14,650	CV	CV	98.99	–
[14]	Zone	SVM	12000	0	3000	93.33	253.2
[19]	Zone	SVM	12000	0	3000	94.1	527.83
[23]	Wavelet	MLP, MV	6000	1000	5000	97.16	–
[24]	Kirsh, PCA	SVM, MV	6000	0	10000	95.05	–

Bhattacharya and Chaudhuri have proposed a multi-resolution wavelet analysis and majority voting approach for handwritten numeral recognition of Bangla script [23]. They have extracted wavelet filtering features in three resolutions and multi-layer perceptron (MLP) neural network for classification on each resolution. Then classification results are combined by majority vote (MV). They have used 6000 Bangla numeral samples for training, 1000 Bangla numeral samples for validation and 5000 Bangla numeral samples for testing. They achieved recognition rate of 97.16 %.

Wen et al. have proposed two approaches for handwritten Bangla numeral recognition [24]: first one is based on image reconstruction error in principal subspace and second one is based on Kirsh gradient (four masks for four orientations), dimensionality reduction is achieved by principal component analysis (PCA) and classification is achieved by SVM. A third approach is based on image PCA and SVM classification. The results of all these three recognizers are combined by majority vote. They have used 6000 Bangla numeral samples for training, 10,000 Bangla numeral samples for testing. They have achieved recognition rate of 95.05 %.

6 Conclusion

In this paper we have proposed the zone based hybrid feature extraction method for the recognition of South Indian scripts. The proposed method is capable of recognizing isolated handwritten numerals with highest recognition rate of 97.55 %. Proposed method provides good recognition results even when certain pre processing steps like filtering, smoothing and slant removing are not considered. Our future work aims to improve classifier to achieve still better recognition rate and also to develop new zone based feature extraction algorithms. In this work we have considered 15000 images from ISI Bangla numeral database. Our future work aims to work on hybrid classifier and also to develop new zone based feature extraction algorithms.

The extensive experiments on different numeral database of Indian scripts and standard database like ISI Bangla are carried out. The recognition results prove the validation of the proposed feature extraction method and also its novelty and robustness.

References

1. Pal, U., Chaudhuri, B.B.: Indian Script Character recognition: A survey. Pattern Recognition 37, 1887–1899 (2004)
2. Anil, K.J., Taxt, T.: Feature Extraction Methods for Character Recognition-A Survey. Pattern Recognition 29(4), 641–662 (1996)
3. Majumdar, A., Chaudhuri, B.B.: Printed and handwritten Bangla numeral recognition using multiple classifier outputs. In: Proceedings of the first IEEE ICSIP 2006, vol. 1, pp. 190–195 (2006)
4. Abdur, R., Shuvabranta, S., Rahman, M., Sattar, A.: Recognition of conjunctive Bangla characters by artificial neural network. In: International Conference on Information and Communication Technology, pp. 96–99 (2007)
5. Pal, U., Sharma, N., Wakabayashi, T., Kimura, F.: Off line handwritten character recognition of Devanagiri Scripts. In: Ninth International Conference on Document Analysis as Recognition (ICDAR 2007), pp. 496–500 (2007)
6. Patil, P.M., Sontakke, T.R.: Rotation scale and translation invariant handwritten Devanagiri numeral character recognition using fuzzy neural network. Elsevier, Pattern Recognition 40, 2110–2117 (2007)
7. Pal, U., Sharma, N., Wakabayashi, T., Kimura, F.: A system for off-line Oriya handwritten character recognition using curvature feature. In: 10th International Conference on Information Technology, pp. 227–229. IEEE, Los Alamitos (2007)
8. Hanmandlu, M., Grover, J., Madasu, V.: Input fuzzy for the recognition of handwritten Hindi numeral. In: International Conference on Informational Technology, vol. 2, pp. 208–213 (2007)
9. Pujari, A.K., Dhanunjaya Naidu, C., Sreenivasa Rao, M., Jinaga, B.C.: An Intelligent character recognizer for Telugu scripts using multi resolution analysis and associative memory. Elsevier, Image vision Computing, 1221–1227 (2004)
10. Suresh, R.M., Arumugam, S.: Fuzzy technique based recognition of handwritten characters. Elsevier, Image Vision Computing 25, 230–239 (2007)
11. Lajish, V.L.: Handwritten character using gray- scale based state-space parameters and class modular neural network. In: IEEE International Conference on signal processing, Comunication and Networking, pp. 374–379 (2008)
12. Pal, U., Wakabayashi, T., Kimura, F.: Handwritten numeral recognition of six popular scripts. In: Ninth International conference on Document Analysis and Recognition ICDAR 2007, vol. 2, pp. 749–753 (2007)
13. Rajput, G.G., Hangarge, M.: Recognition of isolated handwritten Kannada numerals based on image fusion method. In: Ghosh, A., De, R.K., Pal, S.K. (eds.) PReMI 2007. LNCS, vol. 4815, pp. 153–160. Springer, Heidelberg (2007)
14. Rajashekararadhya, S.V., Vanaja, R.: Isolated handwritten Kannada digit recognition: A novel approach. In: Proceedings of the International Conference on Cognition and Recognition, pp. 134–140 (2008)
15. Rajashekararadhya, S.V., Vanaja, R., Manjunath Aradhya, V.N.: Isolated handwritten Kannada and Tamil numeral recognition: A novel approach. In: First International Conference on Emerging Trends in Engineering and Technology ICETET 2008, pp. 1192–1195 (2008)
16. Rajashekararadhya, S.V., Vanaja, R.: Handwritten numeral recognition of three popular South Indian scripts: A novel approach. In: Proceedings of the second International Conference on information processing ICIP, pp. 162–167 (2008)

17. Rajashekararadhya, S.V., Vanaja, R.: Neural network based handwritten numeral recognition of Kannada and Telugu scripts. In: TENCON 2008, Hyderabad, pp. 1–5 (2008)
18. Gonzalez, R.C., Woods, R.E., Steven, L. (Eddins): Digital Image. In: Processing using MATLAB, Pearson Education, Dorling Kindersley, South Asia (2004)
19. Rajashekararadhya, S.V., Vanaja, R.: A novel zone based feature extraction algorithm for handwritten numeral recognition of four Indian scripts. Digital Technology Journal 2009 2, 41–51 (2009)
20. Bhattacharya, U., Chaudhuri, B.B.: Handwritten numeral databases of Indian scripts and multistage recognition of mixed numerals. IEEE Transaction on Pattern analysis and machine intelligence 31(3), 444–457 (2009)
21. Bhattacharya, U., Chaudhuri, B.B.: Databases for research on recognition of handwritten characters of Indian scripts. In: Proceedings of the 8th International conference on document analysis and recognition ICDAR 2005, Seoul, Korea, vol. II, pp. 789–793 (2005)
22. Chaudhuri, B.B.: A complete handwritten numeral database of Bangla – a major Indic script. In: Proceedings of the 10th International workshop on frontiers of handwriting recognition La Baule, France, pp. 379–784 (2006)
23. Bhattacharya, U., Chaudhuri, B.B.: A majority voting scheme for multiresolution recognition of hand-printed numerals. In: Proceedings of the 7th International conference on document analysis and recognition ICDAR 2003, Edinburgh, Scotland, pp. 789–793 (2005)
24. Wen, Y., Lu, Y., Shi, P.: Handwritten Bangla numeral recognition system and its application to postal automation. Pattern recognition 40(1), 99–107 (2007)

Local Subspace Based Outlier Detection

Ankur Agrawal

G.L.A. Institute of Technology and Management, Mathura
111/128, Gurhai Bazar, Mathura, Uttar Pradesh, India
ankuragrawalgla@gmail.com

Abstract. Existing studies in outlier detection mostly focus on detecting outliers in full feature space. But most algorithms tend to break down in high-dimensional feature spaces because classes of objects often exist in specific subspace of the original feature space. Therefore, subspace outlier detection has been recently defined. As a novel solution to tackle this problem, we propose here a local subspace based outlier detection technique, which uses different subspaces for different objects. Using this concept we adopt local density based outlier detection to cope with high-dimensional data. A broad experimental evaluation shows that this approach yields results of significantly better quality than existing algorithms.

Keywords: Outlier detection, Local subspace.

1 Introduction

Outlier mining is an active research field in data mining and has a lot of practical applications in many different domains, such as, financial fraud detection [1], network intrusion detection [2], medical abnormal reactions analysis [3] and signal preprocessing [4] etc.

Hawkins gave the definition of outlier that "An outlier is an observation that deviates so much from other observations as to arouse suspicion that it was generated by a different mechanism" [5]. Based on the definition, researchers have proposed various schemes for outlier mining. Statistical-based approaches were the earliest approaches used for outlier mining [6]. The statistical based approaches assume that the dataset have a distribution and an object having low probability with respect to the distribution may be an outlier [7]-[8]. However, this approach requires a prior underlying distribution of the dataset to compute the outlier scores, which is usually unknown. In order to overcome the limitations of the statistical-based approaches, the distance-based [9] and density-based [10] approaches were introduced to detect outliers, which use k-nearest neighbors (KNN) to compute the similarity between data points.

Many useful outlier detection methods proposed in the last decade compute outlier score of the data points in a complete feature space, i.e. each dimension is equally weighted when computing the distance between points. These approaches are successful for low-dimensional data sets. However, in higher dimensional feature spaces,

S. Ranka et al. (Eds.): IC3 2009, CCIS 40, pp. 149–157, 2009.
© Springer-Verlag Berlin Heidelberg 2009

their accuracy and efficiency deteriorates significantly. The major reason for this behavior is the so-called curse of dimensionality: In high dimensional feature spaces, a full-dimensional distance is often no longer meaningful, since the nearest neighbor of a point is expected to be almost as far as its farthest neighbor [11].

A common approach to cope with high dimensional feature spaces is the application of a global dimensionality reduction technique such as Principal Component Analysis (PCA). A standard outlier detection method can then be used to compute outliers in this subspace. But if different subspaces of the feature space are meaningful for different objects, a global dimensionality reduction will fail.

In this paper, we use the concept of *local subspace* to solve this problem. Our new approach LSOF is founded on the concept of local outlier factor proposed in density-based local outlier detection [10]. In the example given in fig. 1, a local density based algorithm will detect only one outlier i.e. o_1 as shown in fig. 1(a) while local subspace based algorithm will detect two outliers o_1 and o_2 as shown in fig. 1(b). Though this task could have been performed using a global dimensionality reduction technique but in large datasets of high-dimensionality, different subsets of attributes provide better information about the objects. Thus local subspace for each object would have to used.

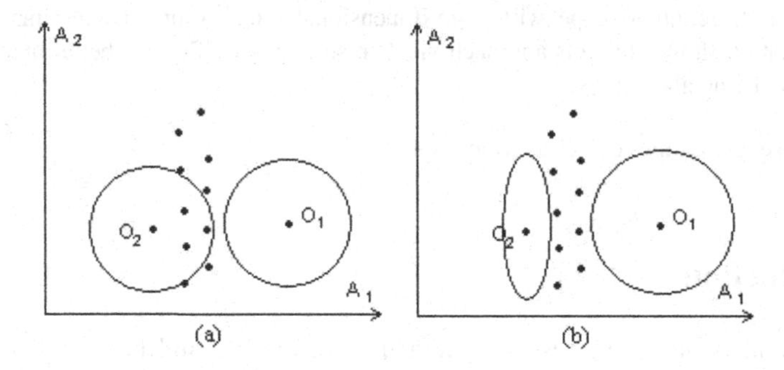

Fig. 1. Outliers according to local density-based outlier detection (a) and according to local subspace based outlier detection (b)

In order to ensure the quality of outlier detection in high-dimensional datasets, we suggest using a weighted Euclidean distance measure to compute outliers in subspaces instead of using each dimension with equal importance. Thus, we build for each point a so-called local subspace weight vector based on the variance in each attribute and use a weighted Euclidean distance measure based on this local subspace weight vector. Using this more flexible model, we propose the algorithm *LSOF (Local Subspace based Outlier Factor)* to efficiently compute exact solutions of the local subspace outlier detection problem. The user can select a parameter β indicating the variance threshold for feature selection in the local subspace of an object. Only those features with a variance of no more than β in the local neighborhood are included in the local subspace.

The remainder of the paper is organized as follows: In Section 2, we discuss related work and point out our contributions. In Section 3, we formalize our notion of local subspace outlier factor to efficiently detect local subspace based outliers. In Section 4, we discuss the time-complexity of our approach. Section 5 contains an extensive experimental evaluation and Section 6 concludes the paper.

2 Related Work

2.1 LOF Algorithm

LOF (Local Outlier Factor) [10] is the first concept of an object which also quantifies how outlying an object is and the LOF value of an object is based on the average of the ratios of the local reachability density of the area around the object and the local reachability densities of its neighbors. The size of the neighborhood of the object is determined by the area containing a user-supplied minimum number of points (*MinPts*). Several concepts and terms to explain the LOF algorithm can be defined as follows.

Definition 1. (k-distance of an object *p*) For any positive integer *k*, the k-distance of object *p*, denoted as k-distance(*p*), is defined as the distance $d(p,o)$ between *p* and an object of *D* such that:
(i) for at least *k* objects $o' \in D\setminus\{p\}$ it holds that $d(p,o') \leq d(p,o)$, and
(ii) for at most k - 1 objects $o' \in D\setminus\{p\}$ it holds that $d(p,o') < d(p,o)$.

Definition 2. (k-distance neighborhood of an object *p*) Given the k-distance of *p*, the k-distance neighborhood of *p* contains every object whose distance from *p* is not greater than the k-distance.

Definition 3. (reachability distance of an object *p* w.r.t. object *o*) Let *k* be a natural number. The reachability distance of object *p* with respect to object *o* is defined as

$$reach-dist_k(p) = \max\{k-distance(o), d(p,o)\} \quad (1)$$

Let *MinPts* be the only parameter and the values $reach\text{-}dist_{MinPts}(p,o)$, for $o \in N_{MinPts}(p)$, be a measure of the volume to determine the density in the neighborhood of an object *p*.

Definition 4. (local reachability density of an object *p*) The local reachability density of *p* is defined as

$$lrd_{Minpts}(p) = 1/\left[\frac{\sum_{o \in N_{MinPts}(p)} reach-dist_{MinPts}(p,o)}{|N_{MinPts}(p)|}\right] \quad (2)$$

Definition 5. (local outlier factor of an object *p*) The local outlier factor of *p* is defined as

$$LOF_{MinPts}(p) = \frac{\sum_{o \in N_{MinPts}(p)} \frac{lrd_{Minpts}(o)}{lrd_{MinPts}(p)}}{|N_{MinPts}(p)|} \quad (3)$$

It is easy to see that the lower $lrd_{MinPts}(p)$ is, and the higher the $lrd_{MinPts}(o)$, $o \in N_{MinPts}(p)$ are, the higher is the LOF value of *p*. The outlier factor of object *p* captures the degree to which we call *p* an outlier. However, the weakness of the LOF algorithm is that it can not detect the outliers existing in a subspace of full feature space.

2.2 DSNOF Algorithm

The DSNOF algorithm [12] considers the density similarity between the object and its neighbors and calculates the DSNOF value. In this algorithm each object in dataset is assigned a density-similarity-neighbor based outlier factor (DSNOF) to indicate the degree of outlier-ness of an object. The algorithm calculates the densities of an object and its neighbors and constructs the similar density series (SDS) in the neighborhood of the object. Based on the SDS, this algorithm computes the average series cost (ASC) of the objects. Finally, it calculates the DSNOF of the object based on the ASC of the object and its neighbors.

2.3 Subspace Outlier Detection in Data with Mixture of Variances and Noise

This algorithm [13] introduced a bottom-up approach to discover clusters of outliers in any m-dimensional subspace from an n-dimensional space. It uses an outlier score function based on Chebyshev (L_∞ norm) distance in order to properly rank the outliers.

First, it uses a method to compute the outlier score for all points in each dimension. It states that if a point is an outlier in a subspace, the score must be high for that point in each dimension of the subspace. It then aggregates the scores to compute the final outlier score for the points in the dataset. It introduces a filter threshold to eliminate the high dimensional noise during the aggregation.

2.4 Our Contributions

In this paper, we make the following contributions: In order to enhance the quality of density-based outlier detection algorithms in high-dimensional space, we extend the well-founded notion of density-based local outliers to ensure high quality results even in high-dimensional spaces. We do not use any sampling or approximation techniques, thus the result of our outlier-mining algorithm is determinate. We propose an efficient method called LSOF which is able to detect all local subspace outliers of a certain dimensionality in a single scan over the database and is linear in the number of dimensions. And finally, we successfully apply our algorithm LSOF to real-world data sets, showing its superior performance over existing approaches.

3 The Proposed Algorithm (LSOF)

In this section, we formalize the notion of local subspace outliers. Let D be a database of d dimensional points ($D \subseteq R^d$), where the set of attributes is denoted by $A = \{A_1, \ldots A_d\}$. The projection of a point p onto an attribute $A_i \in A$ is denoted by $\pi_{Ai}(p)$. Let $dist : R^d \times R^d \rightarrow R$ be a metric distance function between points in D, e.g. one of the L_p-norms. Let $N_\varepsilon(p)$ denote the ε-neighborhood of $p \in D$, i.e. $N_\varepsilon(p)$ contains all points q where $dist(p, q) \leq \varepsilon$. Intuitively, a local subspace outlier is an object whose density is low as compared to its neighbors in a certain local subspace. In order to identify local subspace outliers, we are interested in all sets of points having a small variance along one or more attributes, i.e. a variance smaller than a given $\beta \in R$.

Definition 1 (variance along an attribute)
Let $p \in D$ and $\varepsilon \in R$. The *variance* of $N_\varepsilon(p)$ along an attribute $A_i \in A$, denoted by $Var_{A_i}(N_\varepsilon(p))$, is defined as follows:

$$Var_{A_i}(N_\varepsilon(p)) = \frac{\sum_{q \in N_\varepsilon(p)} (dist(\pi_{A_i}(p), \pi_{A_i}(q))^2}{|N_\varepsilon(p)|} \tag{4}$$

The intuition of our formalization is to consider those attributes in local subspace of an object which have a low variance in its neighborhood. Therefore, we associate each point p with a local subspace vector $\mathbf{w}_p$ which reflects the variance of the points in the ε-neighborhood of p along each attribute in A.

Definition 2 (local subspace similarity measure)
Let $p \in D$, $\beta \in R$ and $\kappa \in R$ be a constant with $\kappa \gg 1$. Let $\mathbf{w}_p = (w_1, w_2, \ldots w_d)$ be the so-called *local subspace weight vector of p,* where

$$w_i = \begin{cases} 1 & if \ Var_{A_i}(N_\varepsilon(p)) > \beta \\ \kappa & if \ Var_{A_i}(N_\varepsilon(p)) \le \beta \end{cases} \tag{5}$$

The *local subspace similarity measure* associated with a point p is denoted by

$$subdist_p(p,q) = \sqrt{\sum_{i=1}^{d} w_i \cdot (\pi_{A_i}(p) - \pi_{A_i}(q))^2} \tag{6}$$

where w_i is the i-th component of $\mathbf{w}_p$.

Let us note, that the subspace similarity measure $subdist_p(p, q)$ is simply a weighted Euclidean distance. The parameter β specifies the threshold for a low variance. As we are only interested in distinguishing between dimensions with low variance and all other dimensions, weighting the dimensions inversely proportional to their variance is not useful. Thus, our weight vector has only two possible values.

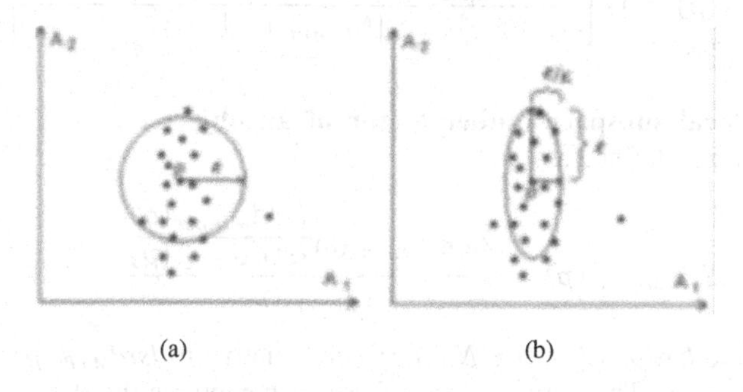

Fig. 2. ε-neighborhood of p according to (a) simple Euclidean and (b) local subspace Euclidean distance

The local subspace similarity measure is visualized in Figure 2. The ε-neighborhood of a 2-dimensional point p exhibits low variance along attribute A1 and high variance along attribute A2. The similarity measure $subdist_p$ weights attributes with low variance considerably lower (by the factor κ) than attributes with a high variance.

After calculating the local subspace distance between different data objects which captures the notion of local subspace for each data object, we now use this distance measure to find out the outlying degree for each object as in LOF algorithm with some modifications.

Definition 3 (local subspace k-distance of an object *p*). For any positive integer *k*, the local subspace k-distance of object *p*, denoted as k-dist$_{sub}$(*p*), is defined as the distance $subdist_p(p,o)$ between *p* and an object of *D* such that:

(i) for at least *k* objects $o' \in D\setminus\{p\}$ it holds that $subdist_p(p,o') \leq subdist_p(p,o)$, and
(ii) for at most *k - 1* objects $o' \in D\setminus\{p\}$ it holds that $subdist_p(p,o') < subdist_p(p,o)$.

Definition 4 (local subspace k-distance neighborhood of an object *p*). Given the local subspace k-distance of *p*, the local subspace k-distance neighborhood of *p* contains every object whose local subspace distance from *p* is not greater than the local subspace k-distance.

Definition 5 (local subspace reachability distance of an object *p* w.r.t. object *o*). Let *k* be a natural number. The local subspace reachability distance of object *p* with respect to object *o* is defined as

$$subreach-dist_k(p,o) = max\,\{\,k-dist_{sub}(o), subdist_p(p,o)\} \tag{7}$$

Let *MinPts* be the only parameter and the values $subreach\text{-}dist_{MinPts}\,(p,o)$, for $o \in N_{MinPts}(p)$, be a measure of the volume to determine the density in the neighborhood of an object *p*.

Definition 6 (local subspace reachability density of an object *p*). The local subspace reachability density of *p* is defined as

$$lsrd_{Minpts}(p) = 1/\left[\frac{\sum_{o\,\in\,\mathrm{N_{MinPts}(p)}} subreach-dist_{MinPts}(p,o)}{|N_{MinPts}(p)|}\right] \tag{8}$$

Definition 7 (local subspace outlier factor of an object *p*). The local subspace outlier factor of *p* is defined as

$$LSOF_{MinPts}(p) = \frac{\sum_{o\,\in\,\mathrm{N_{MinPts}(p)}} \frac{lsrd_{Minpts}(o)}{lsrd_{MinPts}(p)}}{|N_{MinPts}(p)|} \tag{9}$$

The higher the $lsrd_{MinPts}(o)$, $o \in N_{MinPts}(p)$ are relative to $lsrd_{MinPts}(p)$, the higher is the LSOF value of *p*. The outlier factor of object *p* captures the degree to which we call *p* an outlier. Thus, we can find n outliers existing in a local subspace of full feature space by taking the top n objects ordered by their LSOF values.

4 Complexity Analysis

The overall worst-case time complexity of our algorithm based on the sequential scan of the dataset is $O(d \cdot n^2)$. Our algorithm has to associate each point of the dataset with a local subspace weight vector that is used for searching neighbors. The corresponding vector must be computed once for each point. The computation of the local subspace weight vector **w** is based on the result of a Euclidean range query which can be evaluated in $O(d \cdot n)$ time. Then the vector is built by checking the variance of the points in the Euclidean ε-neighborhood along each dimension which requires $O(d \cdot n)$ time. For all points together, this sums up to $O(d \cdot n^2)$.

The local subspace k-distance neighborhood for all points can be calculated in $O(n^2)$ time. Calculation of local subspace reachability distance between all pairs of points requires linear time. The lsrd and LSOF values of all the points can then be calculated in $O(n)$ time. Thus the worst case complexity of our algorithm comes out to be $O(d \cdot n^2)$ with the assumption of no index structure. As this complexity is realistic, our algorithm can be used practically to detect outliers in high-dimensional database.

5 Experimental Evaluation

We conducted all our experiments on a workstation with a Pentium 4 2.6 GHz processor and 1.5 GB of RAM. We implemented all algorithms in Java. In order to test the ability of our algorithm to detect outliers, we used the KDD CUP 99 Network Connections Data Set from the UCI repository [14]. This database contains a standard set of data to be audited, which includes a wide variety of intrusions simulated in a military network environment. The task was to detect the attack connections without any prior knowledge about the properties of the network intrusion attacks. The detection will be simply based on the hypothesis that the attack connections may behave differently from the normal network activities which makes them outliers. We created a test dataset from the KDD original dataset with 1000 connections. Each record has 38 continuous attributes representing the statistics of a connection and its associated connection type, i.e. normal, buffer overflow attack. A very small number of attack connections are randomly selected. There are 22 types of attacks with the size varying from 1 to 4. Totally, there are 43 attack connections in the test dataset.

We used the LOF and DSNOF algorithms as baselines to compare our results as these are well-known outlier detection methods. We ran all the algorithms for different values of input parameters. We found our algorithm performing better in most of the cases with very little dependence on input parameters. Table 1 summarizes the results of running LOF, DSNOF and LSOF algorithms on the test dataset with $MinPts = 10$ and $\beta = 0.0005$ for LSOF.

In table 1, K indicates the total no. of outliers detected (including correct and incorrect detection) and the entries corresponding to LOF, DSNOF and LSOF indicate the actual number of intrusion attacks detected by the algorithms. As we can see from the table, our LSOF algorithm consistently performed well over the LOF and DSNOF algorithms. In figure 3, we have drawn the curve of percentage of correct detection corresponding to the three algorithms.

Table 1. Results of running LOF, DSNOF and LSOF algorithms on the test dataset

No. of outliers reported (K)	No. of correct detection		
	LOF	DSNOF	LSOF
10	6	6	7
20	8	9	10
30	12	13	15
40	14	16	20
50	18	18	25
60	21	20	31
70	24	22	32
80	26	23	33
90	28	25	34
100	30	25	36

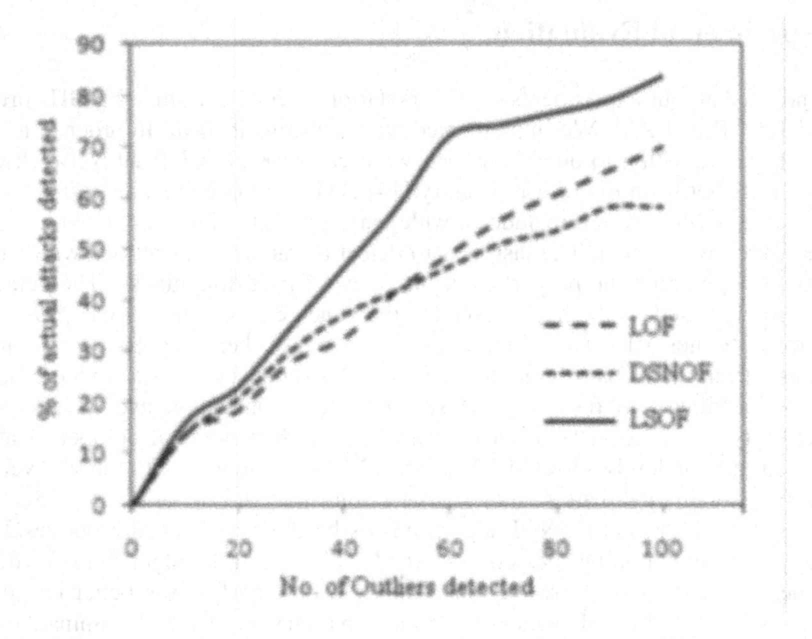

Fig. 3. Percentage of actual attacks detected by LOF, DSNOF and LSOF algorithms

6 Conclusion

In this paper we proposed an enhancement of LOF outlier detection algorithm to cope with the curse of dimensionality by introducing the notion of local subspaces so that outliers existing in specific subspaces of full feature space are detected. We have shown that our approach is superior to the well-known LOF and DSNOF algorithms in detecting outliers in high-dimensional databases. The proposed algorithm LSOF first detects the meaningful attributes for each object by limiting the allowed variance among its neighbors. A parameter β is used to limit the amount of variance within the neighborhood of an object. The future work can be to determine the value of parameter β automatically for better performance for any given data set.

References

1. Yue, D., Wu, X., Wang, Y., Li, Y., Chu, C.: A Review of Data Mining-Based Financial Fraud Detection Research. In: 2007 International Conference on Wireless Communications, Networking and Mobile Computing, Shanghai, P. R. China, pp. 5514–5517 (2007)
2. Zhang, J., Zulkernine, M.: Anomaly Based Network Intrusion Detection with Unsupervised Outlier Detection. In: 2006 IEEE International Conference on Communications, Istanbul, Turkey, pp. 2388–2393 (2006)
3. Podgorelec, V., Heri_ko, M., Rozman, I.: Improving Mining of Medical Data by Outliers Prediction. In: 18th IEEE International Symposium on Computer-Based Medical Systems, Ireland, pp. 91–96 (2005)
4. Näsi, J., Sorsa, A., Leiviskä, K.: Sensor Validation And Outlier Detection Using Fuzzy Limits. In: 44th IEEE Conference on Decision and Control, and the European Control Conference, Seville, Spain, pp. 7828–7833 (2005)
5. Hawkins, D.: Identification of Outliers. Chapman and Hall, London (1980)
6. Hodge, V., Austin, J.: A Survey of Outlier Detection Methodologies. Artificial Intelligence Review, 85–126 (2004)
7. Eskin, E.: Anomaly Detection over Noisy Data Using Learned Probability Distributions. In: 17th International Conference on Machine Learning, Stanford, CA, USA, pp. 255–262 (2000)
8. Yamanishi, K., Takeuchi, J.: Discovering Outlier Filtering Rules from Unlabeled Data-Combining a supervised Learner with an Unsupervised Learner. In: 7th ACM SIGKDD International Conference on Knowledge Discovery and Data Mining, San Francisco, CA, USA, pp. 389–394 (2001)
9. Knorr, E., Ng, R.: Algorithms for mining distance-based outliers in large datasets. In: 24th International Conference on Very Large Data Bases, San Francisco, CA, USA, pp. 392–403 (1998)
10. Breunig, M., Kriegel, H., Ng, R., Sander, J.: LOF: identifying density-based local outliers. In: SIGMOD 2000 International Conference on Management of Data, Dallas, Texas, USA, pp. 93–104 (2000)
11. Hinneburg, A., Aggarwal, C., Keim, D.: What is the Nearest Neighbor in High Dimensional Spaces. In: 26th International Conference on Very Large Databases, Cairo, Egypt, pp. 506–515 (2000)
12. Cao, H., Si, G., Zhu, W., Zhang, Y.: Enhancing Effectiveness of Density-based Outlier Mining. In: 2008 International Symposiums on Information Processing. Moscow, pp. 149–154 (2008)
13. Nguyen, M., Mark, L., Omiecinski, E.: Subspace Outlier Detection in Data with Mixture of Variances and Noise. Report Number GT-CS-08-11, Georgia Institute of Technology, Atlanta, GA 30332, USA (2008)
14. Newman, C., Merz, C.: UCI repository of machine learning databases (1998)

New Robust Fuzzy C-Means Based Gaussian Function in Classifying Brain Tissue Regions

S.R. Kannan[1,2], A. Sathya[2], S. Ramathilagam[1], and R. Pandiyarajan[2]

[1] National Cheng Kung University, Taiwan
[2] Gandhigram Rural University

Abstract. This paper introduces a new fuzzy c-mean objective function called Kernel induced Fuzzy C-Means based Gaussian Function for the purpose of segmentation of brain medical images. It obtains effective methods for calculating memberships and updating prototypes by minimizing the new objective function of Gaussian based fuzzy c-means. The performance of proposed algorithm has been tested with synthetic image and then it has been implemented for segmenting the brain [18] medical images to reduce the inhomogeneities and to allow the labeling of a pixel (voxel) to be influenced by the labels in its immediate neighborhood. Also this paper compares the results of proposed method with the results of existing basic Fuzzy C-Means.

Keywords: Fuzzy C-Means, Clustering, Kernel Function, Gaussian Function, Image Segmentation, MR Imaging.

1 Introduction

Due to the technical limitations, shacking in body, and breaths, during image attainment, MR images have the following problems: (1) low spatial resolution, (2) fragmentations, (3) non-uniformity effect induced by radio- frequency, (4) low contrast, and (5) borders between tissues have not clear. Medical Image segmentation is a very important one before to go for treatment planning based on medical images. Segmentation divides an image into different groups based on the counts of tissues, to have anatomical structures from data acquired via MRI. Initially segmentation was made by manually, but manual segmentation is more difficult, time-consuming, and costly. So researchers have gone for automated processing of segmentation for medical images by a computer system. Automatic brain tumor segmentation from MR images is a not easy task that involves various disciplines covering image analysis, mathematical algorithms [13, 14, 15], and etc. Automated segmentation methods based on artificial intelligence techniques were proposed in (Clark et al., 1998[9]; Fletcher-Heath et al., 2001 [11]). Gering et al. (2002) [12] proposed a method that detects deviations from normal brains using a multi-layer Markov random field framework.

In recent world, many of the segmentation methods are based on the unsupervised clustering algorithms. Unsupervised clustering is a process for ordering objects in such a way that samples of the identical group are more similar to one another than samples belonging to different groups. Recently fuzzy c-means of fuzzy clustering [7] plays main role in unsupervised clustering method for segmenting medical images [21, 22]. In

S. Ranka et al. (Eds.): IC3 2009, CCIS 40, pp. 158–169, 2009.
© Springer-Verlag Berlin Heidelberg 2009

this paper, we propose a system capable to perform segmentation of images [8]. Fuzzy clustering, which produces the idea of partial membership of belonging described by a membership function, fuzzy clustering as a soft segmentation method has been widely studied and successfully applied in image segmentation [1, 16, 17, 19, 22, 23]. The FCM algorithm has also been employed by many researchers. However, drawback of these fuzzy clustering algorithms are; equation in updating prototypes and memberships. The membership of an object has not strong enough or significantly high for a particular cluster, it means that the equation of calculating membership is not an effective. Due to greatly affected by noise in the data, the updated prototypes leads the result of clustering might be incorrect.

The main reason for the underlying drawbacks of above is the objective of fuzzy c-means minimizes based on existed Euclidean distance measures for constructing membership equation and equation for updating prototypes. Further the standard FCM uses the Euclidean-norm to measure similarity between prototypes and data points, it can not be effective in clustering 'non-spherical' clusters and it is only effective in clustering ‘spherical’ clusters.

This paper understood the highlighted drawbacks of using Euclidean distance measures for constructing objective function, so it tries to replace the underline Euclidean – norm in the object function of standard FCM with other similarity measures (metric).

The aims of proposed methods are:

1. To induce a new robust objective function and effective methods for obtaining memberships and updated centers kernel functions for the input space and using the new objective functions to replace nonrobust measure to segment images;
2. To keep ease of computation;
3. To achieve results with less computations and iterations

2 Fuzzy Clustering

Kindly The idea of basic fuzzy clustering called Fuzzy C-Means (FCM) has invented by Bezdek. It provides a method that shows how to group data points that populate some multidimensional space into a specific number of different clusters. *Fuzzy c-means* (FCM) is a data clustering technique wherein each data point belongs to a cluster to some degree that is specified by a membership grade.

2.1 Fuzzy C-Mean Algorithm

Fuzzy c-means (FCM) is a method of clustering which allows one piece of data to belong to two or more clusters. This method (developed by Dunn in 1977 and improved by Bezdek in 1981) is frequently used in pattern recognition. It is based on minimization of the following objective function:

$$\mathrm{J_m} = \sum_{i=1}^{N}\sum_{j=1}^{C} u_{ij}^{m} \left\| x_i - c_j \right\|^2, \qquad 1 \le m < \infty$$

where m is any real number greater than 1, u_{ij} is the degree of membership of x_i in the cluster j, x_i is the ith of d-dimensional measured data, c_j is the d-dimension center of the cluster, and ||*|| is any norm expressing the similarity between any measured data and the center.

Fuzzy partitioning is carried out through an iterative optimization of the objective function shown above, with the update of membership u_{ij} and the cluster centers c_j by:

$$u_{ij} = \frac{1}{\sum_{k=1}^{C}\left[\frac{\left\|x_i - c_j\right\|}{\left\|x_i - c_k\right\|}\right]^{\frac{2}{m-1}}}, \quad c_j = \frac{\sum_{i=1}^{N} u_{ij}^m x_i}{\sum_{i=1}^{N} u_{ij}^m}$$

This iteration will stop when $\max_{ij}\left\{\left|u_{ij}^{(k+1)} - u_{ij}^{(k)}\right|\right\} < \varepsilon$, where ε is a termination criterion between 0 and 1, whereas k is the iteration steps. This procedure converges to a local minimum or a saddle point of J_m.

3 Kernel Induced Fuzzy C-Means Based Gaussian Function [KFCGF]

This paper proposes a modification in basic fuzzy c-means by introducing Kernel function that allows the clustering of objects to be more reasonable. The modified proposed objective function with kernel function is given by

$$J(U,V) = \sum_{i=1}^{n}\sum_{k=1}^{c} u_{ik}^m \left\|\psi(x_i) - \psi(v_k)\right\|^2 \mathbb{k} \tag{1}$$

where ψ stands as map. And the distance function can be expressed using in product space as

$$\left\|\psi(x) - \psi(y)\right\|^2 = \langle\psi(x),\psi(y)\rangle + \langle\psi(y),\psi(y)\rangle - 2\langle\psi(x),\psi(y)\rangle$$

To obtain Kernel induced Fuzzy C-Means based Gaussian Function the distance function can be modified as

$$\left\|\psi(x_i) - \psi(v_k)\right\|^2 = G(x_i, x_i) + G(v_k, v_k) - 2G(x_i, v_k)$$

where $i = 1, 2, \ldots, n$ and $k = 1, 2, .., c$.

Let us express $G(x_i, x_i)$ as Gaussian function

$$G(x_i, v_k) = \exp\left(\frac{-\left\|x_i - v_k\right\|^2}{\sigma^2}\right) \tag{2}$$

Where σ is a parameter which can be adjusted by users.

Using the expression (2) we obtain $G(x_i, x_i) = 1$ and $G(v_k, v_k) = 1$, so the distance function can be rewritten as

$$\|\psi(x_i)-\psi(v_k)\|^2 = 2(1-G(x_i,v_k)) \tag{3}$$

Substituting (3) in (1) Kernel induced Fuzzy C-Means based Gaussian Function is given by

$$J(U,V) = 2\sum_{i=1}^{n}\sum_{k=1}^{c} u_{ik}^{m}.(1-G(x_i,v_k)) \tag{4}$$

3.1 Obtaining Membership

To obtain equation for calculating membership we minimizing the objective function

$$J(U,V) = 2\sum_{i=1}^{n}\sum_{k=1}^{c} u_{ik}^{m}.(1-G(x_i,v_k))$$

subject to the constraints $\sum_{k=1}^{c} u_{ik} = 1$.

We solve the constrained objective function in (4) using Lagrangian multiplier method. The new objective function with Lagrangian multiplier is

$$J(U,V,\lambda) = 2\sum_{i=1}^{n}\sum_{k=1}^{c} u_{ik}^{m}(1-G(x_i,v_k)) - \sum_{i=1}^{n}\lambda_i\left(\sum_{k=1}^{c} u_{ik} - 1\right) \tag{5}$$

where $\lambda = (\lambda_1, \lambda_2, ..., \lambda_n)$ subject to the constraints $\sum_{k=1}^{c} u_{ik} = 1, \forall i$.

Expanding Equation (5) as

$$J(U,V,\lambda) = 2\left[\sum_{i=1}^{n} u_{i1}^{m}(1-G(x_i,v_1)) + \cdots + u_{ic}^{m}(1-G(x_i,v_c)))\right] - \sum_{i=1}^{n}\lambda_i((u_{i1}+u_{i2}+\cdots u_{ic})-1)$$

Taking the derivative of objective function with respect to u and setting the first derivative to zero by the necessary condition of Lagrangian method, we have

$$\Rightarrow u_{i1} = \left(\frac{\lambda_i}{2m(1-G(x_i,v_1))}\right)^{1/m-1}$$

$$\Rightarrow u_{i2} = \left(\frac{\lambda_i}{2m\left(1 - G\left(x_i, v_2\right)\right)} \right)^{1/m-1}$$

$$\circ$$

$$\circ$$

$$\circ$$

$$u_{ic} = \left(\frac{\lambda_i}{2m\left(1 - G\left(x_i, v_c\right)\right)} \right)^{1/m-1}$$

Since our proposed partition matrix in an objective function satisfies the following conditions

$$\sum_{k=1}^{c} u_{ik} = 1, \text{ for } 1 \leq \text{k} \leq \text{n}$$

We Obtain

$$\left(\frac{\lambda_i}{2m} \right)^{\frac{1}{m-1}} = \frac{1}{\left[\sum_{j=1}^{c} \left(\frac{1}{\left(1 - G\left(x_i, v_j\right)\right)} \right)^{1/m-1} \right]} \tag{6}$$

Using equation 6 we obtain the general equation for getting memberships

$$u_{ik} = \frac{\left(\frac{1}{1 - G\left(x_i, v_k\right)} \right)^{\frac{1}{m-1}}}{\sum_{j=1}^{c} \left(\frac{1}{1 - G\left(x_i, v_j\right)} \right)^{\frac{1}{m-1}}} \tag{7}$$

The general equation is used to obtain membership grades for objects in data for finding meaningful groups. how you are to be listed in the author index.

3.2 Obtaining Cluster Prototype Updating

Using Gaussian the objective function can be written as

$$J\left(U,V\right) = 2 \sum_{i=1}^{n} \sum_{k=1}^{c} u_{ik}^{m} \left(1 - \exp\left(\frac{-\left\| x_i - v_k \right\|^2}{\sigma^2} \right) \right) , \tag{8}$$

Taking the derivative of objective function with respect to v , we have the general form of updating center as

$$v_k = \frac{\sum_{i=1}^{n} u_{ik}^m G(x_i, v_k) x_i}{\sum_{i=1}^{n} u_{ik}^m G(x_i, v_k)} \tag{9}$$

The above algorithms can uniformly be summarized in the following steps

Algorithm

Step 1) Fix the number of Clusters and then Select initial centers.

Step 2) Update the partition matrix using (7)

Step 3) Update the centroids using (9)

Step 4) Repeat Steps 2)-3) until the following termination c criterion is satisfied: $\left\| V^{(present)} - V^{(previous)} \right\| < \in$, where $V^{(present)}$ and $V^{(previous)}$ are the vector of cluster prototypes at present iteration and previous iterations, and ς is a

4 Results and Discussions

This section describes some experimental results on synthetic image generated by random data corrupted with noise to show the segmentation performance of proposed method, and then it introduces the proposed method to real data of brain magnetic

Table 1. Random Datra

S.No	Data		S.No	Data	
	x_1	x_2		x_1	x_1
1	1.8	2	11	12	4
2	2	2.2	12	11.5	3.5
3	2	1.8	13	12.5	3.5
4	2	3.5	14	21	10
5	8.8	3	15	21	11
6	9	3.2	16	20.5	10.5
7	9	2.8	17	21.5	10.5
8	9.2	3	18	2	4
9	7	2.8	19	19	20
10	12	3	20	11	12

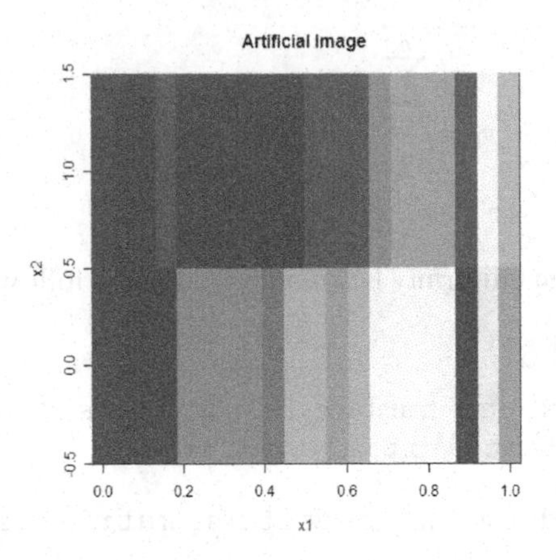

Fig. 1. Image by Random Data

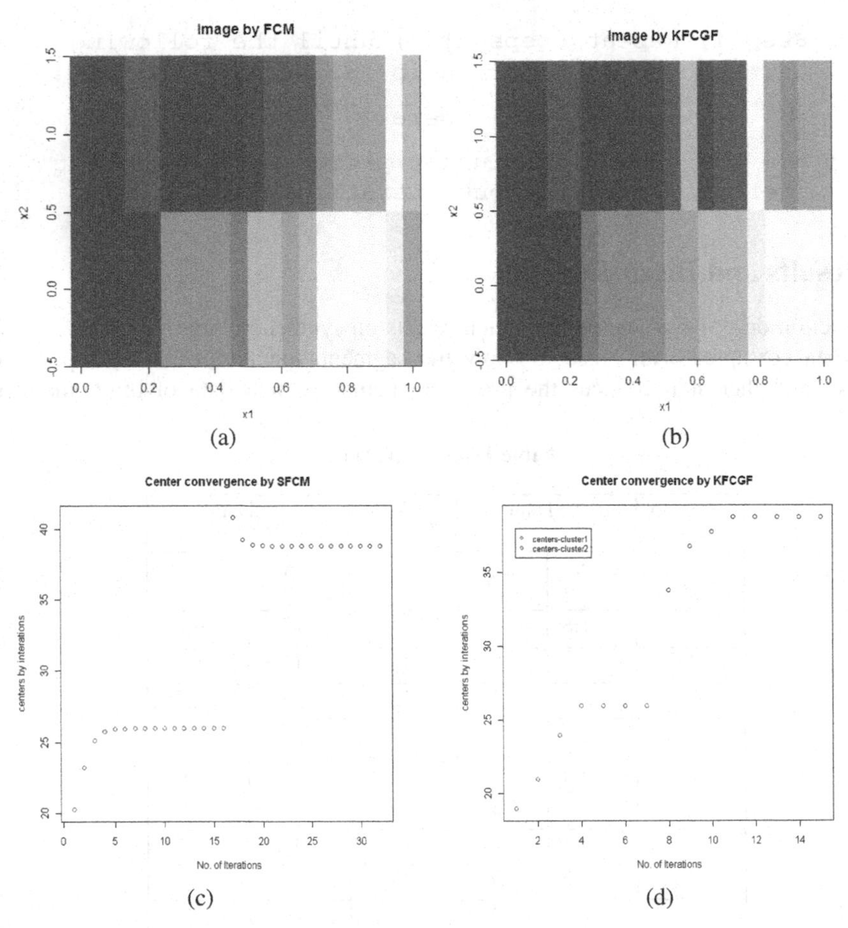

Fig. 2. (a) Segmented Image by FCM. (b) Segmented Image by KFCGF. (c) Updated centers by SFCM. (d) Updated centers by KECGF.

images for segmentation of it. Also it tries to have experimental results for the algorithms of existed FCM to compare the results of proposed method with existed method.

In this section, we describe the experimental results total of three images, i.e., synthetic image; two brains real MR slices our first experiment applies the standard FCM and Proposed algorithms to a synthetic test image. The image with 20 objects includes 3 or 4 classes with intensity values taken 1 to 21.

Table 2. Comparison of Iteration Count

	No. of Iterations	No. of clusters
Standard FCM	16	2
KECGF	7	2

The artificial image is shown in Fig. 1, contains 20 objects taken for the random data, each object of random data is taking different color to construct an artificial image.

Table 3. Final memberships of two clusters of KFCGF method and object allocation

S. No	Data x_1	x_2	Membership for cluster 1	Membership for cluster 2	Appropriate clusters
1	1.8	2	0.986186	0.013814	1
2	2	2.2	0.992940	0.007060	1
3	2	1.8	0.982537	0.017463	1
4	2	3.5	0.989281	0.010719	1
5	8.8	3	0.132239	0.867761	2
6	9	3.2	0.103746	0.896254	2
7	9	2.8	0.117448	0.882552	2
8	9.2	3	0.090034	0.909966	2
9	7	2.8	0.437255	0.562745	2
10	12	3	0.026485	0.973515	2
11	12	4	0.016826	0.983174	2
12	11.5	3.5	0.006171	0.993829	2
13	12.5	3.5	0.034943	0.965057	2
14	21	10	0.432943	0.567057	2
15	21	11	0.442378	0.557622	2
16	20.5	10.5	0.430682	0.569318	2
17	21.5	10.5	0.444227	0.555773	2
18	2	4	0.972505	0.027495	1
19	19	20	0.491037	0.508963	2
20	11	12	0.379073	0.620927	2

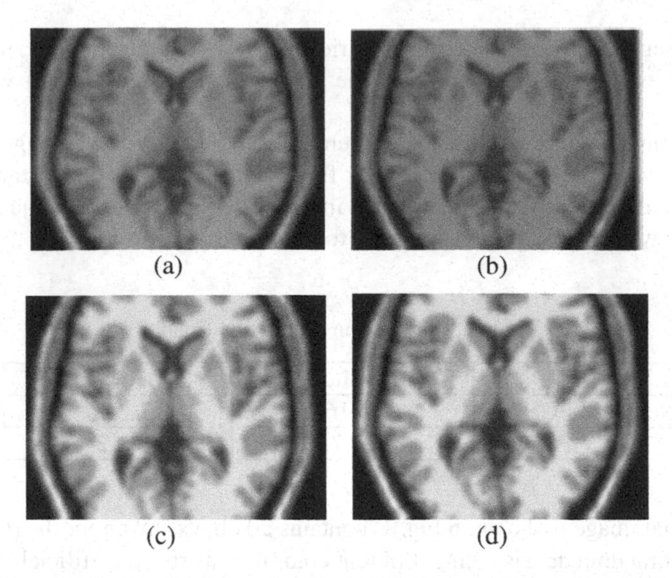

Fig. 3. (a) Corrupted Image, (b) Manual segmented Image, (c) Segmented by FCM, (d) Segmented by KFCGF

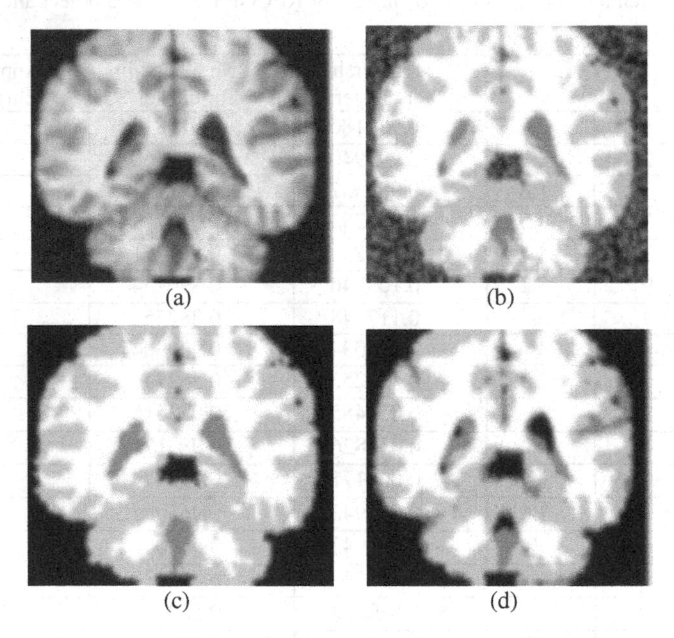

Fig. 4. (a) Corrupted Image, (b) Manual segmented Image, (c) Segmented by FCM, (d) Segmented by KFCGF

Fig. 2(a), 2(b), shows the segmentation results of FCM, KFCGF respectively. Those objects have allocated as two classes using standard FCM and KFCGF. Standard FCM not separated the two classes completely, but KFCGF well separated the data into two classes as shown in Fig. 2(a) and (b). During the experiment of standard FCM on synthetic image, the cluster prototypes or centers were initialized randomly. The standard FCM takes 20 iterations to converge the termination value of algorithm and to achieve optimal result. The proposed method for the experiments of clustering the synthetic image into two clusters has taken seven iterations to converge the value of termination parameter. The convergence speed of cluster centers of Standard FCM, and Proposed method are given in the figures 2 (c) and 3(d) respectively.

Figs. 2 (c) and 3(d) show the center convergence of standard FCM and proposed KECGF, under successive interactions of experiments on the synthetic image corrupted by noise. From Fig. 2 (c), the numbers of updated centers are high under the objective function of Euclidean distance measures and it takes more iteration to converge the termination value of algorithm. According to Fig. 2(d), we know that under the new effective objective function based kernel distance measure, it achieve the termination value with very less iteration with much better performance in getting membership(table 3) than standard FCM. Table 2 gives the number of iterations to achieve the results of cluster on synthetic image by standard FCM and KECGF. It is clear from Figs. 2 (c) and 3(d) and table 2, that our proposed KFCGF is much faster than the standard FCM, and the method is converged fast to termination condition with less run time.

Fig. 3(b-d) presents a comparison of segmentation results between manual segmentation, FCM, and when the proposed method has been introduced on brain image. The segmentation results on three images, it is clear that the proposed method KFCGF has shown a significant improvement over FCM and manual segmentation.

Fig. 4(a) shows the original image with Gaussian noise and Fig. 4(b) shows the manual segmentation result. For comparison with the manual segmentation result, which included four classes, CSF, gray matter, white matter, and others, the cluster number was set to 4. The segmentation results of FCM and KFCGF are shown in Fig. 4(c) and 4(d). KFCGF showed a significant improvement over FCM. From the resulted images using FCM and KFCGF, we note that FCM is much less fragmented than the original image and the incorporation of spatial constraints into the classification has somewhat the disadvantage of blurring of some fine details, but KFCGF had better result than FCM, as shown in Fig. 4(d).

5 Conclusion

This paper proposed a Kernel induced Fuzzy C-Means based Gaussian Function algorithm and applied to brain MR image segmentation. The algorithm was formulated by introducing Kernel, Gaussian function, and Lagrangian methods, with basic objective function of the FCM algorithm to have proper effective segmentation in brain MRI. This paper demonstrated a new KFCGF algorithm with synthetic image generated by random data, before to implement it to real brain MRI. This paper compared the results with standard FCM segmentation. It is clear from our comparison that KFCGF performed well than FCM. Also the proposed KFCGF algorithm produces results with faster convergence than FCM. The results reported in this paper show that the proposed kernel objective function of fuzzy c-means is an effective approach to constructing a robust image segmentation algorithm. And this work hopes that the proposed method can also be used to improve the performance of other FCM-like algorithms based on Euclidean distance functions.

References

1. Ahmed, M.N., Yamany, S.M., Mohamed, N., Farag, A.A., Moriarty, T.: A modified fuzzy c-means algorithm for bias field estimation and segmentation of MRI data. IEEE Trans. on Medical Imaging 21, 193–199 (2002)
2. Al-Sultan, K.S., Selim, S.Z.: A Global Algorithm for the Fuzzy Clustering Problem. Pattern Recognition 26(9), 1357–1361 (1993)
3. Al-Sultan, K.S., Fedjki, C.A.: A Tabu Search-based Algorithm for the Fuzzy Clustering Problem. Pattern Recognition 30(12), 2023–2030 (1997)
4. Bezdek, J.C.: Pattern Recognition with Fuzzy Objective Function Algorithms. Plenum Press, New York (1981)
5. Bezdek, J.C., Pal, S.K. (eds.): Fuzzy Models for Pattern Recognition. IEEE Press, New York (1992)
6. Bezdek, J.C., Hall, L.O., Clarke, L.P.: Review of MR image segmentation techniques using pattern recognition. Med. Phys. 20, 1033–1048 (1993)
7. Brand, M.E., Bohan, T.P., Kramer, L.A.: Estimation of CSF and gray matter volumes in hydrocephalic children using fuzzy clustering of MR images. Comput. Med. Imaging Graph. 18, 25–34 (1994)
8. Carvalho, B.M., Gau, J.C., Herman, G.T., Kong, Y.: Algorithms for Fuzzy Segmentation. Pattern Analysis & Applications 2(1), 73–81 (1999)
9. Clark, M.C., Hall, L.O., Goldgof, D.B., Velthuizen, R., Murtagh, F.R., Silbiger, M.S.: Automatic tumor-segmentation using knowledge-based techniques. IEEE Transactions on Medical Imaging 117, 187–201 (1998)
10. Dunn, J.C.: A Fuzzy Relative of the ISODATA Process and Its Use in Detecting Compact Well-Separated Clusters. Journal of Cybernetics 3, 32–57 (1973)
11. Fletcher-Heath, L.M., Hall, L.O., Goldgof, D.B., Murtagh, F.R.: Automatic segmentation of non-enhancing brain tumors in magnetic resonance images. Artifical Intelligence in Medicine 21, 43–63 (2001)
12. Gering, D.T., Grimson, W.E.L., Kikinis, R.: Recognizing deviations from normalcy for brain tumor segmentation. In: Dohi, T., Kikinis, R. (eds.) MICCAI 2002. LNCS, vol. 2488, pp. 388–395. Springer, Heidelberg (2002)
13. Hall, L.O., Bensaid, A.M., Clarke, L.P., Velthuizen, R.P., Silbiger, M.S., Bezdek, J.C.: A Comparison of Neural Network and Fuzzy Clustering Techniques in Segmenting Magnetic Resonance Images of the Brain. IEEE Trans. Neural Networks. 3(5), 672–682 (1992)
14. Xue, J.-H., Pizurica, A., Philips, W., Kerre, E., Walle, R.V., Lemahieu, I.: An Integrated Method of Adaptive Enhancement for Unsupervised Segmentation of MRI Brain Images. Pattern Recognition Letters 24(15), 2549–2560 (2003)
15. Krishnan, N., Nelson Kennedy Babu, C.V., Joseph Rajapandian, V., Richard Devaraj, N.: A Fuzzy Image Segmentation using Feedforward Neural Networks with Supervised Learning. In: Proceedings of the International Conference on Cognition and Recognition, pp. 396–402
16. Kwon, M.J., Han, Y.J., Shin, I.H., Park, H.W.: Hierarchical fuzzy segmentation of brain MR images. Int. J. Imaging Systems and Technology. 13, 115–125 (2003)

17. Li, X., Li, L., Lu, H., Chen, D., Liang, Z.: Inhomogeneity correction for magnetic resonance images with fuzzy C-mean algorithm. In: Proc. SPIE, vol. 5032, pp. 995–1005 (2003)
18. Li, C.L., Goldgof, D.B., Hall, L.O.: Knowledge-based classification and tissue labeling of MR images of human brain. IEEE Trans. Med. Imag. 12(4), 740–750 (1993)
19. Liew, A.W.C., Leung, S.H., Lau, W.H.: Fuzzy image clustering incorporating spatial continuity. IEE Proc. Visual Image Signal Process. 147, 185–192 (2000)
20. Clark, M.C., Hall, L.O., Goldgof, D.B., Clarke, L.P., Velthuizen, R.P., Silbigger, M.S.: MRI Segmentation using fuzzy clustering Techniques. IEEE Engineering in Medicine and Biology 13(5), 730–742 (1994)
21. Moussaoui, A., Benmahammed, K., Ferahta, N., Chen, V.: A New MR Brain Image Segmentation Using an Optimal Semisupervised Fuzzy C-means and pdf Estimation. Electronic Letters on Computer Vision and Image Analysis 5(4), 1–11 (2005)
22. Noordam, J.C., van den Broek, W.H.A.M., Buydens, L.M.C.: Geometrically guided fuzzy C-means clustering for multivariate image segmentation. In: Proc. Int. Conf. on Pattern Recognition, vol. 1, pp. 462–465 (2000)
23. Pham, D.L., Prince, J.L.: Adaptive fuzzy segmentation of magnetic resonance images. IEEE Trans. Medical Imaging. 18, 737–752 (1999)
24. Runkler, T.A., Bezdek, J.C.: Alternating cluster estimation: A new tool for clustering and function approximation. IEEE Trans. on Fuzzy Systems. 7(4), 377–393 (1999)
25. Ruspini, E.H.: A New Approach to Clustering. Information and Control 15(1), 22–32 (1969)
26. Ruspini, E.H.: Numerical methods for fuzzy clustering. Inform. Sci. 2, 319–350 (1970)

On the Connectivity, Lifetime and Hop Count of Routes Determined Using the City Section and Manhattan Mobility Models for Vehicular Ad Hoc Networks

Natarajan Meghanathan

Jackson State University
Jackson, MS 39217, USA
nmeghanathan@jsums.edu

Abstract. The high-level contribution of this paper is a simulation based analysis of the network connectivity, hop count and lifetime of the routes determined for vehicular ad hoc networks (VANETs) using the City Section and Manhattan mobility models. The Random Waypoint mobility model is used as a benchmark in the simulation studies. Two kinds of paths are determined on the sequence of static graphs representing the topology over the duration of the network session: paths with the minimum hop count (using the Dijkstra algorithm) and stable paths with the longest lifetime (using our recently proposed *OptPathTrans* algorithm). Simulation results indicate that the City Section model provided higher network connectivity compared to the Manhattan model for all the network scenarios. Minimum hop paths and stable paths determined under the Manhattan model have a smaller lifetime and larger hop count compared to those determined using the City Section and Random Waypoint mobility models.

Keywords: Vehicular ad hoc networks, Stable paths, Minimum hop paths, Connectivity, Simulation.

1 Introduction

A mobile ad hoc network (MANET) is a dynamically distributed system of mobile wireless nodes. The network bandwidth is limited. As the medium is shared, the transmitted signals are prone to interference and collisions. The transmission range of a node is limited as nodes operate with reduced battery power. Hence, multi-hop routing is a common feature in MANETs. As nodes move, there is unlikely to be a single fixed route throughout the duration of a source-destination session.

Vehicular ad hoc networks (VANETs) are one of the most promising application areas of MANETs. VANET communication is normally accomplished through special electronic devices placed inside each vehicle so that an ad hoc network of the vehicles is formed on the road. A vehicle equipped with a VANET device should be able to receive and relay messages to other VANET-device equipped vehicles in its neighborhood. In this paper, we use the terms "vehicle" and "node" interchangeably. They mean the same.

S. Ranka et al. (Eds.): IC3 2009, CCIS 40, pp. 170–181, 2009.
© Springer-Verlag Berlin Heidelberg 2009

VANETs resemble MANETs with respect to the dynamically and rapidly changing network topologies due to fast moving vehicles. However, the mobility of the vehicles is normally constrained by predefined roads and speed limitations. Mobility of the vehicles is also affected due to traffic congestion in the roads and the traffic control mechanisms (like stop signs and traffic lights) [1]. Route stability is an important design criterion to be considered in the design of VANET routing protocols. The routing protocols should be able to dynamically adapt to the rapidly changing network topologies while taking into consideration the layout of the roads. The commonly used route discovery approach of flooding the Route-Request (RREQ) packets can easily lead to congestion in the network and also consume the battery charge of the nodes [2]. Frequent route changes can also result in out-of-order packet delivery, causing high jitter in multi-media, real-time applications. For safety applications, it is better to route all the critical data packets through the same path so that the receiver can reassemble the packets and get a consistent view of the network condition.

In [3], we proposed an algorithm called *OptPathTrans* to determine the sequence of stable routes between a given source-destination (*s-d*) pair over the duration of an *s-d* session. Given the complete knowledge of the future topology changes over the entire duration of the communication session between a source *s* and destination *d*, algorithm *OptPathTrans* operates as follows: Whenever an *s-d* path is required at a time instant *t*, choose the longest-living *s-d* path since *t*. The above strategy is repeated over the duration of the *s-d* session. The sequence of such longest living stable paths is called the Stable Mobile Path (SMP). The performance of algorithm *OptPathTrans* has been so far studied only using the commonly used Random Waypoint mobility model [4] for MANETs.

In this paper, we study the performance of algorithm *OptPathTrans* with respect to the commonly used City Section mobility model [5] and the Manhattan mobility model [6] for VANETs. The Random Waypoint mobility model is used as a benchmark to compare and evaluate the relative stability and hop count of routes in a MANET and a VANET. We use the *OptPathTrans* algorithm to compute the sequence of stable paths (the Stable Mobile Path) and the *Dijkstra* algorithm [7] to compute the sequence of minimum hop paths (called the Minimum Hop Mobile Path). For different conditions of network density and node mobility considered, we compute the Minimum Hop Mobile Path (sometimes referred as MHMP) and the Stable Mobile Path (sometimes referred as SMP) under the three mobility models and measure the three critical metrics: (i) network connectivity, (ii) route lifetime and (ii) path hop count. We could not find any such evaluation of the two VANET mobility models in the literature.

The rest of the paper is organized as follows: Section 2 briefly discusses the Random Waypoint, City Section and the Manhattan mobility models. Section 3 provides an overview of the *OptPathTrans* algorithm used to determine the sequence of long-living stable paths in ad hoc networks. Section 4 illustrates the simulation results and compares the three mobility models with respect to the results obtained for network connectivity, route lifetime and hop count. Section 5 concludes the paper.

2 Review of the Mobility Models

All the three mobility models assume that the network is confined within fixed boundary conditions. The Random Waypoint mobility model assumes that the nodes can move anywhere within a network region. The City Section and the Manhattan mobility models assume the network to be divided into grids: square blocks of identical block length. The network is thus basically composed of a number of horizontal and vertical streets. Each street has two lanes, one for each direction (north and south direction for vertical streets, east and west direction for horizontal streets). A node is allowed to move only along the grids of horizontal and vertical streets.

2.1 Random Waypoint Mobility Model

Initially, the nodes are assumed to be placed at random locations in the network. The movement of each node is independent of the other nodes in the network. The mobility of a particular node is described as follows: The node chooses a random target location to move. The velocity with which the node moves to this chosen location is uniformly randomly selected from the interval $[v_{min},\ldots,v_{max}]$. The node moves in a straight line (in a particular direction) to the chosen location with the chosen velocity. After reaching the target location, the node may stop there for a certain time called the *pause time*. The node then continues to choose another target location and moves to that location with a new velocity chosen again from $[v_{min},\ldots,v_{max}]$. The selection of each target location and a velocity to move to that location is independent of the current node location and the velocity with which the node reached that location.

2.2 City Section Mobility Model

Initially, the nodes are assumed to be randomly placed in the street intersections. Each street (i.e., one side of a square block) is assumed to have a particular speed limit. Based on this speed limit and the block length, one can determine the time it would take to move in the street. Each node placed at a particular street intersection chooses a random target street intersection to move. The node moves to the chosen street intersection on a path that will incur the least amount of travel time. If two or more paths incur the same amount of least travel time, the tie is broken arbitrarily. After reaching the targeted street intersection, the node may stay there for a pause time and again choose a random target street intersection to move. Each node independently repeats this procedure.

2.3 Manhattan Mobility Model

Initially, the nodes are assumed to be randomly placed in the street intersections. The movement of a node is decided one street at a time. To start with, each node has equal probability of choosing any of the streets leading from its initial location. After a node begins to move in the chosen direction and reaches the next street intersection, the subsequent street in which the node will move is chosen probabilistically. If a node can continue to move in the same direction or can also change directions, then the node has 50% chance of continuing in the same direction, 25% chance of turning to the east/north and 25% chance of turning to the west/south, depending on the direction of the previous movement. If a node has only two options, then the node has an equal (50%) chance of exploring either of the two options. If a node has only one option to move (this occurs when the node reaches any of the four corners of the network), then the node has no other choice except to explore that option.

3 Algorithm to Determine Optimal Number of Path Transitions

We now briefly review the *OptPathTrans* algorithm, recently proposed by us in [3], to determine the optimal number of path transitions in ad hoc networks. The algorithm uses the notions of mobile graph [8] to record the sequence of network topology changes and mobile path to record the sequence of paths in a mobile graph.

3.1 Mobile Graph

A mobile graph [8] is defined as the sequence $G_M = G_1G_2 \ldots G_T$ of static graphs that represents the network topology changes over some time scale *T*. In the simplest case, the mobile graph $G_M = G_1G_2 \ldots G_T$ can be extended by a new instantaneous graph G_{T+1} to a longer sequence $G_M = G_1G_2 \ldots G_T\ G_{T+1}$, where G_{T+1} captures a link change (either a link comes up or goes down). But such an approach has very poor scalability. In this paper, we sample the network topology periodically for every 0.25 seconds, which could, in reality, be the instants of data packet origination at the source.

3.2 Mobile Path

A *mobile path* [8], defined for a source-destination (*s-d*) pair, in a mobile graph $G_M = G_1G_2 \ldots G_T$ is the sequence of paths $P_M = P_1P_2 \ldots P_T$, where P_i is a static path between the same *s-d* pair in $G_i = (V_i, E_i)$, V_i is the set of vertices and E_i is the set of edges connecting these vertices at time instant t_i. That is, each static path P_i can be represented as the sequence of vertices $v_0v_1 \ldots v_l$, such that $v_0 = s$ and $v_l = d$ and $(v_{j-1},v_j) \in E_i$ for $j = 1,2, \ldots, l$. The timescale of *T* normally corresponds to the duration of a session between *s* and *d*.

3.3 Stable Mobile Path and Minimum Hop Mobile Path

The Stable Mobile Path (SMP) for a given mobile graph and *s-d* pair is the sequence of static *s-d* paths such that the number of route transitions (change from one static *s-d* path to another) is as minimum as possible. In other words, the constituent static paths of an SMP have the largest possible route lifetime. A Minimum Hop Mobile Path (MHMP) for a given mobile graph and *s-d* pair is the sequence of minimum hop static *s-d* paths. The SMP for an *s-d* pair on a given mobile graph is determined by using algorithm *OptPathTrans*. The MHMP for an *s-d* pair on a given mobile graph is determined by repeatedly running the minimum path weight Dijkstra algorithm on the static graphs. We follow the Least Overhead Routing Approach (LORA) [9] for ad hoc networks. Accordingly, a minimum hop *s-d* path determined by running *Dijkstra* algorithm on a static graph G_i is assumed to be used in the subsequent static graphs G_{i+1}, G_{i+2}, …., as long as the path exists in these static graphs.

3.4 Algorithm Description

Algorithm *OptPathTrans* (refer Figure 1) operates on the following greedy strategy: Whenever a path is required, select a path that will exist for the longest time. Let $G_M = G_1G_2 \ldots G_T$ be the mobile graph generated by sampling the network topology at regular instants t_1, t_2, …, t_T of an *s-d* session. When an *s-d* path is required at sampling time instant t_i, the strategy is to find a mobile sub graph $G(i, j) = G_i \cap G_{i+1} \cap \ldots \cap G_j$ such that there exists at least one *s-d* path in $G(i, j)$ and no *s-d* path exists in $G(i, j+1)$. A minimum hop *s-d* path in $G(i, j)$ is selected. Such a path exists in each of the static graphs $G_i, G_{i+1}, \ldots, G_j$. If sampling instant $t_{j+1} \leq t_T$, the above procedure is repeated by finding the *s-d* path that can survive for the maximum amount of time since t_{j+1}. A sequence of such maximum lifetime static *s-d* paths over the timescale of a mobile graph G_M forms the stabile mobile *s-d* path in G_M. The run-time complexity of the algorithm is $O(n^2T)$, where *n* is the number of nodes in the network and *T* is the number of static graphs in a mobile graph (*T* is thus a measure of the timescale of the network communication session between the source *s* and destination *d*).

Input: $G_M = G_1G_2 \ldots G_T$, source *s*, destination *d*
Output: P_S // Stable-Mobile-Path
Auxiliary Variables: *i*, *j*
Initialization: i=1; j=1; $P_S = \Phi$

Begin *OptPathTrans*

1 **while** ($i \leq T$) do
2 Find a mobile graph $G(i, j) = G_i \cap G_{i+1} \cap \ldots \cap G_j$ such that there exists at least one *s-d* path in $G(i, j)$ and {no *s-d* path in $G(i, j+1)$ or $j = T$}
3 $P_S = P_S$ U {minimum hop *s-d* path in $G(i, j)$ }
4 $i = j + 1$
5 **end while**
6 **return** P_S

End *OptPathTrans*

Fig. 1. Pseudo code for algorithm *OptPathTrans*

4 Simulations

The network dimensions are 1000m x 1000m. For the Random Waypoint mobility model, we assume the nodes can move anywhere within the network and $v_{min} = v_{max}$. For the City Section and Manhattan mobility models, we assume the network is divided into grids: square blocks of length (side) 100m. The network is thus basically composed of a number of horizontal and vertical streets. Each street has two lanes, one for each direction (north and south direction for vertical streets, east and west direction for horizontal streets). A node is allowed to move only along the grids of horizontal and vertical streets. The wireless transmission range of a node is 250m. The node density is varied by performing the simulations with 25 (low density) and 50 (high density) nodes. The node velocity values used for each of the three mobility models are 5 m/s (about 10 miles per hour), 15 m/s (about 35 miles per hour) and 30 m/s (about 65 miles per hour), representing scenarios of low, moderate and high node mobility respectively.

We obtain a centralized view of the network topology by generating mobility trace files for 1000 seconds under each of the three mobility models. The network topology is sampled for every 0.25 seconds to generate the static graphs and the mobile graph. Two nodes are assumed to have a bi-directional link at time *t*, if the Euclidean distance between them at time *t* (derived using the locations of the nodes from the mobility trace file) is less than or equal to the transmission range of the nodes. Each data point in Figures 2 through 7 is an average computed over 5 mobility trace files and 20 randomly selected *s-d* pairs from each of the mobility trace files. The starting time of each *s-d* session is uniformly distributed between 1 to 20 seconds.

The following performance metrics are evaluated:

- Percentage Network Connectivity: The percentage network connectivity (refer Figure 2) denotes the probability of finding an *s-d* path between any source *s* and destination *d* in the network for a given density and mobility model. Measured over all the *s-d* sessions of a simulation run, this metric is the ratio of the number of static graphs that have the *s-d* paths to the total number of static graphs in the mobile graph.
- Average Route Lifetime: The average route lifetime is the average of the lifetime of all static paths of an *s-d* session, averaged over all *s-d* sessions.
- Average Hop Count: The average hop count is the time averaged hop count of a mobile path for an *s-d* session, averaged over all *s-d* sessions. The time averaged hop count for an *s-d* session is measured as sum of the products of the number of hops per static *s-d* path and the lifetime of the static *s-d* path divided by the number of static graphs which had a static *s-d* path.

4.1 Percentage Network Connectivity

The Random Waypoint mobility model provided the maximum connectivity among nodes for both low-density (connectivity close to 90%) and high-density networks (connectivity close to 99.5%). The two VANET mobility models provided relatively lower network connectivity for both low and high network density scenarios, but especially for low network density. This is due to the constrained motion of the nodes only along the streets of the network. For a given node velocity, the network connectivity provided by all the three mobility models improved significantly as we increase the network density from low to high. In low-density networks, for a given node velocity, the City Section model provided a network connectivity of about 76-78% and the Manhattan model provided a network connectivity of about 58-62%. The relatively lower connectivity for the Manhattan mobility model can be due to the probabilistic nature of direction selection after reaching each street intersection. In high-density networks, both the City Section and Manhattan models provided network connectivity greater than 97%. As more nodes are added to the streets, the probability of finding source-destination routes at any point of time increases significantly. For a given network density, the network connectivity provided by each of the three mobility models almost remained the same for different values of node velocity. Hence, network connectivity (refer Figure 2) is mainly influenced by the number of nodes in the network and their initial random distribution. The randomness associated with the mobility models ensure that node velocity is not a significant factor influencing network connectivity.

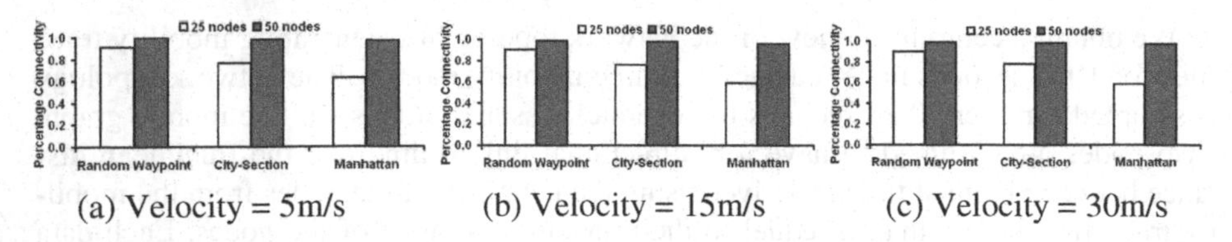

(a) Velocity = 5m/s (b) Velocity = 15m/s (c) Velocity = 30m/s

Fig. 2. Percentage Network Connectivity

4.2 Average Route Lifetime

We now discuss the average lifetime of routes observed for the three mobility models as we determine the Minimum Hop Mobile Path (refer Figure 3) and the Stable Mobile Path (refer Figure 4).

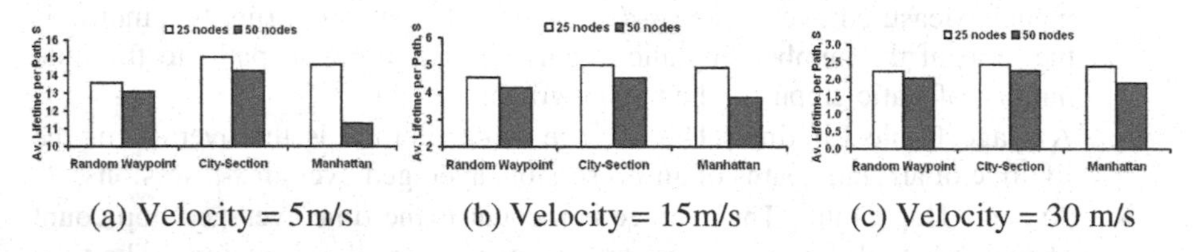

(a) Velocity = 5m/s (b) Velocity = 15m/s (c) Velocity = 30 m/s

Fig. 3. Average Route Lifetime per Static Path for the Minimum Hop Mobile Path

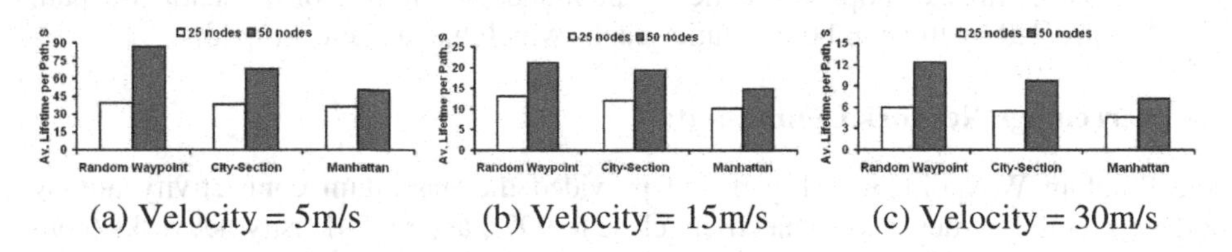

(a) Velocity = 5m/s (b) Velocity = 15m/s (c) Velocity = 30m/s

Fig. 4. Average Route Lifetime per Static Path for the Stable Mobile Path

When we aim for minimum hop count in the paths and determine the Minimum Hop Mobile Path by repeated application of the *Dijkstra* algorithm on the static graphs, the minimum hop paths determined under the City Section mobility model are relatively more stable (i.e., have a larger route lifetime) compared to the minimum hop paths determined under the Manhattan and Random Waypoint mobility models. For a given network density and node mobility, the average lifetime of the minimum hop paths determined under the City Section model is about 10% more than the lifetime of the minimum hop paths determined under the Random Waypoint model. In low-density networks, for a given node mobility, the average lifetime of the minimum hop paths determined under the City Section model is only about 2-3% more than the lifetime of the minimum hop paths determined under the Manhattan model. In high density networks, the average lifetime of the minimum hop paths determined using the City Section model is about 18-25% more than the lifetime of the minimum hop paths determined using the Manhattan model. The relatively high stability of minimum hop routes under the City Section model compared to the Manhattan model can be attributed to the larger average hop count per mobile path determined under the latter model. For the street networks, the larger the number of hops in a minimum hop path, the smaller the lifetime of the path. This is because the physical distance between the constituent nodes of every hop in a minimum hop path is about 80% of the transmission range of the nodes during the time of path discovery and hence all the hops (links in the path) of a minimum hop path in the street networks have a high probability of failure. For each of the three mobility models and for a given network density, the average route lifetime in conditions of high node mobility is about 16% of the average route lifetime in conditions of low node mobility and is about 49% of the average route lifetime in conditions of moderate node mobility.

When we aim for the maximum route lifetime and determine the Stable Mobile Path using algorithm *OptPathTrans*, routes determined under the Random Waypoint model are the most stable for all the simulation conditions. This is due to the unconstrained mobility of the nodes in all directions and algorithm *OptPathTrans* makes best use of this feature. The City Section model leads to relatively more stable routes compared to the Manhattan model, especially with increase in network density. While using the Random Waypoint model and the City Section model, with increase in network density, algorithm *OptPathTrans* gets more options to explore the different routes in the networks and choose the most stable route. However, the random directions chosen for node movement at each of the street intersections makes the Manhattan model to yield relatively less stable routes. For a given node mobility, the average lifetime of the stable paths determined under the City Section model is about 7-18% and 30-35% more than the average lifetime of the stable paths determined under the Manhattan model in low and high density networks respectively. For a given node mobility, the average lifetime of the stable paths determined under the Random Waypoint model is about 10-30% and 40-70% more than the average lifetime of the stable paths determined under the Manhattan model in low and high density networks respectively. For a given node mobility, the average lifetime of the stable paths determined using the Random Waypoint model is about 3-10% and 10-28% more than the average lifetime of the stable paths determined under the City Section model in low and high density networks respectively.

4.3 Average Hop Count per Path

We now discuss the time averaged hop count per path observed for the three mobility models when we determine the Minimum Hop Mobile Path (refer Figure 5) and the Stable Mobile Path (refer Figure 6).

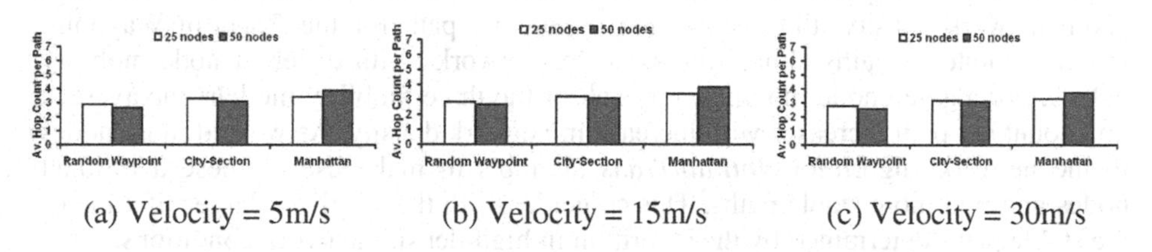

(a) Velocity = 5m/s (b) Velocity = 15m/s (c) Velocity = 30m/s

Fig. 5. Average Hop Count per Static Path for the Minimum Hop Mobile Path

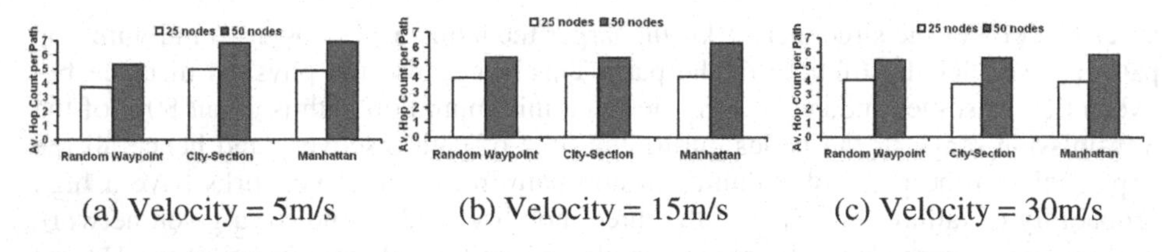

(a) Velocity = 5m/s (b) Velocity = 15m/s (c) Velocity = 30m/s

Fig. 6. Average Hop Count per Static Path for the Stable Mobile Path

The average hop count per minimum hop path determined for the two VANET mobility models is considerably higher than the hop count per minimum hop path determined under the Random Waypoint model. The relatively higher hop count can be attributed to the constrained mobility of the nodes in the street networks. The minimum hop paths in the street networks are most likely not to exist on a straight line between the source and destination. For a given node velocity, the average hop count per path under the City Section model and Manhattan model is respectively about 14-18% and 19-45% more than that incurred for the Random Waypoint model. With increase in network density, the average hop count per path for the Random Waypoint model and the City Section model decreases (by a factor of 5 – 10%). On the other hand, with increase in network density, the average hop count per path for the Manhattan model increases by 12-14%. This can be attributed to the significant increase in network connectivity observed for the Manhattan model with increase in network density. Given the extremely constrained and random nature of node movement under the Manhattan model, in order to connect the source and destination, more intermediate nodes have to be accommodated in the source-destination paths.

As we aim for highly stable routes, algorithm *OptPathTrans* makes use of the less dynamic nature of the low and moderate node mobility street networks and determines paths with relatively larger lifetime. For street networks, the physical distance between the constituent nodes of every hop in the stable paths determined for the City Section and Manhattan models is only 60-70% of the transmission range of the nodes at the time of route determination. As a result, more intermediate nodes with a relatively larger link lifetime have to be accommodated in order to connect the source and destination. Thus, in street networks of low and moderate mobility, paths with larger lifetime are determined at the expense of hop count. This illustrates the tradeoff between route stability and hop count. But, in high node mobility street networks, the chances of finding long-living stable paths by accommodating intermediate nodes with larger link lifetime decreases. Hence, we observe a relative decrease in the hop count of the stable paths determined for high node mobility street networks. For a given network density, the average hop count per path for the Random Waypoint mobility model remains almost the same for networks with different node mobility values. For a given node mobility, for each of the three mobility models, the average hop count per path increases with increase in network density. As we add more nodes to the network, algorithm *OptPathTrans* attempts to make use of these additional nodes and find more stable paths. This coincides with the relatively larger lifetime of the stable paths determined by the algorithm in high-density network conditions.

For a given node mobility, for each of the three mobility models, the average hop count per path increases with increase in network density. As we add more nodes to the network, algorithm *OptPathTrans* attempts to make use of these additional nodes and find more stable paths. This coincides with the relatively larger lifetime of the stable paths determined by the algorithm in high-density network conditions. With increase in the network density from 25 to 50 nodes, the average hop count of the stable paths determined by algorithm *OptPathTrans* increases by a factor of 35-45%, 35-50% and 42-56% for the Random Waypoint, City Section and Manhattan mobility models respectively. In networks of low mobility, for a given density, the average hop count per path incurred with the City Section and the Manhattan models is about 28-35% more than that incurred under the Random Waypoint model. On the other hand, in networks of moderate and high node mobility, for a given network density, the average hop count per path for the City Section and Manhattan models is not more than 20% of the value incurred under the Random Waypoint model.

4.4 Route Lifetime – Hop Count Tradeoff

We observe a tradeoff between the objectives of optimizing route lifetime and hop count for all the three mobility models. Both these performance metrics cannot be optimized at the same time. For a given simulation condition of network density and node mobility, the average hop count of a Minimum Hop Mobile Path (MHMP) is smaller than the average hop count of a Stable Mobile Path (SMP); the average route lifetime of a SMP is more than the average route lifetime of a MHMP.

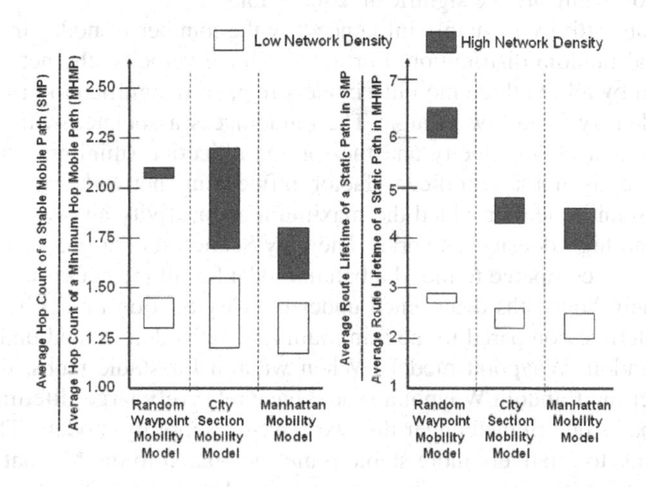

Fig. 7. Route Lifetime – Hop Count Tradeoff

Overall, the general trend of the lifetime-hop count tradeoff is that for a given node mobility condition, when we increase the network density, the static paths of an SMP have a relatively larger route lifetime at the expense of an increase in the hop count. Figure 7 captures this tradeoff for each of the three mobility models by illustrating the range of the ratio of the average hop counts of the SMP and MHMP and the range of the ratio of the average route lifetime of the static paths in SMP and MHMP. As observed in the figure, the ratio of the hop counts is lowest for the Manhattan model for both low and high density networks. The hop count ratios of the City Section and Random Waypoint models are relatively larger. The ratio of the route lifetimes is the lowest under the Manhattan model and is the highest under the Random Waypoint model. The ratio of the route lifetimes for the City Section model is close to that obtained for the Manhattan model.

We thus conclude that the Manhattan mobility model incurs the relatively smallest increase in the hop count when we aim for stable paths vis-à-vis minimum hop paths. In the case of the City Section mobility model, even though the increase in the route lifetime is slightly higher compared that incurred with the Manhattan mobility model, there is a substantial increase in the hop count of the stable paths. In the case of the Random Waypoint mobility model, the increase in the lifetime of the stable paths when compared to the minimum hop paths is substantially high, but this is achieved at a relatively larger hop count for the stable paths.

5 Conclusions

The high-level contribution of this paper is a simulation based analysis of the network connectivity, hop count and lifetime of the routes determined for vehicular ad hoc networks using the City Section and Manhattan mobility models. The MANET Random Waypoint mobility model is used as a benchmark in the simulation studies. Two kinds of routes are determined: routes with minimum hop count and routes with longest lifetime. Following are the significant conclusions:

Network connectivity is mainly influenced by the number of nodes in the network and their initial random distribution. For a given node velocity, the network connectivity provided by all the three mobility models improved significantly as we increase the network density from low to high. The randomness associated with the mobility models ensure that node velocity and the routing objective (minimum hop count or longest lifetime) is not a significant factor influencing network connectivity. The Random Waypoint model provided the maximum connectivity among nodes for both low-density and high-density networks. The City Section model provided higher network connectivity compared to the Manhattan model for all the network scenarios.

The minimum hop paths determined under the City Section model have relatively larger route lifetime compared to the minimum hop paths determined under the Manhattan and Random Waypoint models. When we aim for stable paths, the paths determined under the Random Waypoint model have relatively larger lifetime compared to the stable paths determined under the two VANET mobility models. The City Section model leads to relatively more stable routes compared to the Manhattan mobility model, especially with increase in network density. The random directions chosen for node movement at each of the street intersections makes the Manhattan model to yield relatively less stable routes.

The average hop count per minimum hop path determined for the two VANET mobility models is considerably higher than the hop count per minimum hop path determined under the Random Waypoint model. The minimum hop paths in the street networks are most likely not to exist on a straight line between the source and destination. As a result, more intermediate nodes with a relatively larger link lifetime have to be accommodated in order to connect the source and destination. Thus, in street networks of low and moderate mobility, paths with larger lifetime are determined at the expense of hop count. But, in high node mobility street networks, the chances of finding long-living stable paths by accommodating intermediate nodes with larger link lifetime decreases.

We thus observe a tradeoff between the objectives of optimizing route lifetime and hop count for all the three mobility models. Both of these performance metrics cannot be optimized at the same time. For a given simulation condition of network density and node mobility, the average hop count of a Minimum Hop Mobile Path is smaller than the average hop count of a Stable Mobile Path; average route lifetime of a Stable Mobile Path is more than the average route lifetime of a Minimum Hop Mobile Path.

References

1. Taleb, T., Ochi, M., Jamalipour, A., Nei, K., Nemoto, Y.: An Efficient Vehicle-Heading Based Routing Protocol for VANET Networks. In: International Wireless Communications and Networking Conference, pp. 2199–2204. IEEE, Las Vegas (2006)
2. Siva Ram Murthy, C., Manoj, B.S.: Ad Hoc Wireless Networks: Architectures and Protocols. Prentice Hall, Upper Saddle River (2004)
3. Meghanathan, N., Farago, A.: On the Stability of Paths, Steiner Trees and Connected Dominating Sets in Mobile Ad hoc Networks. Elsevier Ad hoc Networks 6(5), 744–769 (2008)
4. Bettstetter, C., Hartenstein, H., Perez-Costa, X.: Stochastic Properties of the Random-Way Point Mobility Model. Wireless Networks 10(5), 555–567 (2004)
5. Camp, T., Boleng, J., Davies, V.: A Survey of Mobility Models for Ad Hoc Network Research. Wireless Communication and Mobile Computing 2(5), 483–502 (2002)
6. Bai, F., Sadagopan, N., Helmy, A.: IMPORTANT: A Framework to Systematically Analyze the Impact of Mobility on Performance of Routing Protocols for Ad hoc Networks. In: International Conference on Computer Communications, pp. 825–835. IEEE, San Francisco (2003)
7. Cormen, T.H., Leiserson, C.E., Rivest, R.L., Stein, C.: Introduction to Algorithms, 2nd edn. MIT Press/ McGraw Hill, New York (2001)
8. Farago, A., Syrotiuk, V.R.: MERIT: A Scalable Approach for Protocol Assessment. Mobile Networks and Applications 8(5), 567–577 (2003)
9. Abolhasan, M., Wysocki, T., Dutkiewicz, E.: A Review of Routing Protocols for Mobile Ad hoc Networks. Elsevier Ad hoc Networks 2(1), 1–22 (2004)

On the Privacy Protection of Biometric Traits: Palmprint, Face, and Signature

Saroj Kumar Panigrahy, Debasish Jena,
Sathya Babu Korra, and Sanjay Kumar Jena

Department of Computer Science & Engineering
National Institute of Technology Rourkela, 769 008, Orissa, India
{skp.nitrkl,debasishjena}@hotmail.com, {ksathyababu,skjena}@nitrkl.ac.in

Abstract. Biometrics are expected to add a new level of security to applications, as a person attempting access must prove who he or she really is by presenting a biometric to the system. The recent developments in the biometrics area have lead to smaller, faster and cheaper systems, which in turn has increased the number of possible application areas for biometric identity verification. The biometric data, being derived from human bodies (and especially when used to identify or verify those bodies) is considered personally identifiable information (PII). The collection, use and disclosure of biometric data — image or template, invokes rights on the part of an individual and obligations on the part of an organization. As biometric uses and databases grow, so do concerns that the personal data collected will not be used in reasonable and accountable ways. Privacy concerns arise when biometric data are used for secondary purposes, invoking function creep, data matching, aggregation, surveillance and profiling. Biometric data transmitted across networks and stored in various databases by others can also be stolen, copied, or otherwise misused in ways that can materially affect the individual involved. As Biometric systems are vulnerable to replay, database and brute-force attacks, such potential attacks must be analysed before they are massively deployed in security systems. Along with security, also the privacy of the users is an important factor as the constructions of lines in palmprints contain personal characteristics, from face images a person can be recognised, and fake signatures can be practised by carefully watching the signature images available in the database. We propose a cryptographic approach to encrypt the images of palmprints, faces, and signatures by an advanced Hill cipher technique for hiding the information in the images. It also provides security to these images from being attacked by above mentioned attacks. So, during the feature extraction, the encrypted images are first decrypted, then the features are extracted, and used for identification or verification.

Keywords: Biometrics, Face, Palmprint, Signature, Privacy Protection, Cryptography.

S. Ranka et al. (Eds.): IC3 2009, CCIS 40, pp. 182–193, 2009.
© Springer-Verlag Berlin Heidelberg 2009

1 Introduction

The idea of biometric identification is very old. The methods of imprints, handwritten signatures are still in use. The photographs on the identification cards are still an important way for verifying the identity of a person. But developing technology is paving the way for automated biometric identification and is now a highly interested area of research. Biometric techniques are more and more deployed in several commercial, institutional, and forensic applications to build secure and accurate user authentication procedures. The interest in biometric approaches for authentication is increasing for their advantages such as security, accuracy, reliability, usability, and friendliness. As a matter of fact, biometric traits (*e.g.*, fingerprints, palmprints, face), being physically part of the owner, are always available to the user who is therefore not afraid of losing them. However, compared to passwords, biometric traits cannot be strictly considered as "secrets" since often they can be inadvertently disclosed: fingerprints are left on a myriad of objects such as door handles or elevator buttons; pictures of faces are easily obtained without the cooperation of the subjects. Moreover, if they are captured or if their digital representations are stolen, they cannot be simply replaced or modified in any way, as it can be done with passwords or tokens [1]. These aspects have limited so far the number of applications in which biometric authentication procedures were allowed by privacy agencies in several countries. In addition to this, users often perceive the potential threat to their privacy and this reduces the user acceptance of biometric systems, especially on a large scale.

In a typical biometric authentication system, trusted users provide the authentication party with a sample of a biometric trait (e.g., a fingerprint scan). A digital representation of the fingerprint is then stored by the party and compared at each subsequent authentication with new fingerprint scans. The party is then in charge of protecting the database where digital representations of fingerprints are stored. If an intruder gained access to the database, she could prepare fake fingerprints starting from each of the digital images. To limit such a possibility, images of biometric traits are not stored explicitly, rather they can be stored in encrypted form. Only a mathematical description of them is used (the parameters of a model or relevant features). Such a mathematical characterisation is generally called *template* and the information contained in it is sufficient to complete the authentication process. Templates are obtained through *feature extraction* algorithms.

In addition to the potential illegitimate access by imposters, biometric systems raise issues including unintended functions, unintended applications and template sharing.

- **Unintended Functions:** Our biometric traits contain rich private information, which can be extracted from biometrics for non-authentication purposes. DNA containing all genetic information including sex, ethnicity, physical disorder and mental illness can be employed for discrimination. Certain patterns in palm lines also associate with mental disorders such as Down syndrome and schizophrenia.

- **Unintended Applications:** Some biometric traits can be collected without user cooperation. Face and iris are two typical examples. Governments and organizations can employ them for tracking.
- **Template Sharing:** Biometric templates in databases of authorized agents are possible to be shared by unauthorized agents.

1.1 Risks in Biometric Systems

Although biometric authentication approaches are much more secure than the traditional approaches, they are not invulnerable. Biometric systems are vulnerable to many attacks including replay, database and brute-force attacks. Comparing verification, fusion and identification, only limited works are related to palmprint security [2]. Fig. 1 shows a number of points, Points 1-8, all being vulnerable points as identified by Ratha *et al.* [3]. The potential attack points are between and on the common components of a biometric system, input sensor, feature extractor, matcher and database and are especially open to attack when biometric systems are employed on remote, unattended applications, giving attackers enough time to make complex and numerous attempts to break in.

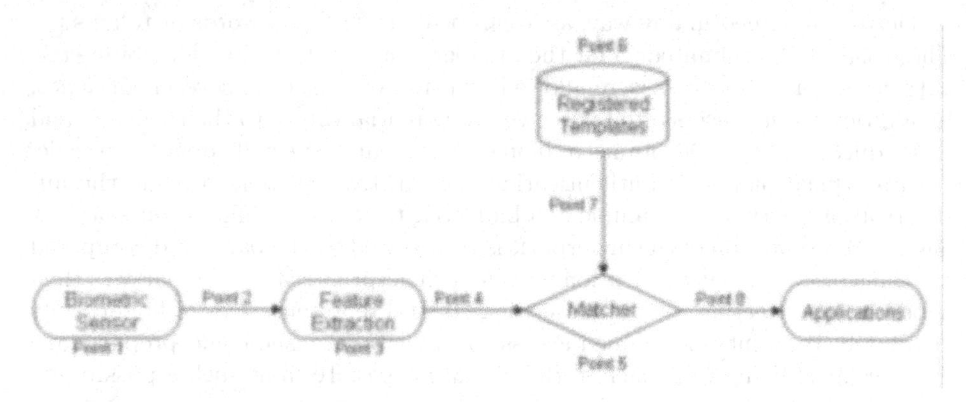

Fig. 1. Potential attack points in a biometric system

At Point 1, a system can be spoofed using fake biometrics such as artificial gummy fingerprints and face masks. At Point 2, it is possible to avoid liveness tests in the sensors by using a pre-recorded biometric signal such as a fingerprint image. This is a so-called replay attack. At Point 3, the original output features can be replaced with a predefined feature by using a Trojan horse to override the feature extraction process. At Point 4, it is possible to use both brute-force and replay attacks, submitting on the one hand numerous synthetic templates or, on the other, prerecorded templates. At Point 5, original matching scores can be replaced with preselected matching scores by using a Trojan horse. At Point 6, it is possible to insert templates from unauthorised users into the database or to modify templates in the database. At Point 7, replay attacks are once again possible. At Point 8, it is possible to override the system's decision output and to collect the matching scores to generate the images in the registered database.

Significant privacy (and operational) concerns arise with unrestricted collection and use of more and more biometric data for identification purposes. To begin with, the creation of large centralized databases, accessible over networks in real-time, presents significant operational and security concerns. If networks fail or become unavailable, the entire identification system collapses. Recognizing this, system designers often build in high redundancy in parallel systems and mirrors (as well as failure and exception management processes) to ensure availability. However, this can have the effect of increasing the security risks and vulnerabilities of the biometric data. Large centralized databases of biometric PII, hooked up to networks and made searchable in a distributed manner, represent significant targets for hackers and other malicious entities to exploit. It is also a regrettable reality that large centralized databases are also more prone to function creep (secondary uses) and insider abuse. There are also significant risks associated with transmitting biometric data over networks where they may be intercepted, copied, and actually tampered with, often without any detection. It should be evident that the loss or theft of one's biometric image opens the door to massive identity theft if the thief can use the biometric for his or her own purposes.

Both Cryptographic techniques and cancellable biometrics can be used for encryption. The difference between these two approaches is that cancellable biometrics performs matching in transform domains while cryptographic techniques require decryption before feature extraction and/or matching. In other words, decryption is not necessary for cancellable biometrics. When matching speed is an issue, e.g., identification in a large database, cancellable biometrics is more suitable for hiding the private information. And when privacy and security of palmprint database is required then cryptographic techniques can be used for encrypting the palmprint images in the database.

Various strategies have been presented to address the problem of supporting personal verification based on human biometric traits, while preserving privacy of digital templates [4]. Most approaches depend on jointly exploiting the characteristics of biometrics and cryptography [5,6]. The main idea is that of devising biometric templates and authentication procedures which do not disclose any information on the original biometric traits, for example replicating the usual approach adopted in password-based authentication system. Similarly, biometric templates are generated by using suited cryptographic primitives so as to protect their privacy and ensure that an attacker cannot retrieve any information on the original biometric trait used for the generation of the template. In this way, user's privacy is guaranteed. Moreover, even if a template is compromised (stolen, copied, etc.) it is always possible to generate a novel template by starting from the same original biometric trait.

In this paper, we propose an encryption technique to encrypt the images of palmprints , faces and signatures by a advanced version of Hill cipher algorithm for maintaining privacy, before storing in database for feature extraction. Our scheme resists the brute-force attacks and database attacks. The rest of the paper is organized as follows. Section 2 deals with description of Hill Cipher encryption technique. In Section 3 our proposed advanced Hill cipher encryption algorithm is explained. Results are discussed in the Section 4. Section 5 gives the concluding remarks.

2 Hill Cipher

Hill ciphers are an application of linear algebra to cryptology. It was developed by the mathematician Lester Hill. The Hill cipher algorithm takes m successive plaintext letters and substitutes m ciphertext letters for them. The substitution is determined by m linear equations in which each character is assigned a numerical value ($a = 0, b = 1, ..., z = 25$). Let m be a positive integer, the idea is to take m linear combinations of the m alphabetic characters in one plaintext element and produce m alphabetic characters in one ciphertext element. Then, an $m \times m$ matrix K is used as a key of the system such that K is invertible *modulo* 26 [7]. Let k_{ij} be the elements of matrix K . For the plaintext block $P = (p_1, p_2, ..., p_m)$ (the numerical equivalents of m letters) and a key matrix K, the corresponding ciphertext block $C = (c_1, c_2, ..., c_m)$ can be computed as follows.

For Encryption: $E_K(P) = KP = C$, i.e.,

$$\begin{pmatrix} c_1 \; c_2 \; ... \; c_m \end{pmatrix} = \begin{bmatrix} k_{11} & k_{12} & ... & k_{1m} \\ k_{21} & k_{22} & ... & k_{2m} \\ ... & ... & ... & ... \\ k_{m1} & k_{m2} & ... & k_{mm} \end{bmatrix} \begin{pmatrix} p_1 \; p_2 \; ... \; p_m \end{pmatrix} \tag{1}$$

For Decryption: $D_K(C) = K^{-1}C = K^{-1}KP = P$, i.e.,

$$\begin{pmatrix} p_1 \; p_2 \; ... \; p_m \end{pmatrix} = \begin{bmatrix} k_{11} & k_{12} & ... & k_{1m} \\ k_{21} & k_{22} & ... & k_{2m} \\ ... & ... & ... & ... \\ k_{m1} & k_{m2} & ... & k_{mm} \end{bmatrix}^{-1} \begin{pmatrix} c_1 \; c_2 \; ... \; c_m \end{pmatrix} \tag{2}$$

If the block length is m, there are 26^m different m letters blocks possible, each of them can be regarded as a letter in a 26^m-letter alphabet. Hill's method amounts to a mono-alphabetic substitution on this alphabet [8].

2.1 Use of Involutory Key Matrix

Hill Cipher requires inverse of the key matrix while decryption. infact that not all the matrices have an inverse and therefore they will not be eligible as key matrices in the Hill Cipher scheme [9]. If the key matrix is not invertible, then encrypted text cannot be decrypted. In order to overcome this problem, we suggest the use of self-invertible or involutory matrix while encryption in the Hill Cipher. If the matrix used for the encryption is self-invertible, then, at the time of decryption, we need not to find inverse of the matrix. Moreover, this method eliminates the computational complexity involved in finding inverse of the matrix while decryption. A is called involutory matrix if $A^{-1} = A$. The various methods for generation of self-invertible matrix are proposed in [10].

2.2 Image Encryption Using Hill Cipher

We note that Hill cipher can be adopted to encrypt grayscale and color images. For grayscale images, the modulus will be 256 (the number of levels is considered as the number of alphabets). In the case of color images, first decompose the color image into (RGB) components. Second, encrypt each component (R-G-B) separately by the algorithm. Finally, concatenate the encrypted components together to get the encrypted colour image [11].

3 Proposed Advanced Hill Cipher Encryption Algorithm

Saeednia S. has proposed a symmetric cipher that is actually a variation of the Hill cipher. His scheme makes use of "random" permutations of columns and

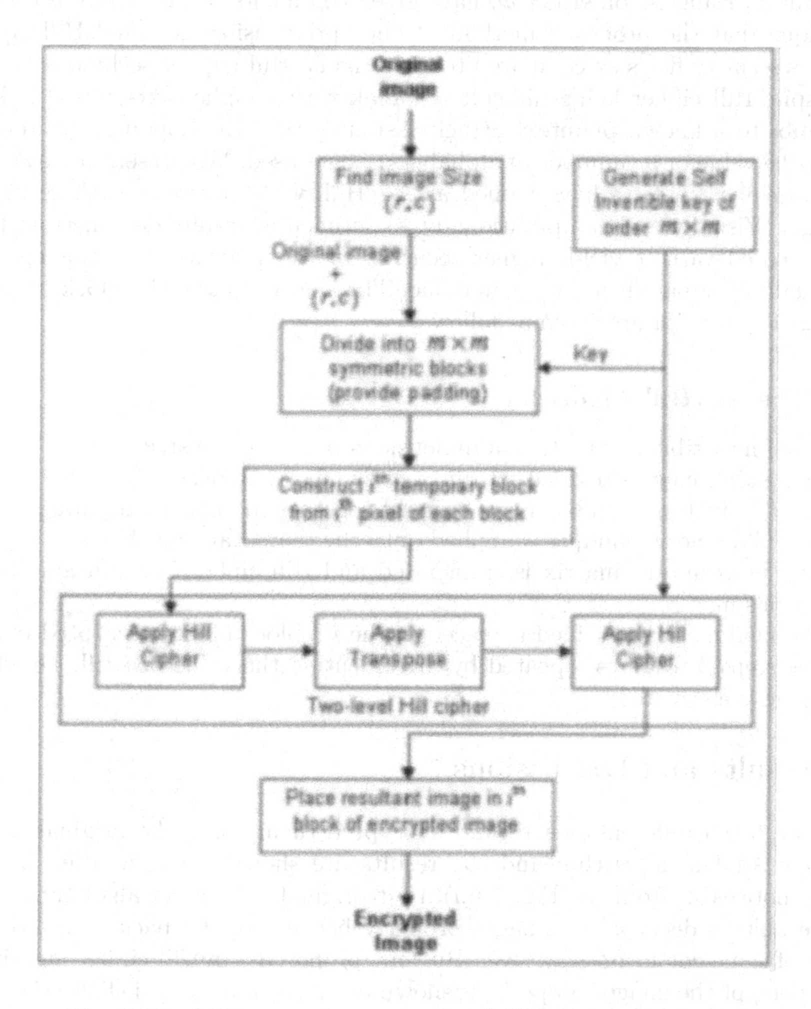

Fig. 2. Block Diagram for proposed AdvHill Cipher Encryption

rows of a matrix to form a "different" key for each data encryption. The cipher has matrix products and permutations as the only operations which may be performed "efficiently" by primitive operators, when the system parameters are carefully chosen [12].

A main drawback of Hill Cipher algorithm is that it encrypts identical plaintext blocks to identical ciphertext blocks and cannot encrypt images that contain large areas of a single colour [13]. Thus, it does not hide all features of the image which reveals patterns in the plaintext. Moreover, it can be easily broken with a known plaintext attack revealing weak security. So, Ismail et al. have proposed a variant of the Hill cipher that overcomes these disadvantages [14]. The proposed technique adjusts the encryption key to form a different key for each block encryption. It is mentioned in the paper that their proposed variant yields higher security and significantly superior encryption quality compared to the original one. But Y. Rangel-Romero *et al.* have given comments on the above proposed technique that the proposed method of encryption using modified Hill cipher still has security flaws as compared to the original Hill Cipher technique [15].

Despite Hill cipher being difficult to break with a ciphertext-only attack, it succumbs to a known plaintext attack assuming that the opponent has determined the value of p (number of alphabets) being used. We present a variant of the Hill cipher that we have named as AdvHill, which overcomes these disadvantages. Visually and computationally, experimental results demonstrate that the proposed variant yields higher security and significantly superior encryption quality compared to the original one. The algorithm and the block diagram (Fig. 2) for AdvHill are given as follows.

3.1 The AdvHill Algorithm

1. A self-invertible key matrix of dimensions m x m is constructed.
2. The plain image is divided into m x m symmetric blocks.
3. The ith pixels of each block are brought together to form a temporary block.
   (a) Hill cipher technique is applied onto the temporary block.
   (b) The resultant matrix is transposed and Hill cipher is again applied to this matrix.
4. The final matrix obtained is placed in the ith block of the encrypted image.
5. The steps 3 to 4 are repeated by incrementing the value of i till the whole image is encrypted.

4 Results and Discussions

We have taken different images and encrypted them using the original cipher and our AdvHill algorithm and the results are shown below in Fig. 3. It is clearly noticeable from the Fig. 3(h,i), that original Hill cipher algorithm could not be able to decrypt the images properly because of the background of the image of same colour or gray level. But our proposed AdvHill cipher algorithm could decrypt the images properly as shown in Fig. 3(m,n). Fig. 4 shows the time

Fig. 3. Original images (a-e), corresponding encrypted images by Hill Cipher (f-j) and by AdvHill Cipher Algorithm (k-o)

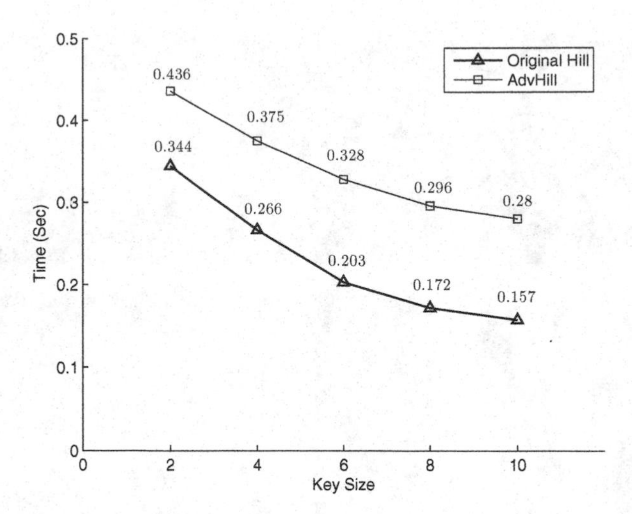

Fig. 4. Encryption time test of Lena image

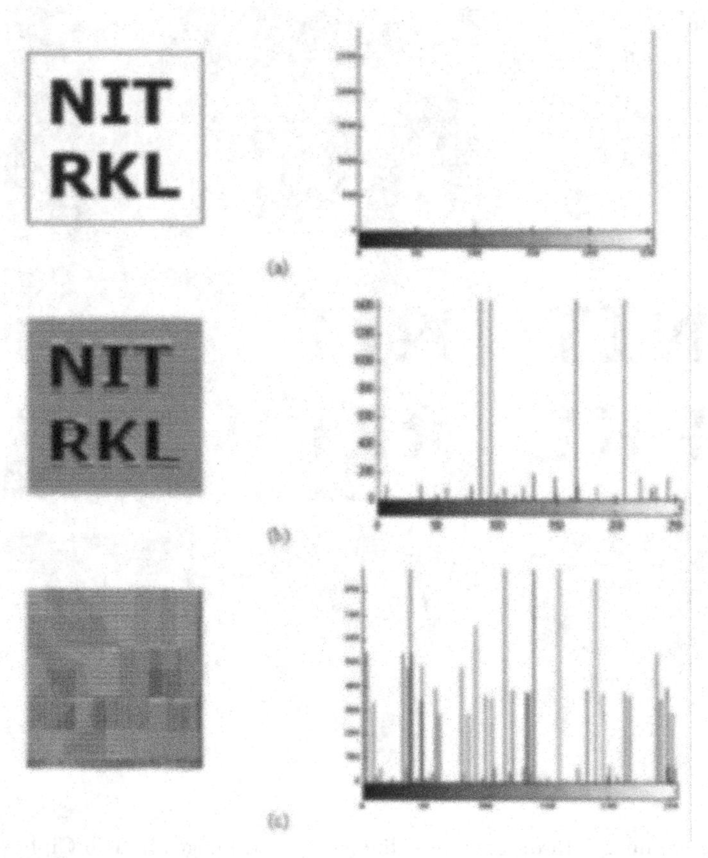

Fig. 5. Histograms of (a) original NITRKL image (b) image encrypted by original Hill cipher and (c) image encrypted by AdvHill cipher

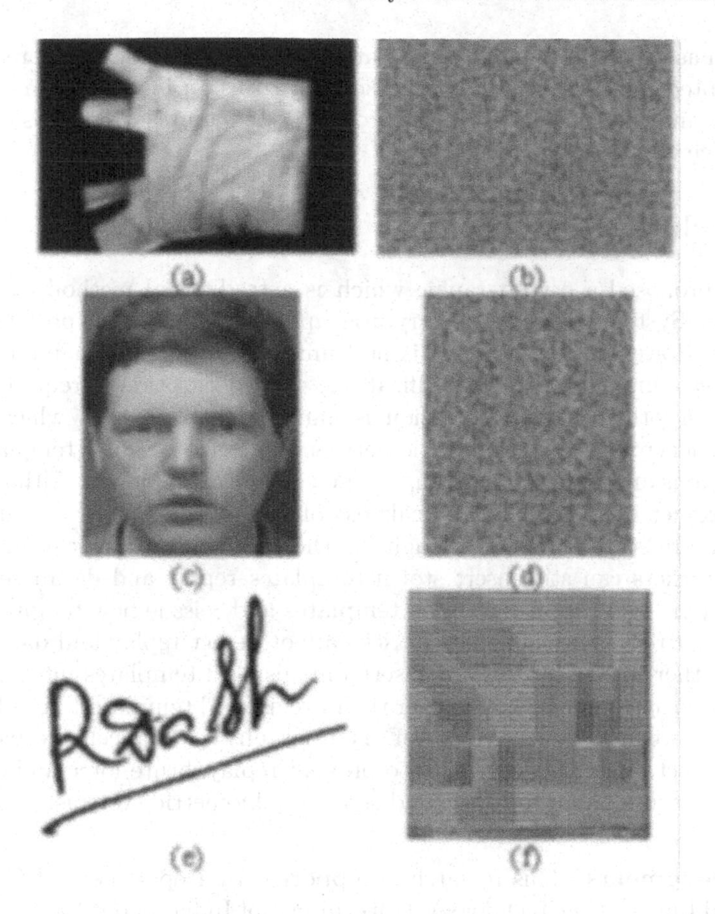

Fig. 6. Encryption of biometric images by AdvHill cipher (a) palmprint image (b) face image (c) signature

analysis for Lena image encryption by original Hill and AdvHill algorithms. It is clear that our AdvHill takes more time than that of original Hill to encrypt the image but with stronger security. Fig. 5 shows how our AdvHill algorithm is capable of decrypting the image as in the histograms it introduces more gray levels which leads to failure of frequency analysis by attackers. This shows that our image encryption scheme is stronger than the original Hill cipher and is also resistant to known-plaintext attack.

4.1 Palmprint, Face and Signature Image Encryption

From the previous section, it is clear that our AdvHill cipher algorithm works well for image encryption compared to that of original Hill Cipher algorithm. We can apply our AdvHill to any type of images for encryption. So, to provide security and maintain privacy of the biometric traits, i.e., palmprint, face and signature images of the users, we encrypt the images and then store in databases for feature extraction as shown in Fig. 6. Before feature extraction, these encrypted images are decrypted by the same key as the key matrix is an involutory matrix. The palmprint and face images are taken from PolyU palmprint database [16] and AT&T Face database [17] respectively.

5 Conclusion

We have proposed a cryptographic which is a traditional method of encrypting images. Systems protected by cryptography store only encrypted images in databases. However, cryptography is not suitable for speed-demanding matching, e.g. real-time large-scale identification, since decryption is required before matching. Another potential solution is cancellable biometrics which can be used for encryption. Cancellable biometrics transform original templates into other domains and perform matching in the transformed domain. Although cancellable biometrics overcome the weakness of cryptography, current cancellable biometrics are still not secure enough for the palmprint identification. For example, attackers can still insert stolen templates replay and database attacks before systems can cancel the stolen templates and reissue new templates. Furthermore, current cancellable biometrics cannot detect replay and database attacks. In other words, if attackers insert unregistered templates into data links or databases, systems cannot discover the unregistered templates. To solve these problems, we can take advantages of cryptography and cancellable biometrics to design a set of security measures to prevent replay, brute force and database attacks for secure palmprint, face and signature biometric systems.

Acknowledgments. This research is supported by Department of Communication and Information Technology, Government of India, under Information Security Education and Awareness Project and being carried out at department of Computer Science and Engineering, National Institute of Technology Rourkela, Orissa, India.

References

1. Schneier, B.: The uses and abuses of biometrics. Communication of the ACM 42(8), 136 (1999)
2. Kong, A.W.K., Zhang, D., Kamel, M.: Analysis of brute-force break-ins of a palmprint authentication system. IEEE Transactions On Systems, Man and Cybernetics-Part B: Cybernetics 36(5), 1201–1205 (2006)
3. Bolle, R.M., Connell, J.H., Ratha, N.K.: Biometric perils and patches. Pattern Recognition 35, 2727–2738 (2002)
4. Uludag, U., Pankanti, S., Prabhakar, S., Jain, A.: Biometric cryptosystems: Issues and challenges. In: Proceedings of the IEEE, Special Issue on Enabling Security Technologies for Digital Rights Management, June 2004, vol. 92, pp. 948–960 (2004)
5. Juels, A., Wattenberg, M.: A fuzzy commitment scheme. In: Proceedings of the 6th ACM conference on Computer and communications security (CCS 1999), pp. 28–36. ACM Press, New York (1999)

6. Hao, F., Anderson, R., Daugman, J.: Combining cryptography with biometrics effectively. Technical Report UCAMCL-TR 640, Computer Laboratory, University of Cambridge, United Kingdom (July 2005)
7. Petersen, K.: Notes on number theory and cryptography (2002), `http://www.math.unc.edu/Faculty/petersen/Coding/cr2.pdf`
8. Menezes, A.J., Oorschot, P.V., Stone, S.V.: Handbook of Applied Cryptography. CRC Press, Boca Raton (1996)
9. Schneir, B.: Cryptography and Network Security, 2nd edn. John Wiley & Sons, Chichester (1996)
10. Acharya, B., Patra, S.K., Panda, G., Panigrahy, S.K.: Novel methods of generating self-invertible matrix for hill cipher algorithm. International Journal of Security 1(1), 14–21 (2007)
11. Li, S., Zheng, X.: On the security of an image encryption method. In: Proceedings of the IEEE International Conference on Image Processing (ICIP 2002), vol. 2, pp. 925–928 (2002), `http://www.hooklee.com/Papers/ICIP2002.pdf`
12. Saeednia, S.: How to make the hill cipher secure. Cryptologia 24(4), 353–360 (2000)
13. Panigrahy, S.K., Acharya, B., Jena, D.: Image encryption using self-invertible key matrix of hill cipher algorithm. In: Proceedings of the 1st International Conference on Advances in Computing (ICAC 2008), Chikhli, India, pp. 1–5 (2008)
14. Ismail, I.A., Amin, M., Diab, H.: How to repair the hill cipher. Journal of Zhejiang Univ. Science A 7(12), 2022–2030 (2006)
15. Rangel-Romero, Y., et al.: Comments on how to repair the hill cipher. Journal of Zhejiang University SCIENCE A, 1–4 (2007)
16. Palmprints. Hongkong PolyU Palmprint Database, `http://www.comp.polyu.edu.hk/~biometric`
17. Faces, Cambridge AT&T Lab Face Database, `http://www.cl.cam.ac.uk/research/dtg/attarchive/facedatabase.html`

Indexing Iris Biometric Database Using Energy Histogram of DCT Subbands

Hunny Mehrotra[1], Badrinath G. Srinivas[2], Banshidhar Majhi[1], and Phalguni Gupta[2]

[1]National Institute of Technology Rourkela, Orissa 769008, India
{hunny,bmajhi}@nitrkl.ac.in
[2]Indian Institute of Technology Kanpur, UP 208016, India
{badri,pg}@iitk.ac.in

Abstract. The key concern of indexing is to retrieve small portion of database for searching the query. In the proposed paper iris database is indexed using energy histogram. The normalised iris image is divided into subbands using multiresolution DCT transformation. Energy based histogram is formed for each subband using all the images in the database. Each histogram is divided into fixed size bins to group the iris images having similar energy value. The bin number for each subband is obtained and all subands are traversed in Morton order to form a global key for each image. During database preparation the key is used to traverse the B tree. The images with same key are stored in the same leaf node. For a given query image, the key is generated and tree is traversed to end up to a leaf node. The templates stored at the leaf node are retrieved and compared with the query template to find the best match. The proposed indexing scheme is showing considerably low penetration rate of 0.63%, 0.06% and 0.20% for CASIA, BATH and IITK iris databases respectively.

Keywords: Indexing, Energy Histogram, DCT, Multiresolution Subband Coding, B Tree, Iris.

1 Introduction

Any identification system suffers from an overhead of more number of comparisons in the large database. As the size of database increases the time required to declare an individual's identity increases significantly [15]. In addition to this the number of false positives also increases with the increase in the database size [16]. Thus there are two ways to improve the performance of a biometric system. First one is by reducing the number of false positives and other is by reducing the search space [2]. The search space can be reduced by using classification, clustering and indexing approaches on the database. Applying some traditional database binning approaches does not yield satisfactory results. The reason behind is that biometrics does not possess any natural or alphabetical order. As a result, any traditional indexing scheme cannot be applied to reduce the search time. Thus the query feature vector is compared sequentially with the all templates in the database. The retrieval efficiency in sequential search depends upon

S. Ranka et al. (Eds.): IC3 2009, CCIS 40, pp. 194–204, 2009.
© Springer-Verlag Berlin Heidelberg 2009

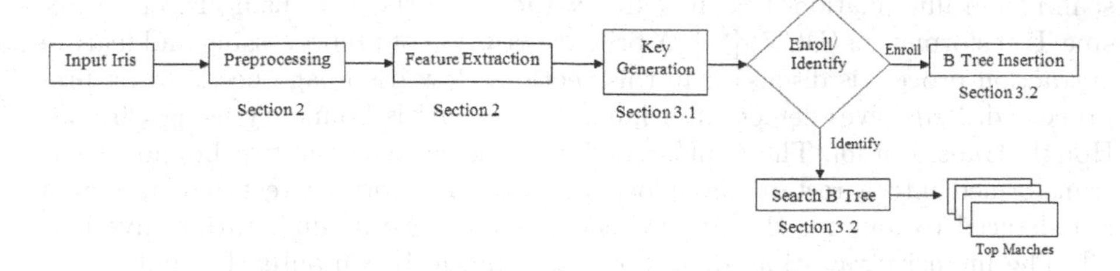

Fig. 1. Block diagram of the proposed indexing scheme

the database size. This leaves behind a challenge to develop a non-traditional indexing scheme that reduces the search space in the large biometric database. The general idea of indexing is to store closely related feature vectors together in the database at the time of enrollment. During identification, the part of the database that has close correspondence with query feature vector is searched to find a probable match.

There already exists some indexing schemes to partition the biometric database. Indexing hand geometry database using pyramid technique has been proposed in [2]. The authors claim to prune the database to 8.86% of original size with 0% FRR. In [1] an efficient indexing scheme for binary feature template using B+ tree has been proposed. In [3] the authors have proposed the modified B+ tree for biometric database indexing. The higher dimensional feature vector is projected to lower dimensional feature. The reduced dimensional feature vector is used to index the database by forming B+ tree. Further, an efficient indexing technique that can be used in an identification system with large multimodal biometric database has been proposed in [4]. This technique is based on Kd-tree with feature level fusion which uses the multi-dimensional feature vector. In [5] two different approaches of iris indexing have been analysed. First one uses the iris code while second one is based on features extracted from iris texture.

In the proposed paper an efficient indexing scheme based on energy histogram on iris database has been studied. The acquired iris image is preprocessed and transformed into a rectangular block. Energy features are extracted from the rectangular block using multiresolution subband coding of DCT coefficients. The energy histogram on extracted features are used to form keys. This key is used to define the B tree and store the iris templates at the leaf that shares similar texture information. The database construction process along with the searching strategy is given in Section 3. The block diagram of system modules along with the section references is given in Fig. 1. The experimental results of the proposed system on three different databases are discussed in Section 4. Conclusions are given in the last section.

2 Preprocessing and Feature Extraction

During preprocessing, the iris circle is localised and transformed into rectangular block known as strip. This transformation makes iris strip invariant to rotation, scaling and illumination. Features are extracted on the strip using Discrete Cosine Transformation (DCT) [11]. A brief description of preprocessing and feature extraction process is discussed in this section. Raw iris image needs to be preprocessed. It involves detection of inner and outer iris boundary using Circular Hough Transform [6]. The annular region lying between the two boundaries is transformed into a rectangular block [7]. The transformed rectangular region is enhanced to improve the texture details and make it illumination invariant [8]. The intensity variation along the whole image is computed by finding the mean of 16×16 block. It helps to estimate the background illumination. The mean image is further expanded to the size of the original image using bi-cubic interpolation. Finally the background illumination is subtracted from the original image. The lightening corrected image is further enhanced using adaptive histogram equalisation approach.

The features are extracted from the preprocessed image using multiresolution DCT. DCT has strong energy compaction property and its coefficients represent some dominant gray level variations of the image. Thus it is the most promising approach for texture classification. The input iris strip is divided into non-overlapping 8×8 pixel blocks which are transformed to generate DCT coefficients. The reason behind using block based DCT approach is that it extracts local texture details of an image. It has been observed that multiresolution decomposition provides useful discrimination between texture. Each block of the computed DCT coefficients has to be reordered to form subbands like 3 level wavelet decomposition. The block of size 8×8 is reordered to transform coefficients into multiresolution form. For a coefficient $D(u, v)$ of the block, ordering is done and stored in S_i where i is defined by

$$i = \begin{cases} 0 & \text{for } m = 0 \\ (m-1) \times 3 + (a/m) \times 2 + (b/m) & \text{otherwise} \end{cases} \tag{1}$$

Let $m = max(a, b)$ for $2^{a-1} \leq u \leq 2^a$ and $2^{b-1} \leq v \leq 2^b$, a and b are the integer values and i ranges from 1 to 10. After reordering, the coefficients $D(1, 1), D(1, 2)$, $D(2, 1)$ and $D(2, 2)$ are stored in subband S_1, S_2, S_3 and S_4 respectively. The multiresolution subband ordering for 8×8 block is shown in Fig. 2.

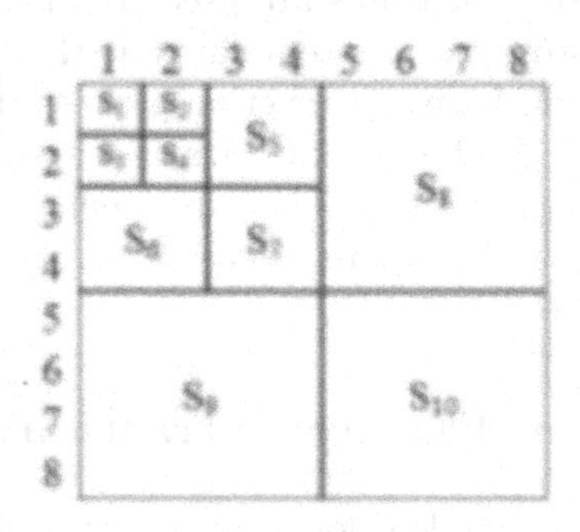

Fig. 2. Multiresolution reordering of 8 × 8 DCT coefficients

After reordering all the DCT blocks, the coefficients from each block belonging to a particular subband are grouped together. Energy value E_i of each subband S_i is obtained by summing up the square of coefficients as

$$E_i = \sum S_i(x, y)^2 \tag{2}$$

Note that the sum of square increases the contribution of significant coefficients and suppresses insignificant coefficients. The feature vector consists of different energy values obtained from 10 subbands.

3 Indexing Scheme

It is expected that the query response time should depend upon the templates similar to the query template and not the total number of templates in the database. Thus the database should be logically partitioned such that images having similar texture patterns are indexed together. To search the large visual databases, content based image indexing and retrieval mechanism based on energy histogram of wavelet coefficients has been proposed in [9]. The scheme provides fast image retrievals. Similar approach has been proposed by considering the energy histogram of reordered DCT coefficients [10]. In the proposed approach biometrics database is indexed using energy histogram of reordered coefficients as given by [10]. The steps involved in indexing are given as follows:

3.1 Key Generation

The feature vector obtained from each image comprises of 10 different energy values one from each subband. The energy histogram (H_i) is build for each subband (S_i) using all the images in the database. This presents the distribution of energy for each subband. Fig. 3 shows the histogram for region S_{10} using all images in the database.

The histogram generated from each subband (H_i) is divided into bins to form logical groups. The texture details of iris strip that have similar energy values (E_i) are placed together in the same bin to have more accurate matches. The size of the bin can be fixed or variable. Here the size of the bin is fixed for experiments. The bins are enumerated in numerical order starting from 1 as shown in Fig. 4. The images falling under each bin are represented on each bar of the histogram. Each image falls under a particular bin of the histogram H_i. This bin number is used to form a global key for indexing. Image key consists of bin number corresponding to each subband. The bin numbers for each subband are combined together using Morton order traversal which places low-frequency coefficients before high-frequency coefficients. The schematic diagram for Morton order traversal is shown in Fig. 5. For example the image I using Morton order forms the key as (3-5-7-8-2-1-4-5-6-7). Similarly all the images in the database obtains keys.

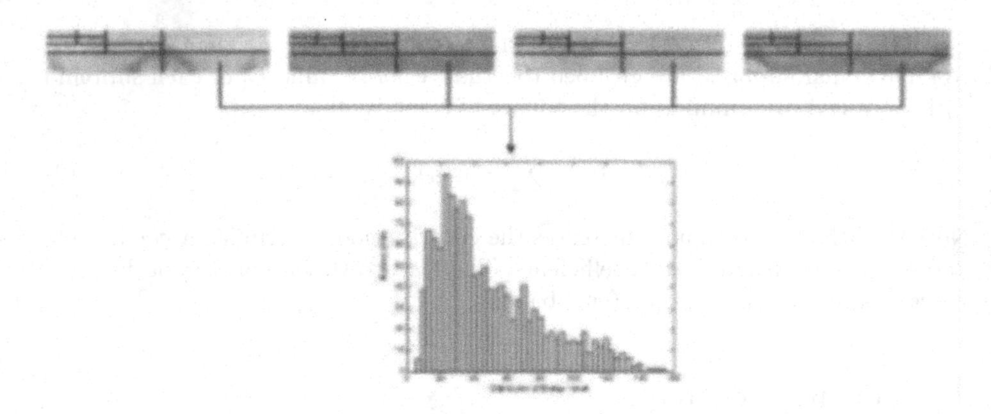

Fig. 3. Energy Histogram of S_{10} region

3.2 Database Creation and Searching

The key is used for inserting an image in the database during enrollment. To store an iris template B tree data structure is used. The degree of the tree is total number of bins that has been constructed for each subband. The height of tree is the number of subbands i that has been taken into consideration. The root node of the tree represents subband S_0 with bins as children that are formed using energy histogram. The leftmost branch represents the first bin and then the next branch represents the second bin and so on. Each node in the second level of the tree corresponds to the immediate following subband. To insert a template in the database, B tree is traversed using the image key generated in Section 3.1. After reaching at the leaf node the template is inserted in the database. Each leaf node in the tree is denoted as a class that contains iris templates. The tree structure used for indexing is given in Fig. 9. Thus more the number of classes lesser will be the retrieval time. The algorithm for inserting an image in the database is given below:

Algorithm 1. database_indexing(n: total number of iris strips)

1. For each image I in the database
2. Find DCT for each 8×8 block
3. Reorder the coefficients using subband coding
4. Find total energy value of each subband (S_i)
5. Construct energy histogram (H_i) for each subband using n images
6. Divide each histogram into bins and enumerate them
7. Obtain a key for I using bin numbers of each (H_i)
8. Traverse the tree using key
9. Store the image at the leaf node

The best match for query strip is obtained by searching the database using the key. Each block of the image is divided into subbands using multiresolution reordering of coefficients. The coefficients of each subband is used to compute

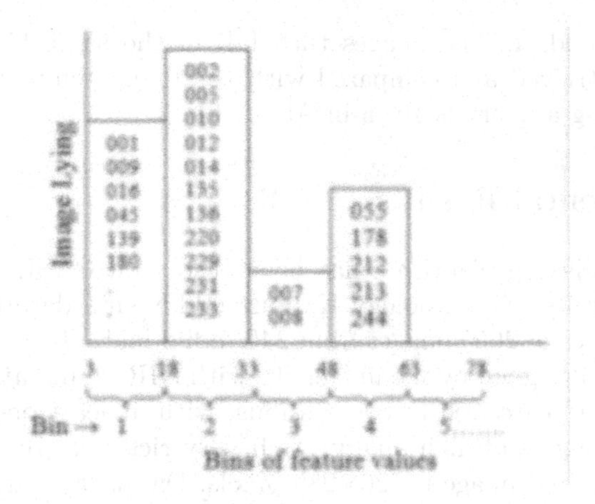

Fig. 4. Logical Grouping of Energy Histogram

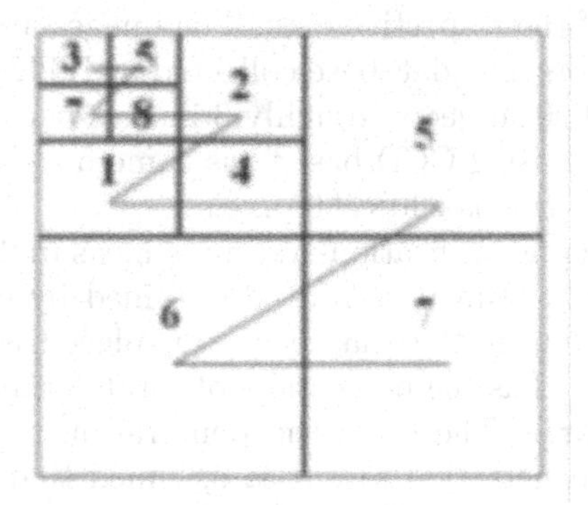

Fig. 5. Global key formation using Morton order traversal

Algorithm 2. searching(q: query strip)

---

1. Find blockwise DCT coefficient of q
2. Reorder the coefficients using subband coding
3. Find energy value of each subband (S_i)
4. Construct query image key using histogram bins
5. Traverse the tree using complete/partial key
6. Retrieve all K images stored in a class
7. Perform comparisons of q with K
8. Find the probable match

---

the energy values. The key for query image is calculated by finding bin number of each subband using bin allocation scheme given in Section 3.1. This key is used for traversing the tree to arrive at the leaf node and retrieve the images stored in a particular class. The query image is compared with the retrieved images to find a suitable match. However if the complete key is used for traversing the tree then the probability of finding exact match becomes less. Thus partial key is used that is constructed from first B subbands where B is less than total number of subbands i. The images that fall in the same bin for the first B subbands are retrieved and compared with the query template. The step-wise process for finding a query is given in Algorithm 2.

4 Experimental Results

The proposed indexing algorithm has been tested on CASIA [13], BATH [14] and Indian Institute of Technology Kanpur (IITK) iris databases. CASIA iris database comprises of 2655 images from 249 individuals. The camera for acquiring images is self designed by the university with NIR illumination scheme. Most of the images were captured in two sessions, with at least one month interval. The acquired image is of high quality with very rich texture details. The resolution of the acquired image is 320×280 pixels. Database available from BATH university comprises of 2000 iris images of 50 subjects each from left and right eye. Within each folder there are two subfolders - L (left) and R (right), each containing 20 images of the respective eyes. The images are in grayscale format with 1280×960 resolution. The database collected at IITK consists of over 1800 iris images taken from 600 subjects (roughly 3 images per person) from left eye. The images are acquired using CCD based iris camera along with uniform light source. The image resolution is 640×480 pixels.

The performance of an identification system is measured in terms of bin miss rate and penetration rate. Bin miss rate is obtained by counting the number of genuine biometric samples that has been mis-placed in a wrong class [12]. Penetration rate is defined as the percentage of total database to be scanned on an average for each search. The lower the penetration rate, more efficient the system. In estimating penetration rate it is assumed that the search does not stop on finding the match but continues through the entire partition.

A comparative study on performance rates is done by changing the number of subbands. The number of subbands determines the length of the key. To find

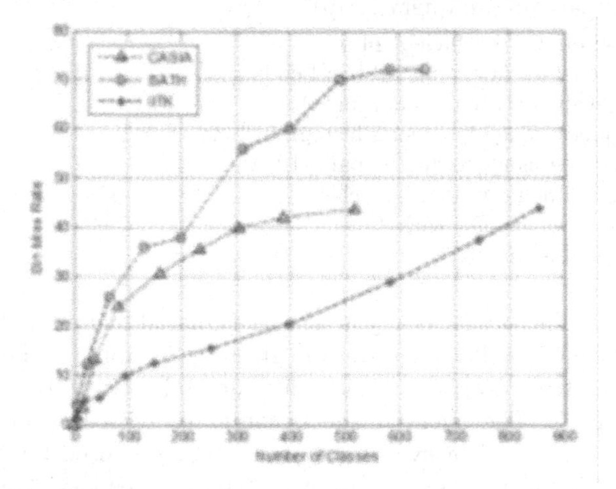

Fig. 6. Bin miss rate for change in number of classes

Table 1. Performance rates for change in the number of classes for CASIA, BATH and IITK datasets. BM: Bin Miss Rate in (%), PR: Penetration Rate in (%), #: Number.

	CASIA			BATH			IITK		
Subband(#)	Class(#)	BM%	PR%	Class(#)	BM%	PR%	Class(#)	BM%	PR%
1	2	0.00	99.69	5	04	26.14	5	1.5	41.44
2	5	1.60	35.96	23	12	7.69	19	5.0	17.21
3	16	3.60	22.70	66	26	3.04	46	5.5	9.24
4	39	13.2	10.23	130	36	1.42	93	10.0	4.77
5	82	24.0	6.12	197	38	0.92	148	12.5	3.25
6	158	30.8	3.46	313	56	0.49	252	15.5	1.56
7	233	35.6	2.63	399	60	0.30	396	20.5	0.92
8	304	40.0	1.77	492	70	0.16	584	29.0	0.50
9	387	42.0	1.22	583	72	0.09	744	37.5	0.27
10	519	43.6	0.63	648	72	0.06	856	44.0	0.20

an exact match the tree is traversed using all the subbands. However to obtain similar matches the tree traversal will stop before reaching the leaf and images having the same partial key is retrieved to find a match. The large set of images will be obtained using partial match which in turn increases the penetration rate. For database construction, an input image is divided into 10 subbands using 8×8 block. Further energy histogram of each subband is divided into 5 bins. Thus every node in B tree is of degree 5. For the sake of convenience fixed number of bins are taken into consideration.

The bin miss rate and penetration rate is obtained by varying the number of subbands. With the change in the number of subbands the number of classes formed at leaf node also changes. Table 1 shows the number of classes, penetration rate and bin miss rate by varying the number of subbands for CASIA, BATH and IITK databases. From the table it has been observed that with increase in the number

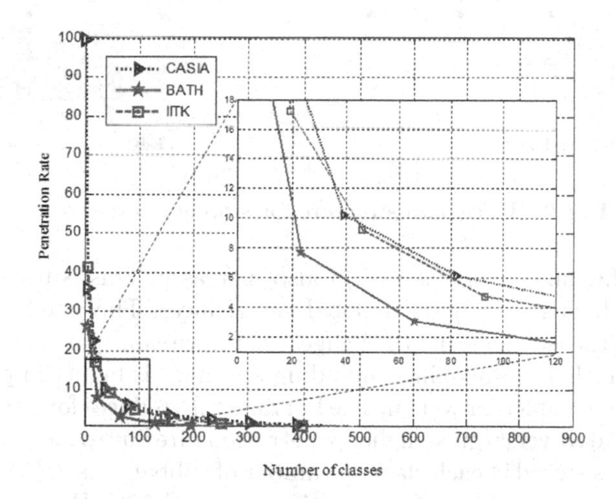

Fig. 7. Penetration rate for change in number of classes with enlarged view of the selected portion

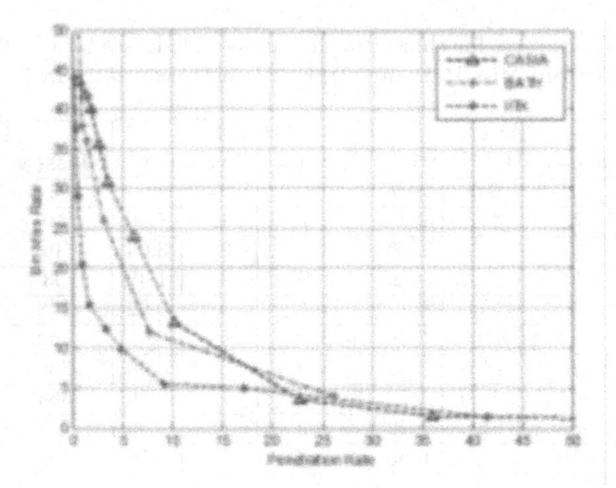

Fig. 8. Graph showing relationship between Penetration Rate versus Bin Miss Rate

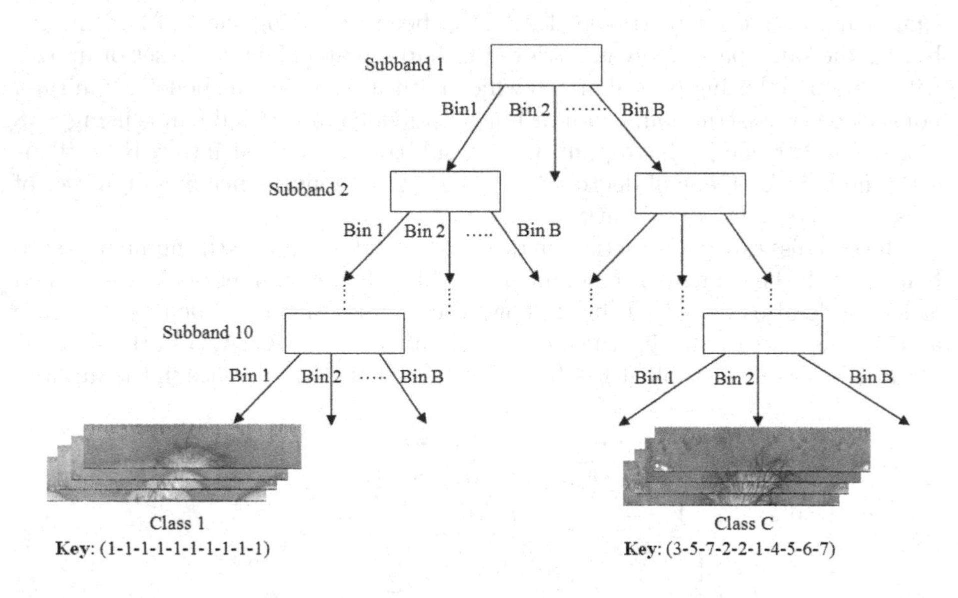

Fig. 9. B tree data structure for storing iris templates

of subbands the number of classes (#) also increases. This is because with less number of subbands the length of global key reduces. The tree is not traversed completely till the leaf node and the images that have same partial key are used to find the match. Hence probability of finding an image is higher in partial traversal compared to complete traversal. The bin miss rate reduces for partial traversal. However, partial traversal gives higher penetration rate due to increase in the number of templates stored in each class. If number of subbands is 2, CASIA database shows bin miss rate of 1.60% and penetration rate of 35.96%. However if number of subbands is 10, the penetration rate reduces significantly to 0.63%. Similar results are obtained for BATH and IITK databases (Table 1). Thus there exists a trade off between the two evaluation rates. The number of subbands used for traversal should be chosen carefully so that both bin miss rate and penetration rate are optimal. Fig. 6 shows change in bin miss rate for change in number of classes. The graph is plotted for all the three databases. Similarly penetration rate is plotted for different number classes as shown in Fig. 7. Fig. 8 represents the relationship between the penetration rate and bin miss rate.

5 Conclusion and Future Work

In the proposed paper, an effort has been made to reduce the search time of iris identification system. A non-traditional indexing scheme has been applied using energy histogram of DCT coefficients. The results are obtained on three available databases. The databases are collected taking various important factors into consideration like difference in time, transformations etc. The number of classes varies depending upon the length of the key. For partial key, the bin miss rate is 1.5% with penetration rate of 41%. For complete key, the bin miss rate increases to 44% while the penetration rate reduces significantly to 0.20%. The two error rates have inverse relationship to each other and is greatly dependent on the number of subbands that are taken into consideration for forming the key. The length of key has to be chosen depending upon the application context and level of security. From the results it can be inferred that the system can be deployed for filtering the database using partial keys. This reduces the penetration rate of the system by grouping the irides with similar texture information. Further, an efficient matching strategy can be applied on the filtered database for finding the exact match. In future the performance of proposed indexing scheme can be extended in the context of invariance to scale and rotation.

References

1. Gupta, P., Sana, A., Mehrotra, H., Hwang, C.J.: An efficient indexing scheme for binary feature based biometric database. In: Proc. SPIE, vol. 6539, p. 653909 (2007)
2. Mhatre, A., Chikkerur, S., Govindaraju, V.: Indexing Biometric Databases using Pyramid Technique. Audio and Video-based Biometric Person Authentication (2005)
3. Jayaraman, U., Prakash, S., Devdatt, G.P.: An Indexing technique for biometric database. In: International Conference on Wavelet Analysis and Pattern Recognition, vol. 2, pp. 758–763 (2008)
4. Jayaraman, U., Prakash, S., Gupta, P.: Indexing Multimodal Biometric Databases Using Kd-Tree with Feature Level Fusion. Information Systems Security, 221–234 (2008)
5. Mukherjee, R., Ross, A.: Indexing iris images. In: 19th International Conference on Pattern Recognition, pp. 1–4 (2008)
6. Kerbyson, D.J., Atherton, T.J.: Circle detection using Hough transform filters. In: Fifth International Conference on Image Processing and its Applications, pp. 370–374 (1995)

7. Daugman, J.: How iris recognition works. IEEE Transactions on Circuits and Systems for Video Technology 14, 21–40 (2004)
8. Ma, L., Tan, T., Wang, Y., Zhang, D.: Efficient Iris Recognition by Characterising Key Local Variations. IEEE Transactions on Image Processing 13(6), 739–750 (2004)
9. Albuz, E., Kocalar, E., Khokhar, A.A.: Scalable Image Indexing and Retrieval using Wavelets (1998)
10. Wu, D., Wu, L.: Image retrieval based on subband energy histograms of reordered DCT coefficients. In: 6th International Conference on Signal Processing, vol. 1, pp. 26–30 (2002)
11. Khayam, S.A.: The Discrete Cosine Transform (DCT): Theory and Application. Tutorial Report, Michigan State University (2003)
12. Wayman, J.L.: Error rate equations for the general biometric system. IEEE Robotics & Automation Magazine 6(1), 35–48 (1999)
13. Center for Biometrics and Security Research, `http://www.cbsr.ia.ac.cn/IrisDatabase.htm`
14. University of Bath Iris Image Database, `http://www.bath.ac.uk/elec-eng/research/sipg/irisweb/`
15. Bolle, R., Pankanti, S.: Biometrics, Personal Identification in Networked Society. Kluwer Academic Publishers, Norwell (1998)
16. Jain, A.K., Maltoni, D.: Handbook of Fingerprint Recognition. Springer, New York (2003)

Secured Communication for Business Process Outsourcing Using Optimized Arithmetic Cryptography Protocol Based on Virtual Parties

Rohit Pathak[1] and Satyadhar Joshi[2]

[1] Acropolis Institute of Technology & Research
[2] Shri Vaishnav Institute of Technology & Science,
Indore, Madhya Pradesh, India
{rohitpathak,satyadhar_joshi}@ieee.org

Abstract. Within a span of over a decade, India has become one of the most favored destinations across the world for Business Process Outsourcing (BPO) operations. India has rapidly achieved the status of being the most preferred destination for BPO for companies located in the US and Europe. Security and privacy are the two major issues needed to be addressed by the Indian software industry to have an increased and long-term outsourcing contract from the US. Another important issue is about sharing employee's information to ensure that data and vital information of an outsourcing company is secured and protected. To ensure that the confidentiality of a client's information is maintained, BPOs need to implement some data security measures. In this paper, we propose a new protocol for specifically for BPO Secure Multi-Party Computation (SMC). As there are many computations and surveys which involve confidential data from many parties or organizations and the concerned data is property of the organization, preservation and security of this data is of prime importance for such type of computations. Although the computation requires data from all the parties, but none of the associated parties would want to reveal their data to the other parties. We have proposed a new efficient and scalable protocol to perform computation on encrypted information. The information is encrypted in a manner that it does not affect the result of the computation. It uses modifier tokens which are distributed among virtual parties, and finally used in the computation. The computation function uses the acquired data and modifier tokens to compute right result from the encrypted data. Thus without revealing the data, right result can be computed and privacy of the parties is maintained. We have given a probabilistic security analysis of hacking the protocol and shown how zero hacking security can be achieved. Also we have analyzed the specific case of Indian BPO.

Keywords: Business Process Outsourcing (BPO), Secure Information, Algorithm.

1 Introduction

Indian firms account for 80% share of the global market for cross border BPO services related to Finance and Accounting, Customer Interaction, and Human Resources Administration. According to the National Association of Software and Service

S. Ranka et al. (Eds.): IC3 2009, CCIS 40, pp. 205–215, 2009.
© Springer-Verlag Berlin Heidelberg 2009

Companies (NASSCOM), the revenues generated by ITES-BPO exports from India were US$2.5 billion during 2002, which increased to US$3.6 billion in 2003. In 2004, export of ITES-BPO to India generated revenues of US$5.2 billion. In addition to exports, the domestic market for BPO services has also grown. The domestic BPO market was US$0.2 billion in 2002, US$0.3 billion in 2003, US$0.6 billion in 2004, and US$0.86 billion in 2005. In India, the value of Human Resources operations outsourced during 2004 was US$165 million, as compared to just US$75 million during 2003. In 2005, the Indian ITES-BPO industry recorded an annual growth rate of 37% to reach a value of US$6.3 billion. Knowledge Process Outsourcing (KPO) and Finance and accounting outsourcing are among the emerging segments of the BPO industry in India [1].

India has become one of the most favored destinations across the world for BPO operations within a span of over a decade. According to NASSCOM, the ITES-BPO exports from India in 2003-04 was US$ 3.1 billion was estimated to be US$ 6.3 billion by 2005-06. It is clearly seen that India has rapidly achieved the status of being the most preferred destination for BPO for companies located in Europe and US because of the availability of appropriate infrastructure, a large English speaking population, and low cost skilled manpower in India.

1.1 Problem Statement and Security Analysis

Security and privacy are the two major issues needed to be addressed by the Indian software industry to have an increased and long-term outsourcing contract from the US. Another important issue is about sharing employee's information to ensure that data and vital information of an outsourcing company is secured and protected. This issue also calls for an immediate action. Given the increasing manpower cost in India coupled with the demand for multi-locational operations from the US companies, countries like the Philippines, Ireland and Mexico may pose a threat to India in due course of time. Even countries such as China, Russia and Hungary are also gearing up on the IT front in a big way [2].

What does security actually mean, in the context of human resources outsourcing, and how is security different from privacy? Is it even different? Are security concerns any different for a multinational company or a company that is considering offshore outsourcing, and is the customer or the service provider ultimately held responsible in the event of failure? One thing's for sure – as corporations continue to broadly adopt outsourcing as a strategy to manage their HR processes and capabilities, concerns around security take on an entirely different dimension [3].

Security has been a human resources concern since long before HR departments began to IT-enable their processes and capabilities and it is not just a consequential incident that results from proliferation of information technology. Prior to IT-enabled HR, the definition of security was more passive – the state of being safe or secure. Usually the types of security topics & questions discussed in the context of HR are who physically saw what records, where the records were stored and how they were transported there were usually. Today's IT-enabled HR systems have amplified and broadened security needs to the extent that security concerns now overarch all IT-enabled HR processes as well.

The burgeoning outsourcing business in India is not restricted to call centers anymore with data processing units coming up in large numbers but the absence of proper data security and cyber laws is hampering their business prospects. Information security concerns surrounding global sourcing will take centre stage alongside public concern over job losses, but will not prevent enterprises from outsourcing. Although security issues will lengthen the sales cycles of global delivery, it will not stop enterprises from adopting global sourcing models. Enterprises and service providers should start an informed dialogue to address security early and to perform due diligence throughout the outsourcing life cycle, it said. The security exposure that both clients and service providers have to deal with, as global sourcing becomes more strategic and complex, increases by orders of magnitude. Service providers are unable to provide standard security solutions because regulations, legislation and consequently risk vary vastly between industries and geographies [4].

There is also tremendous hype and a lack of understanding of the issues surrounding security. The most significant security issues revolve around the protection of data in one manner or another. There are, however, other issues that are not well understood, vague and based on emotion rather than fact. Instances of data thefts and frauds have attracted worldwide attention and become a major cause for concern among the industry players and associations in India. Some of the information security and data privacy challenges that Indian BPOs face include lack of stringent data protection laws, use of portable devices such as laptops by employees to store confidential business information, rising data security costs due to increased employee background checks, training employees in maintaining data security, ensuring compliance with security policies implemented in the company, and systemic plugging of any loopholes through employee activity monitoring procedures.

To ensure that the confidentiality of a client's information is maintained, BPOs need to implement data security measures, which can be classified into measures taken at the recruitment level and measures taken at the operational level. The Indian government is evaluating the possibility of reviewing the Information Technology Act of 2000 to bring various computer crimes relating to information privacy under its purview. NASSCOM announced plans to establish a self-regulating organization (SRO) to deal with information security issues related with outsourcing to India and introduced the National Skills Registry (NSR). NASSCOM also proposed that the Indian government should establish a special court to speed up the trial process of cases related to information/data security and other cyber crimes booked under the Information Technology Act 2000 [1].

1.2 Security Issues

Privacy is the extent to which the data or information of an individual can be shared. In the case of BPO, privacy includes all the data of the client and its' customers. BPO companies have to maintain confidentiality of information and shall use this data only for the purposes by its owner. They achieve this requirement through operational security like software security, physical security, Technology, policies etc. This may include non disclosure of Social security numbers, passport details, bank details, Permanent Account Number (PAN) of Income tax, Health information, financial/loan details etc. An employee can sell information on accounts, passports, credit cards etc. for money to an undercover reporter. People can play around with others bank accounts! Information of the names and credit card numbers of employees or persons can be stolen. In banking security breach consumer accounts of clients at different banks can be attacked. We can have cases of racketeering and disclosing data from databases. The suspects can manually built database accounts using names and Social Security numbers obtained by employees. Attendance recording system must be in place. Every employee logs in to their systems. Email system shall take care of all SPAM and open port issues to stop others exploiting your open SMTP ports, if any.

Data must be checked when it is received from the customer. Ensure that the data is received intact and not tampered with. Record if there are any flaws or deviations. Once data is received, the onus is on the company to maintain the data integrity. The customer data is dynamically backed-up and mirrored frequently at different physical locations. Data should be exchanged over the broadband through the secure server. All entry/exit points are secure and all movement is logged. Some companies seem to have banned internet access in the entire office or the computer in the working area may be secure against data duplication or the computer systems that agents work may not be provided with hard disks or floppy drives. Strong antivirus procedure shall be implemented. While the virus may or may not steal information, they may corrupt the database or the server itself. Ensure that the servers and client machines are protected properly.

Generally the BPO may not need a web page through public domain for a client. Virtual Private Network between the supplier-customer enables better secure communication. Ensure that any transaction/ communication are logged and tracked. Provide a Firewall of repute. Do not compromise. The firewall to be configured to the servers & ports identified with the customer. Intranet server and the data server handling client information shall not be on the same server.

If we have to perform a calculation which includes data from many organizations, than the safety of the data of the organization is the prime concern. Suppose a statistical calculation is to be performed among several organizations. This calculation includes information related to various person's related to the organization, may it be employees working for the organization or the customers of the organization such as customers of a bank. In this case, information of every person is to be kept secure so as to keep privacy of every individual.

2 Recent Works

Yao described millionaires' problem and gave the solution by using Deterministic Computations and introduced a view of SMC (Secure Multi-Party Computation) [5]. Mikhail et al. presented a collaborative benchmark problem and a proposed solution [6]. Later he provided privacy-preserving solutions to collaborative forecasting and benchmarking that can be used to increase the reliability of local forecasts and data correlations, and to conduct the evaluation of local performance compared to global trends [7]. A new model that allows partial information disclosure was proposed by Wenliang *et al.* which was a practical solution to the SMC problem [8]. In our previous work we have proposed the VPP (Virtual Party Protocol) which used virtual parties for computation on encrypted data [9].

Wenliang *et al.* proposes the privacy preserving cooperative linear system of equations problem and privacy-preserving cooperative linear least-square problem [10]. Ran *et al.* has shown how uncorrupted parties may deviate from the case where even protocol by keeping record of all past configurations [11]. Mikhal *et al.* have given a protocol for sequence comparisons in which neither party reveals anything about their private sequence to the other party [12]. A Secure Supply-Chain Collaboration (SSCC) protocols that enable supply-chain partners to cooperatively achieve desired system-wide goals without revealing the private information of any of the parties, even though the jointly-computed decisions require the information of all the parties is proposed by Atallah *et al.* [13]. The problem of defining and achieving security in a context where the database is not fully trusted, i.e., when the users must be protected against a potentially malicious database is discussed by Ueli *et al.* [14]. We have seen building a decision-tree classifier from training data in which the values of individual records have been perturbed, and reconstruction procedure to accurately estimate the distribution of original data values has been described [15]. We have already seen the Anonypro Protocol, which had a good concept to make the incoming data of anonymous identity [16-18]. Anonypro Protocol assumed the connection between the party and anonymizer to be secured.

In this paper we have shown how VPP can be used safely for BPO's to ensure the privacy of the organizations a whole by not revealing the right information. In this technique we will create some fake data and some virtual parties. We encrypt the data and create modifier tokens correspondingly. This modified data is mixed with fake data. These modifier tokens are related to the modification done in the data and will be used in the final computation to obtain the correct result. Now this modified data and the modifier tokens are distributed among the virtual parties. These parties will randomly dispense their data to anonymizers. The anonymizers will send this data to Trusted Third Party for computation. TTP will use the data and the modifier tokens to compute the result. The modifier tokens will aid to bring the result obtained by the encrypted data values. The modifier tokens in any manner will not reveal the identity of the party or such. The modifier is information token which is used in the final computation to ensure the right result. The method of encryption, modifier tokens, encrypted data and the method of computation all arc interdependent. With proper configuration zero hacking security can be achieved with this protocol.

3 Proposed Protocol

3.1 Informal Description

We have to compute the function $f(a_1, a_2, a_3..., a_n)$. There are n organizations O_1, O_2, O_3..., O_n. Each O_i has data X_{i1}, X_{i2}, X_{i3}..., X_{im}. There are z number of anonymizers A_1, A_2, A_3..., A_z. All layers and the data flow is shown in Fig. 1.

For every organization (party) O_i we will create some fake trivial data entries F_{i1}, F_{i2}, F_{i3}..., F_{iq}, where q is the total number of fake entries. The total number of fake entries q may be different for every organization but for the sake of simplicity in explanation it is kept same for every party. The fake data is generated in a manner that it doesn't effects the overall result. We will group this data with original data entries X_{i1}, X_{i2}, X_{i3}..., X_{im}. Thus the new group of data having $m+q$ total number of data items, i.e. D_{i1}, D_{i2}, D_{i3}..., $D_{i(m+q)}$. The value of each data D_{i1}, D_{i2}, D_{i3}..., $D_{i(m+q)}$ is encrypted to obtain the encrypted data E_{i1}, E_{i2}, E_{i3}..., $E_{i(m+q)}$. The whole process is shown in Fig. 2. For every organization O_i we will create k virtual parties O_{i1}, O_{i2}, O_{i3}..., O_{ik}.

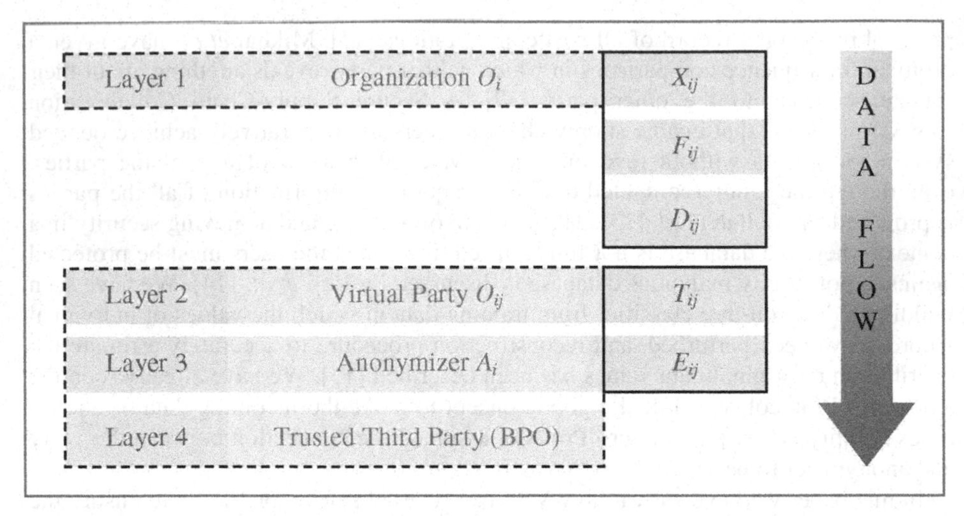

Fig. 1. Different layers and data types of the protocol in a flow

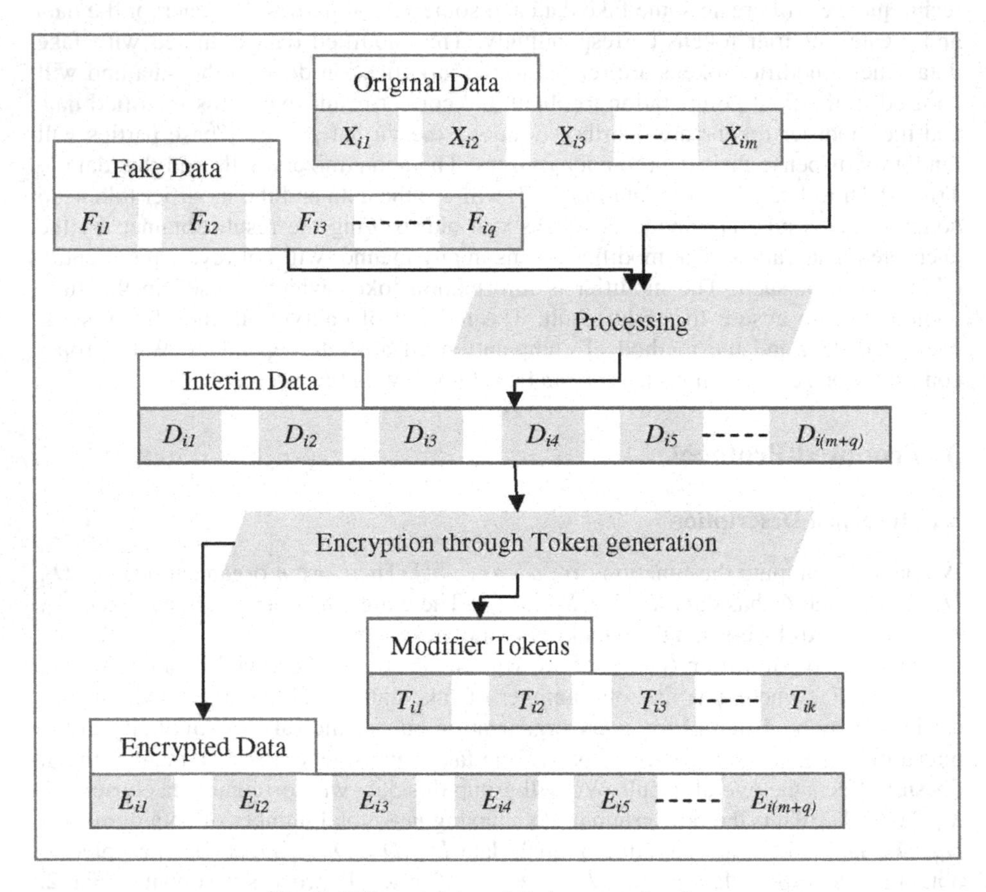

Fig. 2. Different data and their conversion flow

Encrypted data $E_{i1}, E_{i2}, E_{i3}..., E_{i(m+q)}$ is distributed randomly among the virtual parties $O_{i1}, O_{i2}, O_{i3}..., O_{ik}$. Modifier tokens $T_{i1}, T_{i2}, T_{i3}..., T_{ik}$ are generated for every organization O_i. These modifier tokens are randomly distributed among the virtual parties $O_{i1}, O_{i2}, O_{i3}..., O_{ik}$ such that every virtual party gets one modifier token. Now the virtual parties $O_{i1}, O_{i2}, O_{i3}..., O_{ik}$ distributes their data and modifier tokens randomly among the anonymizers $A_1, A_2, A_3..., A_z$. Anonymizers can take data from multiple parties. Anonymizers don't store data they just redirect it. The data of the anonymizers is sent to third party. The function *h()* uses the encrypted data and the modifier tokens to compute the right result. Function *h()*, will vary for different types of computation and will depend highly on *f()*. Third party will compute the value of function $h(E_{11}, E_{12}, E_{13}..., E_{1j}...E_{i1}, E_{i2}, E_{i3}..., E_{ij}, T_{11}, T_{12}, T_{13}..., T_{1j}..., T_{i1}, T_{i2}, T_{i3}..., T_{ij})$ which is the desired result, same as the result computed by the function $f(X_{11}, X_{12}..., X_{1m}, X_{21}, X_{22}..., X_{2m}, X_{31}, X_{32}..., X_{3m}..., X_{n1}, X_{n2}..., X_{nm})$, and this result is declared publicly. The whole scenario can be seen in Fig. 3.

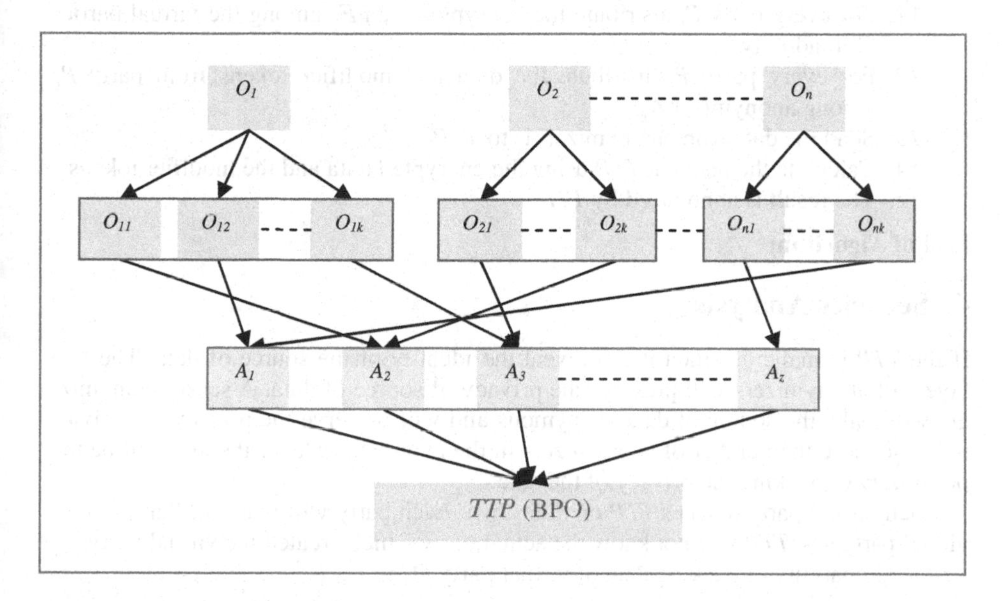

Fig. 3. Party layer, Virtual Party Layer, Anonymizer layer and Computation (Trusted Third Party) layer, from beginning to end respectively

3.2 Formal Description

3.2.1 Algorithm

Identifier List:

P_i – Parties where *i* ranges from *1* to *n*
X_{ij} – Data of party P_i where *j* ranges from *1* to *m*
F_{ij} – Fake data of party P_i where *j* ranges from *1* to *q*
D_{ij} – Total data including the fake and the original data
P_{ij} – Virtual Party of party P_i where *j* ranges from *1* to *k*

E_{ij} – Encrypted data associated with party P_i where j ranges from 1 to $m+q$
A_y – Anonymizer, where y ranges from 1 to z
TTP – Third party

Start

1. Initialize party P_i
2. Create k virtual parties P_{ij} for every party P_i
3. Initialize F_{ij} for every party P_i
4. For every party P_i generate fake data into F_{ij}
5. For every party P_i initialize D_{ij}
6. For every party P_i group fake data F_{ij} with original data X_{ij} to get D_{ij}
7. For every party P_i initialize E_{ij}
8. For every party P_i encrypt data D_{ij} to get E_{ij}
9. For every party P_i initialize T_{ij}
10. Generate modifier tokens T_{ij} for every party P_{ij}
11. For every party P_i distribute the encrypted data E_{ij} among the virtual parties P_{ij} randomly
12. For every party P_i distribute the data and modifier tokens from party P_{ij} among anonymizer A_y
13. Send the data from anonymizer A_y to *TTP*
14. Calculate the result at *TTP* using the encrypted data and the modifier tokens
15. The result is announced by *TTP*

End of Algorithm

4 Security Analysis

If the *TTP* is malicious then it can reveal the identity of the source of data. The two layers of anonymizers will preserve the privacy of source of data. A set of anonymizers will make the source of data anonymous and will preserve the privacy of individual. The more the number of anonymizers in the anonymizer layer the less will be the possibility of hacking the privacy of the data.

Each virtual party reaches *TTP* on their own. Each party will reach *TTP* as an individual party and *TTP* will not know the actual party which created the virtual party.

The probability of hacking data of virtual party O_{ir} is

$$P(VO_{ir}) = \frac{1}{\sum_{i=1}^{n} k_i} \tag{1}$$

When party O_i has k_i number of virtual parties, the probability of hacking data of any virtual party of party O_r is

$$P(VO_r) = \frac{k_r}{\sum_{i=1}^{n} k_i} \tag{2}$$

Even if the data of virtual party is hacked it will not breach the security as this data is encrypted. Probability of hacking the data of any party O_r is calculated as

$$P(O_r) = \frac{k_r}{\sum_{i=1}^{n} k_i} \times \frac{k_r - 1}{\sum_{i=1}^{n} k_i - 1} \times \cdots \times \frac{1}{\sum_{i=1}^{n} k_i - k_r} \tag{3}$$

The graph between number of virtual parties k vs. the probability of hacking $P(O_r)$ for $n=4$ is shown in Fig. 4. which clearly depicts that probability of hacking is nearly zero when the number virtual parties is three or more. Also the graph between number of parties and probability of hacking for $k=8$ is shown in Fig. 5. As the number of virtual parties is eight the probability of hacking is in the order of 10^{-5} or we can say nearly zero.

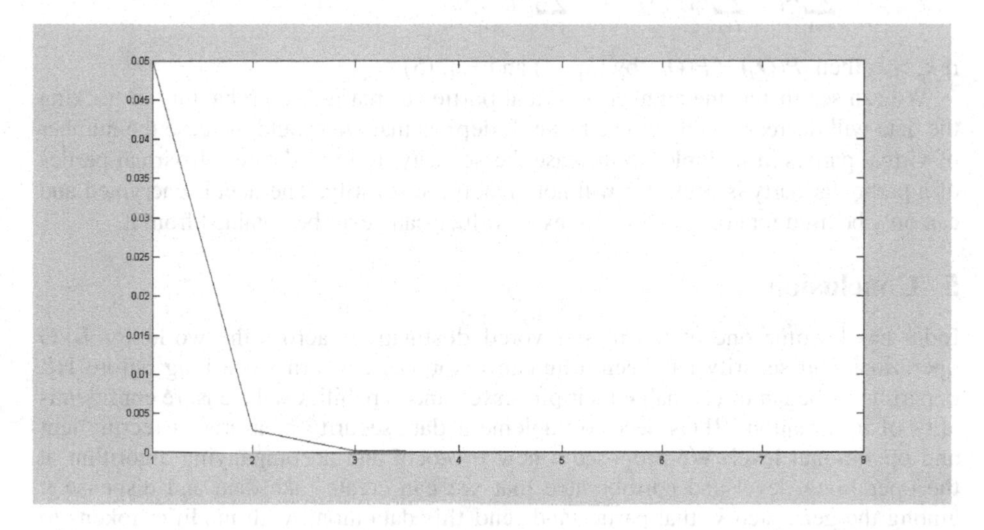

Fig. 4. Graph between number of Virtual Parties (x axis) vs Probability of hacking (y axis)

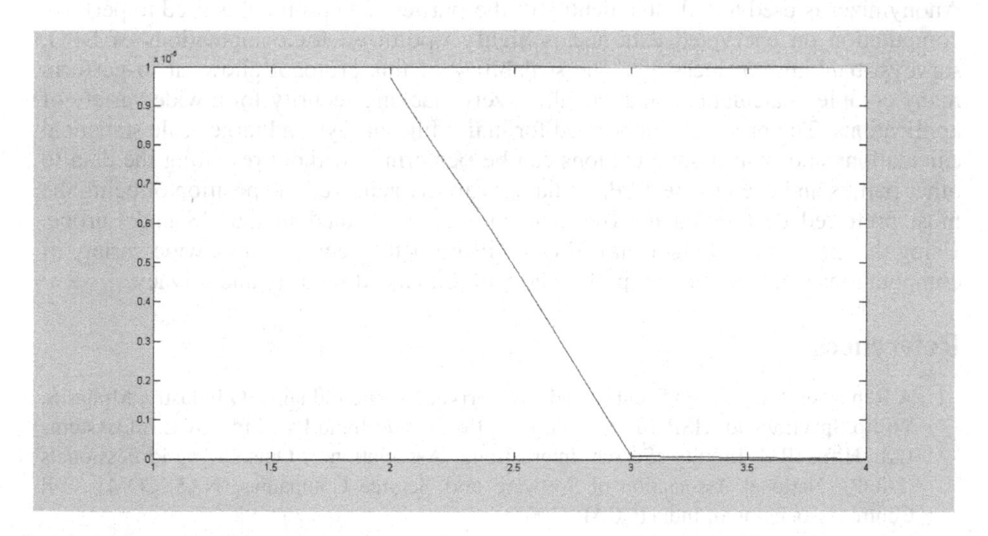

Fig. 5. Graph between number of Parties (x axis) vs Probability of hacking(y axis)

Suppose that the number of virtual parties is k_a then

$$P(O_a) = \frac{k_a}{\sum_{i=1}^{n} k_i} \times \frac{k_a - 1}{\sum_{i=1}^{n} k_i - 1} \times \cdots \times \frac{1}{\sum_{i=1}^{n} k_i - k_a} \quad (4)$$

For k_b number of virtual parties we have

$$P(O_b) = \frac{k_b}{\sum_{i=1}^{n} k_i} \times \frac{k_b - 1}{\sum_{i=1}^{n} k_i - 1} \times \cdots \times \frac{1}{\sum_{i=1}^{n} k_i - k_b} \quad (5)$$

if $k_a > k_b$ then $P(O_a) < P(O_b)$ by Eq. (4) and Eq. (5).

We can see that as the number of virtual parties increases the probability of hacking the data will decrease by harmonic mean. It depicts that we should increase the number of virtual parties in multiples to increase the security. Even if data of all virtual parties of a particular party is hacked it will not breach the security. The data is encrypted and can only be used for computation and exact values can never be obtained from it.

5 Conclusion

India has become one of the most favored destinations across the world for BPO operations and security has been a human resources concern since long before HR departments began to IT-enable their processes and capabilities. To ensure confidentiality of information, BPOs need to implement data security measures at recruitment and operational level. We proposed a new protocol and accompanying algorithm at the operational level and corroborated that we can create fake data and dispense it among the generated virtual parties and send this data along with modifier tokens to carry out computations on encrypted data using an improvised computation method. Anonymizer is used to hide the identity of the parties. The protocol is used to perform computation on encrypted data and is highly optimized for computations of BPO, surveys, banking, business etc. The scalability of this protocol allows it to perform many complex calculations and can allow zero hacking security for a wide variety of applications. The protocol can be used for many big surveys and large scale statistical calculations and with it computations can be performed without revealing the data to other parties and even to the TTP. India has rapidly achieved the position of being the most preferred destination for BPO for companies located in the US and Europe. Using this protocol and algorithm BPO's will be better secured and a wide variety of computations can be optimally performed with enhanced security and privacy.

References

1. A Report on Information Security and Data Privacy in the Indian BPO Industry. MphasiS, Wipro Spectramind, HSBC Electronic Data Processing India Pvt. Ltd., HCL Infosystems Ltd., Hifn, IBM, Infinity eSearch, International Association of Outsourcing Professionals (IAOP), National Association of Software and Service Companies (NASSCOM), Call Centre Association of India (CCAI) (2007), http://www.icmrindia.org/casestudies/catalogue/Business%20Reports/BREP035.htm

2. Security and privacy key issues in US-India BPO relations. The Financial Express (2004), http://www.financialexpress.com/news/security-and-privacy-key-issues-in-usindia-bpo-relations/120451/
3. Security: HR outsourcing deal-maker or deal-killer?. Human Resources Outsourcing Association, http://www.hrmreport.com/article/Issue-1/HR-BPO/Security-HR-outsourcing-deal-maker-or-deal-killer/
4. Security concerns hit BPO firms in India (May 2005), http://www.rediff.com/money/2005/may/25bpo.htm
5. Yao, A.C.: Protocols for secure computations. In: Proc. of 23rd Annual Symposium Foundations of Computer Science, pp. 160–164
6. Atallah, M., Bykova, M., Li, J., Frikken, K., Topkara, M.: Private collaborative forecasting and benchmarking. In: Proc. of the 2004 ACM workshop on Privacy in the Electronic Society (2004)
7. Atallah, M., Bykova, M., Li, J., Frikken, K., Topkara, M.: Private collaborative forecasting and benchmarking. In: Proc. of the 2004 ACM workshop on Privacy in the electronic society, pp. 103–114 (2004)
8. Du, W., Zhan, Z.: A practical approach to solve secure multi-party computation problems. In: Proc. of the New Security Paradigms Workshop (2002)
9. Pathak, R., Joshi, S.: Secure Multi-party Computation Using Virtual Parties for Computation on Encrypted Data. In: Proc. of The First International Workshop on Mobile & Wireless Networks (MoWiN 2009) in Conjunction with The Third International Conference on Information Security and Assurance (ISA 2009). LNCS. Springer, Heidelberg (2009)
10. Du, W., Atallah, M.J.: Privacy-preserving cooperative scientific computations. In: Proc. 14th IEEE Computer Security Foundations Workshop, June 2001, pp. 273–282 (2001)
11. Canetti, R., Feige, U., Goldreich, O., Naor, M.: Adaptively secure multi-party computation. In: Proc. The 28th annual ACM symposium on Theory of computing
12. Atallah, M.J.: Secure and Private Sequence Comparisons. In: Proc. The 2003 ACM workshop on Privacy in the electronic society (2003)
13. Atallah, M.J., Elmongui, H.G., Deshpande, V., Schwarz, L.B.: Secure supply-chain protocols. In: Proc. IEEE International Conference, E-Commerce (2003)
14. Maurer, U.: The role of cryptography in database security. In: Proc. The 2004 ACM SIGMOD international conference on Management of data (2004)
15. Agrawal, R., Srikant, R.: Privacy-Preserving Data Mining. In: Proc. The ACM SIGMOD Conference on Management of Data (2000)
16. Mishra, D.K., Chandwani, M.: Extended protocol for secure multi-party computation using ambiguous identity. WSEAS Transactions on Computer Research 2(2), 227–233 (2007)
17. Mishra, D.K., Chandwani, M.: Arithmetic cryptography protocol for secure multi-party computation. In: Proceeding of IEEE SoutheastCon 2007: The International Conference on Engineering – Linking future with past, Richmond, Virginia, USA, March 2007, pp. 22–24 (2007)
18. Mishra, D.K., Chandwani, M.: Anonymity enabled secure multi-party computation for Indian BPO. In: Proceeding of the IEEE Tencon 2007: International conference on Intelligent Information Communication Technologies for Better Human Life, Taipei, Taiwan, October-November 2007, pp. 52–56 (2007)

Timing Analysis of Passive UHF RFID - EPC C1G2 System in Dynamic Frame

Chandan Maity[1], Ashutosh Gupta[1], and Mahua Maity[2]

[1] Ubiquitous Computing Group,
Centre for Devlopment of Advanced Computing Noida, India
{chandanmaity,ashutoshgupta}@cdacnoida.in
[2] Department of Computer Science,
Raj Kumar Goel Insitute of Technology Ghaziabad, India
mahua21sarkar@gmail.com

Abstract. With the growing application field of RFID, the passive RFID in high speed and dynamic region is the topic of interest of many researchers globally. The major challenges in this area are the reader-tag reading time and anti-collision. After the several readers to tag intercommunications reader is able to process the UII value of the RFID tag, another important consideration in passive RFID system is RFID tags need sufficient time to store energy. The combination of these time quantum results in a quite long time interval which can not be negligible and creates difficulty in reading tags which are moving at a very high speed. In this paper we have done a brief study to realize the minimum required time quantum to read a passive RFID tag and the maximum allowable speed is calculated with the reader parameters.

Keywords: RFID: Radio Frequency Identification, UHF: Ultra High Frequency, UII- Unique Item Identifier, CW: Continuous Power, R=>T: Reader to tag communication, T=>R: Tag to reader communication.

1 Introduction

The real life application of Passive UHF RFID in high-speed domain is the next research challenge in RFID domain. In active RF system, it has been solved and quite a high data transmission is possible, but in passive RFID field there are certain limitations, and particularly when multiple antennas are connected with a single RFID reader system to minimize the cost with comparison of reading area, the time complexity increases and the chances of missing tag also increase, According to the existing architecture of passive RFID network, the passive tag collects power from transmitted RF wave by reader and simultaneously the tag replies back its response data. This whole process takes a specific time and for that period of time, reader must activate one particular antenna. In multi antenna system, the other antennas must sleep during that time when any particular one antenna activated or populating the tags. According to the existing readers with multi port antenna facility, the antennas are switched sequentially and the tags are populated respectively. So, in the sleeping time period, there is a large probability of loosing tags if the tags are in significantly

S. Ranka et al. (Eds.): IC3 2009, CCIS 40, pp. 216–227, 2009.
© Springer-Verlag Berlin Heidelberg 2009

high-speed domain. This paper reveals the realization of high speed domain by calculating the minimum time quantum for the reliable processing and populating the tag, as a result, the tag reading speed (Tag/sec) can be increased significantly which will help to deploy passive UHF RFID not only in high speed domain but also supply chain management with very first huge data acquiring and processing.

2 Reader Tag Communication

To realize the minimum required time quantum to populate the basic Tag ID information, the command, "Query" has been chosen to analyze its minimum time complexity to get only the Tag-ID or UII of a single Tag.

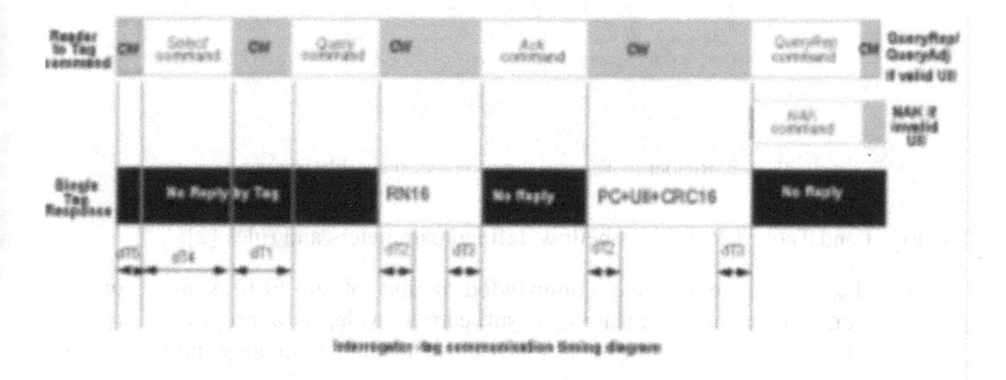

Fig. 1. Timing value of reader – tag inter communication

Fig. 1 shows the timing diagram of Interrogator or reader to tag communication for inventory of a single tag. Where Table 1 indicates the different time duration value T1, T2, T3, T4.

Table 1. Timing parameter for reader – tag inter communication

Parameter	Minimum	Nominal	Maximum
T_1	$MAX(RTcal, 10T_{pri}) \times (1 - \lvert FT \rvert) - 2\mu s$	$MAX(RTcal, 10T_{pri})$	$MAX(RTcal, 10T_{pri}) \times (1 + \lvert FT \rvert) + 2\mu s$
T_2	$3.0T_{pri}$		$20.0T_{pri}$
T_3=T_1	$MAX(RTcal, 10T_{pri}) \times (1 - \lvert FT \rvert) - 2\mu s$	$MAX(RTcal, 10T_{pri})$	$MAX(RTcal, 10T_{pri}) \times (1 + \lvert FT \rvert) + 2\mu s$
T_4	2.0 RTcal		

Fig. 2. shows the Reader to Tag (R=>T) preamble where RTcal and TRcal and Tari values are defined where a Tari is defined as the time period to send data '0'.

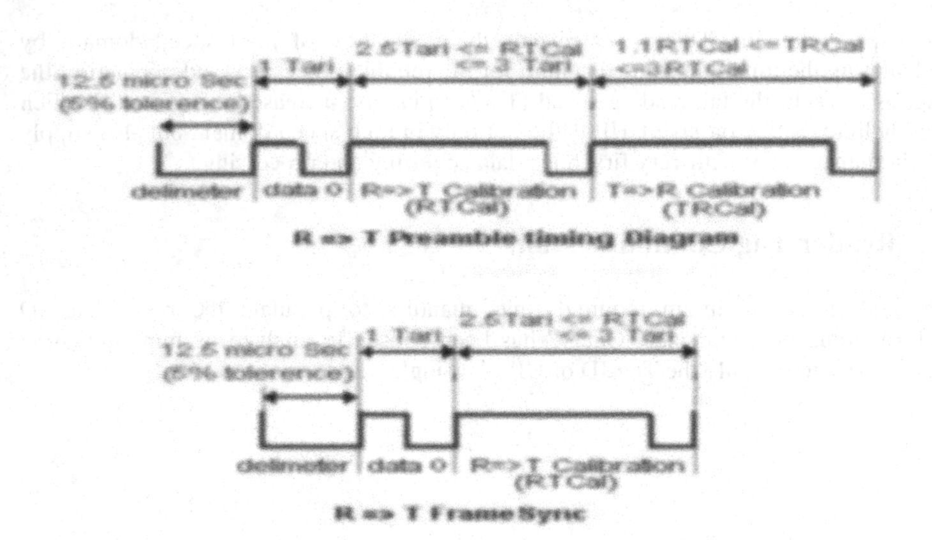

Fig. 2. Reader to Tag and tag to reader preamble frames

The Fig. 1 and Table 1 follows as bellow defined parameters and rules [2]

1. T_{pri} denotes either the commanded period of an FM0 symbol or the commanded period of a single sub-carrier cycle, as appropriate. T_{pri} = 1/ (T=>R link period) i.e. 1/BLF, BLF is the link Frequency. and T_{pri} value follows Table 3.
2. A tag may exceed the maximum value for T_1 when responding to commands that write to memory.
3. The maximum value for T_2 shall apply only to tags in the reply or acknowledged states. For a tag in the reply or acknowledged states, if T_2 expires (i.e. reaches its maximum value):

   — Without the tag receiving a valid command, the tag shall transition to the arbitrate state,
   — During the reception of a valid command, the tag shall execute the command,
   — During the reception of an invalid command, the tag shall transition to arbitrate upon determining that the command is invalid.
   — In all other states the maximum value for T_2 shall be unrestricted. A Tag shall be allowed a tolerance of $20.0T_{pri} \leq T_{2(max)} \leq 26T_{pri} + 2\mu s$ in determining whether T_2 has expired.

4. An interrogator may transmit a new command prior to interval T_2 (i.e. during a tag response). In this case the responding tag is not required to demodulate or otherwise act on the new command, and may undergo a power-on reset.
5. FT is the frequency tolerance at –25 degree to 40 degree Celsius.
6. T_1+T_3 shall not be less than $T_{4.0.}$

Table 2. Back Scatter link Frequency table

DR: Divide Ratio	Trcal[a] (μs +/– 1%)	BLF: Link Frequency (kHz)	Frequency Tolerance FT (-25 °C to 40 °C)
64/3	33.3	640	+ / – 15%
	33.3 < TRcal < 66.7	320 < BLF < 640	+ / – 22%
	66.7	320	+ / – 10%
	66.7 < TRcal < 83.3	256 < BLF < 320	+ / – 12%
	83.3	256	+ / – 10%
	83.3 < TRcal ≤ 133.3	160 ≤ BLF < 256	+ / – 10%
	133.3 < TRcal ≤ 200	107 ≤ BLF < 160	+ / – 7%
	200 < TRcal ≤ 225	95 ≤ BLF < 107	+ / – 5%
8	17.2 ≤ TRcal < 25	320 < BLF ≤ 465	+ / – 19%
	25	320	+ / – 10%
	25 < TRcal < 31.25	256 < BLF < 320	+ / – 12%
	31.25	256	+ / – 10%
	31.25 < TRcal < 50	160 < BLF < 256	+ / – 10%
	50	160	+ / – 7%
	50 < TRcal ≤ 75	107 ≤ BLF < 160	+ / – 7%
	75 < TRcal ≤ 200	40 ≤ BLF < 107	+ / – 4%

Table 3. Reader to Tag communication FM modulation schemes

M: Number of sub-carrier cycles per symbol	Modulation type	Data rate (kbit/s)	T_{pri} = 1/BLF
1	FM0 base-band	BLF	1/BLF
2	Miller sub-carrier	BLF/2	2/BLF
4	Miller sub-carrier	BLF/4	4/BLF
8	Miller sub-carrier	BLF/8	8/BLF
16	Miller sub-carrier	BLF/16	16/BLF
32	Miller sub-carrier	BLF/32	32/BLF
64	Miller sub-carrier	BLF/64	64/BLF

2.1 The Tag Population

To analyze the time complexity, a session, "read one single tag Identity" has been chosen. The simplified diagram of timing value of reader – tag inter communication is described in Fig. 3, In this analysis the QueryRep, QueryAdust and other commands are eliminated carefully to concern only the time quantum to get the UII value of one single tag.

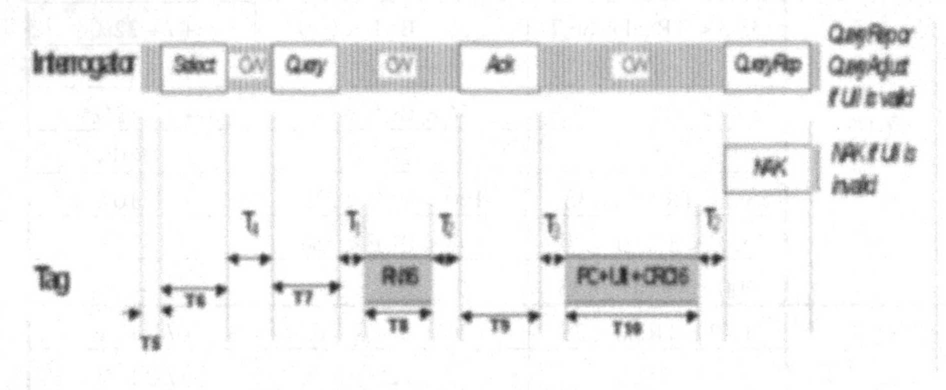

Fig. 3. Internal timing diagram of TCON

To read the Tag Identity or UII, the Interrogator to tag intercommunication follows the internal timing diagram of TCON. As this paper describes only the timing performance of Reader-tag intercommunication, hence all other description of individual command and response has been excluded except the timing complexity of data frame packets.

An interrogator shall begin all R=>T signalling with either a preamble or a frame-sync, both of which are shown in Fig. 2. A preamble shall precede a *Query* command and denotes the start of an inventory round. All other signalling shall begin with a frame-sync.

Delimiter: Refer Fig. 2; Delimiter is a fixed time period of 12.5 μs sent at the header of every *R=>T preample* or *R=>T Frame_sync.*

R=>T frame-sync: Refer Fig. 2; the frame-sync is being sent as the header with all R=>T signalling except the 'Query' command. The frame-sync has following structure:

Delimiter: 12.5 micro-second
Data 0: 1 Tari
RTcal : 2.5 Tari <= RTcal <= 3 Tari

R=>T Preamble: Refer Fig. 2; the preamble is being sent at the header of 'Query' command. The preamble has following structure:

Delimiter: 12.5 micro-second
Data 0: 1 Tari
RTcal : 2.5 Tari <= RTcal <= 3 Tari
TRcal: 1.1 RTcal<= TRcal <=3 RTcal

T=>R preamble: The tag responds with a preamble as it is defined in *Query* command the 'TRext' value. The shortest preamble domes when TRext = 0. The structure of preamble is defined in Fig. 4.

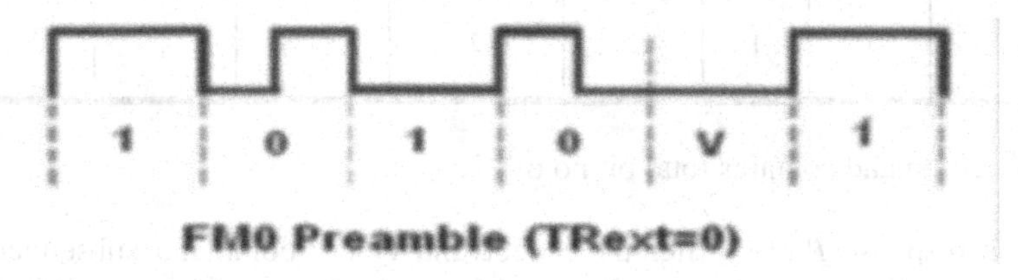

Fig. 4. Tag to Reader preamble frame structure

End of FM0 signaling: Refer Fig. 5; All the data communication between R=>T or T=>R ends with a signal which carries 2 bits when it has only FM0 with no miller sub-carrier.

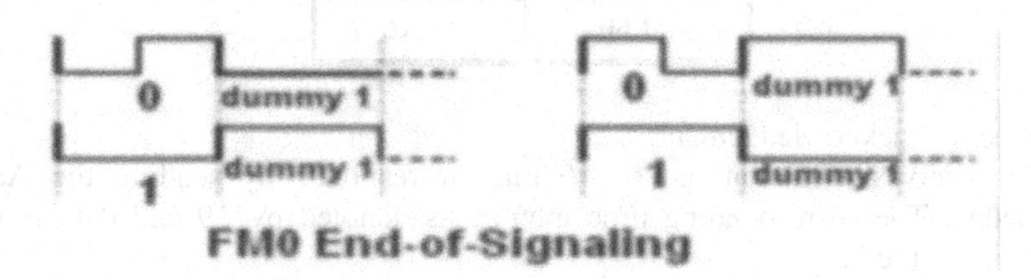

Fig. 5. FM0 end of signalling frame structure

Tag *Power-up* time: Tag needs a time greater or equal to 1500 micro-second to power up. After powering up time at the presence of CW, it must ready to accept and process any command send by reader.

R=>T: The *Select* command: Refer Fig. 3; the corresponding time interval for Select command is denoted by "T6", the Select command is being sent by Reader that contains greater or equal to 46 bits. The data frame structure of the select command is described in Table 4.

Table 4. Data Frame Structure of 'Select' command

	Command	Target	Action	Mem Bank	Pointer	Length	Mask	Truncate	CRC-16
# of bits	4	3	3	2	>=8	8	>=1	1	16

So, the minimum no of bits in Select command is 46.

R=>T: The *Query* command: Refer Fig. 3; the corresponding time interval for Query command is denoted by "T7", the Query command is also being sent by reader to tag. The data frame structure of the query command is described in Table 5.

Table 5. Data Frame Structure of 'Query' command

	Command	DR	TRext	Sel	Session	Target	Q	CRC-5
# of bits	4	1	1	2	2	1	4	5

The Query command contains total bit no of 22.

T=>R: The response *RN16*: After the Select and Query command subsequently, if there is any tag it replies by sending a 16 bit random number i.e. RN16. The corresponding time interval is denoted by T8 and data frame is described in Table 6.

Table 6. Data Frame Structure of RN16

	Response
# of bits	16

R=>T: The *Ack* acknowledgement:
If reader receives a random no of 16 bits, it responds by sending the Ack or acknowledge. The corresponding time interval is denoted by T9 and data frame is described in Table 7.

Table 7. Data Frame Structure of Ack

	Command	**RN**
# of bits	2	16

The 'Ack' response consists of 18 bits.

T=>R: The reply *[PC+UII+CRC16]*:
After getting acknowledge 'Ack' from reader, the tag replies '[PC+UII+CRC16]'. The corresponding time interval is denoted by T10 and data frame is described in Table 8.

Table 8. Data Frame Structure of PC + UII + CRC16

	PC	**UII**	**CRC16**
# of bits	4	96	2

The reply contains minimum bits of 102 in case of 96 bit UII, for 128 bit of UII it will be 134.

2.2 Timing Analysis

As per Fig 2, after getting UII value the reader can terminate the current session or can continue with further command to read memory or write memory or other operation. In the present scope, as only the timing performance to read the UII value of any tag has been considered.

Let total time to read a tag is Tr

$$Tr = \sum_{i=0}^{10} T_i$$

The minimum time required to read any one single tag Min [Tr].

$$\text{Min [Tr]} = \min \left[\sum_{i=0}^{10} T_i \right]$$

$= \min [T_1 + T_2 + T_3 + T_4 + T_5 + T_6 + T_7 + T_8 + T_9 + T_{10}]$
$= \min[T_1] + \min[T_2] + \min[T_3] + \min[T_4] + \min[T_5] + \min[T_6] + \min[T_7] + \min[T_8] + \min[T_9] + \min[T_{10}]$ (1)

T_1: The minimum value of T_1 i.e.

$\text{Min} [T_1] = \text{MIN} [\text{MAX} (\text{RTcal}, 10T_{pri}) \times (1 - |FT|) - 2\mu s]$
$= \text{MAX} (\text{MIN} [\text{RTcal}], \text{MIN} [10T_{pri}]) \times (1 - \text{MIN} [|FT|]) - 2\mu s$ (2)

Now, RTcal = 2.5 tari < = RTcal < = 3 Tari

MIN [RTcal] = 2.5 tari

1 Tari = the reference interval between the falling edges of two successive pulses representing the symbol '0' as transmitted by the interrogator.

According to EPC/ISO180006C tag, Tari value varies between 6.25 μS to 25 μs.

Hence, Min [Tari] = 6.25 μs

MIN [RTcal] = 2.5 × MIN [Tari] = 2.5 × 6.25 μs = 15.6 μs (3)

According to EPC/ISO180006C protocol, 1.1 × RTcal < = TRcal < = 3 × RTcal

MIN [TRcal] = 1.1× MIN [RTcal] = 17.2 μs
As per Table 2, the corresponding BLF of TRca l= 17.2 μs is 320KHz.
i.e. MAX [BLF] = 320 kHz.

T_{pri} = 1/ BLF, MIN [T_{pr}] = 1/MAX[BLF] =1/320000 Sec = 3.125 μs

Hence 10 T_{pr} =10× 3.125 μs = 31.25 μs (4)

FT is defined in Table 2. Let FT ≈ 0% for a stable interrogator and have no temperature tolerance.

Hence resolving Eq. 2 by Eq. 3 and Eq. 4

$$\text{Min}\,[T_1] = \text{MAX}\,(\text{MIN}\,[\text{RTcal}], \text{MIN}\,[10T_{pri}]) \times (1-\text{MIN}\,[|FT|]) - 2\mu s$$
$$= \text{MAX}\,(15.6\ \mu s, 31.25\ \mu s) \times (1-0) - 2\mu s = (31.25\text{-}2)\ \mu s = 29.25\ \mu s \quad (5)$$

T_2: The minimum value of T_2 i.e.

$$\text{Min}\,[T_2] = 3.0T_{pri} = 3.0 \times \text{MIN}\,[T_{pr}] = 3.0 \times 3.125\ \mu s = 9.3\ \mu s \quad (6)$$

T_3: The minimum value of T_3 i.e.

$$\text{Min}\,[T_3] = \text{Min}\,[T_1] = 29.25\ \mu s \quad (7)$$

T_4: The minimum value of T_4 i.e.

$$\text{Min}\,[T_4] = 2.0\ \text{RTca} = 2.0 \times \text{MIN}\,[\text{RTcal}] = 2.0 \times 15.6\ \mu s = 31.2\ \mu s \quad (8)$$

T_5: The minimum value of T_5 is the tag power-up time, i.e.

$$\text{Min}\,[T_5] = T_5 = 1500\ \mu s = 1.5\ \text{ms} \quad (9)$$

Although 1.5 ms is the maximum time allowed for powering up the tag. The minimum time depends on tag manufacturer.

T_6: The minimum value of T_6 i.e. time required to send *select* command.

MIN $[T_6]$ = MIN [Time for Frame_sync] + MIN [time to send bits in *Select* packet] + End of signaling
= [Delimiter+Data 0+RTcal] + Time to send 46 bit data at max data rate i.e. 128KHz in FM0 with no miller sub carrier + Time to send 2 bit at same condition

$$= [12.5 + 6.25 + 15.6]\ \mu s + 48000/128\ \mu s$$

$$\text{MIN}\,[T_6] \approx 410\ \mu s \quad (10)$$

T_7: The minimum value of T_7 i.e. time required to send *Query* command.

MIN $[T_7]$ = MIN [Time for Frame_preamble] + MIN [time to send bits in *Query* packet] + end of signaling

= [Delimiter + Data 0 + Rtcal + TRcal] + Time to send 22 bit data at max data rate i.e. 128KHz in FM0 with no miller sub carrier + Time to send 2 bit at same condition.
$$= [12.5 + 6.25 + 15.6+17.2]\ \mu s + 24000/128\ \mu s$$

$$\text{MIN}\,[T_7] \approx 239\ \mu s \quad (11)$$

T_8: The minimum value of T_8 i.e. time required to send *RN16* by tag at a frequency of MAX [BLF] i.e.320kHz.

MIN [T_8] = MIN [T=>R preamble] + MIN [time to send bits in *RN16* packet] at BLF = 320 kHz + Time to send 2 bit at same condition.
= Time to send [6 bit +16 bit+2] bit data at max data rate i.e. 320 KHz in FM0 with no miller sub carrier = 24000/320 μs

MIN [T_8] ≈ 75 μs (12)

T_9: The minimum value of T_9 i.e. time required to send *Ack* command by reader of 18 bits.

MIN [T_9] = MIN [Time for ame_sync] + MIN [time to send bits in *Ack* packet] + end of signaling
= [Delimiter+Data 0+RTcal] +Time to send 18 bit data at max data rate i.e. 128KHz in FM0 with no miller sub carrier + Time to send 2 bit at same condition.
= [12.5 + 6.25 + 15.6] μs + 20000/128 μs ≈ 190.6 μs

MIN [T_9] ≈ 190.6 μs (13)

T_{10}: The minimum value of T_{10} i.e. time required to send *[PC+UII+CRC16]* command by tag of 102 bits.
MIN [T_{10}] =MIN [T=>R preamble] + MIN [time to send bits in *PC+UII+CRC16* packet at a frequency of MAX [BLF] i.e.320kHz] + Time to send 2 bit at same condition.
= Time to send [6+102+2] bit data at max data rate i.e. 320 KHz in FM0 with no miller sub carrier = Time to send [110 bit] = 110000/320 μs

MIN [T_{10}] ≈ 343.75 μs (14)

Now, Equation Eq. 2.2.1 can be solved by Eq. 2.2.4 to Eq. 2.2.13 and the result found,

Min [Tr] = min[T_1] + min[T_2] + min[T_3] + min[T_4] + min[T_5] + min[T_6] + min[T_7] + min[T_8] + min[T_9] + min[T_{10}]

≈ 29.25 μs + 9.3 μs + 29.25 μs + 31.2 μs + 1500 μs + 410 μs + 239 μs + 75 μs + 190.6 μs + 343.75 μs ≈ 2857.35 μs

Min [Tr] ≈ 2.86 ms (15)

Hence to read a RFID passive tag at least 2.8 ms time is required where as 1.5 ms taken as tag power-up time with reader parameters, data rate of 128 KHz in FM0 with no miller sub carrier and tag responds with BLF of 320 kHz.

3 Speed Estimation in Dynamic Frame

To realize the maximum allowable speed of tag, let assume certain parameters:

L= Maximum avg. linear distance of reading tag when it is placed at 0 degree or 180-degree angle with the reader antenna.
Ø = Angle of transmitted RF wave by reader antenna.
T = Minimum time to read a single tag.
V_{max} = Maximum speed of the tag.

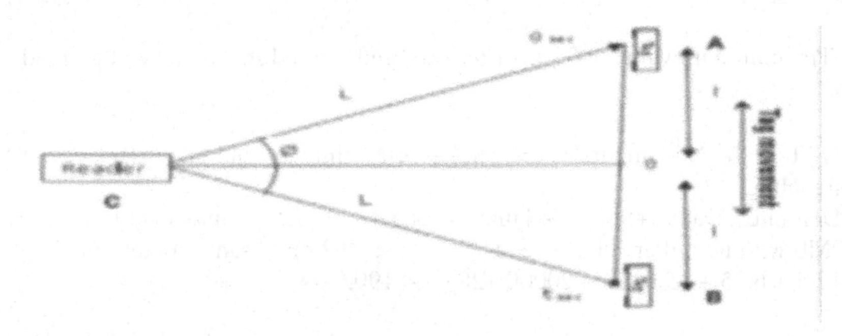

Fig. 6. Setup modelling for speed calculation

Consider an object with RFID enabled tag is moving from point A towards the destination B, the reader is placed at point C. The distance between C and A is equal to distance between C and B and A, C, B creates the angle Ø.

AC = BC = L, Angle < ACB = Ø.

The distance between A and B will be, AB = 2LSin Ø/2 which is being covered by the time t. Hence the maximum speed of the tag will be (2LSin Ø/2)/t

$$V_{max} = (2L\sin Ø/2)/t \quad (16)$$

By solving Eq. 15 and Eq. 16,

$$V_{max} = (2L\sin Ø/2)\times 10^3/2.8 \text{ meter/sec}$$

Let say, a commercial reader having a reading distance of 5 meter with an antenna of 40 degree radiation angle. It will read ant C1G2/ISO180006C tag passing at a maximum velocity of

$$= (2\times 5\times \sin 40/2)\times 10^3/2.8 \text{ meter/sec} = 121.1 \text{ meter/sec.}$$

4 Conclusion

This paper shows the reference mathematical modeling to realize the passive tag reading time and to realize the speed. Changing the parameters L, Ø by choosing different RFID reader, the system can be set up to use passive RFID in high speed dynamic domain.

Acknowledgment

The authors wish to thank Dr. P.C Jain, HOD School of Electronics and Dr. P. R Gupta, HOD School of Information Technology C-DAC Noida for the encouragement. Dr. George Varkey, Executive Director C-DAC Noida to give enough space and freedom to cultivate and nurture the research areas in RFID domain.

References

1. Information technology - Radio frequency identification device conformance test methods - Part 6: Test methods for air interface communications at 860 MHz to 960 MHz
2. Class 1 Generation 2 UHF Air Interface Protocol Standard, Gen 2, `http://www.epcglobalinc.org`
3. Jin, C., Cho, S.H., Jeon, K.Y.: Performance Evaluation of RFID EPC Gen2 Anti-collision Algorithm in AWGN Environment. In: Mechatronics and Automation, August 5-8, pp. 2066–2070 (2007)
4. Engels, D.W., Sarma, S.E.: The reader collision problem. In: Proceedings of the IEEE International Conference on Systems, Man and Cybernetics (SMC 2002), Hammamet, Tunisia, October 2002, vol. 3, pp. 641–646 (2002)

Secure Receipt-Free Sealed-Bid Electronic Auction

Jaydeep Howlader[1], Anushma Ghosh[2], and Tandra DebRoy Pal[1]

[1] National Institute of Technology, Durgapur, India
[2] Microsoft India (R&D) Pvt. Ltd. Hyderabad, India

Abstract. The auction scheme that provides *receipt-freeness*, prevents the bidders from bid-rigging by the coercers. Bid-rigging is a dangerous attack in electronic auction. This happen if the bidder gets a receipt of his bidding price, which proves his bidding prices, from the auction protocol. The coercers used to force the bidders to disclose their receipts and hence bidders lose the secrecy of their bidding prices. This paper presents a protocol for a receipt-free, sealed-bid auction. The scheme ensures the receipt-freeness, secrecy of the bid, secrecy of the bidder and public verifiability.

1 Introduction

Auction has become a major phenomenon of electronic commence in the recent years. Auction is a game for trading some items. There is a competition among the buyers to get their bids as the winning bid. Whereas, another tussle is between the auctioneer and bidders trade the items in a reasonable price. The most common auctions are English auction, Dutch auction and Sealed-Bid auction. Sealed-Bid auction is a mechanism for establishing the price of goods/items and are widely used in the real world. However, Sealed-Bid auction suffers from a serious attack, that is, bid-rigging by coercers. Bid-rigging is a phenomenon where the coercers cheat by ordering other buyers to bid very low prices so that he can win the auction by bidding an unreasonably low price. If bid-rigging occurs, the auction fails to establish the appropriate price. The game of auction goes in favor of the coercers, so it is important to prevent bid-rigging. The traditional paper-based Sealed-Bid auction does not encounter the bid-rigging. There is no receipt issued by the auctioneer on submission of bids. The sealed bids are kept secret and are not opened before the schedule time of the auction. Most of the existing electronic Sealed-Bid auction schemes are subject to be bid-rig, as they issue receipts to the bidders on submission of their bidding prices, which are exploited by the coercers to verify whether the bidders obey their order. The coercers are also able to identify the bidders who have broken the collusion. To enforce the bid-rigging, the coercers reward the colluded bidders and punish the bidders who have broken the collusion.

There are different varieties of Sealed-Bid auction. Vickrey [1] presented a second-price Sealed-Bid auction where the bidder with the highest price won,

S. Ranka et al. (Eds.): IC3 2009, CCIS 40, pp. 228–239, 2009.
© Springer-Verlag Berlin Heidelberg 2009

but he only paid the second highest price. Another Sealed-Bid auction is $M+1^{st}$ price auction. The $M+1^{st}$ Sealed-Bid is used to sell M units of a single kind of item where M bidders win and they pay the $M+1^{st}$ price. Vickrey's auction can be formulated as a $M+1^{st}$ auction for $M=1$.

The receipt-freeness was first introduced by Benaloh and Tuinstra [2] in a voting protocol to overcome the problem of vote-buying. Their protocol was based on two assumptions: a voting booth that physically separated the voters from the others during the vote casting, and a secure communication channel between the trusted authorities (single or multiple) and the voting booth. Later on Sakurai and Miyazaki [3] pointed out the problem of bid-rigging in their electronic auction scheme. The scheme was developed to hide the identity of the winning bidder by using convertible group signature. However the scheme was not a receipt-free, hence it could not prevent the bid-rigging in presence of coercers. Hiding the loser bidders identity was also discussed in [4,5], but they were again not receipt-free. So, hiding the identity of the bidders is not sufficient to solve the problem of bid-rigging. In addition, the auction scheme should be receipt-free so that the coercers are not able to verify whether the bidders obey their order or not. Some other electronic auction schemes were proposed in different literatures. Suzuki *et al.* [6] using chain of hash functions to speed up the computation of auction procedure. In [7] Franklin *et al.* described a secure auction service that guaranteed secrecy of the bids, verifiability of the winning price, anonymity of the bidders. The problem of those schemes is: they do not prevent the bid-rigging in presence of the coercers.

Sealed-Bid auction is the basic mechanism to establishing the price of goods and materials for procurement and is widely used by the governmental/non-governmental organizations. The auction schemes [3,5,6,7] were designed for sealed-bid auctioning. Kikuchi *et al.* presented an anonymous sealed-bid auction in [8] using encrypted bid vector. A time dependent cryptographic key based sealed-bid auction procedure was presented in [9] by Michiharu. Abe and Suzuki [10] proposed a $M+1^{st}$ sealed-bid auction using homomorphic encryption. The above schemes did not provide the receipt-freeness. The first receipt free sealed-bid auction scheme was proposed by Abe and Suzuki [11] to prevent bid-rigging. The scheme was based on multiple auctioneer threshold trust model, i.e., a majority of the auctioneers were assumed to be honest. In their scheme the bidders chose their secret seeds for bidding. A *t-out-of-n* secret share is shared among the auctioneers. During the opening, all auctioneers published their secret shares and recovered the secret seed to compute the winning price. However, a dishonest auctioneer might disclose the secret seeds of the bidders to the coercers after opening, so that coercers could know all the secret information of victim bidders beforehand. If the winning price was not in the favor of the coercer, he could identify the victim bidders who had not followed the order. So Abe and Suzuki's scheme could not prevent the bid-rigging completely. Recently, Chen *et al.* [12] presented a receipt-free auction scheme using homomorphic encryption. Chen *et al.* introduced another entity called *seller*. The scheme was based on single auctioneer and single seller model. The bidders chose their secret seed and constructed their encrypted bid-vectors. The encrypted bid-vectors were sealed by the seller and sent back to the bidders. The bidders checked the validity of their seals, but they could not prove their bidding-prices from the sealed bid-vectors. After verifying the seal, sealed-bids were placed in a bulletin board. On the time of opening the auctioneer opened the bids and declared the winning price. The scheme suffers from two major problems. Firstly, the auctioneer may open the bids before the schedule time and evaluates the winning price. Then he can provide the information to his agents to bid accordingly. Secondly, if the seller is dishonest, he may provide the encrypted bid-vectors to the coercers. The encrypted bid-vectors are all committed messages. The coercers would exploit the committed messages to verify whether the bidders obey his order. In this paper we present a receipt-free auction scheme with multiple sealer and single auctioneer. Our scheme is based on multiple sealer trusted model, assuming that some of the sealers are honest. We ensure that the auctioneer can not open the bids before the schedule time. We also ensure that, any blinded message provided by the dishonest sealers to coercers, in any intermediate stage of auction, does not reveal any information. We assume an anonymous channel between the bidders and the sealers. The anonymous channel hides the correspondences between the sender and receiver.

The rest of the paper is organized as follows: section 2 presents the basic properties and requirements of receipt-free sealed-bid auction. Section 3 describes the auction scheme. The auction scheme is divided in two phases: bidding phase and opening phase. In section 4 we present the security and verifiability analysis. Section 5 presents the efficiency and complexity of the proposed auction scheme. The conclusion the work is in section 6.

2 Receipt-Free Sealed-Bid Auction: Properties and Security Requirements

The sealed-bid auction is an auction mechanism where the bidders submit their bids in sealed envelope. The bids are kept secret during the bidding period. There is a predetermined schedule for bid opening. No bids are accepted after the scheduled time. The bids are opened by the auctioneer at the time of opening and the winning price is determined according to the auction rule. The sealed-bid auction procedure can be formulated in two phases: Bidding phase and Opening phase. To achieve a fair auction, the sealed-bid auction protocol must satisfy the following properties:

2.1 Properties of Receipt-Free Sealed-Bid Auction

Secrecy of Bids: All the bids must be kept secret during the bidding phase. No one even the auctioneer can not open the bids before the time of opening. All the bids except the winning bid must be kept secret even after the auction is over. Hiding the losers' price [4,5] is also important.

Verifiability: Anyone must be able to verify the correctness of the auction. As it is mentioned above that, all the prices except the winning price are kept secret after the opening of auction. So any one should able to verify whether the winning price declared by the auctioneer is indeed the correct price or not.

Anonymity: The bidder must bid anonymously. The encrypted bids should not contain the identity of the bidders, nor would the bidders able to prove their bidding prices from the sealed bids. It is important to keep bidders identity as secret, even after the opening of the bids to prevent bid-rigging.

Efficiency: The computation and communication overhead of auction should be reasonable. The bidders should not be heavily computation intensive related to the auction authorities. Moreover, the bidders should not necessarily be present during the opening.

Non-repudiation: The auctioneer only declares the winning price. The bidder, who has bid the winning price, may not respond or may repudiate. The auction protocol should able to identify the winning bidder, if it is necessary.

Receipt-freeness: The auction scheme should be receipt-free to avoid bid-rigging. Anyone, even the bidder himself, should not be able to prove his bidding price from the sealed bid.

2.2 Physical Requirements for Receipt-Free Sealed-Bid Auction

The physical requirements, to ensure security and prevent bid-rigging, are as follows:

Public Board: The *public board* is a public channel with memory, where the entities can write but no one can delete any information from the public board. The public board is available to every entity.

Anonymous Channel: An *anonymous channel* hides the correspondences between the sender and the receivers. Chaum [17] described a computationally secure anonymous channel called *MIX-net* using public board. The anonymous channel hides *who sends to whom*. We assume an anonymous channel between the bidders and the sealers. Bidders compute their encrypted bids and send the encrypted bids to the sealers through the anonymous channel. The sealers do not know the source of the encrypted bids. They perform the sealing operation and write the sealed bids on the public board.

2.3 Entities of Receipt-Free Sealed-Bid Auction

The entities of the auction protocol are as follows:

Auctioneer: who wants to sell some items or conducts the auction. There is a single auctioneer. The auctioneer publishes a price list $P = \{p_1, p_2, \ldots, p_n\}$, in ascending order.

Bidder: who wants to buy the items and bids for the items. There are m bidders $B = \{B_1, B_2, \ldots, B_m\}$. The number of bidders is not fixed. The bidder B_i selects his bidding price $p_j \in P$ and constructs an encrypted bid-vector with his random seed.

Sealer: who plays a role between the bidders and the auctioneer. The sealer does the sealing operation on the encrypted bid-vectors generated by the bidders. There are t sealers $S = \{S_1, S_2, \ldots S_t\}$. Generally, $t < m$. To avoid collusion we have introduced multiple sealers. The bid-vectors are sealed by all the t sealers.
The auction scheme is based on the following assumptions:

1. $B = \{B_1, B_2, \ldots, B_m\}$ is the set of m bidders, out of them, there may be some coercers.
2. $S = \{S_1, S_2, \ldots, S_t\}$ is the set of sealers, out of them, some of the sealers may collude with the coercive bidders. We assume that all the sealers are not colluded, there are some honest sealers.
3. There is a single auctioneer, A. The auctioneer along with the t sealers open the bids on the schedule time. Then the auctioneer computes and declares the winning price. The scheme ensures that the auctioneer can not open the bids without the cooperation of all the sealers.
4. Coercers are empowered enough to get hold all the stored keys of the bidders. Coercers can enforce the bidders to disclose the plaintext for any committed ciphers that the coercers get. But coercers are not able to influence the bidding process and they are not able to observe the bidders to bid.

3 Receipt-Free Sealed-Bid Auction Procedure

Consider a subgroup G_q of order q from $\mathbb{Z}_p^*$, where p and q are large primes and $q|p-1$. Let $g \in G_q$ is a generator of the group G_q. G_y and G_n are two independent generators of the group G_q, which indicate *'I bid'* and *'I do not bid'* respectively. The keys used in the auction protocol are:

- Bidder B_i's private key is x_{B_i} and public key is $h_{B_i} = g^{x_{B_i}}$.
- Auctioneer A chooses his secret key x_A and published his public key $h_A = g^{x_A}$.
- Sealer S_i has a private key x_{S_i} and publishes his public key $h_{S_i} = g^{x_{S_i}}$.
- $h_S = \prod_{i=1}^{t} h_{S_i}$ is the shared public key of t sealers.
- $h_{S/S_i,S_j,\ldots,S_k} = h_S/(h_{S_i}h_{S_j}\ldots h_{S_k})$ is the shared public key of sealers excluding $S_i, S_j, \ldots, S_k$

3.1 Bidding Phase

The bidder B_i selects his bidding price from the list P and generates an encrypted *bid-vector*. The bidder sends the encrypted bid-vector to the sealers through an anonymous channel. The sealers $S_1, S_2, \ldots S_t$ perform the sealing operation and write the receipt-free sealed bid-vector in a public board. The sealing operation is performed by all the t sealers. The bidder verifies his seal bid-vector. Following is the detail description of the bidding process.

1. Bidder B_i decides his bidding price $p_j \in P$ from the price list published by the auctioneer and encrypts the price as follows:

$$_i\Gamma_j = ({}_iX_j, {}_iY_j) = \begin{cases} g^{{}_ir_j}, h_A^{{}_ir_j} h_S^{{}_ir_j} G_y & \text{if } p_j \text{ is the } j^{th} \text{ price in the list } P \\ g^{{}_ir_j}, h_A^{{}_ir_j} h_S^{{}_ir_j} G_n & \text{otherwise} \end{cases}$$

for $1 \leq j \leq n$, ${}_ir_j \in_R \mathbb{Z}_q$ is randomly selected by bidder B_i. The notation ${}_i\Gamma_j$ denotes the jth encrypted price of the bidder B_i.
B_i also computes a blinded commitment vector as:

$$_iC_j = h_A^{{}_ir_j H({}_i\Gamma_j)\ x_{B_i}}$$

for $1 \leq j \leq n$, H is a one-way hash function. The bidder B_i constructs the encrypted bid-vector as a two tuples $\langle {}_i\Gamma_j, {}_iC_j \rangle$ and sends the encrypted bid-vector to any of the t sealer through the anonymous channel. Let B_i sends the encrypted bid-vector to S_k.

2. Sealer S_k receives the encrypted bid-vector and engraves his random seed ${}_ir_{j,s_k}$ and computes the partially sealed-bid vector as $\langle {}_i\Gamma_{j,k}, {}_iC_j, {}_i\mathbb{R}_j \rangle$ where ${}_i\Gamma_{j,k} = \{{}_iX_{j,k}, {}_iY_{j,k}\}$:

$$\begin{aligned} {}_iX_{j,k} &= {}_iX_j \ . \ g^{{}_ir_{j,s_k}} \\ &= g^{{}_ir_j} \ . \ g^{{}_ir_{j,s_k}} \\ &= g^{{}_ir_j + {}_ir_{j,s_k}} \\ {}_iY_{j,k} &= {}_ir_{j,s_k} \ . \ {}_iY_j \ . \ h_A^{{}_ir_{j,s_k}} \ . \ h_{S/S_k}^{{}_ir_{j,s_k}} \ . \ ({}_iX_j)^{-x_{S_k}} \\ &= {}_ir_{j,s_k} \ . \ h_A^{{}_ir_j} h_S^{{}_ir_j} \ . \ h_A^{{}_ir_{j,s_k}} \ . \ h_{S/S_k}^{{}_ir_{j,s_k}} \ . \ h_{S_k}^{-{}_ir_j} \ . \ G \\ &= {}_ir_{j,s_k} \ . \ h_A^{{}_ir_j + {}_ir_{j,s_k}} \ . \ h_{S/S_k}^{{}_ir_j} \ . \ h_{S/S_k}^{{}_ir_{j,s_k}} \ . \ G \\ &= {}_ir_{j,s_k} \ . \ h_A^{{}_ir_j + {}_ir_{j,s_k}} \ . \ h_{S/S_k}^{{}_ir_j + {}_ir_{j,s_k}} \ . \ G \end{aligned}$$

for $1 \leq j \leq n$, ${}_ir_{j,s_k} \in_R \mathbb{Z}_q$ is randomly selected by sealer S_k, G is either G_n or G_y. Sealer S_k computes a response ${}_i\mathbb{R}_j = \langle {}_iR_j, {}_i\alpha_j, {}_i\beta_j \rangle$. The response is used by the bidders to verify the seal. Sealer S_k selects random numbers ${}_iw_{j,k}$ for $1 \leq j \leq n$ and computes the following:

$$\begin{aligned} {}_iR_j &= {}_iw_{j,k} + {}_iC_j \ . \ {}_ir_{j,s_k} \\ {}_i\alpha_j &= g^{{}_iw_{j,k}} \\ {}_i\beta_j &= ({}_ir_{j,s_k})^{-{}_iC_j} \ . \ h_A^{{}_iw_{j,k}} \end{aligned}$$

The sealer S_k sends the $\langle {}_i\Gamma_{j,k}, \ {}_iC_j, \ {}_i\mathbb{R}_j \rangle$ to some sealer S_l randomly selected from $S - \{S_k\}$.

3. S_l receives the partially sealed bit vector $\langle {}_i\Gamma_{j,k}, \ {}_iC_j, \ {}_i\mathbb{R}_j \rangle$ from S_k and does the sealing operation to produce the partial sealed bit vector as $\langle {}_i\Gamma_{j,l}, \ {}_iC_j, \ {}_i\mathbb{R}_j \rangle$,

where ${}_i\Gamma_{j,l} = \{{}_iX_{j,l}, {}_iY_{j,l}\}$:

$$\begin{aligned}
{}_iX_{j,l} &= {}_iX_{j,k} \,.\, g^{{}_ir_{j,s_l}} \\
&= g^{{}_ir_j + {}_ir_{j,s_k}} \,.\, g^{{}_ir_{j,s_l}} \\
&= g^{{}_ir_j + {}_ir_{j,s_k} + {}_ir_{j,s_l}} \\
&= g^{{}_ir_j + \sum^{a=k,l} {}_ir_{j,s_a}} \\
{}_iY_{j,l} &= {}_ir_{j,s_l} \cdot {}_iY_{j,k} \,.\, h_A^{{}_ir_{j,s_l}} \,.\, h_{S/S_k,S_l}^{{}_ir_{j,s_l}} \cdot {}_iX_{j,k}^{-xs_l} \\
&= {}_ir_{j,s_l} \cdot {}_ir_{j,s_k} \,.h_A^{{}_ir_j + {}_ir_{j,s_k}} .h_{S/S_k}^{{}_ir_j + {}_ir_{j,s_k}} .h_A^{{}_ir_{j,s_l}} .h_{S/S_k,S_l}^{{}_ir_{j,s_k}} .h_{S_l}^{-({}_ir_j + {}_ir_{j,s_k})} .G \\
&= {}_ir_{j,s_l} \cdot {}_ir_{j,s_k} \,.\, h_A^{{}_ir_j + {}_ir_{j,s_k} + {}_ir_{j,s_l}} \,.\, h_{S/S_k,S_l}^{{}_ir_j + {}_ir_{j,s_k} + {}_ir_{j,s_l}} \,.\, G \\
&= \Big(\prod^{a=k,l} {}_ir_{j,s_a} \Big) \,.\, h_A^{{}_ir_j + \sum^{a=k,l} {}_ir_{j,s_a}} \,.\, h_{S/S_k,S_l}^{{}_ir_j + \sum^{a=k,l} {}_ir_{j,s_a}} \,.\, G
\end{aligned}$$

for $1 \le j \le n$, ${}_ir_{j,s_l} \in_R \mathbb{Z}_q$ is randomly selected by bidder S_l. Sealer S_l updates the response ${}_i\mathbb{R}_j$ with his random numbers ${}_iw_{j,l}$ for $1 \le j \le n$ as follows:

$$\begin{aligned}
{}_iR_j &= {}_iR_j + ({}_iw_{j,l} + {}_iC_j \,.\, {}_ir_{j,s_l}) \\
&= \sum^{a=k,l} {}_iw_{j,a} + {}_iC_j \sum^{a=k,l} {}_ir_{j,s_a} \\
{}_i\alpha_j &= {}_i\alpha_j \,.\, g^{{}_iw_{j,k}} \\
&= g^{\sum^{a=k,l} {}_iw_{j,a}} \\
{}_i\beta_j &= {}_i\beta_j \,.\, ({}_ir_{j,s_l})^{-{}_iC_j} \,.\, h_A^{{}_iw_{j,l}} \\
&= \Big(\prod^{a=k,l} {}_ir_{j,s_a} \Big)^{-{}_iC_j} \,.\, h_A^{\sum^{a=k,l} {}_iw_{j,a}}
\end{aligned}$$

The sealer S_l sends the partially sealed bid-vector $\langle {}_i\Gamma_{j,l},\ {}_iC_j,\ {}_i\mathbb{R}_j \rangle$ to some sealer S_m, randomly selected from $S - \{S_k, S_l\}$. Sealer S_m receives the partially sealed bid-vector, engraves his random seed and updates the response vector and sends the bid-vector to some other sealer.

4. In this way, the last sealer, say S_t receives the bid-vector and computes the final sealed bid vector as $\langle {}_i\Gamma_{j,t},\ {}_iC_j,\ {}_i\mathbb{R}_j \rangle$ where ${}_i\Gamma_{j,t} = \{{}_iX_{j,t}, {}_iY_{j,t}\}$:

$$\begin{aligned}
{}_iX_{j,t} &= g^{{}_ir_j + \sum_{a=1}^{t} {}_ir_{j,s_a}} \\
{}_iY_{j,t} &= \Big(\prod_{a=1}^{t} {}_ir_{j,s_a} \Big) \,.\, h_A^{{}_ir_j + \sum_{a=1}^{t} {}_ir_{j,s_a}} \,.\, h_{S/S_1,S_2,\ldots,S_t}^{{}_ir_j + \sum_{a=1}^{t} {}_ir_{j,s_a}} \,.\, G \\
&= \Big(\prod_{a=1}^{t} {}_ir_{j,s_a} \Big) \,.\, h_A^{{}_ir_j + \sum_{a=1}^{t} {}_ir_{j,s_a}} \,.\, G
\end{aligned}$$

Sealer S_t updates the response vector ${}_i\mathbb{R}_j$ as:

$$
\begin{aligned}
{}_iR_t &= \sum_{a=1}^{t} {}_iw_{j,a} + {}_iC_j \sum_{a=1}^{t} {}_ir_{j,s_a} \\
{}_i\alpha_j &= g^{\sum_{a=1}^{t} {}_iw_{j,a}} \\
{}_i\beta_j &= \Big(\prod_{a=1}^{t} {}_ir_{j,s_a}\Big)^{-{}_iC_j} \,.\, h_A^{\sum_{a=1}^{t} {}_iw_{j,a}}
\end{aligned}
$$

Sealer S_t puts the sealed bid-vector $\langle {}_i\Gamma_{j,t},\ {}_iC_j,\ {}_i\mathbb{R}_j \rangle$ in the public board.

5. The bidder B_i checks the validity of the sealed bid vector, that is, B_i checks whether ${}_i\Gamma_{j,t}$ is indeed the sealed bid vector of ${}_i\Gamma_j$. The verification is done as follows:

   - B_i identifies his sealed bid from the challenge vector ${}_iC_j$ and verifies the following relations:

$$
\begin{aligned}
g^{{}_iC_j\ {}_ir_j + {}_iR_j} &=^{?} {}_i\alpha_j ({}_iX_{j,t})^{{}_iC_j} \\
h_A^{{}_iC_j\ {}_ir_j + {}_iR_j}\ G^{{}_iC_j} &=^{?} {}_i\beta_j\ ({}_iY_{j,t})^{{}_iC_j}
\end{aligned}
$$

   for $1 \le j \le n$. G is either G_y or G_n depending on the B_i bidding price. If the bidder B_i fails to verify his bidding price, he raises the complain.

3.2 Opening Phase

After successfully executing the bidding phase, the bids are opened as per the scheduled time. All sealers S_k, for $1 \le k \le t$, compute $V_{j,s_k} = \prod_{i=1}^{m}({}_ir_{j,s_k})$, the product of all random seeds ${}_ir_{j,s_k}$ used for sealing the jth price value and send privately to the auctioneer A. After receiving all V_{j,s_k} $(1 \le k \le t)$ from t sealers the auctioneer A opens the bid-vectors and declares the winning price. The opening of bids is as follows:

1. Auctioneer opens the bids in descending order, that is, auctioneer first opens the bids for P_n, then P_{n-1} and so on until the winning price is not found. The opening of the jth price value is done as follows:
   - A computes

$$
\begin{aligned}
\mathbb{Y}_j &= \frac{\prod_{i=1}^{m} {}_iY_{j,t}}{\prod_{i=1}^{m} ({}_iX_{j,t})^{x_A}} \\
&= \Big(\prod_{i,k=1}^{m,t} {}_ir_{j,t}\Big)\ .\ G \qquad\qquad G \text{ is either } G_n \text{ or } G_y \\
\mathbb{G}_j &= \frac{\mathbb{Y}_j}{\prod_{k=1}^{t} V_{j,s_k}} \\
&= G_n^{m-l} G_y^{l}
\end{aligned}
$$

   $\mathbb{G}_j = G_n^{m-l} G_y^{l}$ for $l \ge 0$. Auctioneer declares the winning price as p_j, for the j where $l \ge 1$ appears first.

2. After declaring the winning price the bidder B_i claims his victory with the encrypted bid-vector $\langle {}_i\Gamma_j, {}_iC_j \rangle$. The auctioneer verifies the claim and the winner is decided by the auctioneer as follows:
   – Auctioneer computes

$$\begin{aligned} {}_i\mathbb{C}_j &= {}_iC_j^{-H({}_i\Gamma_j)x_A} \\ &= {}_iX_j^{x_{B_i}} \end{aligned}$$

   – The bidder B_i and auctioneer A execute an interactive *zero-knowledge proof* protocol to verify that ${}_i\mathbb{C}_j$ and h_{B_i} have the common exponent as x_{B_i}.

     • B_i selects δ_j for $1 \leq j \leq n$ randomly and computes $a_j = g^{\delta_j}$, $b_j = {}_iX_j^{\delta_j}$ and sends (a_j, b_j) to the auctioneer.
     • Auctioneer A selects random challenges c_j for $1 \leq j \leq n$ and sends to the Bidder B_i.
     • B_i computes $\gamma_j = \delta_j + c_j x_{B_i}$ and replies to the A.
     • Auctioneer verifies the following relations:

$$\begin{aligned} g^{\gamma_j} &=^? a_j \,.\, h_{B_i}^{c_j} \qquad \text{for } 1 \leq j \leq n \\ {}_iX_j^{\gamma_i} &=^? b_j \,.\, {}_i\mathbb{C}_j^{c_i} \end{aligned}$$

4 Security Analysis

The proposed scheme satisfies the following security properties:

Security: The bidders are allowed to generate their encrypted bid-vectors and they send the encrypted bid-vectors through an anonymous channel to some sealer. The anonymous channel hides the correspondences between the sender and the receiver. So, if the coercers eavesdrop the communication, they would not get any information about *whose bid is what.* Even if the coercers capture some encrypted bid-vectors from some colluded sealers, they would not able to make any correspondences between the bidders and the encrypted bid-vectors. However, the encrypted bid-vector contains the identity of the bidder engraved in the blind commitment ${}_iC_j$. In that case, the auctioneer can only verify the identity of the bidder B_i, after honestly executing an interactive zero-knowledge proof by auctioneer and bidder.

The encrypted bid-vectors are sealed by t sealers so that the auctioneer can not open the bids without the help of all the sealers. We assume that there are some honest sealer who will not provide $V_{j,s_k} = \prod_{i=1}^{m}({}_ir_{j,s_k})$ before the schedule time of opening. So it is ensured that the auctioneer is unable to open the bids before the schedule time.

Receipt-freeness: ${}_i\Gamma_j = ({}_iX_j, {}_iY_j)$ contains the encrypted price value of the bidder B_i. ${}_i\Gamma_j$ is sealed by t sealers as ${}_i\Gamma_{j,t}$ and written in the public board. The sealed bid-vector contains the random seeds of t sealers. The bidder B_i can verify the seal on his bid, but he can not prove to any third party that ${}_i\Gamma_{j,t}$ is the sealed bid of ${}_i\Gamma_j$, as the bidder does not know the random seeds used by the t sealers. Even a subset of colluded sealers are not able to make the correspondence between the ${}_i\Gamma_{j,t}$ and ${}_i\Gamma_j$. Moreover, initial encrypted price value $({}_iX_j, {}_iY_j)$ generated by the bidder B_i is not require during the validity checking.

Validity: The result of the auction can be verified by any one. If any one wants to check the correctness of the result, the auctioneer publishes all the $\mathbb{Y}_j$ and the V_{j,s_k} $(w \leq j \leq n$ where p_w is the winning price), and the correctness can be verified.

Anonymity: The anonymous channel between the bidders and the sealers provides the bidders to bid anonymously. Moreover, the result of the auction does not reveal the bidder's identity, but the winning price. The winner claims his victory and the winner's claim is verified by the auctioneer. The identity of the bidder present in the blind commitment ${}_iC_j$ is only verified by the auctioneer by executing an interactive *zero knowledge proof* with the bidder. The coercers can not make any relation among the bids and the bidders. So, the anonymity of the bidder is maintained in this auction scheme.

Non-repudiation: The auction scheme does not declare the winner, it declares the winning price. In that case, the winning bidder may repudiate or he may keep silent. The auction scheme should able to identify the winning bidder after the auction, if he repudiates. This is done as follows:

Let p_w, the wth entry of the price list is the winning price. Also let sealer S_k receives the encrypted bid-vector $\langle {}_i\Gamma_j, {}_iC_j \rangle$. The auctioneer asks the sealers to write the initial encrypted price value (not signed by any sealer) $\langle {}_i\Gamma_w, {}_iC_w \rangle$ $(1 \leq i \leq m)$, on the public board. Then every sealer computes $h_{S_k}^{ir_w} = {}_iX_w^{x_{S_k}}$ for $1 \leq i \leq m$, $1 \leq k \leq t$ and writes on the public board. So every entry in the public board look like $\langle {}_i\Gamma_w, {}_iC_w, h_{S_1}^{ir_w}, h_{S_2}^{ir_w}, \dots h_{S_t}^{ir_w} \rangle$. Auctioneer computes

$$G_i = \frac{{}_iY_w}{{}_iX_w^{x_A} \prod_{k=1}^{t} h_{S_k}^{ir_w}} \qquad \text{for } 1 \leq i \leq m$$

and gets the bidders' marks on the price value p_w. The mark is either G_y, that is *'I bid'* or G_n, means *'I do not bid'*. The entry in the public board for which auctioneer computes $G_i = G_y$ is the winner. The identity of the winner is engraved in the blind commitment ${}_iC_w$ maid by the bidder. Auctioneer asks all the bidders to sign ${}_iX_w$ as ${}_i\hat{C}_w = {}_iX_w^{x_{B_i}}$ and asks all the bidders to proof that ${}_i\hat{C}_w$ and h_{B_i} have the same exponent as x_{B_i}. If the bidders do the signature along with the proof, auctioneer can identify the repudiated bidder B_i for which ${}_i\hat{C}_w = {}_iC_w$.

5 Efficiency

The entities of the auction scheme are: multiple bidders, multiple sealers and single auctioneer. In this section we present the computation and communication complexity of the scheme. Let n,m and t represent the number of bidding

Table 1. Communication complexity of the auction scheme

operation	*communication*	*rounds*	*volume*
Bidding	$B_i \rightarrow S_j$	1	$O(n)$
Sealing	$S_j \rightarrow S_k \dashrightarrow S_t$	t	$O(n)$
Verification	PublicBoard $\rightarrow B_i$	1	$O(n)$
Opening	PublicBoard $\rightarrow A$	1	$O(mn)$
	$\{S_1, S_2, \ldots S_t\} \rightarrow A$	t	$O(m)$
Winner Verification	$B_i \leftrightarrow A$	3	$O(n)$
Repudiation	$\{S_1, S_2, \ldots S_t\} \rightarrow$ PublicBoard	t	$O(1)$
	$\{S_1, S_2, \ldots S_t\} \rightarrow$ PublicBoard	mt	$O(1)$
	PublicBoard $\rightarrow A$	m	$O(t)$

prices, number of bidders and number of sealers respectively. Table 1 presents the communication patterns, number of rounds and the volume of data in every round.

6 Conclusion

In this paper we have presented a receipt-free sealed-bid auction scheme to prevent bid-rigging with single auctioneer and multiple bidders. The auction scheme is designed for 1st price auction. However, the scheme can be modified to $M+1^{st}$ price auction. The scheme ensures the receipt-freeness and prevents bid-rigging. The auction scheme allows the bidders to *bid and go*, the bidders need not be present during the opening phase. The scheme provides receipt-freeness, anonymous bidding, bid secrecy and non-repudiation.

References

1. Vickrey, W.: Counterspeculation, Auctions, and Competitive Sealed Tenders. Journal of Finance 16(1), 8–37 (1961)
2. Benaloh, J., Tuinstra, D.: Receipt-free secter-ballot election (extended abstract). In: Proc. 26th ACM Symposium on the Theory of Computing (STOC), pp. 544–553. ACM Press, New York (1994)
3. Sakurai, K., Miyazaki, S.: An Anonymous Election Bidding Protocol Based on a New Convertible Group Signature Scheme. In: Clark, A., Boyd, C., Dawson, E.P. (eds.) ACISP 2000. LNCS, vol. 1841, pp. 385–399. Springer, Heidelberg (2000)
4. Sakurai, K., Miyazaki, S.: A Bulletin-Board Based Digital Auction Scheme with Bidding Down Strategy-Towards Anonymous Electronic Bidding without Anonymous Channels nor Trusted Centers. In: Proc. International Workshop on Cryptographic Techniques and E-Commerce (CryTEC 1999), pp. 180–187 (1999)
5. Sako, K.: An Auction Protocol which Hides Bids of Losers. In: Imai, H., Zheng, Y. (eds.) PKC 2000. LNCS, vol. 1751, pp. 422–432. Springer, Heidelberg (2000)
6. Suzuki, K., Kobayashi, K., Morita, H.: Efficient Sealed-bid Auction using Hash Chain. In: Won, D. (ed.) ICISC 2000. LNCS, vol. 2015, pp. 183–191. Springer, Heidelberg (2001)

7. Franklin, M.K., Reiter, M.K.: The Design an Implementation of a Secure Auction Service. IEEE Trans. Softw. Eng. 22(5), 302–312 (1996)
8. Kikuchi, H., Harkavy, M., Tygar, J.D.: Multi-round Anonymous Auction Protocol. In: Proc. 1st IEEE Workshop on Dependable and Real-Time E-Commerce Systems, pp. 62–69 (1998)
9. Kudo, M.: Secure Electronic Sealed-Bid Auction Protocol with Public Key Cryptography. Trans. IEICE trans. on Fundamentals of Electronics, Communications and Computer Sciences E81-A(1), 20–27 (1998)
10. Abe, M., Suzuki, K.: $M + 1^{st}$ Price Auction using Homomorphic Encryption. In: Naccache, D., Paillier, P. (eds.) PKC 2002. LNCS, vol. 2274, pp. 115–124. Springer, Heidelberg (2002)
11. Abe, M., Suzuki, K.: Receipt-Free Sealed-Bid Auction. In: Chan, A.H., Gligor, V.D. (eds.) ISC 2002. LNCS, vol. 2433, pp. 191–199. Springer, Heidelberg (2002)
12. Chen, X., Lee, B., Kim, K.: Receipt-Free Electronic Auction Scheme using Homorphic Encryption. In: Lim, J.-I., Lee, D.-H. (eds.) ICISC 2003. LNCS, vol. 2971, pp. 259–273. Springer, Heidelberg (2004)
13. Harkavy, M., Tygar, J.D., Kikuchi, H.: Electronic Auction with Private Bids. In: Proc. 3rd USENIX Workshop on Electronic Commerce (1998)
14. Canetti, R., Dwork, C., Naor, M., Ostrovsky, R.: Deniable Encryption. In: Kaliski Jr., B.S. (ed.) CRYPTO 1997. LNCS, vol. 1294, pp. 90–104. Springer, Heidelberg (1997)
15. Chaum, D., Pedersen, T.P.: Wallet Database with Observers. In: Brickell, E.F. (ed.) CRYPTO 1992. LNCS, vol. 740, pp. 89–105. Springer, Heidelberg (1993)
16. Shamir, A.: How to Share a Secret. Communications of the ACM 22(11), 612–613 (1979)
17. Chaum, D.L.: Untraceable Electronic Mail, Return Addresses, and Digital Pseudonyms. Communications of the ACM 24(2), 84–88 (1981)

An Architecture for Handling Fuzzy Queries in Data Warehouses

Manu Pratap Singh[1], Rajdev Tiwari[2], Manish Mahajan[3], and Diksha Dani[4]

[1] Institute of Computer and Information Science
Dr. B.R. Ambedkar University
Khandari, Agra U.P., India
manu_p_singh@hotmail.com
[2] Inderprastha Engineering College
Ghaziabad, U.P., India
rajdevtiwari@yahoo.com
[3] ABES Institute of Technology
Ghaziabad, U.P., India
mahajan_manish@rediffmail.com
[4] Inderprastha Engineering College
Ghaziabad, U.P., India
dikshadani@yahoo.com

Abstract. This paper presents an augmented architecture of Data Warehouse for fuzzy query handling to improve the performance of Data Mining process. The performance of Data Mining may become worst while mining the fuzzy information from the large Data Warehouses. There are number of preprocessing steps suggested and implemented so far to support the mining process. But querying large Data warehouses for fuzzy information is still a challenging task for the researchers' community. The model proposed here may provide a more realistic and powerful technique for handling the vague queries directly. The basic idea behind the creation of Data Warehouses is to integrate a large amount of pre-fetched data and information from the distributed sources for direct querying and analysis .But the end user's queries contain the maximum fuzziness and to handle those queries directly may not yield the desired response. So the model proposed here will create a fuzzy extension of Data warehouse by applying Neuro-Fuzzy technique and the fuzzy queries then will get handled directly by the extension of data warehouse.

Keywords: Data Warehouse Architecture, Data Mining, Fuzzy Query, ANN, Fuzzy Logic, α-cut operation.

1 Introduction

The growth of computers has facilitated the way for quick generation and collection of data. Advancements in the DBMS technologies provide fast data access and information exchange all over the globe. The extended use of computers has resulted in an overwhelming amount of data being collected. The human genome database project has collected gigabytes of data on the human genetic code. The NASA Earth Observing

S. Ranka et al. (Eds.): IC3 2009, CCIS 40, pp. 240–249, 2009.
© Springer-Verlag Berlin Heidelberg 2009

System has generated 50 gigabytes of data on the earth per hour [1]. This has been made possible because of the major improvements in, database technologies including database management system, data warehousing which has resulted in much better facilities to store & search large database very quickly and easily. But in spite of all the improvements, Information Retrieval and consequently knowledge discovery from the vast collection of raw data is a bit challenging and require much involvement of human expertise [2].

Data warehouses are aimed to direct query handling and the architecture being used is therefore tuned for the same. But direct queries are much prone to the vagueness and may carry a lot of vague terms that may not be resolved by data warehouse. Though, we have already a number of techniques implemented and suggested so far to handle the fuzzy queries, still there is a lot of scope for the improvement. This paper suggests architecture for data warehouse which shows the significant improvement over the existing methods of fuzzy query handling. The architecture proposed in this paper is basically the extended version of the three tier architecture currently being used. Relational schemas of data warehouse having crisp valued attributes are extended and some new schemas are designed for fuzzy attributes. Fuzzy query server directly handles the fuzzy query with help of this fuzzy extension of data warehouse without any translation of the query from fuzzy to crisp or so. Underlying techniques used here are ANN to map a crisp record into a fuzzy set and Fuzzy logic to create the fuzzy sets and generating rules fuzzy query handling.

The paper is organized as follows: in the next section review of the related work done so far is done. Section 3 briefly describes the architecture and method suggested and implemented in this paper for fuzzy query handling. This section also explains how α-cut operation is applied on the fuzzy sets and on the union & intersection of them to classify the crisp records into the fuzzy classes. Section 4 elaborates the method of designing appropriate ANN topology, training of which using Back Propagation and using them for prediction. The conclusion section contains the final words about this paper and acknowledgement section includes thanks to the researchers and authors whose work paved for this paper.

2 Related Works

The increase in the number of companies seeking Data Warehousing solution in order to gain significant business advantages has created need for a decision aid approach in choosing appropriate Data Warehouse system [3]. Owing to the vague concepts frequently represented in decision environment, we have proposed a model to help in enhancing the query handling features of Data Warehouse.

Several earlier proposals exist to serve the purpose but most of them concentrate on how to mine the fuzzy information for web based applications where they use fuzzy logic to match the index of documents where the information probably may exist.

As far as vague query handling is concerned at the data warehouse level, approaches suggested so far are mostly based on the query translation, where vague queries are translated into crisp queries e.g. if a query “list the young sales man who has good selling record for house hold goods in the north of England”[4], is posted ; it will get translated into “list salesman under 25 years who have sold more than $20000 of goods in the category of household in the region north England". Problem with this approach is that it will not list the salesman of 26 who has made a real killing, or the one who sold $19000 etc. In [6] Shyi-Ming Chen and Woei- Tzy Jong presented a new method for fuzzy query translation based on the α-cuts operations of fuzzy numbers. For instance consider the following fuzzy query:

```
SELECT Ename
from Emp
where Age=very young OR old
```

The above query will get translated to the crisp query as mentioned below:

```
SELECT Ename
from Emp
where A>= U_w AND A≤ U_x OR
B>=U_y AND B≤U_x
```

Here A is degree of belongingness of an individual's age to fuzzy set *very young* and B is degree of belongingness of an individual's age to fuzzy set *old*. U_w ,U_x ,U_y are hard boundaries of sets.

In another approach a model was suggested where a fuzzy dimensions to the relational crisp databases were created to handle the fuzzy query [5] e.g. consider the database Student as shown in the Table 1 below:

Table 1.

Name	Age	Course	Percentages	Absence
Joe	22	MCA	83	13
David	17	B.Tech.	80	9
Tsewang	21	MBA	77	23

Fuzzy dimensions of student database will look like the Table 2 shown below. The crisp values are now transformed to vague terms and stored in the databases.

Table 2.

Name	Age	Course	Percentages	Absence
Joe	Old	MCA	Very good	Low
David	Middle	B.Tech.	Good	High
Tsewang	Old	MBA	Average	Very low

3 Proposed Architecture

The proposed architecture has close resemblance with the traditional one except the left half portion of the bottom, middle and top tier as shown below. In the bottom tier, Enterprise wide data, Data mart and Metadata is maintained for the same purpose as in traditional DWs. Apart from that, it also maintains fuzzy extension of data marts for the purpose of fuzzy query handling. In the middle tier along with OLAP server, there is a Fuzzy Query Server to take care of the fuzzy queries. In the top tier,

Query/Report interface of the traditional architecture has been segregated to Fuzzy and Crisp query.

In the following sub sections, first it describes how fuzzy sets are created then how α-cut operations are applied on the fuzzy sets. Lastly it elaborates, how fuzzy extensions are created corresponding to the every numeral valued attributes the database.

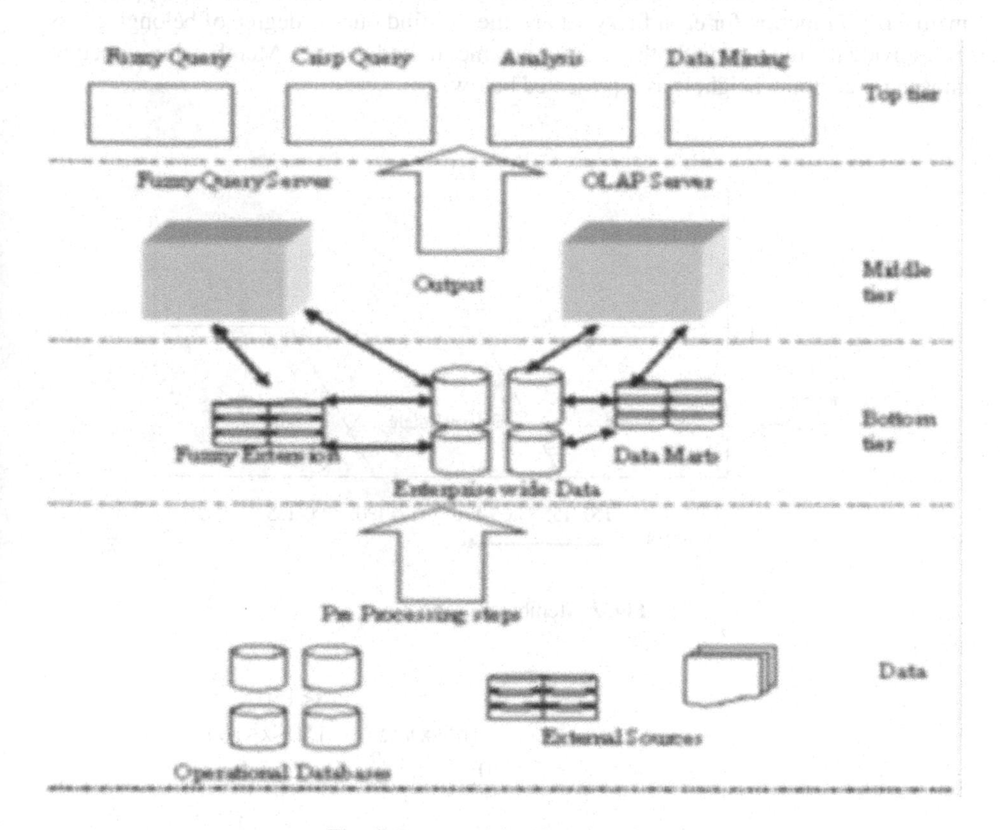

Fig. 1. Proposed Architecture of DW

3.1 Crisp Records to Fuzzy Sets

Consider the relation shown in the table 3. The attributes having numeral values are Height(Ht), Chest, Weight(wt) and Hemoglobin.

Table 3.

ID	Name	Ht (cm)	Chest (cm)	Wt (kg)	Teeth Caries	Hemog lobin
01	X1	170	75	63	Filled cavities	15.2
02	X2	175	72	55	Normal	14.2
03	X3	185	85	80	Normal	16.0
04	X4	152	80	85	cavities	13.0
05	X5	195	72	85	Normal	14.2
06	X6	160	85	60	Normal	16.0

Let us consider the height attribute first, corresponding to which fuzzy sets that can be formed are set of short height students(S1), set of medium height students(S2) and set of tall students(S3). Similarly there are three fuzzy sets corresponding to every attribute mentioned above. What is to be done here is to map the records(i.e. students) into these sets. In other words we can say that we have to choose appropriate membership function for each fuzzy set and then to find out the degree of belongingness of individual students with the fuzzy sets mentioned above. Membership function chosen for attribute height is as represented below:

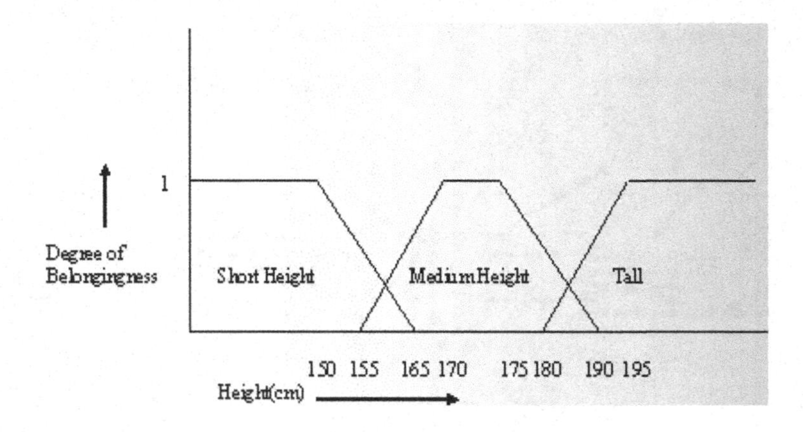

Fig. 2. Membership function

$$S1(x)=\begin{cases} 1 & x<150 \\ (165-x)/15 & 150 \le x \le 165 \\ 0 & x>165 \end{cases}$$

$$S2(x)=\begin{cases} 1 & 170<x<175 \\ (x-155)/15 & 155 \le x \le 170 \\ (190-x)/15 & 175 \le x \le 190 \\ 0 & x<155 \text{ OR } x>190 \end{cases} \qquad (1)$$

$$S3(x)=\begin{cases} 1 & x>195 \\ (165-x)/15 & 180 \le x \le 195 \\ 0 & x<180 \end{cases}$$

An appropriate topology of ANN is designed and trained by Back Propagation algorithm to find the membership of the individual record. Then the primary key of individual record and its membership is represented as shown in table 4.

Table 4.

ID	Ht	S1	S2	S3
01	170.00	0.00	1.00	0.00
02	175.00	0.00	1.00	0.00
03	185.00	0.00	0.33	0.33
04	152.00	0.87	0.00	0.00
05	195.00	0.00	0.00	1.00
06	160.00	0.33	0.33	0.00

3.2 Support and α-Cut Operations

The support of a fuzzy set A is the set of elements whose degree of membership in A is greater than 0. Let A be a fuzzy set in the universe of discourse U. We can formally define support as follows:[7]

$$Spt(A)=\{ x \in U \mid \mu_A(x)>0\} \tag{2}$$

For instance, the supports of the fuzzy sets in table 4 are

Spt(S1)=(152,160)
Spt(S2)=(160,170,175,185)
Spt(S3)=(185,195)

Tabular representation of support for the fuzzy sets of table 4 will look like as shown in table 5.

Table 5.

ID	Ht	S1	S2	S3
01	170.00	0.00	1.00	0.00
02	175.00	0.00	1.00	0.00
03	185.00	0.00	1.00	1.00
04	152.00	1.00	0.00	0.00
05	195.00	0.00	0.00	1.00
06	160.00	1.00	1.00	0.00

The α-cut is more general than that of support. Let α_0 be a number between 0 and 1. The α-cut of a fuzzy set A at α_0 , denoted as $A_{\alpha 0}$, is the set of elements whose degree of membership in A is no less than α_0. Mathematically, the α-cut of a fuzzy set A in U is defined as

$$A_{\alpha 0} = \{ x \in U \mid \mu_A(x)> \alpha_0\} \tag{3}$$

Taking α_0=0.8 and applying the α-cut on table 4 result in table 6 shown below.

Table 6.

ID	Ht	S1	S2	S3
01	170.00	0.00	1.00	0.00
02	175.00	0.00	1.00	0.00
03	185.00	0.00	0.00	0.00
04	152.00	1.00	0.00	0.00
05	195.00	0.00	0.00	1.00
06	160.00	0.00	0.00	0.00

3.3 Fuzzy Extension of Relational Views

Fuzzy extension can now be made on the basis of support and α-cut operations as discussed above. Fuzzy extensions based on these two operations are shown in the table 7 and table 8 respectively. Table 7 is having more than one value in a few of the cells that violets the condition of first normal form (1NF). And table 8 does not have any fuzzy value for height attribute corresponding to certain records. Therefore none of the approaches are serving the purpose hundred percent.

Though table 8 is normalized one and can be used for the fuzzy query handling. But what about the students who have not been mapped to any of the fuzzy sets? To rectify this problem if-then fuzzy rules are derived by considering other attributes also and then with the help of these rules, final shape to the fuzzy extension is given.

Table 7.

ID	Name	Height
01	X1	Medium Height
02	X2	Medium Height
03	X3	Medium Height, Tall
04	X4	Short Height
05	X5	Tall
06	X6	Short Height, Medium Height

Table 8.

ID	Name	Height
01	X1	Medium Height
02	X2	Medium Height
03	X3	None
04	X4	Short Height
05	X5	Tall
06	X6	None

For instance if the degree of membership of a student to all three sets(i.e. S1,S2 & S3) is less than α_0 and the degrees of membership of the student to the sets light weight and slim chest are greater than α_0 then the student will be classified as tall otherwise as medium height.

After applying the fuzzy rules missing values of table 8 is now filled and table 9 gives the final fuzzy extension corresponding to the attribute height. Fuzzy rules are initially derived fro the known set of data and then are used to train the neurons for the further prediction. Fuzzy rules used here are of the form as mentioned below.

If Chest is medium and Weight is heavy and height may be Medium Height, may be Tall then Height is medium.

If Chest is medium and Weight is medium height may be Short height, may be Medium Height, then Height is medium.

Table 9. Fuzzy height

ID	Name	Height
01	X1	Medium Height
02	X2	Medium Height
03	X3	Medium Height
04	X4	Short Height
05	X5	Tall
06	X6	Medium Height

4 Designing of ANN Topologies, Training and Prediction

Artificial neural networks are analytical systems that address problems whose solutions have not explicitly formulated. They contrast to classical computer problems which are designed to solve problems whose solutions have been made explicit. Concel System's BrainCom, a neural network simulator is used here to simulate ANNs as classifiers. Various topologies for appropriate ANN have been designed and tested and the final ones are listed in the table 10. ANN used for the classification of the weight into light weight, medium weight and heavy weight is a two layer feed forward Neural Network consisting of 7 neurons in the first layer and 8 neurons in the second layer. Diagrammatic representation of one of the three ANNs (i.e. ANN used as weight classifier) is shown in figure 3.

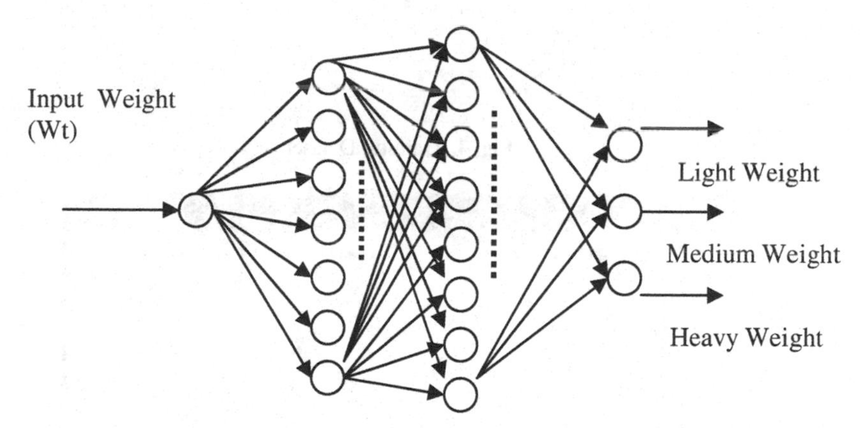

Fig. 3. ANN topology for Weight classifier

The status of the neural networks designed for height, weight and chest is shown in the table 10 below.

Table 10.

Network	Input	Output	Neurons in 1st Layer	Neurons in 2nd Layer	Iteration	% correct
Ht	1	3	13	3	10000	67
Wt	1	3	7	8	5000	72
Ct	1	3	7	8	10000	70

Training parameters and activation function is same for all three networks. Momentum= 0.2, Rate = 0.3 and activation function used is Logistic.

Screen shots of training set, input and prediction for the height is shown below.

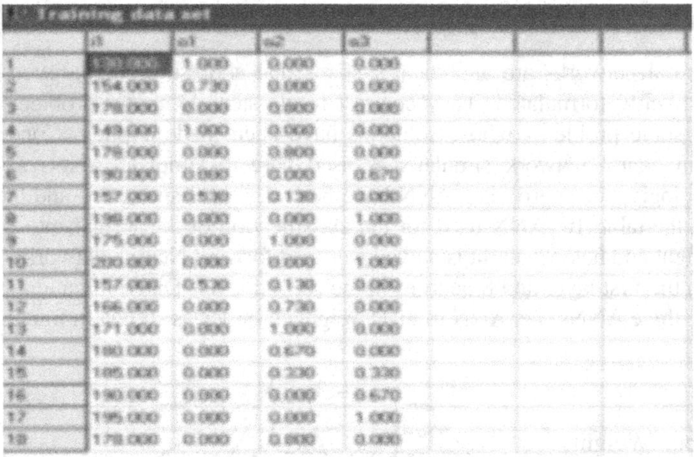

Training data set

	i1	o1	o2	o3
1	130.000	1.000	0.000	0.000
2	154.000	0.730	0.000	0.000
3	178.000	0.000	0.800	0.000
4	149.000	1.000	0.000	0.000
5	179.000	0.000	0.800	0.000
6	190.000	0.000	0.000	0.670
7	157.000	0.530	0.130	0.000
8	198.000	0.000	0.000	1.000
9	175.000	0.000	1.000	0.000
10	200.000	0.000	0.000	1.000
11	157.000	0.530	0.130	0.000
12	166.000	0.000	0.730	0.000
13	171.000	0.000	1.000	0.000
14	180.000	0.000	0.670	0.000
15	185.000	0.000	0.330	0.330
16	190.000	0.000	0.000	0.670
17	195.000	0.000	0.000	1.000
18	178.000	0.000	0.800	0.000

Fig. 4. Training Dataset

Input data set

	i1
1	130.000
2	200.000
3	185.000
4	176.000
5	160.000
6	179.000

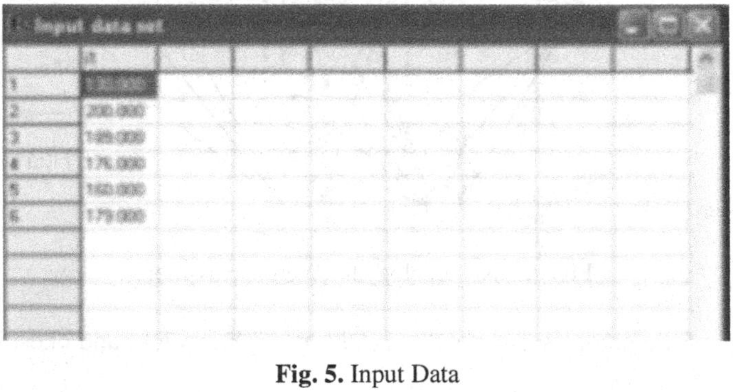

Fig. 5. Input Data

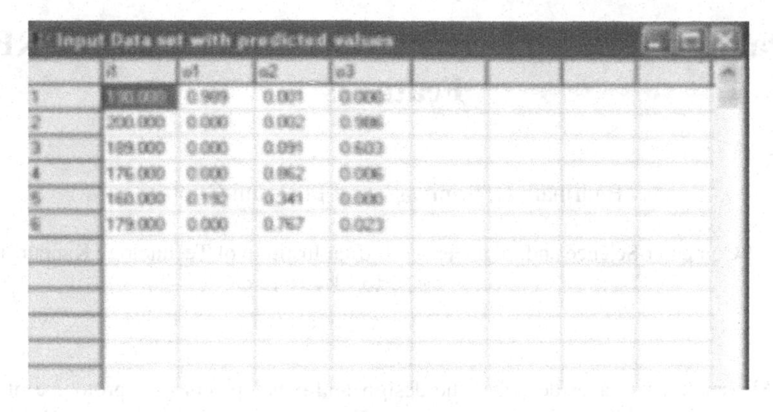

	i1	o1	o2	o3
1	[illegible]	0.989	0.001	0.000
2	200.000	0.000	0.002	0.986
3	189.000	0.000	0.091	0.603
4	176.000	0.000	0.862	0.006
5	168.000	0.192	0.341	0.000
6	179.000	0.000	0.767	0.023

Fig. 6. Prediction by ANN

5 Conclusion

The method presented for fuzzy query handling emphasizes on friendliness and flexibility for the inexperienced users. It is highly compatible with SQL since fuzzy extensions created here are in the form of relations and SQL queries can directly be used to retrieve the fuzzy information. Complete GUI based application is under development to demonstrate the fuzzy query handling using this method.

Acknowledgement

The authors would like to thank to Concel System for its neural network simulator, "The BrainCom" which reduces the effort of simulation to a large extent. We would also like to thanks to all the researchers and authors whose works helped us a lot in our research.

References

1. Kim, M.W., Lee, J.G., Min, C.: Efficient Fuzzy rule generation based on fuzzy decision tree for data mining. In: IEEE International Fuzzy System conference proceedings Seoul, Korea (1999)
2. Fayyad, U.M., Piatetsky-Shapiro, G., Smyth, P.: Data mining to Knowledge Discovery: An overview. In: Advances in Knowledge Discovery and Data Mining, pp. 1–34. MIT Press, Cambridge (1996)
3. Krishna, P.R., De Kumar, S.: A fuzzy approach to build an intelligent data warehouse. Journal of Intelligent and fuzzy system (2001)
4. Gains, B.R.: Logical foundation for database System. Intern. J. of Machine Studies (1979b)
5. Chiang, D., Chow, L.R., Hsien, N.: Fuzzy information in extended fuzzy relational databases. Fuzzy sets and Systems 92 (1997)
6. Chen, S.-M., Jong, W.T.: Fuzzy query translation for Relational Database Systems. IEEE Transactions on Systems, Man, and cybernetics-part B: Cybernetics 27(4) (1997)
7. Yen, J., Langari, R.: Fuzzy Logic, Intelligence, control and Information, pp. 57–84. Pearson Education, London (2003)

Palmprint Based Verification System Using SURF Features

Badrinath G. Srinivas* and Phalguni Gupta

Dept. of Computer Science and Engineering, Indian Institute of Technology Kanpur, India
{badri,pg}@iitk.ac.in

Abstract. This paper describes the design and development of a prototype of robust biometric system for verification. The system uses features extracted using Speeded Up Robust Features (SURF) operator of human hand. The hand image for features is acquired using a low cost scanner. The palmprint region extracted is robust to hand translation and rotation on the scanner. The system is tested on IITK database of 200 images and PolyU database of 7751 images. The system is found to be robust with respect to translation and rotation. It has FAR 0.02%, FRR 0.01% and accuracy of 99.98% and can be a suitable system for civilian applications and high-security environments.

Keywords: Robust, Rotation, Translation, Scanner, Online.

1 Introduction

Biometrics establishes identity of a person by physiological and/or behavioral characteristics. A system based on biometrics has found its applications widely in commercial and law enforcement applications. Fingerprint provides the most widely used biometric features while iris generates the most reliable biometric features [22]. Using human hand as biometric feature is relatively a new approach. A palmprint region between fingers and wrist contains useful features like principle lines, wrinkles, ridges, minutiae points, singular points, and texture pattern for its representation [20].

Palmprint based biometric systems has many advantages. The features of the human hand are relatively stable and unique. At the same time expected co-operation of users for data acquisition is low. Also devices used to acquire images are economic. It is a relatively simple method that can use low resolution images and provides high efficiency. Furthermore, verification/identification systems based on hand features are the most acceptable to users [8].

Nevertheless, limited work has been reported on palmprint identification and verification, despite the importance of palmprint features. Some of the recent research efforts in [1-3], [11] address the problem of palmprint recognition in large databases, and achieve very low error rates. Palmprint features can be extracted using various transforms such as Fourier Transform [10], Discrete Cosine Transform [3], Karhunen-Loeve transform [1], [11], Wavelet transform [2], [11] Fisher Discriminant Analysis [16],

* Corresponding author.

S. Ranka et al. (Eds.): IC3 2009, CCIS 40, pp. 250–262, 2009.
© Springer-Verlag Berlin Heidelberg 2009

Gabor filtering [2] and Independent component Analysis [15]. In [4], identification system based statistical signatures which include spatial dispersivity, center of gravity, density, and texture energy is presented. In [6], verification system based on the datum points of palmprint is presented. Furthermore, in [12] multimodal system which fuse features from from face, hand-geometry and fingerprint at the matching score level is described.

In [6], [7], systems using ink marking to capture the palmprint patterns have been presented. The systems are not widely accepted due to considerable attention and high cooperation is required in providing a biometric sample. In recent papers on palmprint based recognition system, palmprint images are captured from a digital camera [2]. Users hand is placed in constrained environment using pegs for palmprint image acquisition.

This paper proposes a novel method for palmprint matching using Speeded Up Robust Features (SURF) [14] as a means of key point extraction and number of matching points carried out using a nearest neighbour ratio method. SURF extracts the feature, which describes the local features of the image. SURF feature generation process is described in the following section. To increase the speed of matching process a novel sub-image matching strategy is proposed. System parameters are optimized to give lowest equal error rates (EER) on two data sets. Palmprint verification is done on images obtained using low cost flat bed scanner, and images are accepted in constraints free environment without pegs. The system is designed to be robust to translation and rotation on the scanner, and highly accurate, at reasonable price so that it is suitable for civilian applications and high end security applications.

The rest of the paper is organized as follows: Section 2 explains the method of feature extraction. In the next section the proposed system, and palmprint extraction from acquired hand image is described. Section 4 describes the proposed sub-image and original palmprint matching using nearest neighbour ratio technique and verification based on threshold. The experimental results for verification are reported in Section 5. Finally, the conclusions are presented in Section 6.

2 Speeded Up Robust Features

The SURF [14] is recently emerged cutting edge methodology for pattern recognition, and has been used in general object recognition and for other machine vision applications [17], [18]. Features are found to be very much distinctive and stable, key-points are efficiently detected and local pattern around can be used as descriptor.

The extracted features of the image are invariant to scaling, rotation, and translation of the image. Thus extracted feature from an image can be matched correctly with high probability against features from a large image database. The following are the major stages for computing the descriptors of the image.

2.1 Key-Point Detectors

The first stage of computation is to find the key-points. The SURF key-point detector is based on hessian matrix. Given a point $P = (x, y)$ in an image I, the hessian matrix $H = (x, \sigma)$ in P at scale σ is defined as follows

$$H(P,\sigma) = \begin{bmatrix} L_{xx}(P,\sigma) & L_{xy}(P,\sigma) \\ L_{xy}(P,\sigma) & L_{yy}(P,\sigma) \end{bmatrix} \quad (1)$$

where $L_{xx}(P,\sigma)$ is the convolution of the Gaussian second order derivative $\frac{\partial^2}{\partial x^2} g(\sigma)$ with the image I at the point P, and similarly for $L_{xy}(P,\sigma)$ and $L_{yy}(P,\sigma)$. The second order Gaussian derivatives are approximated using box filters shown in Fig. 1. Image convolutions with box filters are computed rapidly using integral images [19]. The integral image $I_{\Sigma}(P)$ at location $P = (x, y)$ represents the sum of all pixels in the input image I of a rectangular region formed by the origin and P.

$$I_{\Sigma}(P) = \sum_{i=0}^{i \le x} \sum_{j=0}^{j \le y} I(i,j) \quad (2)$$

Key-points are localized in scale and image space by applying a non maximum suppression in a $3 \times 3 \times 3$ neighbourhood. The local maxima can be approximated by Hessian matrix determinant are interpolated in scale and image space.

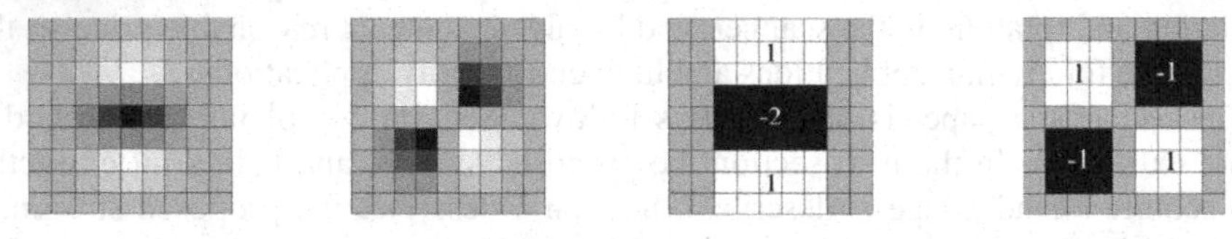

Fig. 1. Left to right: the (discretised and cropped) Gaussian second order partial derivatives in y-direction and xy-direction, and the approximations thereof using box filters. The gray regions are equal to zero. (from [14]).

2.2 Key-Point Descriptor

For the better performance of the system, the key-points extracted are described so that it is distinct compared to other key-points. SURF describes the key-point based on the dominant orientation of the local region around key-point. Following are sub-stages of computing the descriptor of key-point.

2.2.1 Orientation Assignment

In this sub-stage, a circular region is constructed around the extracted key-points. Dominant orientation is computed based on the information from circular region around key-point. The orientation is computed using the Haar wavelet responses in both x and y directions. The dominant orientation is estimated by summing the wavelet response within a rotating wedge covering an angle of 60^0 in wavelet response space. The resulting maximum is considered as dominant orientation and used to describe the key-point. Since the dominant orientation is rotation invariant, the descriptor becomes invariant to image rotation.

2.2.2 Descriptor

In this sub-stage a square region is constructed around the key-point and aligned along the dominant orientation. The descriptor is extracted from the aligned squared region. The Square region is partitioned into smaller sub-regions of size 4 × 4. Haar wavelet responses in vertical and horizontal directions are computed for each sub-region. The sum of the wavelet response d_x and d_y for each sub-region are used as feature values. Furthermore, the absolute values $|d_x|$ and $|dy|$ are summed to obtain the polarity of the image intensity changes. The feature vector of the sub-image is given by

$$V = \left\{ \sum d_x, \sum d_x, \sum |d_x|, \sum |d_y| \right\} \quad (4)$$

Thus summing up the descriptor vectors from all 4x4 sub-regions, descriptor vector of length 64 is obtained. The descriptor vector of length 64 for the key-point is called SURF descriptor.

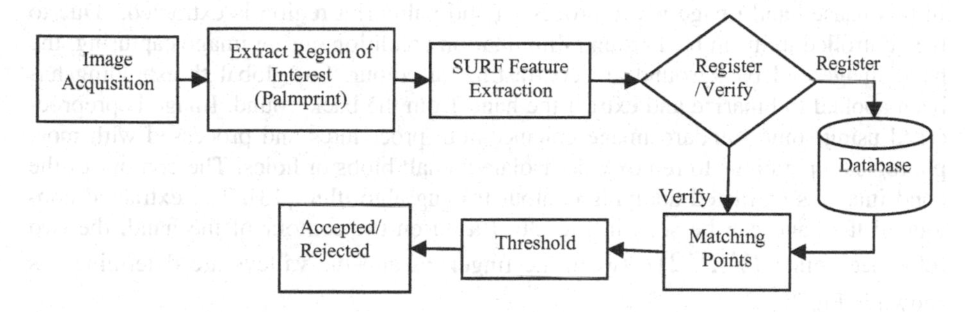

Fig. 2. Block diagram of the steps involved in building an palmprint based biometric system for verification using SURF features

3 Proposed System

The block diagram of the proposed robust biometric system for verification using SURF features of palmprint is shown in Fig. 2. The process starts with acquiring the input image of hand using low cost scanner. In the region of interest module image is pre-processed using some standard image enhancement procedures and normalised. Then the region of interest is extracted. In the following module, features are extracted using SURF. In the subsequent modules feature vectors of live image and enrolled image in database are matched. Matching decision is done based on threshold.

3.1 Image Acquisition

Hand images are obtained using a flatbed scanner at a spatial resolution of 200 dots per inch in gray scale. To avoid non-uniform background illumination/reflection the shutter of the scanner is not closed. The device is constraints (pegs) free, so placing hand is rotation independent relative to line of symmetry of the working surface of scanner. Typical gray level image obtained from the scanner is shown in Fig. 3a.

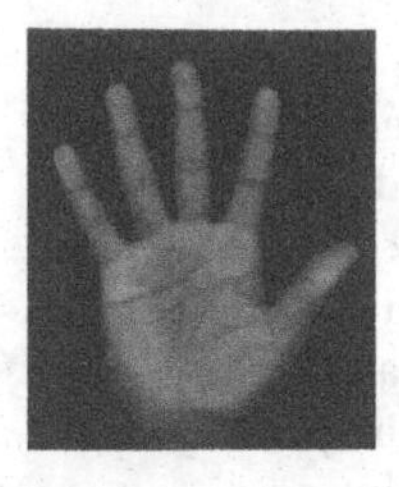
(a) Scanned Image

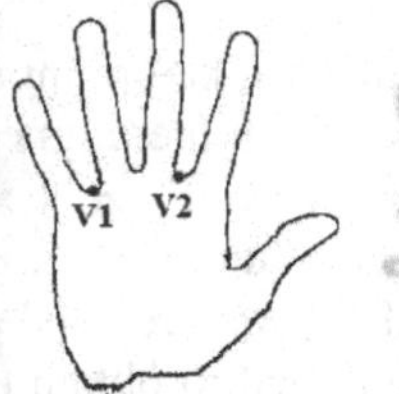

(b) Hand Contour and reference point

(c) Relevant points and palmprint

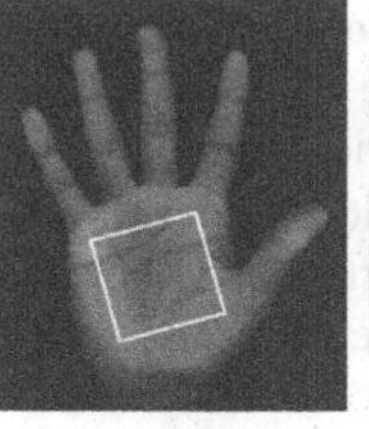
(d) Palmprint in Gray scale image

(e) Extracted palmprint

Fig. 3.

3.2 Pre-processing and Region of Interest Extraction

In this phase hand image is pre-processed and palmprint region is extracted. Due to the controlled uniform background illumination condition during image capturing, the hand image and background are contrasting in colour. So, global thresholding has been applied to binarize and extract the hand from the background. Image is preprocessed using some standard image enhancement procedures and processed with morphological operations to remove any isolated small blobs or holes. The contour of the hand image is extracted applying contour-tracing algorithm [13]. The extracted contour of the hand can be seen in Fig. 3b. Based on the contour of the hand, the two reference points $(V1, V2)$ between the fingertips and the valleys are determined as shown in Fig 3b.

The two reference points are selected:

1. Reference point *V1* is the valley between the little finger and the ring finger, and
2. Reference point *V2* is the valley between the index finger and the middle finger.

Region of interest or palmprint is the square area as shown in Fig. 3c, with two of its corners placed on the middle points of the line segments *C1-V1* and *V2-C2*. The line segments *C1-V1* and *V2-C2* are inclined at an angle of 45^0 and 60^0 respectively to the line joining *V1* and *V2*. The region of interest in gray scale image is shown in Fig 3d.

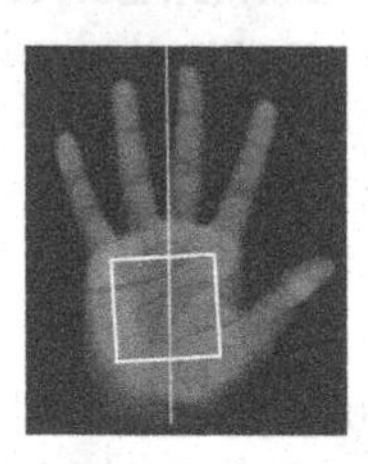
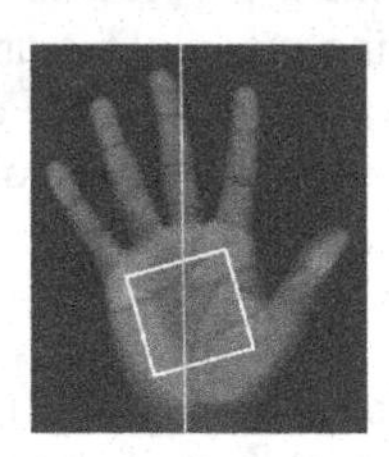
(a) Images of same subject with different orientation of placement relative to the symmetry (Yellow Line) of work surface

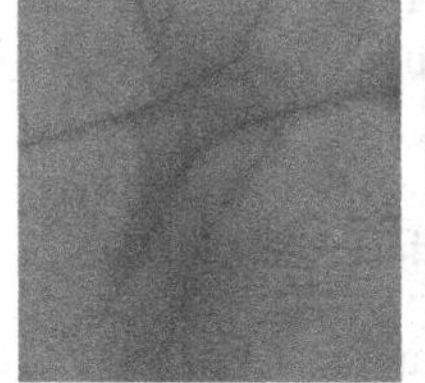
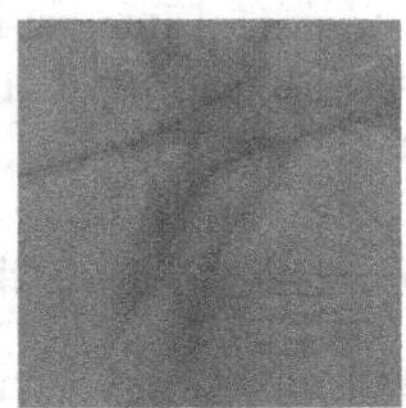
(b) Extracted region of interest for images shown in Fig. 4a

Fig. 4.

The extracted palmprint of the original gray scale image is shown in Fig. 3e. Since the placement of palm on the scanner is pegs free, orientation of placement would vary for every incident. Two images of same subject with different orientations of placement are shown in Fig. 4a. The extracted region of interest is relative to the valley points $V1$ and $V2$ which are stable for the subject. So the extracted palmprint region from different incidents with different orientation of placement for the same subject remains the same as shown in Fig. 4b (corresponding hand image is shown in Fig. 4a). Hence the proposed extraction procedure of the system for extracting region of interest makes the system robust to rotation. The empirical results show that the system is robust to rotation for about ±35 degrees. The extracted palmprint varies in size from subject to subject, because the area of the palm is different for each subject. SURF approach is used to extract features and it is scale invariant, so images need not to be normalized to exactly the same size.

3.3 Palmprint Image Feature Extraction

After hand image pre-processing, and region of interest or palmprint extraction, feature should be extracted for later recognition. SURF is used to extract a feature vector which provides good discrimination ability and has been used for palmprint based verification. The detected SURF key-points for the palmprint image is shown in Fig. 5.

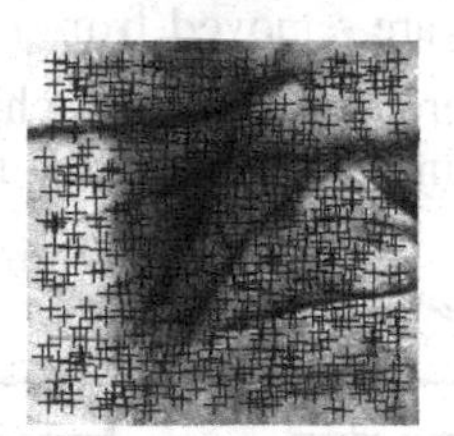

Fig. 5. Detected SURF key-points of the Palmprint

4 Matching

To verify the live palmprint, the SURF features computed for the enrolled image should be matched with SURF features of the live palmprint image. In this section original SURF's staregry and the proposed matching strategy are presented. The proposed matching statergy is different from the original SURF's method and is simpler in a sense less number of matching.

4.1 Matching Method in SURF

This sub-section discusses the method being used to match the live palmprint against those in the database using SURF operator. For any palmprint, features are initially found through SURF operator. To verify a palmprint, the SURF descriptor of each key-point computed in the live image is matched to descriptor of every SURF key-point of the enrolled image in the database.

Let L and E be vector arrays of key-points of live and enrolled images respectively obtained through SURF as:

$$L = \{l_1, l_2, l_3, \ldots l_m\} \quad (5)$$

$$E = \{e_1, e_2, e_3, \ldots e_n\} \quad (6)$$

The descriptor array l_i of key-point i in L and descriptor array e_j of key-point j in E are assumed to be matched if the Euclidean distance $\|l_i - e_j\|$ between them is less than a specified threshold T. Threshold based matching results in several number of matching points. To avoid multiple matches, the key-points pair with minimum descriptor distance and less than threshold is considered. This results in a single matching pair, and is called as nearest neighbor method.

In SURF, the matching method applied is similar to the nearest neighbor matching, except that the thresholding is applied to the distance ratio between the first and the second nearest neighbor. The method used in SURF is called as nearest neighbor ratio method. Thus, the key-points are matched if

$$\|l_i - e_j\| / \|l_i - e_k\| < T \quad (7)$$

where, e_j is the first nearest neighbor and e_k is the second nearest neighbor of l_i.

The matching points (l_i, e_j) are removed from L and E respectively. The matching process is continued until there is no more matching points. Based on the number of matching points between live image L and enrolled image E, a decision is taken.

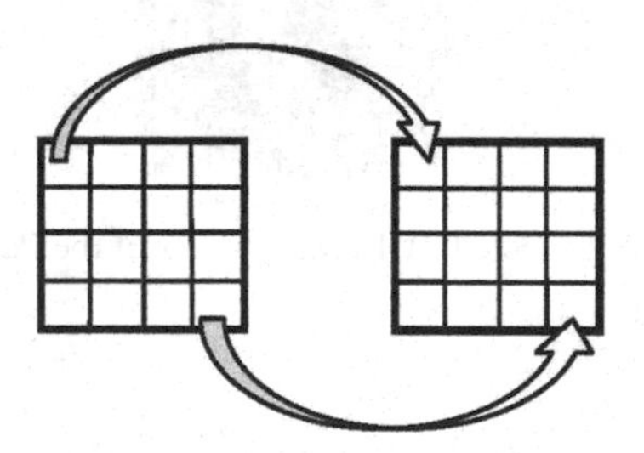

Fig. 6. Block diagram for sub-image based matching

The number of matching points between the palmprint images L and E gives similarity between them. The idea behind matching is that the greater the number of matching points between two images, the greater is the similarity between them.

4.2 Sub-image Matching

This sub-section proposes an efficient matching method which improves the speed of the matching process. In the original proposed paper, the SURF descriptors of each key-point are compared with every key-point of enrolled image. In this proposed method the intuitive fact that the key-points of the corresponding sub-images/locations only match are used to increase the speed of matching process. Block diagram for the sub-image matching is shown in Fig 6. If the palmprint images are registered, location based matching can be performed. A small overlapping region with adjacent sub-images is

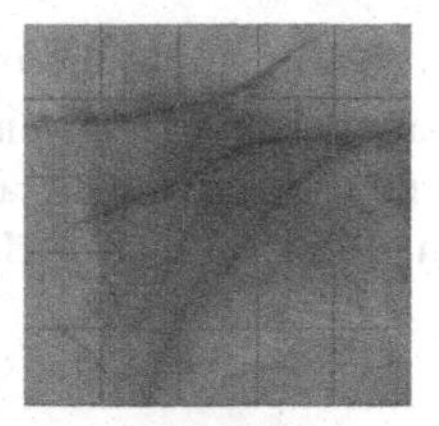

(a) Palmprint image partitioned into sub-images

(b) Overlapping region with adjacent sub-images

Fig. 7.

considered to mitigate the impact of the image not being registered. In this paper only matching methodology is analyzed for extracted image.

The stable valley points between the fingers are used to extract the palmprint. Hence the palmprint is invariant to rotation and sub-image/location dependent matching could be performed. The extracted palmprint is partitioned into sub-images as shown in Fig 7a. The descriptors of corresponding sub-images are compared. To overcome the problem of registration the overlapping between adjacent sub-images are considered. In Fig 7b the area between dashed lines (yellow) in the partitioned image is the overlapping region between the adjacent sub-images.

The SURF descriptors of the key-points are matched from the corresponding sub-images. The number of matching points between the corresponding sub-images is determined.

Formally, let the partitioned palmprint sub-images of enrolled and live palmprint images are given by E_1, $E2$, $E3$,..., E_m and L_1, L_2, L_3,..., L_m respectively. The total matching points is obtained by summing up matching points of corresponding sub-images

$$NoMPs = \sum_{i=1}^{n} MP(E_i, L_i) \tag{8}$$

where $MP(E_i, L_i)$ gives the number of matching points between the ith sub-images in E and L respectively. Empirically it is observed that the sub-image of size 1/5 the width and 1/5 the height gives good performance. The overlapping range with the adjacent sub-image is set to 10%.

5 Experimental Results

The proposed system is tested on two sets of image databases. They are the Indian Institute of Technology Kanpur (IITK) database, and The Hong Kong Polytechnic University (PolyU) [9] database. The execution speed of the feature extraction and matching for SURF is studied.

5.1 IITK Database

IITK has collected a database comprised of 200 images from 100 subjects. Each hand image has been collected at a spatial resolution of 200 dots per inch using a low cost flat bed scanner, and 256 gray levels. The user is independent to rotate his hand around $\pm 35°$ symmetric to working surface of scanner. To illustrate the matching of extracted palmprint using SURF operator for IITK database, an example is shown in Fig 8. Matching between palmprint images of different users is called as imposter matching. An example of imposter matching between users IITK_87_01 and IITK _21 _ 01 in IITK database is shown in Fig 9.

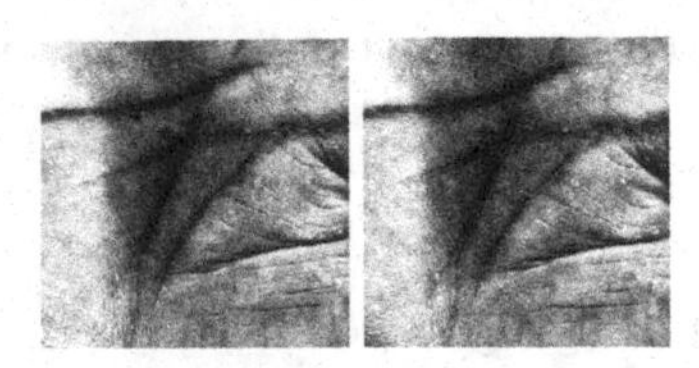

(a) Extracted palmprint from IITK Database (87_00 and 87_01) as input pair

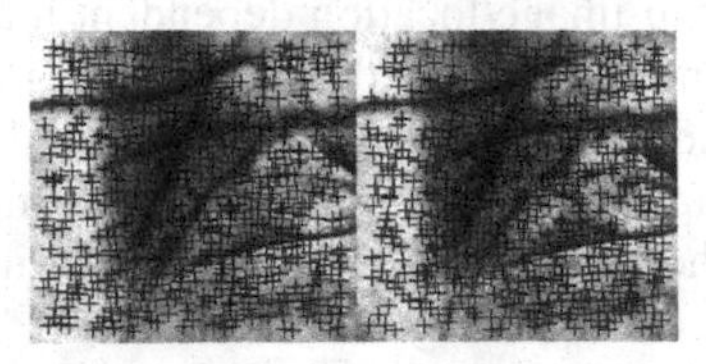

(b) 627 and 607 Key-points detected for the input pair

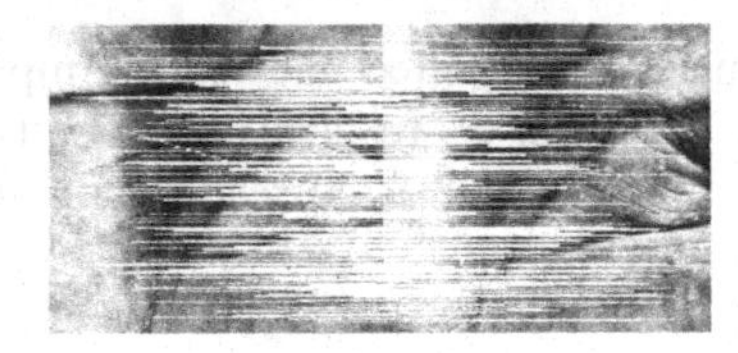

(c) 210 SURF Matching Key-points obtained for the input pair

Fig. 8. Illustration of extracted palmprint matching for IITK database using SURF operator

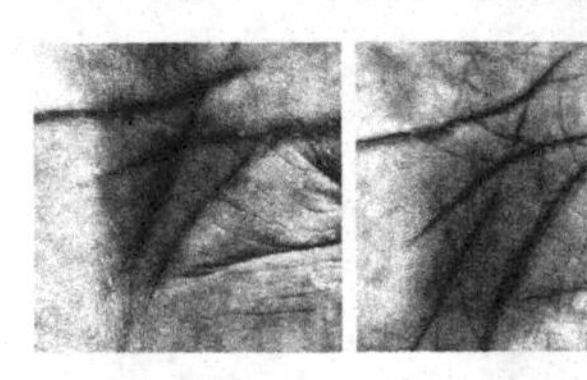

(a) Extracted palmprint from IITK Database (87_00, and 21_01) as input pair

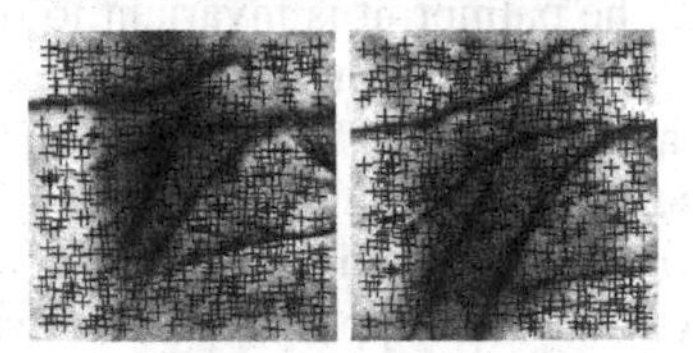

(b) 627 and 656 Key-points detected for the input pair

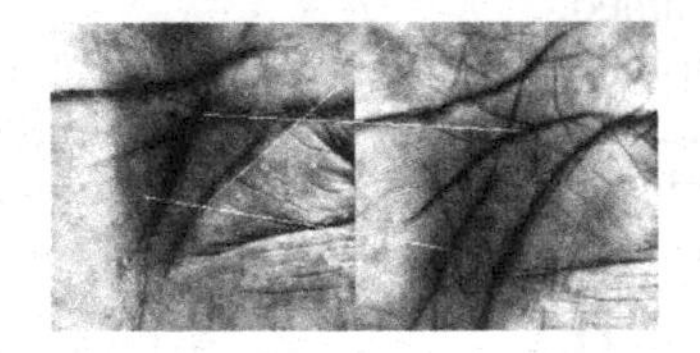

(c) 3 SURF Matching key-points obtained for the input pair input pair

Fig. 9. Illustration of Imposter matching for IITK database

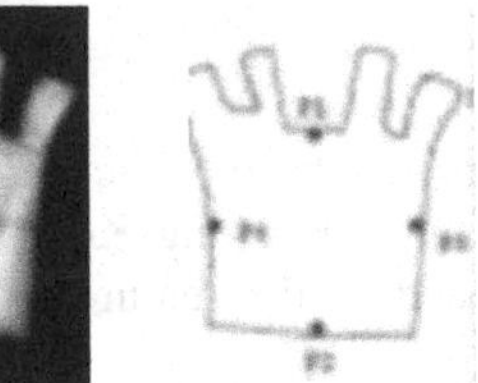

(a) PolyU sample image

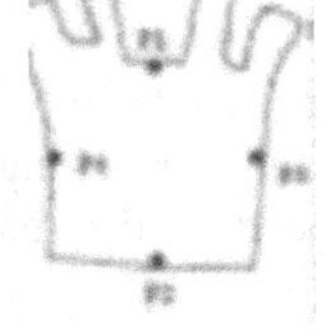

(b) Reference points

(c) Region of interest in gray scale image

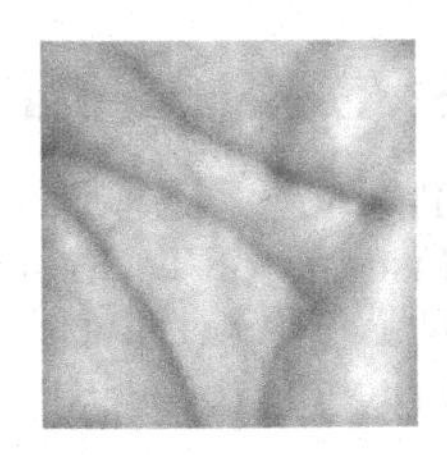

(d) Extracted palmprint

Fig. 10.

5.2 PolyU Database

The proposed system has also been tested on the database from PolyU [9] which consists of 7752 grayscale images corresponding to 386 different palms. Around 17 images per palm are collected in two sessions. The images are collected at spatial resolution of 75 dots per inch, and 256 gray levels using CCD [2]. Images are captured placing pegs. Fig. 10a shows the sample of database.

In order to extract region of interest for PolyU database following method is proposed. Four reference points $P1$, $P2$, $P3$ and $P4$ are located on the contour of palm as shown in Fig. 10b. In gray-scale image extract 200 x 200 pixels palm area with its center coinciding with intersecting point of line segments $P1$–$P2$ and $P3$ – $P4$. Fig. 10c shows the region of interest in gray scale image, and the extracted palmprint image is shown in Fig. 10d.

5.3 Speed

The SURF features have been computed with Bay's source code [5]. The source code is implemented using C++. The system configuration used to run the process of feature extraction and matching is an embedded dual core AMD athlon processor (2 x 1000MHz) PC with 2 GB of RAM.

The total verification time (*VT*) per person consists of a constant time for acquire hand-image, pre-processing, and extracting palmprint and a variable time for feature extraction and matching process. Thus

$$VT = Constant\ Time + E_i \times (Feature\ Extraction + One\text{-}to\text{-}One\ matching\ time)$$

where E_i is the number of enrolled images of user i. For example, if the user is enrolled with 3 images, time needed becomes

$$VT = Constant\ Time + 3 \times (AvgExt_{Time} + Avg\ Mch_{Time})$$

The average time taken to extract SURF features is computed using

$$AvgExt_{Time} = \frac{\sum_{i=1}^{k} T(I_i)}{k} \tag{9}$$

where $T(I_i)$ is the time taken to extract features from the image i in the database, and k is the total number of images in the database. The average matching time between images i and j is computed using

$$AvgMch_{Time} = \frac{\sum_{i=1}^{k} \sum_{j=i+1}^{k} M(I_i, I_j)}{k * (k-1)/2} \tag{10}$$

where $M(I_i, I_j)$ is the time taken to match the SURF features of image i and j, and k is the total number of images in the database. For average matching time, all possible matches are considered, that includes both genuine and imposter cases.

The experiment is conducted as follows: Every image of the user i from the database is matched for false acceptance with every image of remaining 99 other users, results are noted. Process of false acceptance is conducted for all the hundred users and the results are noted. For false rejections every image of user *i* from the database is matched with all other images of the same user *i* in the database, results are noted. This process of false rejection is conducted on all hundred users and results are noted.

Table 1. Accuracy, FAR and FRR of the proposed system and [11]

	IITK database			PolyU Database		
	FAR	FRR	Accuracy	FAR	FRR	Accuracy
Eigenpalm [11]	9.65	9.43	90.45	6.44	8.38	92.58
Haarpalm [11]	5.00	13.8	90.57	4.56	12.8	91.28
Fusion [11]	4.73	5.66	94.80	4.17	8.10	93.85
Proposed	**0.02**	**0.01**	**99.98**	**2.02**	**3.14**	**94.91**

Table 2. Time taken by SURF for extraction and Matching

	Database	
	IITK (ms)	PolyU (ms)
Extraction	523.142857	501.663366
Matching	32.202116	42.1552
Total	555.3450	543.8186

The conducted experiments are more appealing than rank-one recognition [21] method because in rank-one, only inter-class (between testing-set and training-set) comparisons are done for false acceptance and false rejections. The intra-class (within training-set/testing-set) comparisons are not performed for false acceptance and false rejections.

The experiment has been performed for both the datasets using SURF. The Receiver Operating Characteristic (ROC) curves of the proposed system for IITK database and PolyU database is shown in Fig. 11a. The probability distributions for genuine and imposter are estimated by the correct and incorrect matching's respectively. The Genuine and imposter distributions for IITK database is shown in Fig. 11b. Table 1 shows how the accuracy, FAR and FRR of the proposed and previous [11] systems for both IITK and PolyU database. Table 2 gives the execution time for both IITK and PolyU database.

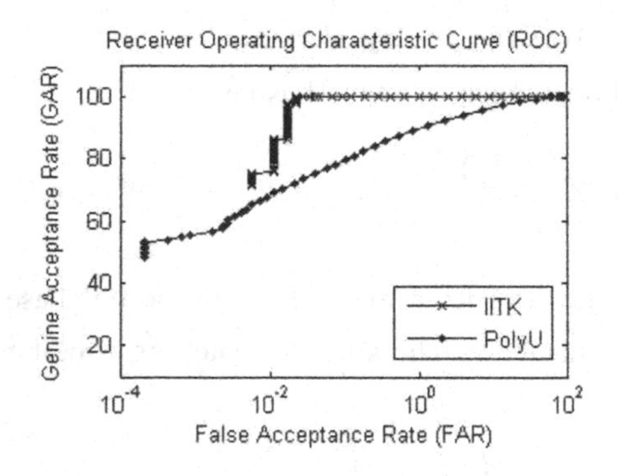

(a) ROC curve for IITK and PolyU database

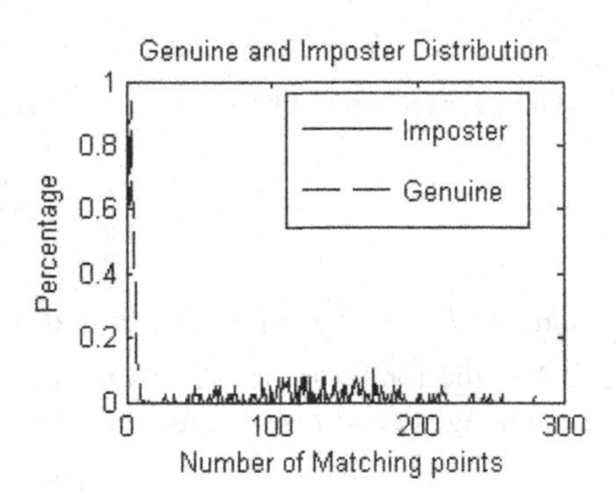

(b) Genuine and Imposter Distribution for IITK database

Fig. 11.

From Table 1, it can be said that the proposed system performs better than the earlier known system which fuses the matching scores of the haarwavelet and eigenpalm classifiers. The total time needed by the system for both feature extraction and matching suggests that the system can be used for online verification.

6 Conclusions

In this paper the use of SURF features in the context of palmprint verification has been investigated. A technique to extract palmprint from hand-image image which is robust to rotation, translation on the scanner surface has been presented. The extracted palmprint is found to be invariant to orientation and translation of palm on scanner, which makes the system robust to rotation and translation. Sub-image based matching technique have been proposed for increased the speed of verification. The proposed system is tested on IITK and PolyU database. The proposed key-point descriptors extracted using SURF outperforms the earlier known system which fuses the matching scores of the haarwavelet and eigenpalm classifiers [11]. The system speed for extraction and verification appeals for the online security environment. The recognition accuracy of the system is 99.98%, along with FAR 0.021% and FRR 0.01%. Thus the design of the system with robustness, performance, speed and use of low cost scanner for acquisition of palm image have demonstrated the possibility of using this system for online applications, high-end security applications, and civilian applications.

References

1. Ribaric, S., Fratric, I.: A biometric identification system based on Eigenpalm and Eigenfinger features. IEEE Trans. on Pattern Analysis Machine Intelligence 27(11), 1698–1709 (2005)
2. Zhang, D., Kong, W.-K., You, J., Wong, M.: Online palmprint identification. IEEE Transaction on Pattern Analysis and Machine Intelligence 25(9), 1041–1050 (2003)
3. Kumar, A., Zhang, A.: Personal recognition using hand shape and texture. IEEE Transaction on Image Processing 15(8), 2454–2461 (2006)
4. Zhang, L., Zhang, D.: Characterization of Palmprints by Wavelet Signatures via Directional Context Modeling. IEEE Transaction on Systems, Man, and Cybernetics 34(3), 1335–11347 (2004)
5. The SURF source code, `http://www.vision.ee.ethz.ch/~surf/`
6. Zhang, D., Shu, W.: Two novel characteristics in palmprint verification: Datum point invariance and line feature matching. Pattern Recognition 32(4), 691–702 (1999)
7. Han, C.-C., Cheng, H.-L., Lin, C.-L., Fan, K.-C.: Personal authentication using palmprint features. Pattern Recognition 36, 371–381 (2003)
8. International Committee for Information Technology Standards. Technical Committee M1-Biometrics (2005), `http://www.incits.org/tc_home/m1.htm`
9. The PolyU palmprint database, `http://www.comp.polyu.edu.hk/~biometrics`
10. Wenxin, L., Zhang, D., Xu, Z.: Palmprint Identification by Fourier Transform. Intl. Journal of Pattern Recognition and Artificial Intelligence 16(4), 417–432 (2002)
11. Badrinath, G.S., Gupta, P.: An Efficient Multi-algorithmic Fusion System based on Palmprint for Personnel Identification. In: Intl. Conf. on Advanced Computing, pp. 759–764 (2007)
12. Ross, A., Jain, A.K.: Information fusion in biometrics. In: Pattern recognition letters, pp. 2115–2125 (2003)
13. Pavlidis, T.: Algorithms for graphics and image processing. Springer, Heidelberg (1982)

14. Bay, H., Tuytelaars, T., Van Gool, L.: SURF: Speeded up robust features. In: Ninth European conference on computer vision, pp. 404–417 (2006)
15. Lu, G., Wang, K., Zhang, D.: Wavelet based independent component analysis for palmprint identification. In: Intl. conf. on Machine Learning and Cybernetics, pp. 3547–3550 (2004)
16. Wang, Y., Ruan, Q.: Kernel Fisher Discriminant Analysis for Palmprint Recognition. In: 18th Intl. Conf. on Pattern Recognition, pp. 457–460 (2006)
17. Bay, H., Fasel, B., Van, L.: Interactive museum guide: Fast and robust recognition of museum objects. In: First Intl. workshop on mobile vision (2006)
18. Murillo, A.C., Guerrero, J.J., Sagues, C.: SURF features for efficient robot localization with omnidirectional images. In: IEEE Intl. Conf. on Robotics and Automation, pp. 3901–3907 (2007)
19. Viola, P., Jones, M.: Rapid object detection using a boosted cascade of simple features. In: IEEE Conf. on Computer Vision and Pattern Recognition, pp. 511–518 (2001)
20. Shu, W., Zhang, D.: Automated Personal Identification by Palmprint. Optical Engineering 37(8), 2359–2362 (1998)
21. Liu, X., Bowyer, K.W., Flynn, P.J.: Experiments with an Improved Iris Segmentation Algorithm. In: Fourth IEEE Workshop on Automatic Identification Advanced Technologies, pp. 118–123 (2005)
22. Independent Testing of Iris Recognition Technology Final Report. Int'l Biometric Group (2005)

A Personnel Centric Knowledge Management System

Baisakhi Chakraborty and Meghbartma Gautam

National Institute of Technology, Durgapur
713209, West Bengal, India
baisakhichak@yahoo.co.in

Abstract. A Knowledge Management System (KMS) is designed to serve as an effective tool for the proper extraction, utilization and dissemination of knowledge. Traditional KMS models incur cost overhead on the extraction of tacit knowledge and conversion to explicit knowledge. The proposed model in this paper takes the concept of mining the tacit knowledge and using it in the KMS instead of following conventional KMS norms. Through interactions and socialization of the personnel participating in the system, the tacit knowledge is extracted, converted to explicit knowledge and preserved in the Knowledge Management System through proper maintenance of knowledge repository. Our model is based on the technology that encourages active participation and sharing of tacit knowledge through interactions of individuals in the knowledge environment. The model builds a database of queries based on user feedback and the database is enhanced and maintained through creation of tags that makes the KMS dynamic and easily maintainable.

Keywords: Knowledge Management System, Personnel, Knowledge Worker, Tags, Feedback, Query, Response.

1 Introduction

A Knowledge Management System (KMS) [1] [2] refers to a system for managing knowledge in organizations, supporting creation, capture, storage and dissemination of information. Knowledge management (KM) can be defined as in [3], "a range of practices and techniques used by organizations to identify, represent and distribute knowledge, know-how, expertise, intellectual capital and other forms of knowledge for leverage, reuse and transfer of knowledge and learning across the organization". Knowledge can be classified roughly into two parts, explicit and tacit as in [4]. Tacit knowledge is the knowledge resident in brain that is hard to formalize and difficult to communicate. It is highly personal, context-specific and therefore difficult to articulate. It is also difficult to quantify and measure tacit knowledge. Explicit knowledge is codified, and can be precisely and formally articulated. As in [5], explicit knowledge resides in an organization in terms of reports, documents, manuals, procedures etc. They are easy to communicate and share in comparison to tacit knowledge. Its ready accessibility has lead to many ways of using it as a management tool. An organization can become learning or innovative one because of its intellectual assets and knowledge workers as per SECI model of [5]. Through interactions and socialization, the tacit knowledge needs to be extracted, converted to explicit knowledge and preserved

S. Ranka et al. (Eds.): IC3 2009, CCIS 40, pp. 263–272, 2009.
© Springer-Verlag Berlin Heidelberg 2009

in the Knowledge Management System through proper maintenance of knowledge repository. Though technology is a key factor of KMS, over dependence on technology may not assist knowledge creation and enhancement until and unless the organization has a culture of knowledge exchange and socialization. Technology just provides a means to the Knowledge Management (KM) program. Our model is based on the technology that encourages active participation and knowledge sharing of tacit assets of individual interactions in the knowledge environment.

2 Related Work

Knowledge Management is seen as the next great tool for improving and optimizing the functioning of any organization. Different organizations are emphasizing on need of having knowledge management systems (KMS). The have been several models of organizational knowledge, one of the earliest being the SECI (Socialization, Externalization, Combination, Internalization) model (Nonaka, 1991) that classifies organizational knowledge as tacit and explicit. Several organizational Knowledge models have been proposed and discussed from Software modeling, domain ontology and semantic ontology perspectives as in [6]. Different models of KMS are formulated depending on how the knowledge is generated in the organization, what is the application area of knowledge, how it is shared and disseminated, how it is re-used and who are the potential users of the system; whether the system is applicable only for internal operations or there are lot of external interfaces. For large, complex KMS where the knowledge interaction, transaction and distribution is heavy, multi-agent models are used employing software agent technology as in [7]. A weblog based KM model has been discussed in [8] for knowledge creation and sharing. Architectural models of KMS have been discussed in [9] using Service-Oriented Architecture methodology on open and distributed environment. Explicit knowledge has to interact with tacit knowledge with various knowledge creation processes and drive the KM program. Explicit knowledge without tacit insight finally loses its meanings. Research document is created by means of interactions between tacit and explicit knowledge rather than from tacit or explicit alone. Leif Edvinsson of Skandia (1991) elaborated on the components of tacit knowledge in terms of intellectual capital in his Model of intellectual capital. The model proposed in this paper is very simple and centered on the premise that tacit or intellectual knowledge is not easily acquired and isolated. It is very difficult to isolate acquired knowledge resident in human brain due to personal learning and experience through interaction with the environment. There is usual inclination towards pressuring and technically utilizing explicit knowledge. The central framework of this model is based on extraction of tacit knowledge possessed by individuals and its conversion to explicit knowledge to be accessible to all. The knowledge query-retrieval in the model is tag-based instead of being based on keyword, ontology or CBR for faster response and for keeping the size of database of the model small.

Section 3 discusses the model and its components, Section 4 elaborates on the database required for the model and search algorithms, Section 5 discusses database modification and maintenance aspects, Section 6 mentions the advantages of the model, Section 7 gives the relevance of proposed model to knowledge pyramid model, Section 8 illustrates application of the model in e-governance, Section 9 winds up with the conclusion and future scope.

3 The Model

The entire KMS model has three components

- Principle
- Framework and
- Feedback

This allows the KMS to propagate its governing principles which it provides as well as serve as a platform for other similar systems.

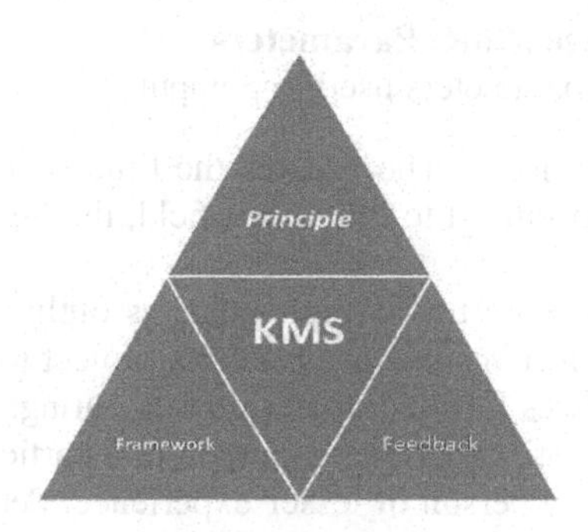

Fig. 1. The three components of the Knowledge Management System

3.1 Principle

Tacit Knowledge is very difficult to isolate and preserve. The model proposes a system by which the knowledge needed in any type-of-work is sought Ad-Hoc. Several individuals work as "Knowledge Workers" (KW) in this KMS who respond to the queries posted to the KMS. The personnel (KW) associated with the Knowledge Management System will be tagged according to the type of work that they have been associated with and the degree of relevance of each KW will be marked according to several factors like the time spent on that particular field, the objectives being fulfilled, the funding being made available for the project in which the KW is working and the educational background of the person in question. The decision to direct the query will be made by the KMS and will involve internal algorithms for evaluation and dynamic decision making as in [10].

3.2 Framework

There will be a central database that houses data on the users involved in the KMS, their Degrees of relevance (DoR), Access points and KMS rating. The entire KMS will be divided into a "Query" (Q) side, a "Response" (R) side and a database to store the existing responses. Q will only be a basic Search interface that waits for a query and returns results ranked according to a Combination Function (C) that combines both the KMS ratings and the DoR. Combination Function, is expressed as:

$$C = f\,(\text{KMS rating, DoR}).$$

The Combination Function is based on the Guidance Function discussed in [11]. The Search result page will only give the Names, their Access Points, KMS rating and DoR.

3.2.1 KMS Rating

Quick and useful replies to queries will bring about an increase in the Rating of the individuals participating in the KMS based on the KM models discussed in [12]. The rating of response to queries is evaluated in steps of 6 hours. There are only 3 windows in increments of 6 i.e. 6, 12 and 18 hours. The quicker the query is responded to, the higher the KMS rating.

3.2.2 Degree of Relevance (DoR) Parameters

The exact break-up of the parameters used to compute the DoR within the KMS are:

1. Time spent on Project – The greater the time spent on any particular field that is, the project related to that exact field, the higher the rating on this parameter.
2. Objectives – These are the explicit goals as outlined by the organization in form of list of objectives that are set for a project to achieve. The greater the number of objectives fulfilled, the higher the rating.
3. Experience – A more experienced person in a particular field will be given a higher rating than a person of lesser experience. People who have worked in diverse fields will, however only be judged on field-wise experience and not on their absolute experiences.
4. Funding – This is considered a direct and formal endorsement by the company and is viewed as a fair representation of the person's credentials. The higher the budget allocation, the more is the rating.
5. Background – The formal education and/or a staggered field-work of that person in question will also be considered as a criterion for ranking.

3.2.3 Response

R will contain the queries that have been received, which have to be responded. The KW can choose one or more ways of replying to the query. It can include:

a. Internal Link – A link to any helpful material within the Intranet of the organization.
b. External Link – Any link outside the local intranet of the organization that may be useful.
c. Referral – It is done to redirect the query to any other user within or outside the local intranet.
d. Post – This is any direct answer to the query. .

3.3 Feedback

3.3.1 Basic Parameters

At the time the query is received, there is a provision to term the query as "Too general" and send it back asking for more specific details. At the time of accepting the response to any query, there is a provision to mark the response as useful or not useful. A useful click will result in a "Merit" and a 'not useful' click will result in a "De-merit".

3.3.2 Rating System

There is a simple 3 star merit and 2 de-merit system in place to rate the responses. 3 stars will result in a reward of few hours off from logging on to the KMS. 2 De-merits will result in a report to the webmaster, warnings and repeat of De-merits will result in summon to the Project/ Department/HR Head.

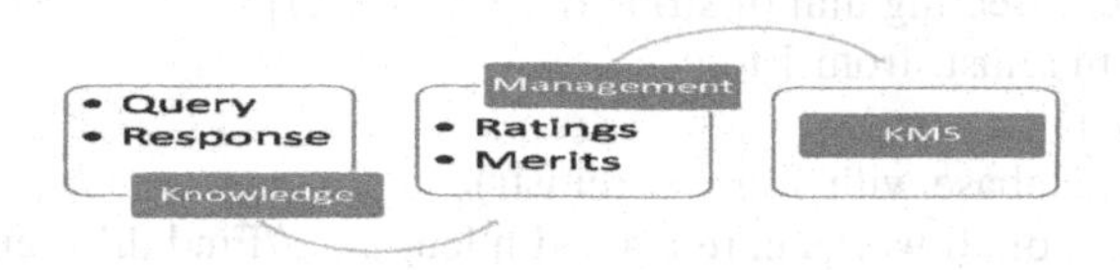

Fig. 2. Proposed Knowledge Management System and its renewal

4 Database Creation

This is done by the generation of tags as governed by the user's query and involves automatic storage of keywords and phrases into the Database of the KMS which serves as a data warehouse. This task is made easier by the Algorithm as described below:

4.1 Algorithm

```
| Questions = {What, Who, When, Where, How, Are, Is, By, Can, Do, Does,.......};
| Articles = {A, an, the};
| Conjunctions={for, and, nor, but, or, yet, and so.......};
| Prepositions={ of, to, in, for, and, on, at........};
|Pronouns={I,you,we,him,her,our,their,them,us,those,that,it,me,mine,this,that..};
|
|str=INPUT 1..n ;                    //Input query of size n taken as the string
|
|For str=1..n
|       Delete words == Questions[ ];
|       Remaining string in order = str;
|
|For str=1..m                        // Numerically, m < n as words are
                                        getting deleted
|       Delete words == Articles[ ];
|       Now str in order = new str;
|
|For str=1..l                        // l < m by the same logic
|       Delete words == Conjunctions[ ];
|       Now str in order = new str;
|
|For str=1..k                        // k < l
|       Delete words == Prepositions[ ];
|       Now str in order = new str;
|
```

```
|For str=1..j                                   // j < k
|       Delete words == Pronouns[ ];
|       Now str in order = new str;
|
|Publish str[1..i];           //After all deletion operations, length = i, where
                               i<j<k<l<m<n
|Smallest Meaning Bearing unit of str[1..i] = remstr[1..i]
|Number words in remstr from 1 to i;
|For(n=i ;  ; n=n-1)
|     Search in database with Tag == (remstr)_n ;
|     Permutation of all words in remstr with length n;//Find different subsequences
                                                        of tags
|     Search database with Tag==(remstr)_n
|If match found in database with n>1
|       If multiple matches found with n>1
|               Return Tag of largest n in order of processing;
|
|If no match found then do
|       For all words of remstr 1..i
|               Number each word from 1 to i
|               Do
|                       Permutations of all the tags from 1 to i;
|                       Compare with the existing tags in database;
|               If match found then return (tag)
|               Else
|                       Make new Tag of original remstr and add to database;
```

4.2 Example of the Algorithm

Query : What is preferred mode of payment for Income Tax?

Algorithm:

First Pass: ~~What is~~ preferred mode of payment for Income Tax
First Pass: preferred mode of payment for Income Tax
Second Pass: preferred mode of payment for Income Tax
Third Pass: preferred mode of payment ~~for~~ Income Tax
Third Pass: preferred mode of payment Income Tax
Fourth Pass: preferred mode ~~of~~ payment Income Tax
Fourth Pass: preferred mode payment Income Tax
Fifth Pass: preferred mode payment Income Tax
str : preferred mode payment Income Tax

remstr:preferred mode pay Income Tax …………Tag

Tag-matching Query: How do I pay Income Tax?

Algorithm:

First Pass: ~~How do~~ I pay Income Tax
First Pass: I pay Income Tax

Second Pass: I pay Income Tax
Third Pass: I pay Income Tax
Fourth Pass: I pay Income Tax
Fifth Pass:~~I~~ pay Income Tax
Fifth Pass: pay Income Tax
str: pay Income Tax

<u>remstr</u>: pay Income Tax..............................New Tag

Permutation and Comparison = Match (3/5)

5 Database Modification and Maintenance

Since the Database of the Knowledge Management System (KMS) is dynamic, it has to be continuously evaluated and kept at an optimum performance point. This involves removal of obsolete entries, quick and easy addition of new ones and general maintenance of the database. An algorithm has been designed to help with this as follows:

5.1 Algorithm

```
| Tags numbered from 1 to n;
| Each tag Timestamped and stored in database
| ts_limit1 is Time-limit for splitting tags
| ts_limit2 is Timelimit for deleting the Tags
| On expiry of ts_limit1 do
|               Break up Tag as
|               for(tag_length =i)
|                       Split Tag into Tags of length 1;
|                       Compare Split Tags == Tag in database;
|                       If match (delete split tag);
|                       else add Split Tags to Tags;
| On expiry of ts_limit2 do          // ts_limit1<ts_limit2
|               Delete Tag;
```

5.2 Description of Algorithm

This algorithm takes into account that full phrases may not be required and in spite of exact queries being used as tags, it is far better to have smaller discrete queries that will offer bits of new information to the existing database. Thus, after one smaller time limit elapses, a Tag is split into its constituent members making mono-tags. After the elapsing of a larger time frame, the tag is deemed useless and deleted.

6 Advantages of the Model

a. There is less overhead of maintenance of knowledge repositories than in conventional KMS.
b. Knowledge creation and sharing are taken care of by the users of the model.

c. There is no explicit need of classification and categorization of knowledge as the tagging is taken care of by the algorithm itself.
d. There is active feedback system which encourages active participation of the users to improve the model.

7 Relevance of Proposed Model to the Knowledge Pyramid Model in E-Government

In the knowledge pyramid model as in [13], there are four layers of data, information, knowledge and wisdom. Wisdom lies in the apex of the pyramid that implies application of knowledge for decision making that assists in taking correct decision. Our model is suggestive of such wise system to be derived from the tacit knowledge of the participants of the system that grows with experience, that is, more the system be used, more shall be its knowledge and wisdom. In the wisdom layer, the queries are expected to be extremely fast. Management ability not only depends on certain management heuristics but also wisdom gathered from the experience gained by handling similar situations and human intuition and mental ability. We term the combination of experience and mental ability to take a correct decision for a situation as wisdom. With the reinforcement of the tags and the feedback system in our proposed KMS model, it is the ultimate wisdom level of the knowledge pyramid that our proposed model hopes to induce into the personnel comprising the KMS.

8 Applications of the KMS Model in E-Governance and E-Government Services

Knowledge management in e-governance and e-government services is essential for enhancement in quality of service (QoS) to citizens. Use of KMS in e-government is aimed towards increased productivity. It is a management tool for government decision makers and its program implementers as mentioned in [13]. Government has been the principal user of knowledge. Primary function of government is decision-making and e-government provides unique support to decision-making as explained in [13]. Government also has largest repositories of information and databases and e-government helps in their efficient management. The inclusion of Knowledge Management in e-government services would convert the existing economy to a knowledge economy which shall accelerate economic growth and development. As mentioned in [13], India has taken the unique initiative among developing economies of setting up a national knowledge commission for leveraging knowledge for economic development (Misra 2006). Emergence of Finland as a leading knowledge economy, which was earlier facing economic crisis, is a success story of leveraging knowledge for economic development. The Knowledge Management System discussed tries to maintain and enhance knowledge base through extraction of tacit knowledge from personnel through interaction and exchange of individual intellectual property that is a part of socialization. Through repeated queries, the tacit knowledge would be added as explicit knowledge to the KMS which shall be free for access to the users of the system. Repeated use of the system shall convert the knowledge base to wisdom base as the newer queries shall be promptly responded with an exact match each and every time a query is submitted as if the system itself has a brain to always take the correct decision. This would surely bring about an increase in the efficiency of queries which may be implemented through several Knowledge Management Systems in the e-Governance services. The citizens, on placing a query to e-government services would receive response to queries in lesser time as compared to queries submitted earlier. The proposed KMS model as well as relevant KMS tools may be essential in assisting and speeding up government decisions and policy maters. Our model of KMS, if in place at Government installations, would improve Quality of Service (QoS) of e-government services to citizens, bureaucrats, policy makers and others in:

1. Emergency or Policy Decision Matters - handling of exceptional conditions like floods or tsunamis , emergency hospital services and so on without having to waste time waiting for critical decisions.
2. Citizen Involvement – Every common user can participate in decision making and give their opinions or feedbacks. This would increase citizen participation in e-governance with respect to national policy decisions.
3. As the model discussed in the paper has the component of feedback, it shall be easier to rate the QoS of individual e-government services with the help of feedback provided by the citizens who are the recipients of the e-government services provided at state or national level.

This would result in an enhanced quality of service of e-government services to citizens. The ultimate aim of technology lies in human development. With use of our proposed model, an overall improvement of the quality of e-government services to the citizens is envisaged which will bring an enhanced quality of life to the citizens at the macro-level.

9 Conclusion and Future Works

The applications of this model are aimed towards enhancement of QoS to citizens in e-governance with respect to public utility and welfare services. As knowledge is a key component for strategic governance, KM based systems shall assist in improving the response to citizens queries. There is scope of active citizen participation which is highly desirable for effective e-governance. KM Systems encourage creation, sharing, and distribution of knowledge among policy makers, service providers and citizens that helps in the ability to analyze problems and provide assistance in decision taking.

The proposed model is a simple logical and effective representation of query and retrieval of a knowledge base with response rating factor. The implementation is envisaged with SQL Server as database backend and Microsoft Dot Net/Visual basic as front end for codification of the algorithms. The codifications shall also consist of database management and knowledge base management systems, which through repeated queries shall improve to wisdom management.

References

1. Knowledge Management System, `http://en.wikipedia.org/wiki/knowledge_management_System`
2. KM System, `http://www.allkm.com/`

3. Knowledge Management, http://en.wikipedia.org/wiki/knowledge_management
4. Nonaka, I., Takeuchi, H., Umemoto, K.: A theory of Organizational Knowledge Creation. Int. J. Technology Management, 833–845 (1996)
5. Knowledge creating process, SECI Model, http://www.allkm.com/km-basics/knowledge-process.php
6. Majid, A.H.M., Lee, S.P., Salwah, S.: An Ontology-based Knowledge Model for Software Experience Management. Int. J. of the Computer, the Internet and Management 14(3), 79–88 (2006)
7. Soto, J.P., Vizcaino, A., Portillo-Rodriguez, J., Piattini, M.: Agents that Help to Detect Trustworthy Knowledge Sources in Knowledge Management Systems. In: ICSOFT/ENASE 2007, CCIS, vol. 22, pp. 297–309. Springer, Heidelberg (2008)
8. Li, J.: Sharing Knowledge and Creating Knowledge in Organizations: The Modelling, Implementation, Discussion and Recommendations of Weblog-based Knowledge Management. In: IEEE International Conference on Service Systems and Service Management, pp. 1–6 (2007)
9. Deng, Z., Hu, X.: Discussion on Models and Services of Knowledge Management System. In: IEEE ISITAE 2007, pp. 114–118 (2007)
10. Wang, C., Leong, T.: Knowledge Based Formulation of Dynamic Decision Models. In: Lee, H.-Y. (ed.) PRICAI 1998. LNCS, vol. 1531, pp. 506–517. Springer, Heidelberg (1998)
11. Kanai, A., Niwa, K.: Knowledge Management System Including Viewpoints as External Constraints. In: IEEE SMC 1999, Japan, vol. 4, pp. 147–152 (1999)
12. Misra, D.C.: Ten Guiding Principles for Knowledge Management in E-government in Developing Countries. In: First international Conference on Knowledge Management for Productivity and Competitiveness, New Delhi, pp. 1–13 (2007)
13. Koanantakool, T.: Struggling Towards a Knowledge-based Society. In: International Symposium on Information Technology and Development Cooperation, Japan, pp. 1–9 (2000)

A Protocol for Energy Efficient, Location Aware, Uniform and Grid Based Hierarchical Organization of Wireless Sensor Networks

Ajay Kr. Gautam[1] and Amit Kr. Gautam[2]

[1] Department of Computer Engineering
M.M. Engineering College, Mullana, Ambala-133 203, Haryana, India
ajaygautam123@yahoo.com
[2] Department of Information Technology
Sankalchand Patel College of Engineering Visnagar-384 Gujarat, India
mailamitgautam@yahoo.com

Abstract. Most of the contemporary clustering protocols require frequent re-clustering in order to rotate the role of cluster heads, CHs, among sensors to avoid the "hot spot" problem. Also, most of the existing, next CH selection strategies are either randomized or complex and the clustering protocols create non uniform clusters. Finally, the clusters are location unaware or even if some kind of location awareness is there, it's either cost ineffective, highly complex or inaccurate.

In this paper we present, design and implementation of a protocol for Energy efficient, Location Aware, Uniform and Grid based Hierarchical organization (E-LAUGH) of Wireless Sensor Networks, WSNs. It provides uniform cluster size enabling an even load distribution in the network and thus provides energy efficiency. The protocol also saves dynamic clustering overheads by allowing a One-Time setup of clusters. The CH selection is in a round robin manner from a list generated by Base Station, BS. This is simpler and better than the randomized or probabilistic approach used by many others [6, 8, 10& 14]. E-LAUGH also provides location awareness to WSN by logically dividing the network into grids of desired granularity.

Keywords: E-LAUGH, WSN, location aware, cluster.

1 Introduction

Many Wireless Sensor Network, WSN, applications are characterized by dense deployment of Sensor Nodes, SNs, in the field to be monitored. Their use in battle fields, volcanic planes etc. implicitly demand redundant nodes to be deployed in order to provide sufficient coverage even if certain subset of nodes gets destroyed. This provides fault tolerance to the system. There can be many other application specific reasons for dense deployment. This density inherently brings about scalability and communicational efficiency issues. Clustering is a proficient rescue to these problems.

In clustering the SNs are logically arranged into groups having their respective focal point called cluster head (CH). The members of cluster transmit the sensed data

S. Ranka et al. (Eds.): IC3 2009, CCIS 40, pp. 273–283, 2009.
© Springer-Verlag Berlin Heidelberg 2009

to their CH in a single-hop [10] or multihop [11] fashion. This data is aggregated at CH and is then transmitted to base station (BS), again in a single-hop [10] or multihop [8] manner. Preventing the requirement of direct communication of all network nodes with BS provides energy efficiency. This also reduces traffic towards BS, thereby providing communicational efficiency and ability to BS to handle more numbers of clusters, thereby enhancing scalability.

Almost all clustering protocols suffer from a few major limitations: non-uniform clusters, frequent re-clustering, complex CH selection process and location unawareness. Non uniform clusters lead to non uniform load distribution over network nodes. This causes certain nodes in network to deplete faster as compared to others, bringing further non-uniformity in network deployment and formation of 'holes' in routing paths. Re-clustering is required to prevent both the inter-cluster and intra-cluster 'hot-spot' problems. But, frequent re-clustering in order to rotate the roles of SNs leads to energy expenditure and sometimes needs lot of maintenance of current network topology. The CH selection too is complex in many protocols, as sometimes being based on residual energy of each SN requiring flooding of control packets and thus, again leading to high energy expenditure. In others it is either random [6] or based on certain probability function [10].

Certain applications demand the location information to be provided along with the data being transmitted to the BS, such as, where the enemy movement is taking place, where the fire is in the forest or where the humidity is undesirable in a farm field. These applications indeed don't require any pin point location information of an individual SN; rather they require the location of the region where the event is taking place. This is where the location aware clustering is required.

In this paper we propose E-LAUGH, a clustering protocol with a location aware data sensing approach. The BS inherently knows the location of the region or cluster whose data it receives. The protocol doesn't require any GPS device or complex uncertain logics to determine this location; rather it uses a grid based, BS controlled approach to determine the location of a region. Moreover the protocol saves all the overheads incurred in dynamic cluster formation and CH selection, by providing a static cluster formation and CH selection from a list generated by BS. The clusters formed are uniform in nature leading to load balance in WSN and thus providing energy efficiency. Finally, each round consists of more than one steady phases unlike LEACH, which requires clustering and CH selection for each round. This also saves a lot of energy and time.

Rest of the paper is organized as follows: Section 2 summarizes related works. In Section 3 the useful system model is introduced. In Section 4 we present the clustering protocol in detail. Section 5 shows effectiveness of E-LAUGH protocol via simulations. Finally, we conclude our paper and draw directions for future work in Section 6.

2 Related Work

Recent years have witnessed many clustering protocols [6, 8, 9, 10, 11, 13, 14, 15], LEACH being ancestor of most of them. LEACH randomly selects a few SNs as CH, based on certain probability function and rotates their role to balance the energy dissipation of the sensors in the network. This rotation is done after every round. This repetitious set-up processes results in unnecessary energy consumption and delay. Its randomized nature creates clusters with non uniform sizes leading to an uneven load distribution.

TEEN [9], a LEACH based protocol was developed for time critical applications. It defined two parameters ST and HT to further reduce the overall transmission towards BS. These parameters also provide a reactive behavior to the network in contrast to many others who require periodic transmission of data. HEED [13] incorporates communication range limits and intra-cluster communication cost information for the decision of selection of CH.

ELCH [14] uses repeated flooding of voting messages in order to determine the suitability of a node to become CH, whereas VAP-E [15] uses heterogeneous SNs requiring extra cost. Protocol proposed in [8] removed the problem of formation of redundant CHs in same locality but suffers from frequent re-clustering and uses LEACH like probability based function for CH selection. Also, none of the above protocol addresses the location awareness problem. Even if certain protocols provide location awareness they either increase the hardware cost or require uncertain and complex methods [3, 4] to localize the sensors.

We propose a grid based location aware clustering protocol for WSNs. E-LAUGH provides this location awareness not at cost of heavy and expensive GPS devices with each SN; rather it uses the BS created grids to uniquely identify the cluster whose data sink receives. The granularity of the region to be sensed can be set at the setup time itself. Thus, this protocol provides a complete control over the number of clusters to be created. It creates uniform sized clusters thus enabling an even load distribution over the network and hence increases network lifetime. The cluster formation is not required before every steady phase as in LEACH; instead it uses a One-Time cluster formation approach thus saving a lot of overheads and energy. The CH selection is done in a simple round robin manner from a BS generated list. The complete clustering process incurs negligible load over a SN as compared to most of other protocols.

3 System Model

In E-LAUGH we have assumed that the sensors are distributed in a uniformly randomized manner throughout a square field and the network has the following properties:

1. There exists a unique BS
2. SNs have unique identity and are stationary
3. Network is homogeneous i.e. all SNs are equivalent, having same computing and communication capacity
4. The network is location unaware i.e. physical location of nodes is not known in advance.
5. The transmitter can adjust its amplifier power based on the transmission distance

The first assumption considers a mobile BS. The sink mobility is desirable [1, 12] in many applications in order to provide energy efficiency. Moreover in E-LAUGH, the mobility is required only once, i.e. immediately after the deployment of SNs. But, because of any reason, if mobility of BS is a big restriction, the protocol provides an alternative to this. The second assumption of lack of mobility of SNs is typical for WSNs employing some clustering or grouping [2] methodology for network organization. SNs with rapid mobility in network degrade the cluster quality, because they frequently alter the organization of cluster. Assumptions like node homogeneity and advance location unawareness are rather advantageous when hardware costs and resource requirements are key issues. In this paper we use the same communicational energy dissipation model as discussed in [7].

The clustering algorithm must divide the network into disjoint clusters, i.e. if G (V, E) is network deployment graph, C_i the ith cluster and N_c the number of clusters, then,

$$G(V,E) = \bigcup_{i=1}^{Nc} c_i(V_i, E_i) where, V_i \subseteq V, E_i \subseteq E \tag{1}$$

We propose E-LAUGH, an energy efficient protocol for location aware grid clustering in wireless sensor networks. The protocol forms clusters in the form of logical grids which are uniform in size, thus providing an even load distribution over the network. This prevents certain parts of network from depleting faster than others leading to network partition. The cluster formation is static in nature, i.e. once the clusters are formed no reclustering is required. This prevents lots of overheads as compared to other clustering protocols [9, 10 &13] who require frequent reclustering requiring high setup energy and time. Moreover the CH selection is very simple and is done from a centrally generated list. The scheduling is done in a round robin manner from this list and hence avoids the complex or the probabilistic nature [10] of the CH selection process.

Above all, we provide location awareness in the clusters, which is extremely rare in the present day clustering protocols. Even if some protocols provide location information in the sensor networks it either requires costly hardware (GPS) with all SNs or is very complex and uncertain in nature. E-LAUGH only requires one-time mobility of BS at initial setup phase. Sink mobility is not a big demand, as many applications require it and it also provides energy efficiency in multi-hop routing by preventing formation of 'hot spots' in network. In time critical applications direct CH-BS communication is preferable and sink mobility provides shorter distances for communicating with CHs, thus providing energy efficiency.

The grid size for clusters can be set by BS according to the granularity of the area to be monitored. More is the granularity more are the number of clusters, Nc. Thus a proper trade off must be there between grid size, Gs, and Nc. A very big value of Nc is not desirable since it affects both energy and communicational efficiency. An optimal approach is to have as many as 5% of SNs as CHs. But in E-LAUGH it may depend on resolution of location awareness required. The clustering algorithm incurs a complexity of O |1| at node level as compared to O |n| or even worse in some cases. There is a negligible load over the individual node for the cluster formation; rather our protocol exploits the resources at the BS for most of the clustering maneuver.

Again E-LAUGH is mainly a clustering protocol and not a routing protocol. We are assuming direct communication with BS by CH as in [9, 10 &11] for the purpose of simplicity. But it is not a restriction; rather we can assume any routing protocol at abstract level that is generally considered for clustered WSNs.

4 Implementation

This paper uses a simple model for radio hardware energy consumption [5]. Thus, for transmitting an l-bit message through the distance d, the energy that radio spends is:

$$E_{tx}(l,d) = \begin{cases} lE_{elec} + l\varepsilon_{fs}d^2, d < d_0 & (2) \\ lE_{elec} + l\varepsilon_{mp}d^4, d \geq d_0 & (3) \end{cases}$$

and for receiving this message, the radio expends:

$$E_{RX}(l) = lE_{elec} \quad (4)$$

The electronics energy, $Eelec$, depends on factors such as the digital coding, modulation, whereas the amplifier energy, $\varepsilon_{fs}d^2$ or, $\varepsilon_{mp}d^4$ depends on the transmission distance and the acceptable bit-error rate. Like any other hierarchical routing protocol E-LAUGH also has a setup phase and a steady phase. Since E-LAUGH is more of a clustering protocol we will emphasize more on setup phase and then use LEACH like steady phase for simplicity.

Once the nodes are deployed, the BS will locate itself at a distance, B_d, from one edge of the field. It then broadcasts a signal with strength, B_s, sufficient to reach all SNs. On receiving this signal the nodes calculate their distance from the BS based on the received signal strength using the free energy dissipation model [15]:

$$P_r(d) = P_{tx} \times \frac{\varepsilon}{d1^2} \quad (5)$$

Where, the power of received signal is P_r, $d1$ is distance and ε the attenuation coefficient.

Form (5) $d1$ can be calculated as:

$$d1 = \frac{r}{\sqrt{p_r}} \quad (6)$$

Where, r is constant.

The BS repeats the same process after moving to an adjacent edge of the field, i.e. broadcasts a signal with strength, B_s from a distance, B_d from the field. The receiving SNs calculate the distance, $d2$ from BS using the same relation as above. Now, if the sink mobility is a big issue, because of any reason, then an effective alternative is to use a SN with strong transmission capacity instead of BS for second transmission on adjacent edge. This will solve the same purpose. The purpose of keeping the BS far from the field if to have a lesser curvature of the arc, i.e. grid formed by broadcasted signal on the field. This created more rectangular like and thus uniform kind of clusters.

Now the network nodes will send the two calculated values, $d1$ and $d2$ respectively, to the BS in the form of a Distance Information Packet, DIP. The header of DIP will be the ID of transmitting SN. The BS on receiving the DIP from SNs will form the clusters using the following method:

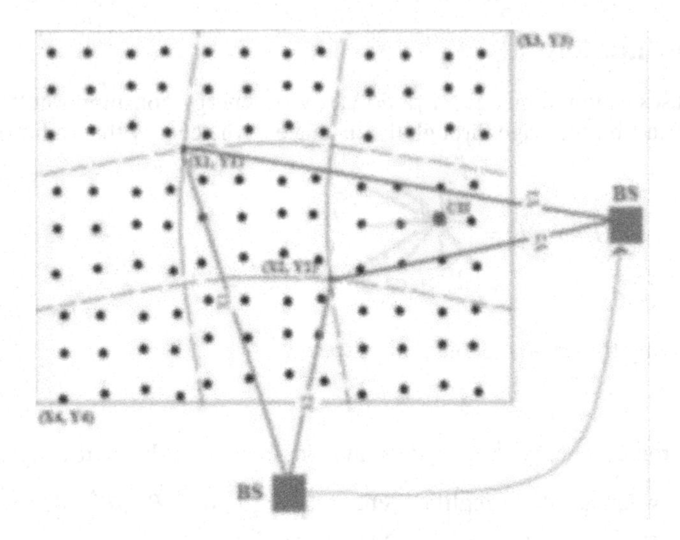

Fig. 1. Grid Formation and Distance Coordinate Formation

BS first forms logical grids on the field according the granularity of the location awareness required. If the data needs to be collected from small chunks of the field then more granularity is required, i.e. more number of clusters will be formed. The BS draws the curved grid lines on the field as shown in Fig. 1 with radii such that they divides the field into equal parts, so as to form uniform sized clusters. The grid lines are drawn with respect to both the adjacent edges from where signal was broadcasted. The curve of grid lines is the curvature of the circle formed by the signal broadcasted by BS. Then the distance coordinates of intersection of grid lines and also for the field vertices are calculated.

The BS then checks the DIPs for the values of $d1$ and $d2$ with respect to the gird line intersection and field vertices coordinates such that:

$$X_1 \leq d1 \leq X_2 \quad \&$$
$$Y_1 \leq d2 \leq Y_2$$

i.e. compares $d1$ and $d2$ with respect to top left, (X_1, Y_1) and bottom right, (X_2, Y_2) corners of the logical grids and accordingly decides the exact cell or cluster in which the SN lies. Based on this information the BS creates a Cell Allocation Packet, CAP, which contains the list of all nodes which fall within a cluster. The header of this packet is the Cluster ID, CID. Now the BS is equipped with location of all clusters, their respective members and sequence of nodes in the CAP. This sequence plays an important role in the CH selection process and scheduling for SN-CH communication as described later.

The BS then broadcasts the CAPs to the network. The receiving nodes check for their own ID in the CAP and thus become aware of their cluster members and CID. The member list and CID are maintained for throughout the lifetime of SNs. The step

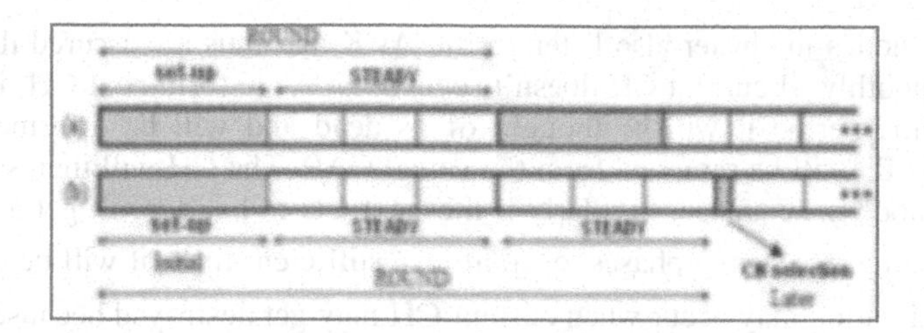

Fig. 2. Time-Line a) LEACH b) E-LAUGH

after the cluster formation is to determine the CH. This is very simple; the order of node IDs in the CAP will determine the sequence in which the nodes will take role of CH. Moreover this order also determines the scheduling for the steady phase, i.e. it tells the succession in which the cluster members will send data to the CH.

Initially the node first in list will become the CH. There is one more modification that E-LAUGH brings about here, i.e. unlike certain protocols [9, 10] it has more than one steady phase in a round. This means that the CH formation will not take place after every steady phase; rather it takes place after every n_s rounds.

$$OneRound = SetupPhase + n_S * Steadyphase \tag{7}$$

This saves energy and prevents the delay induced, as in LEACH, before every steady phase. After the communication proceeding for long time, if a new CH finds its residual energy (E_{rem}) below a threshold (E_{th}), i.e. $E_{rem} < E_{th}$, it reduces the n_S . This must be done at the beginning of a round to prevent unexpected termination of communication at the middle of a round. E_{th} is the energy sufficient to carry out n_S rounds. The CH sends this reduced n_S value to BS in a packet field while sending its first data. The BS then sends an ACK including CID and new n_S to the cluster thus enabling its entire members to know the new number of steady phases. This setting of n_S based on E_{rem} provides fault tolerance to the system and ensures that a CH doesn't die while working just because of lack of energy and also prevents abrupt failure of network.

Another fault tolerance approach is to have a handshake between the current CH and the next in order. This is done by exchanging a REQ and ACK packets at the time of role transfer between CHs after n_S steady phases. These packets include CID, their respective IDs and role taken over, i.e. 0 (in REQ by current CH) for becoming normal node and 1 (in ACK by next CH) for becoming CH. Energy for this role transfer in later stages is very small:

$$E_{Setup}(Later) = e_t(REQ) + e_r(ACK) \tag{8}$$

CID	Self ID	Role

ACK / REQ

All other nodes in cluster also listen to the ACK and thus are assured that work is going on smoothly. If current CH doesn't receive an ACK form next CH, in a certain sufficient time period it will be thought of as dead and will be informed to other members. Its ID will be removed from the saved CAP. The CH will then send a REQ to the next node in sequence. Similarly if the next CH in list doesn't get a REQ from current CH after n_S steady phases, even after a sufficient time, it will be declared as dead. This situation may occur when current CH may get destroyed because of certain reason. The next node takes over by locally broadcasting an ACK.

In the steady phase the CH sends data packet to BS with CID as its header. This enables the BS to know the exact region from where the aggregated data has arrived and can take the required action if needed. This location awareness of sensed data without use of any GPS device or any complex and uncertain logics is the powerful feature of E-LAUGH.

5 Performance Evaluations

The following simulation parameters are considered for the implementation of EEPUSH:

- The distance between the BS and the network is taken as 100m.
- the size of data packet is *512* bytes
- the electronic power is *50* nJ/bit
- free space attenuation coefficient is *12* pJ/bit/m^2
- multipath attenuation coefficient is *0.0012* pJ/bit/m^4
- nodes' initial energy is *6.0* J
- A node is treated as dead when its remaining energy is less than *0.002* J

For simplicity, error free communication links are assumed. We assume a square network field.

First we will see how the selection of number of grid lines along each of the adjacent edges affects the number of clusters formed. The relation is simple; if n_a is number of grid lines along one edge and n_b along the other then the number of clusters, Nc formed are:

$$Nc = (n_a + 1) * (n_b + 1) \tag{9}$$

More number of clusters is generally not desirable as they increase average energy consumption per cluster but in E-LAUGH it depends on the granularity of the location awareness required.

We define a metric called average clustering degree, avClstDeg, as the average number of SNs per cluster in a WSN. This is calculated as:

$$avClstrDeg = \left(\sum_{i=1}^{N_c} ClstrDeg_i\right) / N_c \tag{10}$$

Where, $clstrDeg_i$ is the number of nodes in a particular cluster. Now we will see how Nc affects the avClstrDeg. This reason for this consideration is the fact that generally intra-cluster communication takes place in TDMA manner. Thus if time slot for one node is t_S, then turn of one node comes after approximately,

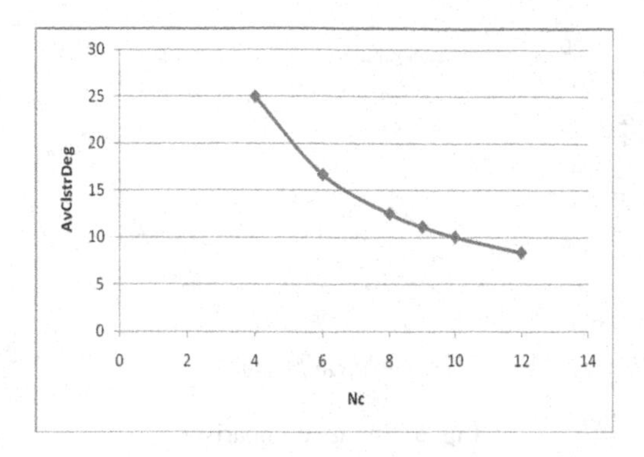

Fig. 3. AvClstrDeg Vs N_c

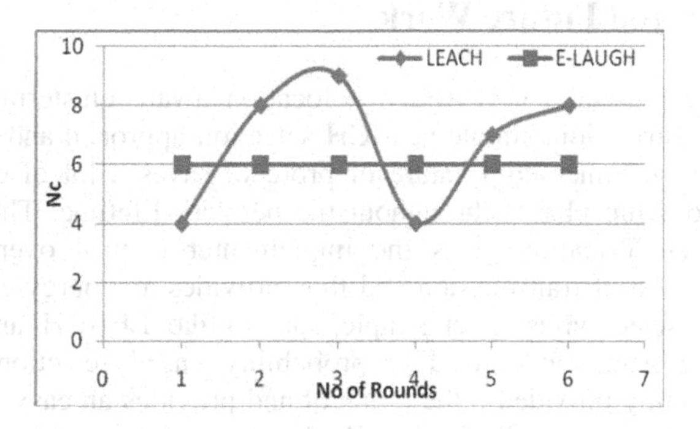

Fig. 4. N_c Vs Number of Rounds

$$T(turn) = t_S * (AvClstrDeg - 1) \quad \text{time.} \quad (11)$$

The bigger the value of $avClstrDeg$, bigger is $T(turn)$. This gives more time to a node for rest between two alternate turns and saves much energy. The Figure clearly shows that more is the Nc lesser is the avClstrDeg and thus lesser is the energy efficiency.

Non-uniform clusters and re-clustering are major issues in any clustering protocol, but our protocol saves the entire headache by providing One-Time cluster formation with uniform load distribution. Fig. 4. gives a comparison of Nc for first six rounds of LEACH and E-LAUGH. LEACH constantly changes the logical network configuration whereas E-LAUGH shows a consistent behavior, thus providing energy efficiency.

Now we consider the setup energies incurred for different number of rounds of both E-LAUGH and LEACH.

The figure clearly depicts that E-LAUGH outperforms the LEACH protocol and is highly energy efficient. This can be attributed to static and uniform clustering in E-LAUGH and least overheads in the next CH selection process that too not after each round but n_s rounds.

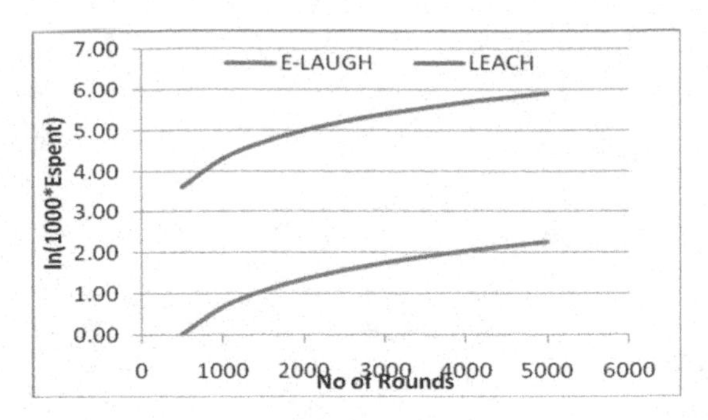

Fig. 5. Energy comparison

6 Conclusion and Future Work

In this paper we presented E-LAUGH, a location aware clustering protocol with uniform cluster distribution, simple next CH selection approach and no re-clustering overheads. The One-Time setup nature of protocol saves a lot of energy and time spent in repeated setup phases throughout the network lifetime. The uniform sized grid based cluster formation gives the implementer control over the maximum distance for intra-cluster transmission and thus provides an energy efficient network model. The CH selection is quiet simple and, unlike LEACH and certain other protocols, is free from randomized or probability based selection process. Fault tolerance is inherently provided n the protocol and provides an easy recovery in case of present CH failure or next CH failure. The location awareness of the sensed data is a unique feature of the proposed protocol. Finally we gave various simulation results and comparisons to prove the effectiveness of protocol.

Future direction in this protocol can be for providing security and intrusion detection system, IDS to the established network model. It would be quiet simple since BS is equipped with a list of all node IDs of a cluster. Some work can be done towards providing a multihop data transmission method towards BS.

References

1. Wang, B., Xie, D., Chen, C., Ma, J., Cheng, S.: Deploying Multiple Mobile Sinks in Event-Driven WSNs. In: IEEE International Conference on Communications, May 19-23, pp. 2293–2297 (2008), doi:10.1109/ICC.2008.437
2. Prasad, D., Patel, R.B., Gautam, A.K.: A Reconfigurable Group Aware Network Management Protocol for Wireless Sensor Networks. In: IEEE Proceeding, IACC 2009, March 6-7, pp. 434–441 (2009), doi:10.1109/IADCC.2009.4809050
3. Li, J., Jannotti, J., De Couto, D.S.J., Karger, D.R., Morris, R.: A scalable location service for geographic ad hoc routing. In: Proc. of ACM MobiCom (2000)
4. Hightower, J., Borriello, G.: Location systems for ubiquitous computing. IEEE Computer 34(8), 57–66 (2001)
5. Krause, A., Guestrin, C., Gupta, A., Kleinberg, J.: Near-optimal sensor placements: Maximizing information while minimizing communication cost. In: Proceedings of Information Processing in Sensor Networks (2006)

6. Wen, C.-Y., Sethares, W.A.: Automatic Decentralized Clustering for Wireless Sensor Networks. EURASIP Journal on Wireless Communications and Networking 5, 686–697 (2005)
7. Krause, A., Guestrin, C., Gupta, A., Kleinberg, J.: Near-optimal sensor placements: Maximizing information while minimizing communication cost. In: Proceedings of Information Processing in Sensor Networks (2006)
8. Kang, T., Yun, J., Lee, H., Lee, I., Kim, H., Lee, B., Lee, B., Han, K.: A Clustering Method for Energy Efficient Routing in Wireless Sensor Networks. In: Proceedings of the 6th WSEAS Int. Conf. on Electronics, Hardware, Wireless and Optical Communications, Corfu Island, Greece, February 16-19 (2007)
9. Manjeshwar, A., Agarwal, D.P.: TEEN: a routing protocol for enhanced effciency in wireless sensor networks. In: 1st International Workshop on Parallel and Distributed Computing Issues in Wireless Networks and Mobile Computing (April 2001)
10. Heinzelman, W., Chandrakasan, A., Balakrishnan, H.: Energy-Effcient Communication Protocol for Wireless Microsensor Networks. In: Proceedings of the 33rd Hawaii International Conference on System Sciences (HICSS 2000) (January 2000)
11. Lindsey, S., Raghavendra, C.: PEGASIS: Power-Effcient Gathering in Sensor Information Systems. In: IEEE Aerospace Conference Proceedings, vol. 3(9-16), pp. 1125–1130 (2002)
12. Khalid, Z., Ahmed, G., Khan, N.M.: Impact of Mobile Sink Speed on the Performance of Wireless Sensor Networks. Journal of Information & Communication Technology 1(2), 49–55 (2007)
13. Younis, O., Fahmy, S.: HEED: A Hybird, Energy-Efficient Distributed Clustering Approach for Ad-hoc Sensor networks. IEEE Trans. On Mobile computing 3(4), 660–669 (2004)
14. Lotf, J.J., Bonab, M.N., Khorsandi, S.: A Novel Cluster-based Routing Protocol with Extending Lifetime for Wireless Sensor Networks. In: 5th IFIP International Conference on Wireless and Optical Communications Networks, May 5-7, pp. 1–5. IEEE, Los Alamitos (2008)
15. Wang, R., Liu, G., Zheng, C.: A Clustering Algorithm based on Virtual Area Partition for Heterogeneous Wireless Sensor Networks. In: Proceedings of the IEEE International Conference on Mechatronics and Automation, Harbin, China, August 5-8 (2007)

Swarm Intelligence Inspired Classifiers in Comparison with Fuzzy and Rough Classifiers: A Remote Sensing Approach

Shelly Bansal[1], Daya Gupta[1], V.K. Panchal[2], and Shashi Kumar[2]

[1] Computer Science Department, Delhi College of Engineering, Delhi
[2] Defence Terrain Research Lab, Metcalfe House, Delhi
bansal.shelly@yahoo.com, dgupt@dce.ac.in, vkpans@ieee.org, shashi120@gmail.com

Abstract. In recent years the remote sensing image classification has become a global research area for acquiring the geo-spatial information from satellite data. There are soft computing techniques like fuzzy sets and rough sets for remote sensing image classification. This paper presents some optimized approach of image classification of satellite multi spectral images which produces comparable results with the fuzzy or rough set based approach. Here we are presenting a comparison of the image classification by fuzzy set and rough set with the swarm techniques as Ant Colony Optimization(ACO) and Particle Swarm Optimization(PSO). The motivation of this paper is to use the improved swarm computing algorithms for the finding more accuracy in satellite image classification.

Keywords: ACO, PSO, Remote Sensing, Rough Set, Fuzzy Set.

1 Introduction

The earth observation satellite image is one of the main source for capturing the geo-spatial information [18]. The main concept in remote sensing with multi spectral satellite imagery is that different features/objects constituting the land cover reflect electro-magnetic radiations over a range of wavelengths in its own characteristics way according to its chemical composition and physical state. A multi-spectral remote sensing system is operated in a limited number of bands and measures radiations in series of discrete spectral bands. The spectral response is represented by the discrete digital number (DN) spectral signatures of an object may be used for identification much like a fingerprint [14].

The ground truth data may itself contain redundant / conflict information. Recently the soft computing mechanism of Rough set and Fuzzy set has emerged as n effective measure to resolve imprecise knowledge, analysis of conflicts and generating rules.

Simultaneously there is a new wide range of computational algorithms for the optimization that have emerged from the behaviour of social insects. Social

S. Ranka et al. (Eds.): IC3 2009, CCIS 40, pp. 284–294, 2009.
© Springer-Verlag Berlin Heidelberg 2009

insects are usually characterized by their self organization and with the minimum communication or the absence of it. This knowledge elicitation is known as Swarm Intelligence [8]. We can found these features in nature such as ant colonies, bird flocking, fish schooling etc. The two most widely used swarm intelligence algorithms are Ant Colony Optimisation (ACO) and Particle Swarm Optimisation (PSO) [9][11].

We are showing here that these swarm intelligence algorithms depicts a competitive accuracy in comparison with the Fuzzy and Rough set approach. For that purpose we have used the cAnt Miner and Hybrid PSO-ACO2 algorithm for the rule extraction for classification. Ant miner was the first data mining algorithm for the classification based on ACO given by Parpinelli et. all [18]. cAnt Miner was the improved version of Ant Miner algorithm which can cope with the continues data also [5]. Unlike a conventional PSO the hybrid PSO-ACO algorithm can directly cope with the nominal attributes, without converting nominal values into numbers in a pre-processing phase. The hybrid PSO-ACO given by Nicholas and Frietas uses sequential covering approach for rule extraction [12]. After that they also proposed a new modified version PSO-ACO2 directly deals with both the continuous and nominal attribute-values [13].

The results with these different versions of the swarm techniques are also discussed and present significant information about the land covering problem.

The paper is organised as follows: section 2, 3 describes fuzzy and Rough set approach for rule generation, section 4,5 presents the swarm based rule generators cAnt Miner, Hybrid PSO/ACO2 Algorithm, section 6,7,8,9 provides approach, experiments, results and case study/application and section 10 conclusion of the paper.

2 Rough Set Theory

Mathematician Z. Pawlak first proposed the theory of Rough Sets (RS) in 1980s. this rough set theory been used in many fields dealing with vagueness and inaccuracy, works well in artificial intelligence and analysis for decision-making. The rough set philosophy is founded on the assumptions that with every object of the universe of discourse we associate some information. Any set of indiscernible (similar) objects is called elementary set, and form a basic granule of knowledge about the universe. Any union of some elementary sets is referred to as crisp (precise) set, otherwise it is rough (imprecise, vague). Each rough set has boundary line cases, i.e. objects which cannot be with certainty classified as members of the set or of its complement. Vague concepts, in contrast to precise concepts, cannot be characterized in terms of information about their elements. So in rough set theory it is assumed that any vague concept is replaced by a pair of precise concepts- called the lower and the upper approximation of the vague concept. This is clearly showed in figure.1. All objects which surely belong to the concept are consist in lower approximation and the upper approximation constitutes the boundary region of the vague concept [17].

The basic concept of the rough sets theory is the notion of approximation space, which is an ordered pair A=(U,R)[26].

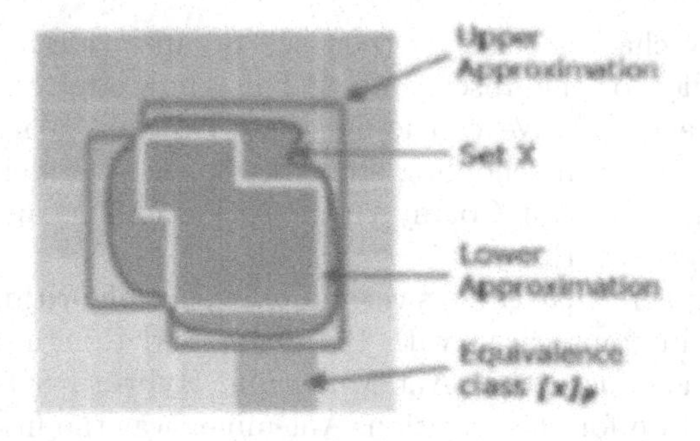

Fig. 1. Concept of Rough Set[17]

Where, U: nonempty set of objects, called universe.
R: equivalence relation (sometimes represented by IND) on U, called indiscernibility relation. If x, y $\in$ U and xRy then x and y are indistinguishable in A.

Each equivalence class induced by R, i.e., each element of the quotient set U/R, is called an elementary set in A. For x $\in$ U let [x]R denote the equivalence class of R, containing x. For each X $\in$ U, X is characterized in A by a pair of sets - its lower and upper approximation in A, defined respectively as:

$A_low(X)$={x$\in$U|$[x]_R$$\in$X}
$A_upp(X)$={x$\in$U||$[x]_R$$\cap$X$\neq\phi$}.

2.1 Dependancy of Attributes

Another important issue in data analysis is discovering dependencies between attributes.Pawlaj and Skowron defined it by means of following equations. Intuitively, a set of attributes D depends totally on a set of attributes C, denoted $C^{\Rightarrow k}$ D, if all values of attributes from D are uniquely determined by values of attributes from C. In other words, D depends totally on C, if there exists a functional dependency between values of D and C. Formally dependency can be defined in the following way. Let D and C be subsets of A.So, D depends on C in a degree k ($0 \leq k \leq 1$), denoted $C^{\Rightarrow k}D$, if,

$$k=\gamma(C,D)= \frac{|POS_c(D)|}{|U|} \text{ [17]},$$

$$POS_c(D)= \bigcup \underline{C}(X), \text{ where X}\in \text{U/D [17]},$$

$POS_c(D)$ called a positive region of the partition U/D with respect to C, is the set of all elements of U that can be uniquely classified to blocks of the partition U=D, by means of C: So,

$$\gamma(C,D)= \sum \frac{|\underline{C}(D)|}{|U|}, \text{ where X}\in \text{U/D [17]},$$

If $k = 1$ we say that D depends totally on C, and if $k < 1$, we say that D depends partially (in a degree k) on C. The coefficient k expresses the ratio of all elements of the universe, which can be properly classified to blocks of the partition U/D, employing attributes C and will be called the degree of the dependency[17].

3 Fuzzy Set Approach

Fuzzy logic, through seminal papers was proposed in 1965 by Lotfi A. Zadeh, professor for computer at the University of California in Berkeley[23]. Basically, fuzzy logic is a multivalued logic that allows intermediate values to be defined between conventional evaluation like true /false, yes/no etc. in order to apply a more human like way of thinking in the programming of computers. Fuzzy system is an alternative to traditional notations of set membership and logic. Fuzzy logic defines no sharp crisp boundaries but just a membership of the element in the given set[26].

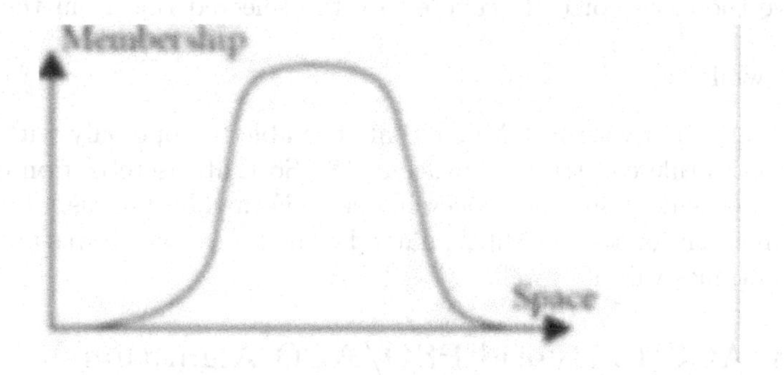

Fig. 2. Fuzzy Set[*17*]

This membership can be full or partial i.e. the membership lies between 0 and 1. This membership of the element is given in terms of the membership function. Membership function is defined as how much that element is contained in our defined set or rangeas shown in figure-2. The natural description of problems,in linguistic terms, rather than in terms of relationships between precise numerical values is the major advantage of this theory[26].

4 cAnt Miner Algorithm

Parpinelli, Lopes and Freitas were the first to propose Ant Colony Optimization (ACO) for discovering classification rules, with the system Ant-Miner. They find out that an ant-based search is more flexible, robust and optimized than traditional approaches. Their method uses a heuristic value based on entropy measure. The goal of Ant-Miner is to extract classification rules from data (Parpinelli et al., 2002). Ant Miner follows a sequential covering approach to discover a list of classification rules from the given data set. It covers all or almost all the training cases. Each classification rule has the form IF <term1 AND term2 AND... > Then <CLASS>. Ant miner requires the discretization method as a pre-processing method and it is suitable only for the nominal attributes. Mostly real-world classification problems are described by nominal or discrete values and continuous attributes.

The algorithm presented by Parepinelli can be viewed as follows –

1. Training set = all possible training cases;
2. While (No. of cases in Training set > maximum uncovered cases)
3. count=0;
4. Repeating count=count+1;
5. Ant count incrementally constructs a classification rule;
6. Prune this constructed rule;
7. Update the pheromone trail followed by Ant count;
8. till (count = No of Ants) or (Ant count construct the same rule as the previous No of Rules Converging-1 Ants)
9. Select the best rule among all constructed rules;
10. Remove the cases correctly covered by the selected rule from the training set;
11. End of while

There is a limitation with Ant-Miner that it is able to cope only with nominal attributes in its rule construction process [18]. So that discretization of continuous attributes is done in a preprocessing step. Fernando, Freitas, and Johnson proposed an extension to Ant-Miner, named cAnt Miner, which was able to cope with the continues values as well.

5 From ACO to Hybrid PSO/ACO Algorithm

The Ant Miner and cAnt Miner has already been a significant approach for data mining, but an extremely large amount of computation is required with the problem of unusually large amount of attributes and classes. The "standard" binary/discrete PSO algorithm [2] does not deal with categorical values in a natural fashion when compared to ACO. In particular, the standard PSO for coping with binary attributes represents a particle by a bit string, where each binary value such as true or false is encoded as 1 or 0. Sousa et al. extended the standard binary PSO to cope with multi-valued categorical attributes [11], developing a Discrete PSO (DPSO) algorithm for discovering classification rules. Nicholas and Freitas proposed several modifications to the original PSO/ACO algorithm [12]. It involves the changes in the splitting of the rule discovery process into two separate phases. In the first phase a rule is discovered using nominal attributes only. In the second phase the rule is potentially extended with continuous attributes. This further increases the ability of the PSO/ACO algorithm in treating nominal and continuous attributes in different ways. Both the original PSO/ACO algorithm and the new modified version PSO/ACO2 uses a sequential covering approach to discover one classification-rule-at-a-time. The new version given by Nicholas and Freitas can be understand as follows-

1. Initially RuleSet is empty(ϕ)
2. For Each class of cases Trs = {All training cases}
3. While (Number of uncovered training cases of class A > Maximum uncovered cases per class)
4. Run the PSO/ACO algorithm for finding best nominal rule
5. Run the standard PSO algorithm to add continuous terms to Rule, and return the best discovered rule BestRule
6. Prune the discovered BestRule
7. RuleSet = RuleSet$\cup$BestRule
8. Trs = Trs – {training cases correctly covered by discovered rule}
9. End of while loop
10. End of for lop
11. Order these rules in RuleSet by descending Quality

It is necessary to estimate the quality of every candidate rule (decoded particle). A measure must be used in the training phase in an attempt to estimate how well a rule will perform in the testing phase. Given such a measure it becomes possible to optimise a rule's quality (the fitness function) in the training phase and this is the aim of the PSO/ACO2 algorithm. In PSO/ACO [12] the Quality measure used was Sensitivity * Specificity (Equation 1) [4]. Where TP, FN, FP and TN are, respectively, the number of true positives, false negatives, false positives and true negatives associated with the rule [2][18].

$$\text{Sensitivity}\quad \text{Specificity} = TP / (TP + FN) \quad TN / (TN + FP)$$

Equation 1: Original Quality Measure [12]

Later it is modified as (Equation 2)

$$\text{Sensitivity}\quad \text{Precision} = TP / (TP + F7) \quad TP / (TP + FP)$$

Equation 2: Quality Measure on Minority Class [12]

This is also modified with using Laplace correction as;

$$\text{Precision} = 1 + TP / (1 + k + TP + FP)$$

Equation 3: New Quality Measure on Minority Class[12]

Where k is the number of classes.

So, PSO/ACO1 attempted to optimise both the continuous and nominal attributes present in a rule antecedent at the same time, whereas PSO/ACO2 takes the best nominal rule built by PSO/ACO2 and then attempts to add continuous attributes using a standard PSO algorithm.

6 Algorithm

There are several standard approaches for image classification like Minimum distance method and Maximum Likelihood classification. We are using the approach of Rough set, fuzzy set, ACO and PSO for the classification of LISS III

EOS image. The algorithm which we are proposing for the image classification is as follows:-

1. Retrieve the 3-band images in .tiff format.
2. Prepare the training data set by these images in .xls format using the ERDAS.
3. Use the Fuzzy method or Rough set approach or cAnt Miner or the Hybrid PSO/ACO2 algorithm on this training data set for rule extraction.
4. Code these rules in matlab file.
5. On executing this matlab file the classified.

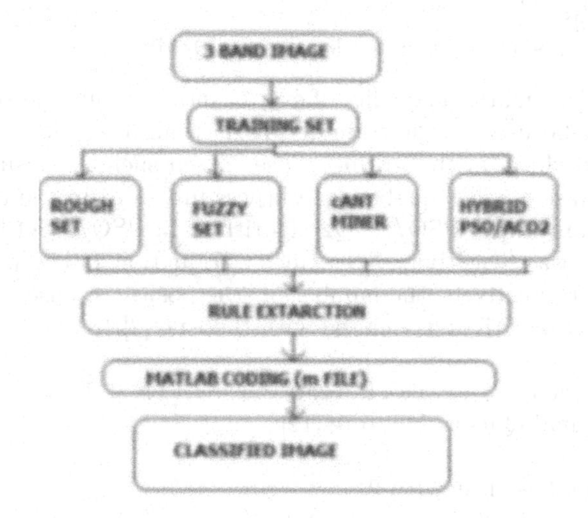

Fig. 3. Process Diagram for Image Classification

7 Experimental Study

We are proposing swarm algorithms as an efficient Landcover classifier for satellite image. We have taken a multi-spectral, multi resolution and multi-sensor

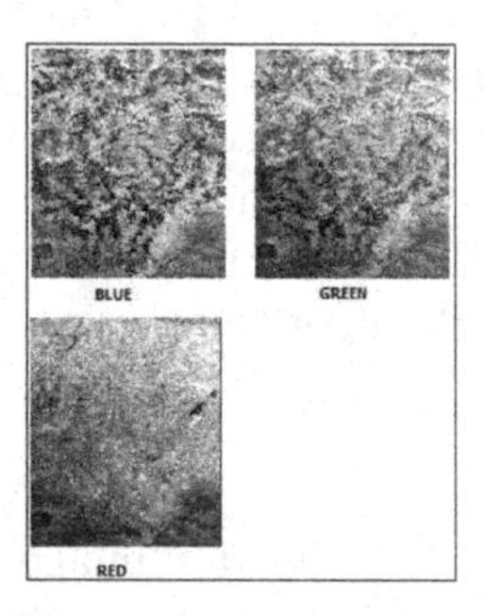

Fig. 4. 3-Band Images

image of Shivpuri area in Madhya Pradesh. The satellite image for three different bands is taken shown in figure-4. These bands are Red, Green, Blue. We are having spectral signatures set from seven bands. This data set provided by the experts in the form of digital numbers (intensity value pixel in a digital image) which were taken with help of Eradas [23]. These sets are taken by carefully selecting the areas (pixel by pixel) from all the images and noting the DN values of the pixels.

8 Results

In remote sensing k, the kappa coefficient is very important and prevalent as a measure of accuracy assessment of Land cover classification. k can be derived from the confusion matrix or error matrix. Hence,

$k \propto$ knowledge

if k=0 then the classifier is inconsistent and does not represent the requisite knowledge

if k=1 then the classifier is consistent and represent the requisite knowledge

Here k, the kappa coefficient for the two probabilistic classifiers i.e., Fuzzy set and Rough set classifier and Swarm Classifier are compared in table1.The classified images by the four proposed ways are shown in figure-5.

Table 1. Classification Comparison

Fuzzy Set Classifier	Rough Set Classifier	cAnt Miner Classifier	PSO/ACO2 Classifier
k=0.785	k=0.847	k=0.964	k=0.975

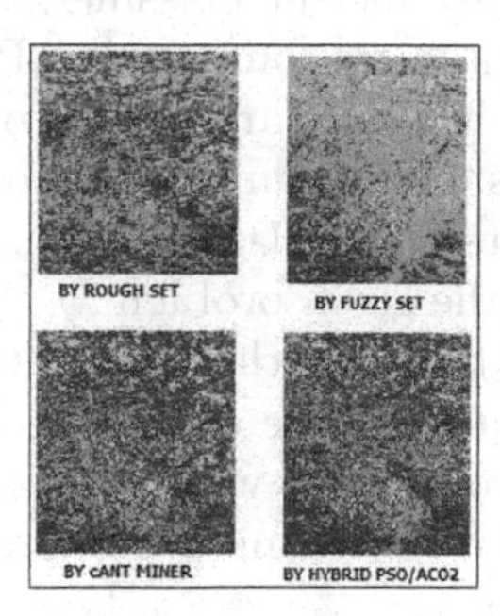

Fig. 5. Classification Results

9 Application

For the analysis of natural resource management/ land cover, accuracy of image classification is very important factor. The very low spatial resolution LU/LC

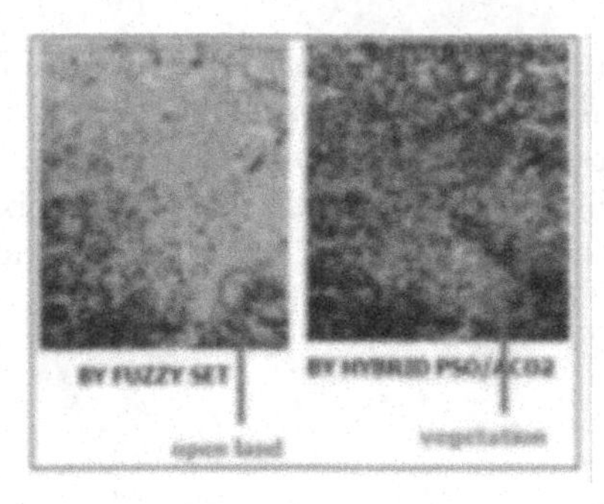

Fig. 6.

image with good numbers of classified accuracy will able to gives land cover analysis result similar to high resolution imageries. So, with lower cost and high degree of classification accuracy will able to replace high resolution high cost satellite imageries.

Here, we have used cAnt Miner ant PSO/ACO2 algorithm for LISS III image (Shivpuri region of Madhya Pradesh), which gives good degree of classification accuracy as compare to the Fuzzy and Rough set methods. In the figure-6 we can see the comparison between the accuracy of ACO-PSO classification and Fuzzy classification . Here, in Fuzzy classified image the encircled region shows that there is open land nearby the water, which is not correctly interpreted. It is correctly given by PSO classified image, where it is shown as vegetation.

10 Conclusion and Future Work

In this paper we have shown the comparative results of algorithms- cAnt-Miner and PSO/ACO2 inspired from social insect behaviour with Fuzzy and Rough set. The results presented are preliminary and there is a lot of scope for improvement to develop these algorithms as efficient classifier. From results it can be seen cAnt miner's efficiency is slightly less compared to PSO/ACO2. PSO introduces a great degree of robustness when compared to the other gradient based learning techniques. This paper details the implementation of the improved social insects inspired techniques for satellite image classification. The results show that these techniques are promising for the given problem.

Future research will focus on using these algorithms together such that the strengths of both the techniques can be exploited. The classifiers that perform better for a particular land cover class will be considered more reliable during conflict resolution. So these swarm techniques can also be combined with fuzzy-rough approach also in future.

References

1. Colorni, A., Dorigo, M., Maniezzo, V.: Distributed Optimization by Ant Colonies. In: Proceedings of the First European Conference on Artificial Life (ECAL 1991), pp. 134–142 (1991)

2. Bratton, D., Kennedy, J.: Defining a Standard for Particle Swarm Optimization. In: Proceedings of the 2007 IEEE Swarm Intelligence Symposium, Honolulu, Hawaii, USA (April 2007)
3. Kaichang, D.I., Deren, L.I., Deyi, L.I.: Remote Sensing Image Classification With Gis Data Based On Spatial Data Mining Techniques. Proceedings of IEEE (2007)
4. Hand, D.J.: Construction and Assessment of Classification Rules. Wiley, Chichester (1997)
5. Otero, F.E.B., Freitas, A.A., Johnson, C.G.: cAnt-Miner: Ans Ant Colony Classification Algorithm to Cope with Continuous Attributes. Springer, Heidelberg (2008)
6. Witten, I.H., Frank, E.: Data Mining: Practical machine learning tools and techniques, 2nd edn. Morgan Kaufmann, San Francisco (2005)
7. Witten, I.H., Frank, E.: Data Mining–Practical Machine Learning Tools and Techniques with Java Implementations. Morgan Kaufmann, San Francisco (1999)
8. Kennedy, Eberhart, R.C., Shi, Y.: SwarmIntelligence. Morgan Kaufmann/Academic Press, San Francisco (2001)
9. Kennedy, J., Mendes, R.: Population structure and particle swarm performance. In: Proceedings of the IEEE Congress on Evolutionary Computation (CEC 2002), Honolulu, Hawaii, USA (2002)
10. Long III, W., Srihann, S.: Geoscience and Remote Sensing Symposium, Unsupervised and supervised classifications. In: Land cover classification of SSC image: unsupervised and supervised classification using ERDAS Imagine, IGARSS 2004. Proceedings, September 20-24, vol. 4 (2004)
11. Dorigo, M., Stuetzle, T.: Ant Colony Optimization. MIT Press, Cambridge (2004)
12. Holden, N., Freitas, A.A.: A hybrid particle swarm/ant colony algorithm for the classification of hierarchical biological data. In: Proc. 2005 IEEE Swarm Intelligence Symposium (SIS 2005), pp. 100–107. IEEE, Los Alamitos (2005)
13. Holden, N., Freitas, A.A.: Hierarchical Classification of GProtein-Coupled Receptors with a PSO/ACO Algorithm. In: Proc. IEEE Swarm Intelligence Symposium (SIS 2006), pp. 77–84. IEEE, Los Alamitos (2006)
14. Omkar, S.N., Manoj, K., Dipti, M.: Urban Satellite Image Classification using Biologically Inspired Techniques. Proceedings of IEEE (2007)
15. Panchal, V.K., Naresh, S., Shashi, K., Sonam, B.: Rough-Fuzzy Sets Tie-Up for Geospatial Information. In: Proceedings of International Conference on Emerging Scenarios in Space Technology and Applications (ESSTA2008) (SSTA 2008), Chennai, India, vol. I (2008)
16. Pawlak, Z.: Rough Set Theory and its Applications to Data Analysis. Cybernetics and Systems 29(7), 661–688 (1998)
17. Pawlak, Z., Skowron, A.: Rudiments of rough sets. Inf. Sci. 177(1), 3–27 (2007)
18. Parpinelli, R.S., Lopes, H.S., Freitas, A.A.: Data Mining with an Ant Colony Optimization Algorithm. IEEE Trans. On Evolutionary Computation, special issue on Ant Colony algorithms 6(4), 321–332 (2002)
19. Lillesand, T.M., Kiefer, R.W.: Remote Sensing and Image Interpretation, 6th edn. Wiley, Chichester (2008)
20. Sousa, T., Silva, A., Neves, A.: Particle Swarm based Data Mining Algorithms for classification tasks. Parallel Computing 30, 767–783 (2004)
21. Fayyad, U.M., Piatetsky-Shapiro, G., Smyth, P.: From data mining to knowledge discovery: an overview. In: Advances in Knowledge Discovery and Data Mining, pp. 1–34. AAAI/MIT, Cambridge (1996)
22. Zadeh, L.A.: Fuzzy sets. Information and Control 8(3), 338–353 (1965)

23. Zadeh, L.A.: Fuzzy logic and its application to approximate reasoning. In: Information processing 74, Proc. IFIP Congr., pp. 591–594 (1977)
24. ERDAS, `http://www.ERDAS.com`
25. MATLAB, `http://www.mathworks.com/`
26. Jensen, R., Shen, Q.: Computational Intelligence and Feature Selection- Rough and Fuzzy Approaches. IEEE press, Los Alamitos

CDIS: Circle Density Based Iris Segmentation

Anand Gupta, Anita Kumari, Boris Kundu, and Isha Agarwal

Department of Information Technology,
Netaji Subhas Institute of Technology,
Delhi University, New Delhi, India
anand@coe.nsit.ac.in, anita.kumari@nsitonline.in,
boriskundu@nsitonline.in, isha.agarwal@nsitonline.in

Abstract. Biometrics is an automated approach of measuring and analysing physical and behavioural characteristics for identity verification. The stability of the Iris texture makes it a robust biometric tool for security and authentication purposes. Reliable Segmentation of Iris is a necessary precondition as an error at this stage will propagate into later stages and requires proper segmentation of non-ideal images having noises like eyelashes, etc. Iris Segmentation work has been done earlier but we feel it lacks in detecting iris in low contrast images, removal of specular reflections, eyelids and eyelashes. Hence, it motivates us to enhance the said parameters. Thus, we advocate a new approach CDIS for Iris segmentation along with new algorithms for removal of eyelashes, eyelids and specular reflections and pupil segmentation. The results obtained have been presented using GAR vs. FAR graphs at the end and have been compared with prior works related to segmentation of iris.

Keywords: Non Ideal Iris segmentation, Circular Hough filter, Sobel filter, CASIA, Eyelash Removal, Specular Reflections, Pupil Segmentation, Canny Edge, Eyelid Detection, Eyelid Removal, Non-ideal iris images.

1 Introduction

Biometrics in its literal sense means the collection, synthesis, analysis and management of biological data. But more recently, it is being used to uniquely identify humans based upon one or more intrinsic physical or behavioural traits such as fingerprints, eye retinas and irides, voice patterns, etc.

Iris recognition uses pattern recognition techniques based on high-resolution images of the irides of an individual's eyes. Iris patterns are thought to be unique and Iris recognition system is safe as it is not intrusive; it does not require physical contact with a scanner but images are obtained through a video-based image acquisition system. Iris recognition method uses the iris of the eye which is the coloured area that surrounds the pupil. But for the systems to work optimally, the segmentation of the iris should be accurate whether obtained under ideal or non-ideal conditions.

S. Ranka et al. (Eds.): IC3 2009, CCIS 40, pp. 295–306, 2009.
© Springer-Verlag Berlin Heidelberg 2009

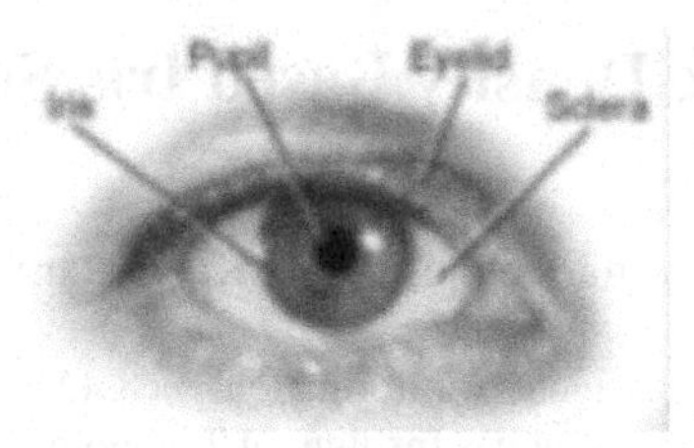

Fig. 1. Image of a human iris

Summing it up from [13], we can say that Iris recognition system has some advantages over others:

1. While fingerprint patterns tend to change after years of manual labour, iris being an internal organ is well protected against damage and wear by a transparent and sensitive membrane called the cornea.
2. The iris has a fine texture that is determined randomly during embryonic gestation itself. Even genetically identical individuals have completely independent iris textures, whereas DNA is not unique for about 1.5% of the human population which includes those who have a genetically identical monozygotic twin.
3. An iris scan is similar to taking a photograph and can be performed from about 10 cm to a few meters away. There is no need for the person to touch any equipment that has recently been touched by a stranger, thereby eliminating an objection that has been raised in some cultures against finger-print scanners, where a finger has to touch a surface.
4. Some medical and surgical procedures can affect the colour and overall shape of the iris but the fine iris texture remains remarkably stable over many decades. Some iris identifications have succeeded over a period of about 30 years also.

Earlier works like region segmentation are less tolerant to noises and some other methods [5], [2] did not work on noise removal. Since, noises not only change the features of noisy regions but also influence the features of neighbouring regions. Hence, we have dealt with noise removal for effective iris segmentation.

To explain CDIS, we have organised the paper as follows: Section 2 comprises of the prior work done till now, Section 3 explains our motivation to use this approach and our contribution in this. Section 4 explains the entire model, Section 5 gives the experimental results achieved, and Section 6 concludes the paper and describes the future work that can be done.

2 Prior Work

A significant amount of work has been done in the area of detecting the limbic boundary. Daugman [5] proposed an Integro-differential operator to find both the pupil and the iris contour. The Integro-differential function is defined by the function

$$max_{(r,\, x0,\, y0)} \left| G_\sigma(r) * \frac{\partial}{\partial r} \oint_{r,x0,y0} \frac{I(x,y)}{2\pi r} ds \right| \tag{1}$$

where $I(x, y)$ = raw image, r = radius of pupil, $(x0, y0)$ = center of pupil, $G_\sigma(r)$ = Gaussian smoothing function with scale σ. But according to [1] and [3] it does not effectively deal with specular reflection and eyelash removal. However, Ross and Shah in [2] described a novel iris segmentation scheme that employs Geodesic Active Contour, popularly known as Snakes, to extract the iris that avoided eyelashes itself. But it is a non-geometric model that depends on the intrinsic properties of the contour. Youmaran et. Al. [3] also proposed a method in which eyelashes were classified as separable and multiple eyelashes, however, the images used have been obtained under constrained conditions and thus it turns out to be image dependent. Zhang et. Al. [4] developed a method for removing eyelashes and restoring the underlying iris texture as much as possible. V.I. Uzunova [6] proposed a method for removing eyelids using parabolic functions on images of the eye.

3 Motivation and Contribution

3.1 Motivation

Most of the previous methods used in this particular area required images with high contrast whereas the images in the databases used did not have much contrast difference between foreground and the background. As discussed earlier, we can conclude that in methods like Geodesic Active Contours,

1. It is difficult to find the radius and centre of the iris accurately.
2. Specular reflections also affect the contour to be developed

Hence, it motivates us to develop a method which mitigated both these drawbacks.

3.2 Contribution

To overcome the above mentioned drawbacks, this paper has been contributed with:

1. A new method for segmentation of iris using CDIS which is based on image contours
2. New algorithms for specular reflection and eyelash removal
3. A new algorithm for eyelid detection
4. Experimental results on CASIA v.3 images [10]
5. Graph-based comparisons with some previous methods [4].

4 The System

4.1 Introduction

Most systems built till now had no provision for removing the specular reflections encountered in iris images, which posed a huge problem in iris segmentation, which laid the foundation of our motivation to remove the same. Moreover, some of the earlier methods removed only eyelids occluding the iris, not the eyelashes which also have been taken care of in CDIS. In discussing the system, various terms and notations used are explained as below.

4.2 Terms and Notations Used

1) I_{min} – Minimum pixel value in image I (of dimensions m and n) in reference to image binarization discussed in Section 4.4.1
2) *K* – Initial Stopping Function in reference to Section 4.5.1
3) *K'* – Final Stopping Function in reference to Section 4.5.1
4) *Val* – Value used to set threshold in reference to image binarization discussed in Section 4.4.1
5) *CD* – Circle Density which detects circles on the basis of density of required intensity pixels on its circumference

4.3 System Flow Model

Elaborating the system flow model shown in Figure 2:

1a. *Image binarization* – takes image from the database [10] as input and outputs a binarized image.
1b. *Pupil Segmentation* – takes binarized image as input and outputs image with pupil segmented.

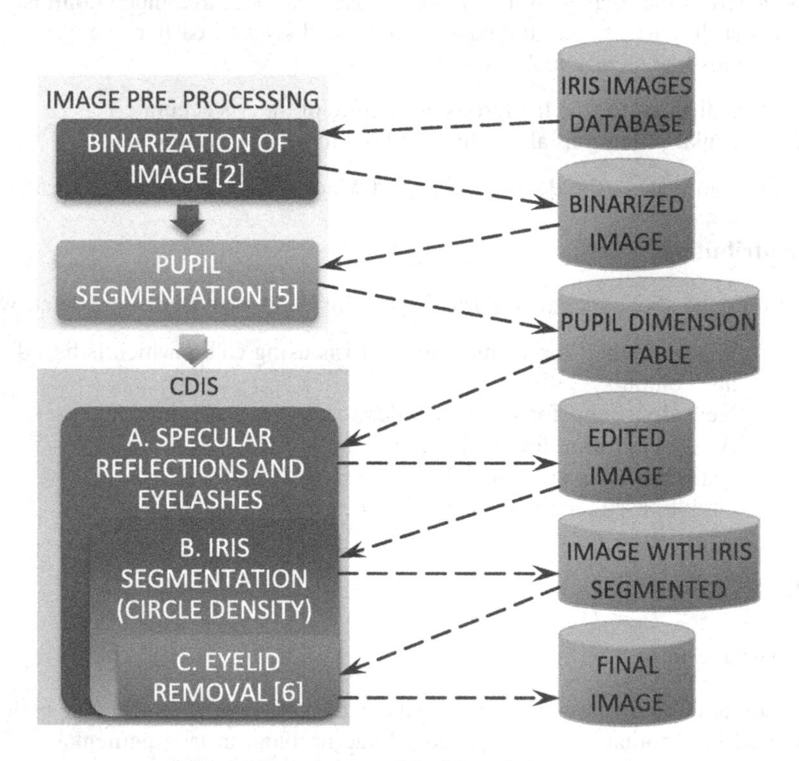

Fig. 2. Work flow model of the whole system

2a. *Specular Reflections and eyelashes removal* – takes image with pupil segmented as input and generates image with specular reflections and eyelashes removed.
2b. *Iris Segmentation* – takes specular reflection removed image as input and generates image with iris segmented.
2c. *Eyelid removal and eyelashes masked* – takes image with iris segmented as input and removes eyelids and masks eyelashes.

Details of each phase are given in following sections.

4.4 Image Pre-processing

4.4.1 Image Binarization

The image is read as an unsigned 8-bit integers' matrix from the database used (CASIA v.3), as shown in Figure 3. As mentioned in [2], the minimum pixel value of the image, I_{min}, is determined. Then, depending on the type of the image, a provisional value, *val*, is set. For example, for CASIA images of size 320 x 280, *val* is set to 70. The image is then binarized using the threshold value $val + I_{min}$. Figure 4 shows the image obtained after binarization.

4.4.2 Pupil Segmentation

From the binary image obtained from binarization (as explained earlier), the minimum and maximum possible pupil radius is determined, which is the minimum and maximum possible length of vertically contiguous white pixels. Circular Hough Transform is then applied on this binary image and the biggest circular area of radius within this range is extracted as the pupil as shown in red. The centre coordinates and final radius are also determined.

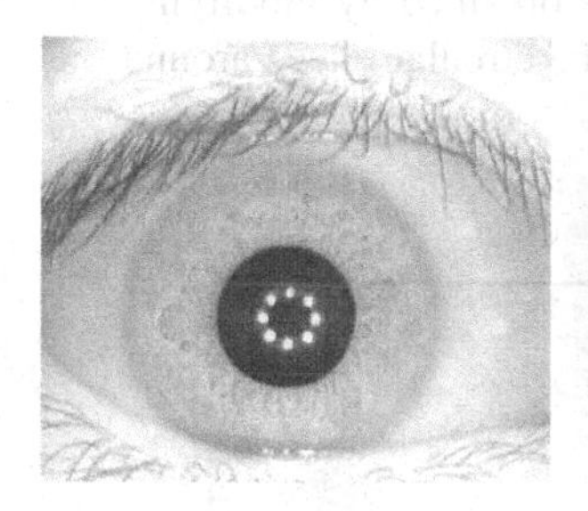

Fig. 3. Original Image

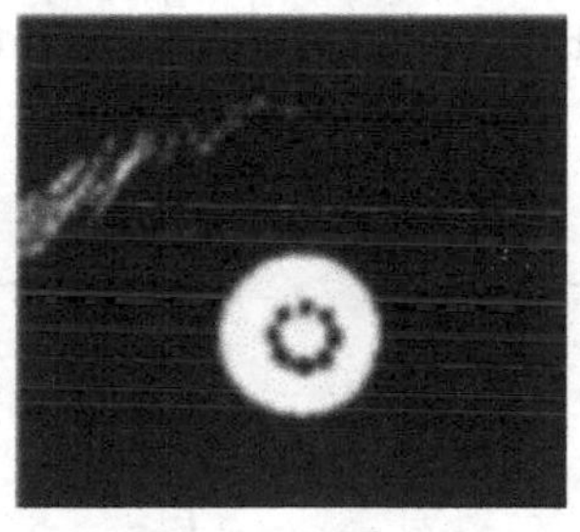

Fig. 4. Binary Image after thresholding

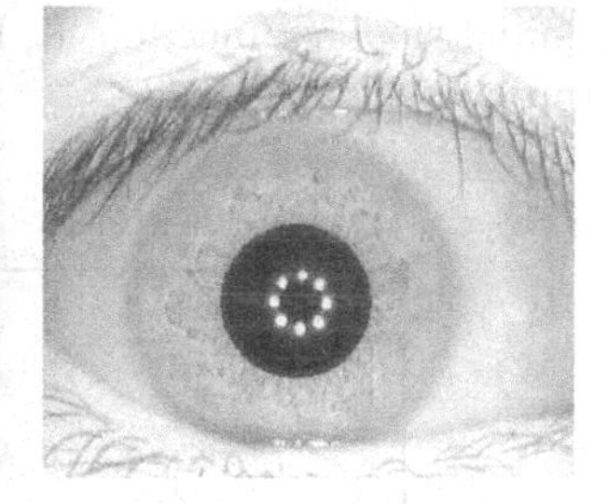

Fig. 5. Pupil Segmented from the image

4.5 Circle Density Based Iris Segmentation (CDIS)

CDIS consists of 3 modules: Specular Reflection removal, Eyelash removal and Iris segmentation along with eyelid removal as shown in Figure 6.

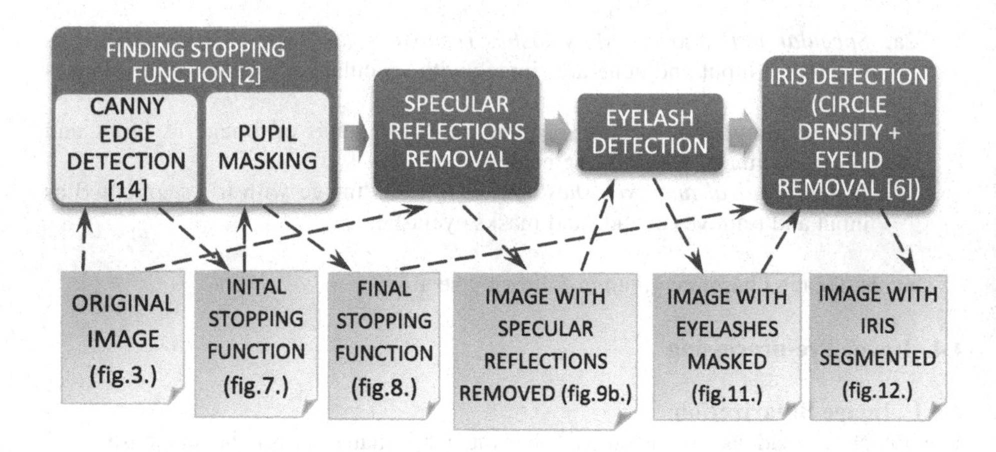

Fig. 6. Flow chart of CDIS

4.5.1 Determining the Stopping Function

4.5.1.1 Finding the Initial Stopping Function

The initial stopping function, K, as shown in Figure 7 is obtained from the image by applying Canny Edge Filter [2] which uses two thresholds, to detect strong and weak edges, and includes the weak edges in the output only if they are connected to strong edges. This method has better resistance to noise and thus, detects true weak edges which generally correspond to the outer limbic boundary.

4.5.1.2 Finding the Final Stopping Function

The final stopping function, K' as shown in Figure 8, is obtained by modifying the initial stopping function, K, by removing the pupil and the circular edges around the pupillary boundary.

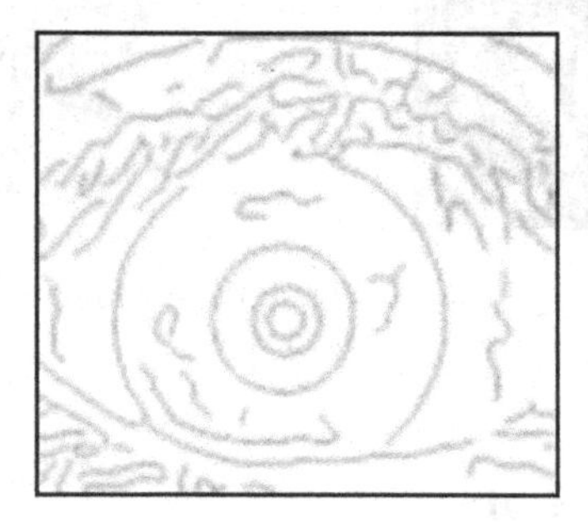

Fig. 7. Initial Stopping Function

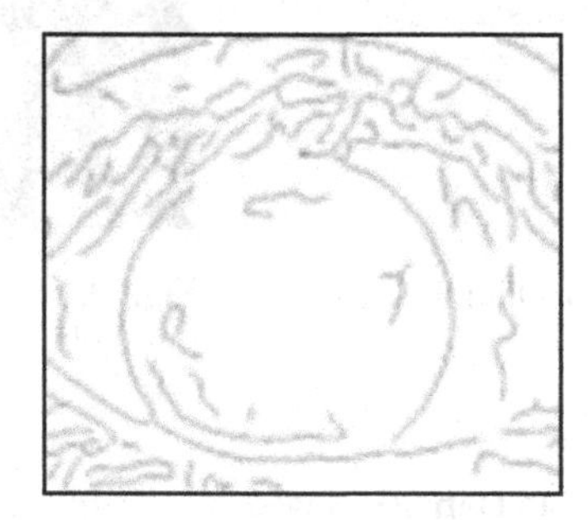

Fig. 8. Final Stopping Function

4.5.2 Specular Reflections Removal

Using the neighbourhood concept, specular reflections are removed from the image. It involves selecting pixels of high intensity value and unpainting these pixels by replacing them with the average of their low-intensity-neighbour values. The neighbours are then further assessed if they also fall within the required range. The results have been highlighted in fig 9(b).

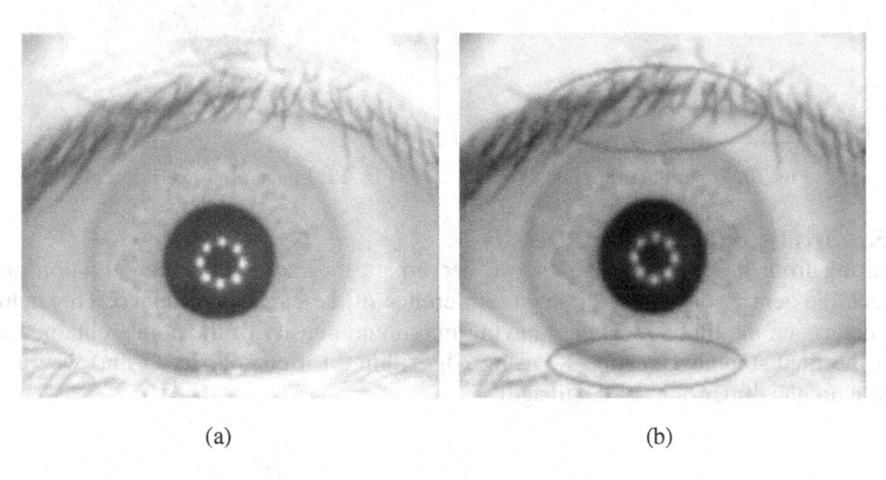

(a) (b)

Fig. 9. (a) Original Image, and (b) Image obtained after contrast enhancement and specular reflection removal (marked)

4.5.3 Eyelash Detection

Firstly, the Sobel filter is applied on the image, which detects the eyelash and eyelid boundaries as shown in Figure 10. Then out of the pixels thus obtained, those pixels are extracted, and their values unpainted, whose value lies below two-thirds of the mean of all pixels obtained by applying the filter. Simultaneously, using the neighbourhood concept, if the neighbours of the so obtained pixels also lie within the required range, their values are also unpainted in a similar fashion. Eyelashes are removed using Algorithm1 (of complexity O(m:width of image)).

Algorithm 1. Eyelash and Specular Reflections Removal

Input : Image I, pupil centre and pupil radius

Output : Updated image

```
1.  J ← Apply Sobel Filter on Image I
2.  FOR all points on edges in J
3.      Find the pixel's neighbours
4.      Update the current pixel value with average
        of high intensity neighbours
5.      Include low intensity neighbours for
        processing
6.  END FOR
```

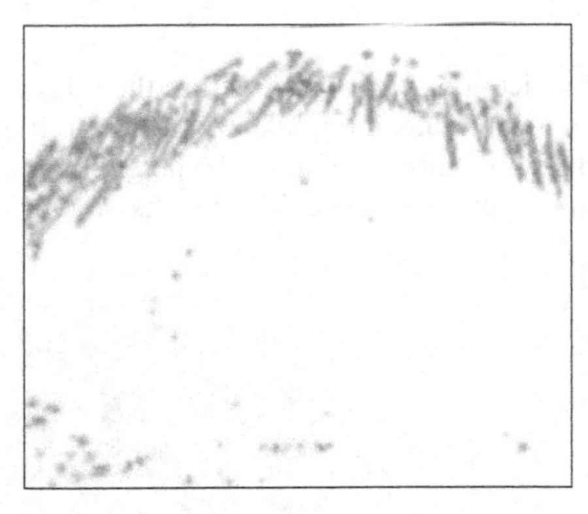

Fig. 10. Sobel Filter applied

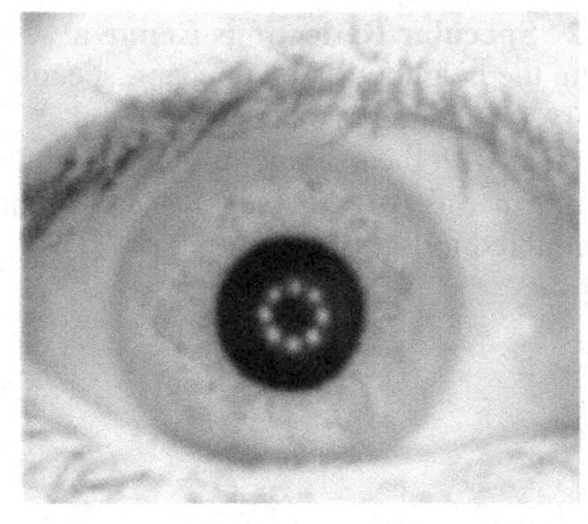

Fig. 11. Image after eyelash removal

4.5.4 Iris Detection by Circle Density

Starting from the image boundary, we decrement the radius by 1 per iteration and check the density of points on the circumference of the circle so obtained. This helps us determine the limbic boundary having maximum density of the points obtained as per the final stopping function K' using Algorithm 2 (of complexity O(mn) where m and n are the dimensions of the image).

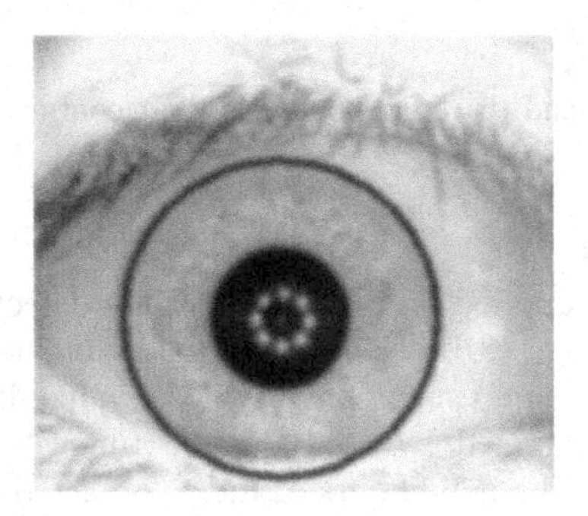

Fig. 12. Limbic boundary marked as circle

Algorithm 2. CDIS

Input : Image I, pupil centre and radius, K'

Output : Iris Radius - *radius*

```
1.  Divide the image into 4 quadrants
2.  Max = 0
3.  FOR all quadrants
4.      FOR radius = image boundary to pupil's radius
5.          Count ← density of black points on circle
6.      End FOR

7.      IF Max < Count
8.          Max ← Count
```

```
9.          Record ← radius
10.     End IF
11. End FOR
```

4.5.5 Eyelids Removal

According to the algorithm given in the thesis [6] by V.I. Uzunova, Arslan Brömme, the various flanks are calculated and accordingly the upper eyelid edge and lower eyelid edge coordinates are determined using Algorithm 3 (of complexity O(n:height of image)), but this requires the subject to hold his head straight. Then a parabola fitting is done to determine the eyelid boundaries, where the latus rectum is taken as the distance between the iris center and the edges, and the vertex as the edge.

Algorithm 3. Eyelid Removal

Input : Image I, pupil centre, pupil radius and iris radius

Output : Upper and lower eyelid borders, latus rectums

```
1. Mask pupil as grey
2. Calculate the horizontal integral projection and
   variance projection function [6]
3. Find the upper and lower edges starting from the
   middle of the image
4. Using algorithm in Listing 1 in [6] the upper
   and lower eyelid borders are calculated
5. Latus rectum ← ( eyelid border - pupil center )
6. Draw Parabola(edge coordinates, Latus rectum)
```

4.5.6 Eyelashes Masking and Final Iris Segmentation

The coordinates of the eyelashes detected in 4.3 were stored in a matrix. In the image of the iris so obtained after slicing it from the original image, the eyelash coordinates are masked as black to remove the unpainted pixels (by neighbourhood method) from the image leaving the user with the final iris image as shown in Figure 13.

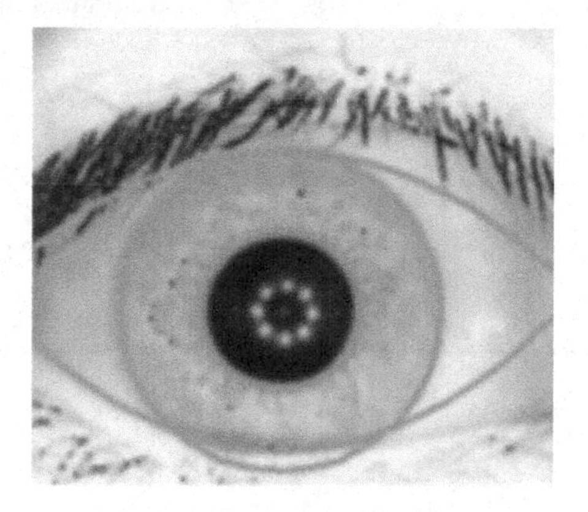

Fig. 13. Final Iris Segmented

5 Experimental Results

The algorithm was tested on 160 iris images taken from the CASIA database [10]. It consists of 2,916 images. The images were scanned at 96 dpi and 256 gray scales. All the steps were carried out on Dell Inspiron 1525, Intel(R) Core(TM) 2 Duo Processor, 2 GB RAM, Windows Vista Home Basic in Matlab 7.5.0. The comparisons of our proposed approach with other algorithms like Libor Masek ([7], [15], [16]) and Region Segmentation [18] are presented in Table1.

Table 1. Comparisons with other algorithms

Method	Accuracy (ideal images)	Accuracy (noisy images)
Libor Masek [7]	83.0%	20.0%
Region Segmentation [18]	Not specified	47.0%
CDIS	99.9%	98.11%

Since the other mentioned original codes were tested on CASIA version 1 images and tests by us were done on CASIA version 3, hence the reason for discrepancy in results.

The circle obtained around the iris is used to slice out the iris from the image and is stored in an array from which the eyelids are removed and eyelashes are masked with black pixels excluding the iris areas occluded by eyelashes and eyelids in the image.

The accuracy of the above approach comes out to be 98.11%. On the above mentioned configuration, the average time for CDIS is 42.5 seconds, for Libor Masek is 47.55 seconds and for regions segmentation 27.15 sec. Its False Accept Ratio (FAR) and Genuine Accept Ratio (GAR) are as shown in Figure 14.

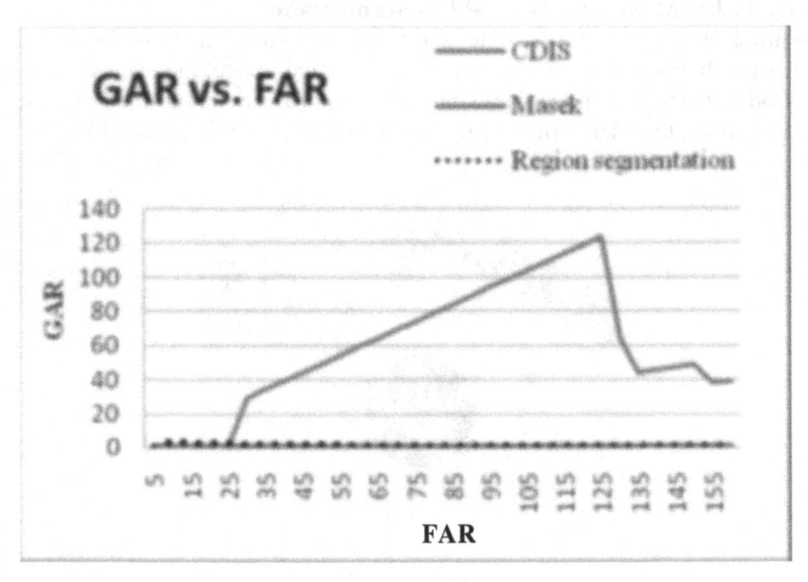

Fig. 14. Genuine Accept Ratio (GAR) vs. False Accept Ratio (FAR) chart as per our dataset

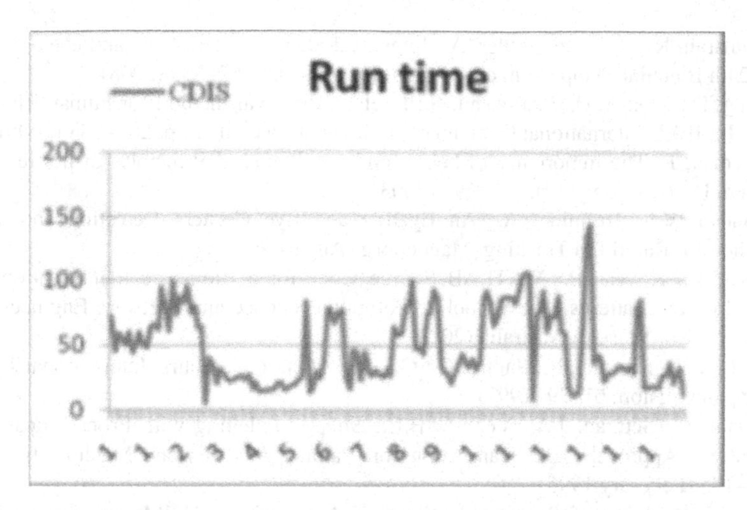

Fig. 15. Graph showing the runtimes for a number of CASIA images on our dataset

The graph shows how the false accept ratio and the true accept ratio varies with the number of database samples. Low GAR vs. FAR value shows that the probability of an error occurring in detecting the iris is little and a linearly growing graph line with high slope shows high accuracy in detecting iris.

The Figure 15 displays the graph showing the run times for various images considered for the experiment.

6 Conclusion and Future Work

The method explained in the paper implements a new crucial and efficient application for offline iris images' segmentation using CDIS algorithms. Specular reflections and eyelashes, which otherwise prove as hindrance in initial phase of all iris segmentation methods, have been accurately removed with a new algorithm based on the local image statistics and block intensity. To make the iris texture analysis phase more efficient, eyelid removal is a necessary step. Since eyelid boundaries obscure the iris in most cases, we extract the visible iris portion effectively using a new algorithm included in CDIS. Hence, we have removed most of the noises from iris images which have not been eliminated in any previous work collectively and thus, accurately segment the iris region for further processing. The experimental results we have presented show that the new algorithm can be a promising enabling tool in time critical areas of security.

Currently our work is being tested for iris segmentation on images with spectacles and lenses.

References

[1] Bowyer, K.W., Hollingsworth, K., Flynn, P.J.: Image Understanding for Iris Biometrics: A Survey. Computer Vision and Image Understanding 110(2), 281–307 (2008)

[2] Ross, A., Shah, S.: Segmenting non-ideal irises using Geodesic Active Contours. In: Biometric Consortium Conference on Biometrics Symposium, pp. 1–6 (2006)

[3] Youmaran, R., Xie, L.P., Adler, A.: Improved identification of iris and eyelash features. In: 24th Biennial Symposium on Communications, pp. 387–390 (2008)

[4] Zhang, D., Monro, D.M., Rakshit, S.: Eyelash Removal method for human iris recognition. In: IEEE International Conference on Image Processing, pp. 285–288 (2006)

[5] Daugman, J.: The importance of being random: Statistical Principles of iris recognition. Pattern Recognition 36(2), 279–291 (2003)

[6] Uzunova, V.I., Brömme, A.: An Eyelids and Eye Corners Detection and Tracking Method for Rapid Iris Tracking, Magdeburg (August 2005)

[7] Masek, L., Kovesi, P.: MATLAB Source Code for a Biometric Identification System Based on Iris Patterns. The School of Computer Science and Software Engineering, The University of Western Australia (2003)

[8] Caselles, V., Kimmel, R., Sapiro, G.: Geodesic Active Contours. International Journal of Computer Vision, 61–79 (1997)

[9] Malladi, R., Sethian, J.A., Vemuri, B.C.: Shape Modelling with Front Propagation: A Level Set Approach. IEE Transactions on Parallel Analysis and Machine Intelligence, 158–175 (Feburary 1995)

[10] Chinese Academy of Sciences Institute of Automation, CASIA, iris image database, `http://www.cbsr.ia.ac.cn/Databases.htm`

[11] Iris Recognition by F. Cheung, Department of Computer Science and Electrical Engineering, The University of Queensland (B.Tech. Project thesis)

[12] Imaging wiki - An Introduction to Smoothing, `http://imaging.mrc-cbu.cam.ac.uk/imaging/PrinciplesSmoothing`

[13] Wikipedia, `http://www.wikipedia.org`

[14] Matlab Online Help, `http://www.mathworks.com`

[15] Wildes, R.: Iris Recognition: An Emerging Biometric Technology. Proceedings of the IEEE 85, 1348–1363 (1997)

[16] Camus, T.A., Wildes, R.: Reliable and fast eye finding in close-up images. In: IEEE 16th Int. Conf. on Pattern Recognition, vol. 1, pp. 389–394 (2002)

[17] Access Excellence – The Eyes have it, `http://www.accessexcellence.org/WN/SU/irisscan.php`

[18] Lankton, S., Tannenbaum, A.: Localizing Region-Based Active Contours. IEEE Transactions on Image Processing 17(11), 2029–2039 (2008)

[19] Papandreou, G., Maragos, P.: Multigrid Geometric Active Contour Models. IEEE Transactions on Image Processing 16(1), 229–240 (2007)

Text and Language-Independent Speaker Recognition Using Suprasegmental Features and Support Vector Machines

Anvita Bajpai[1] and Vinod Pathangay[2]

[1] DeciDyn Systems,
Bangalore, India
[2] Dept. of Computer Science and Engineering,
Indian Institute of Technology Madras, India
anvita@mailcity.com, vinod@cse.iitm.ac.in

Abstract. In this paper, presence of the speaker-specific suprasegmental information in the Linear Prediction (LP) residual signal is demonstrated. The LP residual signal is obtained after removing the predictable part of the speech signal. This information, if added to existing speaker recognition systems based on segmental and subsegmental features, can result in better performing combined system. The speaker-specific suprasegmental information can not only be perceived by listening to the residual, but can also be seen in the form of excitation peaks in the residual waveform. However, the challenge lies in capturing this information from the residual signal. Higher order correlations among samples of the residual are not known to be captured using standard signal processing and statistical techniques. The Hilbert envelope of residual is shown to further enhance the excitation peaks present in the residual signal. A speaker-specific pattern is also observed in the autocorrelation sequence of the Hilbert envelope, and further in the statistics of this autocorrelation sequence. This indicates the presence of the speaker-specific suprasegmental information in the residual signal. In this work, no distinction between voiced and unvoiced sounds is done for extracting these features. Support Vector Machine (SVM) is used to classify the patterns in the variance of the autocorrelation sequence for the speaker recognition task.

Keywords: Speaker recognition, Suprasegmental features, Linear prediction residual, Hilbert envelope, Support Vector Machines.

1 Introduction

In this era of automation, secure and easy access of resources and information is highly important. Most of us must have faced difficulties in remembering passwords and PINs or carrying identity cards. Hence, the biometric-based person authentication system is desirable. Speech being the primary means of communication between humans, it is always desirable to have systems which follows voice commands for the authorized individual and the authorization is also voice

S. Ranka et al. (Eds.): IC3 2009, CCIS 40, pp. 307–317, 2009.
© Springer-Verlag Berlin Heidelberg 2009

based. The existing infrastructure supports the transfer the voice of individuals over telephone and internet, which facilitates remote person authentication also at optimal cost. The objective of automatic speaker recognition task is to recognize a person solely based on his voice by a machine. Speaker characterization also helps in improving synthesized speech quality by adding the natural characteristics of voice individuality, and also to aid in the adaptation of speech recognizers [1] and for supplying metadata for audio indexing and searching.

Speech is a composite signal that contains both speech message and the speaker information. Speaker-specific variations in speech signal are partly due to the anatomical differences in speech-producing organs, and partly due to idiosyncrasies of the speaker, such as speaking habits and emotional state. Both forms of variations are important for automatic speaker recognition task. Speaker-specific anatomical differences are manifested in the two components of speech production mechanism, namely, the excitation source and vocal tract system. Variation in shape of the vocal tract (segmental features) are captured in the form of resonance, anti-resonance and spectral roll-off characteristics, while the excitation source (subsegmental features) is characterized by voice pitch, glottal variations. Features extracted from short-time analysis techniques provide reasonable approximation for speakers speech production mechanism and are relatively easy to extract. Hence, they are widely used for speaker recognition.

For humans, the information about the speaker identity is perceived by listening to a longer segment of the speech signal. This variation in speaking style is related to suprasegmental features such as intonation, duration, stress and co-articulation. However, the suprasegmental features, that is the variation of the speech signal over long duration (typically 50 ms to 200 ms), are difficult to characterize and represent with existing techniques. These behavioral characteristics of the speaker are perceived in the Linear Prediction (LP) residual [2] of speech signal also. Sometimes this difference may not be noticed in the residual waveform, but can be perceived while listening to the signal. The residual of a signal may be less affected by channel degradations as compared to the spectral information [3]. Hence, it is worthwhile to explore these features for speaker identification task. This paper emphasizes the importance of the speaker-specific suprasegmental information present in the LP residual of the speech signals. The excitation peaks can further be enhanced using the Hilbert envelope [4] of the residual signal. The gap between the excitation peaks corresponds to the pitch period in the case of speech. The pattern in these excitation peaks leads to a pattern in the peaks in the autocorrelation sequences of the Hilbert envelope. It further leads to different statistical distribution of autocorrelation peaks for different speakers. This emphasizes the presence of the speaker-specific suprasegmental information in the LP residual signal. For humans, speaking style is a major clue to identify a speaker. Speaking style is a combination of voiced and unvoiced units in an spoken utterance and observed by listening to a longer duration of signal. Hence, no distinction between voiced and unvoiced sounds is done in this work for extracting the speaker-specific suprasegmental information. Support Vector Machines (SVMs) are used to classify the pattern in the variance of the autocorrelation sequence for the speaker recognition task. A speaker recognition system based on suprasegmental features, if combined [5] with existing systems based on the segmental [6], [7] and subsegmental [8], [9] features, can give a better performing speaker recognition system.

Section 2 gives an overview of related work done in the area of speaker recognition. Section 3 discusses the presence of the suprasegmental information in the LP residual signal. The excitation peaks that can be noticed in the LP residual signal are seen clearly in the Hilbert envelope of the LP residual. Section 4 discusses the Hilbert envelope, and also discusses the presence of the speaker-specific suprasegmental information in the Hilbert envelope. The methods to extract suprasegmental information from the Hilbert envelope are discussed in Section 5. In Section 6, the use of SVMs for speaker recognition has been discussed. Section 7 presents the experimental results. Section 8 concludes the paper.

2 Related Work

The first attempts for automatic speaker recognition were made in the 1960s, at Bell Labs [10] by using filter banks and correlating two digital spectrograms for a similarity measure. Li et al. [11] developed it by using linear discriminators. Doddington [12] replaced filter banks by formant analysis. For the purpose of extracting speaker features independent of the phonetic context, various parameters like instantaneous spectral covariance matrix [13], spectrum and fundamental frequency histograms [14], linear prediction coefficients [15], and long-term averaged spectra [16]. As a nonparametric model, vector quantization (VQ) was investigated [17], [18]. As a parametric model, hidden Markov model (HMM) was investigated. Poritz [19] proposed using an ergodic HMM. Reynolds [20] used Gaussian mixture model (GMM), as a robust parametric model. Intra-speaker variation of likelihood (similarity) values is one of the most difficult problems in speaker verification. Variations arise from the speaker himself, from differences in recording and transmission conditions, and from noise. Likelihood ratio and a *posteriori* probability-based techniques were investigated [21], [20]. In order to reduce the computational cost for calculating the normalization term, methods using cohort speakers or a world model were proposed. Score normalization is also done by subtracting the mean and then dividing by standard deviation, both terms having been estimated from the (pseudo) imposter score distribution. Different possibilities are available for computing the imposter score distribution: Znorm, Hnorm, Tnorm, Htnorm, Cnorm and Dnorm [7]. High-level features such as word idiolect, pronunciation, phone usage, prosody, etc. have been successfully used in text-independent speaker verification. Typically, high-level feature recognition systems produce a sequence of symbols from the acoustic signal and then perform recognition using the frequency and co-occurrence of symbols. In Doddingtons idiolect work [22], word unigrams and bigrams from manually transcribed conversations were used to characterize a particular speaker in a traditional target/background likelihood ratio framework. The LP residual is explored to extract subsegmental features using neural networks models for speaker recognition [8], [9], [5].

3 Suprasegmental Features in the LP Residual Signal

3.1 Computation of LP Residual from Speech Signal

The first step is to extract the LP residual from the speech signal using linear prediction (LP) analysis [2]. In the LP analysis each sample is predicted as a linear weighted sum of the past p samples, where p represents the order for prediction. If $s(n)$ is the present sample, then it is predicted by the past p samples as,

$$s'(n) = -\sum_{k=1}^{p} a_k s(n-k) \tag{1}$$

The difference between the actual, and predictable sample value is termed as prediction error or residual, given by,

$$e(n) = s(n) - s'(n) = s(n) + \sum_{k=1}^{p} a_k s(n-k) \tag{2}$$

The linear prediction coefficients $\{a_k\}$ are determined by minimizing the mean squared error over an analysis frame. It has been known that the LP order used for extracting the residual plays a crucial role on the performance of audio, speech and speaker recognition system [23]. The study shows that the optimal range of the LP order for the speaker-specific studies is in the range $8 - 16$ for speech signal sampled at 8kHz.

3.2 Suprasegmental Features in the LP Residual Signal

The behavioral characteristics of one speaker differ from that of the other. In Fig. 2 (a), (b) and (c), the LP residual signals for three speakers are shown. It can be noticed that there are some differences in the patterns in the residual signals of three speakers. Sometimes these differences may not be noticed in the waveform, but could be perceived while listening to the residual signal. Hence, it is worthwhile to explore these features for speaker recognition task.

Patterns in the LP residual signal are in the form of a sequence of excitation peaks. These excitation peaks can be considered as event markers. The sequence of these events contain important perceptual information about the source of excitation and behavioral characteristics of speaker. By listening to the residual signals, one can distinguish between speakers. Excitation peaks can further be enhanced by taking the Hilbert envelope of residual signal. The Hilbert envelope computation removes the phase information present in the residual, thereby leading to better identification of the excitation peaks.

4 The Hilbert Envelope of the LP Residual Signal

4.1 Computation of the Hilbert Envelope from Residual Signal

The residual signal is used to compute the Hilbert envelope, where the excitation peaks are manifested in a better way. The Hilbert envelope is defined as,

$$h_e(n) = \sqrt{e^2(n) + h^2(n)} \tag{3}$$

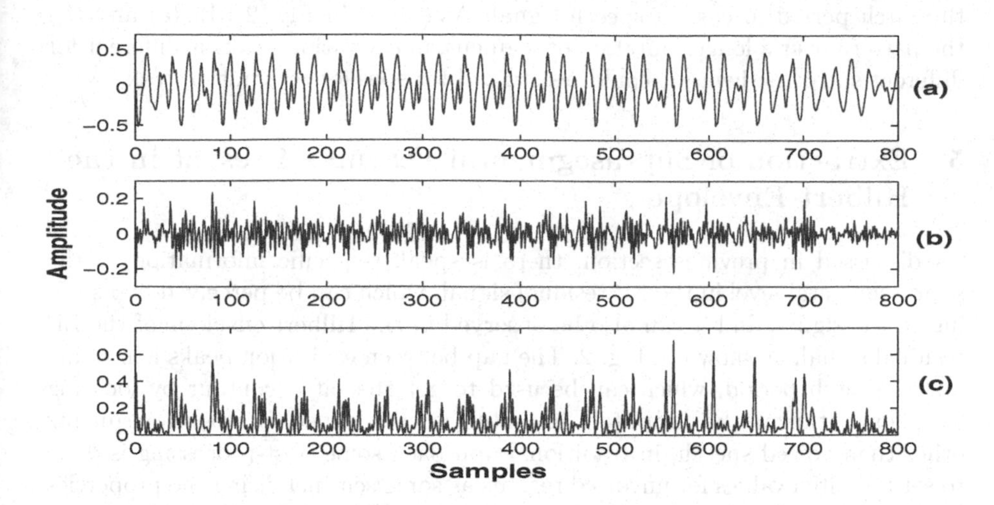

Fig. 1. (a) The speech signal, (b) LP residual and (c) modified Hilbert envelope of the residual signal for a speech segment

where $h_e(n)$ is the Hilbert envelope, $e(n)$ is the LP residual and $h(n)$ is the Hilbert transform of the residual. The Hilbert transform of a signal is the 90^0 phase shifted version of the original signal. Therefore, the Hilbert envelope represents the magnitude of the analytic signal,

$$x(n) = e(n) + ih(n) \tag{4}$$

where $x(n)$ is the analytic signal, $e(n)$ is the residual and $h(n)$ is the Hilbert transform of the residual.

The Hilbert envelope computation removes the phase information present in the residual. This leads to emphasis of the excitation peaks. The excitation peaks are further emphasized by using the neighborhood information of each sample in the Hilbert envelope. The modified Hilbert envelope is computed as,

$$h_{em}(n) = \frac{h_e^2(n)}{\sum_{k=n-l}^{k=n+l} \frac{h_e(k)}{(2l+1)}} \tag{5}$$

where $h_{em}(n)$ is the modified Hilbert envelope, $h_e(n)$ is the Hilbert envelope and l is the number of samples on either side of the neighborhood of current sample n.

4.2 Presence of Suprasegmental Features in the Hilbert Envelope

Fig. 1 shows the speech waveform, the residual and the modified Hilbert envelope for a speech segment. It can be noticed that the excitation peaks are clearly visible in the modified Hilbert envelope[1]. The gap between excitation peaks is

[1] For this study modified Hilbert envelope is considered. So in following text whenever Hilbert envelope is mentioned, it actually refers to the modified Hilbert envelope.

the pitch period in case of speech signal. As shown in Fig. 2 (d), (e) and (f), the pattern over a longer duration of segment, in excitation peaks is different for different speakers, hence it could be utilized for speaker recognition task.

5 Extraction of Suprasegmental Features Present in the Hilbert Envelope

As discussed in previous section, there is speaker-specific information at the suprasegmental level in the LP residual signal, which can be perceived by listening to the signal, and it can also be observed in the Hilbert envelope of the LP residual signal, as shown in Fig. 2. The gap between excitation peaks is nothing but the pitch period, which can be used to get the pitch contour by locating these excitation peaks accurately. As pitch is not defined for speech segments other than voiced speech, in traditional approach some post-processing is done to set the pitch values for unvoiced regions as some sentinel, using the properties like, pitch period cannot vary in a very drastic manner. Once the pitch contour is obtained, it is used to extract the desired clip level features for the task.

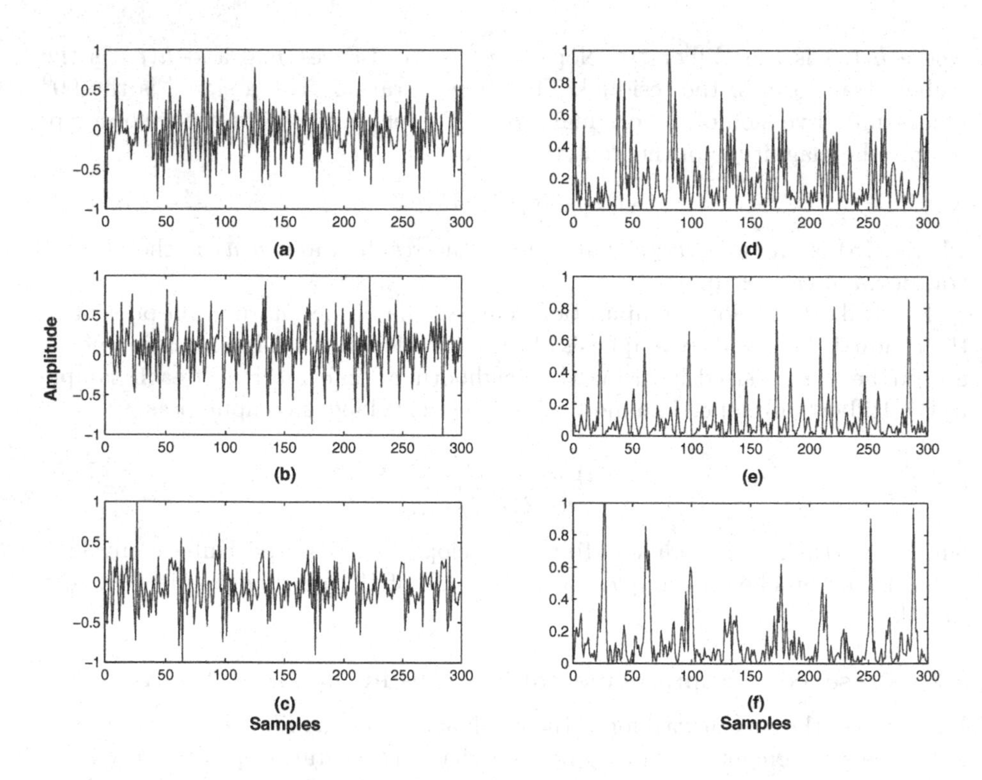

Fig. 2. The LP residual signals (a), (b) and (c), and corresponding Hilbert envelope signals (d), (e) and (f) for the speech segments of three speakers (rows 1 to 3 show residual and Hilbert envelope for speaker 1, speaker 2 and speaker 3 respectively)

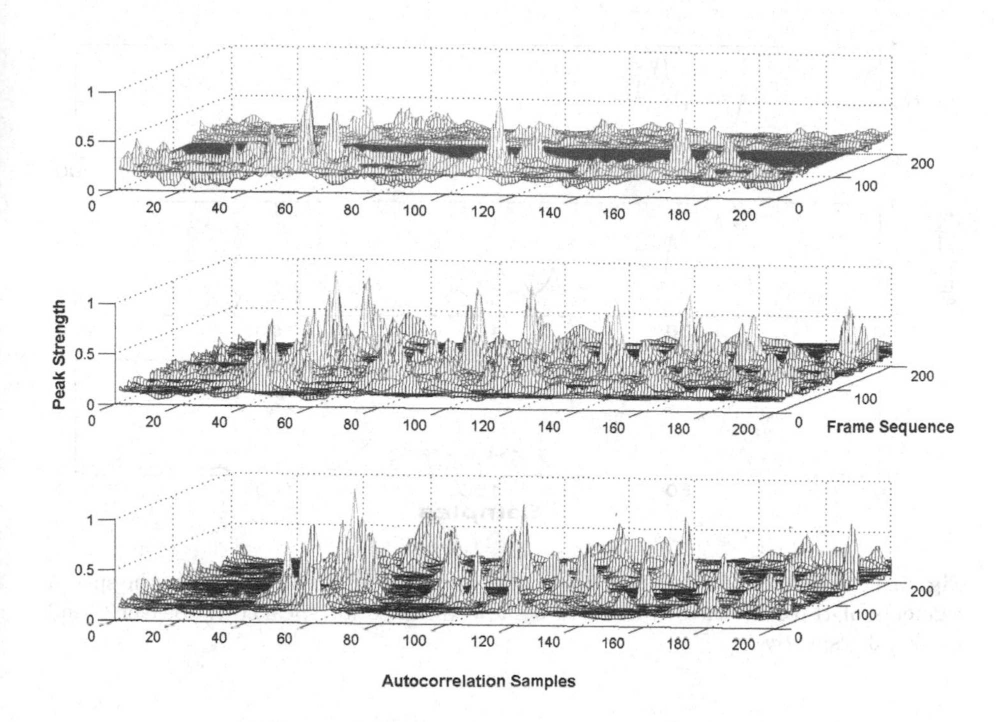

Fig. 3. Autocorrelation sequences of the Hilbert envelope of the residual signal for the speech segments of three speakers (rows 1 to 3 show autocorrelation sequences for speaker 1, speaker 2 and speaker 3 respectively)

However, for humans the speaking style of an individual is captured by listening to a longer duration of the speech, which contains voiced and unvoiced regions both. Hence, in this work a longer duration of window is considered for extraction of suprasegmental features, and the distribution of excitation peaks for the speech segment is considered as it is (without doing any post processing that sets the pitch value for unvoiced regions). One method to capture this pattern in the Hilbert envelope is by taking the autocorrelation of the Hilbert envelope. For a segment of 100 ms of the Hilbert envelope the autocorrelation sequence is calculated. The reason for choosing 100 ms window size for calculation of the autocorrelation is that the long term characteristics of the speech signal are of interest. The window is further shifted by 50 ms, and calculation of the autocorrelation is repeated till whole length of the speech clip is considered. These autocorrelation sequences (starting from 3rd sample from the center peak to 202th sample, normalized with respect to the central peak) are plotted in Fig. 3 for speech segments for three speakers, which shows speaker-specific pattern in the plot.

The variance of autocorrelation sequences along frame sequence axis for each of 3rd to 202th sample are calculated. The variance for the three speaker utterances, is plotted in Fig. 4. This pattern in variance of autocorrelation sequences is

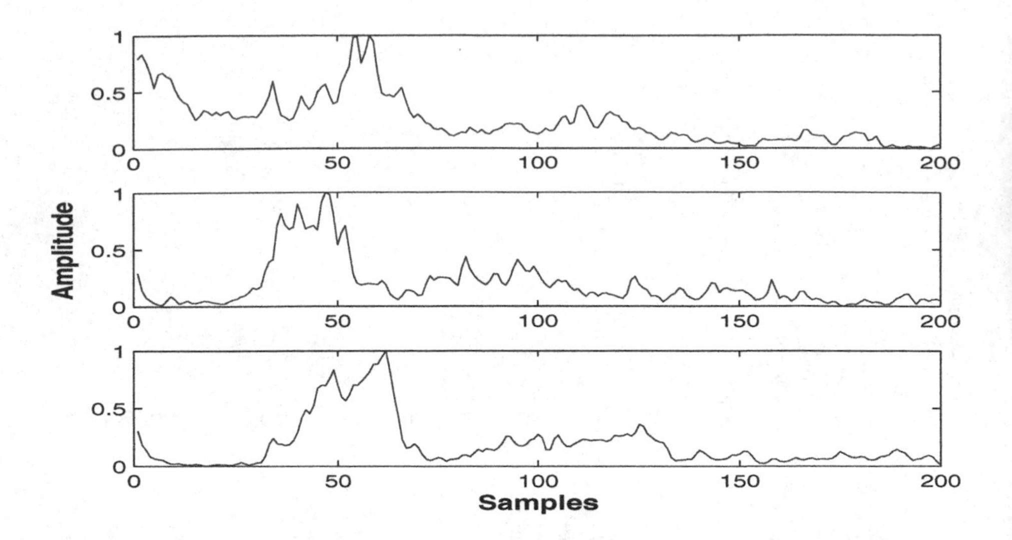

Fig. 4. The variance of autocorrelation sequences of the Hilbert envelope for the speech segments of three speakers (rows 1 to 3 show variance for speaker 1, speaker 2 and speaker 3 respectively)

utilized for speaker modeling and recognition using SVMs. SVMs are well-known for their good generalization performance [24].

6 Support Vector Machines for Speaker Recognition

Support vector machines [24] for pattern classification are built by mapping the input patterns into a higher dimensional feature space using a nonlinear transformation (kernel function), and then optimal hyperplanes are built in the feature space as decision surfaces between classes. Nonlinear transformation of input patterns should be such that the pattern classes are linearly separable in the feature space. According to Cover's theorem, nonlinearly separable patterns in a multidimensional space, when transformed into a new feature space are likely to be linearly separable with high probability, provided the transformation is nonlinear, and the dimension of the feature space is high enough [25]. The separation between the hyperplane and the closest data point is called the margin of separation, and the goal of a support vector machine is to find an optimal hyperplane for which the margin of separation is maximized. Construction of this hyperplane is performed in accordance with the principle of structural risk minimization that is rooted in Vapnik- Chervonenkis (VC) dimension theory [26]. By using an optimal separating hyperplane the VC dimension is minimized and generalization is achieved. The number of examples needed to learn a class of interest reliably is proportional to the VC dimension of that class. Thus, in order to have a less complex classification system, it is preferable to have those features which lead to lesser number of support vectors. The performance of the pattern classification problem depends on the type of kernel function chosen. In this work, we have used radial basis function as the kernel, since it is empirically observed to perform better than the other types of kernel functions. In this work, by classifying the test utterance to one of the known speaker classes (each individual speaker is mapped to a speaker class), the speaker recognition is mapped to the pattern classification problem.

7 Experimental Results

The database considered for study was collected in lab environment for 18 Indian speakers (12 male and 6 female speakers), using two channels - microphone and mobile phone, at 8KHz sampling frequency. Each speaker was asked to give speech samples in two languages, one English and other native language (one of the Indian languages, like Hindi, Kannada, Telugu etc). About 4 min of multilingual data per speaker/channel is used to train channel-dependent SVMs. 504 speech utterances belonging to the 18 speakers (28 multilingual utterances per speaker) are used for testing the models.

Table 1. Speaker recognition accuracy (%) for increasing duration of test utterance, based on suprasegmental features using SVM

	% Accuracy w.r.t. the duration of the test utterance					
Channel used for experiments	10 sec	20 sec	30 sec	40 sec	50 sec	60 sec
Microphone	87.81	89.62	93.52	93.97	95.04	95.23
Mobile	79.34	80.32	81.36	81.89	82.95	83.84

The experimental results of speaker recognition based on suprasegmental features present in Hilbert envelope of LP residual using SVM are shown in Table 1. The variance samples sequence, derived using autocorrelation sequences of Hilbert envelope for a test utterance, is used as feature vector for SVMs. Two similar experiments are conducted, one using microphone data for model building and testing and other using mobile phone data for the same. The accuracy is calculated in percentage of number of utterances correctly recognized out of total test utterances. It can noticed in Table 1, that in both experiments percentage accuracy increases with increasing the duration of the test utterance. It can be concluded from results that about 30 sec is the optimal duration of the test utterance that can be considered for the task. These results are for text-independent and language-independent speaker identification, where voiced and unvoiced regions are equally considered for suprasegmental feature extraction. Future directions of this study are given in the next section.

8 Conclusions and Future Work

The information present in speech signal can be categorized at three levels - subsegmental, segmental and suprasegmental. In this paper the presence of speaker-specific suprasegmental features in the LP residual signal is discussed as an additional evidence for speaker recognition task. The pattern in the excitation peaks in the LP residual for different speakers is enhanced by taking the Hilbert envelope of the residual signal. The statistics of the peak distribution in the autocorrelation sequences of the Hilbert envelopes are noticed to be different for different speakers. The variance of autocorrelation sequences is utilized for speaker recognition using SVMs. In this work, no distinction between voiced and unvoiced sounds is used for extraction of these features. For future work, the study needs to be extended to a larger number of speakers using an standard database and also the effect of channel variation and noise on the system needs to be studied. Further study is needed to explore the combination of features from the residual and spectrum to obtain significantly better performance.

References

1. Furui, S.: Speaker-independent and speakeradaptive recognition techniques. In: Furui, S., Sondhi, M.M. (eds.) Advances in Speech signal processing, pp. 597–622. Marcel Dekker (1991)
2. Makhoul, J.: Linear Prediction: A Tutorial Review. Proc. IEEE 63(4), 561–580 (1975)
3. Yegnanarayana, B., Prasanna, S.R.M., Rao, K.S.: Speech Enhancement using Excitation Source Information. In: Proc. IEEE Int. Conf. Acoust., Speech, and Signal Processing, Orlando, FL, USA (May 2002)
4. Ananthapadmanabha, T.V., Yegnanarayana, B.: Epoch Extraction from Linear Prediction Residual for Identification of Closed Glottis Interval. IEEE Trans. Acoust., Speech, Signal Processing ASSP-27(4), 309–319 (1979)
5. Yegnanarayana, B., Prasanna, S.R.M., Zachariah, J.M., Gupta, C.S.: Combining Evidence from Source, Suprasegmental and Spectral Features for a Fixed-Text Speaker Verification System. IEEE Trans. Speech and Audio Processing 13(4) (July 2005)
6. Campbell, J.P.: Speaker recognition: A tutorial. Proc. IEEE 85(9), 1436–1462 (1997)
7. Bimbot, F., et al.: A tutorial on text-independent speaker verification. EURASIP Journal on Applied Signal Processing 4, 430–451 (2004)
8. Yegnanarayana, B., Reddy, K.S., Kishore, S.P.: Source and System Features for Speaker Recognition using AANN Models. In: Proc. IEEE Int. Conf. Acoust., Speech, and Signal Processing, Saltlake City, Utah, USA (May 2001)
9. Prasanna, S.R.M., Gupta, C.S., Yegnanarayana, B.: Autoassociative Neural Network Models for Speaker Verification using Source Features. In: Proc. Int. Conf. Cognitive and Neural Systems, Boston, USA (May 2002)
10. Pruzansky, S.: Pattern-matching procedure for automatic talker recognition. J. Acoust. Soc. Amer. 35, 354–358 (1963)
11. Li, K.P., et al.: Experimental studies in speaker verification using a adaptive system. J. Acoust. Soc. Amer. 40, 966–978 (1966)
12. Doddington, G.: A method of speaker verification. J. Acoust. Soc. Amer. 49, 139 (A) (1971)
13. Li, K.P., Hughes, G.W.: Talker differences as they appear in correlation matrices of continuous speech spectra. J. Acoust. Soc. Amer. 55(4), 833–837 (1974)

14. Beek, B., et al.: An assessment of the technology of automatic speech recognition for military applications. IEEE Trans. Acoust., Speech, Signal Processing 25, 310–322 (1977)
15. Sambur, M.R.: Speaker recognition using orthogonal linear prediction. IEEE Trans. Acoust., Speech, Signal Processing 24, 283–289 (1976)
16. Furui, S., Itakura, F., Satio, S.: Talker recognition by long-time averaged speech spectrum. Electron Commun., Jap. 55-A, 54–61 (1972)
17. Soong, F.K., Rosenberg, A.E., Rabiner, L.R., Juang, B.H.: A vector quantization approach to speaker recognition. In: Proc. IEEE Int. Conf. Acoust., Speech, and Signal Processing, pp. 387–390 (1985)
18. Rosenberg, A.E., Soong, F.K.: Evaluation of a vector quantization talker recognition system in a text independent and text dependent modes. In: Proc. IEEE Int. Conf. Acoust., Speech, and Signal Processing, pp. 873–876 (1986)
19. Poritz, A.B.: Linear predictive hidden markov models and the speech signal. In: Proc. IEEE Int. Conf. Acoust., Speech, and Signal Processing, pp. 1291–1294 (1982)
20. Reynolds, D.A.: Speaker identification and verification using gaussian mixture models. Speech Comm. 17, 91–108 (1995)
21. Higgins, A.L., Bahler, L., Porter, J.: Voice identification using nonparametric density matching. In: Lee, C.H., Soong, F.K., Paliwal, K.K. (eds.) Automatic Speech and Speaker Recognition, pp. 211–232. Kluwer Academic, Boston (1996)
22. Doddington, G.R.: Speaker recognition based on idiolectal differences between speakers. In: Eurospeech, pp. 2521–2524 (2001)
23. Prasanna, S.R.M., Gupta, C.S., Yegnanarayana, B.: Source Information from Linear Prediction Residual for Speaker Recognition. Communicated to J. Acoust. Soc. Amer. (2002)
24. Collobert, R., Bengio, S.: Svmtorch: Support vector machines for large-scale regression problems. Journal of Machine Learning Research 1, 143–160 (2001)
25. Haykin, S.: Neural Networks: A Comprehensive Foundation. Macmillan College Publishing Company, New York (1994)
26. Vapnik, V.: Statistical Learning Theory. John Wiley and Sons, New York (1998)

Face Recognition Using Fisher Linear Discriminant Analysis and Support Vector Machine

Sweta Thakur[1], Jamuna K. Sing[2,*], Dipak K. Basu[2], and Mita Nasipuri[2]

[1] Department of Information Technology,
Netaji Subhas Engineering College, Kolkata, India
[2] Department of Computer Science & Engineering, Jadavpur University, Kolkata, India
jksing@ieee.org

Abstract. A new face recognition method is presented based on Fisher's Linear Discriminant Analysis (FLDA) and Support Vector Machine (SVM). The FLDA projects the high dimensional image space into a relatively low-dimensional space to acquire most discriminant features among the different classes. Recently, SVM has been used as a new technique for pattern classification and recognition. We have used SVM as a classifier, which classifies the face images based on the extracted features. We have tested the potential of SVM on the ORL face database. The experimental results show that the proposed method provides higher recognition rates compared to some other existing methods.

Keywords: Fisher's Linear Discriminant Analysis (FLDA), Support Vector Machine (SVM).

1 Introduction

Face recognition is one of the important areas in the field of pattern recognition and artificial intelligence. It has been used in wide range of applications such as biometrics, law enforcement, identity authentication and surveillance. The image data are always high dimensional in the face recognition area, and it require considerable amount of computing time for recognition. That's why the feature selection is very important for improving classifier's accuracy and reducing the running time for classification. Principal component analysis (PCA) and Fisher's linear discriminant analysis (FLDA) are two powerful methods used for data reduction as well as feature extraction in appearance-based face recognition approaches. The PCA method is based on linearly projecting the image space into a low dimensional feature space for dimensionality reduction [1]. It yields projection directions that maximize the total scatter across all classes, i.e., across all images of all training faces. Thus, PCA retains unwanted variations due to lighting and facial expression [2]. The FLDA technique projects the face images from high-dimensional image space to a relatively low-dimensional space linearly by maximizing the ratio of between-class scatter to that of within-class scatter. It is generally believed that, when it comes to solving problems

* Corresponding author.

S. Ranka et al. (Eds.): IC3 2009, CCIS 40, pp. 318–326, 2009.
© Springer-Verlag Berlin Heidelberg 2009

of pattern classification, FLDA-based algorithms perform better than PCA-based ones, since the former optimizes the low-dimensional representation of the objects with focus on the most discriminant feature extraction while the latter achieves simply object reconstruction [2].

Support Vector Machine (SVM) is a popular classification tool, which is applied for pattern recognition as well as computer vision domains recently [3]. A SVM is used to find the hyperplane that separates the largest fraction of points of the same class on the same side, while maximizing the distance from the either class to the hyperplane [4]. This hyperplane is called Optimal Separating Hyperplane [10], which minimizes the risk of misclassification in the training as well as unknown test set. In this new method, to achieve higher performance in face recognition, we basically focused on two criteria. Firstly, the features are extracted from a facial image. This feature extracting technique gives the best discriminant features among classes rather than data. Secondly, we choose a classifier, which basically trained and learned those face image and then finally classify test face images based on the extracted features. This classifier helps to achieve better generalization capability.

2 Proposed Method

The schematic diagram of the proposed method is shown in Fig. 1. The features are extracted from the face image using the FLDA technique. The FLDA method extracts the most discriminant features among the classes. Finally, a multiclass SVM is used to classify the face images based on the extracted features.

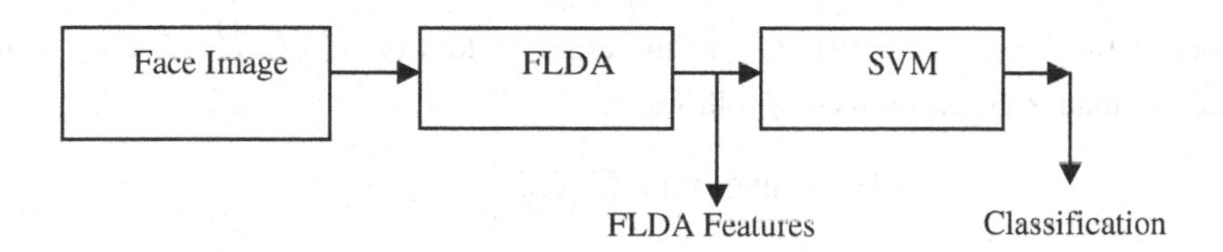

Fig. 1. Schematic diagram of the proposed method

3 Feature Extraction Using FLDA

The facial features are extracted from the face images using the Fisher's linear discriminant analysis (FLDA) technique. The FLDA gives importance to those vectors in the underlying space that best describe the best discriminate among classes rather than best describing the data. It projects the face images from high dimensional image space to a low-dimensional image space linearly by maximizing the ratio of between-class scatter to that of within-class scatter matrix. Let, there are R face images of C individuals (classes) in the training set and each image X_i is a 2-dimensional array of size $m \times n$ of intensity values. An image X_i is converted into a vector of D ($D=m \times n$) pixels, where, $X_i = (x_{i1}, x_{i2}, \ldots, x_{iD})$. Define the training set of R images by $X = (X_1, X_2, \ldots, X_R) \subset \Re^{DxR}$. The between-class scatter matrix is defined as follows:

$$C_B = \sum_{c}^{C} N_c \left(\mu_c - \overline{X}\right)\left(\mu_c - \overline{X}\right)^T \tag{1}$$

where N_c is the number of samples in class c. μ_c and $\overline{X}$ are the mean images of the c class and the training images, respectively. They are defined as follows:

$$\mu_c = \frac{1}{N_c} \sum_{i \in c}^{R} X_i \tag{2}$$

$$\overline{\mathrm{X}} = \frac{1}{\mathrm{R}} \sum_{\mathrm{i}}^{\mathrm{R}} \mathrm{X}_{\mathrm{i}} \tag{3}$$

The within-class scatter matrix is defined as follows:

$$C_W = \sum_{c}^{C} \sum_{i \in c}^{R} (X_i - \mu_c)(X_i - \mu_c)^T \tag{4}$$

Then the Fisher's criteria is defined as follows:

$$F(Q) = \frac{\left|QC_B Q^T\right|}{\left|QC_W Q^T\right|} \tag{5}$$

If the C_W is a nonsingular matrix then this ratio is maximized when the column vectors of the projection matrix Q, are the eigenvectors of $C_B C_W^{-1}$. The optimal projection matrix Q_{opt} is defined as follows:

$$\begin{aligned} Q_{opt} &= \arg\max_{Q} \left|C_B C_W^{-1}\right| \\ &= [q_1, q_2, \ldots, q_m] \end{aligned} \tag{6}$$

where $\{q_i \mid i=1, 2, \ldots, m\}$ is the set of normalized eigenvectors of $C_B C_W^{-1}$ corresponding to m largest eigenvalues $\{\lambda_i \mid i=1, 2, \ldots, m\}$. Each of the m eigenvectors is called a *fisherface.*

Now, each of the face images of the training set $\mathbf{X}_i$ is projected into the lower-dimensional fisherface space spanned by the m normalized eigenvectors to obtain its corresponding FLDA-based features $\mathbf{Z}_i$, which is defined as follows:

$$\boldsymbol{Z}_i = \boldsymbol{Q}_{opt}^T \boldsymbol{Y}_i, \boldsymbol{i} = 1,2,\ldots, \boldsymbol{m} \tag{7}$$

where Y_i is mean-subtracted image of X_i and defined as follows:

$$\mathrm{Y}_i = \mathrm{X}_i - \overline{\mathrm{X}}_i \tag{8}$$

In order to recognize the test images, features are extracted from each of the test images by projecting into the fisherface space using the equation (7) and then fed to the SVM networks as inputs for classification.

4 Support Vector Machine (SVM)

SVM has been used successfully in a wide range of applications due to its high generalization ability. The basic idea behind SVM is that it maximizes the margin around the separating hyperplane between two classes. It basically maps the input space into a high dimension feature space using non-linear transformation, which is defined by the inner product function. This helps the linear classifier to be built in this space by making the samples linearly separable.

4.1 Basic Theory of SVM for Two Classes

The basic idea is to separate the two classes by a function. The Fig. 2 shows the two-dimensional example with the cases completely separated. One category of the target variables is represented by ovals are in the upper side while the other category is represented by rectangles are in the lower side.

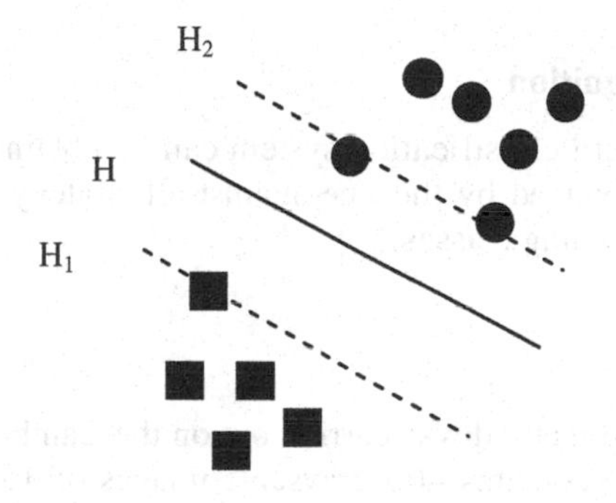

Fig. 2. SVM for two-class problem

The SVM analysis attempts to find a 1-dimensional hyperplane (i.e. a line) that separates the cases based on their target categories. There are an infinite number of possible lines; two candidate lines are shown above. The optimum separation hyperplane (OSH) is the linear classifier with the maximum margin for a given finite set of learning patterns. The Fig. 2 shows that the dashed lines (H1 & H2) drawn parallel to the separating line mark the distance between the dividing line and the closest vectors to the line and H is the hyperplane. The distance between the dashed lines is called the *margin*. The vectors (points) that constrain the width of the margin are the *support vectors*.

Consider the classification of two classes $(x_1, y_1), \ldots, (x_n, y_n)$, where $x_i \in S^m$, $y_i \in \{-1,+1\}$ of patterns that are linearly separable. The equation of the hyperplane is $w.x + b = 0$, where a separating hyperplane must satisfy the following constraints [4]:

$$y_i(w \cdot x_i + b) \geq +1 \tag{9}$$

or

$$y_i(w \cdot x_i + b) - 1 \geq 0 \tag{10}$$

Each such hyperplane (**w**, *b*) is a classifier here that correctly separates all patterns from the training set because we have considered the case of linearly separable classes, where the margin is 2 / || w || .The hyperplane which satisfies equation (9) and minimizes the value of $\| w \|^2 / 2$ is called the optimal hyperplane. According to Vapnik [10], the classification function is defined as follows:

$$F(a) = \sum_{i=1}^{n} a_i - 1/2 \sum_{i,j=1}^{n} y_i y_j a_i a_j (x_i . x_j) \tag{11}$$

where the constraints are $\sum_{i=1}^{n} a_i y_i = 0, a_i \geq 0, i, \ldots\ldots\ldots, n$.

4.2 SVM for Multiclass Recognition

An SVM-based multi-class pattern classification system can be obtained by combing two class SVMs. This can be realized by the one-against-all strategy to classify between each class and all the remaining classes.

5 Experimental Results

The performance of the proposed method was carried out on the Cambridge ORL face database [6]. The ORL database contains 400 grayscale images of 40 persons. Each person has 10 images, each having a resolution of 112x92 and 256 gray levels. Images of the individuals have been taken varying light intensity, facial expressions (open/closed eyes, smiling/not smiling) and facial details (glasses/no glasses). All the images were taken against a dark homogeneous background, with tilt and rotation up to 20° and scale variation up to 10%. Sample face images of a person are shown in Fig. 3. To reduce the computational complexity, each image is down sampled into a size of 16x16.

The recognition rate is defined as the ratio of the total number of correct recognition by the method to the total number of images in the test set for a single experimental run. Therefore, the average recognition rate, R_{avg}, of the method is defined as follows:

$$R_{avg} = \frac{\sum_{i=1}^{q} n_{cls}^{i}}{q * n_{tot}} \tag{12}$$

where q is the number of experimental runs. The n_{cls}^{i} is the number of correct recognition in the ith run and n_{tot} is the total number of faces under test in each run.

Fig. 3. Sample images of a person from the ORL database

5.1 Randomly Partitioning the Database

In our first experiments, we partition the database randomly to form *training* and *test* sets. We have selected randomly *s* images per person from the database to form the *training set*. Remaining images are selected to form the corresponding *test set*. It should be noted that there is no overlap between the training and test images. In this way ten (q=10) different training and test sets has been generated for each value of *s*.

Table 1 shows the average, maximum and minimum recognition rates for different numbers of features using s=5. From the table 1, it may be noted that the best average recognition rate (97.95%) is obtained when the features are considered to be 90. We considered 90 features for all of our experiments in the case of s=4 & 6.

Table 2 summarizes the recognition rates of the proposed method for s=4, 5 and 6 using 90 features. The average recognition rates are found to be 96.30%, 97.95% and 97.31% for s=4, 5 and 6, respectively.

We have compared the average recognition rates (in %) of the proposed method with the FLDA-based methods reported in [4], [7]-[9] and [11], which are also evaluated on the ORL database. The comparisons are shown in Table 3. It can be seen from the table that the proposed method is superior to those reported in [4], [7]-[9] and

Table 1. Recognition rates of the proposed method by varying the number of features using s=5

# features	# average support vectors	R_{avg} (%)	Maximum (%)	Minimum (%)
50	180	95.55	96.50	94.50
60	186	96.70	97.50	96.00
70	183	97.00	97.50	96.50
80	179	97.40	98.50	96.50
90	182	97.95	100.00	97.50
100	179	97.50	97.50	97.50
110	180	97.45	98.50	97.00
120	171	97.55	99.00	96.50
130	174	97.70	99.00	97.00
150	176	97.25	98.50	96.50

Table 2. Recognition rates of the proposed method for *s*=4, 5 and 6

s	4	5	6
# training samples	160	200	240
# avg. support vectors	183	182	176
R_{avg} (%)	96.30	97.95	97.31
Maximum (%)	97.50	100.00	98.12
Minimum (%)	95.00	97.50	96.87

Table 3. Comparison of the average recognition rates (in %) of different FLDA-based approaches on the ORL database

s	4	5	6
DLDA [7]	91.77	94.55	95.97
2DFLD [7]	93.46	95.40	96.30
PCA+FLD [7]	89.42	90.98	92.97
LDA [8]	-	88.87	-
LDA+RBF [8]	-	94.00	-
FLDA [9]	96.75	97.80	96.38
SVM+SGFS [11]	-	95.00	-
SVM+PCA [4]	-	97.00	-
Proposed Method	96.30	97.95	97.31

[11]. In [9] the number of features considered were 150 while in our proposed method we comparatively considered very less number of features i.e. 90.We have performed our evaluation on ten experimental runs while in [4] the result obtained based on four experimental runs.

5.2 N-Fold Cross Validation Test

In this case, the ORL database is randomly divided into (N) ten-folds, taking one image of a person into a fold. Therefore, each fold consists of 40 images, each one corresponding to a different person. For ten-folds cross validation test, in each experimental run, nine folds are used to train the SVM and remaining one fold is used for testing. Therefore, training and test sets consist of 360 and 40 images, respectively in a particular experimental run. The average recognition rate using 90 features in ten-folds cross validation tests is shown in Table 4. In this case, the average recognition rate is found to be 98.75% with maximum and minimum recognition rates of 100% and 97.50%, respectively.

Table 4. Recognition rates using N-fold cross validation test on the ORL database

# training samples	360
# avg. support vectors	157
R_{avg} (%)	98.75
Maximum (%)	100.00
Minimum (%)	97.50

6 Conclusion

In this paper, we have presented a new method for face recognition based on FLDA and an SVM classifier. The facial features are extracted using the FLDA and the SVM is used as a classifier, which helps us in learning and classifying the images based on these extracted features. The average recognition rates we obtained with our proposed method are 97.95%, 96.30% and 97.31% for s=4, 5 and 6, respectively. The experimental results on the ORL database show that the proposed method achieves higher recognition rates in comparison with the other FLDA-based methods reported in the literature.

Acknowledgments. This work was partially supported by the CMATER and the SRUVM projects of the Department of Computer Science & Engineering, Jadavpur University, Kolkata, India. The author, S. Thakur would like to thank Netaji Subhas Engineering College, Kolkata for providing computing facilities and allowing time for conducting research works. The author, D. K. Basu would also like to thank the AICTE, New Delhi for providing him the Emeritus Fellowship (F.No.: 1-51/RID/EF(13)/2007-08, dated 28-02-2008).

References

1. Turk, M., Pentland, A.: Eigenface for recognition. J. Cognitive Neuroscience 3, 71–86 (1991)
2. Belhumeur, P.N., Hespanha, J.P., Kriegman, D.J.: Eigenfaces versus Fisherfaces: recognition using class specific linear projection. IEEE Trans. Pattern Anal. Mach. Intell. 23, 711–720 (1997)
3. Cao, L.J., Chua, K.S., Chong, W.K., Lee, H.P., Gu, Q.M.: A comparison of PCA, KPCA and ICA for Dimensionality Reduction in Support Vector Machine. J. Neurocomputing 55, 321–336 (2003)
4. Guo, G.D., Li, S.Z., Chen, K.L.: Support Vector Machine for Face Recognition. J. Image and Vision computing 19, 631–638 (2001)
5. Chellapa, R., Wilson, C., Sirohey, S.: Human and machine recognition of faces: a survey. J. IEEE 83(5), 705–741 (1995)
6. ORL face database. AT&T Laboratories, Cambridge, U. K., `http://www.uk.research.att.com/facedatabase.html`

7. Xiong, H., Swamy, M.N.S., Ahmad, M.O.: Two-dimensional FLD for face recognition. J. Pattern Recognition 38, 1121–1124 (2005)
8. Pan, Y.Q., Liu, Y., Zheng, Y.W.: Face recognition using kernel PCA and hybrid flexible neural tree. In: International Conference on Wavelet Analysis and Pattern Recognition, China, pp. 1361–1366 (2007)
9. Thakur, S., Sing, J.K., Basu, D.K., Nasipuri, M.: Face recognition by integrating RBF neural networks and a distance measure. In: International Conference on Computer,Communication, Control and Information Technology, India, pp. 264–269 (2009)
10. Vapnik, V.N.: Statistical learning theory. John Wiley & Sons, New York (1998)
11. Wang, L., Sun, Y.: A new approach for face recognition based on SGFS and SVM. Proc. IEEE, 527–530 (2007)
12. Weihong, L., Gong, W., Liang, Y., Chen, W.: Feature selection based on KPCA, SVM and GSFS for face recognition. In: Singh, S., Singh, M., Apte, C., Perner, P. (eds.) ICAPR 2005. LNCS, vol. 3687, pp. 344–350. Springer, Heidelberg (2005)

Threshold Signature Cryptography Scheme in Wireless Ad-Hoc Computing

Sandip Vijay[1] and Subhash C. Sharma[2]

[1] Wireless Computing Research Lab., Electronics & Computer Discipline, DPT, IIT, Roorkee,Saharanpur Campus, Saharanpur, UP-247001, India
snvijdpt@iitr.ernet.in

[2] Associate Professor, Wireless Computing Research Lab., Electronics & Computer Discipline, DPT, IIT, Roorkee, Saharanpur Campus, Saharanpur, UP-247001, India
scs60fpt@iitr.ernet.in

Abstract. Identity-based systems have the property that a user's public key can be easily calculated from his identity by a publicly available function. The bilinear pairings, especially Tate pairings, have high performance in cryptography. With the foundation of above two properties, we have proposed a new ID-Based (t,n) threshold signature scheme from Tate pairings. The scheme is proved secure that it can resist attacks including plaintext attack, recovery equation attack, conspiracy attack and impersonation attack. Furthermore, performance analysis shows that the proposed scheme is simple, efficient so that it will be suitable for an environment of finite bandwidth and low capability equipment.

Keywords: Threshold signature, identity-based, Tate pairing.

1 Introduction

In 1984, Shamir introduced the concept of identity-based (ID-based) systems to simplify key management procedures of CA-based Public Key Infrastructure (PKI) [1]. Since then, several ID-based signature schemes have been proposed [2-4]. ID-based systems can be a good alternative for CA-based systems from the viewpoint of efficiency and convenience. ID-based systems have a property that a user's public key can be easily calculated from his identity by a publicly available function, while his private key can be calculated by a trusted Key Generation Center (KGC). They enable any pair of users to communicate securely without exchanging public key certificates, without keeping a public key directory, and without using online service of a third party, as long as a trusted KGC issues a private key to each user when he first joins the network.

Ever since threshold signature was first proposed by Desmedt and Frankel [5], several threshold signature schemes from bilinear pairings have been proposed. A. Boldyreva proposed a robust and proactive threshold signature scheme which works in any Gap Diffie-Hellman (GDH) group [6]. Baek and Zheng formalized the concept of identity-based threshold signature and gave the first provably secure scheme [7].

S. Ranka et al. (Eds.): IC3 2009, CCIS 40, pp. 327–335, 2009.
© Springer-Verlag Berlin Heidelberg 2009

Chen proposed an ID-based threshold signature scheme without a trusted KGC[8]. Cheng et al. proposed an ID-based signature from m-torsion groups of super-singular elliptic curves or hyper-elliptic curves [9]. It is proved that the time of Tate pairing operations is a half of that of Weil pairings [10]. In this paper, we give a secure ID-based signature scheme from Tate pairings.

The rest of this paper is organized as follows. In Section 2, we briefly introduce related properties of the Tate pairing. We propose an ID-based signature scheme from Tate pairing in Section 3. Section 4 and section5 carries out security and performance analysis of the proposed scheme. Conclusion is drawn in the last section.

2 Tate Pairing

Let E be an elliptic curve over a finite F_q. We write O_E for the point at infinity on E. Let l be a positive integer which is coprime to q. In most applications l is a prime and $l \mid \# E(F_q)$. Let k be a positive integer such that the field F_{q^k} contains the lth roots of unity (in other words, $l \mid (q^k - 1)$). Let $G = E(F_{q^k})$ and write $G[l]$ for the subgroup of points of order l and G / lG for the quotient group (which is also a group of exponent l). Then the Tate pairing is a mapping:

$$t : G[l] \times G / lG \to F_{q^k}^* / (F_{q^k}^*)^l$$

The Tate pairing satisfies the following properties:

a. *Bilinearity*:

$\forall P, P_1, P_2 \in G[l]$, $\forall Q, Q_1, Q_2 \in G / lG$,
$t(P_1 + P_2, Q) = t(P_1, Q) t(P_2, Q)$
and $t(P, Q_1 + Q_2) = t(P, Q_1) t(P, Q_2)$;

$\forall a, b \in Z_q$, we have

$t(aP, bQ) = t(bP, aQ) = t(P, Q)^{ab}$;

b. *Non-degeneracy*:
If $t(P,Q) = 1 \ \forall Q \in G[l]$, then $P = O_E$. Conversely, for each $P \neq O_E$ $\exists Q \in G[l]$ so that $t(P,Q) \neq 1$;

c. *Well-defined*:
$(O_E, Q) \in (F_{q^k}^*)^l$ for all $Q \in G$ and $(P,Q) \in (F_{q^k}^*)^l$ for all $P \in G[l], Q \in lG$

3 ID-Based (t,n) Threshold Signature Scheme from Tate Pairings

In this section, we shall present a new group-oriented threshold signature scheme. It consists of four algorithms: *System setup, Private key extraction, Signature generation and signature verification.* The ID-based (t,n) threshold signature is described as follows.

(1) *System setup*: Let P is the generator of G. Our ID-based signature scheme is based on G. The trust KGC randomly chooses $a_0, a_1, a_2, \cdots, a_{t-1} \in Z_q^*$, $P_{pub} = a_0 P$. It constructs a polynomial of degree $t-1$:

$$f(x) = (a_0 + a_1 x + a_2 x^2 + \cdots + a_{t-1} x^{t-1}) \bmod q.$$

For $i = 1,2,\cdots,n$, $1 \le j \le t$, it computes and publishes $P_{pub}^{(i_j)} = f(ID)P$, where ID is the public identifier of each P_{i_j}. Before requesting his private share, each player can check that $\sum_{j=1}^{t} L_{i_j} P_{pub}^{(i_j)} = P_{pub}$ for any subset $B = \{P_{i_1}, P_{i_2}, \cdots P_{i_t}\}$ of the player set $A = \{P_1, P_2, \cdots P_n\}$, where L_{i_j} denotes the Lagrange coefficient:

$$L_{i_j} = \prod_{k=1, k \ne j}^{t} \frac{0 - ID_{i_k}}{ID_{i_j} - ID_{i_k}} \bmod q.$$

(2) *Private key extraction and distribution*: The group secret key of P_i can be set by $f(0) = a_0$ and the corresponding group public key $Y_i = f(0)P \bmod l = a_0 P \bmod l$. For the purpose of security, the KGC defines two one-way hash functions $H : \{0,1\}^* \to G^*$ and $H' : \{0,1\}^* \to Z_q^*$ and makes it public. Given an identity ID, the KGC plays the role of the trusted dealer. It computes a secret key publishes $S_{ID}^{(i_j)} = f(ID) Q_{ID}$ for each player P_{i_j}, where $Q_{ID} = H(ID)$ is the public key associated with the public identifier ID of P_{i_j}. P_{i_j} accepts $S_{ID}^{(i_j)}$ as his private key if $t(P, S_{ID}^{(i_j)}) = t(P_{pub}^{(i_j)}, Q_{ID})$; otherwise, he shows his complains to the KGC. In summary, the parameters are those listed in Table1.

Table 1. The parameters of the proposed scheme

Participants	Secret Parameters	Public Parameters
KGC	$f(x)$	$G, P, P_{pub}^{(i_j)}, P_{pub}, H, H'$
Signer	$S_{ID}^{(i_j)}$	ID, Q_{ID}
Signer group	$f(0)$	Y_i

(3) *Partial signature generation*: We assume that $B = \{P_{i_1}, P_{i_2}, \cdots P_{i_t}\}$ is the set of the t players designated to join the signing. each player P_{i_j} $(1 \le j \le t)$ randomly chooses $k_j \in Z_q^*$ computes two values u_j and r_j as

$$u_j = k_j P_{pub} \bmod l,$$
$$r_j = k_j Y_i \bmod l.$$

Then each P_{i_j} transmits (u_j, r_j) to the other $(t-1)$ signers via a secure channel.

Upon receiving all (u_j, r_j), each P_{i_j} computes u, r and v as follows:

$$u = \sum_{j=1}^{t} u_j \bmod l, \tag{1}$$

$$r = \sum_{j=1}^{t} r_j \bmod l, \tag{2}$$

$$h = H'(u, r, M), \tag{3}$$

$$v_j = L_{i_j}(P_{pub}^{(i_j)} + hS_{ID}^{(i_j)}) \bmod q. \tag{4}$$

The partial signature on message M given by player P_{i_j} is $\sigma_j = (u_j, v_j)$.

(4) *Threshold signature generation*: Anyone in $B = \{P_{i_1}, P_{i_2}, \cdots P_{i_t}\}$ can be designated to reconstruct the partial signature. After having received the partial signatures, the designated player (DP) first verifies the validity of each partial signature. σ_j is accepted if $t(v_j, P) = t(P + hQ_{ID}, P_{pub}^{(i_j)})^{L_{i_j}}$. Without lose of

generality, we assume that the partial signatures are all valid. DP computes $u=\sum_{j=1}^{t}u_j$ and $v=\sum_{j=1}^{t}v_j$. Then $\sigma=(u,v)$ is the signature on message M .

(5) *Threshold signature verification*: After receiving $\sigma=(u,v)$, the verifier computes $h=H'(M,u)$ and accepts the signature if

$$t(v,P)=t(P+hQ_{ID},P_{pub}). \tag{5}$$

Correctness of the partial signature:

$$\begin{aligned}
t(v_j,P) &= t(L_{i_j}(P_{pub}^{(i_j)}+hS_{ID}^{(i_j)}),P)\\
&= t(L_{i_j}P_{pub}^{(i_j)},P)t(L_{i_j}hS_{ID}^{(i_j)},P)\\
&= t(L_{i_j}f(ID_{i_j})P,P)t(L_{i_j}hf(ID_{i_j})Q_{ID},P)\\
&= t(L_{i_j}P,f(ID_{i_j})P)t(L_{i_j}hQ_{ID},f(ID_{i_j})P)\\
&= t(L_{i_j}P,P_{pub}^{(i_j)})t(L_{i_j}hQ_{ID},P_{pub}^{(i_j)})\\
&= t(L_{i_j}P+L_{i_j}hQ_{ID},P_{pub}^{(i_j)})\\
&= t(P+hQ_{ID},P_{pub}^{(i_j)})^{L_{i_j}}
\end{aligned}$$

Correctness of the threshold signature:

$$\begin{aligned}
v &= \sum_{j=1}^{t}v_j=\sum_{j=1}^{t}(L_{i_j}(P_{pub}^{(i_j)}+hS_{ID}^{(i_j)}))\\
&= \sum_{j=1}^{t}L_{i_j}P_{pub}^{(i_j)}+\sum_{j=1}^{t}L_{i_j}hS_{ID}^{(i_j)}\\
&= P_{pub}+hQ_{ID}\sum_{j=1}^{t}L_{i_j}f(ID_{i_j})
\end{aligned}$$

$$\begin{aligned}
t(v,P) &= t(P_{pub}+hQ_{ID}\sum_{j=1}^{t}L_{i_j}f(ID_{i_j}),P)\\
&= t(P_{pub},P)t(hQ_{ID}\sum_{j=1}^{t}L_{i_j}f(ID_{i_j}),P)\\
&= t(P_{pub},P)t(hQ_{ID},\sum_{j=1}^{t}L_{i_j}f(ID_{i_j})P)\\
&= t(P_{pub},P)t(hQ_{ID},\sum_{j=1}^{t}L_{i_j}P_{pub}^{(i_j)})\\
&= t(P_{pub},P)t(hQ_{ID},P_{pub})\\
&= t(P+hQ_{ID},P_{pub})
\end{aligned}$$

4 Security Analysis of Our Threshold Scheme

The security of the proposed scheme is based on the well-known difficulty of computing the one-way hash function and the cryptographic assumption of discrete logarithms. In the following paragraphs, we shall consider some attacks against our proposed scheme. We shall demonstrate that our proposed scheme can successfully withstand those attacks.

(1) Plaintext attacks

An adversary tries to expose a signer's secret key $S_{ID}^{(i)}$ from the corresponding public key Q_{ID}. However, it is as difficult as breaking the discrete logarithms to obtain the user's secret key from the associated public key, the hardness of which depends on the hardness assumption of the discrete logarithm problem (**DLP**) in $E(F_q)[l]$ [11]. Similarly, assume that the adversary attempts to get P_i's group secret key $f(0)$ from their corresponding group public key $Y_i = f(0)P \bmod l$. The adversary will also have to face the intractability of the same problems as deriving P_i's group secret key.

(2) Recovery equation attacks

An intruder tries to derive the signer's secret key $S_{ID}^{(i_j)}$ from the individual signature v_j by Equation (4). Given a message M and a signature v_j, it is difficult to determine $S_{ID}^{(i_j)}$ because Equation (4) has two unknown parameters. The values $S_{ID}^{(i_j)}$ are kept secret and the commitment values u and r are only known to the signers. Moreover, if one message with its related signature is added, the number of unknown parameters is also increased by one. The number of secret parameters is always greater than the number of equations available. Consequently, the intruder cannot succeed in recovering the equation and breaking the scheme.

(3) Conspiracy attacks

Any $(t-1)$ or less signers in B attempt to reconstruct the secret polynomial $f(x)$ to reveal the other signers' secret keys $S_{ID}^{(i_k)}$ and the group secret key $f(0)$. By using the Lagrange interpolating polynomial, with the knowledge of t or more signers' secret parameters $f(ID)$, the $(t-1)$ th degree polynomial $f(x)$ can be uniquely determined as

$$f(x) = \sum_{j=1}^{t} f(ID_{i_j}) \prod_{k=1,k\neq j}^{t} \frac{x - ID_{i_k}}{ID_{i_j} - ID_{i_k}}$$

Therefore, any $(t-1)$ or less malicious signers cannot conspire to derive the secret polynomial $f(x)$. Then, they cannot obtain any other signer's secret key and the group secret key. Thus conspiracy attacks can be successful.

(4) Impersonation attacks
Now let's discuss some possible impersonation attacks as below:

(a) An adversary attempts to impersonate a signer P_{i_j}. However, she/he cannot create a valid individual signature (u_j, r_j) to satisfy Equation (4) because of the lack of the secret key $S_{ID}^{(i_k)}$.

(b) An adversary tries to forge a valid threshold signature (u, v) of chosen message M to satisfy Equation (5). First, the adversary has to randomly choose u and r and find v to satisfy Equation (5), which is as difficult as solving discrete logarithms. In another similar approach, given u, v, finding r to satisfy Equation (5) is as difficult as solving the one-way hash function and the discrete logarithms. Therefore, the adversary cannot successfully forge the valid threshold signature.

(c)An adversary tries to collect a preciously valid threshold signature (u, v) on the message M and the associated value r to forge the signature of an arbitrary message M' [12]. First, the adversary selects a random $k_j' \in Z_q^*$ and calculates two values u' and r' as follows:

$$u_j' = k_j' P_{pub} \bmod l,$$

$$r_j' = k_j' Y_i \bmod l.$$

Then she/he computes

$$u_j' = u_j H'(u, r, M)^{-1} H'(u', r', M') \bmod l \quad (6)$$

$$v_j' = v_j H'(u, r, M)^{-1} H'(u', r', M') \bmod q.$$

Finally, the adversary sends the signature (u', v') for the message M' to verifiers. The validity of the threshold signature can be checked by Equation (5). Since

$$t(v', P) = t(P + H'(u', r', M')Q_{ID}, P_{pub})$$

The threshold signature (u', v') is valid for the message M'. However, it is hardly possible for the adversary to determine the value u' that satisfies Equation (6). Hence, the proposed scheme is secure against the impersonation attacks.

5 Performance Analysis

The performance of our partial signature is determined by those dominant cost operations. One of dominant operations in our scheme is scalar multiplication. Another dominant operation is the Tate pairing defined in Section 2. Comparing with scalar multiplication or the Tate pairing, point addition and hash functions can be ignored. Performance estimations of our partial signature with reference to those corresponding popular signature schemes are shown in Table 2(see the appendix).

It is believed that 1024-b RSA and 160-b elliptic curve cryptosystem are offering more or less the same level of security [13]. In this case, if our scheme pre-computes

(u_j, r_j), the timing of signing will be much shorter for two scalar multiplication operations reduced. On the other hand, our scheme uses the Tate pairing instead of Weil pairing because the Weil pairing takes longer than twice the running time of the Tate pairing for the cryptographic applications [10]. Obviously, our scheme has high performance.

6 Conclusion

In this paper, we have proposed a new ID-Based (t, n) Threshold Signature Scheme from the Tate pairing. According to our discussions, none of the possible attacks including plaintext attack, equation attack, conspiracy attack and impersonation attack can break our scheme. Performance analysis shows that it is more efficient and is more applicable to systems where signatures are sent over a finite bandwidth channel and with low capability equipment.

References

1. Shamir, A.: Identity-based cryptosystems and signature schemes. In: Blakely, G.R., Chaum, D. (eds.) CRYPTO 1984. LNCS, vol. 196, pp. 47–53. Springer, Heidelberg (1985)
2. Tsuji, S., Itoh, T.: An ID-based cryptosystem based on the discrete logarithm problem. IEEE Journal of Selected Areas in Communications 7(4), 467–473 (1989)
3. Boneh, D., Franklin, M.: Identity Based Encryption from the Weil Pairing. In: Kilian, J. (ed.) CRYPTO 2001. LNCS, vol. 2139, pp. 213–229. Springer, Heidelberg (2001)
4. Yi, X.: An identity-based signature scheme from the Weil pairing. IEEE Communications Letters 7(2), 76–78 (2003)
5. Desmedt, Y., Frankel, Y.: Shared Generation of Authenticators and Signatures. In: Feigenbaum, J. (ed.) CRYPTO 1991. LNCS, vol. 576, pp. 457–469. Springer, Heidelberg (1992)
6. Boldyreva, A.: Threshold Signatures, Multisignatures and Blind Signatures Based on the Gap-Diffie-Hellman-Group Signatures Scheme. In: Desmedt, Y.G. (ed.) PKC 2003. LNCS, vol. 2567, pp. 31–46. Springer, Heidelberg (2002)
7. Baek, J., Zheng, Y.L.: Identity-Based Threshold Signature Scheme from the Bilinear Pairings. In: ITCC 2004, pp. 124–128. IEEE Computer Society, Los Alamitos (2004)
8. Chen, X.F., Zhang, F.G., Konidala, D.M., Kim, K.: New ID-Based Threshold Signature Scheme from Bilinear Pairings. In: Canteaut, A., Viswanathan, K. (eds.) INDOCRYPT 2004. LNCS, vol. 3348, pp. 371–383. Springer, Heidelberg (2004)
9. Cheng, X.G., Liu, J.M., Wang, X.M.: An Identity-Based Signature and Its Threshold Version. In: Advanced Information Networking and Applications-AINA 2005, pp. 973–977. IEEE Computer Society, Los Alamitos (2005)
10. Galbraith, S.D., Harrison, K., Soldera, D.: Implementing the Tate Pairing. In: Fieker, C., Kohel, D.R. (eds.) ANTS 2002. LNCS, vol. 2369, pp. 324–337. Springer, Heidelberg (2002)
11. Boneh, D., Franklin, M.: Identity Based Encryption from the Weil Pairing. In: Kilian, J. (ed.) CRYPTO 2001. LNCS, vol. 2139, pp. 213–229. Springer, Heidelberg (2001)
12. Lee, N.Y.: The security of the improvement on the generalization of threshold signature and authenticated encryption. IEICE Transactions on Fundamentals E85-A(10), 2364–2367 (2002)
13. Lenstra, A.K.: Selecting cryptographic key sizes. J. Cryptology 14(4), 255–293 (2001)

Appendix

Table 2. Performance Comparison

	partial signing	partial signature verification	signature verification
Baek and Zheng's scheme[7]	**1 Weil pairing** **1 scalar multiplication**	**2 Weil pairing** **2t+1 integer exponentiations** **2t+1 scalar multiplications**	**2 Weil pairings** **1 integer exponentiation** **1 scalar multiplication**
Cheng and Liu's[9]	**4 scalar multiplications**	**3 Weil pairings** **3 scalar multiplications**	**2 Weil pairings** **1 scalar multiplication**
Our scheme	**5 scalar multiplications**	**2 Tate pairings** **2 scalar multiplications**	**2 Tate pairings** **1 scalar multiplication**

Vehicular Traffic Control: A Ubiquitous Computing Approach

Naishadh K. Dave and Vanaraj B. Vaghela

EC Department, Sankalchand Patel College of Engineering, Visnagar-384315, India
naishadhiisc@gmail.com, vanaraj79548@yahoo.co.in

Abstract. Vehicular traffic is a major problem in modern cities. We collectively waste huge amounts of time and resources while travelling through traffic congestion. Significant savings of fuel and time could be achieved if traffic control mechanism could be effectively discovered. With the advent of increasingly sophisticated traffic management systems, such as those incorporating dynamic traffic assignments, more stringent demands are being placed upon the available real time traffic data.

To address the problem of real time traffic data availability and processing needs, we propose a method which uses novel concept of ubiquitous computing (ubicomp) which uses ubiquitous database and intelligent agents for traffic data management. The concept of ubicomp is shifting computing paradigm from machines in a room to the augmented contexts in the real world. Ubiquitous database will make data everywhere available automatically, and it augments object that manages information about itself. The proposed system uses URA (Unique Routing Agent) to handle the distribution of database, route discovery and route maintenance. The method has been simulated for the measurement of traffic related parameters like traffic load, occupancy and trip time.

Keywords: URA (Unique Routing Agent), Ubiquitous computing (Ubicomp), Ubiquitous database.

1 Introduction

There is a growing need for the improvement of the efficiency of urban traffic in order to ensure the sustain-ability of modern cities. With the existing vehicular traffic control systems following transportation needs or issues are identified:

- Lack of real-time traffic information.
- Lack of access to travel information and 24 hour real-time alternate route information.
- Better alternate route guidance.
- Lack of readily available transit information to increase ridership.

To alleviate the problem of real time traffic data availability, we present the concept of ubiquitous computing and intelligent agents for vehicular traffic monitoring and management. The goal of *ubiquitous computing* is to place computers everywhere in the

S. Ranka et al. (Eds.): IC3 2009, CCIS 40, pp. 336–348, 2009.
© Springer-Verlag Berlin Heidelberg 2009

real world environment, providing ways for them to interconnect, talk and work together. Ubicomp shifts the computing paradigm from machines in a room to the augmented contexts in the real world. That leads to the new concept of real time database which is ubiquitous database. The ubiquitous database places data everywhere, in turn this helps in access and management of traffic information. Intelligent agents have generated particular research interest to many distributed applications because agents are a powerful, natural metaphor for conceptualizing, designing, and implementing many complex, distributed applications [1].

1.1 Ubiquitous Database

A set of the small-embedded database attached to each real-world entity, such as goods, materials and persons is known as ubiquitous database [2]. A data object on a DBMS can move to another electronically. Furthermore, the DBMS itself moves in the real world as the corresponding entity moves. An external application system interacts with the DBMSs, when each DBMS is moved in the position that can be accessed from the application system. That is, the ubiquitous database enables different organizations to share electronic data through the DBMSs attached to mobile entities.

Every real-world object originally has information such as properties and its historical changes. Traditional databases collect such information to manage in a central manner. However, the central management, although it is highly efficient, does not meet the demand of real time data availability, as traffic control system, electronic commerce and digital libraries demand. The ubiquitous database augments object that manages information about itself. A database-augmented object enables data application integrations through the movement of object in the real world. An augmented product moving from one place to another can carry electronic updating records at the same time.

1.2 Proposed Method

We propose the concept of ubiquitous database and Mo-bile Agents (MAs) for access and management of traffic information. Information relevant to traffic are hosted on static data centers where the URAs are placed at the intersections of streets. URAs keep database table of roadmap, weather information and traffic related information. The URA a specialized node capable of generating MA for route discovery, route maintenance and to handle distribution of traffic information.

1.3 Organization of the Rest of the Paper

The rest of the paper is organized as follows, section 2 presents the research work carried out in the field of vehicular traffic control using database approach, section 3 illustrates our vehicular traffic control system and also describes the working of URA, agents for information dissemination and creation of ubiquitous database. Simulation and results of the work are presented in Section 4, and finally section 5 draw conclusions.

2 Existing Works

There are several works reported that deals with the database approach in vehicular traffic control. Some of the works are as follows. The concept of distributed databases in traffic event detection and management is discussed in [3]. A work described in [4] uses traffic cookies placed on in-vehicle computers to maintain the state (current trip) of vehicles, that in turn be useful for estimation of vehicle trip table in real time. The method leverages the vehicles themselves to store their own travel data, and then physically carry that data around the network. Acknowledge based traffic control architecture with centralized database and agents proposed in [5]. The concept of map database for navigation and driver assistance is proposed in [6]. In [7], authors developed the distributed shared memory system to provide real time traffic data and the range of information services for distributed traffic monitoring. Traffic View, that is a device that can be embedded in the vehicles to provide the drivers with a real-time view of the road traffic is given in [8]. A framework of on-line vehicle routing in a distributed traffic information system based on vehicle to vehicle information sharing architecture is described in [9].

3 Proposed Vehicular Traffic Control System

In our technique, we propose an intelligent agent based ad hoc network. All vehicles moving in the network considered to be part of the network for traffic monitoring and control. The core of the system is creation of ubiquitous database and sustenance of intelligent nodes (URAs). URAs are static nodes placed at intersections and along the streets. Which maintains the database of traffic information and routing information. Road side sensors monitor the traffic situation and it is provided to nearby URAs. Fig. 1 shows the layout of road network which contains 7 intersections and 10 streets. Following subsections describe the working of URAs, proposed algorithms for vehicular traffic control, vehicle tracing and creation of ubiquitous database.

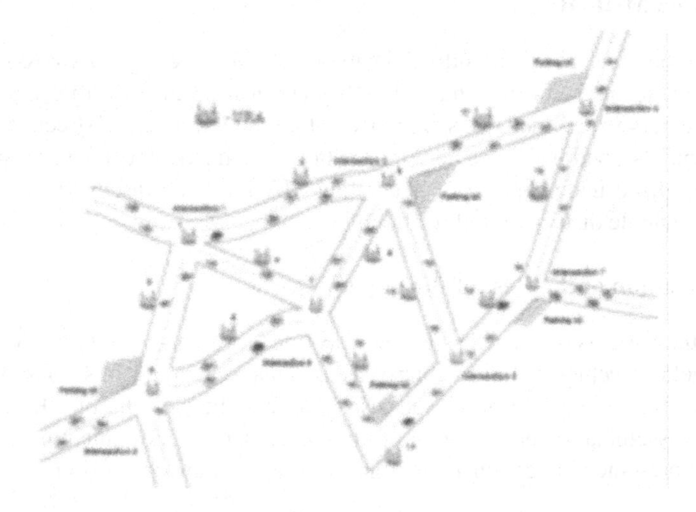

Fig. 1. VANET layout of a road network

3.1 URAs in the Network

The network consists of mobile as well as static nodes in an ad-hoc environment. The URA nodes are strategically placed such that URA node is well connected to at least one neighboring URA node. Desirable characteristics of URA node are:

- Processing power to sustain distribution of database and node management.
- More than average normalized link capacity
- URA to URA and URA to mobile node connectivity and reliability
- Large buffer capacity to maintain ubiquitous database and routing table.

URAs are responsible for route discovery, route maintenance and distribution of ubiquitous database. For this purpose it uses mobile agents. It keeps the route information and presence information in URA routing base.

1) URA routing base: URAs maintain database of network related information which we call as a URA routing base. Each URA finds the path to neighboring RI-MAs using MAs. MAs collect network behavior information which contain dynamic behavior of network resources like bandwidth and buffer availability at a node and pass it to a URA. URA routing base mainly consists of following tables,

I. *Path Reach Table (PRT)* -It gives information about all neighboring URAs with path(s) to reach them.
II. *Destination Reach Table (DRT)* -It gives information about next URA to be used to reach each possible destination.

The example of this is shown in table 1

Table 1. DRT at RIMA node: Next Hop RIMA node for each destination Node

Destination Node	Next Hop RIMA
15	6
12	10

This information helps URAs in delivering information. PRT also keeps path attributes, i.e., the network information about the path like hop distance and reliability. According to these path attributes URA-to-URA traffic is distributed over all paths that give better utilization. These path at-tributes are also used for selection of best path. In case of regular nodes, only PRT is maintained.This stores information about all URAs reachable within their neighborhood. Here, a regular node keeps few or no path attributes.

2) Finding paths between URAs:: After the placement of URAs in a network, the next task is to form routing table at URA node. To form routing table, paths from a URA node to its neighbors are established. Path discovery is carried out by MAs. A URA node sends Forward Mobile Agents (FMAs) in network to discover the paths between itself and neighboring URAs. since any URA node can be connected to nearest URA node within maximal number of hops. This avoids flooding. Between each pair of

URA node, many paths are discovered and all paths are recorded in routing table. Algorithm 1 explains working of FMA. When FMA reaches at URA node it gives traveled path information along with other collected information to it. After receiving FMA, URA node generates Backward Mo-bile Agent (BMA). BMA traverses the same path as FMA and gives the resource information and path information to source URA node.

Algorithm 1. Forward Mobile Agent (FMA)

1: **if** URA node reached **then**
2: Give all the collected information to URA, i.e., path followed and resources available on that path.
3: Create Backward Mobile Agent with path information.
4: **else if** hop == 1 **then**
5: mobile agent is deleted.
6: **else**
7: decrease the hop of mobile agent by 1.
8: collect the network information needed for routing.
9: flood the mobile agents to neighbor nodes.
10: **end if**

URA node adds the path in its routing table and updates other network information to its database. Algorithm 2 explains the working of BMA.

Algorithm 2. Backward Mobile Agent (BMA)

1: **if** URA reached **then**
2: give all the collected information to URA, i.e., path followed and resources available on that path to update routing table.
3: delete mobile agent.
4: **else**
5: give all the information collected to node so that node will update its PRT with updated information.
6: travel to next hop.
7: **end if**

3.2 Agents for Information Dissemination

Information dissemination model consists of mainly three types of agents like Vehicle Agent, Alert Agent, and Information Discovery Agent. These agents are responsible for collecting and disseminating traffic information. All agents follow the routing path discovered by URAs.

I. *Vehicle Agent (VA):* VA is static agent resides in vehicle which communicates with the URA to get/disseminate the relevant information. VA collects the status (moving or stationary) and location information of vehicle from sensors equipped in a vehicle.

II. *Alert Agent (AA):* AA is a mobile agent that travels around the network by creating its clones to disseminate the critical information during the critical situations. Examples of critical situation are accident, traffic jam, bad weather conditions, tracing a vehicle involved in crime or traffic rule violation etc. It also informs VA and updates the vehicle database. AA is sent by URAs to the vehicles moving in the network.

III. *Information Discovery Agent (IDA):* IDA travels in the network to search for the required information as desired by vehicle user. IDA is sent by the URA in the network on the request issued by user or URA itself to get traffic information.

3.3 Vehicular Traffic Monitoring and Control

To monitor the vehicular traffic, traffic density and travel time are measured at each streets and this context information is stored in nearby ubiquitous database situated in URA. Then this traffic context is disseminated by agents to neighboring URA for finding the alternate routes to avoid congestion. URA finds many routes according to cost of a street and stores minimum cost routes to its ubiquitous database. The cost of a street is decided by predicting travel time as described in following subsection. Local traffic density context LCL–TD–CTX at each street is measured and disseminated to neighboring streets URA by AAs. If traffic density is larger than threshold (THD–TD) value (200) than traffic routing information is changed according to current density value. Algorithm 3 shows the method of traffic control using local traffic density.

Algorithm 3. Vehicle traffic monitoring & control

1: Require: LCL–TD–CTX, travel time information on the streets
2: **repeat**
3: Street (*j,k*)
4: Get current LCL–TD–CTX, Travel time
5: Calculate
Traffic Density = Traffic load / Travelling time for a street
6: Update URA database
7: **until** all streets exhausted
8: At URA:
9: **if** LCL–TD–CTX > THD–TD **then**
10: Create max. No. of AAs
11: AA: Migrate to nearest URA with LCL–TD–CTX and route information
12: submit traffic and route information
13: At new URA & Vehicles: Update database; Display information
14: Take decisions
15: **end if**
16: AA: collect information on the path and retract to origin URA
17: submit information; URA: Update database

3.4 Vehicle Tracing

In the case of finding a trace of vehicle which has been involved in traffic rule violation and crime, the history of such vehicles are registered in ubiquitous database. Those vehicles identity is disseminated around the network. According to the identity, the vehicles are detected by sensors and that data provided to URA. URA plants AA on the identified vehicle and stores the identification of AA and time. After migrating to vehicles, AA collects status and location information from VA. It clones itself and migrate to host URA or the nearest URA with the status and location information of vehicle. URA stores this information to its database with time stamp. For getting the trace of a vehicle, IDA searches the vehicle location from URA database according to algorithm 5. The proposed algorithm for vehicle tracing is given in algorithm 4.

Algorithm 4. Vehicle tracing
1: At URA:
2: Require: vehicle identity, customization information
3: Get vehicle identity
4: create Alert Agents (AAs)
5: **if** Identified vehicle in 1-hop neighborhood **then**
6: Migrate to identified vehicle
7: store AA Id. in ubi-database with time stamp
8: **end if**
9: At Vehicle:
10: Get location & status information from VA
11: **while** Status == moving **do**
12: **if** New URA in 1-hop neighborhood **then**
13: clone itself
14: Migrate to URA
15: **end if**
16: **end while**
17: At URA:
18: Submit vehicle location and status information with time stamp
19: Store vehicle information with time stamp in ubi-database
20: Search of vehicle:
21: call procedure IDA-start

3.5 Creation of Ubiquitous Database

For the creation of ubiquitous database, each vehicle collects, stores, traffic information from the network. To achieve this goal, each vehicle stores its own travel history, by accepting information from URAs. The motivation of this approach is that these travel information can be used to route the vehicles in congestion free route as well as predicting the movement of vehicles in the system, which can in turn, be used for traffic management. As URAs are static, so no constraints put on the buffer capacity and processing power. URAs at intersections maintain database of traffic related information, which contains following database entries:

1. Roadmap & Weather information: This information is used for routing the traffic on the different street in the case of congestion and incidents. Weather forecast information is used to alert the driver to avoid any incident.
2. *Current Status:* Simply stores the current state of the traffic signal. It is used to decide on the changes necessary to move to a new state in the change of traffic signal.
3. *Street Loads:* This is a table of the incoming and outgoing streets from an intersection. These are the streets that the intersection polls to calculate its strategy about the timing of state changes. When the loads of these streets are received they are put into this table.
4. *Occupancy:* Simply the current number of different types of vehicles (e.g. two-wheeler, light, and heavy) that are waiting to cross the intersection and leave particular road segment.
5. *Travel time:* Measured travel time of vehicles from sensors is stored in this table.

Algorithm 5. Information Discoverer Mobile Agent Algorithm

1: Require: Data information, Customization information, Migration Depth
2: procedure: IDA-start
3: begin
4: **if** PATH == null **then**
5: *PATH* ← *local URA*
6: Migrate to the nearest non-visited URA
7: **else**
8: Begin migration along PATH
9: **end if**
10: Follow PATH till last URA is reached
11: Forward Journey AND URA node reached
12: *Migration Depth* ← *Migration Depth* − 1
13: **if** Corresponding Data index found **then**
14: collect Data and Retract to the origin URA node
15: **else**
16: call URA node's Indexing Data
17: **if** Corresponding Data index found **then**
18: Migrate to URA node and retrieve Data information
19: **else if** *Migration Depth* > 0 **then**
20: Migrate to nearest non-visited URA node
21: Retract to the origin URA node
22: **end if**
23: **end if**

URA on the streets maintains following database entries for the management of traffic:

1. Capacity & Road condition: Information like number of vehicles can occupy and the maximum speed of the road are stored in this table. Road condition information is useful for planning a state change and routing of traffic.
2. Downstream Loads: This is a table of the percentage of this street's load that it sends to each of its downstream roads so it could be useful to avoid congestion on street.
3. Upstream Loads: This is a table which holds the current loads which will be sent from streets further Upstream. These are received in messages from those streets and are used to calculate this street's load.
4. Vehicle Count: Simply the current number of different types of vehicles that are waiting to cross and leave this road segment.

URAs at intersections and on the streets maintain the database of vehicles those are stolen, involved in crime and traffic rule violation, and vehicle of special interest. This database helps in tracing a vehicle of particular interest and predicting the movement of that vehicle. The vehicle nodes moving on the roads maintain small database which contains information like owner name, vehicle identification number, location, time, date, travel history and traffic knowledge information. The vehicle is automated vehicle unit which uses the Global Positioning System (GPS) to determine the location of the vehicle also it captures the traffic related alert information from the neighboring URA. Data captured through a automated vehicle unit is stored in the database of vehicle unit.

4 Simulation

4.1 Simulation Environment

We have developed the road network as shown in Fig. 1. There are URAs have been placed at each intersection which will monitor and record the vehicles passing through the intersections.

4.2 Simulation Procedure

We have simulated 10 streets environments with the large, medium, and small vehicles population for 24 hours. We discuss the simulation by considering following cases:

1) Case i: Traffic monitoring at each streets: We have considered traffic of three different types of vehicles for 24 hours at each street. The URA nodes updates the information in database according to the types of vehicles pass through them and pass on the information to its neighboring URAs in which direction the vehicle has proceeded to.

For simulation, we have considered different traffic rate on different streets. Peak time slots for traffic are considerd as 8:30 to 10:30 a.m., 12:30 to 2:30 a.m., and 5:30 to 8:30 a.m.. Different vehicle types are taken as different packet sizes; two wheeler-250 bytes, light-500 bytes, and heavy-1000 bytes. Fig. 2 shows the measured

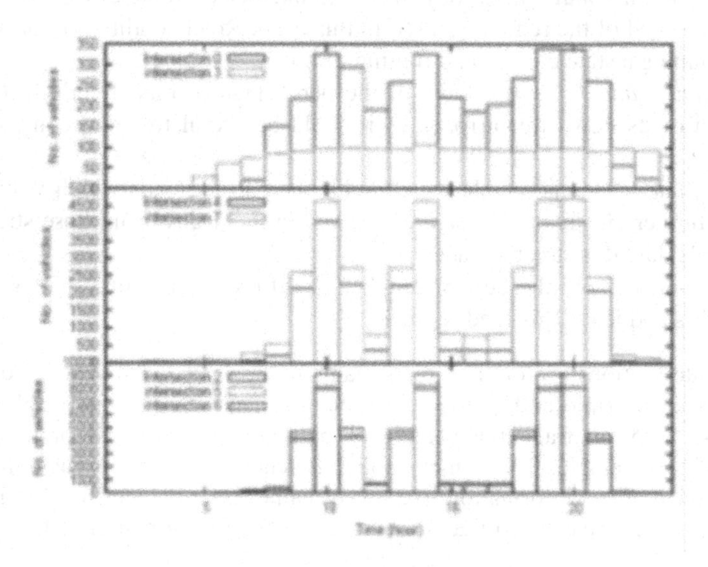

Fig. 2. Trafficload on intersections for 24 hrs

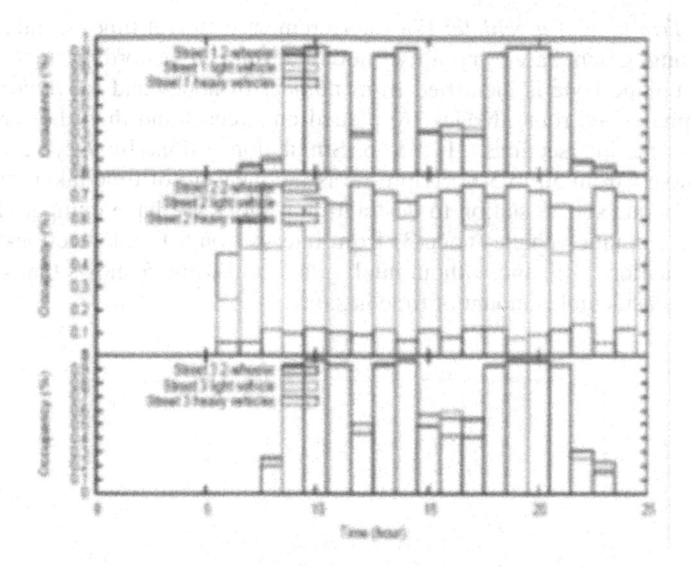

Fig. 3. Different types of vehicle occupancy on the streets 1,2,3

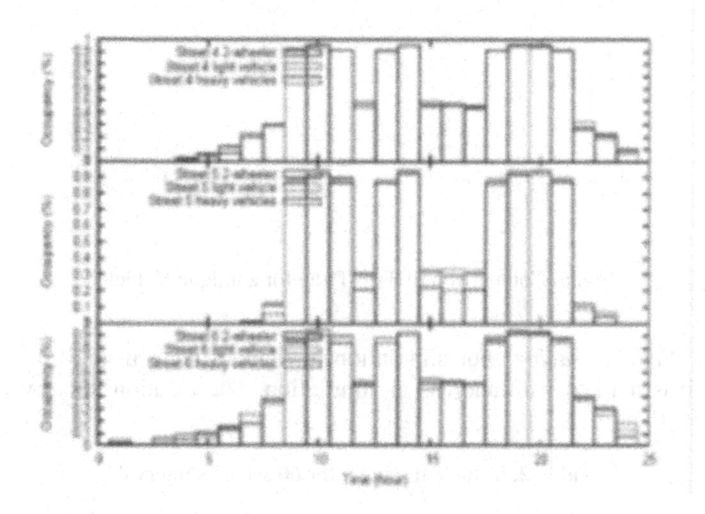

Fig. 4. Different types of vehicle occupancy on the streets 4,5,6

traffic load for 24 hours. Traffic load is the measure of number of vehicles passed from intersections or streets per unit time (one hour). Occupancy is the measure of vehicle count that are waiting to cross the intersection or particular road segment.

The measured percentage occupancy of different types of vehicles (two-wheeler, light and heavy) at different streets is shown in Fig. 3 and Fig. 4 respectively.

2) Case ii: Trip time of a vehicle: For measurement of travel time, simulation is carried out using graph based mobility model developed according to the network (Fig. 1). A unique node is identified from number of nodes and movement of nodes taken in a predefined route. Nodes are placed on street-3 and then they move to the predefined route: intersections - 1, 5 to 6. Simulation is done by varying the number of vehicle nodes from 50 to 500. Trip time is the measure of time taken by a vehicle to move from one source station to destination station. We have compared measured trip time for a unique vehicle (node 3) from intersection-1 to intersection-5 in different traffic situations with and without intelligent agents. Fig. 5 shows that with intelligent agents considerable amount of time is saved.

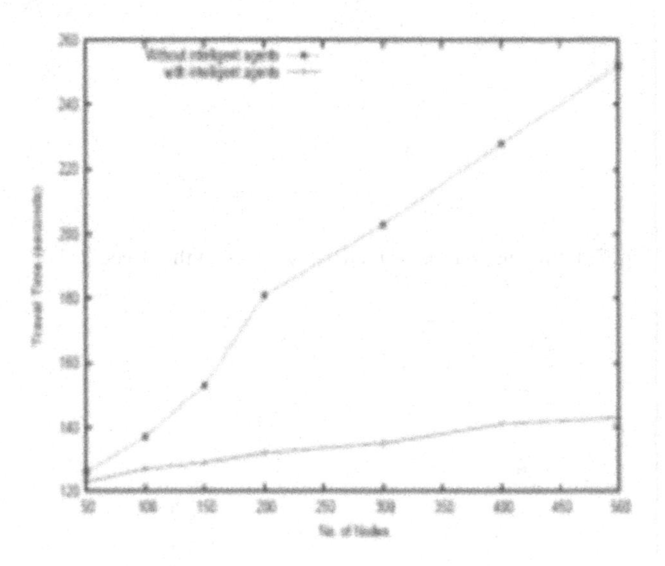

Fig. 5. Comparison of Trip Time for a unique Vehicle

3) Case iii: Vehicle tracing: For simulation, a random node from 100 nodes is selected and it is marked with address in simulation. The location identity is taken as

Table 2. Vehicle tracing after 60 seconds interval

Sr. No.	After Time (seconds)	RIMA No.	Location
1	60	7 & 8	I-6
2	120	6	S-4
3	180	5	I-2
4	240	12 & 14	I-5
5	300	15	I-7

(I= intersection, S= street)

URA identity number and according to that intersection and street number identified. Initially nodes are placed on the street-3 and started its journey on random path for its trip. The simulation has been carried out for 300 seconds and the movement of marked node is stored in nesrest URA database as it traverse the path. Table 2 shows the result of trace of a vehicle after every 60 seconds.

5 Illustration of Ubiquitous Database in Traffic Monitoring

Since ubiquitous databases place data everywhere in the network, database is attached to every entity in the network which stores their status information as well as traffic information. Consider a critical event like traffic jam Occurred at some intersection in the system, sensors deployed. at that intersection detect the event and this information is stored into URA database. URA creates AAs to its neighbors. AA communicates with neighboring URA or VA and informs about the critical event, this information is further propagated by AAs into network according to the route discovered by URAs and in turn URAs and vehicles database is updated. According to updated data this traffic information is displayed on message-board or sign board placed on streets, intersection and in-vehicle.

6 Conclusion

To best of our knowledge, the proposed method for vehicular traffic control using ubiquitous database technique is a new thinking towards dynamic real time data availability application. Really, in the application like vehicular traffic control where data is dynamically updated, ubiquitous database will bring high impact on data integration application. From the simulation results, we believe that ubiquitous database is the attractive solution to potential application like vehicular traffic control.

References

[1] Manvi, S.S., Venkataram, P.: Applications of agent technology in communications: A review. Computer communications 27, 1493–1508 (2004)

[2] Kuramitsu, K., Sakamura, K.: Towards ubiquitous database in mobile commerce. In: MobiDe 2001: Proceedings of the 2nd ACM international workshop on Data engineering for wireless and mobile access, pp. 84–89. ACM Press, New York (2001)

[3] Basu, P., Little, T.D.C.: Database-centered architecture for traffic incident detection, management, and analysis. In: Intelligent Transportation Systems (ITS), pp. 149–154 (2000) ISBN: 0-7803-5971-2

[4] Marca, J.E., Rindt, C.R., Jayakrishnan, R.: A Method for Creating a Real-time, Distributed Travel History Database. Journal of the Transportation Research Board (1972), 69–77 (2006) ISSN:0361-1981

[5] Krogh, C., Irgens, M., Tmtteber, H.: A Novel Architecture for Traffic Control. In: The 3rd International Conference on Vehicle Navigation & Information Systems. IEEE, Los Alamitos (1992)

[6] Martin Rowell, J.: Applying Map Databases to Advanced Navigation and Driver Assistance Systems. Journal of Navigation (2001)
[7] Kosonen, I., Bargiela, A.: A Distributed Traffic Monitoring and Information System. Journal of Geographic Information and Decision Analysis 3(1), 31–40 (1999)
[8] Nadeem, T., Dashtinezhad, S., Liao, C., Iftode, L.: TrafficView: traffic data dissemination using car-to-car communication. In: SIGMOBILE Mob. Comput. Commun. Rev., vol. 8, pp. 6–19. ACM Press, New York (2004)
[9] Yang, X., Recker, W.: Modeling Dynamic Vehicle Navigation in a Self-organizing, Peer-to-peer, Distributed Traffic Information System. Journal of Intelligent Transportation Systems 10(4), 185–204 (2006)

Application of Particle Swarm Optimization Algorithm for Better Nano-Devices

Nameirakpam Basanta Singh[1], Sanjoy Deb[2], Guru P. Mishra[2], Samir Kumar Sarkar[2], and Subir Kumar Sarkar[2]

[1] Department of Electronics & Communication Engineering, Manipur Institute of Technology, Imphal-795004, Manipur, India
[2] Department of Electronics & Telecommunication Engineering, Jadavpur University, Kolkata-700032, India
basanta_n@rediffmail.com, deb_sanjoy@yahoo.com, gurumishra@rediffmail.com, su_sircir@yahoo.co.in, sksarkar@etce.jdvu.ac.in

Abstract. Particle swarm optimization, an intelligent soft computing tool is employed to determine the optimized system parameters of GaAs quantum well for better high frequency performance under hot electron condition at room temperature. The energy loss through LO phonon and momentum loss through LO phonon, deformation acoustic phonon and ionized impurity (both background and remote) are incorporated in the present calculations. For a typical dc biasing field, it is possible to predict the optimum values of system parameters like lattice temperature, well width and two-dimensional carrier concentration for realizing a particular high frequency response characterised by well defined cut-off frequency. Such optimization will make feasible the fabrication of a variety of new quantum devices with desired characteristics.

Keywords: Evolutionary algorithm, Particle swarm optimization, quantum well, mobility.

1 Introduction

A GaAs quantum well (QW) is produced when a thin layer of GaAs is sandwiched between layers of higher band-gap semiconductor such as AlAs [1, 2]. As the thickness of the centre material, called the channel is comparable to the de Broglie wavelength of the carriers, a subband structure is developed. The carriers are then confined in the active layer parallel to the heterojunction interfaces forming a quasi-two-dimensional electron gas. In QW, the density-of-states and scattering rates of the carriers are different from those in the bulk materials. In such structures, mobility of the two-dimensional electron gas is enhanced considerably at low temperature by the modulation doping technique. The carrier mobility in quantum wells can further be enhanced by placing an undoped spacer layer between the doped barrier and undoped channel layer. The spacer layer increases the separation between the carriers and ionized donors thereby increasing the electron mobility because of less Coulomb interaction. High-mobility two dimensional electron gas in QW structures has attracted

S. Ranka et al. (Eds.): IC3 2009, CCIS 40, pp. 349–357, 2009.
© Springer-Verlag Berlin Heidelberg 2009

special attention of researchers because of the possibility of realization of high speed nano-devices [3, 4].

The study of the small-signal carrier transport of the 2D hot electron gas in QWs is of fundamental importance in the realization of high speed and high frequency devices in the microwave and millimetre wave regime. Due to the dependence of the transport properties on the various system parameters, namely, the channel width, carrier concentration, lattice temperature, electron temperature etc., the optimization of these system parameters is very essential for suitable commercial applications with devices of desired characteristics [5, 6]. Optimization provides an important technique in solving some scientific and engineering problems. Among the available optimization techniques, evolutionary algorithm based soft computing tools like Genetic Algorithm, Artificial neural network and Particle swarm optimization (PSO) are very useful techniques to predict optimum system parameters to get better as well as desired performance of nanodevices [7]. This has motivated us to investigate the application of Particle Swarm Optimization Technique for parameter optimization of GaAs quantum wells.

In this work, Particle swarm optimization (PSO) technique is employed to obtain optimized system parameters of GaAs QW. The optimized values of carrier concentration and channel length at a particular dc biasing field are computed at room temperature to get desired mobility and frequency response.

2 Analytical Model

Let us consider a GaAs/AlAs square QW of infinite barrier height. The values of system parameters such as two dimensional (2D) carrier concentration (N_{2D}), channel length (Lz) and lattice temperature (T_L) used here are such that the separation between the lowest and the next higher subband is sufficiently higher than the maximum average electron energy and the carriers populate only the lowest subband. Carrier scattering by polar optic phonon, acoustic deformation potential, remote and background-ionized impurities are considered. Reduced ionized impurity scattering and improved carrier concentration in the QW establish a strong electron-electron interaction, favouring a heated drifted Fermi-Dirac distribution function for the carriers characterized by an electron temperature T_e, and a drifted crystal momentum p_d [8]. The carrier distribution function $f(\vec{k})$ in the presence of an electric field F applied parallel to the heterojunction can be expressed as [8]

$$f(\vec{k}) = f_o(E) + \frac{\hbar \vec{p}_d \vec{k}}{m^*}\left(-\frac{\partial f_o}{\partial E}\right)\cos\gamma \tag{1}$$

where, $f_0(E)$ is the Fermi-Dirac distribution function for the carriers, $\vec{p}_d$ is the drift crystal momentum, ħ is Planck's constant divided by 2π, $\vec{k}$ is the two-dimensional wave vector of the carriers with energy E, m* is the electronic effective mass and γ is the angle between the applied electric field $\vec{F}$ and the two dimensional wave vector

$\vec{k}$. An electric field of magnitude F_1 and the angular frequency ω superimposed on a moderate dc bias field F_0 is assumed to act parallel to the heterojunction interface. The net field is thus given by [8]:

$$F = F_0 + F_1 Sin\omega t \tag{2}$$

The electron temperature and the drift momentum will also have similar components with the alternating ones generally differing in phase as they depend on the field and the scattering processes. Thus

$$T_e = T_0 + T_{1r} \sin \omega t + T_{1i} \cos \omega t \tag{3}$$

$$p_d = p_0 + p_{1r} \sin \omega t + p_{1i} \cos \omega t \tag{4}$$

Where, T_0 and P_0 are the steady state parts, T_{1r} and P_{1r} are real and T_{1i} and P_{1i} are imaginary parts of T_e, and P_d respectively [8]. The energy and momentum balance equations obeyed by the carrier are [8]:

$$ep_d F / m^* + \langle dE / dt \rangle_{scat} = \frac{d\langle E \rangle}{dt} \tag{5}$$

and

$$eF + \langle dp / dt \rangle_{scat} = \frac{dp_d}{dt} \tag{6}$$

Where $-\langle dp/dt \rangle$ and $-\langle dE/dt \rangle$ are the average momentum and energy loss due to scatterings and $\langle E \rangle$ is the average energy of a carrier with charge e.

We insert Eqs. 3 and 4 in Eqs. 5 and 6, retain terms up to the linear in alternating components and equate the steady parts and the coefficients of *sinωt* and *cosωt* on the two sides of the resulting equations following the procedure adopted in Ref. 8. For a given electric field F_0, we solve for P_0 and T_0. The dc mobility μ_{dc} and ac mobility μ_{ac} are then expressed as:

$$\mu_{dc} = \frac{p_o}{m^* F_o} \tag{7}$$

$$\mu_{ac} = \frac{\sqrt{p_{1r}^2 + p_{1i}^2}}{m^* F_1} \tag{8}$$

The phase lag ϕ, the resulting alternating current lags behind the applied field is expressed as

$$\phi = \tan^{-1}\left(-\frac{p_{1i}}{p_{1r}} \right) \tag{9}$$

3 Particle Swarm Optimization

Particle swarm optimization (PSO) is a swarm intelligence based algorithm to find a solution to an optimization problem in a search space. PSO algorithm is inspired by social behaviour of bird flocking or fish schooling [9]. This technique is achieved by mimicking the behaviour of the biological creatures within their swarms and colonies. If one of the particles in entire population discovers a good path to food, then the rest of the swarm will follow it instantly even if they are far away. Swarm behaviour is modelled in multidimensional space that has two characteristics, one being position and the other being velocity. The particles wander around the solution space, remember the best position that they have discovered and communicate to each other to adjust their own position and velocity. A simple flowchart of the PSO algorithm is shown in Fig. 1.

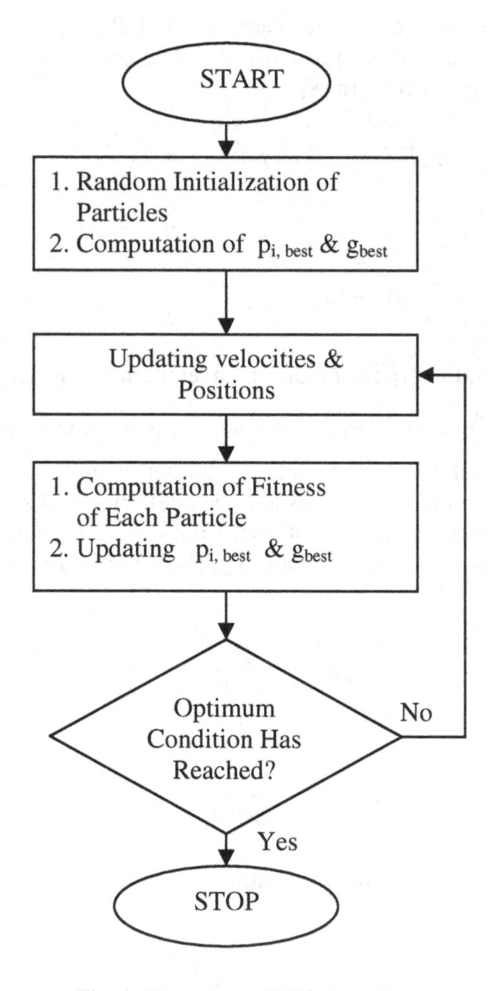

Fig. 1. Flowchart of PSO algorithm

The PSO algorithm begins by initializing a group of random particles. These random particles are the candidate solutions. An iterative process to improve these candidate solutions is set in motion. The particles iteratively evaluate their fitness and remember the location where they had their best success. The individual's best solution is called the local best. Each particle makes this information available to its neighbours. It is also able to see where its neighbours have had success. Movements through the search space are guided by these successes, with the population usually converging, by the end of a trial.

The fitness values of all the particles are evaluated by the fitness function to be optimized. The algorithm stores and progressively replaces local best parameter values of each particle ($p_{best,\ i}$, where i is the index value) as well as the global best (best position found by any particle in the swarm) parameter value evaluated from the fittest particle in the entire population (g_{best}). As the swarm iterates, the velocities and positions of particles are updated according to the following two equations:

$$v_i(t) = wv_i(t-1) + c_1 r_1 (x_{best,\ i}(t) - x_i(t-1)) + c_2 r_2 (g_{best,}\ (t) - x_i(t-1)) \tag{10}$$

$$x_i = x_i(t-1) + x_i(t) \tag{11}$$

Where r_1 and r_2 are two random numbers, which are used to maintain the diversity of the population, and they are uniformly distributed in the interval [0,1], c_1 is a positive constant, called as coefficient of the self-recognition component and c_2 is a positive constant, called as coefficient of the social component. Gradually updating its position and velocity at each generation all the particle will converge to the optimum solution [10, 11].

4 Results and Discussions

The Particle Swarm optimization technique is employed to determine the optimized values of carrier concentration and well width at room temperature for GaAs/AlAs QW. The material parameters are taken from Ref. 12. The particle converges gradually with iterations to reach the optimum values of mobility (both ac and dc) and cut-off frequency within the range of system parameters considered in the present case.

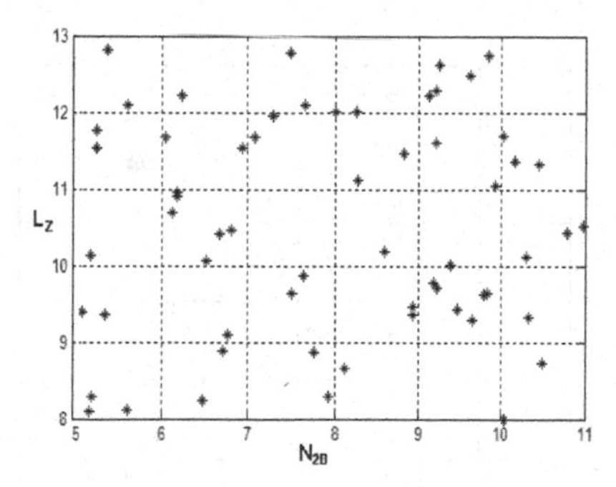

Fig. 2. Position of swarm after 5th iteration

The positions of the particles after 5th, 20th, 40th and 60th iteration are shown in figures 2, 3, 4 and 5. The selection of 5th, 20th and 40th iterations are completely arbitrary with an intention that the readers will feel the intermediate positions during swarm optimization. Other iterations can also be observed. Initially all the particles are randomly distributed over the 2D solution space and their positions are represented by two coordinates, N_{2D} and Lz. As the swarm iterates, the fitness of the global best solution improves and after the 60th iteration, the swarm converges to the optimum solution.

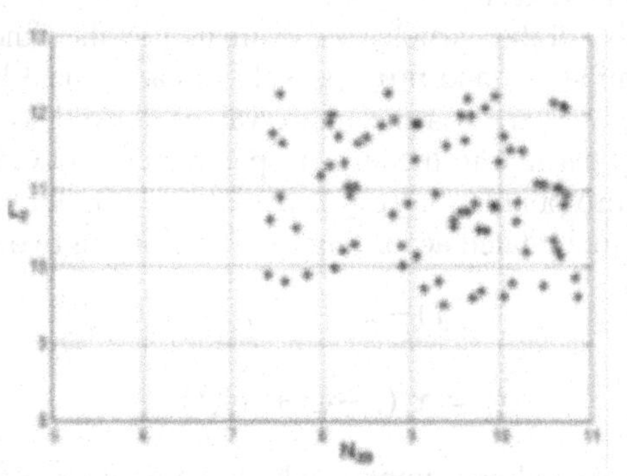

Fig. 3. Position of swarm after 20th iteration

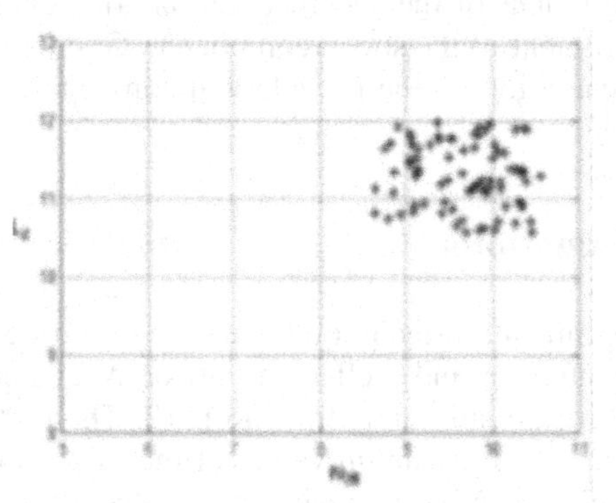

Fig. 4. Position of swarm after 40th iteration

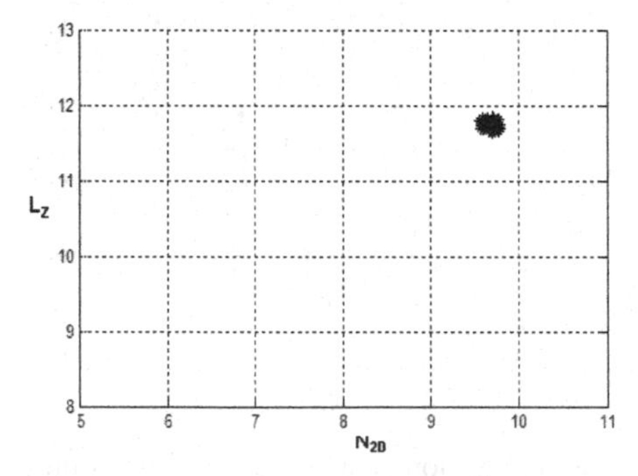

Fig. 5. Position of swarm after 60th iteration

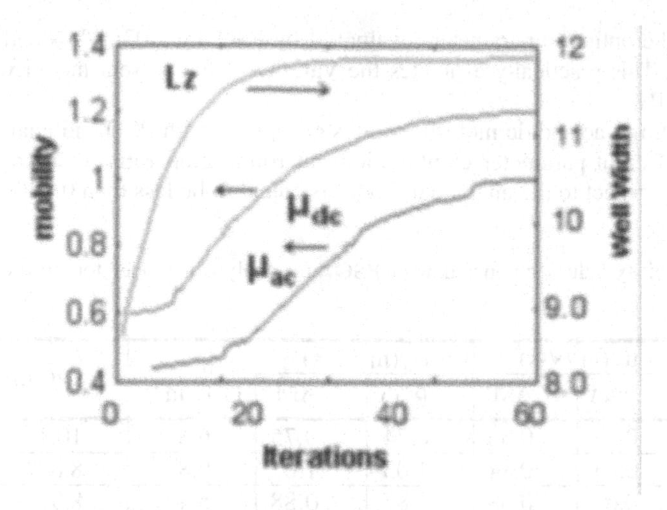

Fig. 6. Plot of μ_{ac}, μ_{dc} and well width (in nm) with iteration

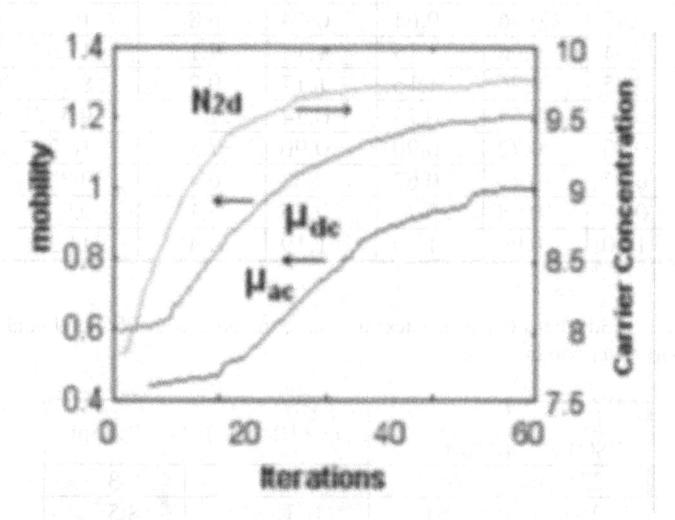

Fig. 7. Plot of μ_{ac}, μ_{dc} and carrier concentration ($10^{15}m^{-2}$) with iteration

The variation of ac mobility, dc mobility and well width with iteration is shown in Fig. 6. The results are obtained at 300 K for a typical ac field frequency of 30GHz. After the 55th iteration, ac and dc mobility values reach their optimum value and the optimized value of the well width is found to be 11.8 nm.

Fig. 7 shows the variation of ac mobility, dc mobility and carrier concentration with iteration. The results are obtained with parameters values similar to those used in Fig. 3. After the 55th iteration, ac and dc mobility values reach their optimum value and the optimized value of the carrier concentration is $9.6{\times}10^{15}m^{-2}$. The ac and dc mobility values with optimum system parameters computed with PSO are 1.01 m^2/V-s and 1.19 m^2/V-s respectively. Calculated ac and dc mobility values with the analytical model

(AM) with the optimum parameters evaluated by PSO are $1.03m^2/V$-s and $1.21m^2/V$-s respectively. This practically evaluates the validity of the present model with an error less than 0.01%.

Table 1 shows ac and dc mobility values computed with PSO and analytical model (AM) for different parameter combinations at room temperature. Computation error of PSO with respect to the analytical model is found to be less then 0.02%.

Table 1. Mobility values computed with PSO and analytical model for different parameter combinations

$\mu_{ac}(m^2/V\text{-}s)$		$\mu_{dc}(m^2/V\text{-}s)$		N_{2D}	Lz(nm)
PSO	AM	PSO	AM	$(10^{15}\ m^{-2})$	
0.55	0.56	0.74	0.75	6.3	10.1
0.83	0.84	1.03	1.01	9.8	8.8
0.67	0.68	0.87	0.88	8.4	8.9
0.42	0.44	0.61	0.63	6.5	10.1
0.45	0.46	0.64	0.64	6.8	10.9
0.94	0.96	1.15	1.14	7.1	11.6
0.95	0.96	1.16	1.17	9.7	8.3
0.54	0.55	0.73	0.74	7.9	9.5
0.70	0.72	0.90	0.90	7.5	10.3
0.47	0.49	0.67	0.68	6.4	9.2
0.57	0.58	0.76	0.75	6.1	9.9
1.00	0.99	1.20	1.19	9.4	11.3

Table 2. Values of 3dB cut-off frequencies in GHz computed with PSO and analytical model for different parameter combinations

f_{3dB} (PSO)	f_{3dB} (AM)	$N_{2D}(10^{15}\ m^{-2})$	Lz(nm)
255	257	8.3	9.3
283	284	6.8	8.5
267	268	7.4	8.8
242	243	8.2	11.2
245	246	7.8	11.9
294	296	8.1	8.4
295	297	6.2	9.4
254	255	8.3	9.4
270	271	8.5	11.1
247	249	8.4	10.5
257	256	8.1	9.8
200	197	9.4	11.3

Optimized system parameters at cut off frequency with the applied small-signal electric field and the dc biasing field of 0.75×10^5 V/m are given in Table 2. For a desired cut-off frequency at a particular dc biasing field, it is possible to predict the optimum values of the system parameters like carrier concentration and quantum well width for realizing a particular high frequency response characterized by a cut-off frequency at room temperature.

5 Conclusion

Application of PSO for parameter optimization in AlAs/GaAs quantum well structure is new and is quite useful to understand the optimum combinations of parameters to get high mobility of this type of nanostructures. In this work, desired values of ac & dc mobility the system parameters are optimized at room temperature which will provide valuable information for the technologists and device fabricators.

Acknowledgement

Sanjoy Deb thankfully acknowledges the financial support obtained from School of Material Science and Nanotechnology, Jadavpur University in the form of UGC Fellowship.

References

1. Akimoto, R., Li, B.S., Akita, K., Hasama, T.: Appl. Phys. Lett. 87, 181104–181108 (2005)
2. Aydogu, S., Akassu, M., Ozbas, O.: Rom Journal, physics 50(9-10), 1047–1053 (2005)
3. Sarkar, S.K., Chattopadhyay, D.: Phys. Rev. B 62, 15331–15335 (2000)
4. Chattopadhyay, D.: Appl. Phys. A 53, 35–38 (1991)
5. Sarkar, S.K., Gosh, P.K., Chattopadhyay, D.: Phys. Stat. Sol. (b) 207, 125–129 (1998)
6. Carlson, J.M., Lyon, S.A., Worlock, J.M., Gossard, A.C., Wiegmann, W.: Bull Amer. Phys. Soc. 29, 213–218 (1984)
7. Sarkar, S.K., Karmakar, A., De, A.K.: Czech. J. Phys. 51, 249–256 (2001)
8. Sarkar, S.K., Gosh, P.K., Chattopadhyay, D.: J. Appl. Phys. 78(1), 283–287 (1995)
9. Kennedy, J., Eberhart, R.: Swarm intelligence. Morgan Kaufmann Publishers, Inc., San Francisco (2001)
10. Angeline, P.J.: In: Proceedings of the 1998 IEEE Congress on Evolutionary Computation, Piscataway, NJ, USA, pp. 84–89. IEEE Press, Los Alamitos (1998)
11. Blackwell, T.M., Bentley, P.J.: In: Proceedings of the 2002 IEEE Congress on Evolutionary Computation, Piscataway, NJ, USA, pp. 1691–1696. IEEE Press, Los Alamitos (2002)
12. Sarkar, S.K.: Elsevier, Computational Materials Science 29, 243–249 (2004)

Measuring the Storage and Retrieval of Knowledge Units: An Empirical Study Using MES

Selwyn Justus[1] and K. Iyakutti[2]

[1] Department of Computer Applications, K.L.N. College of Info. Tech., Madurai, Tamilnadu, India
justus.mku@gmail.com
[2] Department of Mircoprocessor & Computer, Madurai Kamaraj University, Madurai, Tamilnadu, India
iyakutti@gmail.com

Abstract. Computer applications are smart that they require efficient storage and retrieval of data. Object-relational data models are the opted and the widely appreciable approach because of their power in object representation and relational retrieval. Two OR models were designed for representing knowledge units in the Music Expert System and three metrics were proposed to study the storage and retrieval of the knowledge units from the OR schemas. Experiments conducted to asses the storage efficiency and relational retrieval of the objects indicated significant results. The metrics were used to keep in check the size of the objects created during runtime and their relational coupling helped in the retrieval of objects, with minimal disk reads. The empirical results and interpretations concludes the work, focusing on the efficient design of OR schema models which commend the functioning of the system's performance.

Keywords: Knowledge units, Object-Relational Schema, Measurements, Storage & retrieval.

1 Introduction

Computer applications and intelligent systems are demanding efficient storage and intelligent retrieval of the storage units. Object-based and object-oriented storage and retrieval is one of several other approaches. As systems are also approaching object oriented models and design, Object Relational Storage and Retrieval (ORSR) will be the state-of-the-art-technology in representing knowledge units as conceptually related objects [7], [18]. Object relational modeling became widely appreciable because of its power in representing complex data models, those which are impossible with the conventional relational storage mechanisms that are normally available with any relational database management systems. This combination of objects and relations has lead to a new thought on OR modeling of knowledge units, which the database and system designers are approving [14]. Most of today's database management systems support the object relational storage of data. Object-Relational Database Management Systems (ORDBMS) combines the programming power of objects and the storage potential of relational databases [9], [20], [21], [22].

S. Ranka et al. (Eds.): IC3 2009, CCIS 40, pp. 358–369, 2009.
© Springer-Verlag Berlin Heidelberg 2009

There are inadequate measures for assessing the performance of a system in terms of storage and retrieval of objects in relational paradigm. As metrics and measurement are one among the many ways for assuring the internal quality of a product [23], [29], the measuring of several attributes of OR models formally approves them for their applicability [5]. The measurement of ORSR of knowledge units can well be appreciated using two functional models for object representations in the ORDMS, which we have opted for this study. The object-relational (OR) schema used for representing a knowledge unit is assessed for its structural and functional efficiency, in terms of its structural design and performance.

Moreover, the functional trait of the OR schema is studied with the music expert system (MES) proposed in [13]. Using this object based models we were able to obtain knowledgeable chord formation and harmony in the MES. The metrics serve as an indicator in assessing the efficiency of OR schema designed for MES using ORSR architecture.

The paper is organized as follows: Section 2 gives two object-relational schemas based on conceptual representation of a knowledge unit. Section 3 proposes three metrics that are used to validate the two OR models. Section 4 gives the experiment part of the object-relational storage and retrieval and the results and their interpretation. Section 5 concludes the work.

2 Knowledge Based Storage Models

Prior to proposing the metrics, we give here two object-relational models which are used in the representation of a knowledge unit (KU). In our case, the musical notations in MES are represented as KU and the two models are then measured in the aspect of storage and retrieval of KUs. The acquired musical notation is codified into an object-relational storage knowledge unit using object-structured model (OSM) and concept-structured model (CSM). These storage models are discussed in this section.

2.1 The Object-Structured Model (OSM)

The OSM environment is an object-based ontology with knowledge acquisition activities that are widely used for domain modeling [3], [26], [27]. OSM ontology can be represented in four labeled structure

$$OSM = \{Header\ tag, \langle c, s, f, a\rangle\}$$

where c-classes, s - slots, f - facets, and a - axioms.

Fig. 1. Sample musical notation

Classes are the representation of the context. Slots describe properties or attributes of classes. Facets describe constraints associated with a slot. Additional constraints to a relation can be specified with Axioms. The two slots: own slot describes properties of an object represented by that frame and template slots describe properties that an instance of a class can possess. The number of values that can be associated with a slot, restrictions on the value of the slot can be specified using Facets. A class can acquire own slots only by being an instance of a metaclass that has those slots as template slots. Consider a piece of musical notation in figure 1.

This piece of notation given in fig 1 can be represented in OSM as follows: Staff-signature: Treble Clef, three-sharp major, 4/4 time signature; Class: A3; Slots: Crochet, no staccato, no expression; Facets: The scale of 3-sharp major; Axioms: Pre and Post notes, F#minor.

The functional model of OSM representation in the ORDBMS class-type definition is given in figure 2, which is aggregated in a relational table which is used to store a relational tuple of a musical note, here we consider as knowledge unit (KU). The knowledge models given in figure 2 and figure 3 are the object definition language (ODL) schema for the MES. The ODL for creating a User-defined Datatypes (UDT) is based on the Ontology of Oracle 8 [9].

2.2 Concept-Structured Model (CSM)

Knowledge is perceived as a concept, and it is strongly argued and agreed that any knowledge could be represented in the notion of concepts, because knowledge itself is a conceptual understanding of the human brain [1]. Hence a knowledge unit can be represented as a concept, which when linked to related concepts forms a conceptual graph (CG). The conceptual model described in [19], [25] is considered for representing the data by means of a conceptual tree structure. Since a CG representation is in close association with the natural language, practitioners choose this model for representing the entities conceptually. Conceptual graph, according to John F. Sowa, is "A finite, connected, undirected, bipartite graph with nodes of one type called concepts and nodes of the other type called conceptual relations" [26].

```
Create type OSM_Note_t(
        Note_name    varchar(2);
        Note_id      varchar(2);
        Staff_sig    varchar(6);
        Duration     number;
        Position     varchar(4);
        Expression   varchar (10);
  Operations:
        setvalues()    integer;
        getvalues()    integer;
   Relationship:
        OSM_note_t     Pre_note;
        OSM_note_t     Post_note;
        Staff_t         Staff_link
inverse Staff_t::Arc_set;
    );
```

Fig. 2. Class-type Definition of OSM

```
Create type Staff_t(
      Concept_ID        number;
      Concept_type      varchar(15);
      Referent_type     varchar(4);
      Key_type          key_t;
      Signature         sigtime_t;
      Key_Scale         key_scale_t;
   Operation:
      set_values()      integer;
      get_values()      integer;
   Relationship:
      Linked_staff Staff_t
      Arc_set    OSM_Note_t inverse
OSM_Note_t::Staff_link
   );
```

Fig. 3. Class-type Definition of CSM

In our example, a musical staff (the five lines, and the key signatures) are codified as the primary concept, which houses the KUs represented using OSM, discussed in the previous sub-sections. The concept structured functional model for the musical staff in figure 1 is given in figure 3. The allied class-types are defined in figure 4. The CSM representation of a musical staff can be codified in the Object-relational data model by creating a class-type which can be aggregated into a relational database table. The CSM is used to represent each staff of the sheet music along with the staff's properties. The OSM is used to represent all the notes on the stave. Hence these two models themselves are related in context. The metrics we propose in the following section considers these two representation models.

```
Create type Key_t (
    Key_name          varchar(3);
    Key_sharps[12]    varchar(2);
    Key_flats[12]     varchar(2);
);

Create type Key_Scale_t(
    Key_name       varchar(3)
inverse key_t::key_name;
    Scale_notes[12] varchar(2);
);
```

```
Create type Chord_t(
    Chord_name     varchar(12);
    note[7]        varchar(2);
};

Create type sigtime_t (
    Numerator        number;
    Denominator      number;
);
```

Fig. 4. User Data-type definitions used in the OSM and CSM representation models

3 Measuring the ORSR

In this section we propose three simple metrics to assess the structural storage of objects and the efficient retrieval of objects using the OR data model.

3.1 Object Size (OS)

The prime motivation of this metric lies in the understanding of the storage systems, particularly the object-relational storage model. Since the higher-level application code has not been tuned to match the underlying storage mechanisms, it has become a compulsion to measure the nature and enormity of the objects, specifically in ORDBMS [21], [24].

Object Size (OS) metric is the number of bytes the object occupies in the disk storage. An object in object-relational data model is comprised of attributes, interface and relations. This metric is a direct invocation of the sizeof() function, which is supported by any programming language. *Object_Size = sizeof(object_name);*

The metric can normally be measured by the system designers while modeling data, and can check on the object's size so as to lie within the baseline of the metric. The total size of the persistent objects (TOS) that are created dynamically and fetched for retrieval can be defined as the sum of the size of the total number of persistent objects created during runtime.

$$\mu(TOS) = \sum Object\_size(pob)_R \quad (1)$$

Where TOS is total object size and (pob)R is the number of persistent objects created during runtime.

3.2 Relational Coupling (RC)

Class Coupling is a common metric used to measure the degree of interactions in object-oriented systems [2], [4], [16]. In object-relational data modeling, the class-types are modeled in such a way that the objects are related to other objects through relation, the third dimension of OOP [18]. The objects of these class-types are conceptually related through conceptual linking of objects self or non-self class-types [7], thus measuring the internal complexity of the class-types and their relations [8], [10], [15]. Here we measure coupling that arises due to the conceptual relation of an object.

The metric relational coupling is defined as the number of object references created in 'this' object. Object reference is explained as the number of persistent objects, both 'self' and 'non-self' objects created in the class. Consider a class-type has n number of object-references, then the relational coupling (RC) can be formulated as

$$RC = \sum_{i=0}^{n} Ob\_Rf_i \tag{2}$$

where $Ob\_Rf = \sum_{i=0}^{n} SelfOb_i + \sum_{j=0}^{m} NonselfOb_j$

Metric Interpretation:

1. This metric is a static measure which indicates the relatedness of objects with other objects of same class or different class due to conceptual coupling.
2. Higher the metric value, higher the degree of coupling out of conceptual relations. Highly relational-coupled objects are encouraged to be stored in contiguous locations so as to fetch them back in minimal disk read operations.
3. This metric does not assess the interaction of the objects, but indicates the location preference for storage and time optimization for fetching conceptually related objects.

The dynamic nature of this metric is best understood during fetching of objects. In our example, the fetching of related object's fetch is based on the conceptual storage of objects that are codified using OSM and CSM. The above defined metrics are used to assess the performance of the system in the framework of object relational storage and retrieval.

4 Empirical Investigation

After being given the object-relational storage models of knowledge units, we attempt here to evaluate the models using the metrics discussed in the section 3. The two metrics also help us to assess the performance of the system from the storage and retrieval aspects. We derive experimental results for the metrics, OS, TOS and RC, using the Music Expert System (MES) as the experimental system.

4.1 Overview of MES

The Music Expert System is a software developed for composing musical notes and playing the music score. The system also performs an intelligent formation of musical chords for the given progression of musical notes. A Chord is an arrangement of related musical notes and harmony is the output of the intelligent chord selected for a musical score [30]. The most critical part in music composition is the organization of the chords and harmony, which involves creative skills. With an intelligent music system, a music composer will be provided with a set of related chords and few smart suggestions will also be given for producing good harmony.

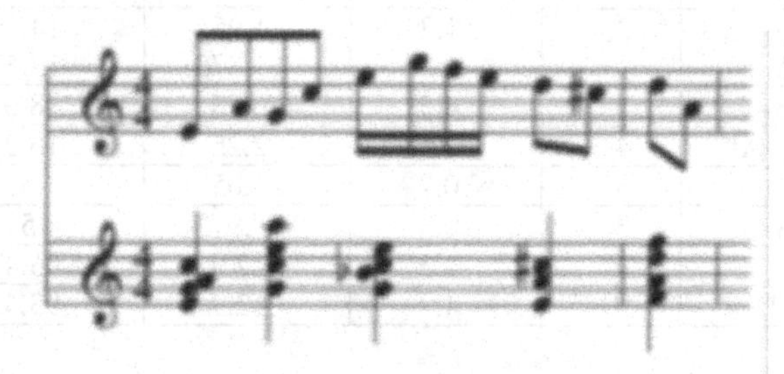

Fig. 5. Sample musical score

The chord formation for are given musical score in a given key-signature (scale) can be found in [31], [32]. Since a chord consists of a set of related notes in the given key-scale, the fetching of these related notes are made intelligent by conceptual storage of the OSM objects. The musical progression given in figure 5 can be humanly interpreted and the related chords are given along with the music score. However, human creative ability is limited to just four chords formation, in this case. The MES gives the following chords, figure 6, for the musical score in figure 5.

The given musical progression is E3-A3-G3-C4-E4-G4-F4-E4-D4-C#4-D4-A3. All of these concepts are constrained in C major, except for C#. Instead of fetching all the chords for Cmajor and related Aminor, the CSM restricts the fetches by retrieving only the chords related to this progression. Three sets of chords are required for the score given in figure 5.

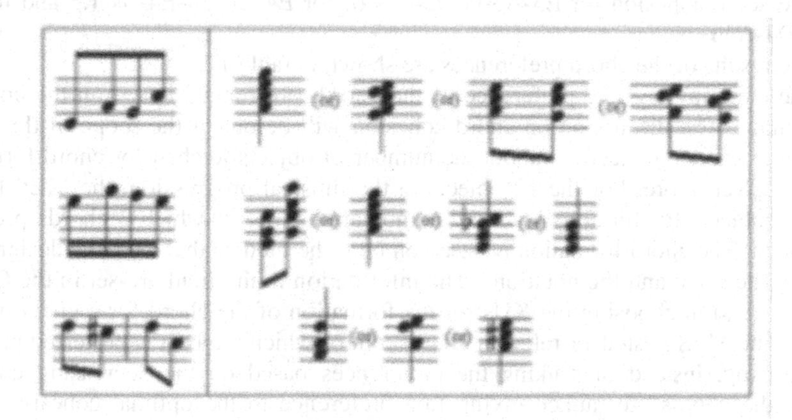

Fig. 6. MES generated Chords

Table 1. Relevant Chords and its preference based on Chord Cohesion

Score	Relevant Chords	KUs (k)	H-Distance (h)	Chord Cohesion (c)	Harmony (%)	Preference
E3-A3-G3-C4-	A-C-E-G	4	4.5	27	27.4	IV
	G-A-C-D-E	5	3.2	19.2	19.5	I
	A-C-E, C-E-G	3, 3	3.7, 2.3	22.2, 13.8	22.5, 14	III
	A-D-E, G-C-D	3, 3	3.7, 2.7	22.2, 16.2	22.5, 16.4	II
E4-G4-F4-E4-	C-E-G, C-E-G-A	3, 4	1.3, 1.25	1.95, 1.875	14, 13.5	II
	A-C-E-G	4	4.5	6.75	48.6	IV
	C-E-G-A#	4	1.5	2.25	16.2	I
	E-G-B	3	0.7	1.05	7.5	III
D4-C#4-D4-A3	D-F-A	3	0.3	0.9	6.5	I
	A-D-E	3	4.3	12.9	93.4	II
	A-C#-E	3	4.3	12.9	93.4	III

4.2 Retrieval of OSM Objects

The retrieval of OSM objects for chord formation is the key performance issue. The MES finds the most suitable set of chords for the progression. This is derived by calculating the H-distance of the OSM objects (KU), and the chord cohesion of the chord.

$$H\text{ - }Distance = D - \frac{N}{K_n} \tag{3}$$

Where D = Distance of the KU from the key signature,
N = Number of semitones between KUs
K_n = number of KUs,
Chord Cohesion = H-Distance × Score Cohesion
where Score cohesion = number of hops × avg of time signature

The score cohesion for E3-A3-G3-C4- is 6; for E4-G4-F4-E4- is 1.5 and for D4-C#4-D4-A3 is 3.

The results of the chord preferences are shown in table 1.

The harmony and the preference of the chords are ranked based on the cohesion percentage. The discussion on chord cohesion will be out of the scope of this work. However, we are concerned about the number of objects fetched for chord formation for the given score. For the 12 objects in the musical progression, the MES fetches just 13 objects for forming 11 chord combinations, out of which 3 chords preferred the much. The chord formation is based on both the models that we have designed for storing the stave and the notations. The information in the field arc-set in the CSM is the key-field in choosing the KUs for the formation of the chord for a given musical score. The MES posted 11 relevant chords, out of which the best is preferred based on the ranking. Instead of ranking the preferences based on the decreasing cohesion value, the chords are ranked giving first preference to the optimal cohesive chord, which may be suitable for the given musical progression.

4.3 Metric Results and Interpretation

The musical notations that are represented in OSM and CSM are collection of persistent objects. Storage of these objects and their retrieval affects the efficiency of the system, in our study, the Music Expert System. The functioning efficiency of the MES is assessed using the two metrics, Object Size and Relational Coupling. Three works of musical scores, 'Peace be Still' (PBS), 'Moonlight' (ML) and 'Then came the Morning' (TCM) are taken for this study.

The chords are formed by analyzing the notation progressions (sequence of OSM objects). While forming a chord for a given notation progression, the *pre_note, post_note* and the *arc_set OSM_note_t* objects are fetched. Since the object size of OSM_note_t is 26 bytes, we preferred to store them in the object-relational database, as persistent objects.

Table 2. Relational fetching of Objects for Chord Formation

Projects	No of Objects Generated	No. of Objects Fetched for Chord formulation	Total chords formulated	Relational fetching of objects (%)
PBS	460	12	234	42
ML	5240	26	2274	68
TCM	2724	18	1306	53

From table 1 it can be noted that MES posted 11 chords for the given musical progression. We also attempted to formulate chords for the three musical pieces, PBS, ML and TCM. From the set of chords suggested by MES 53% of chords were considered by the musicians. Table 2 gives the details on the KUs fetched for chord formation and their relevance with the relations among objects. For each of the musical scores we found that there were instances of new fetches of objects, which is based on the conceptual relations (*Related_links, Arc_set* attributes in *staff_t*), as the score progresses.

A bar within a musical score is defined as the collection of *OSM_note_t* objects whose total count (*OSM_note_t.duration*) equals the time-signature of the musical score. Project PBS consists of 34bars, ML consists of 203 bars and TCM consists of 72 bars. When the fetchings of new *OSM_note_t* objects were plotted against the bar instances, we could observe good performance of MES, in terms of minimal number of object fetches.

The object fetches due to the conceptual relations among the *OSM_note_t* are given in Figures 7(a), 7(b) and 7(c).The number of new objects fetched for chord formation in each new bars of musical progression are assessed based on the number of disk read operations. PBS shows 30% of disk access, ML shows 46% and TCM shows 39% of disc access. These values are directly proportional to the Relational Coupling (RC) metric value. While loading the files, the *OSM_note_t* and *Staff_t* objects are fetched as the progression improves. The steeps in object fetches (figure 7(d)) means that those are bars where new *OSM_note_t* objects are read from the database. Projects ML and TCM show much steeps indicating frequent fetches. The percent of relational fetching of objects is given in table 2.

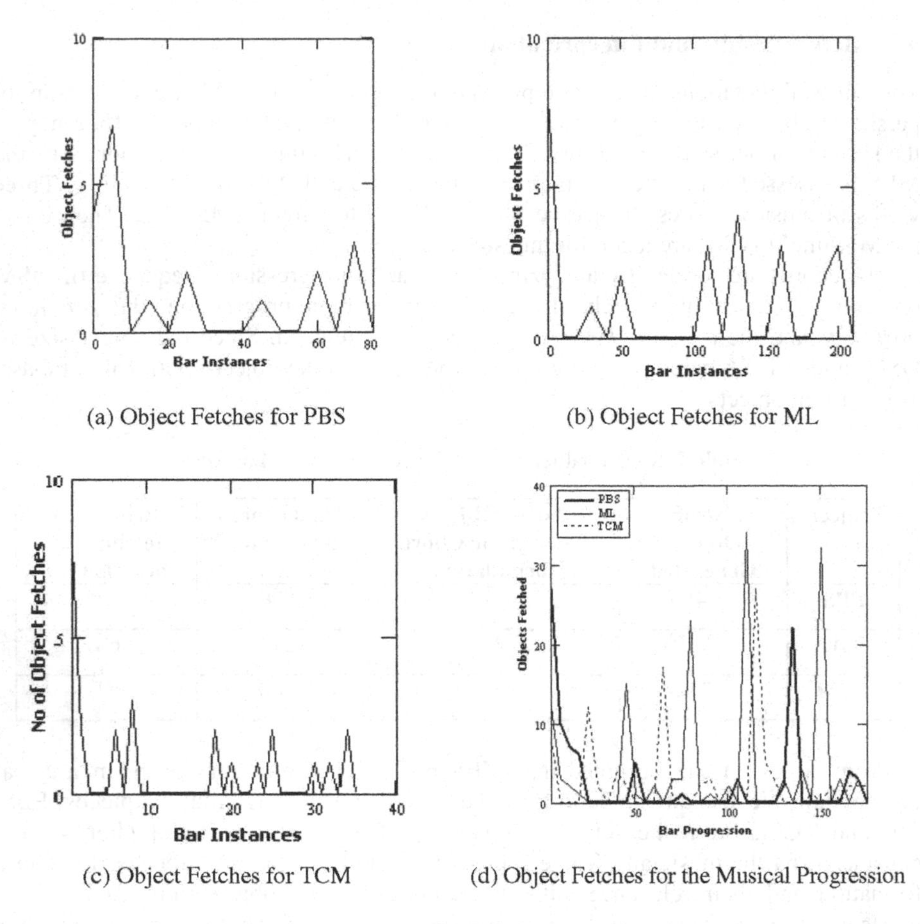

(a) Object Fetches for PBS

(b) Object Fetches for ML

(c) Object Fetches for TCM

(d) Object Fetches for the Musical Progression

Fig. 7.

The metrics object size (OS), total object size (TOS) and relational coupling (RC) give a clear indication of the object fetches and manipulation and thus the performance of the MES is assessed. Number of reads/fetches of objects is an important and one of the significant parameters that commends the performance of a system [5], [8], [30], [31]. Hence, we like to interpret that the metrics we have proposed for the object-relational schemas (*OSM_note_t* and *Staff_t*) signify the performance of the MES and helped us check the efficacy of the design of the schema. Using these metrics, further work is in progress to further refine and redesign the OR schemas. Additional schemas were also considered for storing supplementary information about the knowledge units.

5 Conclusion

The object relational storage and retrieval of knowledge is the requirement of today's computer applications and smart, intelligent systems. Intelligence systems help humans in providing various suggestions and choices and also give the preferences of best results out of them. Hence, the design of such high-end systems has become a significant task. One such example system is the MES that we considered in this work for proposing two OR models to represent knowledge units and thereby assess them using three metrics. The experimental results were encouraging and significant in evaluating the performance of the system from the storage and retrieval points of view. The proposed three metrics were used to quantify the storage and retrieval mechanisms of objects. System throughput is a bright indication for the size of objects, which has to be under check, during design time. Minimal read access to the disk for fetching of objects literally indicates the system's functioning, during runtime.

The object relational storage models proposed in this work will operate well across a wide range of scalable smart systems. By providing support for codifying knowledge, deploying them using conceptually related links, and retrieving the related concepts and posting the required knowledge, the models help in constructing high performance, intelligent systems object-based storage systems. Based on the two OR models discussed in this work, new models could well be designed for the systems that require higher throughput and better performance. Future works on object-relational data modeling focus on their compatibility with the application platform and storage modules. This work may serve as a lead in this area of OR modeling and measuring their efficacy.

References

1. Alagarsamy, K., Justus, S., Iyakutti, K.: Implementation Specification of a SPI supportive Knowledge management Tool. IET Software 2(2), 123–133 (2008)
2. Ammar, H.H., Yacoub, S.M., Robinson, T.: Dynamic metrics for object-oriented designs. In: 5th International Software Metrics Symposium, Boca Raton, Florida, USA, pp. 50–61 (1999)
3. García-Serrano, A., Martínez, P., Teruel, D.: Knowledge-modeling techniques in the e-commerce scenario (2001), `http://www.csd.abdn.ac.uk/~apreece/ebiweb/papers/serrano.doc`
4. Arisholm, E., Briand, L.C., Foyen, A.: Dynamic coupling measures for object-oriented software. IEEE Transactions on Software Engineering 30(8), 491–506 (2004)
5. Baroni, A.L., Calero, C., Abreu, F.B., Piatini, M.: Object relational Metrics Formalization. In: Sixth International Conference on Quality Software (2006), doi:ieeecomputersociety.org/10.1109/QSIC.2006.44
6. Baroni, A.L., Calero, C., Ruiz, F., eAbreu, F.B.: Formalizing Object-Relational Structural Metrics. In: 5th Portuguese Association of Information Systems Conference (2004), `http://ctp.di.fct.unl.pt/QUASAR/Resourses/Paper/2004/baroni5CAPSI.pdf`
7. Calero, C., Ruiz, F., Baroni, A., Brito, A.F., Piattini, M.: An Ontological approach to describe the SQL: 2003 Object-Relational Features. International J. Computer Standards & Interfaces 28, 695–713 (2006)
8. David, P., Kemerer, C.F., Sandra, A.S., James, E.T.: The Structural Complexity of Software: An Experimental Test. IEEE Transactions on Software Engineering 31(11), 982–995 (2005)
9. Elmasri, Navathe, Somayajulu, Gupta: Fundamentals of Database systems, 4th edn. Pearson Education, Dorling Kindersley (2007)
10. Henderson-Sellers, B.: Object-oriented Metrics - Measures of Complexity. Prentice-Hall, Upper Saddle River (1996)

11. Justus, S., Iyakutti, K.: Assessing the Object-level Behavioral Complexity in Object Relational Databases. In: 3rd International Conference on Software Science, Technology and Engineering, Israel, pp. 48–59 (2007), doi:ieeecomputersociety.org/10.1109/SWSTE.2007.6
12. Justus, S., Iyakutti, K.: Object Relational Database Metrics: Classified and Evaluated. In: International Workshop on Software Engineering, Potsdam, Germany, pp. 119–131 (2007) ISBN-10: 3-8322-5611-3
13. Justus, S.: Data Mining for Music Distribution. In: National Conference on Datamining, India (2004)
14. Long, D., Brandt, S., Miller, E., Wang, F., Lin, Y., Xue, L., Xin, Q.: Design and implementation of large scale object-based storage system. Technical Report ucsc-crl-02-35, University of California, Santa Cruz (2002)
15. Michura, J., Capretz, M.A.M.: Metrics Suite for Class Complexity. In: International Conference on Information Technology Coding and Computing (ITCC 2005) (2005)
16. Moris. K.: Metrics for object oriented software development, Masters Thesis, M.I.T Sloan School of Management, Cambridge, MA (1998)
17. Noy, N.F., Fergerson, R.W., Musen, M.A.: The knowledge model of Protégé-2000: combining interoperability and flexibility (2000), http://pms.ifi.lmu.de/mitarbeiter/ohlbach/Ontology/Protege/SMI-2000-0830.pdf
18. Zhang, N., Ritter, N., Härder, T.: Enriched Relationship Processing in Object-Relational Database Management Systems. In: Third International Symposium on Cooperative Database Systems for Advanced Applications (2001)
19. Nguyen, P.H.P., Corbett, D.: A Basic Mathematical Framework for Conceptual graphs. IEEE Transactions on Knowledge and Data Engineering 18(2), 261–271 (2006)
20. Piattini, M., Calero, C., Sahraoui. H., Lounis H.: Object-Relational Database Metrics, L'object (March 2001), http://www.iro.umontreal.ca/~sahraouh/papers/lobjet00_1.pdf
21. Liu, Q., Feng, D., Qin, L.-j., Zeng, L.-f.: A Framework for Accessing General Object Storage. In: International Workshop on Networking, Architectures, and Storages (2006), doi:ieeexplore.ieee.ord/10.1109/IWNAS.2006.8
22. Weil, S.A., Wang, F., Xin, Q., Brandt, S.A., Miller, E.L., Long, D.D.E., Maltzahn, C.: Ceph: A Scalable Object-Based Storage System. Technical Report UCSC-SSRC-06-01, Storage Systems Research Center, Baskin School of Engineering, University of California, Santa Cruz, CA (March 2006)
23. Morasca, S.: Software Measurement. In: Handbook of Software Engineering and Knowledge Engineering - Volume 1: Fundamentals (refereed book), Knowledge Systems Institute, Skokie, IL, USA, pp. 239—276 (2001)
24. Sears, R., van Catherine, I., Jim, G.: To BLOB or not to BLOB: Large Object Storage in a Database or a Filesystem. Technical Report, MSR-TR-2006-45, Redmond (2006)
25. Sowa, J.F.: Knowledge Representation: Logical, Philosophical, and Computational Foundations. Brooks Cole Publishing Co., Pacific Grove (2000)
26. Sowa, J.F.: Conceptual Graphs for a Data Base Interface. IBM J. of Research and Development 20(4), 336–357 (1976)
27. Torgeir, D., Reidar, C.: A Survey of Case Studies of the Use of Knowledge Management in Software Engineering. Intl. J. Software Engineering and Knowledge Engineering 12(4), 391–414 (2002)

28. Han, W.-S., Whang, K.-Y., Moon, Y.-S.: A Formal Framework for Pre-fetching based on the Type-Level Access Pattern in Object-Relational DBMSs. IEEE Transactions on Knowledge and Data Engineering 17(10), 1436–1448 (2005)
29. Zusc, H.: Properties of Software measures. Software Quality J. 1, 255–260 (1992)
30. Wikipedia, http://en.wikipedia.org/wiki/Chord_%28music%29
31. 8notes, http://www.8notes.com/resources/notefinders/piano_chords.asp
32. Bedrockband, http://www.bedrockband.com/CTheory.htm

Implementation of QoS Aware Q-Routing Algorithm for Network-on-Chip

Krishan Kumar Paliwal[1], Jinesh Shaji George[1], Navaneeth Rameshan[1], Vijay Laxmi[1], M.S. Gaur[1], Vijay Janyani[2], and R. Narasimhan[1]

[1] Department of Computer Engineering, Malaviya National Institute of Technology, Jaipur, India

[2] Department of Electronics and Communication Engineering, Malaviya National Institute of Technology, Jaipur, India

{krishan332001,jinesh.s.george,navaneeth.rameshan, gaurms,vlgaur,raghavn86}@gmail.com, vijay.janyani@ieee.org

Abstract. The objective of this paper is to implement QoS aware Q-routing algorithm for providing different level of Quality-of-Service (QoS) such as Best Effort (BE) and Guaranteed Throughput (GT) in Network-on-Chip. In this paper, a novel scheme which contrast the performance of Q-routing with the well known XY routing strategy in context of QoS in Network-on-Chip (NoC) is presented. Simulation study with discrete event, cycle accurate, Network-on-Chip simulator NIRGAM reveals that Q-routing proves to be superior to the non-adaptive routing algorithm for both type of traffic BE and GT. The paper explores the performance of the network for different values of bandwidth reserved for GT traffic.

Keywords: Q-routing, Quality-of-Service, Best Effort and Guaranteed Throughput.

1 Introduction

Recent advances in the VLSI technology, system on chip (SoC) designs have gained popularity. As we are approaching the limits of silicon technology in Deep Sub Micron (DSM), placing more and more components (in terms of gate density) is proving to be a challenging task. One of the problems is interconnecting these gates reliably. As the gate size reduce, interconnection delays gain significance than gate delays. As the design complexity increases, and physical layout approaches its limit, it would be difficult for bus based interconnection techniques to provide communication to the future SoC design. Network-on-Chip (NoC) has emerged as a solution to the problem of non scalability of the bus based interconnection technique [1]. NoC is often described by topology (placement of tiles and inter tile connections), switching techniques (messages are routed from source node to destination), routing mechanism, architecture of router and algorithm to determine actual path from one router to another [2].

S. Ranka et al. (Eds.): IC3 2009, CCIS 40, pp. 370–380, 2009.
© Springer-Verlag Berlin Heidelberg 2009

Numerous topologies including mesh, torus, fat trees and butterflies and different switching mechanisms such as wormhole and virtual-cut through has been proposed [3]. Network-on-chip owes a legacy from the off-chip networks. This architecture consists of routing elements (routers/switches/nodes) connected to processing elements (IP cores/processors/memory elements) and interconnected through channels. A channel is a physical connection between two adjacent routers. Number of channels per router is fixed and depends on the topology; that is, arrangement of routers such as mesh, torus, butterfly etc. Complexity of interconnection network design is reduced to selecting a topology, designing a router and selection of suitable routing algorithm. Scalability simply means adding more routers and channels.

QoS refers to the capability of the network to provide typical services (dedicated bandwidth or reserving resources, control jitter and latency) [4]. To manage the allocation of resources to packets more efficiently, it is useful to divide network traffic into a number of classes. Different classes of packets have different requirements and different levels of importance [5]. The requirements of different traffic classes can be latency sensitivity, jitter intolerance and packet loss intolerance. Broadly speaking, the network traffic can be classified into two traffic classes, Guaranteed Throughput(GT) and Best Effort (BE).

In this paper, we contrast Q-routing with deterministic XY routing in the context of Quality of Service support in NoC. This paper is compiled as follows: Section 2 deals with NoC simulation framework. Section 3 describes experimental setup. Section 4 deals with routing algorithms. Section 5 explains simulation results and analysis. Section 6 presents conclusion and future work.

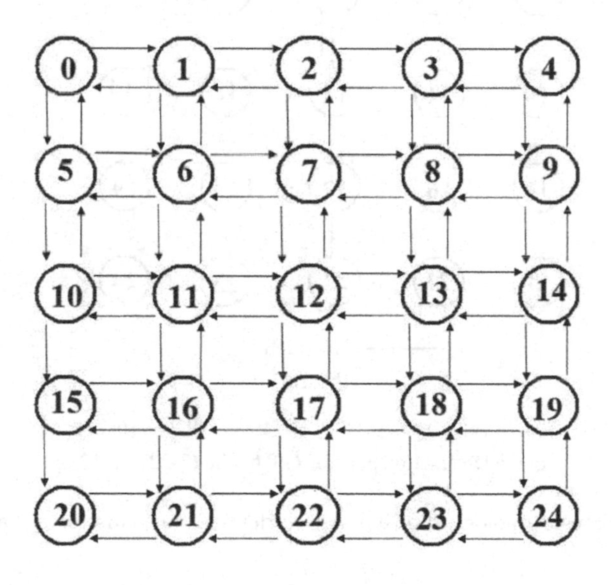

Fig. 1. 5x5, 2D mesh topology

2 NoC Simulation Framework

We have used NIRGAM (a simulator for NoC Interconnect Routing and Application Modeling), a discrete event cycle accurate simulator targeted at NoC research [6]. It allows to experiment with various options available at every stage of NoC design topology, switching techniques, virtual channels, buffer parameters, routing mechanisms, quality of service and applications besides built in capabilities it can be easily extended to include new applications and routing algorithms [7]. Besides built in capabilities and routing algorithms, the simulator can output performance matrix (latency and throughput) for a given set of choices.

3 Experimental Setup

In our experimental setup a 5x5, 2D topology is selected and wormhole switching technique is adopted using two classes of traffic, GT traffic as well as BE traffic. In figure 2, (0,17) and (6,17) shows source destination pairs for BE traffic whereas (3,22) and (9,12) represents source destination pairs for GT traffic.

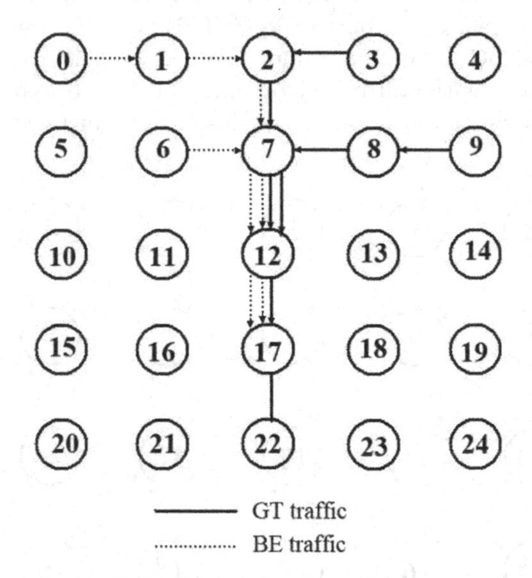

Fig. 2. 5x5, 2D mesh topology showing source destination pairs for BE and GT traffic

With the above setup, we have two types of flits BE which has priority 0 (low) and GT flits having priority 1(high) [8]. A single physical channel is divided into 4 virtual channels. Out of these 4 virtual channels, one virtual channel is reserved for GT traffic implying that bandwidth reserved for GT traffic (BW GT) is 0.25 [9]. When 2 virtual channels out of these 4 channels are reserved for GT, the bandwidth reserved for GT is 0.5. Similarly when 3 virtual channels are reserved for GT traffic the BW GT becomes 0.75. The number of buffers is 32. Flit size is 5 bytes. The network performance is evaluated on the basis of average latency of both GT and BE traffic for Q-routing (for different values of learning rate α) and XY routing strategy.

4 Routing Algorithms

4.1 XY Routing

XY routing is deterministic routing, i.e. route is determined solely by addresses of source and destination nodes and same path is chosen for a given pair of source and destination nodes [10]. In XY routing, first route is determined in X-dimension and then in Y-dimension. XY routing is 2D version of dimension order routing. In latter, a packet is successively routed in each dimension until distance in that dimension is zero. In other words, routing is done in first dimension and then second and then third and so on. Order of picking dimensions is pre-defined. Advantages of XY Routing are: easy to implement, simpler router design leading to reduction in cost and deadlock free routing [11].

4.2 Proposed Adaptation of Q-Routing for QoS

Q-routing is an estimate-based adaptive algorithm, that uses a variant of reinforcement learning algorithm (as mentioned in Algorithm 1). It is a distributed routing algorithm, in which each router acts as an agent that tries to minimize estimated cost involved in routing the packets to its destination.

Any outbound packet by a router can be routed via one of neighbors. A Q-router maintains a m x n table, m is number of destinations and n is number of neighbors. A (p,q) entry in this table is estimated cost of routing a packet to destination q via neighbor p.

In Q-routing, x sends packets to the neighbor y with minimum estimate for a given destination d. The neighbor y then sends back its estimate of cost in routing a packet to destination d. With this new information, the router x updates its 'estimate' for d. With constant updating of cost involved in sending a packet, Q-router adapts to congestion by routing packets onto routes with least estimated cost. With this adaptivity, a router can route for d via all its neighbors whereas XY employs same neighbor always. Thus traffic is distributed much more uniformly in Q-routing resulting delay in onset and building of congestion [12].

Apart from the normal NoC issues, a typical Q-router has to address issues such as updating estimate values, routing decisions, penalizing dropped packets and sending estimate values.

Algorithm 1. Q-routing

```
input: A mesh of N nodes, D nodes marked destination, learning
       parameter α

procedure Cost Initialization:
for x = 1 to N do
    for d = 1 to D do
        for y = 1 to neighborhood(x) do
            Q_x(d,y)=0
        end
    end
end
endprocedure

procedure Q-routing(router x, destination d):
Min=0
Generate a random number p ∈ [0,1]
if p ≤ ∈ then
    y = random neighbor
end
else
    Min = ∞
    for k = 1 to neighborhood(k) do
        if Q_x(d,k) ≤ Min then
            Min = Q_x(d,k)
            y = k
        end
    end
end
if routing node then
    Route packet to y at cost q
    Receive estimate s from y
    if estimate s is received in time t then
        Compute S=s + q
    end
    else
        S = penalty + q
    end
    Update cost: Q_x(d,y) = (1-α)Q_x(d,y) + αS
end
else
    S = Q_x(d,y)
    Send S to previous router node
end
endprocedure
```

In the following sections we will discuss the mechanism used to implement QoS aware Q-routing.

Updating Estimate Values. Each Q-router x maintains an estimate of cost involved in sending packets to a given destination d through each neighbor y, denoted by $Q_x(d,y)$. For routing a packet to d, router x sends it through neighbor y for which $Q_x(d,y)$ is minimum.

Router y sends its minimum estimated cost $Q_y(d)$ back to the router x which updates its own cost $Q_x(d,y)$ according to the following equation:

$$Q_x(d,y) = (1-\alpha)Q_x{}^{old} + \alpha(q+Q_y(d))$$
$$R=Q_x{}^{old}(d,y),\ S = q + Q_y(d),\ \alpha = [0,1)$$
$$Q_x(d,y) = (1-\alpha)R + \alpha S$$

Here q is the cost involved in sending a packet from x to y; α is a positive constant less than 1; 'R' is the old estimate value and 'S' is the received estimate value. α is a measure of the rate of learning. Higher α means more weight-age to the received estimate value S. If α is low, more weight-age is given to the old value 'R'. If α is 1, then the new estimate value will be equal to the received estimate value 'S'. If α is 0, then no learning takes place.

Routing Decision. A router that chooses the best known path is said to use Greedy Policy. For a destination d, cost $Q_x(d,y)$ will change only when x routes packet via neighbor y. In a greedy approach, x continues to choose $y1$ (nearest neighbor for the chosen path) as long as cost $Q_x(d,y)$ remains minimum. As a result, cost for all other neighbors $Q_x(d,y)$, $y \neq y1$ is never updated. To avoid this, near greedy approach such as epsilon-greedy ($\in$-Greedy) can be used [13]. In $\in$-Greedy approach, the router sending a packet will choose neighbor, randomly, with a probability $\in$ and will choose the neighbor with minimum cost in all other cases. In all our simulations we use $\in$-Greedy Method.

Penalizing Dropped Packets. In general there will be no packets drop in NoC, but router routes a packet via a faulty neighboring router, it is possible that the estimate value 'S' is never sent back. As a result there is no change in the estimate. The same can happen if the input channel of the router sending the data packet is faulty. To avoid such situations the router has to have a mechanism for finding packet drops and penalizing the estimate value by making it larger.

Sending Estimate Values. In order to implement Q-routing for NoC, each Q-router has to maintain a table of size m x n for storing estimate values, where m is the number of destinations and n is the number of neighbors. The initial values can be application specific. In some cases, we set the initialization table with a very high value. In such a case once packet takes that path, then most of the time it takes that path only because the Q value for that path is decreased comparatively. But in our case, we are initializing table with zero Q values so that once the packet takes that path the Q value for that path will increase with respect to other paths so next time the packets will choose another path and hence tries to explore all the path in initial learning period. When a router y receives a packet for destination d from x, it has to create an estimate packet for destination value 'S'. The newly created estimate packet can go to one of the following, depending on the implementation.

1. Add required routing information and queue it to the input buffer [14].
2. A separate buffer where the estimate values will be stored before sending it to the output channel.
3. Send it through a separate line for sending estimate values. A separate line will be added to every NoC router.

In our implementation, we have used the third method. We chose it because, in the first approach, not only do estimate packets compete for the output channel, but they also occupy the input buffers. In the second approach, even though they don't occupy the input buffer, they compete for the output channel. Also if there are 4 input channel (say) then there will be 4 times more estimate packets, more chances of overflowing. In both the cases it is quite possible that at higher loads estimate packets may be dropped. This can be problematic if for example, "west" input port for a router is free but output port is not, in such a case estimate packets going to west output port might get dropped. Now the west side router will penalize for this even though the west input channel is free.

All the above issues can be easily handled by adding an extra line. By adding an extra line, the estimate packets no longer compete for output channel, and won't occupy any buffers. Also, there won't be any packet dropping, since packets are sent as soon as they arrive.

5 Results

In order to compare the performances of Q-routing algorithm and XY, we implemented both XY and Q-routing in NIRGAM with the additional feature of QoS (Quality-of-Service).

5.1 Result Analysis

Since XY is deterministic routing algorithm, hence the path flit takes for every source-destination pair is fixed and does not depend on the current status of the network. But, if some routers or links become congested as in our experiment, the performance of XY start decreasing because it still takes the same path while the neighboring router and links are free hence, the total number of packets received by the cores are significantly less, as now it depends only on the capacity of the links which is provided by the XY router and it is fixed. Also it increases the overall latency of the packets.

Q-routing is an adaptive routing algorithm which takes into account the current network status and on the basis of the status, it decides the path for the packets being routed. In our experiment as some links become busy, the time

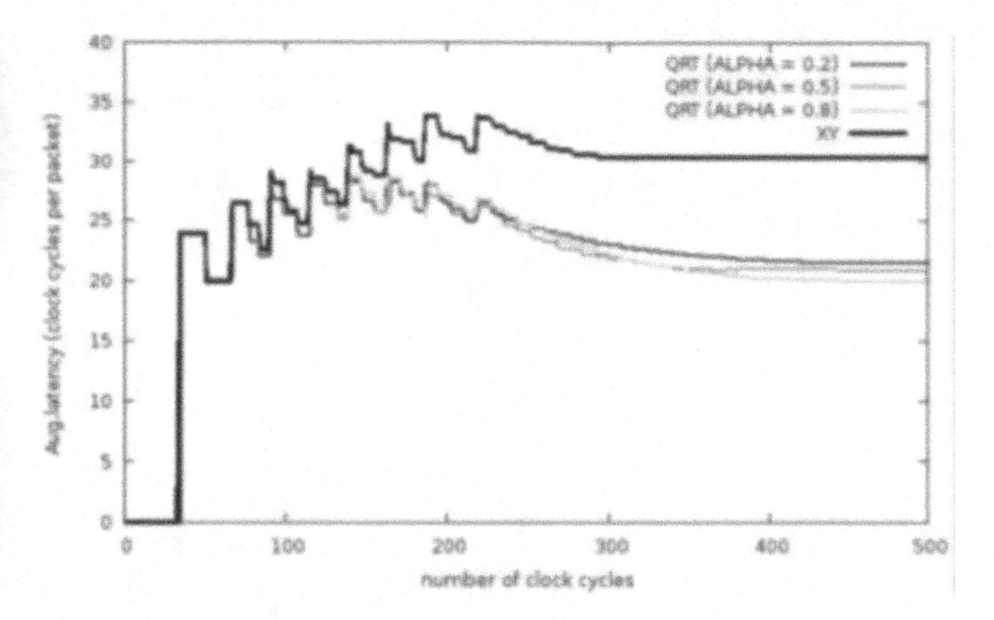

Fig. 3. Average latency of BE traffic vs Clock Cycles(BW GT=0.25)

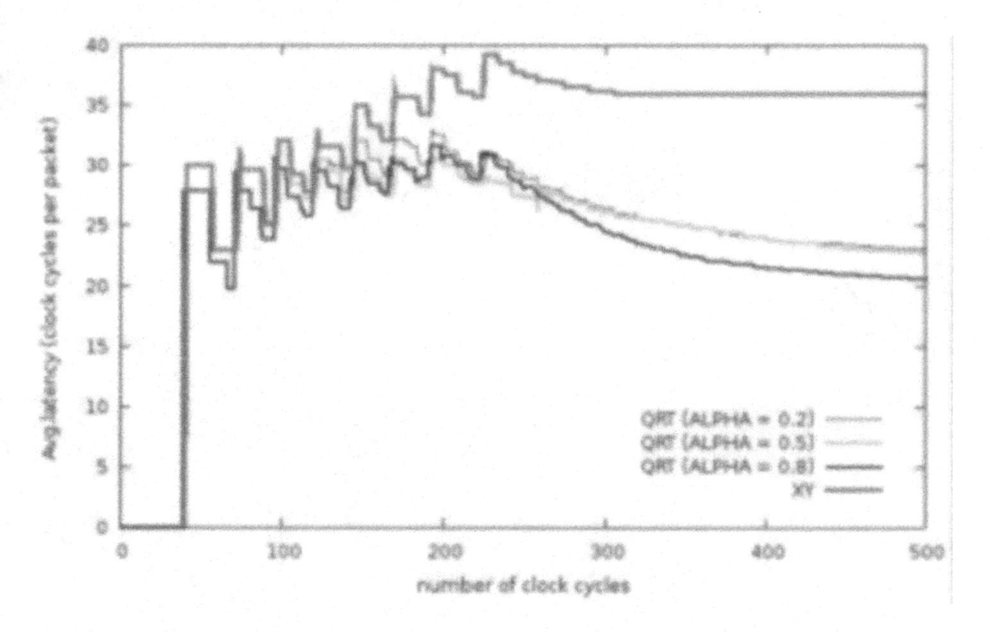

Fig. 4. Average latency of GT traffic vs Clock Cycles(BW GT=0.25)

taken by the packets traveling through these links increases, which Q-routing takes into account and updates its table. While taking the routing decision it sends the packet through some other path which has lower Q value than previously selected. This way it makes use of the neighboring routers and links

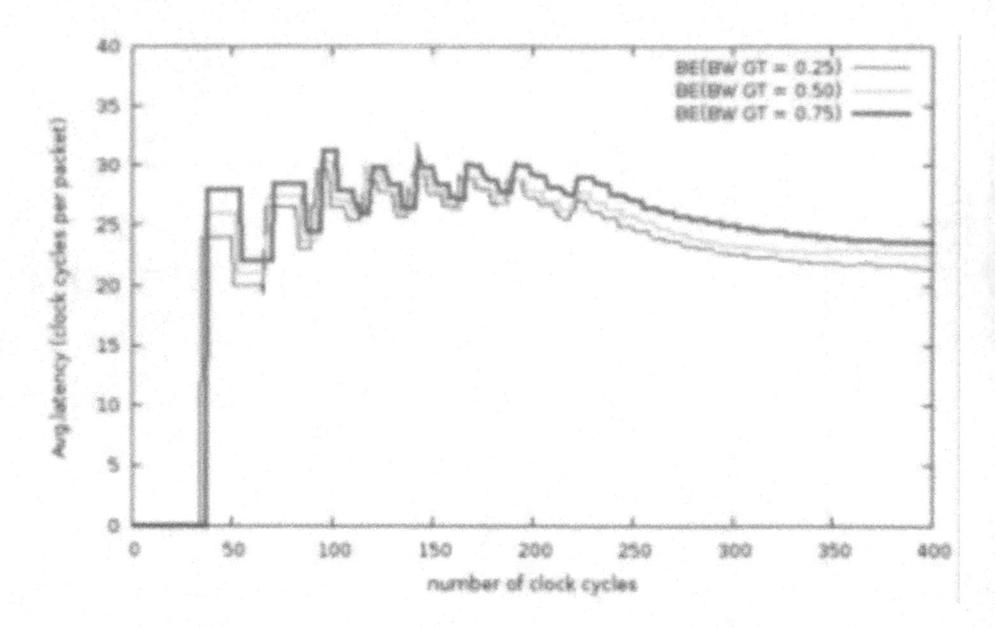

Fig. 5. Average latency of BE traffic vs Clock Cycles(QRT α=0.8)

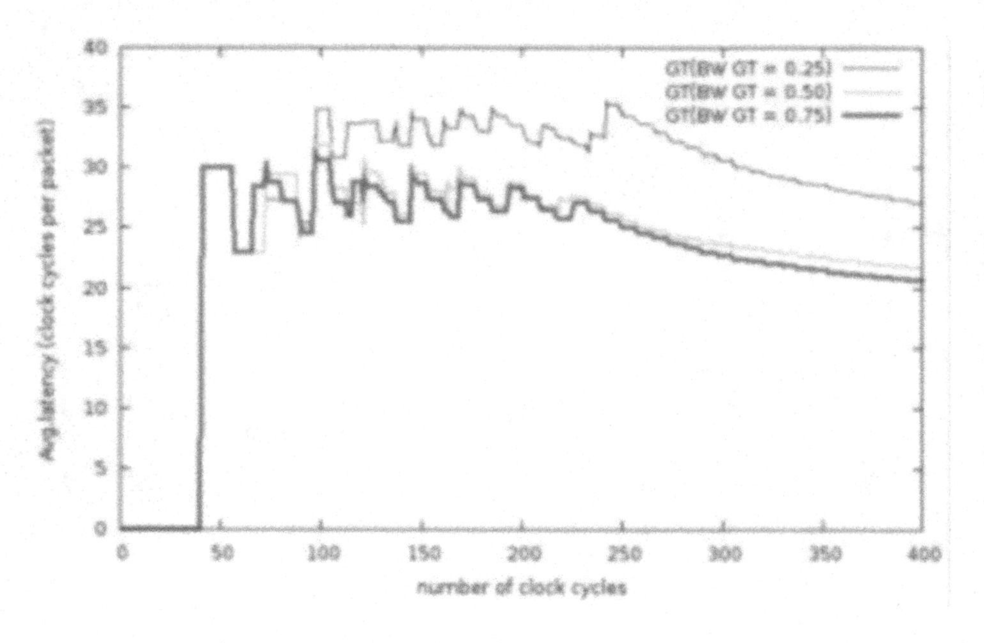

Fig. 6. Average latency of GT traffic vs Clock Cycles(QRT α=0.8)

to distribute the load and decrease the congestion. Also the randomness introduced in Q-routing helps in finding new paths with lower Q values. Without this, routers can go unnoticed after recovering from congestion. Without randomness the router will keep sending the packets to the same link with lower Q values at the time of congestion. So we see that Q routing is highly adaptive to the current status of the network and hence gives a better result from XY.

Figure 3 and 4 shows average latency of BE and GT traffic vs clock cycles for Q-routing (with different learning rate α) and XY routing while the bandwidth reserved for GT traffic (BW GT) equals to 0.25. For less than 150 clock cycles, the results of average latency in case of BE traffic and GT traffic for Q-routing (with α=0.2, 0.5 and 0.8) and XY routing, are comparable. However, for clock cycles greater than 250, the graph in Figure 3 shows that there is an increasing trend in average latencies for Q-routing (α=0.8, 0.5, 0.2) and is highest for XY routing. This might be due to the reason that as the network gets congested Q-Routing adapts itself to the changing network conditions.

As shown in Figure 4, it also exhibits an increasing trend in average latency of GT traffic for Q-routing (α=0.8, 0.5, 0.2) and is highest for XY routing. Comparing Figure 3 and Figure 4, it is observed that for higher value of learning rate (α=0.8), initially the average latency is high due to network congestion and load imbalance but Q-routing responds quickly to the changing network conditions and settles down to a lower value of average latency.

As shown in Figure 5 and 6, average latency of BE traffic and GT traffic is observed for Q-routing algorithm (learning rate α=0.8) and varying the bandwidth reserved for GT (BW GT). It is observed that as BW GT increases, the average latency of BE traffic also increases. However, the average latency of GT traffic decreases.

Q-routing with a lower value of alpha (0.2) does not perform well as it adapts very slowly to the current situation and so it could not give good results. Q-routing with lower values will not shift to a better path immediately and detects congestion slowly. On the other hand, Q-routing with high alpha values learns quickly and so reacts quickly.

Limitation is overhead of maintaining and updating cost values. Some mechanism is needed to check deadlock avoidance. Cycles in routing path may lead to deadlocks. One way is to avoid cycle formation. Virtual Channels (VCs) can be employed for this purpose [15]. By routing on an output VC with less number than input VC, deadlock can be avoided. Our current implementation however does not deal with this and is part of our future work.

6 Conclusion and Future Work

This paper presents the concept of handling the issue of dynamic communication among modules with differing Quality-of-Service requirements which are dynamically placed on a re-configurable device. Using simulation results we can observe that both XY and Q-routing algorithms are suitable for a dynamic QoS aware NoC implementation. Both algorithms have the advantage to depend on local information only. XY routing algorithm is simple and small, whereas Q-routing with QoS presents the learning advantage which quickly adapts to the changing network conditions and brings the system to a state of convergence.

Q-routing algorithm outperforms XY for higher values of learning rates α and for higher values of bandwidth reserved for GT traffic.

In our future work, the performance of the NoC can be evaluated in context of QoS and Q-routing can be contrasted with other routing algorithms like Odd-Even, DyAD and fully adaptive algorithms.

References

[1] Dally, W., Towles, B.: Route Packets, Not Wires: On-Chip Interconnection Networks. In: DAC, June 2001, pp. 684–689 (2001)
[2] Benini, L., De Micheli, G.: Networks on Chip: A new SoC Paradigm. IEEE Computer 35, 70–78 (2002)
[3] Jantsch, A.: Communication Performance in Network-on-Chips. Network on Chip Seminar Linkping, November 25 (2005)
[4] Andreasson, D., Kumar, S.: On improving Best Effort throughput by better utilization of Guaranteed-Throughput channels in an on-chip communication system. In: Proceedings of Norchip conference, November 8-9, pp. 265–268 (2004)
[5] Rostilav, D., Ginosar, R., Kolodny, A.: QNoC Asynchronous Router. Integration VLSI Journal 814, 1–13 (2008)
[6] Jain, L., Al-Hasimi, B.M., Gaur, M.S., Laxmi, V., Narayanan, A.: NIRGAM: A Simulator for NoC Interconnect Routing and Application Modeling. In: Design, Automation and test in Europe 2007 (DATE 2007), Nice, France, April 16-20 (2007)
[7] Dally, W.: Virtual Channel Flow Control. IEEE Transactions on parallel and distributed systems 3, 194–205 (1992)
[8] Andreasson, D., Kumar, S.: Improving BE Traffic QoS using GT slack in NoC Systems. In: NORCHIP Conference, November 21-22, pp. 44–47 (2005)
[9] Bolotin, E., Cidon, I., Ginosar, R., Kolodny, A.: QNoC: QoS Architecture and design process for network on chip. Journal of Systems Architecture, 105–128 (2004)
[10] Ye, T.T., Benini, L., De Micheli, G.: Packetization and routing analysis of on-chip multiprocessor networks. Journal of Systems Architecture 50(2-3) (February 2004)
[11] Dally, W., Towles, B.: Principles and Practices of Interconnection Networks. Morgan Kaufmann publishers, San Fransisco (2004)
[12] Majer, M., Bobda, C., Ahmadinia, A., Teich, J.: Packet Routing in Dynamically Changing Networks on Chip. In: Proceedings of the 19th IEEE International Parallel and Distributed Processing Symposium (IPDPS 2005), Denver, CA, USA, April 4-8, pp. 154b–154b (2005)
[13] Mansour, M., Kayssi, A.: FPGA-Based Internet Protocol Version 6 Router. In: Proceedings of International Conference on Computer Design: VLSI in Computers and Processors (ICCD 1998), Austin, Texas, USA, October 2-5, pp. 334–339 (1998)
[14] Sathe, S., Wiklund, D., Liu, D.: Design of a Switching Node (Router) for On-Chip Networks. In: 5th International Conference on ASIC Proceedings (ASICCON 2003), Beijing, China, October 21-24, vol. 1, pp. 75–78 (2003)
[15] Hansson, A., Goossens, K., Radulescu, A.: A Unified Approach to Mapping and Routing on a Network-on-Chip for both Best-Effort and Guaranteed Service Traffic. VLSI Design, vol. 2007 (2007)

Color Image Restoration Using Morphological Detectors and Adaptive Filter

Anita Sahoo, Rohal Suchi, Neha Khan, Pooja Pandey, and Mudita Srivastava

JSS Academy of Technical Education, Noida
anitasahu1@gmail.com

Abstract. A two phased impulse restoration scheme for color images is presented. In the first phase of the proposed image restoration scheme it detects the pixels corrupted with impulse noise by employing the tools of mathematical morphology and then in the next phase a linear adaptive mean filter attempts to remove those noisy pixels in an efficient manner. Experimental results indicate that the proposed scheme can suppress impulse noise effectively in color images. This provides a better restoration performance than many other filters used for removing impulse noise from color images.

Keywords: Morphological Operations, Adaptive Mean Filter, Alpha Trimmed Mean Filter, Image Restoration, Impulse Noise, Color image in RGB space.

1 Introduction

With the wide use of color in many areas such as multimedia applications, biomedicine, internet and so on, the interest on the color perception and processing has been growing rapidly. Color, as we know is a powerful descriptor and helps us interpret and identify the objects in a picture more clearly than a gray scale image. They provide the picture with an element of closeness with the real world which obviously, isn't black and white. To add to it, there are thousands of shades our eyes can recognize thus elaborating the scope of this paper to all of these instead of restricting it to a few dozen shades of gray.

Impulse noise is characterized by discontinuities in the form of very high or very low values in a dynamic range thus giving it an overall granular kind of appearance. The standard median filter has been widely used for removing impulse noise. The problem with traditional linear or median filters and it's like [5], [7] was that they treated all the pixels even the undisturbed ones in the same manner. Though they achieve noise suppression but at the expense of distorting image features considerably. Improvements are the adapted switching filters [1], [8] [9], the opening closing sequence filter [2], Adaptive switching mean filter with morphological noise detector [3]. Mathematical morphology has been widely used in image processing [6]. Morphological filters are nonlinear signal transformations that locally modify geometric features of signals. It has been shown that morphological filters can be efficiently used to clean noise in binary image and gray-scale image [2], [3], [6].

Deng, Z.F et.al [4] used Median Controlled Adaptive Recursive Weighted Median Filter for restoring color image corrupted by impulse noise. In this paper, a new class

S. Ranka et al. (Eds.): IC3 2009, CCIS 40, pp. 381–388, 2009.
© Springer-Verlag Berlin Heidelberg 2009

of filter for color image processing is developed. The Morphological Operations in RGB color space are introduced, which are an extension of mathematical morphology from gray-scale image to color image. The experimental results have been compared with various other filters indicating the efficiency of the proposed technique.

2 Proposed Method

The proposed method uses a sequence of morphological operations for noise classification. A flat structuring element has been used for morphological noise detection that reduces the complexity of the overall technique. Once the noise has been detected effectively it can be filtered. Generally in the fixed small window size filters, the amount of noise density filtered will be very less, for filtering high density noise the window size of the filter may increase. This may lead to blurring in the output images. In order to overcome this, the adaptive window length filters are designed for filtering high density noises. Here we use an adaptive mean filter that restores the color image.

2.1 Morphological Noise Detection

A color pixel at (x, y) in a given domain of color image can be described by a color vector (R(x, y), G(x, y), B(x, y)) in RGB color space, where $(x, y) \in Z^2$. In order to process the color image, it is first decomposed into its three component images. Now each component image is treated separately. Morphological noise detection is performed in two stages.

2.2 Stage-I of Noise Detection

This involves the application of basic morphological operations of Dilation and Erosion on the image function f(x, y) that is the value of the component under consideration at position (x, y). Say b(x, y) is an appropriate structuring element. Dilation and Erosion operations can be represented by equation (1) and (2) respectively.

$$(f \oplus b)(x, y) = \max\{f(x-s, y-t) + b_n(s,t) \,|\, (x-s),(y-t) \in D_f ; (s,t) \in D_b\} \tag{1}$$

$$(f \Theta b)(x, y) = \max\{f(x+s, y+t) - b_n(s,t) \,|\, (x+s),(y+t) \in D_f ; (s,t) \in D_b\}, \tag{2}$$

where D_f and D_b denote the domain of the image f and the domain of the structuring element b, respectively.

As shown, the above operations find out the minimum and maximum color component pixels in their neighborhood. Now as the salt and pepper impulse noise itself implies significant discontinuities in their neighborhood, they can be detected comparing the results of dilated or eroded image with the original one. If the values after dilation or erosion turn out to be the same as before, the pixel is classified as the noisy one with the noise flag set. Else it is considered a noise free pixel. Using this concept a binary array B(x, y) containing the noise flags can be derived as explained in equation (3).

$$B(x,y)=\begin{cases}1 & f(x,y)=(f\oplus b)(x,y) \quad or \quad f(x,y)=(f\Theta b)(x,y)\\ 0 & otherwise\end{cases} \tag{3}$$

However there are some issues with this. Sometimes the edge pixels or other such sharp discontinuity areas which otherwise form a part of the image itself are also misclassified as being noisy pixels. Thus a second stage of noise classification is also employed.

2.3 Stage-II of Noise Detection

Based on the basic operations of Erosion and Dilation, Opening and Closing operations are defined as given in equation (4) and (5) respectively.

$$(f\circ b)(x,y)=((f\Theta b)\oplus b)(x,y) \tag{4}$$

$$(f\bullet b)(x,y)=((f\oplus b)\Theta b)(x,y) \tag{5}$$

Combining the sequence of opening-closing operators with the closing-opening operators, the local characteristic at position (x, y) is determined as in equation (6).

$$D(x,y)=\left|\frac{((f\circ b)\bullet b)(x,y)+((f\bullet b)\circ b)(x,y)}{2}-f(x,y)\right| \tag{6}$$

If the local characteristic D for a pixel is greater than a predefined threshold value T, it means it differs with its neighborhood significantly. Using this concept the binary array B(x, y) containing the noise flag is revised as described in equation (7).

$$B(x,y)=\begin{cases}1 & B(x,y)=1 \quad and \quad D(x,y)\geq T\\ 0 & otherwise\end{cases} \tag{7}$$

As discussed earlier these morphological operations defined in two stages are to be applied on the three color components of the image separately. The sequence of above described operations is used to realize accurate noise detection. Once the noise is detected we move to the next phase, where the noise will be eliminated by applying an adaptive mean filter.

2.4 Noise Elimination

Adapted switching filters improve the efficiency by freeing the filter from the constraints imposed by a fixed window size. Adaptive switching filters have the capability to alter the window size according to the availability of noise affected pixels. The adaptive mean filter used determines the filtering window size for every detected noise pixel based on the number of its neighboring noise-free pixels. For each detected noise pixel (x, y), the filtering window with the size of $(2L_f+1)*(2L_f+1)$ centered about it is used. Starting with $L_f=1$, the filtering window is iteratively extended outwards by one pixel in its four sides until the number of noise-free pixels within this window is 1 at least. Let W(x, y) denote the noise-free pixels in the filtering window, i.e.

$$\mathrm{W}(x,y)=\{f(x+s,y+t)|B(x+s,y+t)=0,B(x,y)=1,(s,t)\neq 0,-L_f\leq s,t\leq L_f\} \tag{8}$$

The output of the adaptive mean filter applied at position (x, y) is obtained by:

$$h(x, y) = B(x, y)m^{\alpha}(x, y) + (1 - B(x, y))f(x, y), \tag{9}$$

where $m^{\alpha}(x, y)$ is the alpha-trimmed mean value calculated as:

$$m^{\alpha}(x, y) = \frac{1}{p(x, y) - 2\lfloor \alpha p(x, y) \rfloor} \sum_{k=\lfloor \alpha p(x,y) \rfloor + 1}^{p(x,y) - \lfloor \alpha p(x,y) \rfloor} F(k), \tag{10}$$

where p(x, y) is the number of noise pixels within the filtering window, α is the trimming parameter ($0<\alpha<0.5$) and F (k) denoted the kth data item in increasing order samples of W(x, y). After each component of the color image in RGB space have been processed under the above mentioned techniques, they are recombined into one, rendering a single restored color image

3 Experimental Results

The proposed method has been simulated using MATLAB7. For the experimentation the size of the structuring element b is chosen as 5x5 while T and α has been considered to be 25 and 0.3, respectively. A number of color images have been corrupted by impulse noise with the noise density varying from 0.1 to 0.9 and the method has been applied for the restoration. The results have been compared with some of the recent filters [4], [5], [7] using the PSNR value. Table 1 and 2 shows the comparison of PSNR of different filters for Lena.jpg and Zelda.jpg respectively. It is easy to see from the tables and subjective evaluation based on visual comparisons that the proposed method provides significantly higher PSNR values than other compared filters.

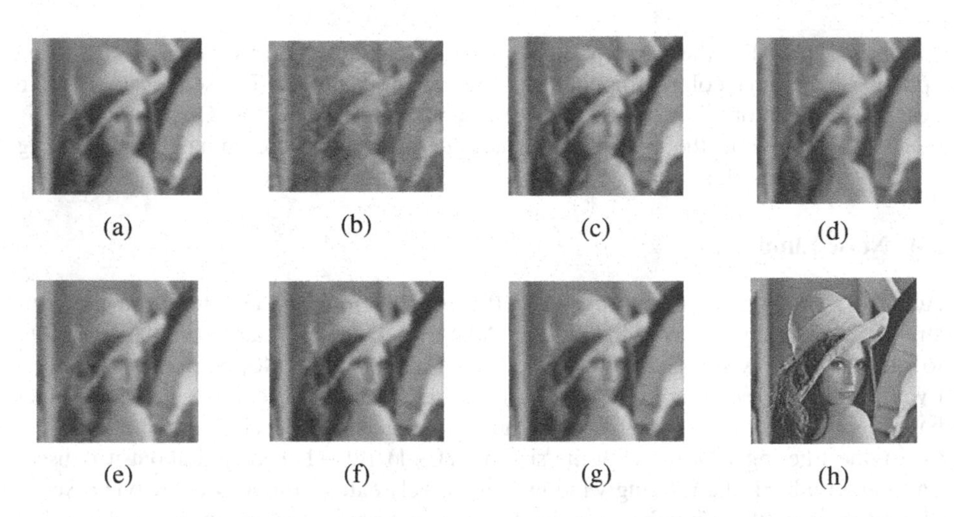

Fig. 1. (a) Original Lena 256 X256 image (b) Noisy image (density 40%) (c) SMF output (d) WMF output (e) RWM output (f) MC LIN's output (g) ARWMF output (h) proposed method's output

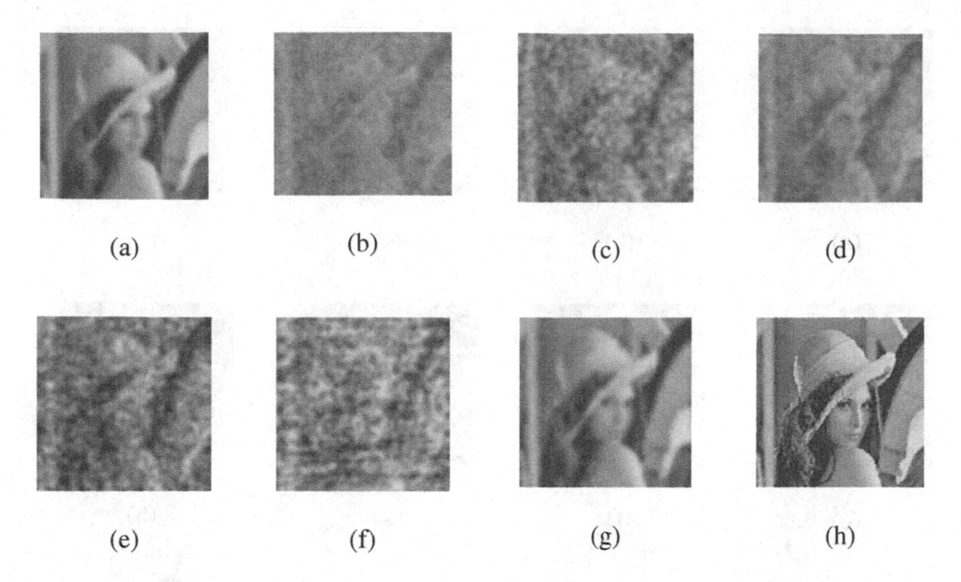

Fig. 2. (a) Original Lena 256 X256 image (b) Noisy image (density 80%) (c) SMF output (d) WMF output (e) RWM output (f) MC LIN's output (g) ARWMF output (h) proposed method's output

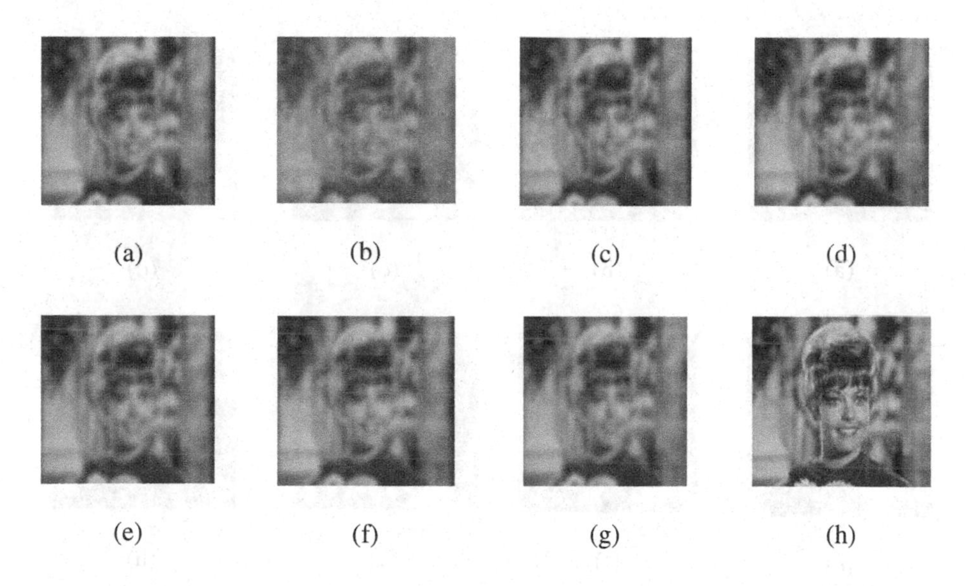

Fig. 3. (a) Original Zelda 256 X256 image (b) Noisy image (density 20%) (c) SMF output (d) WMF output (e) RWM output (f) MC LIN's output (g) ARWMF output (h) proposed method's output

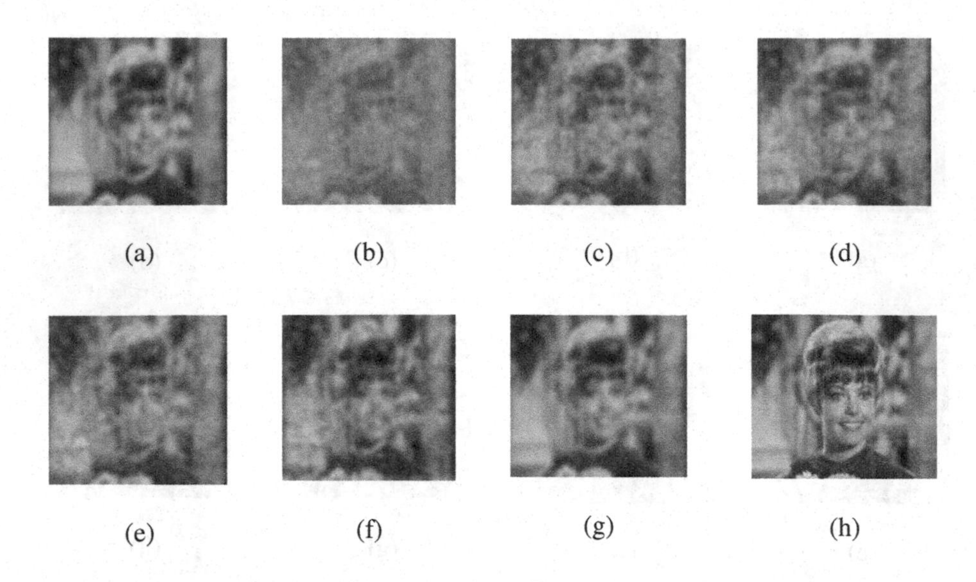

Fig. 4. (a) Original Zelda 256 X256 image (b) Noisy image (density 60%) (c) SMF output (d) WMF output (e) RWM output (f) MC LIN's output (g) ARWMF output (h) proposed method's output

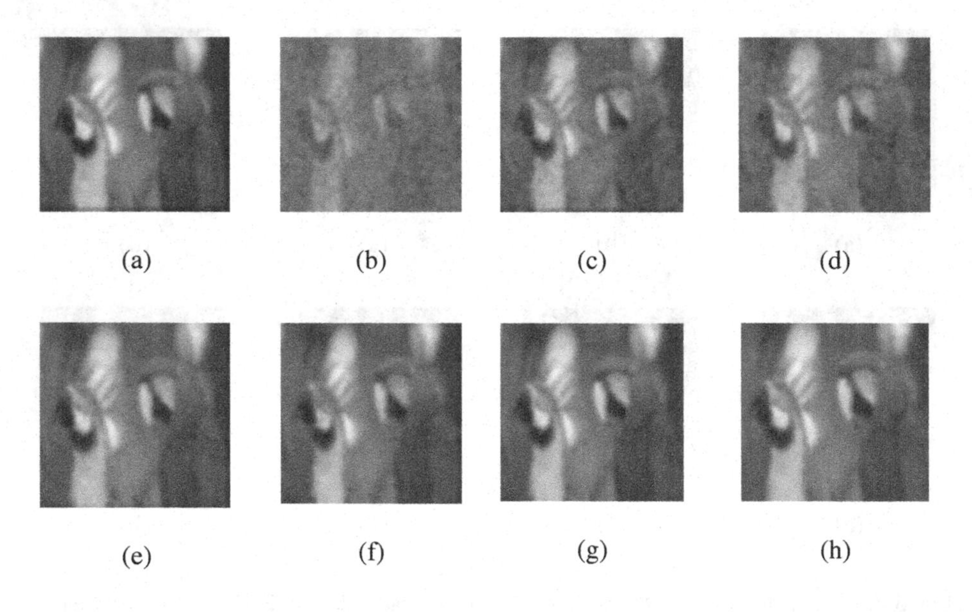

Fig. 5. (a) Original Parrot 512X512 image (b) Noisy image (density 30%) (c) SMF output (d) WMF output (e) RWM output (f) MC LIN's output (g) ARWMF output (h) proposed method's output

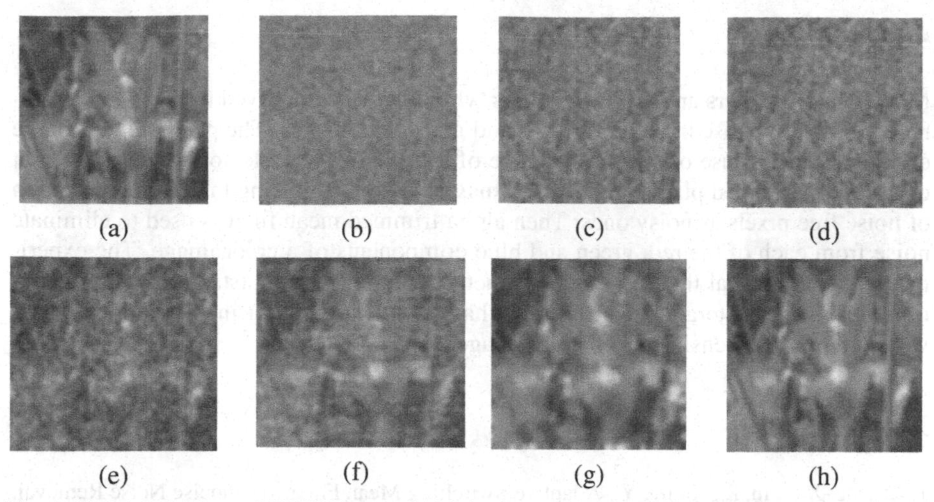

Fig. 6. (a) Original Flower 512X512 image (b) Noisy image (density 90%) (c) SMF output (d) WMF output (e) RWM output (f) MC LIN's output (g) ARWMF output (h) proposed method's output

Table 1. Comparison of Restoration Performance in *PSNR* for various Filters applied on Lena. Image Corrupted with Impulse Noise of different Densities

Noise Density	SMF	WMF	CWM	RWM	MC(lin's)	ARWMF	ASM
10	**33.9**	**35.04**	**34.87**	**32.07**	**33.76**	**30.53**	**36.1**
20	**31.58**	**30.62**	**29.28**	**30.79**	**31.88**	**29.73**	**33.8**
30	**27.42**	**25.58**	**24.53**	**28.63**	**29.32**	**29.18**	**32.6**
40	**23.28**	**21.65**	**20.63**	**26.5**	**25.42**	**28.48**	**31.16**
50	**19.82**	**18.6**	**17.7**	**24.19**	**21.44**	**28.01**	**29.71**
60	**16.89**	**16.12**	**15.38**	**20.8**	**18.01**	**27.43**	**28.39**
70	**14.62**	**14.1**	**13.63**	**17.01**	**15.13**	**26.75**	**26.84**
80	**12.73**	**12.48**	**12.2**	**13.68**	**12.78**	**25.95**	**25.04**
90	**11.24**	**11.11**	**10.9**	**11.69**	**11.17**	**24.1**	**22.19**

Table 2. Comparison of Restoration Performance in *PSNR* for various Filters applied on Zelda.jpg Image Corrupted with Impulse Noise of different Densities

Noise Density	SMF	WMF	CWM	RWM	MC(lin's)	ARWMF	ASM
10	**35.71**	**35.04**	**36.36**	**33.85**	**36.3**	**30.76**	**39.77**
20	**32.52**	**30.62**	**29.18**	**32.28**	**33.74**	**30.52**	**38.88**
30	**27.79**	**25.58**	**24.56**	**29.69**	**30.06**	**30.35**	**38.51**
40	**23.44**	**21.65**	**20.7**	**28.01**	**26.15**	**29.79**	**37.25**
50	**20**	**18.6**	**17.68**	**24.58**	**21.84**	**29.34**	**35.52**
60	**17.01**	**16.12**	**15.44**	**21.12**	**18.27**	**28.96**	**33.84**
70	**14.65**	**14.1**	**13.67**	**16.97**	**15.28**	**28.39**	**31.64**
80	**12.77**	**12.48**	**12.19**	**13.62**	**13.07**	**27.07**	**28.79**
90	**11.74**	**11.11**	**11**	**11.3**	**11**	**26.52**	**23.91**

4 Conclusions

Morphological filters are nonlinear filters which have been proved to be able to eliminate powerfully noise in binary images and gay-scale images. The proposed technique effectively makes use of such a sequence of morphological detectors to detect the set of actually corrupted pixels in the color images, thereby avoiding the misclassification of noise free pixels as noisy ones. Then alpha trimmed mean filter is used to eliminate noise from each of the red, green and blue components of a color image. The experimental results reveal that the proposed method not only has outstanding noise detection and image restoration feat but also has excellent strength in combating a wide variation of noise densities from color image.

References

1. Zhang, X., Yin, Z., Xiong, Y.: Adaptive Switching Mean Filter for Impulse Noise Removal. Congress on Image and Signal Processing 3, 275–278 (2008)
2. Deng, Z.F., Yin, Z.P., Xiong, Y.L.: High probability impulse noise removing algorithm based on mathematical morphology. IEEE Signal Process. Lett. 14(1), 31–34 (2007)
3. Yin, Z.P., Zhang, X.M., Xiong, Y.L.: Adaptive switching mean filter using conditional morphological noise detector. Image and Signal Processing 44(6) (2008)
4. Vijay Kumar, V.R., Manikandan, S., Ebenezer, D., Vanathi, P.T., Kanagasabapathy, P.: High Density Impulse noise Removal in Color Images Using Median Controlled Adaptive Recursive Weighted Median Filter. IAENG International Journal of Computer Science 34(1), IJCS_34_1_2 (2007)
5. Arce, G.: A General Weighted Median Filter Structure Admitting Negative Weights. IEEE Tr. On Signal Proc. 46 (1998)
6. Serra, J. (ed.): Mathematical Morphology and Its Applications to Image Processing. Kluwer Academic Publishers, Boston (1994)
7. Arce, G., Paredes, J.: Recursive Weighted Median Filters Admitting Negative Weights and Their Optimization. IEEE Tr. on Signal Proc. 48 (2000)
8. Wang, Z., Zhang, D.: Progressive switching median filter for the removal of impulse noise from highly corrupted images. IEEE Trans. Circuits Syst. II, Analog Digit. Signal Process. 46(1), 78–80 (1999)
9. Eng, H.L., Ma, K.K.: Noise adaptive soft-switching median filter. IEEE Trans. Image Process. 10(2), 242–251 (2001)

Secure Multi-party Computation Protocol for Defense Applications in Military Operations Using Virtual Cryptography

Rohit Pathak[1] and Satyadhar Joshi[2]

[1] Acropolis Institute of Technology & Research
[2] Shri Vaishnav Institute of Technology & Science
Indore, Madhya Pradesh, India
{rohitpathak,satyadhar_joshi}@ieee.org

Abstract. With the advent into the 20th century whole world has been facing the common dilemma of Terrorism. The suicide attacks on US twin towers 11 Sept. 2001, Train bombings in Madrid Spain 11 Mar. 2004, London bombings 7 Jul. 2005 and Mumbai attack 26 Nov. 2008 were some of the most disturbing, destructive and evil acts by terrorists in the last decade which has clearly shown their evil intent that they can go to any extent to accomplish their goals. Many terrorist organizations such as al Quaida, Harakat ul-Mujahidin, Hezbollah, Jaish-e-Mohammed, Lashkar-e-Toiba, etc. are carrying out training camps and terrorist operations which are accompanied with latest technology and high tech arsenal. To counter such terrorism our military is in need of advanced defense technology. One of the major issues of concern is secure communication. It has to be made sure that communication between different military forces is secure so that critical information is not leaked to the adversary. Military forces need secure communication to shield their confidential data from terrorist forces. Leakage of concerned data can prove hazardous, thus preservation and security is of prime importance. There may be a need to perform computations that require data from many military forces, but in some cases the associated forces would not want to reveal their data to other forces. In such situations Secure Multi-party Computations find their application. In this paper, we propose a new highly scalable Secure Multi-party Computation (SMC) protocol and algorithm for Defense applications which can be used to perform computation on encrypted data. Every party encrypts their data in accordance with a particular scheme. This encrypted data is distributed among some created virtual parties. These Virtual parties send their data to the TTP through an Anonymizer layer. TTP performs computation on encrypted data and announces the result. As the data sent was encrypted its actual value can't be known by TTP and with the use of Anonymizers we have covered the identity of true source of data. Modifier tokens are generated along encryption of data which are distributed among virtual parties, then sent to TTP and finally used in the computation. Thus without revealing the data, right result can be computed and privacy of the parties is maintained. We have also given a probabilistic security analysis of hacking the protocol and shown how zero hacking security can be achieved.

Keywords: Secure Multi-party Computation (SMC), Defense, Information Security, Privacy.

S. Ranka et al. (Eds.): IC3 2009, CCIS 40, pp. 389–399, 2009.
© Springer-Verlag Berlin Heidelberg 2009

1 Introduction

Terrorism has become the biggest issue of concern in the 20th century worldwide. In year 2004 approximately 3,259 attacks occurred which resulted in nearly 7,902 deaths, in 2005 approximately 11,000 attacks which resulted in nearly 14,500 deaths, in 2006 approximately 14,570 attacks occurred which resulted in nearly 20,872 deaths, and in 2007 approximately 14,000 terrorist's attacks occurred which resulted in 22,000 deaths [1, 2]. As we can see there is more that approximately 9 percent increase in deaths every year due to terrorism from year 2004 to 2007 and is expected to increase further if terrorism is not dealt with strictly.

The suicide attack on World Trade Center Twin Towers in United States of America on 11th September 2001 was one of the biggest acts of terrorism in the last decade in which approximately 2,900 died including nationals from 90 different countries and many were injured [3, 4]. The whole operation was carried out by 19 terrorists affiliated with al Quaida which is a terrorist organization lead by Osama bin Laden. Later on 11 March 2004 a series of blasts in Madrid, Spain caused death of 191 people [5, 6]. The official investigation determined that the attacks were directed by an al Quaida inspired terrorist cell. On 7 July 2005, there were a series of suicide bomb blasts in London followed by another series on 21 July of the same year [7-9]. Lately on 26 November 2008 in India, terrorists carried out a series of attacks in Mumbai which lasted for 4 days and nearly 180 were killed and 320 were injured [10, 11]. This attack was carried out by Lashkar-e-Taiba, a Pakistani based militant organization, which is considered terrorist organization by India, US and United Kingdom among others. Such acts of terrorism are carried out by terrorist organizations such as al Quaida, Harakat ul-Mujahidin, Hezbollah, Jaish-e-Mohammed, Lashkar-e-Toiba, etc. [12]. These organizations are carrying out training camps and terrorist operations which are accompanied with latest technology and high tech arsenal as funds are no issue for them.

To counter such terrorism we need military forces equipped with advanced weaponry and defense technology. Secure communication is among one of the chief areas of concern. It has to be made sure that all the communications carried out by military forces is secure and no third party is able to hack the concerned data. Military forces need to keep their private information, statistics, confidential data, etc. secure from terrorist forces. Leakage of concerned data can prove lethal for military, thus privacy preserving and security measures are required. Many such situations can be countered with Secure Multi-Party Computations as there may be a need to perform computations that require data from many military forces, but in some cases the associated forces would not want to reveal their data to other forces.

In this paper, we have shown how our VPP (Virtual Party Protocol) [13] for SMC (Secure Multi-Party Computations) can be used for defense applications in military operations. This protocol can be used to perform computations on encrypted data. Every party encrypts their data in accordance with a particular scheme. This encrypted data is distributed among some created virtual parties. These Virtual parties send their data to the TTP through an Anonymizer layer. TTP performs computation on encrypted data and announces the result. As the data sent was encrypted its actual value can't be known by TTP and with the use of Anonymizers we have covered the identity of true source of data. Modifier tokens are generated along encryption of data which are distributed among virtual parties, then sent to TTP and finally used in the computation. Thus without revealing the data, right result can be computed and privacy of the parties is maintained. We have also shown how the protocol and accompanying algorithm is optimized and have given a probabilistic security analysis of hacking the protocol and shown how zero hacking security can be achieved. The main advantage is that encrypted data is sent to third party, so even if the data is hacked from the TTP it will not breach the privacy and security of the party as the data is encrypted. The decryption complexity is very high and with appropriate settings we can reach nearly zero hacking security.

2 Problem Statement

There is a region occupied by terrorists where they are operating their training camps and carrying out other illegal activities. Some countries want to perform a joint operation against terrorists in this region. Consider the situation that there are terrorists training camps being held in some region of Pakistan. Suppose United States, India and Israel wants to conduct a joint military operation to eliminate these terrorists. At any situation one of the countries wants to launch a missile on a target area with possible terrorist activity. The attacking country has to make sure that the missile poses no threat to the cooperating countries' army/navy/air-force. It has to be made sure that troops/ships/planes of all the countries is out of the radius of impact of missile and it doesn't inflict any of the cooperating countries' army/navy/air-force harmfully. There need to be performed a joint computation which can decide whether the target of the missile is appropriate and safe or not. The information required for such computation would be from all the countries involved in the operation but neither of them would want to reveal their private data of army/navy/air-force to other cooperatives. The information would include coordinates of missile's target, information about the missile, coordinates of troops/ships/planes of cooperating countries, radius of missile's area of impact, strategy information, and other statistics etc. The missile launching country doesn't want to reveal the target location's coordinates and all the countries don't want to reveal the coordinate locations of their troops/ships/planes.

This state of affairs leaves us with a situation with many questions. Where or by whom would this computation be performed? Would it be performed by a third country not in the alliance or would it be performed by one of the countries involved in the operation? How would the computation be performed? What method or technique would be used for the computation? Would it involve direct information from all the cooperatives or will the information be altered in a particular fashion. How the security and privacy would be achieved? What type of security measures would be taken? How would one country protect its data from being revealed to other cooperatives? How would the alliance protect this computation and related information from enemies? What all information from the alliance countries would be used in this computation? Would it involve position coordinates of their troops/ships/planes?

2.1 Description

There are n countries C_1, C_2, C_3..., C_n, in alliance jointly performing a terrorist elimination operation in a region R. C_k wants to launch a missile in an area A_k with missile

M_k and wants to make sure that the target area has no troops/ships/air-planes of the alliance { C_1, C_2, C_3..., C_n }. The alliance {C_1, C_2, C_3..., C_n} needs to perform a computation *f()* which decides whether the selected target area A_k is appropriate and safe or not.

The problems arising from such situation are:

Where to perform the computation?
How to perform the computation?
How to achieve security and privacy?
What information is required in the computation?

This problem can be solved with the proposed SMC protocol.

3 VPP (Virtual Party Protocol)

3.1 Informal Description

There is an alliance of *n* countries C_1, C_2, C_3..., C_n, each having data X_1, X_2, X_3..., X_n respectively, X_1, X_2, X_3..., X_n being the distribution of troops/ships/planes over a

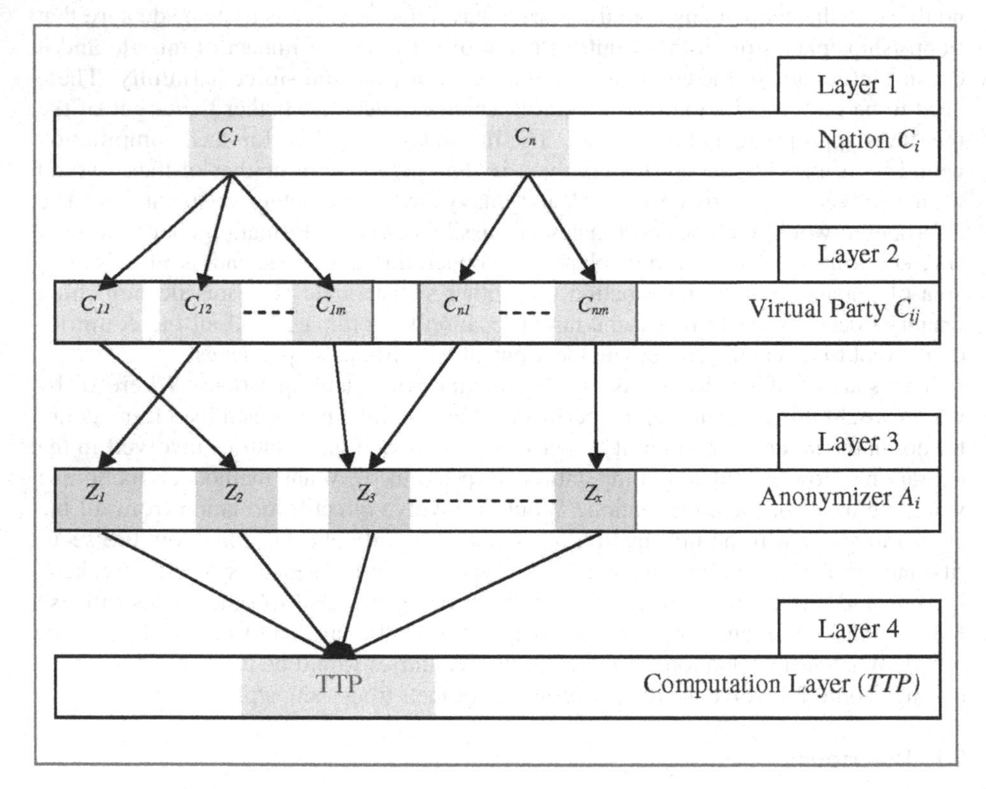

Fig. 1. Various layers in the proposed protocol namely the Party layer (1st), Virtual Party layer (2nd), Anonymizer layer (3rd) and Computation layer (4th)

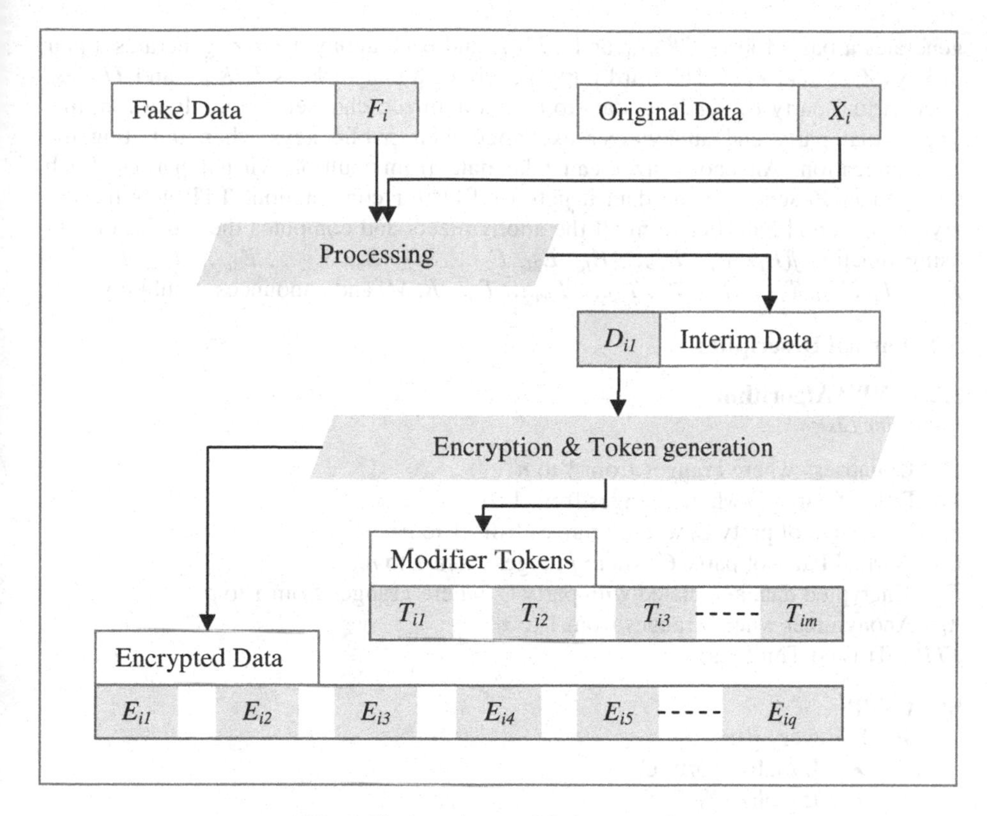

Fig. 2. Various data and their conversion

region *R*. Country C_k wants to launch a missile *M* on an area A_k in region *R* having possible terrorist activity. The function $f(X_1, X_2, X_3..., X_n, R, M)$ computes whether the target of missile is safe and appropriate or not. *M* and *R* are the information of the launched missile and involved region. M also holds the trajectory and other information related to the missile. There are *x* number of anonymizers $Z_1, Z_2, Z_3..., Z_x$. The protocol consists of four layer namely the party layer, virtual party layer, anonymizer layer and computation layer as shown in Fig. 1. The communication between layer three layers is secured using encryption keys.

Every country C_i will generate fake data F_i such that $f(F_i, X_1, X_2, X_3..., X_n, R, M) = f(X_1, X_2, X_3..., X_n, R, M)$. The country C_i will then generate encrypted data $E_{i1}, E_{i2}, E_{i3}..., E_{iq}$ along with *m* modifier tokens $T_{i1}, T_{i2}, T_{i3}..., T_{im}$ such that $f(F_i, X_i, R, M) = f(E_{i1}, E_{i2}, E_{i3}..., E_{iq}, T_{i1}, T_{i2}, T_{i3}..., T_{im}, R, M)$. The data is encrypted in a manner such that it doesn't effects the overall result. The method of encryption is highly dependent on the type of information and computation being performed and may vary for different computations.

Every country/party C_i creates m_i virtual parties. For the sake of simplicity in explanation the number of virtual parties m_i is kept *m* for every party. The country C_i randomly distributes the data's $E_{i1}, E_{i2}, E_{i3}..., E_{iq}$ among the virtual parties $C_{i1}, C_{i2}, C_{i3}..., C_{im}$. The modifier tokens are distributed randomly among the virtual parties such that every virtual party gets strictly one modifier token. Each virtual party C_{ij}

generates a pair of keys CK_{PUBij} and CK_{PRIij} and each anonymizer Z_i generates a pair of keys Z_{PUBi} and Z_{PRIi}. The third party generates a pair of keys TPK_{PUB} and TPK_{PRI}. Each virtual party C_{ij} sends its data to an anonymizer, choosen randomly at runtime. The virtual party and anonymizer exchange their public keys when initiating the communication. An anonymizer can take data from multiple virtual parties. Each anonymizer Z_i sends all the data it gets to TTP for computation. TTP gets the encrypted data and modifier from all the anonymizers and computes the required result using function $f(E_{11}, E_{12}, E_{13}\ldots, E_{1q}, E_{21}, E_{22}\ldots, E_{2q}, E_{31}, E_{32}\ldots, E_{3q}\ldots, E_{nq}, T_{11}, T_{12}, T_{13}\ldots, T_{1m}, T_{21}, T_{22}\ldots, T_{2m}, T_{31}, T_{32}\ldots, T_{3m}\ldots, T_{nm}, R, M)$ and announces it publicly.

3.2 Formal Description

3.2.1 VPP Algorithm

Identifier List:

C_i – Countriess where i ranges from 1 to n
X_i – Data of party C_i where i ranges from 1 to n
F_i – Fake Data of party C_i where i ranges from 1 to n
C_{ij} – Virtual Party of party C_i where j ranges from 1 to m
E_{ij} – Encrypted data associated with party C_i where j ranges from 1 to q
A_i – Anonymizer where i ranges from 1 to x
TTP – Trusted Third party

Start VPP

- For every party C_i
  - Initialize party C_i
  - Initialize X_i
  - Initialize F_i
  - Encrypt (X_i, F_i) to get E_{i1} - E_{iq}
  - Initialize & Generate T_{ij}
  - Initialize m Virtual Parties C_{i1} - C_{im}
  - For every data E_{ij}
    - Generate random number r
    - Send E_{ij} to C_{ir}
  - Distribute all T_{ij} among C_{ij}
- For every virtual party C_{ij}
  - Generate random number r
  - Initiate secure communication with Z_r
  - Send data(C_{ij}) to Z_r
  - Terminate communication
- For every anonymizer Z_i
  - Initialize secure communication with TTP
  - Send data(Z_i) to TTP
  - Terminate communication
- Compute the value of function f() at TTP using encrypted data and modifier tokens
- TTP - Announce the result publicly

End of Algorithm

4 Security Analysis

If the TTP is malicious then it can reveal the identity of the source of data. The two layers of anonymizers will preserve the privacy of source of data. A set of anonymizers will make the source of data anonymous and will preserve the privacy of individual. The more the number of anonymizers in the anonymizer layer the less will be the possibility of hacking the privacy of the data. Each virtual party reaches TTP on their own. Each country will reach TTP as an individual party and TTP will not know the actual country which created the virtual party. The probability of hacking data of virtual party C_{ir} is

$$P(VC_{ir}) = \frac{1}{\sum_{i=1}^{n} k_i} \tag{1}$$

When party C_i has k_i number of virtual parties, the probability of hacking data of any virtual party of party C_r is

$$P(VC_r) = \frac{k_r}{\sum_{i=1}^{n} k_i} \tag{2}$$

Even if the data of virtual party is hacked it will not breach the security as this data is encrypted. Probability of hacking the data of any party r is calculated as

$$P(C_r) = \frac{k_r}{\sum_{i=1}^{n} k_i} \times \frac{k_r - 1}{\sum_{i=1}^{n} k_i - 1} \times \cdots \times \frac{1}{\sum_{i=1}^{n} k_i - k_r} \tag{3}$$

The graph between number of virtual parties k vs. the probability of hacking $P(C_r)$ for $n=8$is shown in Fig. 3. which clearly depicts that probability of hacking is nearly zero when the number virtual parties is three or more. Also the graph between number of parties and probability of hacking for $k=7$ is shown in Fig. 4. As the number of virtual parties is eight the probability of hacking is in the order of 10^{-5} or we can say nearly zero. Suppose that the number of virtual parties is k_a then

$$P(C_a) = \frac{k_a}{\sum_{i=1}^{n} k_i} \times \frac{k_a - 1}{\sum_{i=1}^{n} k_i - 1} \times \cdots \times \frac{1}{\sum_{i=1}^{n} k_i - k_a} \tag{4}$$

For k_b number of virtual parties we have

$$P(C_b) = \frac{k_b}{\sum_{i=1}^{n} k_i} \times \frac{k_b - 1}{\sum_{i=1}^{n} k_i - 1} \times \cdots \times \frac{1}{\sum_{i=1}^{n} k_i - k_b} \tag{5}$$

if $k_a > k_b$ then $P(C_a) < P(C_b)$ by Eq. (4) and Eq. (5). We can see that as the number of virtual parties increases the probability of hacking the data will decrease by harmonic mean.

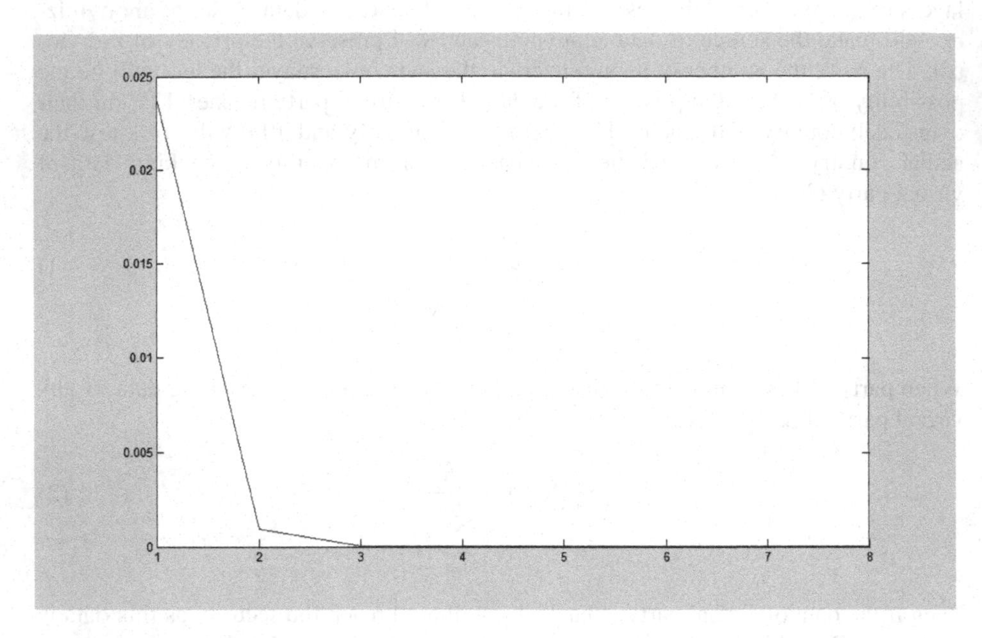

Fig. 3. Graph between number of Virtual Parties (x axis) vs Probability of hacking (y axis)

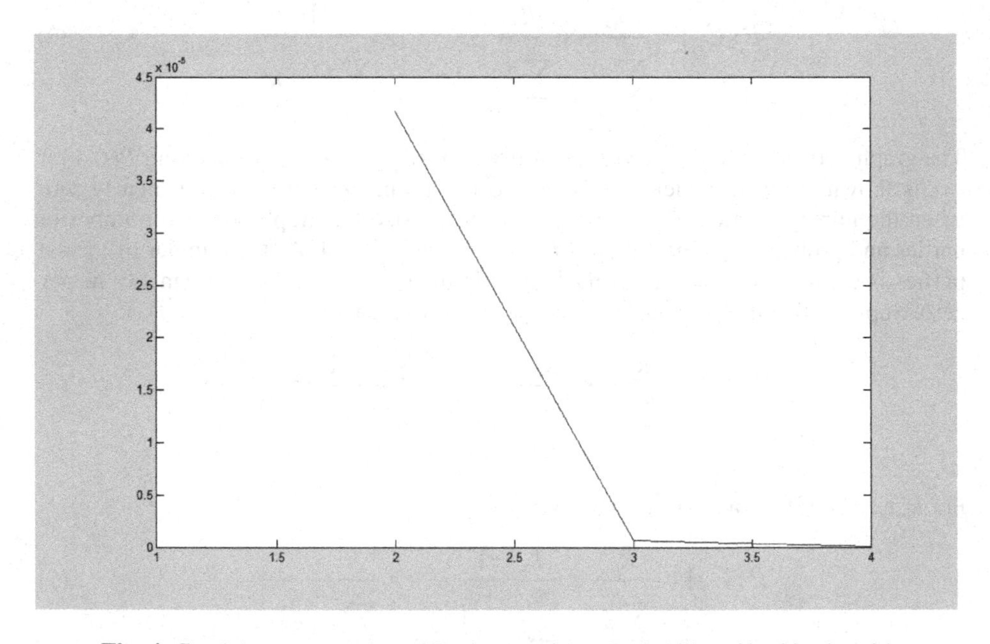

Fig. 4. Graph between number of Parties (x axis) vs Probability of hacking(y axis)

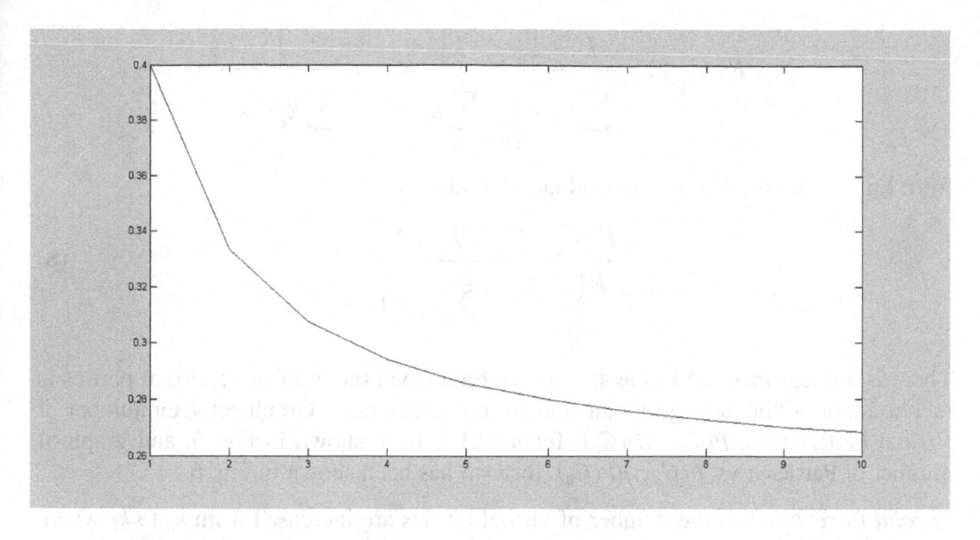

Fig. 5. Graph of number of Virtual Parties k (x axis) vs. $P(C_{a+1})/P(C_a)$ (y axis)

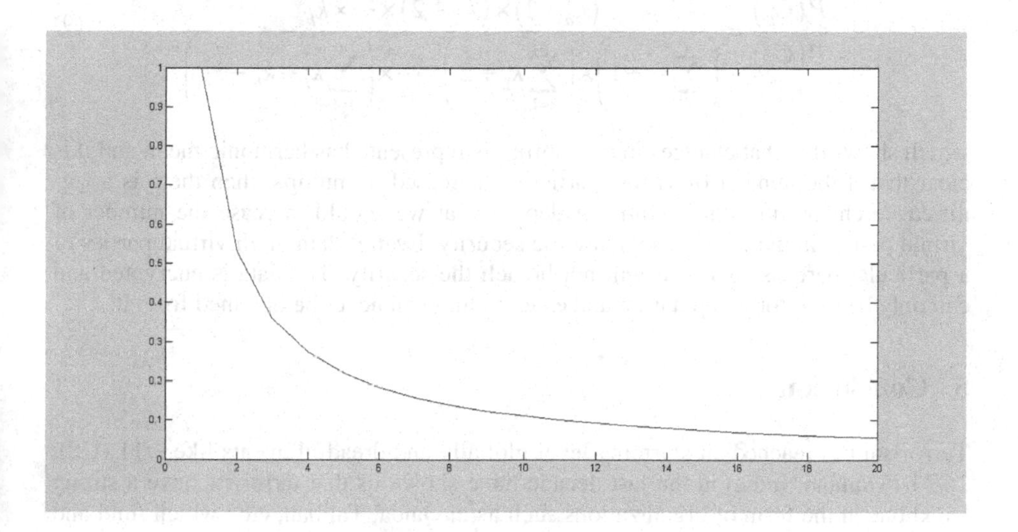

Fig. 6. Graph of number of Parties n (x axis) vs. $P(C_{a+1})/P(C_a)$ (y axis)

Special Case 1: When the number of virtual parties is increased from k_a to k_a+1, the effect in probability of hacking is evaluated as

$$P(C_a) = \frac{k_a}{\sum_{i=1}^{n} k_i} \times \frac{k_a - 1}{\sum_{i=1}^{n} k_i - 1} \times \cdots \times \frac{1}{\sum_{i=1}^{n} k_i - k_a} \tag{6}$$

$$P(C_{a+1}) = \frac{k_a+1}{\sum_{i=1}^{n} k_i + 1} \times \frac{k_a}{\sum_{i=1}^{n} k_i} \times \cdots \times \frac{1}{\sum_{i=1}^{n} k_i - k_a} \tag{7}$$

from Eq. (6) and Eq. (7) we can evaluate the ratio as

$$\frac{P(C_{a+1})}{P(C_a)} = \frac{k_a+1}{\sum_{i=1}^{n} k_i + 1} \tag{8}$$

There is a linear increase in the security of data when the number of virtual parties is increased, providing no significant change in security ratio. Graph between number of Virtual Parties k vs. $P(C_{a+1})/P(C_a)$ for $n=4$ has been shown in Fig. 5. and graph of number of Parties n vs. $P(C_{a+1})/P(C_a)$ for $k=8$ has been shown in Fig. 6.

Special Case 2: When the number of virtual parties are increased from k_a to k_b where $k_b > k_a$ then the security ratio is evaluated as

$$\frac{P(C_b)}{P(C_a)} = \frac{(k_a+1)\times(k_a+2)\times\cdots\times k_b}{\left(\sum_{i=1}^{n} k_i + 1\right)\times\left(\sum_{i=1}^{n} k_i + 2\right)\times\cdots\times\left(\sum_{i=1}^{n} k_i + k_b - k_a\right)} \tag{9}$$

which shows that that changes in probability is represented as harmonic mean and it is clear that if the number of virtual parties is increased in multiple then there is a significance change in security ratio. It depicts that we should increase the number of virtual parties in multiples to increase the security. Even if data of all virtual parties of a particular party is hacked it will not breach the security. The data is encrypted and can only be used for computation and exact values can never be obtained from it.

5 Conclusion

Terrorism has reached an alarming level globally and dreadful events like 9/11 (US), 26/11 (Mumbai, India) in the last decade have shown us that terrorists have a strong backbone in the form of organizations such as al Qaida, Taliban, etc., which fund and supply them with latest weaponry and equipments, which allows them to bring up a better fight every time they perform an act of terror. Our military forces stand against such terrorism, risking their lives to maintain peace and order in the world. To deal with such terrorism, the defense forces must be well advanced in all means to minimize the odds of casualties. Our proposed protocol provides a secure means of joint computations for defense applications in military operations. It can be used to securely perform many computations involving critical data from different military squads/forces etc. preventing any third party from gaining knowledge of the information being processed and the computation being performed. We have corroborated that we can create encrypted data and modifier tokens using original and fake data, and then distribute them among the created virtual parties for further processing. The virtual parties send their data to TTP using anonymizers to hide their identity. Encrypted data along with modifier tokens is used to carry out computations at TTP using an improvised computation method. TTP performs computation on encrypted data and announces result. The protocol and algorithm are highly scalable and optimized for defense and other similar computations. As security is a major concern for military operations, zero hacking configuration can be achieved using this protocol.

References

1. 2005 Report on Terrorism. National Counter Terrorism Center (NCTC). United States of America (2006)
2. 2007 Report on Terrorism. National Counter Terrorism Center (NCTC). United States of America (2008)
3. National Commission Up-on Terrorist Attacks in the United States, http://www.9-11commission.gov/archive/hearing7/9-11Commission_Hearing_2004-01-27.htm (retrieved on 2008-01-24)
4. Bin Laden claims responsibility for 9/11. CBC News, http://www.cbc.ca/world/story/2004/10/29/binladen_message041029.html (retrieved on 2009-01-11)
5. Spanish Indictment on the investigation of March 11, http://www.elmundo.es/documentos/2006/04/11/auto_11m.html
6. Spain furious as US blocks access to Madrid bombing 'chief'. The Times, http://www.timesonline.co.uk/tol/news/world/europe/article1391123.ece (retrieved 2007-02-15)
7. Incidents in London. United Kingdom Parliament, http://www.parliament.the-stationery-office.co.uk/pa/cm200506/cmhansrd/cm050707/debtext/50707-11.htm (retrieved on 2008-07-30)
8. Report into the London Terrorist Attacks on 7 July 2005. Intelligence and Security Committee (May 2006)
9. Indepth London Attacks. BBC News, http://news.bbc.co.uk/1/shared/spl/hi/uk/05/london_blasts/what_happened/html/russell_sq.stm
10. Dossier From India Gives New Details of Mumbai Attacks. The New York Times, http://www.nytimes.com/2009/01/07/world/asia/07india.html (retrieved on 2009-01-06)
11. India terrorist attacks leave at least 101 dead in Mumbai. Los Angeles Times, http://www.latimes.com/news/nationworld/world/la-fg-mumbai27-2008nov27,0,3094137.story (retrieved on 2008-11-27)
12. Foreign Terrorist Organizations List. United States Department of State. USSD Foreign Terrorist Organization, http://www.state.gov/s/ct/rls/fs/2002/12535.htm (retrieved on 2007-08-03)
13. Pathak, R., Joshi, S.: Secure Multi-party Computation Using Virtual Parties for Computation on Encrypted Data. In: Proc. of The First International Workshop on Mobile & Wireless Networks (MoWiN 2009) in Conjunction with The Third International Conference on Information Security and Assurance (ISA 2009). LNCS. Springer, Heidelberg (2009)

An Angle QIM Watermarking in STDM Framework Robust against Amplitude Scaling Distortions

Vijay Harishchandra Mankar, Tirtha Sankar Das, and Subir Kumar Sarkar

Dept. of Electronics and Telecommunication, Jadavpur University, Kolkata, India
vijaymankar@yahoo.com, tirthasankardas@yahoo.com,
sksarkar@etce.jdvu.ac.in

Abstract. Quantization index modulation (QIM) watermarking proposed by Chen and Wornell provides computational efficient blind watermarking based on Costa's dirty paper codes. The limitation of this is its vulnerability against amplitude scaling distortion. The present work is proposed to solve this problem based on angle QIM within spread transform dither modulation (STDM) framework. AQIM embeds the information by quantizing the angle formed by the host-signal vector with respect to the origin of a hyperspherical coordinate system as opposed to quantizing the amplitude of pixel values. It has been shown experimentally that the proposed work not only provides the resistance against this valumetric scaling distortion but also against non-linear, gamma correction and constant luminance change.

Keywords: Quantization index modulation (QIM), STDM watermarking, angle QIM, valumetric scaling, gamma correction.

1 Introduction

The growing use of the Internet and other digital transmission channels made distribution and access of multimedia data with no effort. This has created the problems regarding security and illegal distribution of copyrighted multimedia data as they can easily be copied, modified and redistributed. The watermarking emerges as the potential solution to overcome the copyright infringement and tampering of multimedia data. The watermarking is the process of hiding the useful information into the multimedia data imperceptibly which at the decoder can be used for various applications like proof of authenticity, copyright protection etc [1-4]. Chen and Wornell have recently introduced a class of watermarking methods called quantization index modulation (QIM) for data hiding emerged as a result of information theoretic analysis [1]–[4]. The QIM refers to embedding information by first modulating an index or sequence of indices with the embedded information and then quantizing the host signal with the associated quantizer or sequence of quantizers. The digital watermarking has been modeled as communication with side information [19]. The host signal is known at the embedder. The information of host signal at embedder is used to reduce the host

S. Ranka et al. (Eds.): IC3 2009, CCIS 40, pp. 400–410, 2009.
© Springer-Verlag Berlin Heidelberg 2009

signal interference. In terms of additive noise attacks, these schemes have proven to perform better than traditional spread-spectrum (SS) watermarking because they can completely cancel the host signal interference, which makes them invariant to the host signal. The structured lattice code in high dimensions can be directly and efficiently implemented that has made quantization-based schemes of practical interest with high data capacity [5]. Although QIM have shown a significant improvement in terms of watermark capacity over SS, it has been beaten by even the simplest attacks. This has lead to evolution the Spread Transform Dither Modulation (STDM) scheme, introduced by Chen and Wornell, which gains the capacity achieving properties of quantization based schemes with robustness of SS schemes. The message in STDM is embedded in the subspace spanned by a key-dependent spreading vector, but its value is computed by quantizing the projection of the host image onto this subspace.

Lattice-based schemes are vulnerable to amplitude scale attacks because these attacks introduce mismatch between the encoder and the decoder lattice volumes. The amplitude scale attack is a very common signal processing operation and it occurs whenever the brightness of the image is changed or volume of an audio is changed. Furthermore, amplitude scaling induces a large amount of distortion with respect to the mean squared error but does not cause significant perceptual degradations.

Several researchers have worked for combating amplitude scale attacks. In one of the approaches watermarking codes is designed that are invariant to amplitude scale operations, such as modified trellis codes [6], order-preserving lattice codes [7], and rational dither modulation [8]. Another approach is based on estimating the nonadditive operations and inverted back prior to watermark decoding, using pilot signals [9] or blind estimation [10]–[12]. Oostveen *et al.* [15] improved the robustness of QIM against amplitude scaling by using an adaptive quantization step size based on perceptual model using Weber's law. Lee *et al.* used an EM algorithm to estimate the global scaling factor but with impractical complexity [12]. In [13], an iterative estimation procedure in combination with error-correcting codes was proposed, which performed well even for low watermark-to-noise ratios (WNRs). The advantage of the approach in [9] is the ability to estimate the scaling factor from a small number of signal samples, which makes the estimation procedure applicable in situations where the scaling factor slowly varies. The disadvantage of the method is that the pilot signals consume part of the capacity of the watermarking system and its need for a signal calibration may lead to security weakness. Li *et al.* used the modified Watson's perceptual model to provide resistance against valumetric scaling for QIM watermarking [18]. The method proposed in [11] performs well for low WNR but lacks security, in the sense that an attacker knowing the distortion of the embedder is able to estimate the scaling factors and decode the watermark. The methods based on invariant codes give small probability of error with respect to amplitude scale attacks at the expense of increased probability of error, with respect to additive noise attacks and reduced payload [6-8].

In the proposed paper, we present a novel technique to combat amplitude scaling using angle QIM within the spread transform dither modulation (STDM) framework. The angle QIM was originally proposed by Ourique *et al.* AQIM embeds information by quantizing the angle formed by the host-signal vector with respect to the origin of a hyperspherical coordinate system rather than quantizing the amplitude of pixel values [16].

This paper is organized as follows. Section 2 gives the preliminaries about the QIM and STDM, a framework for embedding the message. The section 3 describes about

AQIM, a technique to embed message by quantizing angle instead of amplitudes of pixel. The experimental result and discussion is given in section 4. The conclusion drawn in the present work is given in section 5.

2 Preliminaries

2.1 Quantization Index Modulation (QIM)

In basic QIM, a quantizer maps a value to the nearest point belonging to a class of pre-defined discontinuous points. The function *round* (.) here denotes rounding value to the nearest integer and the standard quantization operation with step size is defined as $Q_\Delta(x) = round(x/\Delta)\Delta$.

Let Δ be the quantization step size and L represent the length of the host signal x and the message m (we embed one bit per sample). QIM embeds a message by modulating an index or sequence of indices with a watermark to be embedded and then quantizing the host signal with the associated quantizer. For QIM with dithering, we choose d_0 pseudo-randomly with a uniform distribution over [-Δ/2, Δ/2] and

$$d_1 = \begin{cases} d_0 + \Delta/2, & d_0 < 0 \\ d_0 - \Delta/2 & d_0 > 0 \end{cases} \tag{1}$$

Here d_0 or d_1 is used for embedding message bit "0" or "1" respectively. The watermarked signal is given by:

$$\begin{aligned} y_n &= Q_\Delta(x_n, d_m) \\ &= Q_\Delta(x_n + d_m) - d_m \quad m \in [0,1] \quad n=1,2,\ldots,L \end{aligned} \tag{2}$$

where $Q_\Delta(x) = round(x/\Delta)\Delta$ and the function *round* (:) denotes rounding value to the nearest integer.

At the detector the received signal z, possibly a corrupted version of y, is re-quantized with the family of quantizers used while embedding to determine the embedded message bit, i.e.

$$\hat{m} = \underbrace{\arg\min}_{m \in 0,1} \left| z - Q_\Delta(z, d_m) \right| \tag{3}$$

The above description embedded one bit per sample. In practice, we usually spread one message bit into a sequence of Nsamples $\{x_1, \ldots, x_N\}$, to improve the robustness. The detection equation then becomes:

$$\hat{m} = \underbrace{\arg\min}_{m \in 0,1} \sum_{i=1}^{N} \left| z_i - Q_\Delta(z_i, d_m) \right| \tag{4}$$

We have assumed the use of soft decoder as with the same code rate of $1/N$ soft decision decoding usually outperforms hard decision decoding.

2.2 Spread Transform Dither Modulation STDM

The spread transform dither modulation (STDM) consists of quantizing the projection of the host vector **x** along a randomly generated unit-length vector **p** and the resulting value is then quantized before being added to the components of the signal that are orthogonal to **p**. The embedding equation is given by:

$$\mathbf{y} = \mathbf{x} + (Q_\Delta(\mathbf{x}^T\mathbf{p}, \mathbf{d}) - \mathbf{x}^T\, \mathbf{p})\, \mathbf{p} \tag{5}$$

where the superscript T denotes vector transpose. The decoder projects the received data onto direction **p** to decide which of the quantizer was used. The corresponding detection is given by:

$$\hat{m} = \arg \min_{m \in \{0,1\}} \left| \mathbf{z}^T \mathbf{p} - Q_\Delta(\mathbf{z}^T \mathbf{p}, \mathbf{d}) \right| \tag{6}$$

Since the distortion due to embedding takes place in **p** direction only and no other component of **x** is modified, the embedder can allocate the entire distortion in direction **p**, enabling the use of a large quantizer step size. The large quantizer step size offers an increased protection against noise [20].

3 Angle Quantization Index Modulation (AQIM)

The idea of phase modulation schemes to the data hiding problem under amplitude scaling attacks was first identified by Chen in his thesis [17], correlating the phenomenon with communications theory. Under certain circumstances, the carrier's phases are modulated instead of amplitude thereby giving substantial performance improvement. This is very well known by the superior noise performance of FM over AM techniques in analog communications and by the performance of PSK methods over multipath fading environments. The analogy was put forward to combat amplitude scale attack by Ourique *et al.* into practical implementation by introducing AQIM in [16].

Consider a point x in the two dimension Euclidean space. The two samples x_1 and x_2 of $x_i \in \Re$ for i =1, 2 may be viewed as a point in a two dimensional plane. In QIM, the point x would be quantized to the closest centroid of the lattice defined in (3) or (5). Instead of using Cartesian coordinates, here this point x be represented by the tuple (r, θ) in polar coordinates. The angle θ and radius r are given as below

$$\theta = \tan^{-1}\left(\frac{x_2}{x_1}\right) \tag{7}$$

$$r = \sqrt{x_1^2 + x_2^2} \tag{8}$$

Then the angle θ can be quantized using (5) as given below:

$$Q_\Delta(\boldsymbol{\theta}, \mathbf{d}) = \boldsymbol{\theta} + (Q_\Delta(\boldsymbol{\theta}^T\mathbf{p}, \mathbf{d}) - \mathbf{x}^T\, \mathbf{p})\, \mathbf{p} \tag{9}$$

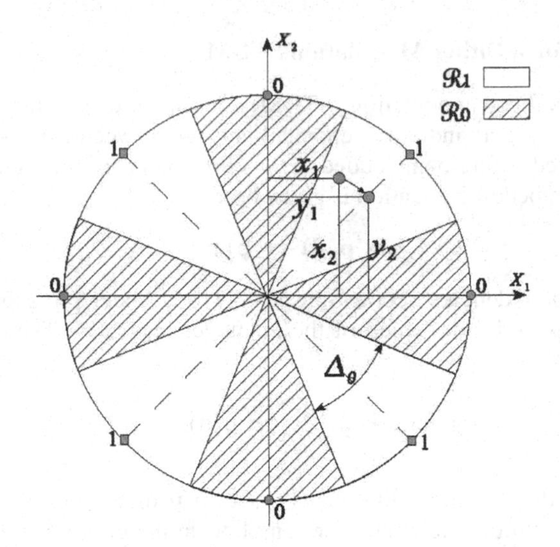

Fig. 1. Angle Quantization Index Modulation (AQIM) for L = 2

The quantized θ along with its radius r i.e. $(r, Q_\Delta(\boldsymbol{\theta}, \mathbf{d}))$ from its polar form is again converted back to its Cartesian coordinate representation which yields the new amplitude values for the pixels, i.e. $y_1 = r \cos \theta$ and $y_2 = r \sin \theta$. This process is shown in fig. 1. In the same way, this 2-Dimensional case can be extended for L-Dimensional case as follows:

Let $\mathbf{x}$ be a vector in the L-dimensional hyperplane with Cartesian coordinates given by $(x_1,\ldots, x_L)$ are converted in hyperspherical coordinates by its radius r and angle vector $\boldsymbol{\theta} = (\theta_1, \theta_2, \ldots \theta_{L-1})$. These quantities can be calculated as under:

$$\theta_1 = \tan^{-1}\left(\frac{x_2}{x_1}\right) \tag{10}$$

$$\theta_i = \tan^{-1}\frac{x_{i+1}}{\left(\sum_{k=1}^{i} x_k^2\right)^{1/2}}, \qquad \forall i = 2,3,\ldots, L-1 \tag{11}$$

$$r = \sum_{k=1}^{L} x_k^2 \tag{12}$$

Then, the quantization in (9) is applied to the components of $\boldsymbol{\theta} = (\theta_1, \theta_2, \ldots \theta_{L-1})$ where the appropriate lattice is chosen according to the value of message m to be embedded.

The radius and the quantized angle vector is mapped back to its Cartesian coordinates yielding the watermarked pixels, as follows:

$$y_1 = \prod_{k=1}^{L-1} \cos\theta_k \tag{13}$$

$$y_i = r \sin \theta_{i-1} \prod_{k=i}^{L-1} \cos \theta_k, \quad \forall i = 2,3,\ldots,L \tag{14}$$

In order the distortion level lies below visual perception, the norm of watermark, *w* needs to be bounded. The watermark distortion is defined as

$$D_w = \frac{1}{L}\|w\|^2 = \frac{1}{L}\|y - x\|^2 \tag{15}$$

By means of this distortion measure, the *Document to Watermark Ratio* (DWR) is defined as

$$DWR = 10\log_{10}\frac{\sigma_x^2}{D_w} \tag{16}$$

where σ_x^2 is the variance of the host. The distortion introduced by the channel is parameterized by the *Watermark to Noise Ratio*, defined as

$$WNR = 10\log_{10}\frac{D_w}{D_c} \tag{17}$$

where *Dc* is the distortion introduced by the channel.

4 Result and Discussions

We have used a set of benchmark images such as Lena, Fishing Boat, Pepper, Baboon, US Air Force, Woman etc. shown in fig. 2 having dimensions 256 X 256. The subjective binary watermark message of 32 X 32 with message length 1024 bits is used for embedding. The proposed work is carried out in spatial domain only. Each bit is embedded into 8 X 8 block of host image. The required DWR is set by changing the quantization steps. Fig. 3 shows bit error rate as a function of AWGN for different scaling factor keeping DWR = 25dB. The performance clearly shows the BER is well below 0.2 even at high distortion level of WNR = -5dB for all practical level of scale factor i.e. 0.5 to 1.5. This result clearly outperforms the results obtained by [15-16]. There is almost no effect of AWGN for WNR $\geq$ 0 dB for all practical range of scaling factor. The adverse effect of AWGN is observed for scaling factor less than 1 only. The sample result is shown in fig. 5 for scaling factor, β = 0.5 with AWGN, WNR = 0 dB, without scaling at WNR = -5 dB and for scaling factor, β = 1.5 with strong AWGN, WNR = -10 dB respectively.

There is a great deal of challenge posed in the literature for robustness of watermarking against non-linear scaling distortion such as gamma correction. We have also tested the proposed algorithm against the gamma correction. Fig. 4 shows the BER for different gamma factor. The results indicate the performance improvement for different values of DWR. There is almost no effect of gamma factor at DWR = 20dB and significant improvement of BER less than 0.2 is observed for gamma range of 0.8 to 1.2. The sample result is shown in fig. 5 for gamma factor, $\gamma = 0.8$ and $\gamma = 1.2$.

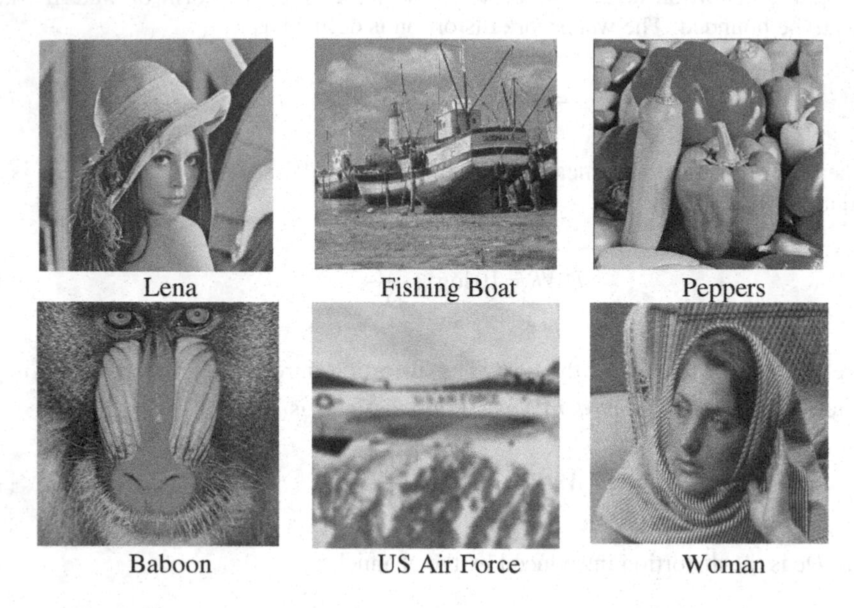

Fig. 2. The sample benchmarking images used for testing the present work

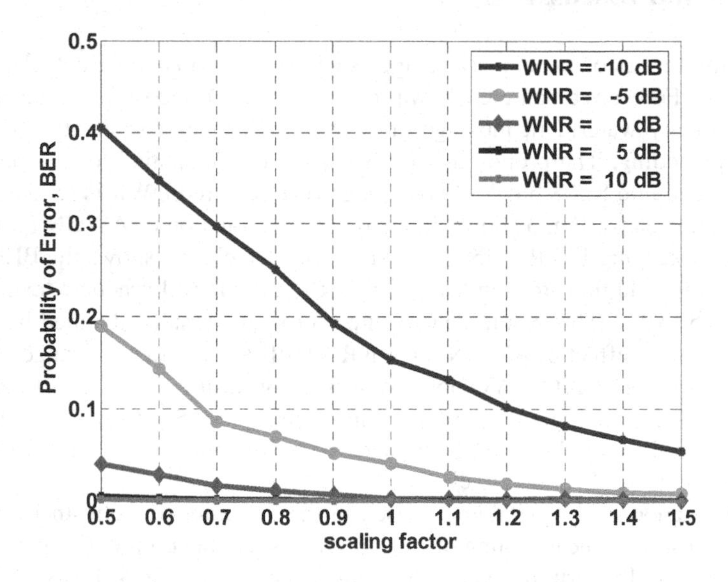

Fig. 3. Robustness versus amplitude scaling with additive white Gaussian noise (AWGN)

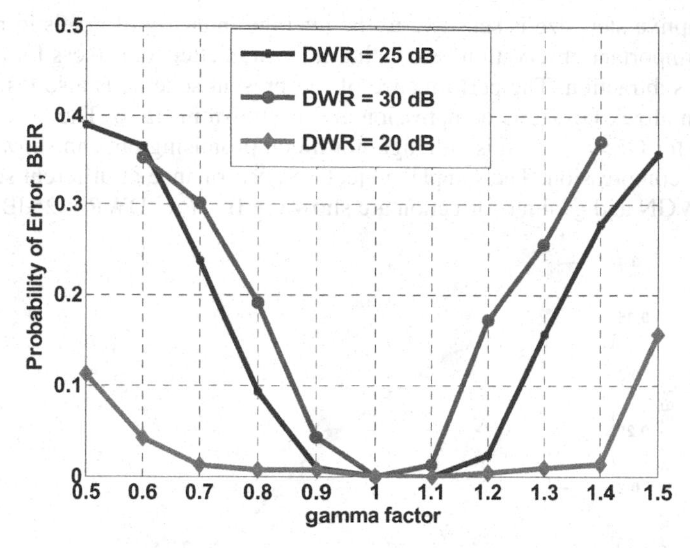

Fig. 4. Bit error rate performance against gamma correction with various strengths of watermarking

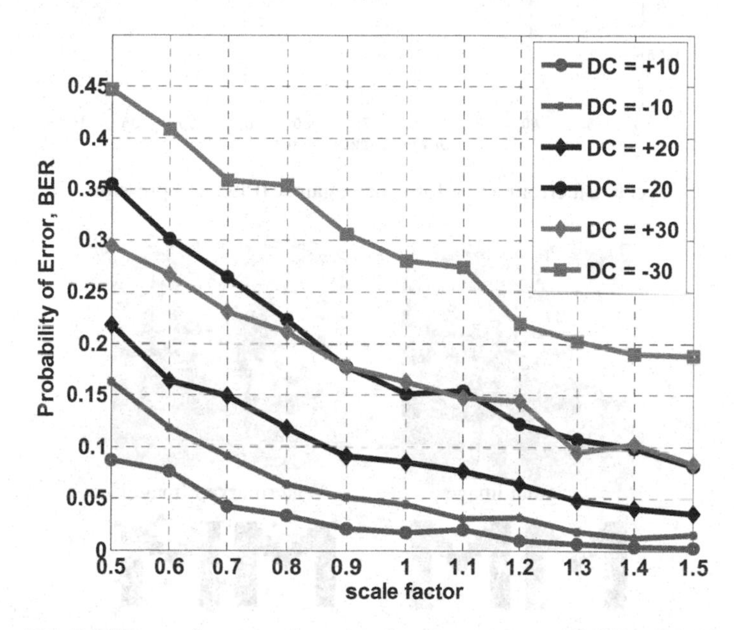

Fig. 5. BER as a function of constant luminance change (DWR = 25 dB)

Fig. 5 shows the sensitivity of our algorithm to the offset attack such as addition/subtraction of a constant luminance value. The experimental results illustrate bit error rate performance to the constant luminance change in the range of ± 30. The results obtained show improvement in robustness in the range ± 10 for all practical scaling factor range and also in the range ± 20 for scaling factor greater than 0.7. The performance clearly outperforms with the scheme [15] which is sensitive to this attack as their adaptive step size is sensitive to the average intensity of pixels in neighbourhood. One important observation here is that it shows better robustness for addition as compare to subtraction. The performance of the present scheme is also tested against quantization attack i.e. JPEG compression and is shown in fig. 6. The bit error rate is below 0.25 for QF>75 which is although not much promising but can show resilience against low compression. The sample subjective performance at different scaling factor with AWGN and gamma correction are shown in fig. 7, at DWR = 25dB.

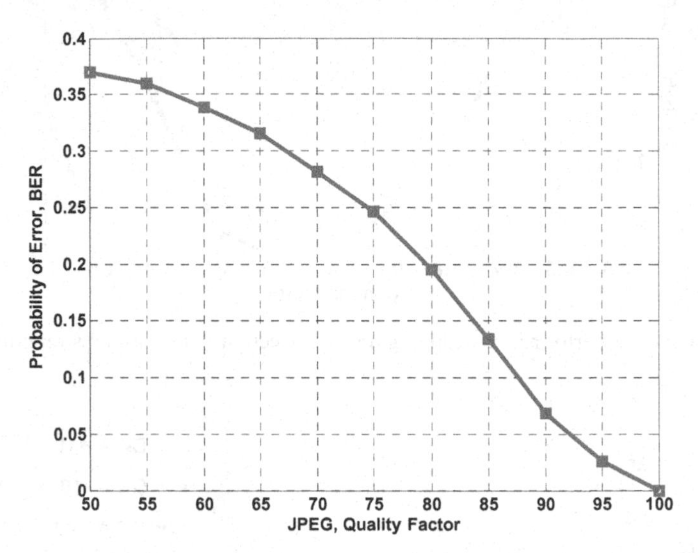

Fig. 6. Bit error rate performance against JPEG compression

Fig. 7. The sample output performance at different scaling factor, β with AWGN and gamma correction, γ (DWR = 25dB)

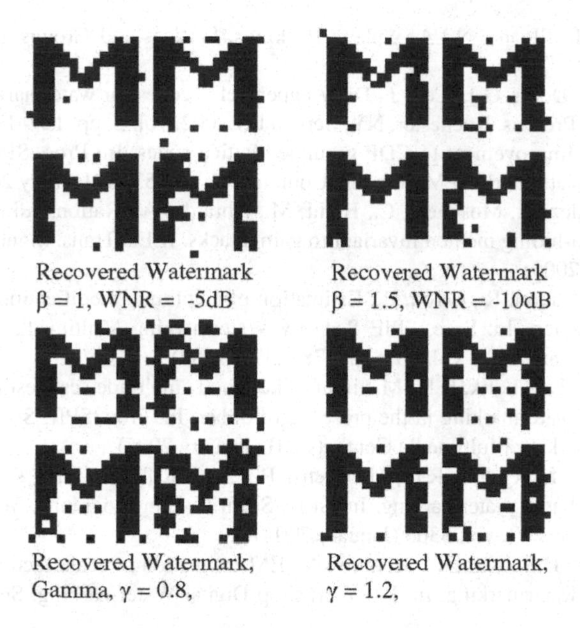

Fig. 7. *(continued)*

5 Conclusion

The present algorithm is proposed to resist linear valumetric scaling and nonlinear gamma correction using AQIM within the STDM framework. The AQIM, which keeps the amplitude of pixels unchanged, is completely invariant to scaling distortion theoretically and it is also validated to good extent practically barring some adverse effect of round and clip operation. We have shown the performance improvement of message decoding under additive white Gaussian noise attack in addition to distortion caused by amplitude scaling. The experimental results show robustness to good extent against gamma correction. The present work is also evaluated against the offset attack i.e. addition/subtraction of constant luminance change. The performance improvement is promising. The performance of the proposed method is compared and found much better with the results obtained in [15], [16]. The present scheme has shown significant resiliency against JPEG compression.

References

1. Moulin, P., O'Sullivan, J.A.: Information-theoretic analysis of information hiding. IEEE Trans. Inf. Theory 49(3), 563–593 (2003)
2. Chen, B., Wornell, G.: Quantization index modulation: A class of provably good methods for digital watermarking and information embedding. IEEE Trans. Inf. Theory 47, 1423–1443 (2001)
3. Gel'fand, S.I., Pinsker, M.S.: Coding for channel with random parameters. Prob. Contr. Inf. Theory 9, 19–31 (1980)
4. Costa, M.H.: Writing on dirty paper. IEEE Trans. Inf. Theory IT-29(3), 439–441 (1983)

5. Conway, J.H., Sloane, N.J.A.: Sphere Packings, Lattices and Groups, 3rd edn. Springer, Berlin (1999)
6. Miller, M.L., Doerr, G.J., Cox, J.: Dirty-paper trellis codes for watermarking. In: IEEE Int. Conf. Image Process, Rochester, NY, September 2002, vol. 2, pp. 129–132 (2002)
7. Bradley, B.: Improvement to CDF grounded lattice codes. In: Proc. SPIE Security, Steganography, Watermarking Multimedia Contents VI, vol. 5306 (January 2004)
8. Perez-Gonzalez, F., Mosquera, C., Barni, M., Abrardo, A.: Rational dither modulation: A high rate data-hiding method invariant to gain attacks. IEEE Trans. Signal Process. 53(10), 3960–3975 (2005)
9. Eggers, J.J., Bauml, R., Girod, B.: Estimation of amplitude modifications before SCS watermark detection. In: Proc. SPIE Security Watermarking Multimedia Contents IV, San Jose, CA, January 2002, vol. 4675, pp. 387–398 (2002)
10. Shterev, I.D., Lagendijk, R.L.: Maximum likelihood amplitude scale estimation for quantization-based watermarking in the presence of dither. In: Proc. SPIE Security, Steganography, Watermarking Multimedia Contents VII (January 2005)
11. Shterev, I.D., Lagendijk, R.L., Heusdens, R.: Statistical amplitude scale estimation for quantization-based watermarking. In: SPIE Security, Steganography, Watermarking Multimedia Contents VI, vol. 5306 (January 2004)
12. Lee, K., Kim, D.S., Kim, T., Moon, K.A.: EM estimation of scale factor for quantization-based audio watermarking. In: Int. Workshop Digital Watermarking, Seoul, Korea (October 2003)
13. Balado, F., Whelan, K.M., Silvestre, G.C.M., Hurley, N.J.: Joint iterative decoding and estimation for side-informed data hiding. IEEE Trans. Signal Process. 53(10), 4006–4019 (2005)
14. Schuchman, L.: Dither signals and their effect on quantization noise. IEEE Trans. Commun. Technol. COM-12(4), 162–165 (1964)
15. Oostveen, J., Kalker, T., Staring, M.: Adaptive quantization watermarking. In: Proceedings of SPIE –IS&T Electronic Imaging, vol. 5306, pp. 296–303 (2004)
16. Ourique, F., Licks, V., Jordan, R., Perez-Gonzalez, F.: Angle QIM: A novel watermark embedding scheme robust against amplitude scaling distortions. In: International Conference on Acoustics, Speech, and Signal Processing, pp. 797–800 (2005)
17. Chen, B.: Design and Analysis of Digital Watermarking, Information Embedding, and Data Hiding Systems, PhD Thesis, Massachusetts Institute of Technology, Massachusetts, USA (June 2000)
18. Li, Q., Cox, I.J.: Using perceptual models to improve fidelity and provide invariance to valumetric scaling for quantization index modulation watermarking. Presented at the IEEE Int. Conf. Acoustics, Speech and Signal Processing, Philadelphia, PA (March 2005)
19. Cox, I.J., Miller, M.L., McKellips, A.: Watermarking as communications with side information. Proc. IEEE 87(7), 1127–1141 (1999)
20. Moulin, P., Koetter, R.: Data-Hiding Codes. Proceedings of the IEEE 93(12) (December 2005)

Parallelization Issues of Domain Specific Question Answering System on Cell B.E. Processors

Tarun Kumar[1], Ankush Mittal[1], and Parikshit Sondhi[2]

[1] Department of Electronics and Computer Engineering,
Indian Institute of Technology, Roorkee, India
tarun.krgtm2002@gmail.com,
ankumfec@iitr.ernet.in
[2] Department of Computer Science,
University of Illinois at Urbana Champaign
sondhi1@illinois.edu

Abstract. A question answering system is an information retrieval application which allows users to directly obtain appropriate answers to a question. In order to deal with an explosive growth of information over internet and increased number of processing stages in answer retrieval, time and processing hardware required by question answering system has increased. The need of hardware is currently served by connecting thousands of computers in cluster. But faster and less complex alternatives can be found as a multi-core processor. This paper presents a pioneer work by identifying major issues involved in porting a general question answering framework on a cell processor and their possible solutions. The work is evaluated by porting the indexing algorithm of our biomedical question answering system, INDOC (Internet Doctor) on cell processors.

Keywords: Biomedical Question Answering system, Indexing Algorithm, Cell Broadband Engine Processor.

1 Introduction

Question Answering systems represent the next step in information retrieval applications as they allow users to directly obtain answers to questions rather than following the search engine style approach of returning a list of documents for queries. A general question answering system framework usually involves steps such as document preprocessing, semantic parsing, indexing, question classification, question keyword weighing, document ranking and answer extraction. Thus there's often a deeper level of document and question processing involved both in the indexing and retrieval stages. While such an extended pipeline of NLP (natural language processing) operations greatly helps in improving accuracy, it also greatly increases the time and processing power required for indexing and retrieval operations. This makes it infeasible to run sophisticated information extraction systems over very large corpora where these operations are really required.

S. Ranka et al. (Eds.): IC3 2009, CCIS 40, pp. 411–421, 2009.
© Springer-Verlag Berlin Heidelberg 2009

Taking the example of biomedical literature domain, where there are currently an estimated 17 million citations in PUBMED [1], the current breed of search engines have been proven to be grossly inadequate [2] as they lack the knowledge of biomedical terminology [3]. As a solution to these problems [4] suggests a biomedical question answering system- INDOC [4], which is designed and developed at IIT Roorkee. It is based on the novel ideas of indexing and extracting the answer to the question posed. The system achieves an accuracy of 76% over first five documents and increases up to 83% for 50 documents retrieved. The drawback of this algorithm is however that it is slow to be used. The algorithm is slow because of the large search space.

One solution to deal with the above problem is to have a large number of computers in a cluster or grid for searching and indexing. However, this is a costly method [5]. An alternative approach is presented in the form of multi-core computers like Cell Processor.

In this paper we present our experience in porting indexing algorithm of INDOC [4] a biomedical domain QA system on cell processors. The major contributions of this paper are:

1. We identified the major issues involved in porting a general question answering framework on a cell processor.
2. We proposed potential solutions to these issues.
3. We evaluated our solutions by porting our biomedical QA system, INDOC on cell processors.
4. Our approach resulted in achieving a performance gain of roughly 23 times over a linear implementation.

Most of the issues and solutions discussed here are also largely applicable to porting search engines.

Paper is organized into 9 sections. Section 2 describes INDOC architecture, section 3 provides features of Cell broadband Engine, section 4 discusses the porting issues of question answering system on Cell BE, section 5 presents the ways in which various porting issues can be overcome, section 6 presents implementation details of Indexing of INDOC on Cell BE, section 7 discuss our general observations during implementation. Section 8 shows the results and comparison analysis and section 9 presents the future work.

2 INDOC

INDOC [4] is a biomedical Question Answering system. Major tasks in it are indexing, question processing, document ranking, clustering and display results. Complete architecture of INDOC is shown in figure. 1.

MMTX, a programming implementation of MetaMap [6] server is an NLP toolkit which is used to map free text into Unified Medical Language System (UMLS) [7] concepts. Question processing and Indexing module use it to find out concepts corresponding to the keywords in the sentences. First the entire document set is processed by the indexing module and an indexed database is prepared. At run time, when the user submits a query, the question processing module recognizes the keywords of question and finds the UMLS concepts corresponding to these keywords from MMTX server. The ranking module searches the indexed database for retrieving the documents and assigning them a rank on the basis of their relevance to the question concepts. Finally the display module displays the documents in decreasing order of their weights.

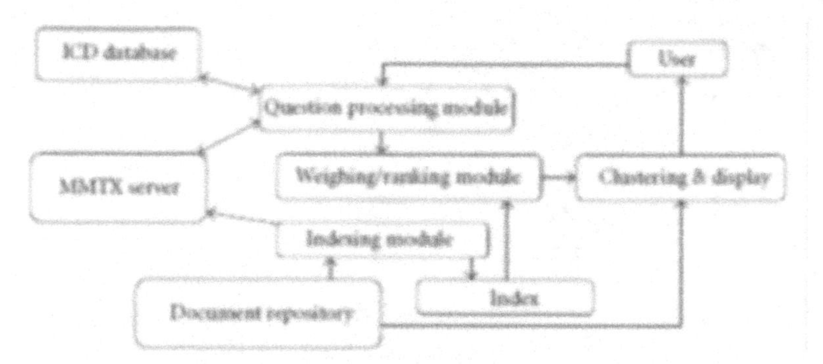

Fig. 1. Complete architecture of INDOC [4]

Indexing module for preparing indexed database not only selects the important keywords and concepts from document; but also represents the entire document in the form of sections. Each section has a section heading and number of sentences in it. Section heading consists of one or more concepts that represent the section. The algorithm to perform the task of indexing is shown in figure 2.

Algorithm in Figure 2 obtains the concepts of title and stores them in file. Beside this, sections are formed on the basis of concept present in the sentences. A new sentence is added to the current section till intersection of concepts of the current section and sentence to be added is not empty. But there are two restrictions on the size of section. If size of the current section < M (a Const.), the sentence is added to the section and section concept will be intersection of concepts of the current section and sentence to be added.

If size of the current section > M, the sentence is added to the current section only if sentence's concept are a subset of concepts of the current section.

If size of the current section < L(minimum number of sentence in a section), then the current section is merged with previous section.

An indexed file containing the concepts of titles of all the documents in the document set is also prepared. This file is used by the ranking module at the time of retrieval of documents corresponding to a question submitted by the user.

3 Sony-Toshiba IBM Cell BE Architecture

The Cell Broadband Engine (Cell B.E) [8] is a heterogeneous multi-core chip that is significantly different from conventional multiprocessor. It consists of a central microprocessor called the Power processing element (PPE) that controls

```
1. Obtain the concepts of the title and store them.
2. Initialize i =1 and j=1 and set all Xi, SCj, XCi to be empty where
        Sj : jth sentence in the document.
        Xi : ith section.
        SCj : set of concepts in jth sentence (concepts in an individual sentence).
        XCi : set of concepts in ith section.
        L : min number of sentences necessary in a section.
        M : minimum number of sentences in a section so that merging is not
        necessary.
3. Formation of Sections
        Set XCi to concepts in the first sentence.
        Define |S| as the number of elements in set S.
        For each sentence Sj left in the document to process
        {
                If(|Xi|==0)
                {
                        Add Sj to Xi
                        Add SCj to XCi
                }
                else
                {
                        if( (|Xi|<M && |Xci∩SCj|>0 ) || XCi==SCj )
                        {
                                Add Sj to Xi
                                Set XCi= XCi∩SCj
                        }
                        else
                        {
                                i=i+1
                                Add Sj to the new section Xi
                                Add SCj to XC
                        }
                }
        }
4. Final Section merging step
for each section Xi
{       If( i>1 && ( |Xi|<L || XCi is a subset of XCi-1) )
        {       Merge Xi with Xi-1
        }
}
```

Fig. 2. Indexing Algorithm of INDOC [4]

eight SIMD co-processing units called synergistic processor elements (SPEs), a high speed memory controller, and a high bandwidth bus interface, all integrated on a single chip. Figure 3 gives an architectural overview of the Cell/B.E.

The Cell operates on the fundamentals of increasing concurrency through the use of multiple processing cores and increasing specialization in execution through non-homogeneous parallelization. For this purpose, it employs 8 SPEs onto which threads of an application can be mapped controlled by PPE.

The SPE offers a high bandwidth interface to a direct memory access (DMA) engine that can transfer 32 GB/sec to and from the 256 KB local memory. The important point to note is that the SPE works only on the data that is in its local memory. However the SPE local storage is a limited resource as only 256 Kbytes is available for program, stack, local buffers and data structures. Rather

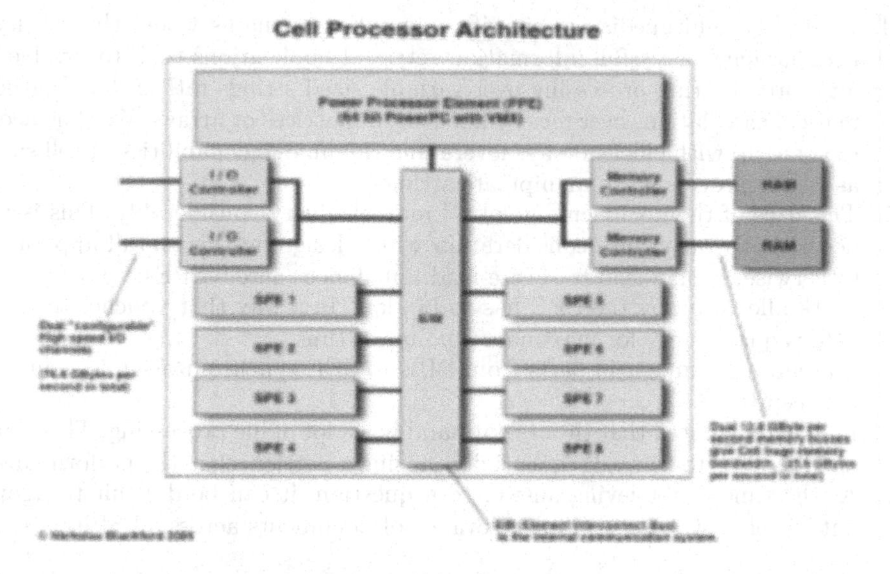

Fig. 3. Architecture of Cell BE Processor [9] showing eight SPUs connected to one PPU through a common bus

than considering cache control and the impact of memory bandwidth, the focus is on structuring data movement within the Cell processor to keep the SPEs busy, and dividing the application into vectorized functions to make efficient use of the SPEs. Cell BE attains computational speed up through several levels of such as SIMD (Single Instruction Multiple Data) or Vector processing, dual-issue superscalar micro architecture, multithreading, multiple execution units with heterogeneous architectures etc [9].

4 Issues in Porting a Typical Question Answering System

We now enumerate various issues that arise during porting of a question answering system on Cell processor.

1. Each SPE of Cell BE has 256KB local store. This local store is shared by code segment and data segment. This limited memory may not allow the entire document to fit on SPE store.
2. The SPEs cannot directly read or write a file. This means that the PPE needs to read files for all SPEs, the SPEs then perform the task of indexing and send back the output to the PPE. Thus it can potentially become a bottleneck if all these operations are not performed efficiently enough.
3. The NLP toolkits such as the MMTX server are not implemented for the cell processor. Porting them is non trivial task. Thus these toolkits need to run on the PPE. This could again make the PPE a bottleneck and put severe limitations on the amount of gain that could be achieved. Moreover the APIs provided by MMTX are in Java which can't be accessed through C/C++.

4. Unlike the multimedia or scientific computing domains where the cell has been largely successful, information retrieval applications tend to involve a lot more of string processing over variable sized strings rather than mathematical calculations over mostly fixed sized matrices or arrays. We thus need to come up with efficient ways leveraging the unique capabilities of cell such as vector processing to manipulate strings.
5. The sizes of the documents involved may also vary considerably. This issue needs to be taken into consideration while designing the overall approach. Otherwise, it may lead to severe load imbalances across SPEs.
6. Work allocation by the PPE has to be done in a way that vouches to keep SPEs equally busy for maximum amount of time.
7. To send a document in parts from PPE to SPE synchronization is required between them.
8. It should be noted that the task at hand has a lot of file processing. This type of data needs to be read sequentially causing a bottleneck in the performance.
9. At the time of retrieving answer to a question, it can be difficult to figure out a global strategy to rank relevance of documents across all SPEs.

5 Potential Solutions to the Issues

This section discusses the solution to the issues arising in indexing the document set.

1. Since limited memory of SPE may not allow entire document to fit at once, the document should be worked upon in such a way that only a part of document is required at a time for indexing purpose. Since SPE cannot directly read the document, PPE is required to read the document and send its contents in parts to the SPEs as and when required.
2. To deal with programming language issues of MMTX toolkit, we can interface the kit programmatically so that they receive input and generate output which is then used for indexing purpose. Thus, cell processor need not worry about Java APIs. In that case SPE is required to have a networking support.
3. A solution to the problem of variable sized documents is to let PPE read many documents and built a sort of pool of read documents. Whenever an SPE finishes off with its current document, it simply requests the PPE for the next. One may not pre-assign a set number of documents to SPEs. Instead, one can just let them request the documents whenever they need it
4. A large number of DMA operations take place, so instead of sequentially reading line and making DMA, these operations are overlapped. This compensates the time of transferring the content from PPE to SPE.
5. Since many files are needed to be opened on PPE for serving many SPE's simultaneously, many threads can be spawned on PPE so that I/O overhead of opening, closing and reading can be minimized.
6. Synchronization between SPE and PPE can be done through mailboxes.
7. In order to deal with the string processing efficiently, the strings are needed to be converted into vectors and then SIMD operations can be applied to these vectors.

6 Design and Implementation of INDOC Indexing Algorithm on CELL BE

This section presents the implementation of some of the solutions provided in previous section by applying them while porting indexing algorithm of biomedical QA system, INDOC on cell processors.

Cell BE offers a number of ways to achieve parallelism viz. SIMD processing, multithreading, shared memory multiprocessing, multiple execution units with heterogeneous architectures. Of all these, we have selected to use multithreaded approach to achieve data level parallelism. The data is partitioned such that entire document set is divided into eight subsets and one thread at PPE corresponding to one SPE, is responsible for assigning one such document subset to that SPE. We have done so because I/O operations in opening, reading and closing files by multiple threads may be overlapped with other computations. The logic used on PPE and SPEs is as follows:

6.1 PPE

The main task of PPE is to read files for the SPEs. PPE creates one thread each for one SPE and has one file allocated per SPE. Once this specific file is indexed completely, PPE picks up the next file for that particular SPE. PPE reads the file character by character until it reads one complete line in a temporary buffer. Later the corresponding SPE is directed to pick up this line of text by using SPE read inbound mailbox. Status variable of mailbox represents the number of free entries in mailbox. Initially its value is 4. A write operation on mailbox by PPE decreases the status value by 1 and read operation by SPE increases the value by 1.

Value of status is checked repeatedly. If the value is 4 then temporary buffer is copied to original buffer whose address is available to the SPE. Status variable is updated to 3 to indicate that buffer is ready to be read by SPE. Status variable is updated with a write operation to the mailbox value. This value is used to indicate the SPE about end of file. SPE reads the line with DMA operation and perform the task of indexing on this line of text. During this time PPE is busy reading the file, constantly generating raw data for the 8 SPEs. Working of PPE is shown in figure 4.

Value of the status variable of mailbox is 4 when it is reset by the SPE after completing DMA. As soon as PPE updates the buffer, it sets the status to 3 to indicate SPE about update of buffer. In this way tasks occur simultaneously both on the PPE and the SPE.

6.2 SPE

SPE has the task of creating sections of the input lines that were sent to it from the PPE. To receive the data from main memory (PPE) to the SPE we use DMA operations. This also allows us to make use of double buffering.

Initially value of status of mailbox is 0 therefore SPE waits to perform DMA until status becomes 1. As soon as a DMA operation of taking one line from

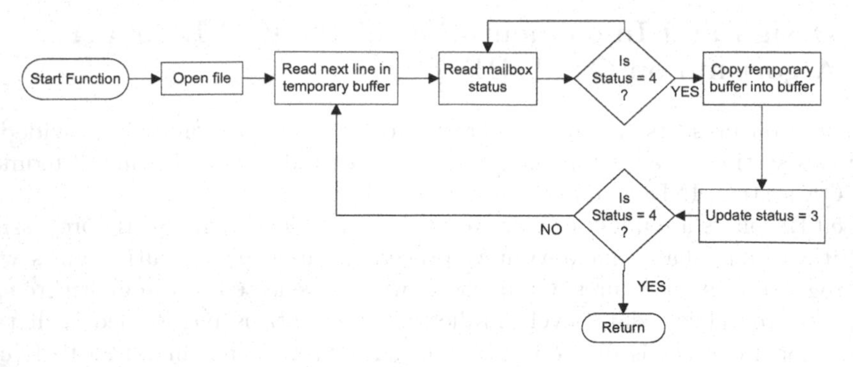

Fig. 4. Working of a thread function on PPE showing how the file is read and thread waits for signal by SPE for next read

PPE to SPE is over, the mailbox which was earlier used by PPE to signal to SPE to indicate the presence of a new line of data is now used by the SPE to ask for next line of data. Here value of status variable represents number of occupied entries in the mailbox. Initially it is 0. A write operation by PPE increases the value of status by 1 and read operation by SPE decreases the value by 1. The value read from mailbox is helpful in deciding whether the document/ file has finished or not. Working of a particular SPE is as shown in figure 5.

Status variable of the mailbox in the SPE remains 0 till the buffer is not updated by PPE. As soon as SPE finds the values of status non zero, it starts DMA to read the buffer. After the DMA operation is over, it resets the status to 0 by reading the mailbox. On the basis of read value from mailbox it is checked whether file ended or not. The line just read is used for making sections. Since DMA is a non blocking call, task of processing current line on SPE overlaps the next DMA operation.

As soon as the line is received, a counter which keeps track number of sentences is incremented. If the sentence is first sentence then first section is initialized with the sentence. Current sentence is added to the current section if either of the following two conditions is satisfied: (1) whether the length of intersection of concepts of current section and current sentence is greater than 0, and length of current section is less than a const and (2) whether the number of concepts

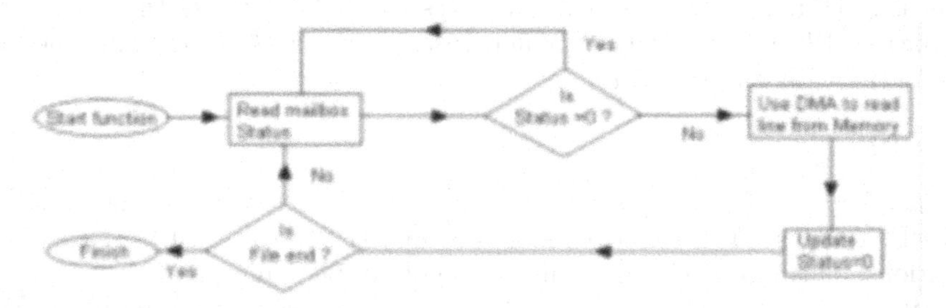

Fig. 5. Showing file read operation by SPE and signaling to PPE for next read

in current section and current sentence are equal. If neither of the condition is satisfied then a new section is made and initialized with current sentence. Here the operations performed are mainly subset finding, string comparison, string matching therefore, vector operations were difficult to be applied.

7 General Observations

Some tasks that were difficult and can result in less speed up are as discussed below:

1. The sizes of the documents were varying greatly. This causes an imbalance of work among the SPE's. For maximum speed up we would ideally want all the SPEs to stay busy for same amount of time. For example, in video compression, the sizes of all video frames are same as against our case.
2. In some of the cases, it was observed that synchronization between the PPE and SPE resulted in some periods of time when either of the machines were idle. For example if we have a long line followed by a small line, in such a case the DMA transfer of the long line shall take considerable more time keeping the PPE idle for some time.
3. While implementing the indexing system for INDOC, it was observed that the issues encountered were likely to be fairly general to a lot of other QA systems.

8 Results and Conclusion

Performance of the code built for Cell BE is evaluated on Georgia Tech Cell Buzz [10]. Number of documents used for mesuring performance has been varied from 8 - 40. Three samples were taken for each observation and an average time was calculated. Results show a comparison with Intel Pentium4 (3.2 Ghz) processors with 1 GB RAM. Table 1 shows the observed comparison. It is found that speedup in execution times on Cell Buzz is about 22+ times against Intel Xeon.

From table 1, It can be observed that there is variation of speedup for different number of documents. The reason behind this is variation in size of documents. Size of documents vary from 1 KB to 95 KB , which actually hinders the speedup because of an imbalance in of work distribution among the SPEs.

On the other hand if we take documents of same size for measuring performance, results are consistent and better because of equal amount of work for all spes. the results are shown in table 2.

9 Future Work

In this paper we implemented the indexing component of a domain specific question answering system on a cell processor. We identified the major issues encountered and proposed and implemented solutions to them. While there were

Table 1. Comparison of execution times (micro sec.)

	8 docs	16 docs	24 docs	32 docs	40 docs
8 SPE	2194	6769	9210	9771	14901
	4270	7749	7229	8738	10814
	3703	7043	8614	9274	10233
Average(t1)	3389	7187	8351	9261	11982
Intel P4 (3.2 Ghz)	86440	161749	216708	259634	310906
	87072	164603	218688	259771	314959
	87030	163615	215714	258467	312666
Average(t2)	86847	163322	217036	259290	312843
Speed up (t2/t1)	25.6	22.7	25.9	27.9	26.1

Table 2. Table showing low variation in speedup when all documents are of same size

	8 docs micro sec	16 docs micro sec	24 docs micro sec	32 docs micro sec	40 docs micro sec
8 SPE	780	1542	2232	3212	4067
	731	1403	2214	3020	3518
	791	1591	2209	3149	4220
Average(t1)	767	1512	2218	3127	3935
Intel P4 (3.2 Ghz)	30239	59741	90173	120283	150184
	29955	60050	90094	120352	150136
	30362	60219	88071	117889	150782
Average(t2)	30065	60003	89446	119508	150367
Speed up (t2/t1)	39.1	39.7	40.32	38.2	38.2

a number of major challenges involved and some of them were only partially dealt with, we still managed to obtain reasonable speedup. This suggests that Cell processor holds considerable amount of potential for information retrieval applications. In future, one can try implementing NLP toolkits like MMTX on the Cell processor so that they will lead better compatibility while used with application designed on cell processor.

References

1. National Library of Medicine, `http://www.nlm.nih.gov/news/pubmed_15_mill.html`
2. Jacquemart, P., Zweigenbaum, P.: Towards a medical question-answering system: a feasibility study. In: Beux, P.L., Baud, R. (eds.) Proceeding of Medical Informatics Europe (MIE 2003). Studies in Health Technology and Informatics, vol. 95, pp. 463–468. IOS Press, San Palo (2003)
3. Schultz, S., Honeck, M., Hahn, H.: Biomedical text retrieval in languages with complex morphology. In: Proceedings of the Workshop on Natural Language Processing in the Biomedical domain, Philadelphia, Pa, USA, July 2002, pp. 61–68 (2002)

4. Sondhi, P., Raj, P., Kumar, V.V., Mittal, A.: Question processing and clustering in INDOC: a biomedical question answering system. EURASIP Journal on Bioinformatics and Systems Biology 2007(3), 1–7 (2007)
5. Barroso, L.A., Dean, J., Hölzle, U.: Web Search for a Planet: The Google Cluster Architecture. IEEE Micro 23(2), 22–28 (2003)
6. http://mmtx.nlm.nih.gov/
7. http://www.nlm.nih.gov/research/umls
8. IBM alphaWorks. Cell BE SDK, http://www.alphaworks.ibm.com/topics/cell
9. Cell Broadband Engine - An Introduction, Cell Programming Workshop, IBM Systems and Technology Group, April 14-18 (2007)
10. User guide, Cell buzz, http://wiki.cc.gatech.edu/cellbuzz/index.php/User_Guide

Security Issues in Cross-Organizational Peer-to-Peer Applications and Some Solutions

Ankur Gupta[1] and Lalit K. Awasthi[2]

[1] Model Institute of Engineering and Technology
Camp Office: B.C Road, Jammu, J&K, India – 180001
ankur_g1@yahoo.com
[2] National Institute of Technology
Hamirpur, Himachal Pradesh, India – 177005
lalitdec@yahoo.com

Abstract. Peer-to-Peer networks have been widely used for sharing millions of terabytes of content, for large-scale distributed computing and for a variety of other novel applications, due to their scalability and fault-tolerance. However, the scope of P2P networks has somehow been limited to individual computers connected to the internet. P2P networks are also notorious for blatant copyright violations and facilitating several kinds of security attacks. Businesses and large organizations have thus stayed away from deploying P2P applications citing security loopholes in P2P systems as the biggest reason for non-adoption. In theory P2P applications can help fulfill many organizational requirements such as collaboration and joint projects with other organizations, access to specialized computing infrastructure and finally accessing the specialized information/content and expert human knowledge available at other organizations. These potentially beneficial interactions necessitate that the research community attempt to alleviate the security shortcomings in P2P systems and ensure their acceptance and wide deployment. This research paper therefore examines the security issues prevalent in enabling cross-organizational P2P interactions and provides some technical insights into how some of these issues can be resolved.

Keywords: Peer-to-Peer Networks/Computing, Peer Enterprises, Security Issues in Cross-Organizational P2P Interactions.

1 Introduction

Peer-to-Peer (P2P) networks [1] and the computations that they facilitate have received tremendous attention from the research community, simply because of the huge untapped potential of the P2P concept – extending the boundaries of scale and decentralization beyond the limits imposed by traditional distributed systems, besides enabling end users to interact, collaborate, share and utilize resources offered by one another in an autonomous manner. Moreover, P2P architectures are characterized by their ability to adapt to failures and a dynamically changing topology with a transient population of nodes/devices, thus exhibiting a high degree of self-organization and fault tolerance. P2P networks therefore have many desirable properties for designing massively distributed applications.

S. Ranka et al. (Eds.): IC3 2009, CCIS 40, pp. 422–433, 2009.
© Springer-Verlag Berlin Heidelberg 2009

With the desirable properties, P2P systems also have some characteristics which pose great challenges in designing robust, reliable and secure applications around them. *Anonymity* of peers poses issues of trust and security, whereas frequent *node transience* does not allow any useful work to be performed apart from best-effort information sharing. Sharing of compute resources leads to vulnerability to a variety of *security attacks*, whereas overlay routing performed by peers raises issues of *data privacy*. *Lack of centralized control* and censorship leads to copyright violations and infringement of intellectual property rights. Thus, P2P systems have not been adopted by organizations even though their desirable properties can be leveraged to meet many organizational requirements effectively. Some organizations and network service providers have resorted to blocking P2P traffic entirely to prevent any potential impact on their operations. These organizations have the same information/data sharing requirements, scalable storage and resource requirements and solutions which can adapt to dynamically changing environments and are extremely fault tolerant and resilient to security attacks. There is therefore no cogent reason why the P2P concept cannot be applied to share information, compute resources and expert human knowledge across organizations.

However, the requirements of an organization are somewhat different than that of an anonymous peer connected to the internet. Organizations need to be completely convinced about the security of their data and confidential information, which can be potentially compromised if its resources are accessed by an external agency. Moreover, organizations do not like to expose their compute resources to the outside world for fear of security attacks which might lead to loss of revenue. Organizations would also not like to interact with an anonymous external agency, unless it can control the nature and scope of interaction. While P2P systems represent decentralization, autonomy and no censorship, organizations require centralized control and some level of censorship to ensure security. How can then the P2P concept be applied successfully to organizations, if at all?

Some researchers have attempted to evaluate the feasibility of applying the P2P concept to the Enterprise. While some have advocated the use of P2P for information sharing and computing purposes within organizations, others have dismissed the thought, describing P2P as being too flaky and tangential to organizational requirements. A report by InformationWeek [2] talks about the P2P Peril and how content sharing P2P applications deployed by employees in an organization inadvertently exposed sensitive data. Another whitepaper by Verso technologies [3] talks about the high cost of P2P on the enterprise listing loss of productivity, degraded network performance and above all a loss of intellectual property amongst the many real threats. With P2P traffic constituting upto 70% of the internet traffic some organizations have blocked P2P traffic entirely. Others have cautiously adopted the P2P concept for specific applications such as sharing real-time market information amongst the sales force, building large scale information access systems, video streaming to keep employees abreast of latest happenings in the organization and the industry. The P2P concept has thus so far not been applied to enable cross-organizational collaborations, motivating the authors to propose the concept of Peer Enterprises [4, 5].

From an organizations perspective the risks involved in deploying P2P applications are:

- Risk of data exposure
- Risk of security attacks
- Risk to organizational compute resources (when allowing outside access)
- Risk of degradation of network performance (due to P2P traffic overload)
- Risk of loss of employee productivity

The potential benefits of enabling cross-organization P2P interactions [4, 5], necessitates that the research community attempt to mitigate the security threats posed by the P2P system model. To enable organizations to open up their resources to prospective partner organizations would require a fool-proof security net around its critical data and compute resources.

The rest of the paper is organized as follows: Section 2 details the security issues prevalent in P2P systems and some of the proposed solutions to counter these threats. The limitations of the existing security schemes are also discussed. The concept of "Peer Enterprises" enabling cross-organization P2P collaboration is presented in section 3, while section 4 proposes research-based solutions to the identified security issues. Finally section 5 concludes the paper.

2 Security Issues in P2P Systems

As always security is a major research area, especially with the anonymous nature of P2P systems offering malicious users the cover to indulge in disruptive acts [6]. In fact, P2P based systems have become notorious for blatant copyright violations and propagation of security threats and distributed attacks. Table 1 summarizes the security issues prevalent in P2P networks/computation and proposed solutions in literature to counter these threats.

Anonymity remains the core issue affecting security for P2P applications. How can a peer know whether its counterpart peer is genuine or malicious? Can the content from a peer be trusted? Can the other peer be allowed to access your compute resources, content without harming your computer? The basis for such interactions is the "Trustworthiness" of a peer, which needs to be established. Several solutions based on trust and reputation management of peers participating in the P2P network have been proposed. EigenTrust [7] and TrustMe [8] are two well-known schemes for establishing the trust ratings of a peer through its interactions with other peers. The sheer scale of P2P systems makes trust computation and ensuring the security of every peer a Herculean task. As such, the solutions proposed by researchers have focused only on certain class of P2P applications like file sharing and storage management. These distributed solutions have focused on identifying trustworthy peers and attempting to isolate malicious peers within the realm of the application under consideration. Trust and reputation values are computed at a per peer basis and then aggregated and communicated to the entire network of peers, which can be time consuming and less than efficient. Over the years several other researchers have proposed schemes for computing and disseminating trust values for peers. A detailed discussion on various trust and reputation-based systems is available in [9].

Table 1. Summary of P2P security issues and applicable solutions

P2P Feature	Resultant Security Issue	Proposed Solutions
Anonymity	Untrusted Content, cannot Share Resources, cover for disruptive activities, no traceability	Centralized identity management, trust and reputation management
Overlay Application Layer Routing	Man-in-the-Middle (MitM) and Distributed-Denial-of-Service (DDos) attacks, sybil attacks, network/content poisoning, eclipse attacks	Strong encryption, trust and reputation management, secure routing
Sharing of Compute Resources/Remote Work Processing	Viruses/malware/spyware, potential damage to peer hosting remote work, cannot trust result of remote computation	Role-Based-Access-Control (RBAC), admission control in peer groups, trust and reputation management
Transience/Churn	Leeching, no guarantee of retrieval of remote stored content, malicious peers can vanish with critical data, incomplete execution of remote work	Fault-tolerance schemes, content replication, trust and reputation management
Censorship-Resistance	Sharing of copyright content/ classified information, pornography	No major research-based solution proposed - legal recourse is only action as in case of Napster.
Social Networking	Exploitation, crime	None exist
Lack of Centralized Control	Not possible to shutdown, not possible to control activities of individual peers	Distributed schemes for building trust and reputation, individualized security policies built into P2P middleware

Security solutions for P2P applications operating within federated domains such as organizations or those relying on setting up specialized communities/groups range from identity management [10] (introduce a trusted authority for managing identities of peers which are repudiation resistant), authentication and authorization [11] (a central authority authenticates the identities of peers based on digital certificates and other credentials), admission control in peer groups [12] (allows only peers meeting certain criteria to join the peer group), role-based access control (RBAC) [13, 14] (allow only trusted peers to access the services offered by peer groups or individual peers) and the like have been proposed. However, these solutions tend to introduce centralized elements for issuing certificates, public/private keys etc, which does not fit well in the decentralized model for P2P networks. Moreover, these represent a single-point-of-failure in the scheme, besides constituting a performance bottleneck. Some attempts are being made by researchers to propose decentralized security models. Some attempts have focused on letting peers define security policies suited to their interactions with other peers, most notably in P2P SL (Security Layer) [15], in which a peer can define security policies for each peer interaction. However, once the remote code is resident on the host peer, this scheme cannot ensure the security of the peer as it does not provide any mechanism for the run-time monitoring of the remote application. Thus, existing solutions are inadequate to cater to the security requirements for cross-organizational P2P applications.

3 Peer Enterprises: Enabling Cross-Organizational P2P Interactions

The Peer Enterprises (PE) framework [4, 5] is a hybrid, P2P application on a two-tier P2P network, enabling:

a. Organized P2P interactions between peers residing in different enterprises, organizations or geographical sites, allowing them to increase utilization levels of their compute resources, thereby increasing return-on-investment (ROI).
b. Small and Medium Businesses (SMBs) to utilize the compute resources of other peer enterprises, without having to invest in them.
c. Creation of collaborative P2P applications over the framework to meet specific requirements of organizations, such as joint R&D, sharing of patent portfolios, collaborative projects etc.
d. Communication, collaboration and exchange of expert human knowledge across enterprises.

The first-tier P2P network is the community of inter-connected peers within an organization, while the second-tier comprises the P2P network formed by representative edge-peers from participating organizations, facilitating the aggregation and sharing of compute resources, content and enabling communication. Once edge-peers negotiate a "contract" with another edge-peer, individual peers from both organizations begin their interactions in a purely decentralized manner. The edge-peer plays no further role during peer interactions.

The system model is depicted in Figure 1. Peer Enterprises interact through designated Edge Peers whose main responsibility is to aggregate the peer resources for a particular organization and try and negotiate the best possible contract with another edge peer. A single edge peer can enter into contracts with multiple other edge peers, leading to the creation of a hierarchical, multi–level P2P network. Hierarchical P2P networks [16, 17] have been proposed earlier primarily to reduce the lookup times for P2P queries. The context in which the hierarchical P2P network is employed in the PE framework is novel and represents a new application domain.

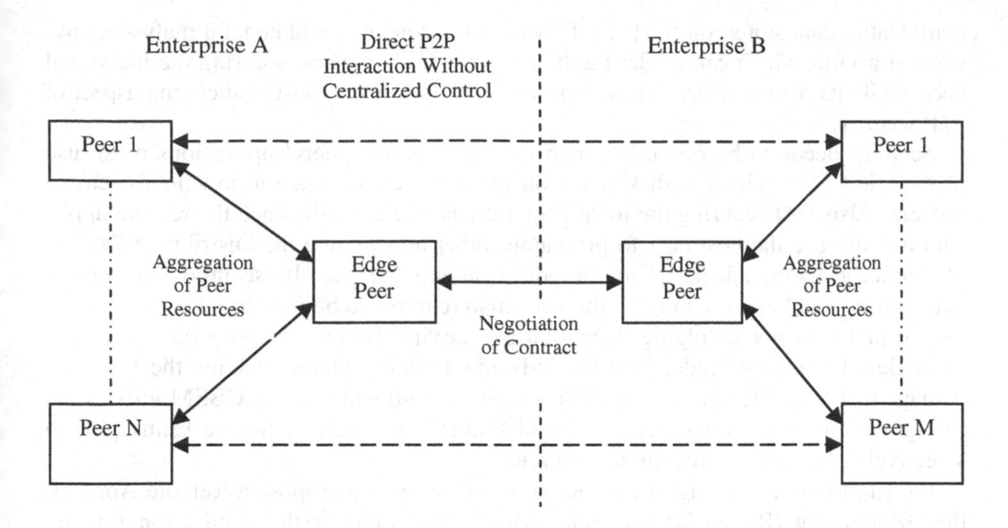

Fig. 1. Conceptual representation of the Peer Enterprises Framework [4]

The plethora of security issues identified in earlier sections are also applicable to frameworks enabling cross-organizational P2P collaborations and need to be resolved before such frameworks can be deemed feasible.

4 Security Solutions

This section discusses the security solutions to alleviate the security concerns prevalent in existing schemes and also proposes some new mechanisms to ensure security in the Peer Enterprises Framework.

4.1 Identity Management

Useful Cross-organizational interactions can only take place when identities of organizations and the peer's belonging to them can be irrefutably established, since organizations would not risk interacting with an anonymous external agency. Towards this end the Peer Enterprises framework requires that each organization first register with a Regional Nodal Authority (RNA), which provides each organization its unique identifier and also its digital credentials (an encrypted digitally signed certificate). The organization is required to supply its digital credentials to other organizations as its identity. Each peer also uses the same digital credentials while interacting with peers belonging to other peer enterprises. The registration process is a one-time affair and does not impact the scalability of the framework. Moreover, the RNA plays no further role in the P2P interactions.

4.2 Ensuring Security of Individual Peers

Since individual peers shall be hosting remote applications from other peers and also downloading content, it makes it vulnerable to various kinds of security attacks. A malicious remote application could cause damage to the host peer besides compromising the confidential data stored on the peer. Downloaded content could contain malware, spyware or a virus which can render the host peer unusable. Thus, securing the individual peer, while its resources are available for external use is the most challenging aspect of P2P security.

Security needs to be provided when multiple remote peers/applications make use of the idle CPU cycles and disk space on the host peer or when it downloads remote content. Also, just securing the local peer may not be enough, since the remote application could use the host peer to propagate other attacks like the Distributed-Denial-of-Service (DDoS) attacks. Thus not only should the peer be secured, the remote application which is executing on the peer also requires to be monitored at run-time to ensure that it is not indulging in malicious activity. Hence, we propose a Containment-Based Security Model [18] for individual peer's which contains the potential damage that a remote application can cause by sand-boxing it. The CBSM utilizes the fine-grained privileges and access-control features provides by Secure Linux [19] to selectively grant access to system resources.

For run-time monitoring of the remote application, we propose a Remote Application Monitoring (RAM) [18] module, which shall monitor the application for any suspicious activity like increased/heavy CPU utilization, increased outbound/inbound network traffic, multiple instances of remote application, increased disk space usage, increased memory utilization, staying active after expiry of allotted time etc. If the application exceeds its pre-defined quotas (quotas could be pre-negotiated based on the local peers characteristics and resource availability or via an organization-wide policy), the RAM promptly terminates the application. By using the access control features provided by SE Linux in conjunction with the custom application monitor (RAM), the local peer can be secured and malicious remote code can be contained without causing serious damage within the P2P network. Figure 2 provides an overview of the system model for the proposed solution.

4.3 Ensuring Security of Organization's Data

4.3.1 Audit Trail of Shared Content

To have additional checks in the system, the PE framework provides an audit trail of all shared content. This audit trail is maintained at each individual peer and keeps track of the peer's activities over a period of time. During times of low network usage, the audit trail gets uploaded to the edge peer (or any other peer designated for accounting purposes), so that organization-wide statistics can be analyzed. Knowing that all shared content gets logged into the audit trail will make a malicious employee think twice about sharing sensitive content knowingly.

4.3.2 Sandboxing the File-Sharing Application

The file-sharing functionality is built into the PE framework. Although the file-sharing application will look only into the designated shared folder/directory while searching for files, for additional security, it can be restricted to not being able to read any additional directories on the host peer. This can be done by specifying a small SE Linux access-control policy for the file sharing module and will prevent any unauthorized access on the host peer. Moreover, if any content which gets downloaded to the shared folder/directory is infected with malware which attempts to access areas outside the shared folder, the access to it shall also be denied.

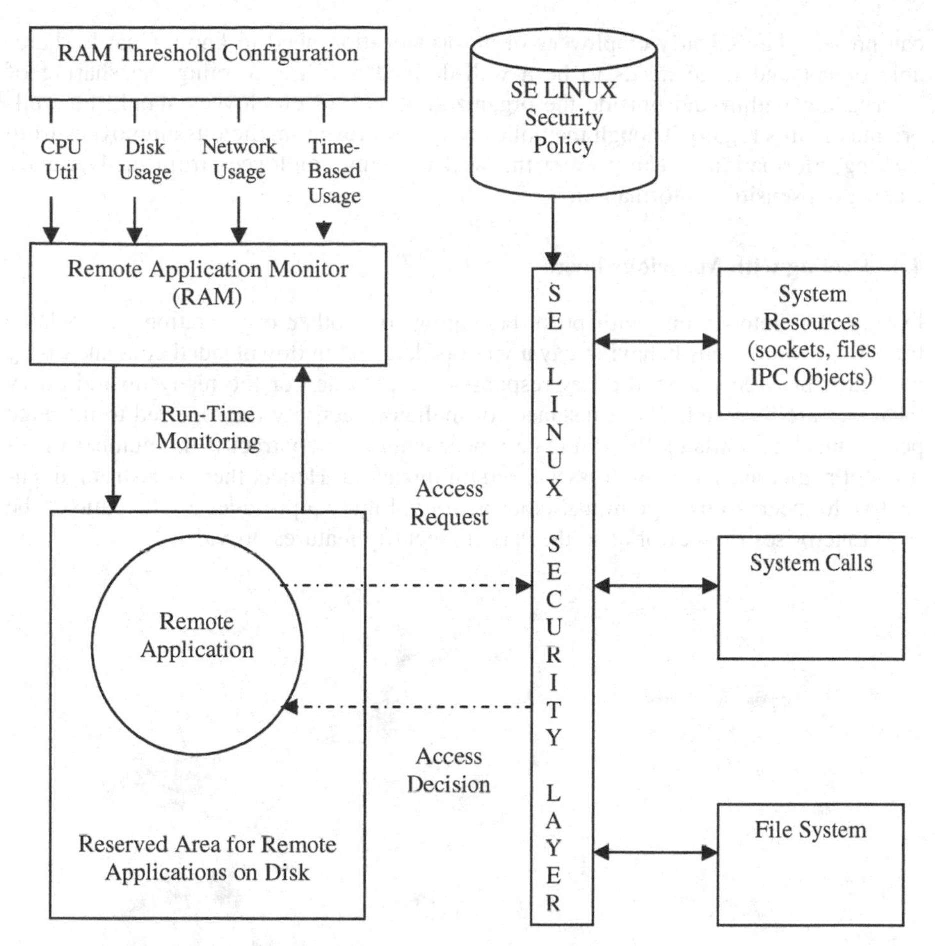

Fig. 2. Containment-Based Security Model; Uses SE Linux Access Control Features with Remote Application Monitor to Ensure Security of Local Peer [18]

4.3.3 Digitally Signed Content to Prevent Sharing

Organizations need to set up certain privileges for shared data access. It should ideally be able to specify which files can be shared and which can't. Within the Peer Enterprises framework, sensitive content can be digitally signed to secure it. To prevent unauthorized access we propose that organizations setup a content server, which makes available digitally signed confidential data for use within its HPN. Peers can use the content server to digitally sign any new sensitive content that they generate. To prevent unauthorized sharing checks in the middleware stalls the transfer of all digitally signed content, thus reducing the risk of sharing confidential data.

4.3.4 Well-Defined Policy on Sharing Confidential Information

The security of any application ultimately depends on its human users. If an employee wants to share a company's sensitive content, there are many means for doing the same, including through email or by carrying a physical hard copy. No technology can prevent that. Clearly employees of an organization need to know what is shareable or not and there needs to be a well-defined policy governing the sharing of information within and outside the organization and all employees should be well-oriented in this regard. Though the policy may not prevent malicious employees from stealing information, it can prevent the well-meaning employees from inadvertently leaking out sensitive information.

4.4 Dealing with Malicious Peers

During their interactions with peers belonging to another organization peer's keep track of any malicious behavior, say a virus is detected in downloaded content, wrong content is provided, invalid query responses are provided or too many invalid query messages are received. These instances of malicious activity are reported to the edge peer, which can unilaterally terminate a peer enterprise contract if the number of reports of malicious activities crosses a certain threshold. Hence, there is a strong disincentive for peers to indulge in malicious activity. Figure 3 provides a schematic of the Peer Enterprises Framework and the various security features provided.

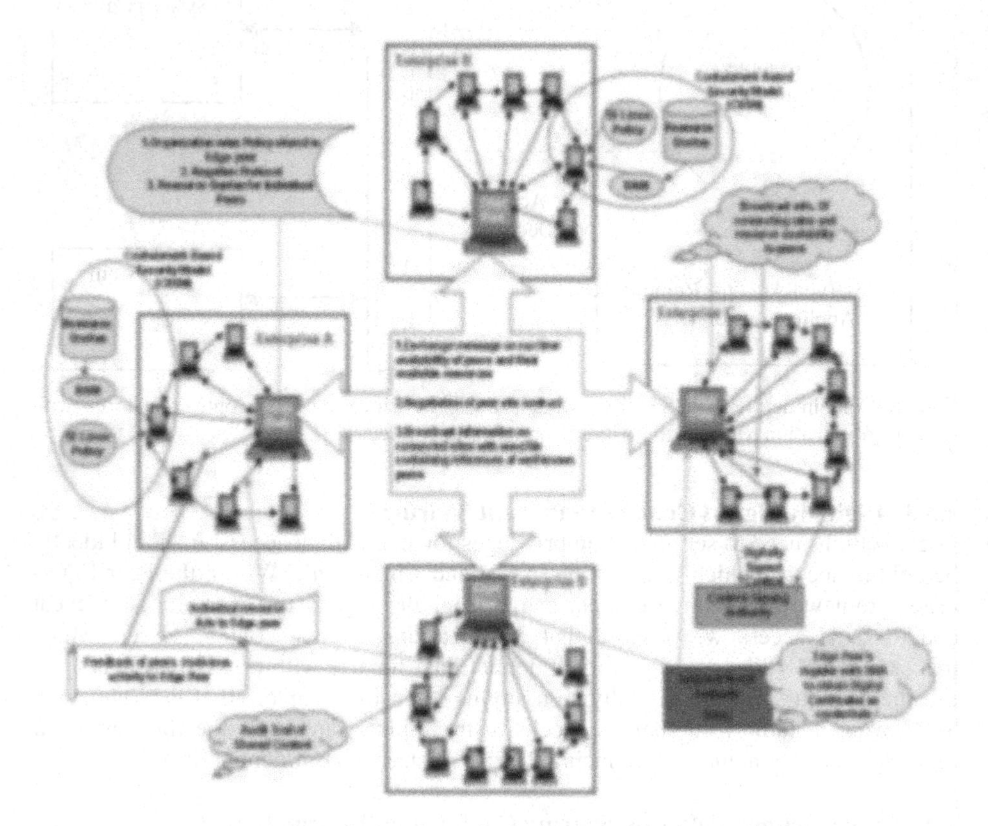

Fig. 3. Schematic of security provisions for cross-organizational applications

5 Conclusions

This paper looks at the specific security issues involved in cross-organizational P2P collaborations, a new application domain. P2P networks have not been proposed for cross-organizational collaborations in the past. To realize the envisaged framework solutions to issues like identity management, individual peer security, security of organization's sensitive data, overcoming malicious peer behavior etc. are presented. Table 2 provides a summary of the security features proposed to address specific security risks.

Table 2. Summary of security features for Cross-Organizational P2P Applications

Organizational Risks Due to P2P Applications	Security Provisions in Peer Enterprises Framework
Risk of data exposure	• Shared data is sandboxed – the file sharing application cannot access any other directory on the host peer. Prevents inadvertent sharing of confidential/critical data. • Audit Trail of all files shared by the peer is maintained. • If central content server digitally signs critical content and shares within organization. Middleware does not allow sharing of signed content outside the organization.
Risk of security attacks	• Containment-based security model. Creates sandbox for remote code. Cannot access critical system services and features. Cannot access anything outside permissible directory. Run-time monitor terminates remote code if it indulges in malicious activity. • Peers report malicious activity to edge peer which can terminate contract and blacklist a peer enterprise.
Risk to organizational compute resources (when allowing access to peers outside the organization)	• Containment-Based Security Model (CBSM) secures individual peers against untrusted or even hostile remote applications.
Risk of degradation of network performance (due to P2P traffic overload)	• Organization-wide policy configured at the edge peer can allow an organization to specify hours of operation, typically during hours of low network utilization. • Run-time monitor prevents remote code from establishing more than allowed number of network connections or sending large amounts of data.
Risk of loss of employee productivity	• Audit Trail of peer activity is reported to the edge peer. Any activity over permissible limits can be detected. • Security provisions ensure compute resource uptime and prevent loss of productivity due to unavailable resources.

Many of the proposed schemes have been validated through implementation and testing, especially the CBSM model and the RAM, which is a novel mechanism for ensuring security of individual peers. The various other schemes like digitally signed content and audit trailing help in preventing inadvertent sharing of sensitive data, besides discouraging users from sharing data outside the organization. These can ensure that security concerns prevalent in existing P2P systems get alleviated to a large extent. We believe that the proposed security mechanisms shall go a long way in ensuring the security of participating organizations and enable them to deploy cross-organizational P2P applications, opening up exciting new technical collaborations and business opportunities.

References

1. Oram, A. (ed.): P2P: Harnessing the Power of Disruptive Technologies. O'Reilly, Sebastopol (2001)
2. InformationWeek Report : P2P Peril (March 2008), `http://www.informationweek.com/news/security/showArticle.jhtml?articleID=206903416`
3. Verso Technologies Report: The High Cost of P2P on the Enterprise (July 2003), `http://jobfunctions.bnet.com/abstract.aspx?docid=312772`
4. Gupta, A., Awasthi, L.K.: Peer Enterprises: Possibilities, Challenges and Some Ideas Towards Their Realization. In: Meersman, R., Tari, Z., Herrero, P. (eds.) OTM-WS 2007, Part II. LNCS, vol. 4806, pp. 1011–1020. Springer, Heidelberg (2007)
5. Gupta, A., Awasthi, L.K.: Peer Enterprises: Enabling Advanced Computing and Collaboration Across Organizations. In: IEEE International Conference on Advanced Computing, pp. 3543–3548. IEEE Press, Los Alamitos (2009)
6. Engle, M., Khan, J.I.: Vulnerabilities of P2P Systems and a Critical Look at their Solutions. Technical Report, Internet and Media Communications Research Laboratories, Kent State University (2006)
7. Kamwar, S.D., Schlosser, M.T., Garcia-Molina, H.: The EigenTrust Algorithm for Reputation Management in P2P Networks. In: 12th International Conference on World Wide Web, pp. 640–651 (2003)
8. Singh, A., Liu, L.: TrustMe: Anonymous Management of Trust Relationships in Decentralized P2P Systems. In: Proceedings of the Third International Conference on Peer-to-Peer Computing, pp. 142–149 (2003)
9. Marti, S., Garcia-Molina, H.: Taxonomy of Trust: Categorizing P2P Reputation Systems. J. Comp. Net. 50(4), 472–484 (2006)
10. Lesueur, F., Me, L., Tong, V.V.T.: A Sybilproof Distributed Identity Management for P2P Networks. In: IEEE Symposium on Computers and Communications, pp. 246–253 (2008)
11. Gupta, R., Manion, T.R., Rao, R.T., Singhal, S.K.: Peer-to-Peer Authentication and Authorization. United States Patent: 7350074 (2008)
12. Kim, Y., Mazzocchi, D., Tsudik, G.: Admission Control in Peer Groups. In: Second IEEE International Symposium on Network Computing and Applications, p. 131 (2003)
13. Tran, H., Hitchens, M., Varadharajan, V., Watters, P.: A Trust based Access Control Framework for P2P File-Sharing Systems. In: Proceedings of International Conference on System Sciences, p. 302 (2005)
14. Park, J.S., An, G., Chandra, D.: Trusted P2P Computing Environments With Role-Based Access Control. Information Security, IET 1(1), 27–35 (2007)

15. Gaspary, L.P., Barcellos, M.P., Detsch, A., Antunes, R.S.: Flexible Security in Peer-to-Peer Applications: Enabling New Opportunities Beyond File Sharing. J. Comp. Net. 51(17), 4797–4815 (2007)
16. Lua, E.K.: Hierarchical Peer-to-Peer Networks Using Lightweight SuperPeer Topologies. In: Proceedings of the 10th IEEE Symposium on Computers and Communications, pp. 143–148 (2005)
17. Peng, Z., Duan, Z., Qi, J., Cao, Y., Lv, E.: HP2P: A Hybrid Hierarchical P2P Network. In: First International Conference on the Digital Society, pp. 8–18 (2007)
18. Gupta, A., Awasthi, L.K.: Secure Thyself: Securing Individual Peers in Collaborative Peer-to-Peer Environments. In: International Conference on Grid Computing and Applications, pp. 140–146 (2008)
19. Linux SE Website, `http://www.nsa.gov/selinux/info/docs.cfm`

Protocols for Secure Node-to-Cluster Head Communication in Clustered Wireless Sensor Networks

A.S. Poornima[1] and B.B. Amberker[2]

[1] Dept. of Computer Science and Engg, Siddaganga Institute of Technology, Tumkur, Karnataka, India
[2] Dept. of Computer Science and Engg, National Institute of Technology, Warangal, Andhra Pradesh, India

Abstract. Cluster based organization is widely used to achieve energy efficiency in Wireless Sensor Networks(WSN). In order to achieve confidentiality of the sensed data it is necessary to have secret key shared between a node and its cluster head. Key management is challenging as the cluster head changes in every round in cluster based organization. In this paper we are proposing deterministic key establishment protocols, which ensures that always there exists a secret key between node and its cluster head. The proposed protocols establishes key in an efficient manner every time a cluster head is changed. The hash based protocol achieves key establishment with very minimal storage at each node and by performing simple computations like one way hash functions and EX-OR operations. Where as the polynomial-based protocol establishes key in every round by using preloaded information without performing any additional communication.

Keywords: Cluster-based WSN, Seed value, Cluster head, Sensor Node, Key Matrix.

1 Introduction

Wireless Sensor Networks (WSN) typically consist of small, inexpensive, battery powered sensing devices fitted with wireless transmitters, which can be spatially scattered. Sensors have the ability to communicate through wireless channels, and their energy, computational power and memory are constrained. WSN's have many advantages. They are easier, faster and cheaper to deploy than wired networks or other forms of wireless networks. They have higher degree of fault tolerance than other wireless networks : failure of one or few nodes does not affect the operation of the network. Also, they are self organizing or self-configuring. These advantages make them very promising in a wide range of applications ranging from health care to warfare. The envisioned growth in using sensor networks is demanding extensive research in securing these networks.

Wireless Sensor networks are used in many applications like battlefield, patient monitoring, emergence response information and environmental monitoring.

S. Ranka et al. (Eds.): IC3 2009, CCIS 40, pp. 434–444, 2009.
© Springer-Verlag Berlin Heidelberg 2009

Providing security for WSNs presents unique challenges. In addition to unknown topography sensors have very high computation, storage and battery constraints. WSNs lack physical protection and are usually deployed in open, unattended environments, which makes them vulnerable to attacks. Cryptographic methods can be used to secure such Wireless Sensor Networks. An important issue to be addressed when cryptographic methods are used to secure WSNs is key distribution. Key pre distribution schemes are the most widely used key distribution methods in WSNs. Many such pre distribution schemes are discussed in literature [3,4,5,6,7,8,9,10,11]. The schemes proposed in the literature consider number of WSN architectures and key distribution methods that is well suited to one architecture is likely not to be the best for another, as different network architectures exhibit different communication patterns.

The available redundancy and inherent energy scarcity of a sensor network encourages the use of aggregation of data. Clustering of the Wireless Sensor Networks (WSN's) improves data aggregation ability. The cluster-based architecture is an effective way to achieve the objective of energy efficiency in Wireless Sensor networks. In a clustered WSN nodes in a neighborhood organize themselves into a cluster, with one node designated as the Cluster Head (CH) [1,2]. The CH collects sensed data from the other nodes in its neighborhood and uses an aggregation scheme to aggregate this information. It then sends the information to a neighboring CH in the direction of the Base Station (BS). Cluster based organization [12] has been proposed for ad hoc networks in general and WSNs in particular. In clustered WSNs rotating cluster heads concept is widely used for energy efficiency. The concept of rotating cluster head is introduced in [12] (which is also called as LEACH- Low Energy Adaptive Clustering Hierarchy protocol). In LEACH protocol to save energy, sensor nodes send their messages to their CH's, which then aggregate the messages and send the aggregate to the BS. To prevent energy drainage of a restricted set of CH's, LEACH randomly rotates CHs among all nodes in the network, distributing aggregation and routing related energy consumption among all nodes in the network.

Security issues in cluster-based sensor networks are addressed in [13,14,15,16]. Bohge et. al. [13] proposed an authentication framework for a concrete 2-tier network organization, in which a middle tier of more powerful nodes between the BS and the ordinary sensors were introduced in order to carry out authentication function. Oliveira et. al. [14] propose solution that relies exclusively on symmetric key schemes and is suitable for networks with an arbitrary number of levels; and Ferreira et.al. [15] proposed F-LEACH where each node has two symmetric keys ; a pairwise key shared with the BS, and the last key of a key chain held by the BS used in authenticated broadcast. In [16] SecLeach, a protocol for securing node-to-CH communication in LEACH based networks is discussed. SecLeach bootstraps security from random key pre distribution scheme which is studied extensively in [3,4,5,6,7,8,9,10,11].

In this paper we are proposing dynamic key establishment protocols to secure node-to-CH communication. Here we are proposing two protocols which deal with establishing secret key between node and its cluster head in order to achieve confidentiality of the transmitted data. Every time when a cluster head is changed the protocols establishes new key between a node and its new cluster head. The protocols discussed in this paper are developed based on a scheme known as Efficient Pairwise Key Establishment and Management in Static Wireless Sensor Networks (EPKEM) proposed by Cheng.et.al in [11]. The scheme in [11] is modified so that it can be used in clustered WSNs in an efficient manner. The problem we are considering here is similar to that of [16] (i.e., establishing key between nodes and cluster head in a clustered WSN dynamically for every round). Hence we compare our protocols with the scheme discussed in [16]. The first protocol we are presenting in this paper uses simple hash and EX-OR operations to establish a key, also the storage at each node is constant. The second protocol is a polynomial-based scheme which requires some additional storage but keys are established without performing any additional communication.

The rest of the paper is organized as follows : in Section 2 we describe our problem statement along with assumptions and attack model, Section 3 we focuses on details of the EPKEM scheme. In Section 4 we present in detail the new proposed deterministic protocols for key establishment. Section 5 we discuss the details of security analysis and performance analysis. We conclude in Section 6.

2 Problem Statement

A. Network Assumptions
In this paper we are considering homogeneous distributed static sensor network. Each node is equal in terms of computation, communication, energy supply and storage capability. The nodes we are considering here are immobile. The nodes are arranged into clusters after deployment using cluster formation procedure as explained in [12]. The clusters considered in this paper are set of small number of sensor nodes. The communication pattern used is : every node send the sensed data to its cluster head, in turn cluster head aggregates the received data and forwards the aggregated data towards the base station.

B. Attacker Model
The type of attacker we are considering in this paper are of two types. First type of attacker is an *outside attacker* who is able to eavesdrop on the communications. Second type of attacker is *inside attacker* a compromised node which is able to get all the secrets.

C. Problem Definition
In order to achieve energy efficiency in clustered sensor networks, cluster heads are changed periodically. We need to provide security to the sensed data that is transmitted from nodes to the cluster head in presence of the attacker model mentioned above and the network assumptions. To achieve confidentiality of the data we need a secret key that is shared between a node and its cluster head. As the cluster head changes periodically, dynamic key establishment protocols are required to establish key between node and its cluster head every time a cluster head is changed. In this paper we are presenting two protocols to achieve dynamic key establishment in the above discussed setup.

3 New Protocols for Dynamic Key Establishment

In this section we explain the proposed protocols which establishes key between any node and its cluster head each time a cluster head is changed. In Section 3.1 we explain the hash based protocol for key establishment and in Section 3.2 polynomial based protocol is explained.

3.1 Hash Based Protocol

In section 3.1.1 modified EPKEM scheme is explained, which presents how EPKEM scheme is modified for our hash based protocol. In Section 3.1.2 key establishment protocol which establishes key efficiently in every round using the modified EPKEM is discussed.

Modified EPKEM Scheme. We propose a protocol which is a modification of EPKEM scheme. The new protocol is suitable for clustered WSNs. The protocol is explained below.

Key Pre Assignment Phase: KDS pre loads each node with a key P_k that it shares with the base station (BS) for confidential communication. For a network of size n, KDS constructs a $m \times m$ key matrix K where $m = \sqrt{n}$. The KDS chooses a random value for each row and column which we call as row seed value rs and column seed value cs. Entries for the matrix K are computed using these seed values and one way hash function h [17,18]. The $(i,j)^{th}$ entry of K is computed as $h^i(rs_i) \oplus h^j(cs_j)$ where rs_i is the ith row seed value and cs_j is the jth column seed value. Here, $h^i(x) = h^{i-1}(h(x))$ for $i = 2, \ldots, m$ and $h^1(x) = h(x)$. Each node is assigned with some row and column of the matrix and instead of storing entire subset of values in that row and column as in [11], here we are storing only the corresponding row seed value and column seed value.

To illustrate let us consider the key matrix K shown in the Table1 : Here rs_1, rs_2, rs_3, and rs_4 are row seed values of corresponding rows and cs_1, cs_2, cs_3 and cs_4 are column seed values. The entries (i.e., k_{11}, $k_{12} \ldots k_{44}$) in the matrix are computed using the respective row and column seed values. For example, entry $k_{11} = h(rs_1) \oplus h(cs_1)$. Similarly $k_{34} = h^4(rs_3) \oplus h^3(cs_4)$.

Common Key Setup Phase: After deployment, nodes will be arranged into clusters and cluster head will be elected using the cluster formation procedure as explained in [12]. Once the cluster is formed, any two nodes S_{ij} and S_{kl}

Table 1. Key matrix K

Key Matrix K	cs_1	cs_2	cs_3	cs_4
rs_1	k_{11}	k_{12}	k_{13}	k_{14}
rs_2	k_{21}	k_{22}	k_{23}	k_{24}
rs_3	k_{31}	k_{32}	k_{33}	k_{34}
rs_4	k_{41}	k_{42}	k_{43}	k_{44}

where i,j,k,l are their row and column indices assigned respectively, can setup a common key. The nodes S_{ij} and S_{kl} can compute the two entries of key matrix K on their own. The entries that the nodes can compute are $h^l(rs_i) \oplus h^i(cs_l)$ and $h^j(rs_k) \oplus h^k(cs_j)$. Each node will send its ID, row and column indices assigned to it. Upon receiving such broadcast message S_{ij} and S_{kl} will setup a common key. If S_{ij} receives message from S_{kl} first, S_{ij} will compute $h^l(rs_i)$ and send it to S_{kl} and S_{kl} will compute $h^i(cs_l)$ and send it to S_{ij}. Once the hashed row and column seed values are exchanged between the nodes then the nodes can compute the common key on their own as $h^l(rs_i) \oplus h^i(cs_l)$. Suppose if node S_{kl} receives the message first, then S_{kl} will compute $h^j(rs_k)$ and send it to S_{ij} and S_{ij} will compute $h^k(cs_j)$ and send it to S_{kl}. Now the common key computed between S_{ij} and S_{kl} is $h^j(rs_k) \oplus h^k(cs_j)$. Each node is required to perform two transmit and two receive operation in order to setup a common key.

Pairwise Key Computation Phase: If common key between nodes S_{ij} and S_{kl} is $h^l(rs_i) \oplus h^i(cs_l)$, then pairwise key is $ID_{S_{ij}} \oplus h^l(rs_i) \oplus h^i(cs_l) \oplus ID_{S_{kl}}$.

Key Establishment for Every Round. Cluster based WSNs are used to address the efficiency and lifetime of the battery by changing the role of cluster head among the available nodes. Hence it is required to design key management protocols such that when a cluster head is changed it should be possible to compute common shared key between any node and its cluster head in an efficient manner. This is accomplished as follows:

Step 1: Compute key matrix K as explained in *key pre assignment* phase of modified EPKEM scheme in Section 3.1.1. Now every sensor is preloaded with seed values (row seed value rs and column seed value cs) of row index and column index of key matrix K. Also, every sensor is loaded with a private secret information P_k to communicate securely with the base station (BS).

Step 2: After deployment, nodes will be arranged into clusters using the cluster formation procedure as explained in [12].

Step 3: After the nodes are arranged into clusters, the nodes will now establish a secret key with the corresponding cluster head. Cluster head S_{ij} will broadcast (i, j) i.e., the row and column indices of key matrix K that is assigned to it. Upon receiving this, each node S_{kl} will compute $h^j(rs_k)$ and send it to cluster head along with its own row and column indices (k, l). Now for each node cluster head computes $h^k(cs_j)$. Message sent after computing hash values is (Node $ID_m||h^m(cs_j)$) for $m = 1, 2, \ldots, t$ where t is the number of nodes in the cluster.

Step 4: After the hash values are exchanged between the cluster head and other nodes in the cluster, each node S_{kl} computes the key that it shares with the cluster head as : $ID_{s_{ij}} \oplus h^j(rs_k) \oplus h^k(cs_j) \oplus ID_{S_{kl}}$.

3.2 Polynomial-Based Protocol

In this section we present our polynomial-based protocol to setup a common key between a node and its cluster head dynamically in every round. The protocol is based on the polynomial-based key pre distribution protocol [19,5] which is discussed in Section 3.2.1. In Section 3.2.2 we discuss how the the schemes in [19,5] can be used to design a deterministic key establishment technique. Section 3.2.3 explains about dynamic key establishment for confidential communication between the nodes and its cluster head in cluster based WSN. The protocol ensures that when the cluster head is changed in every round it is possible to establish a key between every node and its cluster head.

Polynomial-Based Key Predistribution. In this section we discuss poly nomial-based key pre distribution protocol proposed by Blundo for group key setup in [19]. In this scheme key server randomly generates a symmetric bivariate t-degree polynomial $f(x,y) = \sum_{i,j=0}^{t} a_{ij}x^iy^j$ over a finite field F_q, where q is a large prime number. Now server computes a polynomial share $f(x,y)$ i.e., $f(i,y)$ where i is the unique ID of the sensor node and pre loads this share in the memory of node i. Similarly for node j $f(j,y)$ is computed and pre loaded. Common key between nodes i and j (where i and j are ID of node i and node j respectively) is computed as follows: node i can compute the key $f(i,j)$ by evaluating its polynomial share $f(i,y)$ at point j, and node j can compute the same key $f(j,i) = f(i,j)$ by evaluating its polynomial share $f(j,y)$ at point i.

Polynomial-Based Scheme for Key establishment. Pre Assignment Phase: Key Distribution Server (KDS) pre loads each node with a key P_k that it shares with the base station (BS) for confidential communication. For a network of size n, KDS constructs a $m \times m$ matrix K where $m = \sqrt{n}$. Each entry of the matrix K is a symmetric t- degree bivariate polynomial. For eg., the polynomial stored at row i and column j is represented as $f_{ij}(x,y)$. Polynomials stored at i^{th} row of matrix K are labelled as $\{f_{ij}(x,y)\}_{j=1,\ldots m}$, similarly polynomials at j^{th} column are labelled as $\{f_{ij}(x,y)\}_{i=1,\ldots m}$. Every node is preloaded with shares of the polynomials by selecting one row and one column of the matrix K. The shares assigned to node S_{ij} (where i and j are the row and column indices assigned to node S) are $\{f_{kl}(ID_{S_{ij}},y)\}_{k=i,l=1,\ldots m}$ and $\{f_{kl}(ID_{S_{ij}},y)\}_{k=j,l=1,\ldots m}$.

Key establishment phase: After deployment, any two nodes S_{ij} and S_{kl} where i,j,k,l are their row and column indices assigned respectively, can establish a common key. Each node will send its ID, row and column indices assigned to it. Upon receiving such a message nodes S_{ij} and S_{kl} compute common key using the polynomial shares that are assigned and common between the nodes. The shares of the polynomial that are used by node S_{ij} to compute the common key are $f_{il}(ID_{S_{ij}},y)$ and $f_{kj}(ID_{S_{ij}},y)$. The shares with node S_{kl} are $f_{il}(ID_{S_{kl}},y)$ and $f_{kj}(ID_{S_{kl}},y)$. Now node S_{ij} evaluates its common polynomial shares by replacing y term with the ID of the node S_{kl}, therefore key computed by node S_{ij} is $f_{il}(ID_{S_{ij}},ID_{S_{kl}}) \oplus f_{kj}(ID_{S_{ij}},ID_{S_{kl}})$. Similarly node S_{kl} will compute $f_{il}(ID_{S_{kl}},ID_{S_{ij}}) \oplus f_{kj}(ID_{S_{kl}},ID_{S_{ij}})$. Since symmetric polynomials are used both computations will yield same key using which nodes S_{ij} and S_{kl} can perform confidential communication.

The above protocol is simple and using which we can establish key between any two nodes in the network and also no communication cost is involved to set up the common key.

Key Establishment for Every Round. Cluster based WSNs are used to address the efficiency and lifetime of the battery by changing the role of cluster head among the available nodes. Hence it is required to design key management protocols such that when a cluster head is changed it should be possible to compute common shared key between any node and its cluster head in an efficient manner. This section explains the complete protocol to set up key for confidential communication between a node and its cluster head CH. The protocol ensures that key is established between every node and its corresponding cluster head dynamically in every round.

Step 1: Compute key matrix K as explained in *Pre assignment Phase* in Section 3.2.2. Now every node is pre loaded with shares of polynomials again as explained in Section 3.2.2.

Step 2: After deployment, nodes will be arranged into clusters using the cluster formation procedure as explained in [12].

Step 3: After the nodes are arranged into clusters, the nodes will now establish a secret key with the corresponding cluster head. Cluster head S_{ij} will broadcast (i, j) i.e., the row and column indices of key matrix K that is assigned to it. Upon receiving this broadcast message every node in the cluster will communicate the key that it shares with the cluster head for this round. The key is computed by evaluating the shares of the polynomial that the node shares with the cluster head. The common share of the polynomial that a node S_{kl} shares with cluster head S_{ij} are $f_{il}(ID_{S_{kl}}, y)$ and $f_{kj}(ID_{S_{kl}}, y)$, now node S_{kl} evaluates the shares by replacing y term with the ID of the cluster head. The computed by S_{kl} is $f_{il}(ID_{S_{kl}}, ID_{S_{ij}}) \oplus f_{kj}(ID_{S_{kl}}, ID_{S_{ij}})$. The cluster head S_{ij} computes the key by replacing y term of its polynomial shares with the ID of the node S_{kl}. Key computed by cluster head that it shares with the node S_{kl} is $f_{il}(ID_{S_{ij}}, ID_{S_{kl}})$ $\oplus$ $f_{kj}(ID_{S_{ij}}, ID_{S_{kl}})$. Similarly every node in the cluster will compute the a key that it uses for confidential communication with the cluster head by evaluating respective shares of the polynomial. Also the cluster head computes the key by evaluating appropriate shares of the polynomial for every node.

In the above protocol no communication in required to establish a key between a node and its cluster head. By broadcasting the ID's of each other nodes can establish a key with its cluster head that they can use for confidential communication in the current round. Every time when a cluster head is changed without performing any additional communication keys are computed dynamically between the nodes and the cluster head for the current round.

4 Analysis of the Proposed Protocols

In this section we analyze our proposed protocols. In security analysis we discuss how the proposed protocols achieve the security goal in presence of the attacker model as assumed in section 2. In Performance analysis we prototype the overhead with respect to storage, communication and computation of the proposed protocols and compare it with existing scheme.

4.1 Security Analysis

Key Setup for Each Round. In cluster based WSNs the cluster head will change after certain predefined time interval, which we call it as *one round.* In every round the cluster head is elected as explained in[12]. After the cluster head is changed, every node the cluster can establish key with its cluster head by using any of the protocols as discussed in Section 3.1 and Section 3.2. At the end of the successful execution of the protocol each node shares a key with its cluster head. In our proposed protocols every node will establish a secret key with the cluster head at the end of key establishment phase. Hence every communication between node and its cluster head is secure. In the scheme discussed in [16] they use probabilistic key pre distribution method, which may not yield full connectivity. That is some nodes may not have a shared key with the cluster head. But in the proposed protocols this is eliminated and at any round it is possible to establish a secret key between a node and its cluster head.

Node Compromise. In the hash based protocol, if a node is compromised, the secret information revealed are : the P_k, private key that the node shares with BS, cluster key that it shares with cluster head, the row seed value rs and the column seed value cs. When a node is compromised, only the communication between that node and its cluster head is affected as the shared key is revealed. But the row or column seed value of the compromised node will not reveal any other key. In polynomial-based protocol, the secret information revealed are : the P_k, private key that the node shares with BS, cluster key that it shares with cluster head, and preloaded shares of the polynomial. Shares of the polynomial are computed by evaluating the polynomial using the ID of the node for which shares are assigned. Therefore, same polynomial will yield different shares when evaluated using different ID's. When shares are obtained by using this method compromised shares will not reveal any information regarding other shares obtained by the same polynomial. Hence when a node is compromised the revealed shares can not be used to derive the other shares. The scheme in [16] uses key pre distribution method, here when a node is compromised entire key ring on that node is compromised. This may affect communication of non compromised nodes which share some keys same as that of the compromised node.

4.2 Performance Analysis

Storage. In the hash based protocol, at any point of time, each node will store the following four secret values : Private key used to communicate with the BS securely (P_k), cluster key used to communicate with its cluster head for the current round and preloaded row seed value (rs) and column seed value (cs) from which the node computes common key with the cluster head. In the polynomial-based protocol, at any point of time, each node will store the following four secret values : Private key used to communicate with the BS securely (P_k), cluster key used to communicate with its cluster head for the current round and preloaded shares of the polynomial from which the node computes common key with the cluster head. Each node stores $2n$ polynomial shares. The base station stores private keys P_k of all the nodes in the network. Also it stores the entire key matrix K of size $m \times m$ where $m = \sqrt{n}$ and n is number of nodes in the network. Therefore, the total storage in BS is $2n$. The scheme in [16] uses key pre distribution method proposed in [3]. The key pre distribution method [3] requires every node to be preloaded with 150 keys for a key pool of P of size 10,000 in order to achieve a key sharing probability of 0.9. Therefore storage required at each node is large in [16] compared to proposed protocols in which hash based protocol requires only 4 keys to be stored in each sensor node. Also, as our protocols are deterministic, every node can establish a key with its cluster head using the stored information.

Computation Cost. In hash based protocol to compute secret key of a node we use simple one way hash function [17,18] and bitwise EX-OR operation. Each node S_{ij} has to perform $i + j$ number of hash operations and one bitwise EX-OR operation. For a cluster of size t the cluster head computes at the most tm number of hash operations and t EX-OR operations. In the polynomial-based protocol to compute secret key each node performs evaluation of polynomial shares and bitwise EX-OR operation. Each node S_{ij} has to evaluate two polynomial shares and one bitwise EX-OR operation. For a cluster of size t the cluster head performs $2t$ polynomial evaluations and t EX-OR operations.

Communication Cost. In every round, once a cluster head is elected, the communications that are required by hash based protocol to compute common key between node and its cluster are : Each node will perform one transmit and two receive operations whereas the cluster head has to perform t receive operations and two transmit operations, where t is the number of nodes in the cluster. The scheme in [16] also require same number of communications in order to establish common key with the cluster head.

In the polynomial-based protocol no communication in required to establish a key between a node and its cluster head. By broadcasting the ID's of each other nodes can establish a key with its cluster head that they can use for confidential communication in the current round. Every time when a cluster head is changed without performing any additional communication keys are computed dynamically between the nodes and the cluster head for the current round. The scheme in [16] require one transmit and one receive operation in order to establish common key with the cluster head.

5 Conclusion

The paper presents a new protocols for establishing secret key for confidential communication between node and its cluster head in cluster based WSN. The proposed protocols perform the task of establishing common key for node-to-cluster head communication in an efficient manner with respect to communication, computation and storage. In the hash based protocol every time when a cluster head is changed key can be established by performing two transmit and two receive operations and it ensures that key can be computed between every node and its cluster head. Also each sensor node needs to store only four keys at any point of time. Efficiency in Computation is achieved by using simple operations like one way hash functions and bit wise exclusive or operations. Also the polynomial-based key establishment protocol performs the task of establishing common key for node-to-cluster head communication. To establish key in every round no additional communication cost is incurred in the proposed polynomial-based protocol. Also when a node is compromised only the keys of the compromised nodes are affected and these keys and other information revealed will not reveal keys of other nodes.

References

1. Bandyopadhyay, S., Coyle, E.: An energy efficient hierarchical clustering algorithm for wireless sensor networks. In: 22 nd Conference of the IEEE Communication Society, INFOCOM, vol. 3 (2003)
2. Younis, O., Fahmy, S.: Distributed clustering in adhoc sensor networks: a hybrid, energy-efficient approach. In: 23rd Conference of the TEEE Computers and Communication Societies, pp. 629–640 (2004)
3. Eschenauer, L., Gligor, V.D.: A key management scheme for distributed sensor networks. In: Proceedings of the 9th ACM conference Computer and Communications security, November 2002, pp. 41–47 (2002)
4. Chan, H., Perrig, A., Song, D.: Random key pre distribution schemes for sensor networks. In: IEEE symposium on Research in Security and Privacy, pp. 197–213 (2003)
5. Liu, D., Ning, P.: Establishing pairwise keys in distributed sensor networks. In: Proceedings of the 10th ACM conference on Computers and Communication Security (CCS 2003), pp. 52–61 (2003)
6. Du, W., Deng, J., Han, Y.S., Varshney, P.K.: A pairwise key pre distribution scheme for wireless sensor networks. In: Proc. of the 10th ACM conference of Computers and Communication Security (CCS 2003), pp. 42–51 (2003)
7. Zhu, S., Setia, S., Jajodia, S.: LEAP: Efficient Security Mechanisms for Large Scale Distributed Sensor Networks. In: Proc. of 10th ACM Conference on Computers and Communication Security (CCS 2003) (2003)
8. Zhu, S., Xu, S., Setia, S., Jajodia, S.: Establishing pairwise keys for secure Communication in Ad Hoc Networks: A Probabilistic approach. In: 11th IEEE International conference on Network Protocols (ICNP 2003) (2003)
9. Pietro, R.D., Mancini, L.V., Mei, A.: Random Key assignment to secure wireless sensor networks. In: 1st ACM workshop on Security of Ad Hoc and Sensor Networks (2003)
10. Hwang, J., Kim, Y.: Revisiting random key pre distribution schemes for WSN. In: Proc. of the 2nd ACM workshop on Security of ad hoc and sensor networks, pp. 43–52 (2004)

11. Cheng, Y., Agrawal, D.P.: Efficient pairwise key establishment and management in static wireless sensor networks. In: Second IEEE International Conference on Mobile ad hoc and Sensor Systems (2005)
12. Heinzelman, W.R., Chandrakasan, A., Balakrishnan, H.: Energy-efficient Communication protocol for wireless microsensor networks. In: IEEE Hawaii Int. Conference on System Sciences, January 2000, pp. 4–5 (2000)
13. Bohge, M., Trappe, W.: An authentication framework for hierarchical ad hoc sensor networks. In: 2003 ACM workshop on Wireless Security, pp. 79–87 (2003)
14. Oliveira, L.B., Wong, H.C., Loureiro, A.A.F.: Lha=sp: Secure protocols for hierarchical wireless sensor networks. In: 9th IFIP/IEEE International Symposium on Integrated Network Management (IM 2005), pp. 31–44 (2005)
15. Ferreira, A.C., Vilaça, M.A., Oliveira, L.B., Habib, E., Wong, H.C., Loureiro, A.A.F.: On the security of cluster-based communication protocols for wireless sensor networks. In: Lorenz, P., Dini, P. (eds.) ICN 2005. LNCS, vol. 3420, pp. 449–458. Springer, Heidelberg (2005)
16. Oliveira, L.B., Wong, H.C., Bern, M., Dahab, R., Loureiro, A.A.F.: SecLeach: A random key distribution solution for securing clustered sensor networks. In: 5th IEEE International symposium on network computing and applications, pp. 145–154 (2006)
17. N.F.P.180-1. Secure hash standard. Draft, NIST (May 1994)
18. Rivest, R.: The MD5 message-digest algorithm. RFC 1321 (April 1992)
19. Blundo, C., De Santis, A., Herzberg, A., Kutten, S., Vaccaro, U., Yung, M.: Perfectly-secure key distribution for dynamic conferences. In: Brickell, E.F. (ed.) CRYPTO 1992. LNCS, vol. 740, pp. 471–486. Springer, Heidelberg (1993)

Significant Deregulated Pathways in Diabetes Type II Complications Identified through Expression Based Network Biology

Sanchaita Ukil[1], Meenakshee Sinha[1], Lavneesh Varshney[1], and Shipra Agrawal[1,2,*]

[1] Institute of Bioinformatics and Applied Biotechnology
[2] BioCOS Life Sciences Pvt. Limited, G-05, Tech Park Mall,
International Technology Park Bangalore (ITPB),
Whitefield Road, Bangalore, India
shipra@ibab.ac.in

Abstract. Type 2 Diabetes is a complex multifactorial disease, which alters several signaling cascades giving rise to serious complications. It is one of the major risk factors for cardiovascular diseases. The present research work describes an integrated functional network biology approach to identify pathways that get transcriptionally altered and lead to complex complications thereby amplifying the phenotypic effect of the impaired disease state. We have identified two sub-network modules, which could be activated under abnormal circumstances in diabetes. Present work describes key proteins such as P85A and SRC serving as important nodes to mediate alternate signaling routes during diseased condition. P85A has been shown to be an important link between stress responsive MAPK and CVD markers involved in fibrosis. MAPK8 has been shown to interact with P85A and further activate CTGF through VEGF signaling. We have traced a novel and unique route correlating inflammation and fibrosis by considering P85A as a key mediator of signals. The next sub-network module shows SRC as a junction for various signaling processes, which results in interaction between NF-kB and beta catenin to cause cell death. The powerful interaction between these important genes in response to transcriptionally altered lipid metabolism and impaired inflammatory response via SRC causes apoptosis of cells. The crosstalk between inflammation, lipid homeostasis and stress, and their serious effects downstream have been explained in the present analyses.

1 Introduction

Type 2 diabetes is a metabolic disorder, characterized by high blood glucose level due to insulin resistance and relative insulin deficiency in the body. Defective insulin action leads to glucose uptake blockage in the cells. Deregulated glucose and lipid homeostasis, improper control of cellular proliferation, differentiation and protein synthesis are some of the phenotypes manifested due to altered signalling. T2D can give rise to other complications due to the deregulation of normal cascade. Diabetes is recognized as an independent risk factor for cardiovascular morbidity and mortality. Obesity, dyslipidemia and activation of systemic inflammatory cascades with resultant vascular dysfunction have been implicated as risk factors for diabetes mediated cardiovascular diseases. The

* Corresponding author.

S. Ranka et al. (Eds.): IC3 2009, CCIS 40, pp. 445–453, 2009.
© Springer-Verlag Berlin Heidelberg 2009

characteristic features of diabetic dyslipidemia are a high plasma triglyceride concentration, low HDL and high small dense LDL-cholesterol particles concentration [1].

Our study is focused towards identifying significant links between T2D and cardiovascular disease (CVDs) using system biology approach. We have used an integrated functional networks concept to detect the transcriptionally altered pathways and regulations linking T2D with CVDs. Integration of co-expressed gene network with the corresponding protein interaction network has been employed to identify signature pathways and mechanisms. Systematic analysis of high throughput data is used to construct transcriptional networks from differentially expressed genes to identify novel pathways and candidate biomarkers affected during disease process. We describe two signaling routes, which get switched on during disease conditions. In the first pathway, we link stress activating kinases expressed during stress and inflammation leading to scar formation and fibrosis due to increased expression of CTGF induced by P85A-VEGF interaction as highlighted in our network (Figure 1). We describe the signal being transduced from MAPK to CTGF via p85, an important connecting protein, which interacts with VEGF and transmits signal to CTGF. Such a connection is predicted to be a novel path involving stress inflammation and scar formation. The second network module explains the activation of SRC by LDL which further activates NF-kB – beta catenin signaling and their role in apoptosis (Figure 2). The functions of SRC, NF-kB and beta catenin are well documented. However interaction involving the induction of beta catenin gene by NF-kB is assumed to be unique. Through our studies, we present diverse signaling directions, which can contribute to disease phenotypes significantly.

2 Materials and Methods

The microarray expression data has been retrieved from Gene Expression Omnibus datasets. Inflammatory changes drive the progression of various metabolic and cardiovascular dysfunctions. The dataset Inflam_H reports genes from inflammatory cardiomyopathy [2]. Information regarding the data source for the biological network Inflam_H is provided in Table 1. The differentially expressed genes were selected using Dchip software. The criteria for selection of differentially expressed genes were: cut off p-value of $<=0.05$ and fold change value of 2. Pearson's correlation coefficient (r-value) is calculated for the genes of the dataset in order to construct a gene-gene correlation matrix. Cladist software constructs the correlation matrix after computing the correlation coefficient (r-value) for the differentially expressed genes. In our study the r-value threshold for selection of gene pairs was 0.9. The data list was fed into Cytoscape to create coexpression networks. The corresponding protein-protein interactions were imported using the Cytoscape plug-in APID2NET. The coexpression networks and the protein interaction networks were integrated using the implemented plug-in of Cytoscape to merge two or more networks.

Table 1. Details of the datasets on their origin, control and diseased sets

Dataset	Disease condition	Tissue source	Control	Disease
Inflam_H	Inflammatory Dilated Cardiomyopathy	Endomyocardial biopsy	4-(healthy)	8-(inflammatory dilated cardiomyopathy)

Sub networks with significant functional relevance were extracted from these integrated union networks and named as MAPK8_P85A_CTGF and SRC_NFKB_CTNB1 respectively. Genes related to T2D, CVD as well as SNPs related to diabetes were obtained from literature sources, online repositories and T2Ddb (Agrawal S. et al) and denoted significant for our study due to their genetic association with the disease (Figures 1 and 2). Subnetworks created were assessed to validate their biological function. The networks were validated using literature and statistical measures such as node degree distribution, clustering coefficients, topological coefficients, betweeness centrality and shortest path length, closeness centrality and network heterogeneity were calculated using Network Analyser plug-in. Created networks were biologically significant when compared to random networks created which were generated using the plug-in Random networks.

R-squared values are known as coefficient of determination which measure how well the data points fit to a curve to establish hierarchical nature of biological networks. R-squared values are used to quantify the fit to the power line indicating the scale freeness of networks. R-squared values that are closer to 1 indicate higher correlation and a stronger linear relationship between the data variables. The R-squared values for the networks were computed and verified to demonstrate that they are scale free networks. The statistical values were calculated for both the networks as tabulated in the article (Tables 2 and 3). Statistical assessment is carried out in order to strengthen our hypotheses.

3 Results and Discussion

Study of sub networks identified as statistically significant with respect to their functional relevance led to the inference of the following signaling mechanisms, which can play a role in impaired disease conditions.

3.1 Linking Stress Inflammation, Insulin Resistance and CVD Complications via Significant Signaling Molecules Such as P85A

MAPK plays a very integral role in cell signaling, cell growth, differentiation, metabolism and development. In response to stress stimuli we see the induction of stress activated protein kinases such as JNK and p38 isoforms. In the given network, presence of TNF genes/proteins and c-JUN are indicative of inflammation [3]. ATF2 and ELK1 are transcription factors, which coordinate gene expression in response to stress [4]. Phosphorylation of c-JUN implies roles in inflammation, apoptosis and cellular transcription. CDC42 expression contributes to activation of p38 regulatory genes [3]. Presence of inflammation condition is further strengthened by the over expression c-Jun Terminal kinases such as MAPK8 and MAPK9 and also p38MAPK like MAPK14 along with other JNK interacting proteins. MAPK8 appears to be the central hub for all these interactions capable of further participating in other signaling cascades to advance other complications.

In the sub-network, it is seen that MAPK8 directly interacts with P85A, which is the regulatory subunit of Phosphatidylinositol 3-kinase. P85A is a very significant node in this study. From literature it is understood that P85A has a negative effect on insulin actions. It has been reported that monomeric P85 can inhibit IRS mediated signaling and down regulate IGF/insulin signaling. In addition, enhanced expression of P85 has been proposed to have a contradictory effect on glycogen synthase [5]. Our sub-network module also shows a direct link between P85A and CDC42 and Fasl/TNFL6. Tumor necrosis factor alpha (TNF) stimulates stress and also results in negative regulation of IRS-1 tyrosine phosphorylation thereby weakening its association with P85A [6]. These genes play a role in inflammation and their signals converge at P85A. Regulation of inflammatory cytokines by the stress responsive MAPK such as MAPK8 and MAPK9 is shown in the network (Figure 1).

VEGFA is over expressed in response to hypoxia and inflammation [7]. VEGF promotes angiogenesis in vascular endothelium and its expression is up regulated in ischemia where it is observed that it promotes inflammation by causing vascular leakage and mobilizing leukocytes. Direct involvement is seen in case of altered cardiac lipid metabolism and induction of myocardial hypertrophy. P85A is linked to VEGFA via different VEGF receptors.

CTGF is a vital entity playing a role in cardiovascular complications. High expression of CTGF leads to failure to terminate tissue repair resulting in pathological scarring as seen in case of fibrosis [8]. VEGF can cause up regulation of CTGF via p85 [9].

Thus, it is envisaged that inflammation, insulin signaling and cardiac problems, all are related to one other through some very significant genes. VEGFa and MAPK8 are retrieved from co expression network. CTNNB1, GFPT1, PTPN11 and variants of Tumor necrosis factor receptor super family are also present in the co expression network. P85A is an important molecule which can link inflammation with insulin signaling and further cardiac complications. From this analysis, we hypothesize that inflammation can cause the over expression of certain stress responsive kinases such as MAPK8 which can further activate VEGF expression via p85 and finally CTGF over expression. This can be predicted as an alternative pathway tracing its path from MAPK8 to CTGF. We identify P85A and VEGF as important central signaling molecules which carry the effector signal from MAPK8 to CTGF and contribute to fibrosis which occurs due to the action of proinflamotory cytokines.

3.2 SRC as a Link between Type 2 Diabetes and Cardiovascular Phenotype via NF-kB and Beta Catenin Interaction

T2D involves the disruption of glucose and lipid metabolic pathways. Glucose and ldl uptake is affected in adipose cells leading to high blood levels of ldl (low density lipoprotein) and glucose. Disrupted lipid metabolism induces the release of several inflammatory cytokines and up regulation of their receptors on the target cells.

Low density lipoprotein-related protein 1 (alpha-2-macroglobulin receptor), also known as LRP1 or CD91, is a protein forming receptor found in the plasma membrane of human cells involved in receptor-mediated endocytosis and cellular lipid homeostasis. LDL receptors sense high LDL and can activate LRP1. Under normal conditions insulin down regulates LRP1 level [10]. Impairment of this regulation inT2D increases LRP1 activation. LRP1 causes the SRC activation [11]. SRC is a protein tyrosine kinase. It phosphorylates its target molecules to regulate their activity. In T2D SRC integrates the signals induced by several pathophysiological conditions. It activates MAPK pathway and functions as a proto oncogene. High blood glucose induces high VEGFA levels [12]. VEGFA can activate SRC via VGFR as indicated in the sub

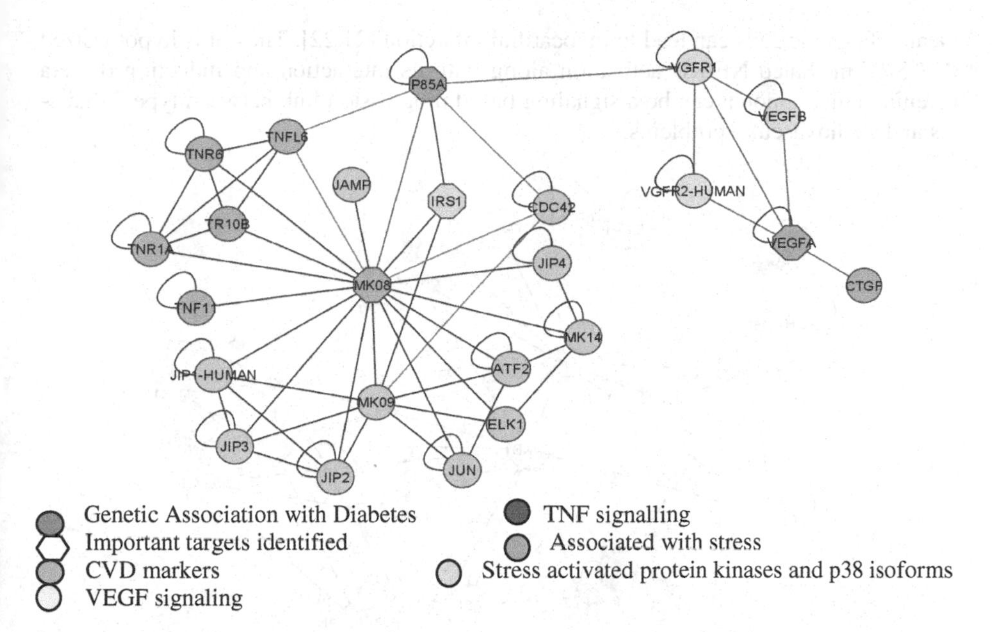

Fig. 1. MAPK8_P85A_CTGF pathway: The network explains the interaction between stress activated kinases with cardiovascular gene CTGF via intermediate genes such as P85A and VEGFA. MAPK8 acts as a hub for various stress stimuli and transmits signal through regulatory gene P85A and VEGF signaling to CTGF to cause fibrosis in cardiac.

network module (Figure 2) Additionally, cytokines such as TNF 11, TNFL6 and their receptors such asTNR6 are also capable of activating SRC [13].

Once activated, SRC can activate IkappaB kinase subunit alpha (IKKA) and IkappaB kinase subunit beta (IKKB) by phosphorylation [14]. Activated IKKA and IKKB in turn phosphorylate IRS1 [15] thus blocks insulin signaling and induces insulin resistance in cardiomyocyte. As a result it can disrupt energy metabolism and contraction in cardiomyocyte. Low glucose uptake can induce ROS production thereby induces apoptosis through caspase activation [16].As observed in the second sub network module; CASP3 and CASP8 along with p53 get activated. Caspases can further activate NF-KB through its interaction with IKKA and IKKB [17]. This further supports the high expression of beta catenin by NF-KB (Figure 2).

NF-KB is found to be bound to an inhibitory protein IkappaB (IkB) which sequesters it in cytoplasm. IKKA and IKKB can phosphorylate IkB and thus release NF-KB, increasing its cytoplasmic concentration and favoring nuclear localization [18]. In this manner SRC can be envisaged to induce NF-KB activation. SRC activation of NF-KB is well reported [19]. In the nucleus NF-KB regulates the expression of its target genes. Our network displays an interaction between NF-KB and beta catenin. Beta catenin gene CTNNB1 is highly expressed in our study. Therefore, we predict that NF-KB acts as a transcriptional activator and induces enhanced expression of beta catenin gene. Beta catenin has been reported to be involved in cardiovascular defects such as cardiac hypertrophy [20]. Beta catenin can further interact with other cardiovascular markers such as FHL2 and ACTN4, which can result in induction of apoptosis in cardiomyocyte.

Death of cardiac cells can lead to myocardial infarction [21,22]. Thus, it is hypothesized that SRC mediated NF-KB activation along with its interaction and induction of beta catenin in high amount can be a signaling based progression link between type 2 diabetes and cardiovascular problems.

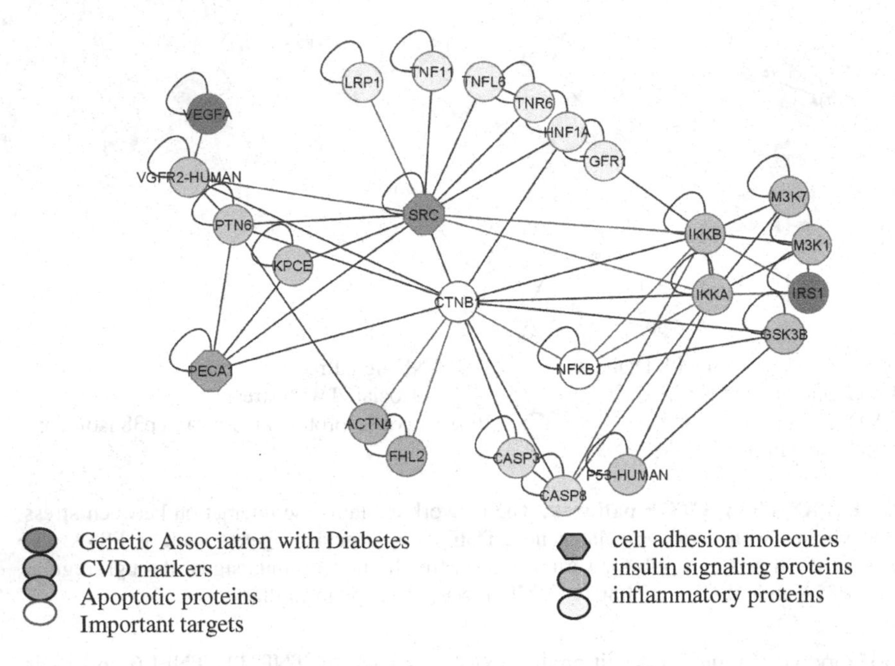

Fig. 2. SRC_NKB_CTNB1 pathway: SRC behaves like an important central signaling node in T2D and subsequently activates other key nodes. The activation of NF-kB following its interaction with CTNB1 is revealed here. High CTNB1 expression is linked to cardiovascular complications through ACTN4 and FHL2 markers.

3.3 Statistical Validation

Statistical analysis was carried out for main network, sub networks and the corresponding random networks, that were generated using Cytoscape plugin. Comparison of topological properties of the biological network with corresponding random network indicates scale free topology by displaying good power law distribution (Table 2 and Table 3). The strong disparity among the values for biological network Inflam_H and random network indicate that the associations are biologically significant and not random.

Clustering coefficients characterize the cohesiveness of the neighborhood of a node. The clustering coefficient for the Inflam_H network was approx 10 times higher than for the random network. R-squared values are used to quantify the fit to the power line. Average clustering coefficient indicating the presence of clusters and modules showed better R-squared values for our network in comparison to the random network (Table 2 and Table 3). Average number of neighbors and network centralization indicate the presence of well connected hubs in a network. The average number of neighbors is almost same for both random and biological networks and indicates good connectivity of the nodes. High network centralization values show that biological network presented here is dominated by one or few highly connected nodes. These nodes act as hubs during signaling process. Node degree distributions for the main network have R-squared value approaching 1, thereby obeying the power law and demonstrating scale free topology [23]. The parameter values for random vs biological networks are summarized in Table 2.

The statistical measure calculations for the two sub networks (named as MAPK8_P85A_CTGF and SRC_NFKB_CTNB1 respectively) are summarized in Table 3. Network heterogeneity signifies the tendency of the networks to form hubs. The biological network with a value 0.8 presents a higher tendency to form hubs than random .Higher R-squared value for average clustering coefficient in MAPK8_P85A_CTGF (0.927) indicate greater level of redundancy and cohesiveness than SRC_ NFKB_ CTNB1 (0.905). Networks SRC_NFKB_CTNB1 and MAPK8_P85A_CTGF have R-squared values of 0.862 and 0.822 respectively for topological coefficients determination. Closeness centrality measures how fast information spreads from a given node to other reachable nodes in the network. Higher range shows rapid signal transfer in the biological network. The closeness centrality distribution for both networks lies in the range 0.45-0.65. Betweeness centrality reflects the capability of a node to communicate between protein pairs. The R-squared values for betweeness centrality lies above 0.5 for the two networks.

Table 2. Comparison between Random and Biological Network

Parameters	Random Network	Inflam_H
Clustering coefficients	0.025	0.326
Avg. number of neighbours	11.612	10.783
network centralization	0.02	0.347
Avg. clustering coefficients	0.331	0.759
Node Degree (R sq value)	0.003	0.82

Table 3. Statistical analyses of sub two network modules

Networks analyzed	**R-squared values**				**Network heterogeneity**	
	Avg. clustering coefficients	**Toplogical coeffcients**	**Closeness centrality**	**Betweeness centrality**	**Random network**	**Biological sub-networks**
MAPK8_P85A_CTGF	0.927	0.822	0.458	0.552	0.43	0.78
SRC_NFKB_CTNB1	0.905	0.862	0.645	0.553	0.33	0.80

A strong power law distribution is indicative of scale free hierarchical architecture. With this report we can conclude that our networks are statistically valid and can be further endorsed by conducting molecular experiments on the same lines. The networks predicted can serve as basis for predicting newer targeted therapies against diabetic and cardiovascular syndromes.

Acknowledgements

We thank Prof. Yathindra, Director, Institute of Bioinformatics and Applied Biotechnology for continuous support during the course of this work. We also acknowledge Department of Information Technology, COE New Delhi, India for providing financial support.

References

1. Mooradian, A.D.: Dyslipidemia in type 2 diabetes mellitus. Nat. Clin. Pract Endocrinol. Metab. 5(3), 150–159 (2009)
2. Wittchen, F., Suckau, L., Witt, H., Skurk, C.: Genomic expression profiling of human inflammatory cardiomyopathy (DCMi) suggests novel therapeutic targets. J. Mol. Med. 85(3), 257–271 (2007)
3. Kyriakis, J.M., Avruch, J.: Protein kinase cascades activated by stress and inflammatory cytokines. Bioessays 18(7), 567–557 (1996)
4. Dunand-Sauthier, I., Walker, C.A., Narasimhan, J., Pearce, A.K., Wek, R.C., Humphrey, T.C.: Stress-activated protein kinase pathway functions to support protein synthesis and translational adaptation in response to environmental stress in fission yeast. Eukaryot Cell 4(11), 1785–1793 (2005)
5. Mauvais-Jarvis, F., Ueki, K., Fruman, D.A., Hirshman, M.F., Sakamoto, K., Goodyear, L.J., Iannacone, M., Accili, D., Cantley, L.C., Kahn, C.R.: Reduced expression of the murine p85alpha subunit of phosphoinositide 3-kinase improves insulin signaling and ameliorates diabetes. J. Clin. Invest. 109(1), 141–149 (2002)
6. Grimble, R.F.: Inflammatory status and insulin resistance. Curr. Opin. Clin. Nutr. Metab. Care 5(5), 551–559 (2002)
7. Jelkmann, W.: Pitfalls in the measurement of circulating vascular endothelial growth factor. Clin. Chem. 47(4), 617–623 (2001)
8. Shi-Wen, X., Leask, A., Abraham, D.: Regulation and function of connective tissue growth factor/CCN2 in tissue repair, scarring and fibrosis. Cytokine Growth Factor Rev. 19(2), 133–144 (2008)
9. Suzuma, K., Naruse, K., Suzuma, I., Takahara, N., Ueki, K., Aiello, L.P., King, G.L.: Vascular endothelial growth factor induces expression of connective tissue growth factor via KDR, Flt1, and phosphatidylinositol 3-kinase-akt-dependent pathways in retinal vascular cells. J. Biol. Chem. 275(52), 40725–40731 (2000)
10. Ceschin, D.G., Sánchez, M.C., Chiabrando, G.A.: Insulin induces the low density lipoprotein receptor-related protein 1 (LRP1) degradation by the proteasomal system in J774 macrophage-derived cells. J. Cell. Biochem. 106(3), 372–380 (2009)
11. Loukinova, E., Ranganathan, S., Kuznetsov, S., Gorlatova, N., Migliorini, M.M., Loukinov, D., Ulery, P.G., Mikhailenko, I., Lawrence, D.A., Strickland, D.K.: Platelet-derived Growth Factor (PDGF)-induced Tyrosine Phosphorylation of the Low Density Lipoprotein Receptor-related Protein (LRP). J. Biol. Chem. 277(18), 15499–15506 (2002)
12. Benjamin, L.E.: Glucose, VEGF-A, and Diabetic Complications. AJP 158(4), 1181–1184 (2001)

13. Wenzel, J., Sanzenbacher, R., Ghadimi, M., Lewitzky, M., Zhou, Q., Kaplan, D.R., Kabelitz, D., Feller, S.M., Janssen, O.: Multiple interactions of the cytosolic polyproline region of the CD95 ligand: hints for the reverse signal transduction capacity of a death factor. FEBS Lett. 509(2), 255–262 (2001)
14. Huang, W.C., Chen, J.J., Inoue, H., Chen, C.C.: Tyrosine phosphorylation of I-kappa B kinase alpha/beta by protein kinase C-dependent c-Src activation is involved in TNF-alpha-induced cyclooxygenase-2 expression. J. Immunol. 170(9), 4767–4775 (2003)
15. Gao, Z., Hwang, D., Bataille, F., Lefevre, M., York, D., Quon, M.J., Ye, J.: Serine Phosphorylation of Insulin Receptor Substrate 1 by Inhibitor _B Kinase Complex. J. Biol. Chem. 277(50), 48115–48121 (2002)
16. Herrera, B., Álvarez, A.M., Sánchez, A., Fernández, M., Roncero, C., Benito, M., Fabregat, I.: Reactive oxygen species (ROS) mediates the mitochondrial-dependent apoptosis induced by transforming growth factor b in fetal hepatocytes. The FASEB Journal 15, 741–751 (2001)
17. Chaudhary, P.M., Eby, M.T., Jasmin, A., Kumar, A., Liu, L., Hood, L.: Activation of the NF-kB pathway by Caspase 8 and its homologs. Oncogene 19(39), 451–460 (2000)
18. Häcker, H., Karin, M.: Regulation and function of IKK and IKK-related kinases. Sci. STKE. 357: re13 (2006)
19. Jalal, D.I., Kone, B.C.: Src Activation of NF-kB Augments IL-1beta-Induced Nitric Oxide Production in Mesangial Cells. J. Am. Soc. Nephrol. 17, 99–106 (2006)
20. Chen, X., Shevtsov, S.P., Hsich, E., Cui, L., Haq, S., Aronovitz, M., Kerkelä, R., Molkentin, J.D., Liao, R., Salomon, R.N., Patten, R., Force, T.: The beta-Catenin/T-Cell Factor/Lymphocyte Enhancer Factor Signaling Pathway Is Required for Normal and Stress-Induced Cardiac Hypertrophy. Mol. Cell. Biol. 26(12), 4462–4473 (2006)
21. Sun, J., Yan, G., Ren, A., You, B., Liao, J.K.: FHL2/SLIM3 Decreases Cardiomyocyte Survival by Inhibitory Interaction With Sphingosine Kinase-1. Circ. Res. 99(5), 468–476 (2006)
22. Triplett, J.W., Pavalko, F.M.: Disruption of alpha-actinin-integrin interactions at focal adhesions renders osteoblasts susceptible to apoptosis. Am. J. Physiol Cell Physiol. 291(5), 909–921 (2006)
23. Albert, R.: Scale-free networks in biology. J. Cell Sci. 118(21), 4947–4957 (2005)

Study of Drug-Nucleic Acid Interactions: 9-amino-[N-2-(4-morpholinyl)ethyl]acridine-4-carboxamide

Rajeshwer Shukla and Sugriva Nath Tiwari

Department of Physics,
D.D.U. Gorakhpur University, Gorakhpur-274 009, India
rajshukla_biop@rediffmail.com, sntiwari123@rediffmail.com

Abstract. 9-amino-[N-2-(4-morpholinyl)ethyl]acridine-4-carboxamide (9AMC) elicits its antitumour activity through intercalative binding with the genetic material, DNA molecule. The binding of this acridine molecule with DNA fragments has been examined using quantum mechanical methods. Second order perturbation theory valid for medium range interactions has been used to obtain binding sites of the acridine drug. Relative stability of various acridine-base pair complexes and preferred molecular associations have been discussed.

Keywords: DNA, Acridine, CNDO/2 Method, Molecular Interactions and Computer Simulation.

1 Introduction

Many of the anticancer drugs employed clinically exert their antitumour effect by inhibiting nucleic acids or protein synthesis. DNA is a well-characterized intracellular target but its large size and sequential nature makes it an elusive target for selective drug action. Binding of low molecular weight ligands to DNA causes a wide variety of potential biological responses. The biological activity of certain low molecular weight antitumour compounds appears to be related to their mode and specificity of interaction with particular DNA sequences. Such small molecules are of considerable interest in chemistry, biology and medicine [1-4]. 9-amino-[N-2-(4-morpholinylethyl] acridine-4-carboxamide (9AMC) is an antibiotic drug with potent antitumour, antimicrobial, amebicidal and chemosterilant activities. It is DNA intercalating agent that forms ternary complexes with mammalian topoisomerases and poison their cleavage and rejoining activities [5,6]. It preferentially binds to GC-rich nucleotide sequences [7,8]. There are tight correlations between ligand structure, cytotoxicity and DNA-binding kinetics for the 9-aminoacridine-4-carboxamide class of compounds [9,10]. The molecule of 9AMC binds to six base pairs of calf thymus DNA, and the involvement of at least three groups on the alkaloid in binding to DNA has been established. 9AMC requires native helical DNA, and, possibly, the presence of all four bases to show the activity of DNA. The structure of the intercalated complex enables a rationalization of the known structure-activity relationships for the inhibition of topoisomerase II activity, cytotoxicity, and DNA-binding kinetics for 9-aminoacridine-4-carboxamides [5-10].

S. Ranka et al. (Eds.): IC3 2009, CCIS 40, pp. 454–460, 2009.
© Springer-Verlag Berlin Heidelberg 2009

In continuation of our earlier studies on interaction of acridine drugs with DNA fragments [11], the present paper deals with the binding mechanism of 9AMC with DNA base pairs namely, G-C and A-T using quantum mechanical methods.

2 Method of Calculation

The molecular geometry of 9-amino-[N-2-(4-morpholinyl)ethyl]acridine-4-carboxamide has been constructed using the crystallographic data from literature and standard values of bond lengths and bond angles [12]. Net atomic charge and corresponding dipole moment components at each of the atomic centres of the molecule have been computed by CNDO/2 method [13]. Modified Rayleigh-Schrodinger second order perturbation theory along with multicentred-multipole expansion technique as developed by Claverie and coworkers has been used to calculate interaction energy between drug molecule and DNA base pairs. According to the energy decomposition obtained by perturbation treatment, the total interaction energy (E_{TOT}) between two molecules is expressed as [14]:

$$E_{TOT} = E_{EL} + E_{POL} + E_{DISP} + E_{REP}$$

where E_{EL}, E_{POL}, E_{DISP} and E_{REP} represent electrostatic, polarization, dispersion and repulsion energy components respectively. The calculation of electrostatic energy has been restricted only up to first three terms namely monopole-monopole, monopole-dipole and dipole-dipole interaction energy [15]. During energy minimization, base pairs are kept fixed throughout the process while both lateral and angular variations are introduced in the acridine molecule in all respects relative to the fixed one and vice versa. Accuracies up to 0.1Å in sliding (translation) and 1^0 in rotation have been achieved. The details of the mathematical formalism and optimization process may be found in literature [11,14,16].

3 Results and Discussion

The molecular geometry of 9-amino-[N-2-(4-morpholinyl)ethyl]acridine-4-carboxamide (9AMC) has been shown in Fig.1. Net atomic charge and dipole moment components corresponding to each atomic center of the molecule are given in Table1.

The variation of stacking energy with respect to change of relative orientation between drug molecule and base pairs, has been shown in Fig. 2(a). Here, the interplanar separation corresponds to 3.1Å in each case. As evident from Fig. 2(a), two minima are exhibited by each energy curve. The energy curve for GC-9AMC complex shows one minima at 160^0 with energy –20.90 kcal/mole and the other at 330^0 with energy –13.45 kcal/mole. The energy curve for AT-9AMC complex also shows one minima at 50^0 with energy –13.53 kcal/mole and the other at 160^0 with energy –19.45 kcal/mole. Similarly, the energy curve for 9AMC-GC complex exhibits two minima, one at 30^0 with energy –13.42 kcal/mole and the other at 200^0 having energy –20.90 kcal/mole. The energy curve for 9AMC-AT complex also shows one minima at 200^0 with energy –19.44 kcal/mole and the other at 310^0 with energy –13.46 kcal/mole. Obviously, In case of GC-9AMC or 9AMC-GC complexes energy curves show an energy difference of nearly 7.5 kcal/mole between the two minima positions while in case of AT base pairs the energy difference between the two minima is reduced to approximately 6.0 kcal/mole.

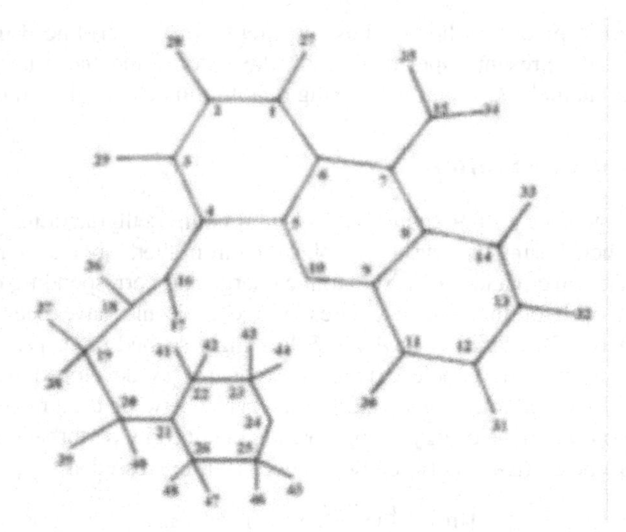

Fig. 1. Molecular geometry of 9-amino-[N-2-(4-morpholinyl)ethyl]acridine-4-carboxamide with various atomic index numbers

It indicates that 9AMC shows strong orientational specificity of stacking interactions in case of both GC and AT base pairs. The minima for GC-9AMC and AT-9AMC correspond to the same orientation and similar is the situation with 9ABC-GC and 9ABC-AT complexes. The minima located by GC-9AMC and 9AMC-GC curves are having more energy as compared to those noticed in case of AT-9AMC and 9AMC-AT energy curves though the energy difference is less than 1.0 kcal/mole.

Similar to that noticed for 9-amino-[N-(2-dimethylamino)ethyl]acridine-4-carboxamide [11], 9AMC also shows strong orientational specificity of stacking interactions in case of both GC and /or AT base pairs. The minima for GC-9AMC and AT-9AMC correspond to the same orientation and similar is the situation with 9AMC-GC and 9AMC-AT complexes. The minima located by GC-9AMC and 9AMC-GC curves are having more energy as compared to those noticed in case of AT-9AMC and 9AMC-AT energy curves.

The variation of stacking energy with interplanar distance between the drug (9AMC) and base pair is shown in Fig. 2(b), which indicates that complexes with GC and AT base pairs are stabilized at 3.1Å. The minimum energy configurations, thus obtained, are depicted in Fig.3 which clearly shows that acridine chromophore of the drug (9AMC) is stacked nearly perpendicularly through the hydrogen bonded regions of base pairs and partially over the purine (guanine and adenine) base of the base pairs. Since the acridine chromophore of the drug molecule possesses functional groups such as amino and carboxamide groups, the stacking patterns further indicate the possibility of formation of hydrogen/covalent bonds between the intercalated drug molecule and the backbone and /or the nucleotide bases of nucleic acid helices. The stacking energy values corresponding to various stacked complexes are shown in Table 2, which implies that the drug (9AMC) like 9-aminoacridine, prefers to bind and intercalate into a dinucleotide unit containing guanine and cytosine bases [11,12,17]. Table 2 indicates the following order of the stability of the stacked complexes:

$$\text{I} \geq \text{III} > \text{II} \geq \text{IV}$$

Table 1. Molecular charge distribution of the 9-amino-[N-2-(4-morpholinyl)ethyl]acridine-4 carboxamide molecule

Atom No.	Atom Symbol	Charge (e.u.)	Atomic dipole components (debye)		
			X	Y	Z
1	C	0.047	0.035	0.074	-0.013
2	C	-0.034	-0.095	0.133	0.014
3	C	0.052	-0.069	-0.036	0.003
4	C	-0.100	0.023	-0.172	0.037
5	C	0.161	-0.015	-0.044	-0.174
6	C	-0.084	-0.128	-0.099	0.132
7	C	0.202	-0.130	-0.230	-0.056
8	C	-0.082	-0.015	-0.144	0.024
9	C	0.158	-0.008	-0.053	0.001
10	N	-0.241	-0.937	-1.551	0.148
11	C	-0.053	-0.123	-0.111	-0.023
12	C	0.038	0.026	-0.106	0.015
13	C	-0.025	0.134	-0.017	-0.052
14	C	0.031	0.055	0.073	0.035
15	N	-0.223	0.058	0.102	0.029
16	C	0.333	0.161	0.061	-0.047
17	O	-0.322	0.619	-1.201	-0.168
18	N	-0.202	-0.478	0.817	-1.078
19	C	0.084	-0.155	-0.101	-0.062
20	C	0.071	0.073	-0.101	-0.079
21	N	-0.159	0.729	1.246	-0.068
22	C	0.073	0.119	-0.085	0.045
23	C	0.135	0.052	-0.200	0.099
24	O	-0.228	0.398	0.930	0.923
25	C	0.143	-0.155	0.147	-0.046
26	C	0.072	-0.095	-0.016	-0.060
27	H	-0.021	0.000	0.000	0.000
28	H	-0.008	0.000	0.000	0.000
29	H	-0.018	0.000	0.000	0.000
30	H	0.006	0.000	0.000	0.000
31	H	-0.009	0.000	0.000	0.000
32	H	-0.010	0.000	0.000	0.000
33	H	-0.019	0.000	0.000	0.000
34	H	0.122	0.000	0.000	0.000
35	H	0.124	0.000	0.000	0.000
36	H	0.100	0.000	0.000	0.000
37	H	-0.009	0.000	0.000	0.000
38	H	0.003	0.000	0.000	0.000
39	H	0.008	0.000	0.000	0.000
40	H	-0.020	0.000	0.000	0.000
41	H	-0.007	0.000	0.000	0.000
42	H	-0.013	0.000	0.000	0.000
43	H	-0.007	0.000	0.000	0.000
44	H	-0.018	0.000	0.000	0.000
45	H	-0.011	0.000	0.000	0.000
46	H	-0.016	0.000	0.000	0.000
47	H	-0.009	0.000	0.000	0.000
48	H	-0.017	0.000	0.000	0.000

(Total energy = -242.68 a.u., Binding energy = -24.85 a.u.,Total dipole moment = 2.41 debyes).

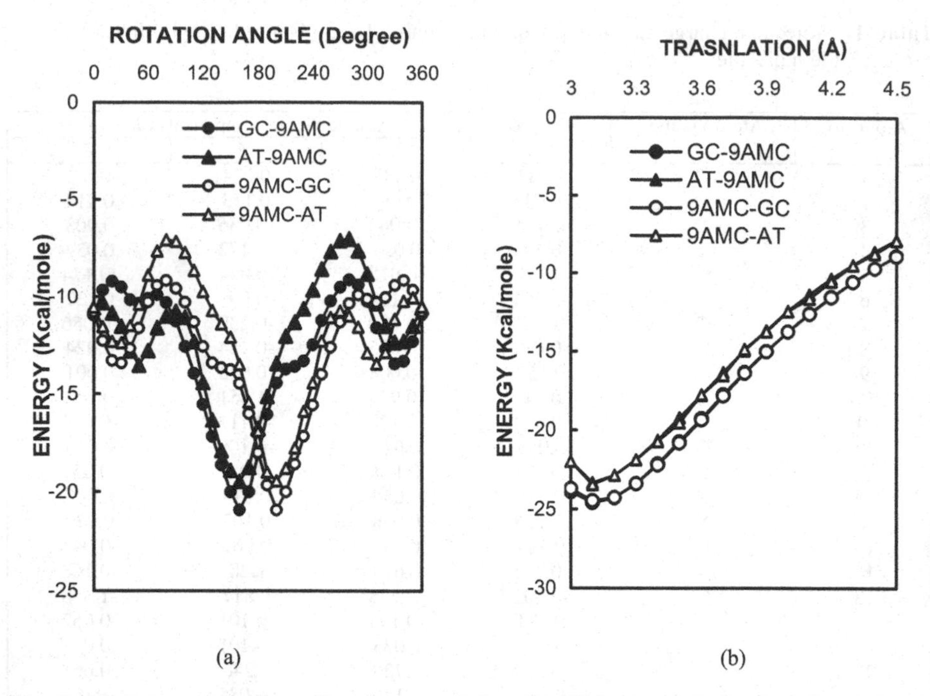

Fig. 2. Variation of total stacking energy of 9-amino-[N-2-(4-morpholinyl)ethyl]acridine-4-carboxamide with various base pairs as a function of (a) angular rotation and (b) interplanar separation

Table 2. Stacking energy of various complexes formed between 9-amino-[N-2-(4-morpholinyl) ethyl]acridine-4-carboxamide and DNA base-pairs

Energy terms (Kcal/mole)	**Stacked Complexes**			
	GC-9AMC (I) (3.1)*	**AT-9AMC (II)** (3.1)*	**9AMC-GC(III)** (3.1)*	**9AMC-AT (IV)** (3.1)*
E_{QQ}	-1.95	-1.41	-1.98	-1.39
E_{QMI}	-3.85	-3.55	-3.95	-3.45
E_{MIMI}	-1.62	-1.86	-1.45	-1.87
E_{EL}	-7.42	-6.82	-7.38	-6.71
E_{POL}	-3.33	-2.90	-3.40	-2.84
E_{DISP}	-28.91	-27.31	-28.72	-27.36
E_{REP}	13.74	12.99	13.68	12.95
E_{TOT}	**-25.92**	**-24.04**	**-25.82**	**-23.96**

*drug-base pair inter-planar separation (Å)

It is apparent that contribution of dispersion energy component plays a dominant role in stabilizing all the complexes. Electrostatic energy has larger contribution (-7.42 k cal/mole) in case of complexes formed with GC base pair (Table 2), which implies that the drug (9AMC) like other acridine drugs prefers to bind/ intercalate into a dinucleotide unit containing guanine (G) and cytosine (C) bases.

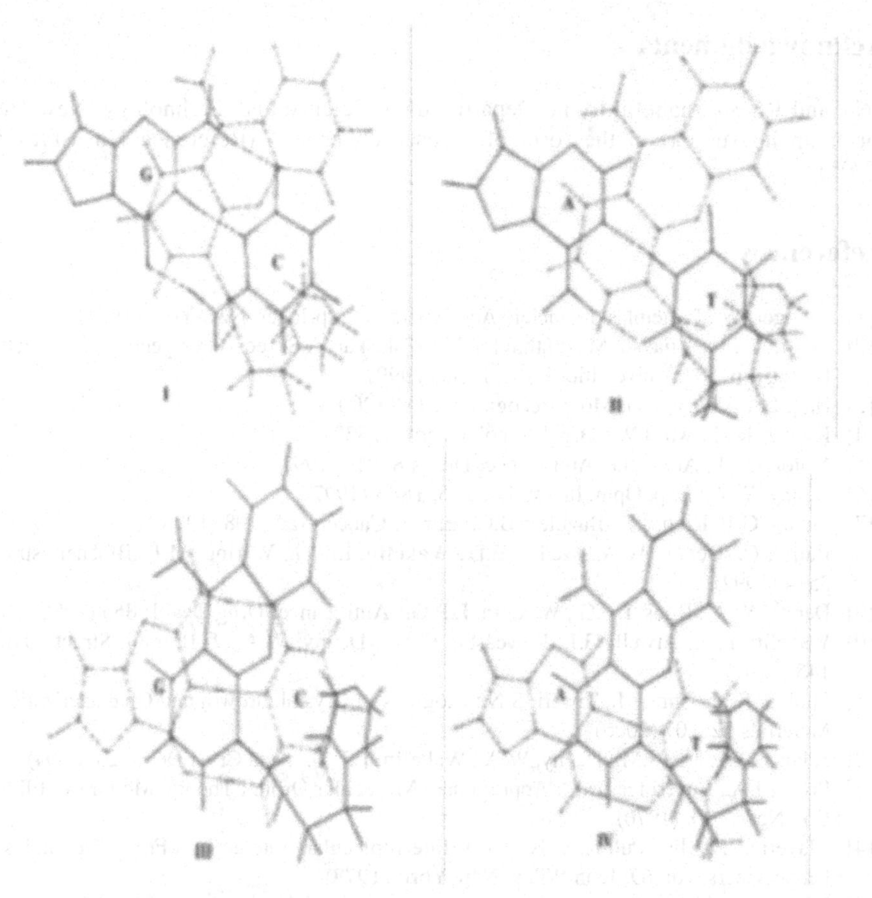

Fig. 3. Stacked minimum energy configurations of 9AMC with DNA base pairs. The geometry shown by dotted lines represents the upper molecule in each case.

4 Conclusion

The present study reveals that binding of 9-amino-[N-2-(4-morpholinyl)ethyl]acridine-4-carboxamide (9AMC) to GC rich region of DNA helices is more preferred and the mode of binding is similar to that of other intercalating acridines. The largest stability contribution is derived from dispersion forces irrespective of the base pairs involved. Also as observed from the stacking patterns, there exists a possibility of the bond formation/ interaction between the functional groups associated with the chromophore of the drug molecule and the backbone of the nucleic acid helices. Therefore, intercalative binding may be held responsible for the pharmacological properties of the drug. Also, it seems probable that 9-amino-[N-2-(4-morpholinyl)ethyl]acridine-4-carboxamide intercalates from the major groove side with their carboxamide groups in the plane of the chromophore and dimethylammonium groups interacting with the O6/N7 atoms of guanine.

Acknowledgments

SNT and RS are thankful to the Department of Science and Technology, New Delhi for financial support in the form of a research project (Reference No. SP/SO/D-12/95).

References

[1] Saenger, W.: Principles of Nucleic Acid Structure. Springer, New York (1984)
[2] Maiti, M.: In: Vijayan, M., Yathindra, N., Kolaskar, A.S. (eds.) Perspectives in Structural Biology, p. 583. Universities Press, India (1999)
[3] Haq, I., Ladbury, J.: J. Mol.: Recognit 13, 188 (2000)
[4] Kumar, R., Lown, J.W.: Org. Biomol. Chem. 1, 3327 (2003)
[5] Malonne, H., Atassi, G.: Anti-Cancer Drugs 8, 811 (1997)
[6] Denny, W.A.: Exp. Opin. Invest. Drugs 6, 1845 (1997)
[7] Finaly, G.J., Riou, J.F., Baguley, B.C.: Eur. J. Cancer 32A, 708 (1996)
[8] Bailly, C., Denny, W.A., Mellor, E.L., Wakelin, L.P.G., Waring, M.J.: Biochemistry 31, 3514 (1992)
[9] Denny, W.A., Roos, I.A.G., Wakelin, L.P.G.: Anti-Cancer Drug Des. 1, 855 (1986)
[10] Wakelin, P.G., Atwell, G.J., Rewcastle, G.W., Denny, W.A.: J. Biomol. Struct. Dyn. 5, 145 (1987)
[11] Shukla, R., Mishra, M., Tiwari, S.N.: Progress in Crystal Growth and Characterization of Materials 52, 107 (2006)
[12] Adams, A., Guss, J.M., Denny, W.A., Wakelin, L.P.G.: Acta Cryst D60, 823 (2004)
[13] Pople, J.A., Beveridge, D.L.: Approximate Molecular Orbital Theory. Mc-Graw Hill Pub. Co., New York (1970)
[14] Claverie, P.: In: Pullman, B. (ed.) Intermolecular Interactions-From Diatomics to Biopolymers, vol. 69, John Wiley, New York (1978)
[15] Rein, R.: In: Pullman, B. (ed.) Intermolecular Interactions-From Diatomics to Biopolymers, p. 307. John Wiley, New York (1978)
[16] Tiwari, S.N., Mishra, M., Sanyal, N.K.: Indian J. Phys. 76B, 11 (2002)
[17] Pritchard, N.J., Blake, A., Peacocke, A.R.: Nature 212, 1360 (1966)

IDChase: Mitigating Identifier Migration Trap in Biological Databases*

Anupam Bhattacharjee[1], Aminul Islam[1],
Hasan Jamil[1], and Derek Wildman[2]

[1] Integration Informatics Laboratory, Department of Computer Science
[2] Center for Molecular Medicine and Genetics
Wayne State University, USA
{anupam,aminul}@wayne.edu, jamil@cs.wayne.edu, dwildman@med.wayne.edu
http://integra.cs.wayne.edu

Abstract. A convenient mechanism to refer to large biological objects such as sequences, structures and networks is the use of identifiers or handles, commonly called IDs. IDs function as a unique place holder in an application for objects too large to be of immediate use in a table which is retrieved from a secondary archive when needed. Usually, applications use IDs of objects managed by remote databases that the applications do not have any control over such as GenBank, EMBL and UCSC. Unfortunately, IDs are generally not unique and frequently change as the objects they refer to change. Consequently, applications built using such IDs need to adapt by monitoring possible ID migration occurring in databases they do not control, or risk producing inconsistent, or out of date results, or even face loss of functionality. In this paper, we develop a wrapper based approach to recognizing ID migration in secondary databases, mapping obsolete IDs to valid new IDs, and updating databases to restore their intended functionality. We present our technique in detail using an example involving NCBI RefSeq as primary, and OCPAT as secondary databases. Based on the proposed technique, we introduce a new wrapper like tool, called *IDChase*, to address the ID migration problem in biological databases and as a general platform.

1 Introduction

Databases and applications refer to biological objects and to each other using identification numbers or IDs for convenience. IDs are usually used to refer to large objects that are non-traditional data, such as DNA (GenBank [4], EMBL [2]) or protein sequences (SWISS-PROT [8]), protein structures (PDB [5]), metabolic pathways (KEGG [3]), gene expression (SGD [7]), and so on. Such objects are often large and complex, and users require interpretive tools to understand what they represent. The complexity involved in their intended use is also a factor in the way they are represented in the public databases, and

* Research supported in part by National Science Foundation grants CNS 0521454 and IIS 0612203.

S. Ranka et al. (Eds.): IC3 2009, CCIS 40, pp. 461–472, 2009.
© Springer-Verlag Berlin Heidelberg 2009

the approach to application design involving such objects. Moreover, different databases use their own unique identifiers to represent what otherwise would be an identical real world object. For example, the gene OAZ1 has the ID 4946 in GenBank while in EMBL it has the ID ENSG00000104904. Finally, the quest for sanitizing the collected data sets introduces another dimension of complexity because of process called curation that corrects the perceived error in the machine generated data by some form of manual consensus. Once an object is cured, it migrates to a more stable state in a separate collection (e.g., from GenBank to RefSeq database [6]) rendering the object before the curation and its ID obsolete which are often discontinued and removed from the databases such as GenBank. But before the final curation takes place, objects undergo several cycles of revisions. Each database that houses them uses different techniques to update IDs to reflect these revisions. Nonetheless, all these changes essentially make the IDs before the change unusable or obsolete.

Regardless of the reason why IDs migrate from one to the next, applications that are developed cannot adapt to these changes easily. This is partly due to the fact that source databases such as GenBank or UCSC do not provide any mechanism for applications to migrate to updated states, neither do they notify applications about the change simply because they do not follow any client-server type of relationship. Applications use public databases in an asynchronous manner and thus, assume the responsibility of staying current. Databases such as GenBank do provide some help in the form of publishing the changes on a periodic basis and maintaining the older versions of the data so that applications using them still remain functional. But applications now must first identify the changes, recognize the objects that changed, and update on their own. Usually such adaptations are manual, and hence, error prone and time consuming. Too long an adaptation period will significantly reduce usability of an application, or even lead to abandoning it altogether. So, solutions addressing the ID migration problem must be comprehensive, automatic and efficient, and preferably incremental so that a batch update is not the only option.

1.1 A Motivating Example

Two issues complicate application design and maintenance in biological databases – materialized views, and autonomous changes in the source databases. These issues have been plaguing biological databases for a long time, and little to no solutions have been found. Since most public data repositories are read only and autonomous, they follow a "use as you find and need" policy where users use the contents without any awareness of the repositories. Consequently, change management becomes the user's responsibility. One such application is OCPAT [14] in which codon-preserved alignment is computed to automatically generate multiple sequence alignments from the coding sequences of any list of human gene IDs and their putative orthologs from genomes of other vertebrates. The tool is designed to perform multiple sequence alignment in ways other tools cannot so that new age phylogenetic analysis can be performed. The result of the tool kit is a set of about 30,000 multiple sequence alignments using each of the human mRNAs found in RefSeq database. For each such mRNAs, an Ensembl entry is found that shows the gene tree of the mRNA. The Ensembl gene trees show the gene evolution in all the species of interest. The final results are collected in the form of a table as shown in figure 1 which includes the human RefSeqID from NCBI RefSeq database, the corresponding Ensembl gene tree IDs, and all the orthologs found, alongwith some other details.

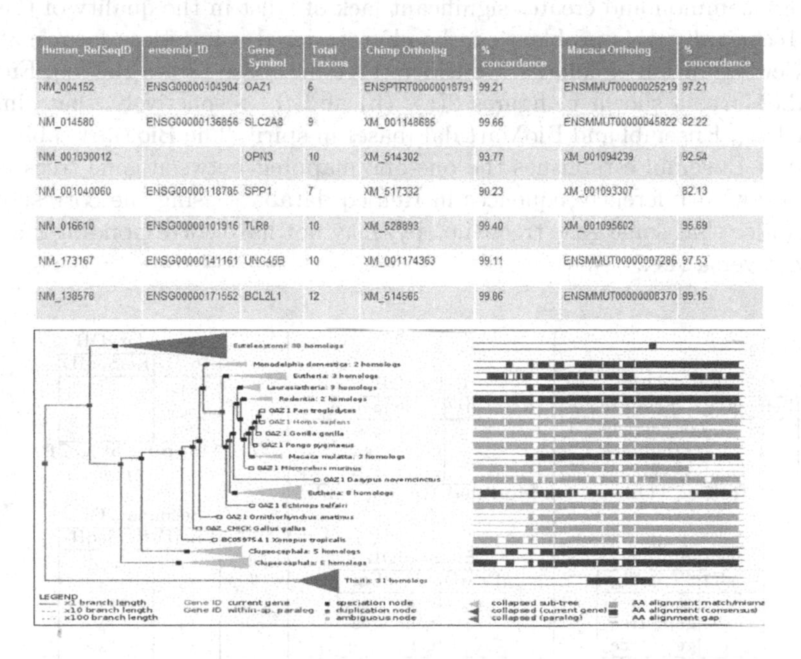

Human_RefSeqID	ensembl_ID	Gene Symbol	Total Taxons	Chimp Ortholog	% concordance	Macaca Ortholog	% concordance
NM_004152	ENSG00000104904	OAZ1	6	ENSPTRT00000018791	99.21	ENSMMUT00000025219	97.21
NM_014580	ENSG00000136856	SLC2A8	9	XM_001148685	99.66	ENSMMUT00000045822	82.22
NM_001030012		OPN3	10	XM_514302	93.77	XM_001094239	92.54
NM_001040060	ENSG00000118785	SPP1	7	XM_517332	90.23	XM_001093307	82.13
NM_016610	ENSG00000101916	TLR8	10	XM_528893	99.40	XM_001095602	95.69
NM_173167	ENSG00000141161	UNC45B	10	XM_001174363	99.11	ENSMMUT00000007286	97.53
NM_138578	ENSG00000171552	BCL2L1	12	XM_514565	99.86	ENSMMUT00000008370	99.15

Fig. 1. OCPAT alignment result and corresponding Gene Tree view for the first RefSeq

The table created by OCPAT follows a defined process discussed in [14]. Following this process, for each RefSeq mRNA, it produces the most likely gene tree from Ensembl database, alongwith all the orthologs using codon preserving multiple alignment. This table is then materialized at the OCPAT site. Users of this site use the table to study evolution of genes and proteins, and from the table they can immediately see how close the codon preserved homologs (i.e., the listed putative orthologs in selected species) are to the human mRNA. Using a secondary evolutionary analysis, they are able to see the evolution of these orthologs relative to each of the human mRNAs, and compare with the most likely Ensembl analysis listed. Although informative and useful, the materialization of the table invites the unwanted view materialization problems, because both Ensembl and RefSeq databases change often, making the RefSeq-Ensembl mapping in the table obsolete and necessitating a revision of the mapping. Fortunately, the newly introduced BioMart database [1] in Ensembl currently maintains the most up to date mappings of RefSeq IDs to Ensembl IDs.

Materialized Views and Database Inconsistency. Although the ID migration apparently does not cause loss of information or inconsistency in OCPAT today, inconsistency still exists in general. Just the way OCPAT materializes the mappings from Ensembl, other applications that do secondary materialization of the mappings from OCPAT, actually become inconsistent. In Life Sciences, such propagation of update inconsistency caused by materialized views is actually very common and creates significant lack of trust in the quality of the data [12]. Technically, this problem can be characterized using ideas from materialized view evolution as follows. Consider three database tables RefSeq, Ensembl and BioMart as shown in figures 2(a), (b) and (c) respectively, that simulate the RefSeq, Ensembl and BioMart databases in spirit. The BioMart table maintained at Ensembl establishes the one-one mapping between gene trees in this database with reference sequences in RefSeq database using the corresponding IDs. Notice that some gene trees (i.e., e_3) may not have corresponding mapping, and vice versa (i.e., r_3).

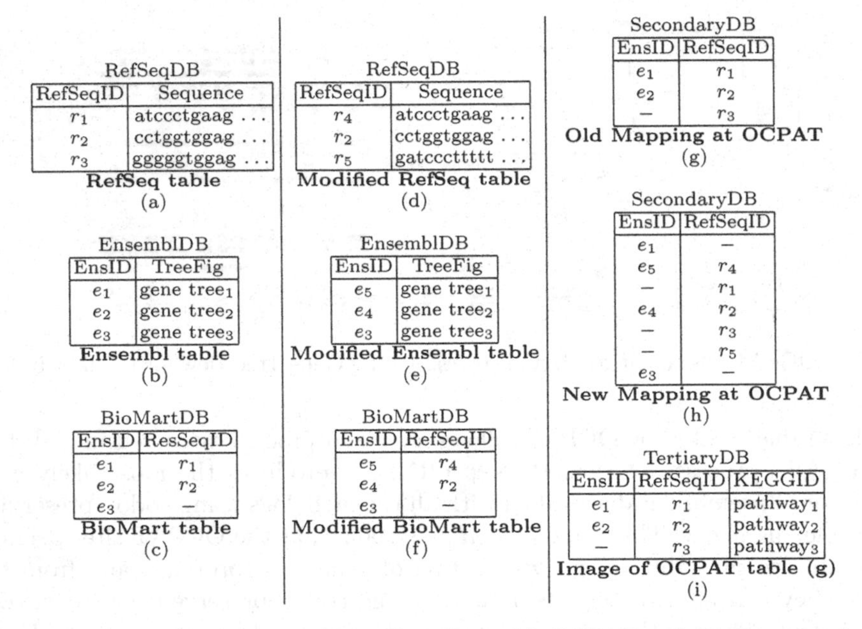

RefSeqDB

RefSeqID	Sequence
r_1	atccctgaag ...
r_2	cctggtggag ...
r_3	gggggtggag ...

RefSeq table
(a)

EnsemblDB

EnsID	TreeFig
e_1	gene tree$_1$
e_2	gene tree$_2$
e_3	gene tree$_3$

Ensembl table
(b)

BioMartDB

EnsID	ResSeqID
e_1	r_1
e_2	r_2
e_3	–

BioMart table
(c)

RefSeqDB

RefSeqID	Sequence
r_4	atccctgaag ...
r_2	cctggtggag ...
r_5	gatcccttttt ...

Modified RefSeq table
(d)

EnsemblDB

EnsID	TreeFig
e_5	gene tree$_1$
e_4	gene tree$_2$
e_3	gene tree$_3$

Modified Ensembl table
(e)

BioMartDB

EnsID	RefSeqID
e_5	r_4
e_4	r_2
e_3	–

Modified BioMart table
(f)

SecondaryDB

EnsID	RefSeqID
e_1	r_1
e_2	r_2
–	r_3

Old Mapping at OCPAT
(g)

SecondaryDB

EnsID	RefSeqID
e_1	–
e_5	r_4
–	r_1
e_4	r_2
–	r_3
–	r_5
e_3	–

New Mapping at OCPAT
(h)

TertiaryDB

EnsID	RefSeqID	KEGGID
e_1	r_1	pathway$_1$
e_2	r_2	pathway$_2$
–	r_3	pathway$_3$

Image of OCPAT table (g)
(i)

Fig. 2. ID migration scenarios at RefSeq, Ensembl, OCPAT and Tertiary databases (using simulated data)

Recall that the databases RefSeq and Ensembl may change any time independently. RefSeq may change IDs of existing sequences – as shown in the modified RefSeq table in figure 2(d), ID r_1 has been changed to r_4, sequence r_3 was completely removed, and r_5 was introduced. Similarly, in the Ensembl database (modified Ensembl table in figure 2(e)), ID e_1 was changed to e_5, and e_2 to e_4. These autonomous changes cause serious loss of information and currency problems in secondary databases such as OCPAT, or tertiary databases such as the one shown in figure 2(i) that was created from OCPAT database. Ensembl being the secondary database, changes in RefSeq trigger update needs in it to keep ID mapping in BioMart database current[1]. In BioMart, the focus is on maintaining Ensembl ID to RefSeq ID mapping, while in OCPAT, we are interested in maintaining RefSeq to Ensembl ID mappings, both one-to-one. It should be apparent that changes in any one of these databases affect OCPAT mappings, i.e., the changes in RefSeq and Ensembl cause more serious problems and synchronization needs in OCPAT[2].

In particular, in OCPAT, each RefSeq sequence is edited (2KB BP flanking region expansion plus deletion of introns) before a multiple sequence alignment is performed to find the orthologs. Hence, the RefSeq sequences in OCPAT are syntactically and semantically very different from their counterparts in RefSeq database. Consequently, from the standpoint of OCPAT, it is not readily apparent from tables 2(d) and (g), the conclusions that can be derived about the relationship between the IDs r_1, and r_4. In other words, it is not simple to conclude that r_1 has changed to r_4 without in depth analysis. This is partly because in OCPAT we do not locally store the original RefSeq sequences (we only store the edited versions), and even if it did, it cannot determine the equality without doing an alignment with all sequences in RefSeq. As a result, the changes in BioMart as shown in figure 2(e), and the RefSeq and Ensembl changes force OCPAT modifications as shown in table 2(g) to (h). It is not possible to guess that we should update e_1 and r_1, and hence are retained as unmatched mappings until a manual curation is conducted. This lag time in synchronization in OCPAT causes a ripple effect throughout the system making the entire system inconsistent as shown in table TertiaryDB in figure 2(i). This table was created from OCPAT in figure 2(g) to maintain a mapping between RefSeq IDs and Ensembl IDs to KEGG IDs. For TertiaryDB, it simply is not possible to maintain the mappings without any knowledge of how RefSeq-Ensembl mappings were established in OCPAT. Only option we have is to decide if the IDs have changed, and if so, to what. Clearly, until some form of system wide synchronization is carried out, all secondary and higher order databases risk being inconsistent. Hence, higher the frequency of update, greater the risk, and the more critical it becomes to find an automated solution that can adapt to the changes rapidly.

2 Chasing ID Migration

As IDs migrate, causing a system wide ripple effect, each application must find its own way of adjusting, because only the application designers are aware of the semantics and objectives. Consequently, the "one size fits all" type of solution may not be entirely applicable, though preferred. Based on this observation,

[1] The method Ensembl follows to update itself is not the subject of this article, and hence not the issue. We just assume that it has a sound process to tackle this issue.

[2] At least, BioMart and RefSeq administrators communicate the changes among themselves and in a way synchronize their databases against changes. But OCPAT does not have that freedom of collaboration with either one of them and hence must do whatever it finds suitable to stay current.

we hypothesize that a solution for OCPAT synchronization by chasing the ID migration history may serve as a model for other database synchronization solutions. Before we introduce our tool, let us examine possible opportunities and pitfalls in this section toward the development of a generalized solution. In the next few sections, we discuss two main technical issues – (i) detecting database inconsistency in materialized views, and (ii) autonomous reconciliation of ID inconsistency from the online resources – that we leverage to propose a solution to the ID migration problem.

2.1 Detecting Database Inconsistency in Materialized Views

While the ID inconsistencies are created in stages as databases copy information from other databases, the solution seemingly lies at the source of the change that started the ripple effect in the first place. Fortunately, these databases, such as RefSeq and Ensembl, have a notification mechanism that users can browse through to search for obsolete IDs to find out their status as well as their new IDs. Unfortunately though, these databases do not offer another table or repository to report the changes. Since these resources are geared for human consumption, the changes are published in human readable bulletin boards that users need to browse, read and digest. It is perhaps necessary to read these bulletins to understand the changes, why they were made and when. But oftentimes, it is just routine to find out the changes and adjust we have discussed. Figure 3 shows one such discontinued RefSeq ID notification. The figure clearly demonstrates that simple database querying will not reveal the fact that the ID NM_001030012.1 has been discontinued.

gi|71999139|ref|NM_001030012.1|

NM_001030012.1 was permanently suppressed because it is a nonsense-mediated mRNA decay (NMD) candidate

Fig. 3. An example of a discontinued RefSeq notification in NCBI

On the other hand, changes in Ensembl objects are reported in complicated reports as shown in figure 4, where plain and traditional text strings are used without any guarantee of finding the target responses in any specific format or location. In such cases, case by case solutions become essential. Examples in figures 3 and 4 highlight the fact that simple text search is not enough and if used could lead to incorrect computation. But a prudent text query could reveal the needed information.

It should be clear from the discussion above and the preceding examples that, as the size of the databases grow exponentially, changes are made more frequently. As the number of databases to be monitored and adjusted increases, manual adaptation is no longer feasible. It can be argued that an autonomous system must take over the role of an expert user capable of detecting changes, selecting IDs needing update and analyzing available information to ascertain possible ID migration, to alleviate this problem and to be efficient. However,

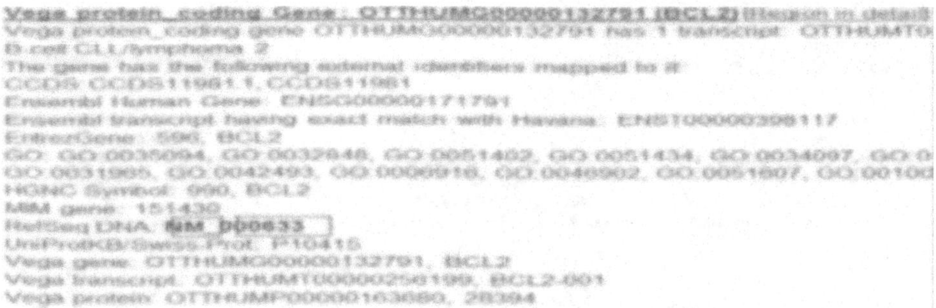

Fig. 4. A sample ID change bulletin at Ensembl database. This example reveals the fact that for the refseq ID NM_000633, BioMart does not have a mapping to an Ensembl ID because the ID has been changed. However, a text search for the RefSeq ID establishes an indirect mapping from NM_000633 to ENSG00000171791 via VEGA OTTHUMG00000132791.

an autonomous system then must follow a well structured approach, must know where to go to conduct an investigation, and be aware of how to find the needed information fully automatically.

2.2 Autonomous Online Reconciliation of ID Inconsistencies

Technically, there are two basic ways to determine the fate of a changed ID and find the new ID, if it exists. The first choice is to visit the source database, and search the bulletins and use text queries or text mining methods. The second option is to use the old sequence (if available) in the database to Blast against the source sequences to find a match. An example of a Blast search sequence in Ensembl database is shown in figure 5 where RefSeq sequence is being collected in Fasta format from GenBank and submitted in Emsembl to gather the set of Ensembl homologs. Here the highest scoring homolog of a RefSeq ID is taken as the corresponding ID in Ensembl. In the former case, the chances of finding the migration or update information are higher because the sources usually publish some form of bulletin board when a change is made, and these bulletins are usually comprehensive. The advantage of this approach is that the current database need not store the sequence and it only needs the ID to query the source.

On the other hand, the latter approach is more complicated and non-definitive, and should be used only if the first option fails. In this case, not only is the current database forced to store the original sequence, it will also need to store the modified sequence such as in the case of OCPAT, and incur higher storage cost. Even then, the outcome is not definitive. For example, if it uses Blast to find the most similar sequence, there can be several outcomes. If it finds a sequence with 100% match but with different ID, it can assume that ID in the source database is the new ID. But, it is possible to have two valid sequence fragments that are identical but with two distinct IDs making the decision process truly complicated. So, in the absence of a 100% match, should we assume that the ID has been discontinued? At what threshold?

Fig. 5. (a) DNA Sequence collected from NCBI for RefSeq mRNA NM_004152 in FASTA format, (b) Fetched DNA Sequence submitted for Ensembl Blast Search, (c) A dynamically assigned ticket to retrieve Blast result, (d) Acquired Gene Information from Ensembl, (e) A mapping warehouse for Ensembl ID - RefSeq ID Mapping, (f) Error message while retrieving a Blast result

Regardless of the merits of the two approaches, it is safe to assume that the individual applications will decide what suits their application the most, and design a process using one or the other, or both. Once a decision on the reconciliation method has been made, one needs to be aware of technological hurdles that might lie ahead since the idea is that an autonomous agent will conduct the reconciliation using the services that exist, for machines or for humans. Among many such difficulties, one that is of interest in the current context is that the Ensembl database interacts with users through a dynamic ticket for every Blast submission, and if a machine or autonomous agent like the ones we are advocating reads these tickets, Ensembl disables the tickets for security reasons (as shown in figure 5(f)) making it impractical to use autonomous agents to gather information from this database. In this case, even though the information possibly is available, we cannot access it automatically without full site cooperation.

3 ID Migration Management Using IDChase

We now present an ID migration management tool, called *IDChase*, as a possible reconciliation mechanism for change management of materialized IDs. We adopt a hybrid strategy in IDChase that explores change related bulletins (such as NCBI or Ensembl sites), and any structured information (such as BioMart) available at the source, and uses homology based techniques for the purpose of validation when the exploration of text query based approach returns ambiguous results.

As a prototype, IDChase is built as an interface where users are able to plug in a set of RefSeq IDs for which IDChase returns the corresponding Ensembl IDs, if they exists. The interface to the current IDChase system is shown in figure 6. The tool essentially functions as a wrapper that encapsulates a set of functionalities mimicking the process we outline in section 3.1. Since the current implementation was meant to serve as a proof of concept, we used a three tier database preference relation $\preceq_\tau$ involving databases NCBI, Ensembl, OCPAT, BioMart, HPRD, UniProt, and DDBJ as follows: {NCBI, Ensembl} $\preceq_\tau$ {OCPAT, BioMart} $\preceq_\tau$ {HPRD, UniProt, DDBJ}. Furthermore, this prototype also uses a user supplied similarity threshold ϵ as opposed to dynamically computing it from the rule base as discussed in section 3.1, and very specific and limited text search.

3.1 IDChase as a Mechanism for Change Management

The discussion above illustrates the fact that ID changes can be tackled and maintained in OCPAT and that a solution exists. Unfortunately, the solution is not robust and may break down with changes in the way the source databases report ID migrations. But this is to be expected – we cannot have full source autonomy and expect to be robust and expect the sources to behave the way we would like. A further observation is that these are not the only two sources that assign, manage and change IDs. So, to be a viable mechanism, we will need to factor in the fact that each site will deal with ID change differently, and that the applications using them will decide what is best. From these observations, it is clear that a dynamically adaptable system would be more suitable to address this problem.

We envision IDChase as a platform in which ideas from several technologies come together – such as continual querying, wrapper generation, ontology, text

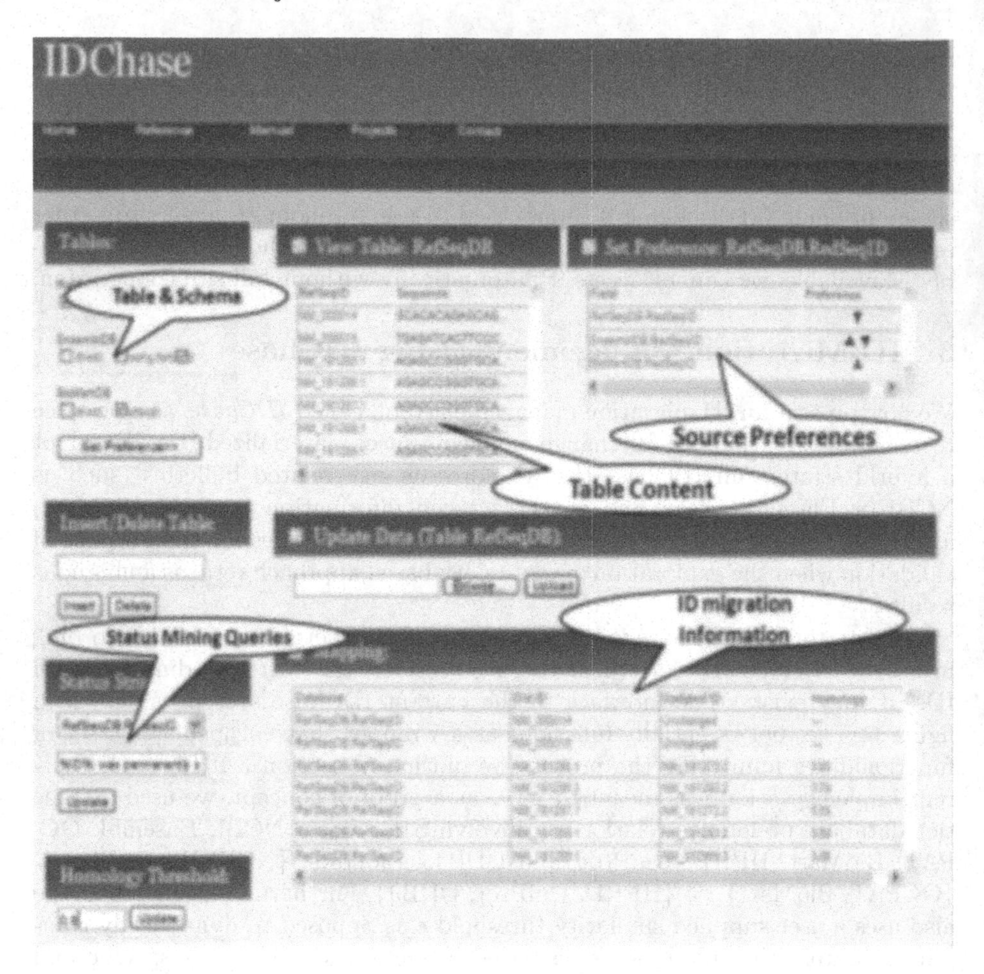

Fig. 6. IDChase interface showing RefSeq-Ensembl mapping

mining, and information retrieval to make it possible to track down ID changes and establish currency from online resources. The steps involved in the process are as follows:

1. Establish an ordering $\preceq_\tau$ among the sources of IDs of type τ, i.e., RefSeq ID.
2. Save ID change rules in an ontology Ω for IDs of type τ.
3. Test if a ID change took place. If true, then,
   (a) Compute the set of IDs Σ that are seemingly obsolete or their status unknown requiring update.
   (b) For each $\sigma \in \Sigma$, go to source U of σ and use text mining queries to extract the status, such that U is the most preferred source in the relation $\preceq_\tau$.
      i. If the status of σ is not definitive in U, then
         A. Establish similarity threshold ϵ for sequence σ using ID change rule in Ω.

B. Submit a homology search in U with sequence of σ and threshold ϵ.
C. If a unique sequence survives the filter condition, return the corresponding ID as a match and go to step 3b for next ID in Σ.
D. Else go to next site U in step 3b in the order $\preceq_\tau$ with the current σ. If no sites are left in the order, return failure and go to step 3b for next ID in Σ.

ii. Else return success, return ID as a match, and go to step 3b for next ID in Σ.

In the above process, preference is given to a text search in the preferred source. Only in the event of a non definitive conclusion, a homology search is carried out to find a one-one match to establish correspondence. If the sites in the preferred order fail to produce a match, an error is reported, otherwise we return the changed IDs as a set of mappings. The important question is how do we build a system that is capable of implementing the above algorithm? It turns out that all the necessary technological ingredients are already available. We briefly discuss the idea below.

In our Integration Informatics Laboratory, we are building a new data integration and workflow query language called BioFlow [13]. BioFlow uses wrapper generation technique FastWrap [9], ontology matching technique OntoMatch [10] and remote site integration technology called *remote user defined function* [11] that makes interacting with remote web sites possible and fairly trivial by treating the sites as remote functions in a relational database. BioFlow also allows integration of local and remote tools in relational database as part of the workflow queries. All the steps described in the above algorithm can be implemented fairly easily in BioFlow, once we can devise a way to implement the ID change rules, and a mechanism to compute the sequence similarity threshold. A number of tools are currently available to compute the text mining queries we have alluded to, and can be integrated in BioFlow as a local system function. We believe, the current IDChase toolkit can be quickly redesigned as a system that can be used as a general mechanism for ID maintenance for biological sequences. With proper fine tuning, it has the potential for a wide usage in Life Sciences databases where IDs are used and potentially changed, even for non sequence data.

4 Summary and Future Research

The primary objective of this paper was to show that although tackling ID migration is not a trivial issue, a collection of emerging technologies can be used in a prudent way to address this issue. It was also our goal to introduce the IDChase prototype we have built to demonstrate the feasibility of our approach as a test case for OCPAT database. Eventually, it is our goal to implement the ideas presented in section 3.1 as a platform for change management in biological databases due to ID migration. The proposed IDChase tool can be found at `http://integra.cs.wayne.edu:8080/IDChase`

Acknowledgement. The authors would like to acknowledge the many helpful discussions they have had with Munirul Islam, the co-developer of the OCPAT database [14], during the course of the development of IDChase.

References

1. BioMart, http://www.ensembl.org/biomart/martview
2. EMBL, http://www.ebi.ac.uk/embl/
3. KEGG, http://www.genome.ad.jp/kegg/
4. NCBI, http://www.ncbi.nlm.nih.gov/
5. PDB, http://www.pdb.org/
6. RefSeq, http://www.ncbi.nlm.nih.gov/RefSeq/
7. SGD, http://www.yeastgenome.org/
8. Swiss-Prot, http://www.ebi.ac.uk/swissprot/
9. Amin, M.S., Jamil, H.: FastWrap: An efficient wrapper for tabular data extraction from the web. In: IEEE IRI, Las Vegas, Nevada (2009)
10. Bhattacharjee, A., Jamil, H.: OntoMatch: A monotonically improving schema matching system for autonomous data integration. In: IEEE IRI, Las Vegas, Nevada (2009)
11. Chen, L., Jamil, H.M.: On using remote user defined functions as wrappers for biological database interoperability. IJCIS 12(2), 161–195 (2003)
12. Gupta, A., Mumick, I.S.: Maintenance of materialized views: Problems, techniques, and applications. IEEE Data Eng. Bull. 18(2), 3–18 (1995)
13. Jamil, H., El-Hajj-Diab, B.: BioFlow: A web-based declarative workflow language for Life Sciences. In: 2nd IEEE Workshop on Scientific Workflows, pp. 453–460. IEEE Computer Society Press, Los Alamitos (2008)
14. Liu, G., Uddin, M., Islam, M., Goodman, M., Grossman, L., Romero, R., Wildman, D.: OCPAT: an online codon-preserved alignment tool for evolutionary genomic analysis of protein coding sequences. Source Code for Biology and Medicine 2(1), 5 (2007)

Multi-domain Protein Family Classification Using Isomorphic Inter-property Relationships

Harpreet Singh[1], Pradeep Chowriappa[1], and Sumeet Dua[1,2]

[1] Data Mining Research Laboratory (DMRL), Department of Computer Science, Louisiana Tech University, Ruston, LA, U.S.A.
[2] School of Medicine, Louisiana State University Health Sciences Center, New Orleans, LA, U.S.A.
{hsi001,pradeep,sdua}@latech.edu

Abstract. Multi-domain proteins result from the duplication and combination of complex but limited number of domains. The ability to distinguish multi-domain homologs from unrelated pairs that share a domain is essential to genomic analysis. Heuristics based on sequence similarity and alignment coverage have been proposed to screen out domain insertions but have met with limited success. In this paper we propose a unique protein classification schema for multi-domain protein superfamilies. Segmented profiles of physico-chemical properties and amino acid composition are created for vector quantization based dimensionality reduction to create a feature profile for rule-discovery and classification. Association rules are mined to identify isomorphic relationships that govern the formation of domains between proteins to correctly predict homologous pairs and reject unrelated pairs, including those that share domains. Our results demonstrate that effective classification of conserved domain classes can be performed using these feature profiles, and the classifier is not susceptible to class imbalances frequently encountered in these databases.

Keywords: Multi-domain proteins, supervised classification, association rules.

1 Introduction

In the post-genomic era, it is increasingly important to analyze the protein sequences obtained from the genome sequencing projects for a variety of drug-design and translational and transformative bioinformatics research. Numerous sequence alignment algorithms have been developed for comparing protein sequences efficiently and reliably and to deduce the function or 3-D structure of polypeptide sequences through homology. However, a considerable research effort is still needed to determine whether two proteins sharing the same sequence homology have the same biological function. If two such proteins do share biological function, how much of such a similarity exists? The current methods yield a low, effective cut-off point for the effectual homology modeling of proteins with ~30% sequence identity, a lower bound at which the computed structure can still accurately depict the arrangement of secondary structure elements or folds in 3-D. Proteins of the same fold often have similar biological functions. However, one encounters many cases of high similarity both in fold and in

S. Ranka et al. (Eds.): IC3 2009, CCIS 40, pp. 473–484, 2009.
© Springer-Verlag Berlin Heidelberg 2009

function that are not reflected in sequence similarity. Such cases are often missed by current search methods, and several interesting investigative challenges remain open.

The degree of sequence conservation varies among protein families. However, homologous proteins almost always have the same fold. According to Pearson [1], homology is a transitive relation, meaning that if homology has been inferred between proteins A and B, and between proteins B and C, then proteins A and C must be homologous, even if they share no significant similarity. However, multi-domain sequences, especially those with promiscuous domains that occur in many contexts, are frequently excluded from genomic analyses due to the lack of a theoretical framework and practical methods for detecting multi-domain homologues [2]. The deduction of homology is particularly difficult. If Protein 1 contains Domains A and B, Protein 2 contains domains B and C, and Protein 3 contains domains C and D, then it is difficult to determine whether Proteins 1 and 3 should also be considered homologous. Multi-domain proteins evolve via both vertical descent (domain duplication) and domain insertions. Figure 1, depicts two proteins (2 and 3) that share a homologous domain through vertical descent. In contrast, Proteins 1 and 2 share a domain because of domain insertions where they share only sequence similarity, which are otherwise unrelated.

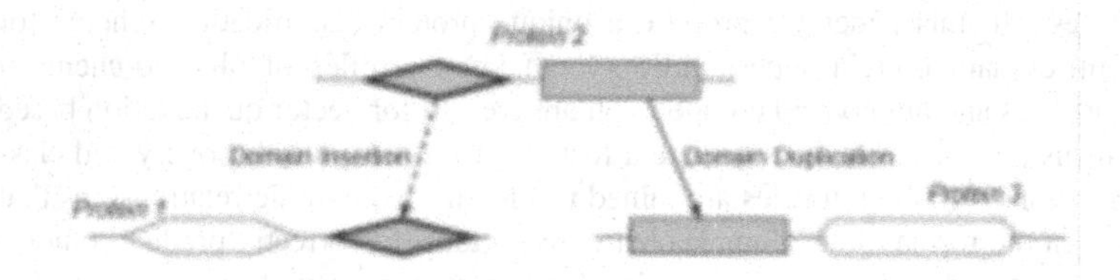

Fig. 1. Evolution of Multi-Domain Proteins

The ability to distinguish multi-domain homologs from unrelated pairs that share a domain is essential to genomic analysis. Heuristics based on sequence similarity and alignment coverage (the fraction of the mean sequence length covered by the optimal local alignment) have been proposed to screen out domain insertions [3]. Since traditional methods are heavily dependent on sequence similarity, the general effectiveness of these methods is questionable, especially in the case of weak sequence similarity, short alignments, and similar combination of shared and unique domains.

We propose a unique homology detection method to analyze multi-domain protein sequences using both physico-chemical properties and amino acid composition, which will identify isomorphic relationships between proteins of the same superfamily. The method employs the principles of association rules to identify isomorphic relationships that govern the formation of domains between proteins to correctly predict homologous pairs and reject unrelated pairs, including those that share domains [4].

1.1 Homology Using Domain Information

We define a protein domain as an independent, evolutionary unit that can form a single-domain protein or be part of one or more multi-domain proteins [5]. Several structural classification databases exist, of which SCOP (Structural Classification of Proteins) [6] and CATH (Class, Architecture, Topology and Homologous super-family) [7] are the most widely used. Both databases use protein domains as the classification unit and progressively group domains into hierarchical levels on the basis of sequence and structural similarity information.

Although these databases share some common features, they differ in the classification procedure and in domain definition. SCOP relies on a purely manual classification method using human expertise, whereas CATH uses a mixture of automated procedure and human intervention. The domain definition of CATH places a strong emphasis on structural compactness. In SCOP, domains are defined as independent evolutionary units that can either form a single-domain protein on their own or recombine with others to form part of a multi-domain protein.

The SCOP database has a four-level hierarchical classification scheme. Starting from the lowest level, all proteins of known structure (all those in the protein database (PDB)) are first split into their constituent domains. Then, protein domains that have clear sequence similarities (that is, greater than 30%) are clustered together into families. The next level of hierarchy is the super-family, which groups domains using structural and functional evidence to show descent from a common ancestor. The highest level of classification is the fold, which brings together domains that have the same major secondary structures with the same chain topology. Although the term protein super-family is widely used, its meaning is not well-defined when applied to multi-domain proteins and does not allow the database to be unambiguously partitioned into super-families.

In this study, we aim to classify multi-domain proteins into their SCOP super-families with different types of evolutionary histories: evolution via gene duplication or via domain insertion. The proposed method captures domain architectural characteristics based on the hydrophobic residue interactions and their corresponding structural characteristics.

1.2 Sequence and Structural Information

As was postulated by Rossmann [8], the polypeptide sequence of a globular protein is folded into two or more structural domains, which are spatially separated and are frequently based on different architectural principles. The conformation of such structural domains is, in general, conserved more than the amino acid sequences. The preservation of structure may originate in the 3-D requirements for the conservation of essential functions.

Using this description as our starting point, we infer that, with the exception of extremely small proteins that are held together by numerous disulfide bridges, each domain will possess a hydrophobic core. Therefore, a domain can be recognized as a series of shorter interactions which are separated by longer interactions from the rest of the molecule. The shorter interactions are dictated by the interactions between residues. Expressions of the hydrophobic effect are palpable in many facades of protein sequence-structure-function dependencies, including (1) the stabilization of the folded conformation of globular proteins in solutions, (2) the subsistence of amphipathic structures in peptides or of membrane proteins at lipid boundaries, and (3) protein-protein interactions associated with protein subunit assembly, protein-receptor binding, and other intermolecular bio-recognition processes [9].

In our proposed methodology, we create a means of integrating the hydrophobic residue information provided by Aboderin [10] along with the secondary structural information provided by the Chou and Fassman [11] method of secondary structure prediction. Our goal is to find isomorphic similarities between regions of proteins that exhibit conserved behavior in amino acid composition and physico-chemical characteristics. More details about the integration mechanism are included in the Feature Extraction part of the Methodology Section.

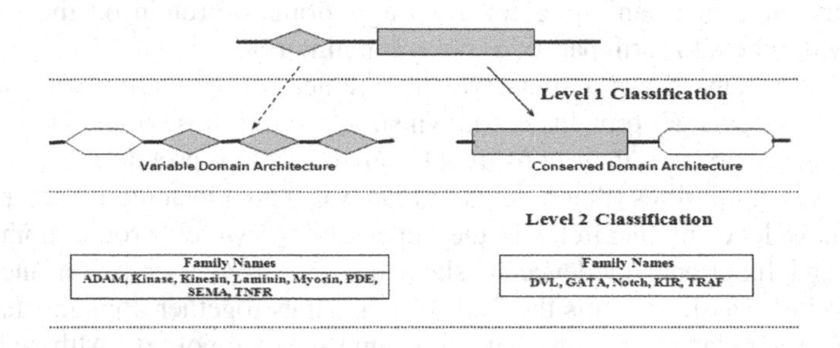

Fig. 2. Different Levels of Classification

1.3 Dataset

Our goal is to correctly identify homologs in multi-domain superfamilies without degrading performance in other types of families. To this end, we perform classification at two different levels (as seen in Figure 2). At level 1, classification is performed between different domain architectures, and level 2 classification is performed among individual families in each domain. The SCOP superfamilies with varied domain architectures represent the primary challenge undertaken in this study. Such families result from domain duplication, domain accretion, and further duplication. Some of these families are defined by a single domain that is unique to the family (e.g., Kinase), while others are characterized by a particular combination of domains (e.g., ADAM) or by a conserved set of domains with variation in domain copy number (e.g., Laminin). Modularity in multi-domain (conserved and non-conserved) can also arise through the presence of sequence motifs, such as localization signals, transactivation sequences (e.g, Tbox), and functional components that confer substrate specificity (e.g., UPS). These motifs can result in matches to unrelated sequences. In addition, promiscuous domains challenge homology identification because they can result in significant sequence similarity but carry little information about gene homology. Promiscuity can confound reliable detection of homologs even in families with conserved domain architectures.

Remote homology detection is a serious challenge that has received widespread attention. In our work, this challenge is represented by FGF, TNF, TNFR, and USP, families that exhibit low sequence conservation. This selection was based on the work of [12]. Since their method is based on a pair-wise match of proteins, with classification results ignoring the first 100k false positives, it is difficult to directly compare our results to theirs. As seen in Table 1, we thus use the classes of proteins only.

Table 1. Individual Classes for Different Multi-Domain Families

Class	N	Class	N
Multi-domain Conserved Architecture		Multi-domain families: Variable Architecture	
DVL	7	ADAM	44
GATA	12	Kinase	40
Notch	8	Kinesin	56
KIR	14	Laminin	22
TRAF	12	Myosin	46
		PDE	44
		SEMA	38
		TNFR	55

2 Methodology

The methodology consists of four parts. First, we extract the features from individual protein properties; second, we perform vector quantization and transform this data into a transaction database format. Third, we generate association rules from this database, and finally we train the classifier using the association rules formed for the classification of previously unseen data.

2.1 Feature Extraction

We take the sequence of a single protein and divide it into M overlapping windows composed of 30 amino acids each. Each window is then shifted by 10 amino acids. If Window 1 starts at Sequence Location 1 and covers amino acids 1 to 30, then Window 2 will start at Location 11 and end at Location 40.

Let P_j represent the protein sequence of jth protein, and let W_i where $i=1...M$ be the M overlapping windows over P_j. Then $\forall W_i \in P_j$ we define the feature vector V_{seq_i} as follows:

$$V_{seq_i} =< X_1, X_2, ..., X_N >$$

Where N= number of amino acids (20), and $X_K : K = 1, ..., N$ is the % composition of Kth amino acid. In addition to the protein sequence, we use three properties $Chou-Fasman\alpha, Chou-Fasman\beta$ and *hydrophobicity* (*Aboderin*). We apply the same windowing procedure to these properties separately and extract the corresponding features vectors as follows:

$$V_{\alpha_i} =< X_1, X_2, X_3, X_4 >, V_{\beta_i} =< X_1, X_2, X_3, X_4 >, \text{and } V_{hy_i} =< X_1, X_2, X_3, X_4 >$$

Where $X_1 = Mean, X_2 = Variance, X_3 = Skewness$ and $X_4 = Kurtosis$

These M individual vectors are combined one below the other to form profile matrices D1 of size Mx20, D2, D3, and D4 each of size Mx4.

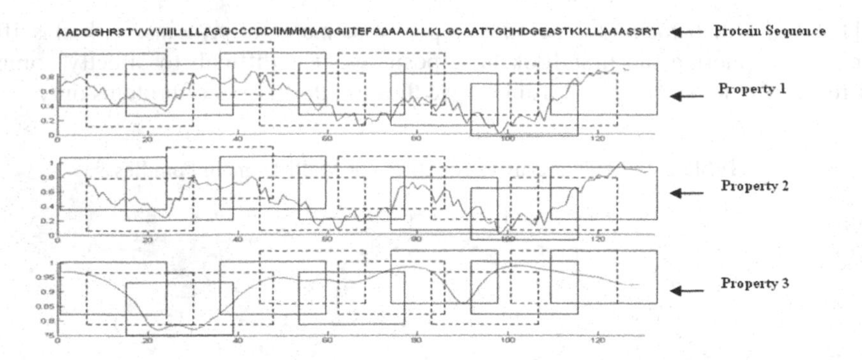

Fig. 3. Feature Extraction by Integrating Protein Properties

Each row of a matrix represents an overlapping window, and each column represents either the amino acid composition (for D1) or one of the four moments of data (for D2, D3, and D4). Figure 3 illustrates the concept of overlapping windows.

2.2 Vector Quantization

After feature extraction each matrix $D_l : l = 1:4$ consisting of M vectors is represented as $D_l = \{V_1, V_2, \ldots\ldots, V_M\}$. Each vector $V_k : k = 1:M$ is of dimension 20 if $l = 1$ and dimension 4 if $l = 2, 3, or\ 4$. The next step in the methodology is to quantize the individual M vectors (rows) in a matrix into N voronoi regions, N<<M, and replace the M vectors by their corresponding region labels. The basic idea of vector quantization is to map the k-dimensional vectors in vector space R^k into a finite set of vectors $V = \{V^j : j = 1,2,\ldots N\}$. Each vector in V is called a codeword and, the complete set of these vectors (V) is called a codebook. Each codeword corresponds to a region where individual regions can contain any of the original M vectors from k-dimensional space [13]. These set of voronoi regions partition the entire space R^k satisfying the condition $\bigcup_{j=1}^{N} V^j = R^k$. Every D_l will have its own codebook *V*. Figure 3 explains the vector quantization of all four matrices. In most cases, a codeword is of the same dimension as the original vector, but we use only the label of the region where the original vector resides. A separate matrix keeps an account of all the original vectors falling in a particular voronoi region which is used later on to quantize matrices from other proteins. We use hierarchical clustering with complete linkage to quantize the vectors into N = 4 regions. In this scheme, initially every point is considered a cluster, and we calculate the pairwise distances between them. Then these N clusters are reduced to N-1 clusters by combining the two clusters which have the minimum pairwise distance. In the next step we calculate pairwise distance among these N-1 clusters. To calculate the distance between a cluster with one point (C1) and the cluster with k points, the pairwise distance between the point from C1 and each of the points in C2 is calculated. Then distance between C1 and C2 is calculated by taking the average of all the k pairwise distances. The quantization stops when we have reached the maximum number of clusters.

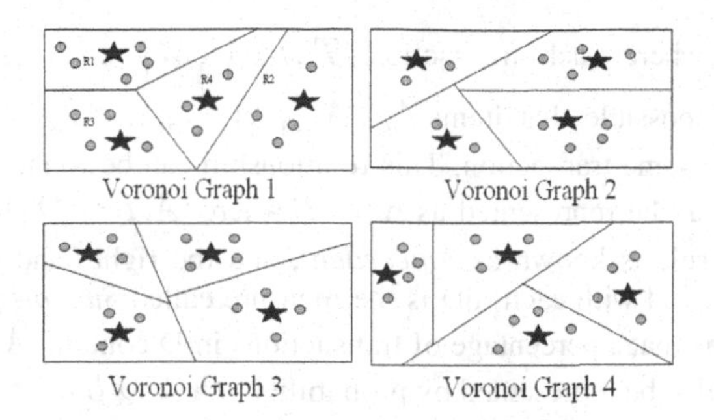

Fig. 4. Vector Quantized Regions for the Amino Acid Profile and the Three Properties

In Figure 4 above, the star represents the voronoi region, and the points represent the vectors falling into this region. The four graphs in Figure 4, represent the amino acid profile and each of the three protein properties. The voronoi region representation is different for all four graphs as the properties have different profiles and hence different vector quantized regions. Each matrix D_l of size MxN is now reduced to size Mx1, where m^{th} row now represents the label of the region where m^{th} vector lies. As such, we get four Mx1 vectors (one for the amino acid sequence and one each for the three properties). We combine these four vectors side-by-side to build a new matrix of length Mx4 where each column now represents a single value (region label of corresponding vector) corresponding to the protein sequence and three protein properties. The first cell of each row represents the label of the region in Voronoi Graph 1 where the amino acid subsequence falls; the second cell represents the label of the region in Voronoi Graph 2 where the second properties' protein profile lies, and so on.

This new matrix is in the format of a transaction database where each row represents a transaction and each column represents items. In our case each of the M overlapping windows is a transaction, and each column is an observation from a multi-domain protein. Every value in each cell is labeled from 1 to 4. In order to distinguish the labels of the first column from the labels of the second, third, and fourth columns, we perform the following preprocessing. First we multiply the value in each column by 100, then multiply the product of column index *i* and add 1000 to each value in order to make a value (e.g. 2) in one column different from the same value (value 2) in another column. The following formula is used for this purpose

$$C(i, j) = (1000 * j) + C(i, j)$$
$$i = 1{:}\ n \quad ; \quad j = 1{:}\ 4$$

Here, $C(i, j)$ represents the value of Column j in Row i.

2.3 Association Rule Mining

Association rules, which present interesting causal relationships among attributes have recently been adapted for finding isomorphisms in multidimensional datasets [4]. They capture information about items which are usually associated. Let I = $A_1, A_2, \ldots, A_N$ be a set of items and D be a database consisting of transactions

$T_1, T_2, \ldots\ldots, T_M$, where each transaction $T_i, i \in (1, M)$ is a set of items such that $T \subseteq I$. It is possible that items $A_K, A_{K+1}, A_{K+2}, \ldots\ldots, A_{K+L}$, where $K+L < N$ could occur in the same transaction. This relationship can be found using association rule mining and can be represented as $A \Rightarrow B$ where $A, B \in D$ [14]. Here the left-hand side of the rule is known as *Antecedent*, and the right-hand side is known as *Precedent.* Associated with each rule is the measure called *Support s* and *Confidence c*. *Support s* means that *s* percentage of transactions in D contain $A \cup B$ (i.e., both *A* and *B*). This can also be represented by probability, $P(A \cup B)$.

$$Support(A \Rightarrow B) = P(A \cup B), \text{ or } \frac{\text{Number of transactions having A and B}}{\text{Total number of transactions}}$$

In *Confidence c*, *c* is the percentage of transactions in C, containing both *A* and *B*. This number can also be represented as conditional probability, *P (B/A)*.

$$Confidence(A \Rightarrow B) = P(B \mid A), \text{ or } \frac{\text{Number of transactions having A and B}}{\text{Number of transactions containing only A}}$$

In our case, we get association rules of the form 1200, 2300=> 4200 with sup=10% and confidence=87%. This rule can be explained as follows: if the value of the first column is 2 and the value of the third column is 3, then the value of the fourth column would be 2 with sup=10% and confidence=87%. If the current amino acid sequence vector profile of the window lies in Region 2 and the vector profile of the corresponding third property lies in voronoi Region 3, then the vector profile of the fourth property would lie in Region 2 with sup=10% and confidence=87%. These types of associations could be present in one class of proteins and might be absent in others and can form the basis for the classification of new proteins.

2.4 Classifier Training

Association rules are extracted for all proteins in every class using the procedure explained earlier. Next, depending on the percentage of training data used, we combine the rules for the proteins in one class to make a class level rule set. This forms the representation of each class. We further combine these class level rule sets to form a global rule set having only unique rules from all the classes. From this set, we calculate a measure of importance of rules, namely the *frequency* of the rule. The frequency of a rule for a class is equal to the percentage of proteins in a class which have that specific rule. For example, say Rule R1 is present in 30% of training proteins in Class 1, 80% of proteins in Class 2, and is not present in any of the other classes. Then the frequency of Rule R1 for Class 1 is .30, for Class 2 is .80 and for all other classes is 0. This frequency measure tells us how important a rule is for a particular class. The cumulative weight of a rule for each class is calculated by taking the product of rule frequency and rule cardinality (the number of items in the rule). The cumulative weight of Rule R1 would be .30xN1 for Class 1, .80xN1 for Class 2 and 0 for other classes. (N1 is the number of items in Rule R1.) Rule weight is calculated in the same way for every rule present in the global rule set. During classification, the rules from a new protein are matched with the rules from the global rule set, and their corresponding weights are calculated. Two rules are said to match if they have the same items in their antecedent and precedent, respectively. The sum of these weights is then combined for each matching rule from the protein, and the protein is classified into the class with the highest weight. A description of the rule weighting can also be found in [4].

3 Results and Discussions

We performed two levels of classification. The first level classification (multi-domain classification) was performed to assess classification between multi-domain families with conserved architecture and multi-domain families with variable architecture. The second level of classification (superfamily classification) was performed to classify individual superfamilies within multi-domain architectures.

3.1 Multi-domain Classification

We have two sets of multi-domain architectures: conserved and variable. A total of 53 proteins belong to five superfamilies in conserved architecture, and 345 proteins belong to eight superfamilies in variable architecture. The data is highly imbalanced, but our classifier is not sensitive to class imbalances. We performed binary classification between conserved and variable architecture classes using 70% data for training and 30% data for testing, respectively. After careful evaluation we decided to use 10% support and 60% confidence to extract association rules from individual proteins. These values of support and confidence were selected based on the notion that each protein should have enough strong representative rules. The classifier performed well, achieving an average accuracy of 100% over 5-fold repetition. This 100% accuracy shows that we can use inter-property causal relationships to classify multi-domain architectures. The confusion matrix for average over 5-folds is shown in Table 2.

3.2 Superfamily Classification

In this level of classification, we performed two experiments. The first experiment was designed to classify the data pertaining to conserved architecture, and the second was used to classify data in variable architecture class. For conserved architecture domain, classification was performed among five families, namely GATA, DVL, Notch, KIR, and TRAF. As in the case of multi-domain classification we used 70% data for training and 30% data for testing from each family. The support and confidence values were also kept the same at 10% and 60%, respectively. The average

Table 2. Confusion Matrix for Multi-Domain Classification

	Conserved	Variable
Conserved	16/16	0/16
Variable	0/104	104/104

class level accuracy over 10 repetitions is shown in Table 3. Except for Notch, all other classes had a high accuracy rate. The overall average accuracy for 10-fold repetition was 87.5%. Table 4 represents the overall accuracy for each of these 10 runs of classification. Table3 also shows exactly what percentage of data from each class is classified into different classes. It can be seen it is KIR classes which is the major cause of misclassification most of the data from classes GATA, Notch and TRAF is misclassified into this class. The corresponding heat map of the confusion matrix can be seen in figure 5. Black portion in the map represents zero.

Table 3. Confusion Matrix showing Class level accuracies

%	GATA	DVL	Notch	KIR	TRAF
GATA	95	0	0	5	0
DVL	0	100	0	0	0
Notch	0	0	35	40	25
KIR	0	0	0	87.5	12.5
TRAF	0	0	0	7.5	92.5

The variable architecture domain has eight families, namely ADAM, Kinase, Kinesin, Laminin, Myosin, PDE, SEMA, and TNFR. In these experiments again we used 70% data for training and 30% for testing from each family and used the same support and and confidence measure as was used for classification in conserved architecture. The overall average accuracy over 10-fold repetition for this architecture was considerably lower, at 61.16%, leading us to believe that the classes in variable architecture have a great deal of overlap in their protein profiles. These results can be seen in Table 5.

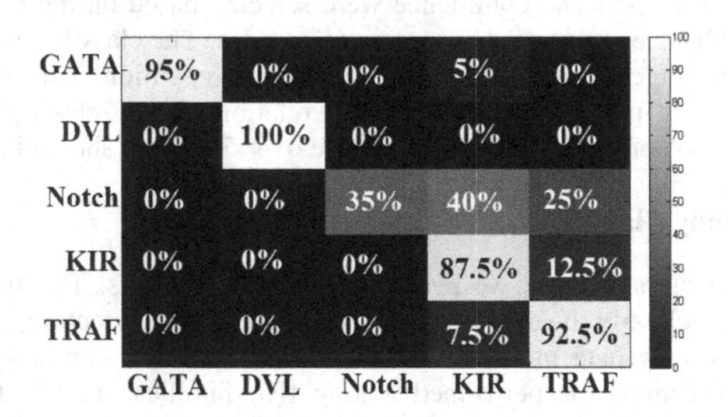

Fig. 5. Heat map of the confusion matrix

3.3 Classification between Single and Multi-domain Proteins

Individual proteins can either be single-domain or Multidomain. So the first set of experiments that we performed was to check the efficacy of classifier in discriminating between these base domains only. These set of experiments are more of precursor to the main goal of the paper which is multi-domain protein classification. We call this classification as Level-0 classification.

Table 4. Results for Conserved Domain

Different Runs	Class. Accuracy
1st	87.50%
2nd	87.50%
3rd	87.50%
4th	87.50%
5th	68.70%
6th	87.50%
7th	68.75%
8th	81.25%
9th	87.50%
10th	87.50%
Average	83.12%

Table 5. Results for Variable Domain

Different Runs	Class. Accuracy
1st	69.56%
2nd	43.00%
3rd	60.80%
4th	57.00%
5th	69.56%
6th	65.00%
7th	58.00%
8th	67.39%
9th	52.00%
10th	69.28%
Average	61.16%

Single domain data consists of seven families: ACSL, FGF, FOX, Tbox, TNF, USP, and WNT having 10, 44, 81, 31, 32, 77, and 38 proteins, respectively.

Table 6. Confusion matrix for Single Domain vs. Multi-Domain Proteins

	Single Domain	Multi Domain
Single Domain	94/94	0/94
Multi Domain	0/120	120/120

The data from variable and conserved domain was combined into one class called multi-domain proteins. Classification was performed among these single domain and multi-domain classes using same support, confidence and training/testing splits used in experiments of Sections 3.1 and 3.2. We achieved an average accuracy of 100% showing that the classifier can distinguish between single and multi-domain families. The results can be seen in Table 6.

4 Conclusion

Homology has traditionally been defined in terms of families that evolve by vertical decent, i.e., by speciation and gene duplication. The ability to distinguish multi-domain homologs from unrelated pairs that share a domain is essential to genomic analysis. Heuristics based on sequence similarity and alignment coverage have been proposed to screen out domain insertions with limited success. In this paper we have presented a new protein homology/domain classification mechanism based on weighted relationships among protein properties. Our test sets include multi-domain families with promiscuous domain that are at risk for domain matches only. The results show that our classifier is not susceptible to class imbalances and can accurately perform both domain and super-family classification.

Acknowledgements. This research project was made possible by National Institutes of Health Grant Number P20 RR16456 from the INBRE Program of the National Center for Research Resources. Its contents are solely the responsibility of the authors and do not necessarily represent the official views of the National Institutes of Health. This project was also partially supported by National Science Foundation (NSF) EPSCoR Award No. EPS-0701491.

References

1. Pearson, W.R.: Effective protein sequence comparison. Methods Enzymol. 266, 227–258 (1996)
2. Song, N., Joseph, J.M., Davis, G.B., Durand, D.: Sequence Similarity Network Reveals. Common Ancestry of Multidomain Proteins 4(5), e1000063 (2004)
3. Wilce, M.C.J., Aguilar, M.-I., Hearn, M.T.: Physicochemical Basis of Amino Acid Hydrophobicity Scales: Evaluation of Four New Scales of Amino Acid Hydrophobicity Coefficients Derived from RP-HPLC of Peptides. Analytical Chemistry 67(7), 1210–1219 (1995)
4. Dua, S., Singh, H., Thompson, H.W.: Associated Classification of Mammograms using Weighted Rules Based Classification. Elsevier Expert System with Applications (in press)
5. Vogel, C., Bashton, M., Kerrison, N.D., Chothia, C., Teichmann, S.A.: Structure, function and evolution of multidomain proteins 14, 208–216 (2004)
6. Hubbard, T.J.P., Murzin, A., Brenner, S., Chotia, C.: SCOP: a structural classification of proteins database. Nucl. Acids Res. 25, 236–239 (1997)
7. Cuff, A.L., Sillitoe, I., Lewis, T., Redfern, O.C., Garratt, R., Thornton, J., Orengo, C.A.: The CATH classification revisited–architectures reviewed and new ways to characterize structural divergence in superfamilies. Nucleic Acids Research (2008)
8. Rossman, M.G., Lijas, A.: Recognition of structural domains in globular proteins. Journal of Molecular Biology 85(1), 177–181 (1974)
9. Song, N., Sedgewick, R.D., Durand, D.: Domain architecture comparison for multidomain homology identification. Journal of Computational Biology 14, 496–516 (2007)
10. Aboderin, A.A.: Mobilities of amino acids on chromatography paper (RF). Int. J. Biochemistry 2, 537–544 (1971)
11. Chou, F.G.: Conformational parameters for amino acids in helical, beta-sheet, and random coil regions calculated from proteins. Biochemistry 13, 211–222 (1974)
12. Vec Quantization, `http://www.geocities.com/mohamedqasem/vectorquantization/vq.html`
13. Agrawal, R., Imielinski, T., Swami, A.N.: Mining association rules between sets of items in large databases. In: Proceedings of the 1993 ACM SIGMOD ICMD, pp. 207–216. ACM, Washington (1993)

IITKGP-SESC: Speech Database for Emotion Analysis

Shashidhar G. Koolagudi[1], Sudhamay Maity[1], Vuppala Anil Kumar[2], Saswat Chakrabarti[2], and K. Sreenivasa Rao[1]

[1] School of Information Technology
[2] G.S. Sanyal School of Telecommunications
Indian Institute of Technology Kharagpur, Kharagpur - 721302, West Bengal, India
koolagudi@yahoo.com, friendsudha@yahoo.com, anil.vuppala@gmail.com, saswat@ece.iitkgp.ernet.in, ksrao@iitkgp.ac.in

Abstract. In this paper, we are introducing the speech database for analyzing the emotions present in speech signals. The proposed database is recorded in Telugu language using the professional artists from All India Radio (AIR), Vijayawada, India. The speech corpus is collected by simulating eight different emotions using the neutral (emotion free) statements. The database is named as Indian Institute of Technology Kharagpur Simulated Emotion Speech Corpus (IITKGP-SESC). The proposed database will be useful for characterizing the emotions present in speech. Further, the emotion specific knowledge present in speech at different levels can be acquired by developing the emotion specific models using the features from vocal tract system, excitation source and prosody. This paper describes the design, acquisition, post processing and evaluation of the proposed speech database (IITKGP-SESC). The quality of the emotions present in the database is evaluated using subjective listening tests. Finally, statistical models are developed using prosodic features, and the discrimination of the emotions is carried out by performing the classification of emotions using the developed statistical models.

Keywords: IITKGP-SESC, Duration, Emotion, Emotion recognition, Energy, Prosody, Statistical models, Pitch.

1 Introduction

Human beings use emotions extensively for expressing their intentions through speech. It is observed that the same message (text) will be conveyed in different ways by using appropriate emotions. At the receiving end, the intended listener will interpret the message according to the emotions present in the speech. In general, it is the fact that the speech produced by human beings is embedded with emotions. Therefore, for developing speech systems (i.e., speech recognition, speaker recognition, speech synthesis and language identification), one should appropriately exploit the knowledge of emotions. But, most of the existing systems are not using the knowledge of emotions while performing the tasks. This is due to the difficulty in modeling and characterization of emotions present in

S. Ranka et al. (Eds.): IC3 2009, CCIS 40, pp. 485–492, 2009.
© Springer-Verlag Berlin Heidelberg 2009

speech. Most of the existing speech systems are developed using the speech data collected under controlled environment by imposing many constraints. For example, speech recognition systems are developed using read speech collected from the news readers [1]. Speech systems developed under these constraints have the applications in the limited domain. Broad applications such as real time speech-to-speech translation and sophisticated human-machine interface, demand the robust speech systems which can work in unconstrained environments.

In this direction to develop robust speech systems, there is a need to analyze and characterize the emotions present in speech. This is very essential, because human beings use the emotions very effectively for conveying the message by incorporating the desired intensions, and perceive the intended message by understanding the implicit emotions conveyed through speech. Therefore, we have focused our research in the area of expressive speech processing [2,3]. The basic issues that need to be addressed in this area are:

1 Exploring the features for discriminating the basic emotions in speech.
2 Exploring the models for capturing the emotion-specific knowledge from speech.
3 Characterization and Acquisition of emotions from speech.
4 Incorporation of emotions in speech synthesis.

To address the above issues, we need to analyze the emotional speech data. Emotion-specific information is embedded in speech at all the levels. Therefore, we need to extract the features from different levels (excitation source, vocal tract system, supra-segmental and linguistic), and develop the emotion specific models. These models can be further used to characterize and discriminate the emotions present in speech. The acquired emotion specific knowledge can be used for developing emotional speech synthesis systems[3].

In practice, it is very difficult to define the emotions in a crisp way, and lot of variability is observed with respect to speakers, language and semantics [4]. Therefore, at the first step we are proposing to analyze the simulated emotions from the speech recorded through the professional artists. Otherwise, collecting the emotional speech data from real and practical situations will be difficult. The emotions present in practical situations have lot of variability, and it is difficult to model the specific emotions. Therefore, we have collected the present database using the simulated emotions from the neutral (emotion free) sentences. Since, the speakers have professional experience, they can simulate the appropriate emotions close to the real and practical situations. Since, all the artists are from the same organization, it ensures the coherence in the quality of the collected speech data.

The proposed speech database is the first one developed in an Indian language Telugu for analyzing the basic emotions present in speech. This database is sufficiently large to analyze the emotions in view of speaker, gender, text and session variability. Before this, we have carried out the study on emotions using the Speech Under Simulated Emotions (SUSE) database at IIT Guwahati [2]. SUSE database is very small. It consists of four emotions (Anger, Compassion, Happy and Neutral), spoken by 30 students. A single sentence *india match odipoinda*, of Telugu language, is uttered in four emotions. In each emotion, the same sentence is uttered 5 times. This gives a total of 600 sentences ($30 speakers \times 4 emotions \times 5 times$). Since, the students have not much experience in simulating the emotions, the quality of the SUSE database may not be very accurate. Therefore, the emotion analysis using the above database may not be appropriate for developing the robust speech systems. Hence, to bridge this gap, we have developed IITKGP-SESC, which will be essential for the analysis of emotions present in speech in the context of Indian languages. There exist several emotion speech databases in the literature in different languages. A brief review of these databases is given in [5].

The details of the IITKGP-SESC are discussed in Section 2. The statistics of the prosodic parameters for various emotions are discussed in Section 3. The discrimination of emotions using prosodic features is also carried out in Section 3. Subjective evaluation of the proposed database (IITKGP-SESC) is carried out in Section 4. The summary of the paper and the future works that can be carried out on this database are given in the final section of the paper.

2 IITKGP-SESC (Indian Institute of Technology Kharagpur Simulated Emotion Speech Corpus)

The proposed database is recorded using 10 (5 male and 5 female) professional artists from All India Radio (AIR) Vijayawada, India. The artists have sufficient experience in expressing the desired emotions from the neutral sentences. All the artists are in the age group of 25-40 years, and have the professional experience of 8-12 years. For analyzing the emotions we have considered 15 Telugu sentences. All the sentences have emotionally neutral in meaning. Each of the artists has to speak the 15 sentences in 8 basic emotions in one session. The number of sessions considered for preparing the database is 10. The total number of utterances in the database is 12000 ($15 sentences \times 8 emotions \times 10 artists \times 10 sessions$). Each emotion has 1500 utterances. The number of words and syllables in the sentences are varying from 3-6 and 11-18 respectively. The total duration of the database is around 7 hours. The eight basic emotions considered for collecting the proposed speech corpus are: Anger, Compassion, Disgust, Fear, Happy, Neutral, Sarcastic and Surprise. The speech samples are recorded using SHURE dynamic cardioid microphone C660N. The speech signal was sampled at 16 kHz, and represented as 16 bit numbers. The sessions are recorded in alternate days to capture the variability of the sessions. In each session, all the artists have given the recordings of 15 sentences in 8 emotions. The recording has done in such a way that each artist has to speak all the sentences at a stretch in a particular emotion. This provides the coherence among the sentences for each emotion category. The entire speech database is recorded using single microphone and at the same location. The recording was done in a quite room, without any obstacles in the recording path.

3 Prosodic Analysis

From the state-of-the-art literature, it is observed that the characteristics of the emotions can be seen at the source level (characteristics of excitation signal and shape of the glottal pulse), system level (shape of the vocal tract and nature of movements of different articulators) and at the prosodic level [6]. Among the features from different levels for characterizing the emotions, prosodic features have attributed to a major role by the existing literature as well as perceptual point of view also [7]. Therefore, in this paper we show the importance of prosodic parameters by categorizing the emotions using the first order statistics of the prosody parameters derived from the database. The prosodic features considered in this study are (1) duration patterns, (2) average pitch, (3) variation of pitch with respect to mean (Standard Deviation(SD)), and (4) average energy.

The duration of each of the speech files is determined in seconds. The mean of the durations is determined for each emotion category. The pitch values of each utterance are obtained from the autocorrelation of the Hilbert envelope of the linear prediction residual [8]. As each utterance has sequence of pitch values according to its intonation pattern, analysis is carried out using mean and standard deviation of pitch values. The energy of speech signal is determined, and the normalized energy for each speech utterance is calculated by dividing the energy with duration (number of samples) of speech utterance. In this paper, we have not used the entire database for deriving the statistics of the prosody parameters. We have considered one male and one female artist's speech files for illustrating the statistics of prosodic parameters. Each artist has uttered 1200 sentences ($15 sentences \times 8 emotions \times 10 sessions$), out of which, 960 sentences are used for deriving the statistical models and the remaining 240 are used for verifying the models. In this study we use the first order statistics (Mean of the distribution) for the analysis of basic emotions. The average values of the prosodic parameters are determined at sentence level. Table 1 shows the mean values of the prosodic parameters for different emotions. In Table 1, the first column indicates the basic emotions considered for the analysis. Columns 2-5 contain the average values of the prosodic features of speech utterances of different emotions for a male artist. Columns 6-9 contain the average prosodic features for a female artist.

From the utterance level features, it is observed that prosodic features are overlapped for some emotions and distinct for other emotions. For illustrating these details, the distribution of the durations for different emotions from one of the female artists is shown in Fig.1. The distribution of duration patterns indicate that the emotion disgust has distribution at lower end, compassion has at higher end and other emotions have the distributions in between them. From this evidence, it is difficult to discriminate emotions using only duration patterns. Combined evidence from other prosodic parameters may improve the discrimination capability.From the analysis of duration patterns of the male artist, it is observed that the surprise and anger emotions have smaller mean durations, whereas compassion, happy, sarcastic and surprise emotions have comparatively larger mean durations (see column 2 of Table 1). For the female artist, disgust

Table 1. Mean values of the prosodic parameters for each emotion

Emotion	Male Artist				Female Artist			
	Duration (Sec)	Pitch (Hz)	SD of pitch (Hz)	Energy	Duration (Sec)	Pitch (Hz)	SD of pitch (Hz)	Energy
Anger	1.76	195.60	48.74	203.45	1.80	301.67	80.51	103.36
Compassion	2.09	204.00	52.70	225.98	2.13	294.33	75.43	86.00
Disgust	1.62	188.05	45.11	11894	1.67	308.62	77.42	90.19
Fear	1.79	210.70	54.93	263.68	1.89	312.07	86.47	144.87
Happy	2.03	198.30	53.99	164.68	2.09	287.78	80.34	83.65
Neutral	1.93	184.37	54.20	160.44	2.04	267.13	77.78	83.42
Sarcastic	2.16	188.44	40.88	120.03	2.20	301.11	83.86	75.26
Surprise	2.05	215.75	62.76	202.06	2.09	300.10	85.63	86.72

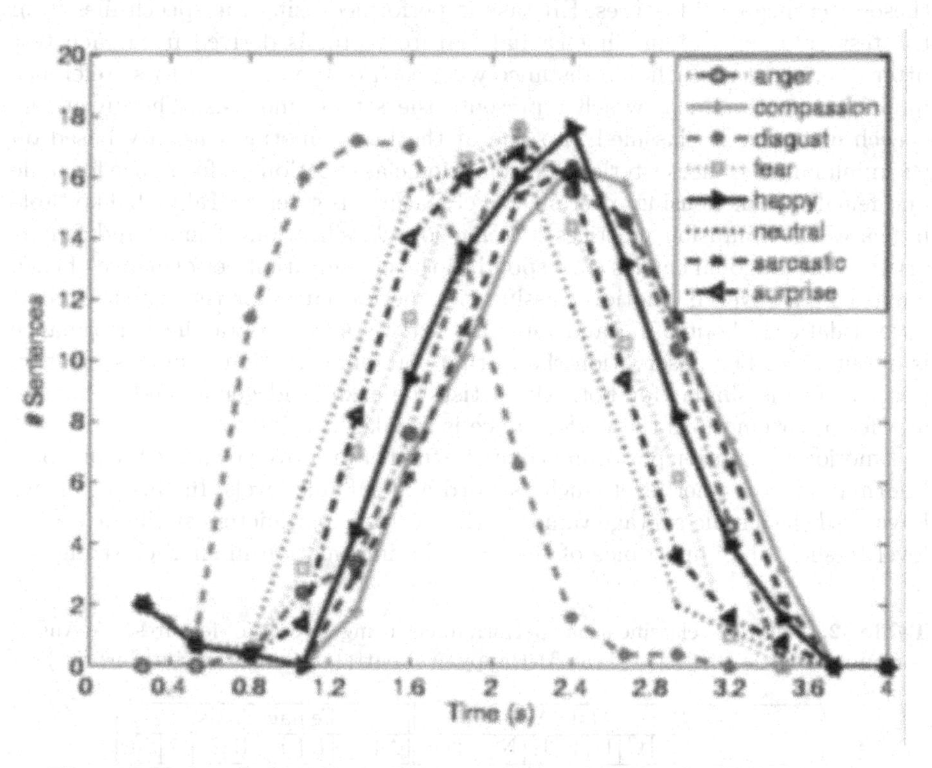

Fig. 1. Distribution of the durations for different emotions of the female artist

and anger emotions have minimum mean durations, and compassion and sarcastic have maximum mean durations (see column 6 of Table 1). Similarly, using the pitch feature, it is observed that neutral, sarcastic and disgust have very close mean pitch values (see column 3 of Table 1). In the case of female artist, the pattern of mean pitch values with respect to emotions is entirely different from the male artist. From the mean values of the prosodic parameters shown in Table 1 indicating that for some emotions like fear, anger and surprise have emotion specific characteristics independent of speaker and gender, for other emotions the emotion specific prosodic characteristics are speaker and gender dependent. In this work, we have explored the prosodic characteristics for analyzing the emotions in view of speaker dependent and independent cases.

In this work we used a simple Euclidian distance measure for performing the Emotion Recognition (ER) task. The mean prosodic feature vector consists of the average values of the prosodic features derived from the whole utterance. These mean vectors correspond to each emotion represent the reference models for the basic emotions. In this paper, we have prepared separate models for male and female artists. The reference models for the emotions of the male artist are shown in columns 2-5 in Table 1. For the female artist the reference models are given in columns 6-9 of Table 1. To evaluate the performance of these utterance level features, ER task is performed using the speech files from the test data set. A four dimensional feature vector is derived from each test utterance, and the Euclidian distance with each of the mean vectors (reference models), is determined, which represents the set of emotions. The given test speech utterance is classified into one of the basic emotion category based on the minimum distance criterion. The emotion classification performance for male and female speakers using utterance level features is given in Table 2. The table indicates the confusion matrices for emotion classification of male and female artists. The diagonal entries correspond to the classification performance of each emotion. The overall emotion classification performance for the female artist's speech data is about 69%, and for male artist's speech data the performance is about 75%. The observation shows that for some emotions the classification performance is similar for both the artists (speaker and gender independent), and for other emotions the performance is speaker specific.

Emotion recognition performance can be further improved by using the prosodic information from finer levels such as word and syllable levels. In this paper, we have used the simple average values of the prosodic parameters at the utterance level to show the importance of the prosodic information in characterizing the

Table 2. Emotion classification performance using prosodic features (A-Anger, C-Compassion, D-Disgust, F-Fear, H-Happy, N-Neutral, S-Sarcastic, Sur-Surprise)

	Male Artist								Female Artist							
	A	C	D	F	H	N	S	Sur	A	C	D	F	H	N	S	Sur
Anger	84	4	4	0	0	0	4	4	68	16	8	0	0	0	0	8
Compassion	0	92	0	8	0	0	0	0	0	100	0	0	0	0	0	0
Disgust	6	0	82	6	0	0	6	0	0	0	76	8	0	0	16	0
Fear	0	8	0	92	0	0	0	0	0	0	0	100	0	0	0	0
Happy	10	0	0	0	74	16	0	0	34	0	0	0	66	0	0	0
Neutral	0	9	0	0	8	66	9	8	0	10	10	0	10	55	5	10
Sarcastic	8	0	8	16	0	8	60	0	16	0	8	16	0	8	52	0
Surprise	14	15	0	0	0	4	14	53	24	17	0	0	0	0	24	35

emotions. This analysis also indirectly indicate the quality of the IITKGP-SESC. The classification performance can be further enhanced using nonlinear models such as neural networks and support vector machines [9].

4 Subjective Evaluation

The quality of the database is evaluated using subjective listening tests. Here, the quality represents how well the artists simulated the emotions from the neutral sentences. The human subjects are used to assess the naturalness of the emotions embedded in speech utterances. This evaluation is carried out by 25 subjects who are pursuing post graduation and research programs in our institute.

The subjective tests are performed on the lines of emotion classification using the first order statistics. The goal of these listening tests is to estimate the capability of the human beings in recognizing the basic emotions. This study will be useful for the analysis of emotions in human versus machine perspective. That is, among the 8 basic emotions, which are easily recognizable and which are confusing.

In this study, 40 sentences (5 sentences from each emotion) from each artist are considered for evaluation. Before taking the test, the subjects have given the pilot training by playing 24 sentences (3 sentences from each emotion) from each artist's speech data, for understanding (familiarizing) the characteristics of basic emotions. The forty sentences used in the evaluation are randomly ordered, and played to the listeners. For each sentence, the listener has to mark the emotion category from the set of 8 basic emotions. The overall emotion classification performance for the chosen male and female artists speech data is given in Table 3. The observation shows that the average emotion recognition rates of male and female artists speech utterances are 61% and 66% respectively. Similar to the statistical models, subjective evaluation is also indicating the speaker dependent and independent emotion characteristics. The results of the listening tests show that the emotion recognition performance is low for humans compared to machines. This is due to confusability among the subset of emotions, which are

Table 3. Emotion classification performance based on subjective Evaluation (A-Anger, C-Compassion, D-Disgust, F-Fear, H-Happy, N-Neutral, S-Sarcastic, Sur-Surprise)

	Male Artist								Female Artist							
	A	C	D	F	H	N	S	Sur	A	C	D	F	H	N	S	Sur
Anger	73	0	17	2	3	4	0	1	69	0	19	3	2	5	0	2
Compassion	0	61	3	16	3	13	4	0	4	52	2	25	1	12	3	1
Disgust	28	0	56	7	0	4	5	0	40	2	44	5	0	3	3	3
Fear	7	19	6	49	0	8	1	10	6	25	8	37	2	7	1	14
Happy	0	9	2	6	62	8	5	6	0	7	4	4	66	10	3	6
Neutral	0	2	5	0	6	86	0	1	1	0	8	1	6	83	1	0
Sarcastic	6	4	5	0	0	0	85	0	4	3	5	0	0	0	88	0
Surprise	0	5	7	5	16	5	7	55	6	16	6	3	17	3	1	48

rarely faced in the daily life. Humans may perform better emotion recognition by exposing to more sentences during training.

5 Summary and Conclusions

In this paper we have proposed an emotional speech corpus (IITKGP-SESC) in Telugu. The basic emotions considered for developing IITKGP-SESC are Anger, Compassion, Disgust, Fear, Happy, Neutral, Sarcastic and Surprise. From the developed speech corpus, the basic emotions were analyzed using prosodic parameters. The importance of the prosodic parameters for discriminating the emotions was shown by performing the emotion classification using the simple statistical parametric models. The proposed statistical models may not capture the complex nonlinear relations present in the prosodic features, which were extracted from the longer speech segments. Hence, nonlinear models can be explored to further improve the recognition performance. The quality of the emotions present in the developed emotional speech corpus is evaluated using subjective listening tests.

The proposed emotional speech database can be further exploited for characterizing the emotions using the emotion specific features extracted from vocal tract and excitation source. The discrimination of emotions can be improved further by combining the evidences from different models developed using features extracted from speech at various levels. The proposed database has wide variety of characteristics in terms of different emotions, speakers and sentences (text). One can perform the systematic study on emotional characteristics in view of speaker and text variability.

References

1. Database for Indian languages. Speech and Vision lab, Indian Institute of Technology Madras, India (2001)
2. Ramamohan, S., Dandapat, S.: Sinusoidal model-based analysis and classification of stressed speech. IEEE Trans. Speech and Audio Processing 14, 737–746 (2006)
3. Sagar, T.V.: Characterisation and synthesis of emotionsin speech using prosodic features. Master's thesis, Dept. of Electronics and communications Engineering, Indian Institute of Technology Guwahati (May 2007)
4. Lee, C.M., Narayanan, S.: Toward detecting emotions in spoken dialogs. IEEEAUP 13(2), 293–303 (2005)
5. Ververidis, D., Kotropoulos, C.: A state of the art review on emotional speech databases. In: Eleventh Australasian International Conference on Speech Science and Technology, Auckland, New Zealand (December 2006)
6. Yang, L.: The expression and recognition of emotions through prosody. In: Proc. Int. Conf. Spoken Language Processing, pp. 74–77 (2000)
7. Cowie, R., Cornelius, R.R.: Describing the emotional states that are expressed in speech. Speech Communication 40, 5–32 (2003)
8. Prasanna, S.R.M., Yegnanarayana, B.: Extraction of pitch in adverse conditions. In: Proc. IEEE Int. Conf. Acoust., Speech, Signal Processing, Montreal, Canada (May 2004)
9. Haykin, S.: Neural Networks: A Comprehensive Foundation. Pearson Education Aisa, Inc., New Delhi (1999)

Detection of Splice Sites Using Support Vector Machine

Pritish Varadwaj, Neetesh Purohit, and Bhumika Arora

Indian Institute of Information Technology, Allahabad
pritish@iiita.ac.in, np@iiita.ac.in, bhumikarora@gmail.com

Abstract. Automatic identification and annotation of *exon* and *intron* region of gene, from DNA sequences has been an important research area in field of computational biology. Several approaches viz. Hidden Markov Model (HMM), Artificial Intelligence (AI) based machine learning and Digital Signal Processing (DSP) techniques have extensively and independently been used by various researchers to cater this challenging task. In this work, we propose a Support Vector Machine based kernel learning approach for detection of splice sites (the *exon-intron* boundary) in a gene. Electron-Ion Interaction Potential (EIIP) values of nucleotides have been used for mapping character sequences to corresponding numeric sequences. Radial Basis Function (RBF) SVM kernel is trained using EIIP numeric sequences. Furthermore this was tested on test gene dataset for detection of splice site by window (of 12 residues) shifting. Optimum values of window size, various important parameters of SVM kernel have been optimized for a better accuracy. Receiver Operating Characteristic (ROC) curves have been utilized for displaying the sensitivity rate of the classifier and results showed 94.82% accuracy for splice site detection on test dataset.

Keywords: Splice site, Support vector machine, Electron-ion interaction potential.

1 Introduction

The successful completion of several genomic projects in recent past has yielded vast amount of sequence data. Rational analysis of these genomic data to extract relevant information can have profound implications on automated annotation and functional motif identification. Identification of genes from sequence data is an important area of research in field of computational biology. The complexity of these gene finding approaches further increases due to the presence of coding regions called *exons* interrupted by non-coding regions called *introns* complemented by *intergenic* regions. The *exon-intron* border is known as donor splice site where as *intron-exon* border is known as acceptor splice site. Identification of *exon, intron* and splice site regions of a gene is not a new problem. There exists classical probabilistic based, artificial intelligence based and Digital Signal Processing based approaches for addressing above problem. Approaches such as Hidden Markov Model (HMM), Dynamic Programming and Bayesian Networks falls in first category while Artificial Neural Network (ANN), Support Vector Machine (SVM) based approaches categorize to artificial intelligence category. Furthermore discrete nature of DNA representation has motivated many signal processing engineers to obtain an equivalent numeric sequence for DNA

S. Ranka et al. (Eds.): IC3 2009, CCIS 40, pp. 493–502, 2009.
© Springer-Verlag Berlin Heidelberg 2009

strands and then apply various Digital Signal Processing (DSP) methods to find some interpretable results. Proposed SVM based method for splice site detection is basically a hybrid version of computational intelligence approach and DSP based approach. Herewith we summarize a brief review of relevant previous work.

Artificial Neural Network (ANN) based approach has been adopted in [1],[2],[3] for gene recognition. A window size of 99 nucleotides has been used and the classification of center nucleotide of the window, either as coding or non-coding region, was done using 9 inputs, 14 hidden layers and 1 output ANN. Genetic algorithm was used for evolving biases and interconnection weights. In [4],[5],[6],[7] many other variations of ANN or rule based system combined with ANN has been used for the purpose of gene identification. Various statistical coefficients and frequency indicators calculated from genomic sequence has been used as input features in [1-7] but unfortunately, none of these methods have satisfactory level of accuracy for different genes and different species. A survey of various computational approaches used in gene identification has been reported in [8]. Support vector machine (SVM) is an alternative method used in machine learning which has much robust theoretical background and often it gives better results than ANN. If the input data belongs to two classes with n common features then each input sample can be represented as a vector in an n-dimensional space and the classification problem reduces in finding a hyperplane in that space which should separate two classes. SVM finds two parallel hyperplanes in the same space with any orientation such that the margin between them could be maximized. The input samples (called vectors) which fall on these parallel hyperplanes are called support vectors. Several kernel functions have been recommended for SVMs but radial basis function (RBF) often gives better performance. The performance of RBF is heavily dependent on two parameters C and γ. The optimum values of these parameters vary from problem to problem this aspect has been ignored by most researchers while using SVM. The SVM has been used [9-13] for several bioinformatics problems; even the problem of splice site detection and also has been recently addressed elsewhere [14-22]. These methods involve feature and model selection (simple or hidden Markov Model) and SVM kernel engineering for splice site prediction using conditional positional probabilities. Above methods are cumbersome as compared to proposed approach which is simple and straight forward. For DSP based approaches, various schemes of converting a character sequence into numeric sequence and applying various DSP techniques for gene finding and other such applications have been summarized in [23],[24]. The binary or Voss representation [25], one, two or three dimensional tetrahedron representation [26], EIIP method [27],[28], paired numeric, paired and weighted spectral rotation (PWSR), Paired spectral content [29] etc are popular conversion techniques. All such techniques utilizes the statistical properties of *exon* and *intron* regions observed in many genes of different species e.g. period-3 property, frequency of particular nucleotides in *exon* or *intron* regions etc. Either time domain methods e.g. correlation structures [30], average magnitude difference function (AMDF) [29] etc. or frequency domain methods e.g. DFT [31],[32],[33], wavelet [27][28], autoregressive [34] etc have been used for exploring above mentioned properties of genes. For overcoming spectral spreading problem of transform method, few filtering approaches have also been suggested [35]. In short, many researchers have done good work by adopting one of the above mentioned three approaches and they have obtained satisfactory results but the dataset used for most of such studies are probably biased towards a single chromosome or single gene hence not giving desired output for other genes of the same species. The hybrid method adopted in proposed scheme is a novel approach which is giving high accuracy identification for a large variety of genes with less preprocessing required. Being natural characteristic feature of nucleotides, EIIP values appear more appealing for obtaining numeric sequences, thus this method has been used in present work. Furthermore we found SVM to be a better classifier than artificial neural network (ANN) at least for aforesaid purpose. Hence this was preferred which resulted in showing higher prediction accuracy. In this work, EIIP mapped numeric dataset of genomic sequences has been prepared for training the SVM and the optimum values of various SVM parameters have been explicitly determined. Use of these optimized values during testing phase has increased the accuracy of results by many folds. This paper has been organized as follows: Detailed methodology has been described in section 2. Results followed by discussion have been reported in section 3. Finally the conclusion and future work appears in section 4 followed by references in section 5.

2 Materials and Methods

We have selected ***Arabidopsis thaliana*** as model species and genomic sequence data were collected across all five chromosomes of the same species. Data were collected from *The Exon-Intron Database* (EID) [36], further the sequences entries were subjected to similarity screening to get the final set of data with less than 23% inter similarities. The selected dataset for this study comprises of 1000 splice site bearing sequences of 12 residues length (500 acceptor splice site and 500 donor splice site) and 1000 non splice site sequences of 12 residues length (500 each, both from *exon* and *intron* region). Further we randomly split the whole dataset of 2000 sequence entries of 12 residues length each, into training and testing set at a ratio of 2:3, i.e. 800 training set comprised of 400 splice site (donor and acceptor) and 400 non-splice site data. Similarly 1200 test sequences were taken which comprised of 600 splice site (donor and acceptor) and 600 non-splice site sequences from *exon* and *intron* region. Training set data was used for training various classifiers, while the testing examples were not exposed to the system during learning, kernel selection and hyper-parameter selection phases. Training dataset and Test dataset can be obtained from http://profile.iiita.ac.in/pritish/svmdata

The genomic character sequence of training and test set were converted into numeric sequence using EIIP values as described below (Table 1).

Table 1. EIIP values of nucleotide residues

Nucleotide Letters	Name	EIIP value
A	Adenine	0.1260
G	Guanine	0.0806
C	Cytosine	0.1340
T	Thymine	0.1335

As described earlier other potential character to numeric mapping schemes appears to be biased in attempt to exploit various statistical properties of gene. EIIP values are natural characteristic feature of nucleotides and probably it carries the chemical information patterns required to be recognized by *spliceosome* [37]. Further it's not only a single EIIP value guided reorganization by *spliceosome* rather it should be an environmental effect of neighborhood residues which makes the specific pattern detection possible. For this reason we have considered sequences of 12 residues length (sequence window size=12) after validating the result with different window length of 6, 8, 10, 12, 14 and 16 residues sequences. The ROC plot of this has been shown in Fig. 1.

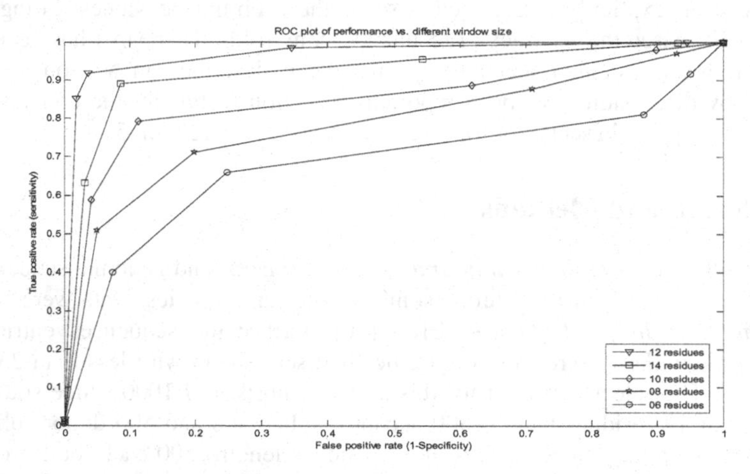

Fig. 1. ROC plot of prediction accuracy with different input vector sequence window length

Thus window size equal to 12 is selected which is moved from one end of a gene (3' / 5') to other(5' / 3') and the presence or absence of splice site is detected for each windows sliding with one residue ahead per move (Fig. 2).

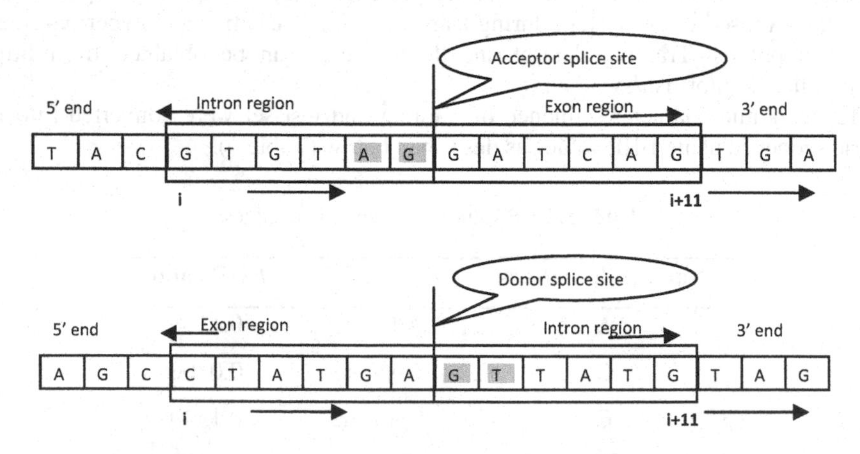

Fig. 2. Window sliding by one residue at a time to create input vector for classifier

In this process there may exist 4 possibilities for the type of nucleotides patterns falling within the window: *a)* sequence pattern belongs *exon* region, *b)* sequence pattern belongs to *intron* region, *c)* sequence pattern belongs to a donor splice site and *d)* sequence pattern belongs to an acceptor splice site. Training data of above four types of sequences were prepared in 2000 such windows of with target labels -1 for *a)* and *b)*; +1 for *c)* and *d)* patterns type respectively. So input vector for training as well as test set has been quantified as: $X^i = (X_1^i, X_2^i, \ldots\ldots\ldots, X_{12}^i)$, each labeled by corresponding $y^i = +1$ or $y^i = -1$ depending on whether it represents a splice site or non-splice site patterns, respectively.

Training set data were subjected to SVM classifier, which involved fixing several hyper-parameters and values of these hyper-parameters determining the function that SVM optimizes and therefore have a crucial effect on the performance of the trained classifier [38]. We have used several kernels: linear, polynomials and radial basis function (RBF). We found RBF as the suitable classifier function (as the number of features is not very large), for which training errors on splice site data (false negatives) outweigh errors on non-splice site data (false positives). The classical Radial Basis Function (RBF) used in this work has similar structure as SVM with Gaussian kernel

$$K(X_i, X_j) = \exp(-\gamma \,||\, X_i - X_j \,||^2), \gamma > 0 \quad (1)$$

This kernel (1) is basically suited best to deal with data that have a class-conditional probability distribution function approaching the Gaussian distribution. It maps such data into a different space where the data becomes linearly separable. To actually visualize this, it is convenient to observe that the kernel (which is exponential in nature) can be expanded into an infinite series, thus giving rise to an infinite-dimension polynomial kernel: each of these polynomial kernels will be able to transform certain dimensions to make them linearly separable. However, this kernel is

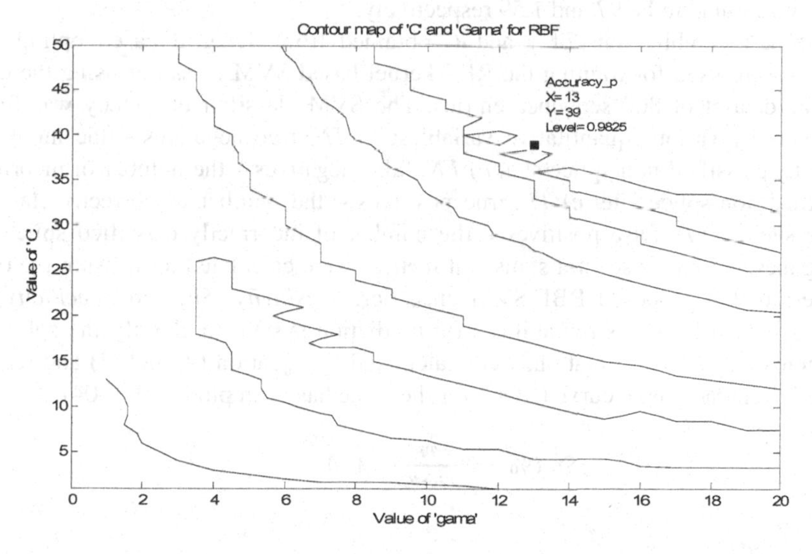

Fig. 3. Contour plot of grid search result showing optimum values of hyper-parameter

difficult to design, in the sense that it is difficult to arrive at an optimum 'γ' and choose the corresponding C that works best for a given problem. Since searching the best hyperplane parameters is associated with the problem of overfitting, a grid parameter search exploring all combinations of C and γ with ten folds cross-validation routine, where γ ranged from 2^{-15} to 2^{4} and C ranged from 2^{-5} to 2^{15}[39] has been implemented. To identify an optimal hyper-parameter set we have performed a two step grid-search on C and γ using 10 folds cross-validation, by dividing training set into 10 subsets of equal size (80 each). Iteratively each subset is tested using the classifier trained on the remaining 9 subsets. Pairs of (C; γ) have been tried and the one with the best cross-validation accuracy has been picked. The best cross-validation performance, for a value of $\gamma = 1.59$ and $C = 97$ was obtained by the RBF kernel, with parameter and cost factor (Fig. 3). The result obtained shows very good classification accuracy 98.25 % during the cross-validation.

3 Result and Discussion

To optimize the SVM parameters γ and C, 10-fold cross-validation has been applied on each of the training datasets bin, exploring various combinations of C (2^{-5} to 2^{15}) and γ (2^{-15} to 2^{4}). In 10-fold cross-validation, the training dataset (800 sequence entries, each of 12 residues length) was spilt into 10 subsets of 80 sequence entries (40 splice site and 40 non-splice site), where one of such subsets was used as the test dataset while the other subsets were used for training the classifier. The process is repeated 10 times using a different subset of corresponding test and training datasets, hence ensuring that all subsets are used for both training and testing. A two fold grid optimization has been considered and result shown (Fig. 3) suggests the optimized C and γ were found to be 97 and 1.59 respectively.

The best combinations of γ and C obtained from the grid based optimization process were used for training the RBF kernel based SVM classifier using the entire training dataset of 800 sequence entries. The SVM classifier efficiency was further evaluated by various quantitative variables: *a) TN*, true negatives – the number of correctly classified non-splice site, *b) FN*, false negatives – the number of incorrectly classified non-splice site, *c) TP*, true positives – the number of correctly classified splice site, *d) FP*, false positives – the number of incorrectly classified splice site. Using these variables several statistical metrics were calculated to measure the effectiveness of the proposed RBF-SVM classifier. *Sensitivity (Sn)* and *Specificity (Sp)* metrics, which indicates the ability of a prediction system to classify the splice site and non-splice site information, were calculated by equation (2) and (3) and receiver operating characteristic curve (ROC) for the same has been plotted (Fig-04)

$$Sn\ (\%) = \frac{TP}{TP+FN} \text{ x } 100 \tag{2}$$

$$Sp\ (\%) = \frac{TN}{TN+FP} \text{ x } 100 \tag{3}$$

To indicate an overall performance of the classifier system; *a)* Accuracy *(Ac)*, for the percentage of correctly classified splice sites and the Matthews Correlation Coefficient *(MCC)* were computed as follows:

$$Ac(\%) = \frac{TP + TN}{TP + TN + FP + FN} \, X\, 100 \tag{4}$$

$$MCC = \frac{(TP\, X\, TN) - (FP\, X\, FN)}{\sqrt{(TN + FP)(TN + FN)(TP + FP)(TP + FN)}} \tag{5}$$

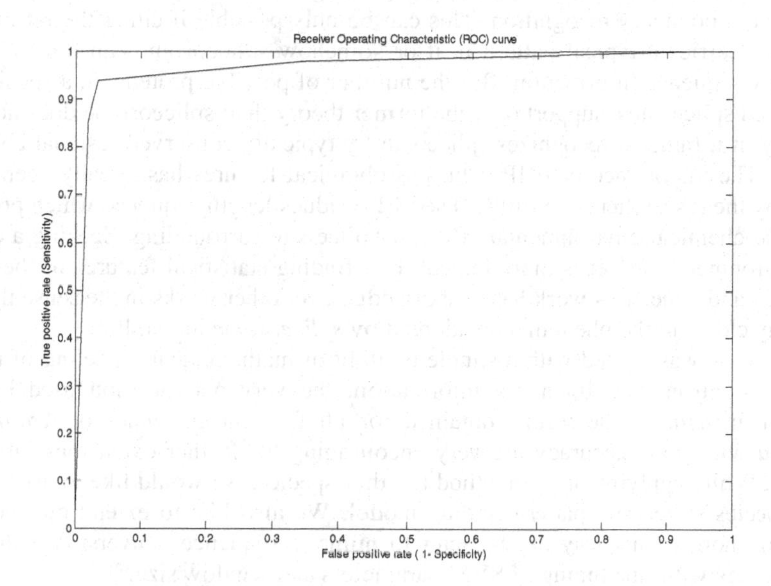

Fig. 4. Receiver operating characteristic (ROC) plot for classifier with optimized values of C and γ

Sensitivity (*Sn*) is found to be 96.55% with false positive proportion (*FP*) 3.45 %, where as Specificity (*Sp*) is found to be 93.10 % with false negative (*FN*) proportion 6.90%. Similarly *Youden's Index* (*Youden's Index= sensitivity + specificity – 1*) is 0.8865 and Matthews Correlation Coefficient *(MCC)* found to be 0.9124. The overall accuracy (*Ac*) is calculated as 94.82% , which is significantly higher than existing methods. Area under ROC curve is found to be 0.98095 with standard error 0.00567. We have chosen the RBF kernel with optimized parameters γ and *C*. Using 10-fold cross-validation, the parameters γ and *C* were optimized at 1.59 and 97 with an overall training datasets classification accuracy of 98.25%, which is reasonably good. While the reported accuracy on the training datasets may indicate the effectiveness of a prediction method, it may not accurately portray how the method will perform on novel, hitherto undiscovered splice sites. Therefore, testing the SVM methodology on independent out-of-sample datasets, not used in the cross-validation is critical. Here, we applied the SVM classifiers, on the entire test datasets, the SVM method obtained an accuracy of 94.82% using the RBF kernel with $\gamma = 1.59$ and C = 97. These findings suggest that the SVM-based prediction of splice site detection might be helpful in identifying potential *exon-intron* boundary hence gene annotation.

4 Conclusion and Future Work

In the process of spliceosome mediated RNA splicing, introns are removed and exons are joined from transcribed pre-mRNA to form the final mRNA, for translation into successive protein. *Spliceosome* recognizes the splice sites (donor and acceptor) on pre-mRNA and it does so for invariably large ranges of pre-mRNA with very accurate intron-exon boundary recognition. This can be only possible if either the intron-exon boundary carries a typical pattern in it or somehow spliceosome can remember the boundary sequence information. But the number of possible proteins in a species and associated splice sites, support only the former theory that; spliceosome does not have memory in it rather it recognizes splice site by typically conserved residual-chemical pattern. The importance of EIIP values as chemical features has already been established by the researchers, we further used 12 residues length sequence which probably carry the chemical environmental effect of splice site surrounding. Seeking a chemical environmental effect is more logical than finding statistical features of the whole sequence and hence this work has an extra edge over other works in the sense that it is probably closer to the phenomenon adopted by *spliceosome* in real life.

This work was started with a simple thought of mathematical modeling of the detection of intron-exon boundary information, the very phenomenon used by *spliceosome* in nature. The results obtained for all five chromosomes of ***Arabidopsis thaliana*** with good accuracy are very encouraging for further extensions into other species. While applying above method to other species, we would like to establish an inter-species splice site pattern finding model. We also like to extend our work for adopting more robust way of character to numeric sequence conversion, enhancing the accuracy with fine tuning of SVM parameters and window size.

References

[1] Uberbacher, E.C., Xu, Y., Mural, R.J.: Discovering and understanding genes in human DNA sequence using GRAIL. Methods Enzymol. 266, 259–281 (1996)

[2] Fickett, J.W., Tung, C.-S.: Assessment of protein coding measures. Nucleic Acids Res. 20, 6441–6450 (1992)

[3] Fogel, G.B., Chellapilla, K., Corne, D.W.: Identification of coding regions in DNA sequences using evolved neural networks. In: Fogel, G.B., Corne, D.W. (eds.) Evolutionary Computation in Bioinformatics, pp. 195–218. Morgan Kaufmann, San Francisco (2002)

[4] Hebsgaard, S.M., Korning, P.G., Tolstrup, N., Engelbrecht, J., Rouze, P., Brunak, S.: Splice site prediction in Arabidopsis thaliana pre mRNA by combining local and global sequence information. Nucleic Acids Res. 24(17), 3439–3452 (1996)

[5] Reese, M.G.: Application of a time-delay neural network to promoter annotation in the Drosophila melanogaster genome. Comput. Chem. 26(1), 51–56 (2001)

[6] Ranawana, R., Palade, V.: A neural network based multi-classifier system for gene identification in DNA sequences. Neural Comput. Appl. 14(2), 122–131 (2005)

[7] Sherriff, A., Ott, J.: Applications of neural networks for gene finding. Adv. Genet. 42, 287–297 (2001)
[8] Bandyopadhyay, S., Maulik, U., Roy, D.: Gene Identification: Classical and computational Intelligence approaches. IEEE Trasaction on systems, man and cybernatics 38(1) (January 2008)
[9] Rätsch, G., Sonnenburg, S., Srinivasan, J., Witte, H., Müller, K.R., Sommer, R., Schölkopf, B.: Improving the C. elegans genome annotation using machine learning. PLoS Computational Biology 3(2), e20 (2007)
[10] Jaakkola, T., Haussler, D.: Exploiting Generative Models in Discriminative Classifiers. In: Kearns, M., Solla, S., Cohn, D. (eds.) Advances in Neural Information Processing Systems, vol. 11, pp. 487–493. MIT Press, Cambridge (1999)
[11] Zien, A., Rätsch, G., Mika, S., Schölkopf, B., Lengauer, T., Müller, K.R.: Engineering Support Vector Machine Kernels That Recognize Translation Initiation Sites. BioInformatics 16(9), 799–807 (2000)
[12] Brown, M.P.S., Grundy, W.N., Lin, D., Cristianini, N., Sugnet, C., Furey, T.S., Ares, J.M., Haussler, D.: Knowledge-based analysis of microarray gene expression data using support vector machines. PNAS 97, 262–267 (2000)
[13] Tsuda, K., Kawanabe, M., Rätsch, G., Sonnenburg, S., Müller, K.: A New Discriminative Kernel from Probabilistic Models. Advances in Neural information processings systems 14, 977 (2002)
[14] Sonnenburg, S., Rätsch, G., Jagota, A., Müller, K.R.: New Methods for Splice-Site Recognition. In: Dorronsoro, J.R. (ed.) ICANN 2002. LNCS, vol. 2415, p. 329. Springer, Heidelberg (2002)
[15] Sonnenburg, S.: New Methods for Splice Site Recognition. Master's thesis Humboldt University (Supervised by Müller, K.-R., Burkhard, H.-D., Rätsch, G.) (2002)
[16] Lorena, A., de Carvalho, A.: Human Splice Site Identifications with Multiclass Support Vector Machines and Bagging. In: Kaynak, O., Alpaydın, E., Oja, E., Xu, L. (eds.) ICANN 2003. LNCS, vol. 2714. Springer, Heidelberg (2003)
[17] Yamamura, M., Gotoh, O.: Detection of the Splicing Sites with Kernel Method Approaches Dealing with Nucleotide Doublets. Genome Informatics 14, 426–427 (2003)
[18] Rätsch, G., Sonnenburg, S.: Accurate Splice Site Detection for Caenorhabditis elegans. In: Schölkopf, B., Tsuda, K., Vert, J.P. (eds.) Kernel Methods in Computational Biology. MIT Press, Cambridge (2004)
[19] Degroeve, S., Saeys, Y., Baets, B.D., Rouzé, P., de Peer, Y.V.: SpliceMachine: predicting splice sites from high-dimensional local context representations. Bioinformatics 21(8), 1332–1338 (2005)
[20] Huang, J., Li, T., Chen, K., Wu, J.: An approach of encoding for predictionof splice sites using SVM. Biochimie 88, 923–929 (2006)
[21] Zhang, Y., Chu, C.H., Chen, Y., Zha, H., Ji, X.: Splice site prediction using support vector machines with a Bayes kernel. Expert Systems with Applications 30, 73–81 (2006)
[22] Baten, A., Chang, B., Halgamuge, S., Li, J.: Splice site identification using probabilistic parameters and SVM classification. BMC Bioinformatics 7(suppl. 5), S15 (2006)
[23] Anastassiou, D.: Genomic signal processing. IEEE Signal Process. Mag. 18(4), 8–20 (2001)
[24] Zhang, X., Chen, F., Zhang, Y., Agner, S.C., Akay, M., Lu, Z., Waye, M.M.Y., Tsui, S.K.: Signal processing techniques in genomic engineering. Proc. IEEE 90(12), 1822–1833 (2002)
[25] Voss, R.F.: Evolution of long-range fractal correlations and 1/f noise in DNA base sequences. Phy. Rev. Lett. 68(25), 3805–3808 (1992)

[26] Silverman, B.D., Linsker, R.: A measure of DNA periodicity. J. Theor. Biol. 118, 295–300 (1986)
[27] Ning, J., Moore, C.N., Nelson, J.C.: Preliminary wavelet analysis of genomic sequences. In: Proc. IEEE Bioinformatics Conf., pp. 509–510 (2003)
[28] deergha Rao, K., Swamy, M.N.S.: Analysis of Genomics and proteomics using DSP Techniques. IEEE Transactions on circuits abd systems 55(1) (Feburary 2008)
[29] Akhtar, M., Epps, J., Ambikairajah, E.: Signal processing in sequence analysis: advances in eukaryotic gene prediction. IEEE journal of selected topics in signal processing 2(3) (June 2008)
[30] Li, W.: The study of correlation structure of DNA sequences: A critical review. Comput. Chem. 21(4), 257–271 (1997)
[31] Anastassiou, D.: Genomic signal processing. IEEE Signal Process. Mag. 18(4), 8–20 (2001)
[32] Tiwari, S., Ramaswamy, S., Bhattacharya, A., Bhattacharya, S., Ramaswamy, R.: Prediction of probable genes by Fourier analysis of genomic sequences. Comput. Appl. Biosci. 13, 263–270 (1997)
[33] Kotlar, D., Lavner, Y.: Gene prediction by spectral rotation measure: A new method for identifying protein-coding regions. Genome Res. 18, 1930–1937 (2003)
[34] Rao, N., Shepherd, S.J.: Detection of 3-periodicity for small genomic sequences based on AR techniques. In: Proc. IEEE Int. Conf. Comm., Circuits Syst., vol. 2, pp. 1032–1036 (2004)
[35] Vaidyanathan, P.P., Yoon, B.-J.: Gene and exon prediction using allpass-based filters. Presented at the IEEE Workshop Genomic Signal Processing and Statistics, Raleigh, NC (2002)
[36] Saxonov, S., Daizadeh, I., Fedorov, A., Gilbert, W.: An exhaustive database of protein-coding intron-containing genes. Nucleic Acids Res. 28(1), 185–190 (2000)
[37] Burge, C.B., et al.: Splicing precursors to mRNAs by the spliceosomes. In: Gesteland, R.F., Cech, T.R., Atkins, J.F. (eds.) The RNA World, pp. 525–560. Cold Spring Harbor Lab. Press (1999)
[38] Cortes, C., Vapnik, V.: Support-Vector Networks. Machine Learning 20(3) (1995)
[39] Chang, C.-C., Lin, C.-J.: LIBSVM: a library for support vector machines (2001), http://www.csie.ntu.edu.tw/~cjlin/libsvm

Gibbs Motif Sampler, Weight Matrix and Artificial Neural Network for the Prediction of MHC Class-II Binding Peptides

Satarudra Prakash Singh[1,2] and Bhartendu Nath Mishra[2,*]

[1] Amity Institute of Biotechnology, Amity University Uttar Pradesh, Lucknow-226010, India
[2] Department of Biotechnology, Institute of Engineering & Technology, U.P. Technical University, Sitapur Road, Lucknow-226021, India
profbnmishra@gmail.com

Abstract. The identification of MHC class-II restricted epitope is an important goal in peptide based vaccine and diagnostic development. In the present study, we discuss the applications of Gibbs motif sampler, weight matrix and artificial neural network for the prediction of peptide binding to sixteen MHC class-II molecules of human and mouse. The average prediction performances of sixteen MHC class-II molecules in terms of Aroc, based on Gibbs motif sampler, sequence weighting and artificial neural network are 0.56, 0.55 and 0.51 respectively. However, further improvements in the performance of software tools for prediction of MHC class-II binding peptide based on various methods largely depends on the size of training and validation datasets and the correct identification of the peptide binding core.

Keywords: MHC, Weight matrix, ANN, Gibbs sampler, Motif, Epitope.

Abbreviations: ANN-artificial neural network, MHC-major histocompatibility complex, Aroc-area under receiver operating characteristic, IEDB- immune epitope database.

1 Introduction

Activation of helper T lymphocytes (HTL) require recognition by $CD4^+$ T-cell receptor of specific peptide bound to MHC class-II molecules on the surface of professional antigen presenting cells [1]. Peptides bound to MHC molecules that trigger an immune response are referred to as T-cell epitopes. The identification of MHC class-II restricted peptide epitope is an important goal in immunological research and therefore, prediction of such MHC class-II binding peptide is the very first step towards vaccine design [2]. X-ray crystallographic studies demonstrated that the MHC class-II epitope binding site consists of a groove and several pockets provided by a β-sheet and two α-helices [3]. Unlike, MHC class-I, the class-II binding groove is open at both ends. As a result, peptides binding to class-II molecules tend to be of variable length (13-25 residues). The

* Corresponding author.

S. Ranka et al. (Eds.): IC3 2009, CCIS 40, pp. 503–509, 2009.
© Springer-Verlag Berlin Heidelberg 2009

9-mer core region of the binding peptide interaction largely determines binding affinity and specificity [4]. In addition, peptide residues immediately flanking the core region have been indicated to make contact with the MHC molecule outside the binding groove, and contribute to MHC-peptide interaction [5].

The establishment of numerous MHC class-II epitope databases such as SYFPEITHI [6], MHCBN [7], AntiJen [8], EPIMHC [9] and IEDB [10] has facilitated the development of a large number of prediction algorithms based on motifs [11] weight matrices [12] and machine learning including hidden markov models [13], support vector machines [14], and artificial neural networks [15] aimed at predicting peptide binding to MHC molecules. Several other methods such as ARB [16] and SMM-align [17] were also based on scoring matrices that evaluate the contribution to binding of different residues in a peptide based on quantitative binding data.

Besides, the successful computational prediction is mainly based on the use of a sufficiently large set of high quality training data. These databases typically combine data from different sources and different experimental approaches, which can complicate the generation of consistent training and evaluation datasets. In this study, we have developed a suitable model for prediction of peptide binding to MHC class-II molecules using two web servers viz: EasyGibbs based on Gibbs sampler [18] and EasyPred based on clustering [19] and ANN [20]. We have also used a large dataset of newly published MHC class-II peptide binding affinities that were experimentally determined under uniform conditions.

2 Materials and Methods

2.1 Wang et al. 2008 Dataset

We have downloaded the assembled dataset of peptides for various MHC class-II molecules from Wang *et al.,* 2008 [21] as given in Table 1. These data comprise of a total 16 human and mouse MHC class-II molecules. The number of unique MHC-peptide affinities measured per molecule varies greatly from 3,882 for HLA-DRB1-0101 to only 39 for H-2-IEd compared to datasets available publicly on the IEDB [10]. Our new dataset significantly expands the number of measured peptide-MHC class-II interactions for a large number of MHC class-II molecules. Based on measured affinities, peptides were classified into binders (IC_{50}<1000 nM) and nonbinders (IC_{50}≥1000 nM) using experimental data. In order to, classify peptides into binders (positive data) and non-binders (negative data), a cutoff of 1000 nM was chosen, keeping in mind a immunologically relevant threshold.

2.2 Algorithms Used for the Prediction of MHC Class-II Binding Peptides

2.2.1 Gibbs Motif Sampler

MHC class-II binding peptides have a broad distribution length complicating the development of suitable prediction methods. The identification of correct alignment of a set of peptide known to bind the MHC class-II molecule leading to prediction of core of MHC class-II binding peptide based on Gibbs motif sampler is a crucial part of the algorithm [18]. Here, we used the default Gibbs sampling parameters to find the 9-mer motif in a set of MHC class-II binding peptide data using EasyGibbs web-server available at http://www.cbs.dtu.dk/biotools/EasyGibbs/

2.2.2 Sequence Weighting

The three different sequence weighting methods i.e. Henikoff and Henikoff 1/nr [19], Hobohm clustering at 62% identity [22] and no clustering are available for use to weight 9-mer peptide sequences. The Henikoff method is fast and the computation time increases linearly with the number of sequences, whereas, in the Hobohm clustering algorithm, computation time increases as the square of the number of sequences. In order to, generate weight matrix for the prediction of MHC binding peptides using Henikoff and Henikoff 1/nr sequence weighting scheme with weight on pseudo count 200, we used the EasyPred web-server available at http://www.cbs.dtu.dk/biotools/EasyPred/

2.2.3 Artificial Neural Network

The neural network architecture used here is a conventional feed-forward network [20] with an input layer with 180 neurons, one hidden layer with 2-10 neurons, and a single neuron output layer. We trained the neural network using EasyPred web server with two hidden neurons, one bin for balanced training, running upto 300 training epochs and the top 80% of the training set was used to train and rest was used to validate.

2.3 Evaluation Parameters

Based on these datasets and algorithms, we have developed a suitable computational model for prediction of binding affinities between MHC molecule and peptide. The efficiency of algorithms was determined using discrimination between binders and nonbinders. A predicted peptide belongs to one of the four categories, i.e. True Positive (TP); an experimentally binding peptide predicted as a binder, False Positive (FP); an experimentally nonbinding peptide predicted as a binder, True Negative (TN); an experimentally nonbinding peptide predicted as a nonbinder and False Negative (FN); an experimentally binding peptide predicted as nonbinder. Here, we used a non-parametric performance measure, area under receiver operating characteristic (Aroc) curve to evaluate the prediction performance of the applied algorithms. The ROC curve is a plot of the true positive rate TP/(TP+FN) on the vertical axis *vs* false positive rate FP/(TN+FP) on the horizontal axis for the complete range of the decision thresholds [25]. The values Aroc>=0.9 indicate excellent, 0.9>Aroc>=0.8 indicate good, 0.8>Aroc>=0.7 indicate marginal and 0.7>Aroc indicate poor predictions.

3 Results and Discussion

We assembled a dataset of peptide binding and nonbinding affinities for sixteen MHC class-II molecules from Peng Wang *et al,* 2008 [21]. The Table 1 gives an overview of the training and validation datasets. Compared to the training datasets publicly available on the IEDB database [10], our dataset significantly expands the number

Table 1. Overview of MHC Class-II binding peptide dataset used in study

Organism	MHC class–II alleles	Number of 15-mer training data	Number of 15-mer positive training data	Number of 15-mer validation data
Human	DRB1-0101	2557	612	1279
	DRB1-0301	315	129	158
	DRB1-0401	320	106	160
	DRB1-0404	299	81	150
	DRB1-0405	283	39	142
	DRB1-0701	315	62	158
	DRB1-0802	163	92	82
	DRB1-0901	275	113	137
	DRB1-1101	315	91	158
	DRB1-1302	171	105	86
	DRB1-1501	315	109	158
	DRB3-0101	269	196	134
	DRB4-0101	163	66	82
	DRB5-0101	315	74	158
Mouse	H2-IAb	326	234	163
	H2-IEd	17	14	9

of measured peptide-MHC interactions. From the experimental data, peptides were classified into binders (IC_{50}<1000 nM) and nonbinders ($IC_{50}\geq$1000 nM) based on measured affinities. From these datasets, the performance of the prediction methods were then measured based on area under ROC curves (Aroc). The calculation of Aroc provides a highly useful measure of prediction quality, which is 0.5 for random prediction and 1.0 for perfect prediction.

The weight matrix performance in term of Aroc is maximum (0.76) for allele DRB1-0101 and the minimum (0.33) for allele H-2-IEd based on Henikoff and Henikoff 1/nr weighting scheme. The prediction performance of the Gibbs motif sampler is maximum (Aroc=0.71) for the allele DRB1-0101 which is lower than the weight matrix method (Figure 1). The average prediction performance of sixteen MHC class-II molecules based on Gibbs sampler (Aroc=0.56) and sequence weighting (Aroc=0.55) are higher than the artificial neural network (Aroc=0.51).

From the above results it is clear that the size of training dataset may be an important factor contributing to better performance of sequence weighting for the prediction of MHC class-II binding peptides. A key difference between MHC class-I and MHC class-II molecule is that the binding groove of class-II molecules is open at both ends. As a result, the length of peptide binding to class-II molecules can vary considerably, typically ranges 13-25 amino acids long. Therefore, a requisite for all MHC class-II binding prediction approaches are the capacity to identify the correct 9-mer core residues within longer sequences that mediate the binding interaction [23]. Thus the ANN and weight matrix approaches are only suited to describe sequence motifs of fixed length e.g. 9-mer amino acids. The weight-matrix approach is only suitable for prediction of a binding event in situations, where the binding specificity can be represented independently at each position in the motif and this assumption can only be

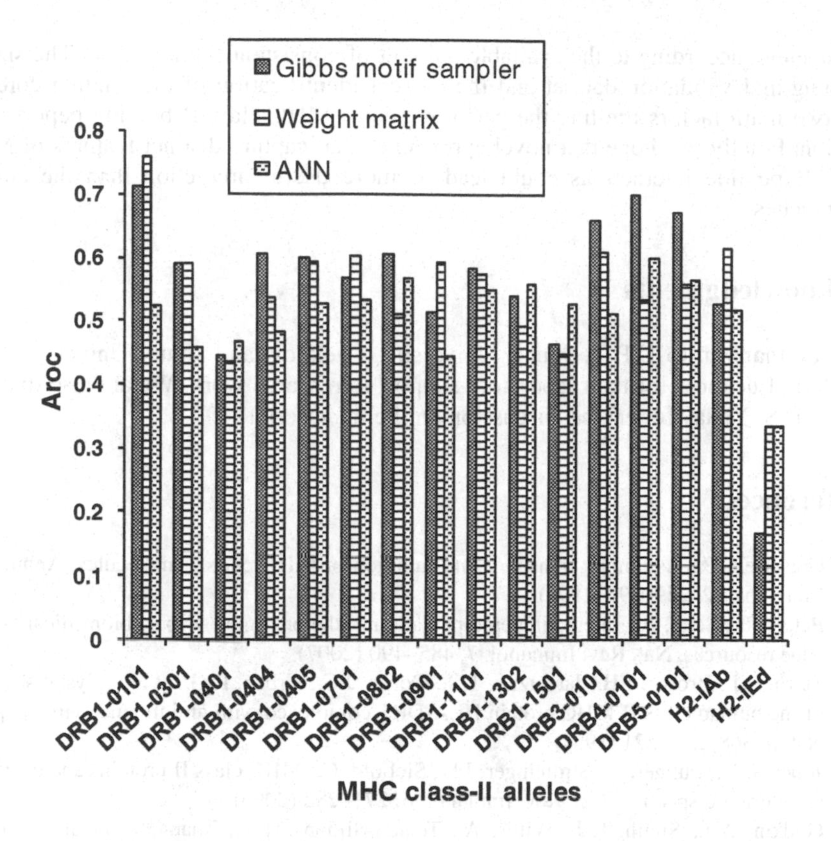

Fig. 1. Prediction performances of Gibbs motif sampler, Weight matrix (Henikoff & Henikoff 1/nr) and ANN (feed-forward back propagation) for the sixteen MHC class-II molecules

considered to be an approximation. In the binding of a peptide to the MHC molecule, the amino acids might for instance compete for the space available in the binding grove. Overall, these data suggest that there is a substantial room to improve the quality of the prediction using novel approaches that capture distinct features of MHC class-II -peptide interactions [24].

4 Conclusions

As, the identification of MHC class-II restricted epitope using wet lab experiment is expensive and time consuming, the computational methods can be used as a fast alternatives. In this paper, MHC class-II pathway was examined with a goal to develop suitable software tool to predict HTL epitope for vaccine designing. For peptide binding to MHC class-II molecule, a new prediction approach was developed based on Gibbs sampling motif, Henikoff clustering weight matrix and feed-forward back-propagation neural network. The superiority of this approach is believed to be the consequence of the flexibility in optimizing the motif, matrix and ANN training parameters, according to the available amount of consistent training data. The size of training and validation dataset and the correct identification of the binding core are the two main factors limiting the performance of MHC class-II binding peptide prediction. Finally, we hope that novel approaches that capture distinct features of MHC class-II peptide interactions could lead to more useful prediction than the current approaches.

Acknowledgments

We are thankful to U.P Technical University, Lucknow and Amity University Uttar Pradesh, Lucknow for their laboratory support to research work. We are also thankful to Dr. D.S. Yadav for critical discussion on the manuscript.

References

1. Cresswell, P.: Assembly, transport, and function of MHC class II molecules. Annu. Rev. Immunol. 12, 259–293 (1994)
2. Peters, B., Sette, A.: Integrating epitope data into the emerging web of biomedical knowledge resources. Nat. Rev. Immunol. 7, 485–490 (2007)
3. Stern, L.J., Brown, J.H., Jardetzky, T.S., Gorga, J.C., Urban, R.G., et al.: Crystal structure of the human class II MHC protein HLA-DR1 complexed with an influenza virus peptide. Nature 368, 215–221 (1994)
4. Jones, E.Y., Fugger, L., Strominger, J.L., Siebold, C.: MHC class II proteins and disease: a structural perspective. Nat. Rev. Immunol. 6, 271–282 (2006)
5. Godkin, A.J., Smith, K.J., Willis, A., Tejada-Simon, M.V., Zhang, J., et al.: Naturally processed HLA class II peptides reveal highly conserved immunogenic flanking region sequence preferences that reflect antigen processing rather than peptide-MHC interactions. J. Immunol. 166, 6720–6727 (2001)
6. Rammensee, H., Bachmann, J., Emmerich, N., Bachor, O., Stevanovic, S.: SYFPEITHI: database for MHC ligands and peptide motifs. Immunogenetics 50, 213–219 (1999)
7. Bhasin, M., Singh, H., Raghava, G.P.: MHCBN: a comprehensive database of MHC binding and non-binding peptides. Bioinformatics 19, 665–666 (2003)
8. Toseland, C.P., Clayton, D.J., McSparron, H., Hemsley, S.L., Blythe, M.J., et al.: AntiJen: a quantitative immunology database integrating functional, thermodynamic, kinetic, biophysical, and cellular data. Immunome Res. 1, 4 (2005)
9. Reche, P.A., Zhang, H., Glutting, J.-P., Reinherz, E.L.: EPIMHC: a curated database of MHC-binding peptides for customized computational vaccinology. Bioinformatics 21(9), 2140–2141 (2005)
10. Peters, B., Sidney, J., Bourne, P., Bui, H.H., Buus, S., et al.: The immune epitope database and analysis resource: from vision to blueprint. PLoS Biol. 3, 91 (2005)
11. Hammer, J., Bono, E., Gallazzi, F., Belunis, C., Nagy, Z., et al.: Precise prediction of major histocompatibility complex class II-peptide interaction based on peptide side chain scanning. J. Exp. Med. 180, 2353–2358 (1994)
12. Reche, P.A., Glutting, J.-P., Reinherz, E.L.: Prediction of MHC Class I Binding Peptides Using Profile Motifs. Human Immunology 63, 701–709 (2002)
13. Mamitsuka, H.: Predicting peptides that bind to MHC molecules using supervised learning of hidden Markov models. Proteins 33(4), 460–474 (1998)

14. Donnes, P., Elofsson, A.: Prediction of MHC class I binding peptides, using SVMHC. BMC Bioinformatics 3(1), 25 (2002)
15. Honeyman, M.C., Brusic, V., Stone, N.L., Harrison, L.C.: Neural network-based prediction of candidate T-cell epitopes. Nat. Biotechnol. 16, 966–969 (1998)
16. Bui, H.H., Sidney, J., Peters, B., Sathiamurthy, M., Sinichi, A., et al.: Automated generation and evaluation of specific MHC binding predictive tools: ARB matrix applications. Immunogenetics 57, 304–314 (2005)
17. Nielsen, M., Lundegaard, C., Lund, O.: Prediction of MHC class II binding affinity using SMM-align, a novel stabilization matrix alignment method. BMC Bioinformatics 8, 238 (2007)
18. Nielsen, M., Lundegaard, C., Worning, P., Hvid, C.S., Lamberth, K., Buus, S., Brunak, S., Lund, O.: Improved prediction of MHC class I and class II epitopes using a novel Gibbs sampling approach. Bioinformatics 20, 1388–1397 (2004)
19. Henikoff, S., Henikoff, J.: Amino acid substitution matrices from protein blocks. Proc. Natl. Acad. Sci. USA 89, 10915–10919 (1992)
20. Nielsen, M., Lundegaard, C., Worning, P., Lauemøller, S.L., Lamberth, K., Buus, S., Brunak, S., Lund, O.: Reliable prediction of T-cell epitopes using neural networks with novel sequence representations. Protein Sci. 12(5), 1007–1017 (2003)
21. Wang, P., Sidney, J., Dow, C., Mothé, B., Sette, A., et al.: A Systematic Assessment of MHC Class II Peptide Binding Predictions and Evaluation of a Consensus Approach. PLoS Comput. Biol. 4(4), e1000048 (2008)
22. Hobohm, U., Scharf, M., Schneider, R., Sander, C.: Selection of representative protein data sets. Protein Sci. 1(3), 409–417 (1992)
23. Singh, S.P., Mishra, B.N.: Prediction of MHC binding peptide using Gibbs motif sampler, weight matrix and artificial neural network. Bioinformation 3(4), 150–155 (2008)
24. Singh, S.P., Mishra, B.N.: Ranking of binding and nonbinding peptides to MHC class–I molecules using inverse folding approach: Implications for vaccine design. Bioinformation 3(2), 72–82 (2008)

Classification of Phylogenetic Profiles for Protein Function Prediction: An SVM Approach

Appala Raju Kotaru and Ramesh C. Joshi

Department of Electronics and Computer Engineering,
Indian Institute of Technology, Roorkee, India
kotaruraju@gmail.com, joshifcc@iitr.ernet.in

Abstract. Predicting the function of an uncharacterized protein is a major challenge in post-genomic era due to problems complexity and scale. Having knowledge of protein function is a crucial link in the development of new drugs, better crops, and even the development of biochemicals such as biofuels. Recently numerous high-throughput experimental procedures have been invented to investigate the mechanisms leading to the accomplishment of a protein's function and Phylogenetic profile is one of them. Phylogenetic profile is a way of representing a protein which encodes evolutionary history of proteins. In this paper we proposed a method for classification of phylogenetic profiles using supervised machine learning method, support vector machine classification along with radial basis function as kernel for identifying functionally linked proteins. We experimentally evaluated the performance of the classifier with the linear kernel, polynomial kernel and compared the results with the existing tree kernel. In our study we have used proteins of the budding yeast saccharomyces cerevisiae genome. We generated the phylogenetic profiles of 2465 yeast genes and for our study we used the functional annotations that are available in the MIPS database. Our experiments show that the performance of the radial basis kernel is similar to polynomial kernel is some functional classes together are better than linear, tree kernel and over all radial basis kernel outperformed the polynomial kernel, linear kernel and tree kernel. In analyzing these results we show that it will be feasible to make use of SVM classifier with radial basis function as kernel to predict the gene functionality using phylogenetic profiles.

Keywords: Protein function prediction, support vector machine, phylogenetic profiles.

1 Introduction

Protein function prediction is one of the key goals in computational biology. The concept of protein function is highly context-sensitive and not very well-defined. In fact, this concept typically acts as an umbrella term for all types of activities

S. Ranka et al. (Eds.): IC3 2009, CCIS 40, pp. 510–520, 2009.
© Springer-Verlag Berlin Heidelberg 2009

that a protein is involved in, be it cellular, molecular or physiological. Another way to define is "function is everything that happens to or through a protein" [1].

Assigning functions to proteins is primarily done through biochemical experimentations. However, irrespective of the details, these approaches have low-throughput because of the huge experimental and human effort required in analyzing a single gene or protein. Rapid advances in genome sequencing technology have resulted in a continually expanding sequence-function gap for the discovered proteins. To minimize the gap various procedures have been invented to investigate the mechanisms leading to the accomplishment of a protein's function.

Traditionally researchers have been using nucleotide sequence in case of genes, or amino acid sequence in case of proteins to determine the function of genes or the corresponding proteins [2]. This approach relies on the fact that a set of genes that have sufficiently similar sequences also perform the same function. The explosive growth of the amount of sequence information available in public databases has made such an approach particularly accurate and it is an indispensable tool towards functional genomics.

Recently numerous high-throughput experimental procedures have been invented to investigate the mechanisms leading to the accomplishment of a protein's function. These procedures have generated a wide variety of useful data such as gene expression data sets [3], phylogenetic data[4] and protein interaction networks[5]. These data offer various insights into a protein's function and related concepts. We considered phylogenetic profiles which come under phylogenetic data for predicting the function.

The phylogenetic profile of a protein is a binary vector whose length is the number of available genomes. The vector contains a 1 in the ith position if the ith genome contains a homologue of the corresponding gene, and 0 otherwise.

In this paper we proposed a method for classification of phylogenetic profiles using supervised machine learning method, support vector machine classification along with radial basis function as kernel for identifying functionally linked proteins. In the implementation of the classifier we compared with linear kernel, polynomial kernel[16] and tree kernel[6]. Our work focuses on the yeast genome and covers a large number of protein functions defined in the Munich Information Centre for protein sequences (MIPS) database [15]. Our experimental results show that the accuracy achieved by the kernel functions widely depend on the function that we try to predict. For certain classes we achieve high accuracy using polynomial kernel and for some classes the high accuracy is obtained using radial basis function.

The rest of this paper is organized as follows. Section 2 presents previous and related work in protein function prediction using phylogenetic profiles. Section 3 explains method used svm classifier and the various kernel functions used. Section 4 describes the sources and the structure of the dataset used in the study, phylogenetic profiles and protein functional class assignment. Section 5 discusses the experimental results. Section 6 provides some concluding remarks.

2 Brief Review

The first study to analyze protein function using phylogenetic profiles was presented by Pellegrini[4]. The underlying hypothesis was that proteins which function together in a pathway or a protein complex are likely to have a similar evolutionary path. To test this hypothesis, phylogenetic profiles were constructed from the fully sequenced genomes of sixteen organisms, other than E.coli, which was the model organism used in this study. Using three E. coli proteins, RL7, FlgL and His5, it was verified that proteins with profiles differing by at most one out of sixteen bits are indeed functionally related as per the SwissProt annotation.

The idea of relaxing phylogenetic profiles is carried forward by [7], particularly for the annotation of bacterial genomes. The modification suggested here is to use the normalized BLAST score [2] denoting the best match for a protein in a genome, instead of using a 0 or 1. Annotation is carried out by finding the statistically dominant class of the MultiFun database in the neighborhood of a protein induced using cosine similarity. Results better than [4] are shown, thus showing the potential of real-valued phylogenetic profiles. This annotation procedure is available via the website of the Phydbac database.

Wu et al. [8] advocate the use of more general measures of similarity for pairs of phylogenetic profiles. Three popularly used measures of similarity, namely the Hamming distance (D), Pearson's Correlation Coefficient (r) and mutual information (MI) are evaluated for this task. It is concluded from the analysis that, although the three measures are strongly related to each other, MI is the most informative measure of profile similarity for inferring functional relationship between two proteins. In addition, it is argued that proteins with complimentary profiles may suggest that they are functionally similar, which is likely to be missed if exact similarity of profiles is required.

Certain amount of research is focused on selection of reference genome for construction of Phylogenetic Profiles. Sun [9] suggested that reference organism should be selected based on genetic distance, rather than the relationship of taxanomy tree, because homology information used in the construction of phylogenetic profiles directly relies on the genetic distance of the sequences. In the paper by Loganantharaj [10] selection of reference organism, for all members in a clade should evolve from a common ancestor and the one far apart from the rest is close to their ancestor. Therefore select the organism that is evolutionarily the farthest apart from the rest of the organisms in that clade essentially selecting an outlier of that clade.

Snitkin[11] explored the application of phylogenetic profiling to eukaryotic genomes. It concluded that the current set of completely sequenced eukaryotic organisms, phylogenetic profiling using profiles generated from any of the commonly used techniques was found to yield extremely poor results.

A new technique, namely Annotating Genes with Positive Samples(AGPS), for defining negative samples in gene function prediction is proposed by Xing[12]. The AGPS algorithm is different from existing methods, which have inappropriate assumptions about genes that have no target annotations. Specifically, this approach does not regard the genes without target annotation as negative samples because one gene generally can have multiple functions and it may indeed have the function even though it is not annotated with the target function currently.

3 Methods

The goal of supervised learning methods, also known as classification methods, is to build a set of models that can correctly predict the class of the different objects. The input to these methods is a set of objects (training set), the classes that these objects belong to (dependent variable), and a set of variables describing different characteristics of the objects (independent variables). Once such a predictive model is built, and then it can be used to predict the class of the objects for which class information is not known a priori. The key advantages of supervised learning methods over unsupervised methods such as clustering, is that by having an explicit knowledge of the classes the different objects belong to, these algorithms can perform an effective feature selection (e.g., ignoring some of the independent variables) if that leads to better prediction accuracy.

Over the years a variety of different classification algorithms have been developed by the machine learning community. Depending on the characteristics of the data sets being classified certain algorithms tend to perform better than others. In recent years, algorithms based on support vector machine have been shown to produce reasonably good results for problems in which the independent variables are continuous and homogeneous. Hence, our study uses primarily this classification algorithm.

3.1 Support Vector Machine

Support vector machine (SVM) is a learning algorithm proposed by [13]. This algorithm was introduced to solve two-class pattern recognition problems using the Structural Risk Minimization principle [13]. Given a training set in a vector space, this method finds the best decision hyper plane that separates two classes. The quality of a decision hyper plane is determined by the distance (referred as margin) between two hyper planes that are parallel to the decision hyper plane and touch the closest data points of each class. The best decision hyper plane is the one with the maximum margin. By defining the hyper plane in this fashion, SVM is able to generalize to unseen instances quite effectively. The SVM problem can be solved using quadratic programming techniques [13]. SVM extends its applicability on the linearly non-separable data sets by either using soft margin hyper planes, or by mapping the original data vectors into a higher dimensional space in which the data points are linearly separable. The mapping to higher dimensional spaces is done using appropriate kernel functions, resulting in efficient algorithms. A new test object is classified by looking on which side of the separating hyper plane it falls and how far away it is from it.

$$k(a,b) = e^{-gamma*\|a-b\|^2} \tag{1}$$

We tuned the classifier for this kernel function and got the gamma value as 1. The kernel functions are conveniently implemented in the software package SVM Light[14] used in this work.

4 Data

In our study we used the yeast Saccharomyces cervisiae genome which is the same dataset as in [6],[16]. Proteins with accurate functional classification were selected. the phylogenetic profiles of 2465 yeast genes selected for their accurate functional classifications were generated by computing the E-value reported by BLAST version 2.0 [2] in a search against each of the 24 selected target genomes. Each bit in the phylogenetic tree was set to 0 or 1 based on whether the E-value for the corresponding organism was larger or smaller than 1 respectively.

Determining the functional classes of the different protein is very much ongoing process and to a large extent one of the key steps in understanding the genomes of the various species. Fortunately, in the case of the yeast genome, there exist extensive annotations for a large fraction of the genes. For our study we used the functional annotations that are available in the MIPS database [15]. For the data we used there are 251 gene functional classes, organized in tree structure. Based on the amount of information that is known for each gene, the MIPS database assigns it to one or more nodes of the functional classes. Genes for whom detailed functional information is known tend to be assigned towards the leaves of the tree, whereas genes for which the information is more limited tend to be assigned at the higher-level nodes of the tree. Consider a gene YHR037W is assigned a function named amino-acid biosynthesis. Amino-acid biosynthesis is a sub-function of the top-level function metabolism, YHR037W has all those functions, amino-acid biosynthesis, amino-acid metabolism, metabolism, a function at a node and all the functions of its path to the top-level node. A gene also may have functions assigned from multiple branches. For the case of YAL001C it has functions from the top level classes transcription, cellular organization and their subcategories. As a result of this functional class assignment, each gene has an average of 3 to 4 functions assigned on the average. The distribution of the number of classes at the different level of the tree is shown in table 1.

Table 1. Number of defined function categories at each level in the tree

Level	Functions
1	16
2	107
3	86
4	40
5	2

Most of the functions are small in their size, which makes functionality prediction difficult. For this reason only functional classes that contain at least 10 genes were extracted which resulted in 133 functional classes. Figure 1 shows the size of the different genes function.

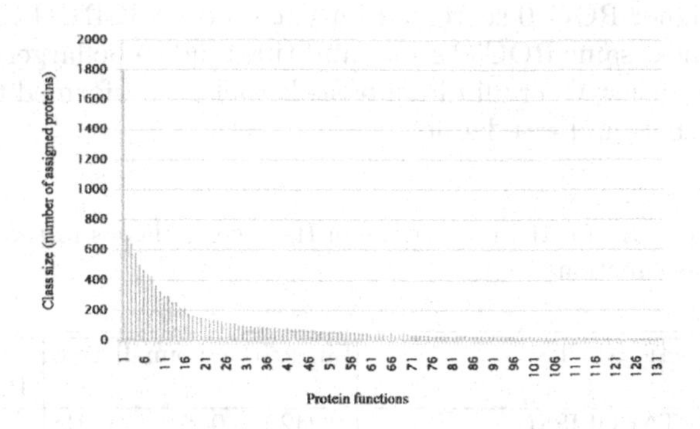

Fig. 1. Distribution of the size of gene functions

A 3-fold cross validation was adopted for the experiments. For each functional class, two third of members are randomly selected as positive training examples, and rest as positive testing examples. Genes not belonging in that class were randomly split into two thirds as negative training and one third as negative testing examples. Now the positive training examples and negative training examples are combined to form training data, and positive testing examples and negative testing examples are combines to form testing data for that particular class.

5 Results and Discussion

SVM is trained for each functional category to predict whether a gene should be assigned to it or not based on phylogenetic profiles. Using 3-fold cross validation repeated 10 times for each class, we compared the performance of proposed kernel function to svm with the linear kernel, polynomial kernel [16] and tree kernel [6] through their receiver operating characteristic (ROC) curves, i.e., the plot of true positives as a function of false positives. After training the svm with training data the output file we get when we test the model with test data contains a score related to the distance between the test example and the linear boundary in the feature space. As each functional class contains a small number of genes learning problem is very unbalanced (there are few positive examples but many negative ones). This issue is handled by giving more weight to the positive examples in svm learning. Moreover this implies that only a small percentage of false positives can be tolerated in real world applications (such as function predictions), so we measured the ROC50 score for each SVM, i.e., the area under the ROC curve up to the first 50 false positives [17].

Table 2 shows the functional categories of the level 1 in functional class tree and their ROC50 scores obtained by SVM using linear kernel, polynomial kernel, radial basis and tree kernel. It show that the performance of the radial basis kernel is similar to polynomial kernel is some functional classes. However for the class TRANSPORT FACILITATION both linear kernel and polynomial kernel has much higher ROC50 score and for class TRANSCRIPTION all the four kernels have almost same ROC50 score, and this tend to be larger and more general classes than other. Over all radial basis kernel outperformed the polynomial kernel linear kernels and tree kernel.

Table 2. ROC50 scores for the predictions of the level 1 classes in the functional class tree using 4 kernel functions

Functional Class	Linear	Polynomial	Tree	RBF (Proposed Kernel)
METABOLISM	0.242	0.272	0.218	0.32316
ENERGY	0.099	0.265	0.105	0.29
PROTEIN SYNTHESIS	0.061	0.274	0.186	0.288
CELLULAR ORGANIZATION	0.169	0.218	0.221	0.229
IONIC HOMEOSTASIS	0.047	0.179	0.105	0.217
TRANSPORT FACILITATION	0.318	0.381	0.273	0.19
CELLULAR TRANSPORT AND TRANSPORT MECHANISMS	0.072	0.109	0.054	0.147
CELL RESCUE, DEFENSE, CELL DEATH AND AGEING	0.054	0.113	0.049	0.141
TRANSCRIPTION	0.114	0.125	0.117	0.122
CELL GROWTH, CELL DIVISION AND DNA SYNTHESIS	0.059	0.08	0.06	0.086
PROTEIN DESTINATION	0.048	0.09	0.04	0.078
CELLULAR COMMUNICATION /SIGNAL TRANSDUCTION	0.057	0.122	0.079	0.062
CELLULAR BIOGENESIS	0.03	0.031	0.034	0.033

We plotted on Figure 2 the ROC50 scores of the 50 largest protein functions in descending order sorted by size. The sum of the ROC50 scores over the 50 functions for the linear, polynomial, radial basis and tree kernel are 4.8, 8.0, 7.5, and 6.0 respectively. Even though radial basis has outperformed all three kernels, the reason for sum being more for polynomial kernel is in classes like amino-acid metabolism polynomial has 0.437, and transport facilitation also it is 0.381 which is much higher than other kernels. So the difference of score is more in the classes where polynomial kernel value is more, so the sum of polynomial is more.

Figures 3 shows the ROC curves up to 50 false positives corresponding to the top three classes with the highest radial basis function ROC score, in order to further study the differences in performance. These plots show that in several cases the RBF kernel significantly outperforms the other three kernels.

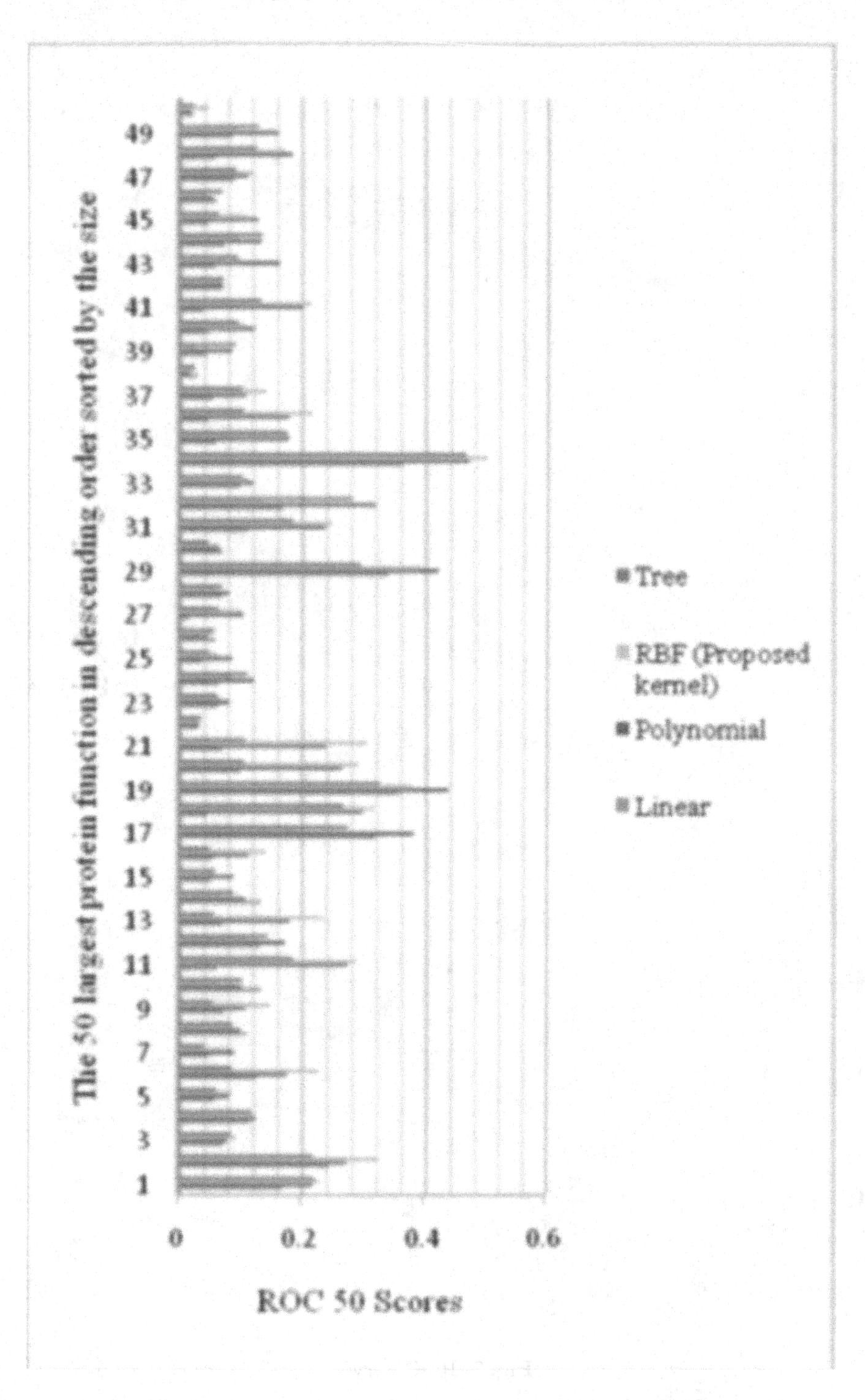

Fig. 2. Comparison of 4 SVM kernel types

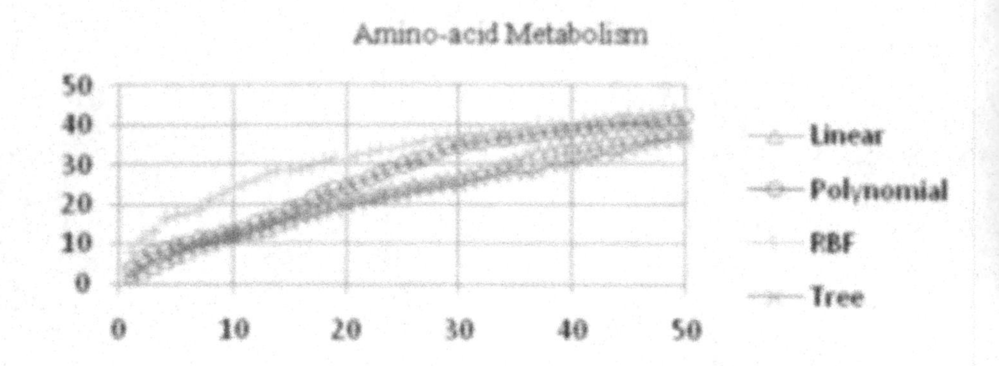

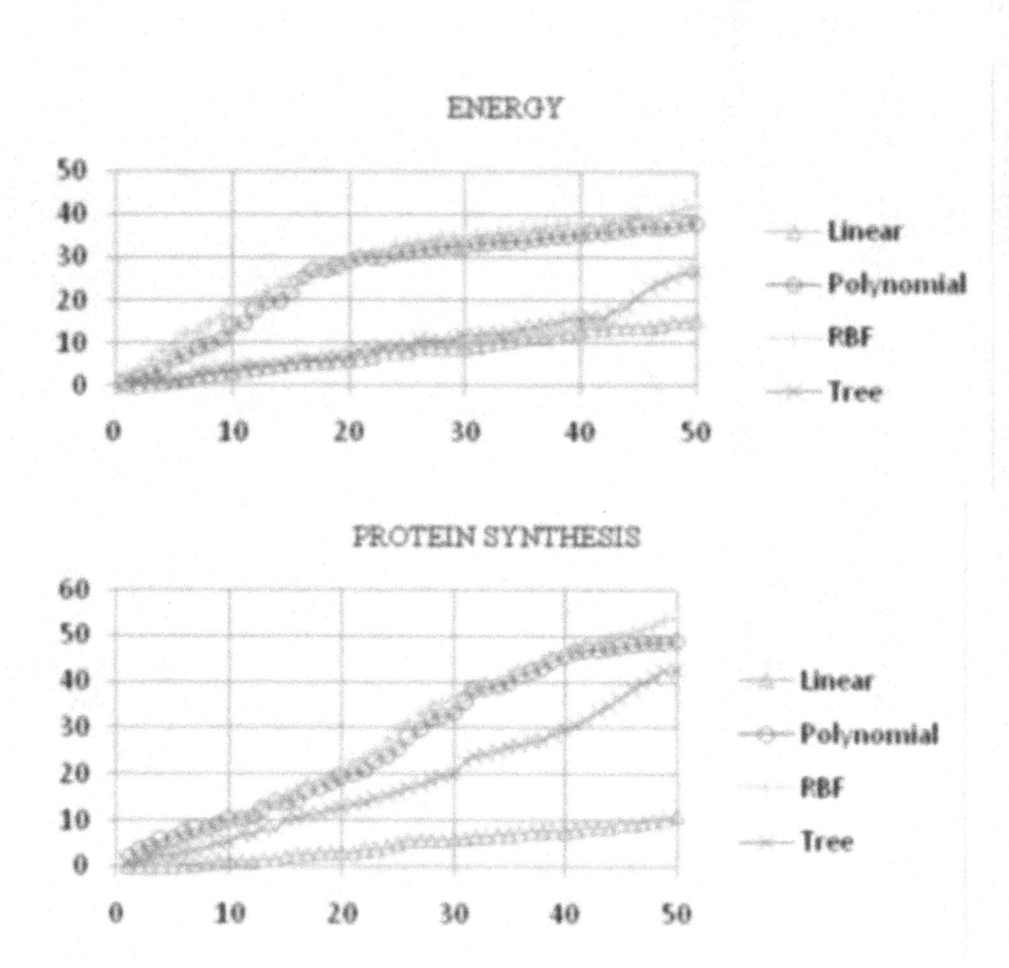

Fig. 3. ROC curves

5.1 Conclusion

In this paper we explored the possibility of using phylogenetic profiles to find the functions of the genes. We proposed a method for classification of phylogenetic profiles using supervised machine learning method, support vector machine classification along with radial basis function as kernel for identifying functionally linked proteins. We tested the algorithm for 2465 annotated genes from the yeast genome. The goal of the classification was to predict functional categories of genes defined in the MIPS database. Because of the functions are small in their size, which makes functionality prediction difficult. For this reason only functional classes that contain at least 10 genes were extracted which resulted in 133 functional classes.

We compared the results with linear kernel, polynomial kernel, and tree kernel. Polynomial kernel and RBF kernel together gave prediction more accurate than linear and tree kernel. Overall RBF kernel outperformed other three kernels. The sum of the ROC50 scores over the 50 functions for the linear, polynomial, radial basis and tree kernel are 4.8, 8.0, 7.5, and 6.0 respectively. Even though radial basis has outperformed all three kernels, the reason for sum being more for polynomial kernel is in classes like amino-acid metabolism polynomial has 0.437, and transport facilitation also it is 0.381 which is much higher than other kernels. So the difference of score is more in the classes where polynomial kernel value is more, so the sum of polynomial is more.

Based on these experiments and the analysis we believe it will be feasible to make use of SVM classifier with radial basis function as kernel to predict the gene functionality using phylogenetic profiles.

References

1. Rost, B., Liu, J., Nair, R., Wrzeszczynski, K.O., Ofran, Y.: Automatic prediction of protein function. Cellular and Molecular Life Sciences 60, 2637–2650 (2003)
2. Altschul, S.F., Madden, T.L., Schffer, A.A., Zhang, j., Zhang, Z., Miller, W., Lipman, D.J.: Gapped BLAST and PSI-BLAST: a new generation of protein database search programs. Nucleic Acids Research 25(17), 3389–3402 (1997)
3. Ben-Dor, A., Shamir, R., Yakhini, Z.: Clustering gene expression patterns. Journal of Computational Biology 6(3-4), 281–297 (1999)
4. Pellegrini, M., Marcotte, E.M., Thompson, M.J., Eisenberg, D., Yeates, T.O.: Assigning protein functions by comparative genome analysis: protein phylogenetic profiles. Proc. Natl. Acad. Sci. U.S.A. 96(4), 4285–4288 (1999)
5. Schwikowski, B., Uetz, P., Fields, S.: A network of protein-protein interactions in yeast. Nature Biotechnology 18(12), 1257–1261 (2000)
6. Vert, J.P.: A tree kernel to analyze phylogenetic profiles. Bioinformatics 18(1), S276–S284 (2002)
7. Enault, F., Suhre, K., Abergel, C., Poirot, O., Claverie, J.M.: Annotation of bacterial genomes using improved phylogenomic profiles. Bioinformatics 19(1), i105–i107 (2003)
8. Wu, J., Kasif, S., Delisi, C.: Identification of functional links between genes using phylogenetic profiles. Bioinformatics 19(12), 1524–1530 (2003)

9. Sun, J., Xu, J., Liu, Z., Liu, Q., Zhao, A., Shi, T., Li, Y.: Refined phylogenetic profiles method for predicting protein-protein interactions. Bioinformatics 21(16), 3409–3415 (2005)
10. Loganantharaj, R., Atwi, M.: Towards validating the hypothesis of phylogenetic profiling. BMC Bioinformatics 8(7), S25 (2007)
11. Snitkin, E.S., Gustafson, A.M., Mellor, J., Wu, J., DeLisi, C.: Comparative assessment of performance and genome dependence among phylogenetic profiling methods. BMC Bioinformatics 7(420) (2006)
12. Zhao, X.-M.: Yong, W., Luonan, C., Kazuyuki, A.: Gene function prediction using labeled and unlabeled data. BMC Bioinformatics 9(57) (2008)
13. Vapnik, V.N.: The nature of stastical learning theory. Springer, New York (1995)
14. Joachims, T.: Making large-Scale SVM Learning Practical. In: Schölkopf, B., Burges, C., Smola, A. (eds.) Advances in Kernel Methods - Support Vector Learning, pp. 169–184. MIT-Press, Cambridge (1999)
15. Mewes, H.W., Fridhman, D., Guldener, U., Mannhaupt, G., Mayer, K., Mokrejs, M., Morgenstern, B., Munsterkoetter, M., Rudd, S., Weil, B.: MIPS: a databse for genomes and proteins sequences. Nucleic Acids Research 30, 31–34 (2002)
16. Narra, K., Liao, L.: Use of extended phylogenetic profiles with e-values and support vector machines for protein family classification. International Journal of Computer and Information Science 6(1), 58–63 (2005)
17. Gribskov, M., Robinson, N.: Use of receiver operating characteristic (roc) analysis to evaluate sequence matching. Computers and Chemistry 20, 25–33 (1996)

FCM for Gene Expression Bioinformatics Data

Dhiraj Kumar, Santanu Kumar Rath, and Korra Sathya Babu

Dept of Computer science and Engineering,
National Institute of Technology Rourkela, India, 769008
{kumardhiraj.nit.rourkela,rath.santanu,
sathyababukorra}@gmail.com

Abstract. Clustering analysis of data from DNA microarray hybridization studies is essential for evaluating and identifying biologically significant co-expressed genes. The K-means algorithm is one of the most widely used clustering technique. It attempts to solve the clustering problem by assigning each gene to a single cluster. However, in practice especially in case of Bioinformatics data, one gene can be found in many clusters simultaneously. To sort out this problem, Fuzzy C-means (FCM) clustering algorithm is applied to microarray data. Two pattern recognition data (IRIS and WBCD data) and thirteen microarray data is used to evaluate performance of K-means and Fuzzy C-means. Improvement of approx. 30 percent clustering accuracy is achieved in case of FCM compared to K-means algorithm. Extensive simulation results shows that the FCM clustering algorithm was able to provide the highest accuracy and generalization results compared to K-means clustering algorithm.

Keywords: K-means clustering, Fuzzy c-means, Microarray, Cancer data, Gene Expression Analysis, Bioinformatics.

1 Introduction

In recent years, the DNA microarray [17], [29], [30], [31], [32], [41], [42], [43], [44] has become an important and widely used technology since it enables the possibility of examining the expressions of thousands of genes simultaneously in a single experiment. A key step in the analysis of gene expression data is the detection of gene groups that manifest similar expression patterns. The main algorithmic problem here is to cluster multi-conditions gene expression patterns. Basically, a cluster algorithm partitions entities into groups based on the given features of the entities, so that the clusters are homogeneous and well separated. Although a number of clustering methods have been studied in the literature [2], [7], [11], [18], [19], [20], [29], [30], [31], [32], [33], [34], [35], [36], [37], [41], [42], [43], [44] to cluster Microarray data, they are not satisfactory in terms of automation, quality, and efficiency. Over the past decade, a number of several variants of FCM clustering algorithm have been also applied to cluster gene expression datasets [33], [34], [35], [36], and [37]. But, none of these papers gave performance evaluation of a clustering algorithm based on percentage

S. Ranka et al. (Eds.): IC3 2009, CCIS 40, pp. 521–532, 2009.
© Springer-Verlag Berlin Heidelberg 2009

accuracy rather they use various cluster validation index [38], [40] to evaluate the performances of a clustering algorithm. The problem here with existing approaches is how to decide about accuracy of a clustering algorithm even if a cluster validation index is optimal [38], [40].

In this paper, we use FCM to identifying and validating clusters in multivariate datasets and apply it to the mining of multi-conditions gene expression data. Through experiments conducted on real gene expression data and machine learning data, the FCM clustering approach is shown to deliver higher clustering quality than K-means clustering algorithms.

The rest of the paper is organized as follows: In section 2, related studies are presented. Methodology is described in section 3. Experiments conducted to evaluate the performance of the k-means and FCM method are presented in section 4, Conclusions in section 5 followed by references.

2 Related Work

In recent years, a number of clustering methods have been proposed, and they can be classified into several different types: partitioning-based methods (e.g., k-means[25], k-medoids, PAM, and CLARA) [24], hierarchical methods (e.g., UPGMA [25], CURE [23]), density-based methods (e.g., CAST [19], DBSCAN [22]), grid-based methods (e.g., CLIQUE [17]), model-based methods (e.g., SOM [26]), etc. Over the past decade, a number of several clustering methods also have been applied to cluster gene expression datasets [11], [18], [19], [20], [21], [29], [30], [31], and [32]. A good review of work about clustering of Bioinformatics data can be found in reference [39], [40].

3 Methodology

In this section, we first define the problem. Then we describe K-means algorithm and FCM approach in detail, including the principles behind it.

3.1 Problem Definition

The problem of multivariate gene expression clustering can be described briefly as follows. Given a set of genes with unique identifiers, a vector Ei = {Ei1, Ei2… Ein} is associated with each gene i, where Eij represents the response of gene i under condition j. The goal of gene expression clustering is to group genes based on similar expressions over all the conditions. i.e. genes with similar corresponding vectors should be classified into the same cluster.

3.2 Clustering

The clustering problem is defined as the problem of classifying n objects into K clusters without any a priori knowledge.

Let the set of n points be represented by the set S and the K clusters be represented by $\{C_1, C_2, \ldots, C_K\}$. Then

$$C_i \neq \acute{\emptyset} \quad \text{for } i = 1,2,\dots,K,$$
$$C_i \cap C_j = \acute{\emptyset} \text{ for } i = 1,\dots,K, \quad j = 1,\dots,K \text{ and } i \neq j$$
$$\text{and } \bigcup_{i=1}^{K} C_i = S.$$

3.3 K-Means Clustering Algorithm

The K-means algorithm [25] is one of the most widely used clustering technique. It attempts to solve the clustering problem by optimizing a given metric (Euclidean distance).

The basic steps of the K-means algorithm has been explained in the following references [23], [24], [25], [26], and [41] It has been shown in Ref. [26] that K-means algorithm may converge to values that are not optimal. Also global solutions of large problems cannot be found with a reasonable amount of computation effort [27]. It is because of these factors that several approximate methods are developed to solve the underlying optimization problem.

3.4 Fuzzy C-Means Clustering Algorithm

The Fuzzy C-Means algorithm (FCM) generalizes the hard c-means algorithm to allow points to partially belong to multiple clusters. Therefore, it produces a soft partition for a given data set. In fact, it generates a constrained soft partition. [27], [28]. The details about FCM can be obtained from following references [7], [23], [24], [25], [33], [34], [35], [36], and [37]. One of the important issues in FCM is the selection of fuzziness parameter "m". Finding an optimal value is a cumbersome process. Recently a little amount of work has been done in this area [7]. But, development of an optimal method for selection of "m" still remains open issue as a research problem. A Little insight on selection of parameter "m" is given in section 4.2.

4 Experimental Work

To validate the feasibility and performance of the proposed approach, Implementation was carried out on MATLAB 7.0(Intel C2D processor, 2.66GHz, 2 GB RAM) and applied it to several datasets whose short descriptions are given below.

4.1 Datasets

This section describes the datasets used for assessing the performance of k-means and FCM, which are listed in Table 1. The first two datasets represent pattern recognition data, while the other represents gene-expression microarray data. For pattern recognition, IRIS [1], [2] and WBCD [2], [16] data were used and for microarray data serum data (Iyer et. al) [3], [4], [8], [14], yeast data (Cho et. al) [5], [6], [7], [8], [14], leukemia data (Golub et. al) [8], [14], [23], breast data (Golub et. al) [9], [15], Lymphoma data (Alizadeh et al.) [9], [10], [11], lung cancer (Bhattacharjee et. al) [9], [12], and St. Jude leukemia data (Yeoh et. al) [9], [13] were used. To identifying common subtypes in independent disease data four different types of breast data (Golub et. al) and four DLBCL (Diffused Large B-cell Lymphoma) data are taken.

4.2 Parameter Selection for FCM

The proper selection of *"m"* is an important and tough task in fuzzy c-means (FCM) clustering algorithm, which directly affects the performance of algorithm and the validity of fuzzy cluster analysis. Lots of work to study the fuzzy exponent *"m"* has been done [10], [11], [12]. The value of *"m"* should be always greater than 1 for FCM to be converging for optimal solution [28]. In the literature about FCM, *"m"* is commonly fixed to 2. 2 is not an appropriate fuzziness parameter for microarray data. For m = 2 for the microarray data sets, it was found that in many of the cases, FCM

Table 1. Comparison of Results for All Fifteen Datasets

Datasets	Primary source	Secondary source	Dimension	# cluster	K-means	FCM		% Increase In accuracy
					# correct	mfuz	# correct	
IRIS	[1]	[2]	[150x4]	3	133	4.2	136	2
WBCD	[16]	[2]	[683x9]	2	429	1.58	654	32.94
Iyer data/Serum data	[4]	[3], [8], [14]	[517x12]	11	268	1.92	252	-3.1
Cho data (yeast data)	[5],[6]	[7], [8], [14]	[386x16]	5	235	2.15	246	2.85
Leukemia (Golub Experiment)	[23]	[8], [14]	[72x7129]	2	43	1.28	40	-4.16
Breast data A	[15]	[9]	[98x1213]	3	71	1.6	86	15.31
Breast data B	[15]	[9]	[49x1024]	4	26	1.15	32	12.24
Breast Multi data A	[15]	[9]	[103x5565]	4	82	1.3	93	10.68
Breast Multi data B	[15]	[9]	[32x5565]	4	17	1.79 5	16	-3.125
DLBCL A	[10],[11]	[9]	[141x661]	3	75	1.40	83	5.669
DLBCL B	[10],[11]	[9]	[180x661]	3	140	1.28 5	136	-2.22
DLBCL C	[10],[11]	[9]	[58x1772]	4	30	1.34 5	44	24.14
DLBCL D	[10],[11]	[9]	[129x3795]	4	55	1.25	65	7.74
Lung Cancer	[12]	[9]	[197x581]	4	142	1.07	150	4.06
St. Jude Leukemia data	[13]	[9]	[248x985]	6	211	1.26	219	3.23

failed to extract any clustering structure underlying within data, since all the membership values became similar. Some literatures [10], [11] illuminated that the best range of fuzzy exponent, m, is [1.5, 2.5]. It was shown in [28] that when m goes to infinity, values of u_{ki} tends to1/K. Thus, for a given data set, there is an upper bound value for m (m_{ub}), above which the membership values resulting from FCM are equal to 1/K. When higher "*m*" values are used, the u_{ki} distribution becomes more spread out. With a lower value of "*m*" all genes become strongly associated to only one cluster, and the clustering is similar to that obtained with K-means. Thus, the selected value for "*m*" appears to be a good compromise between the need to assign most genes to a given cluster, and the need to discriminate genes that classify poorly.

4.3 Results and Discussion

Table 1 summarizes the result of FCM and K-means for all fifteen datasets. It contains the basic information about the various datasets which have used in experimental work. This deals with the comparative analysis of FCM and k-means for all fifteen datasets. Fig.1 shows the comparative results in bar chart. Fig.2 represents the FCM convergence graph for all fifteen datasets. The figure plots the fnorm of Membership partition matrix U_i vs. Number of iterations.

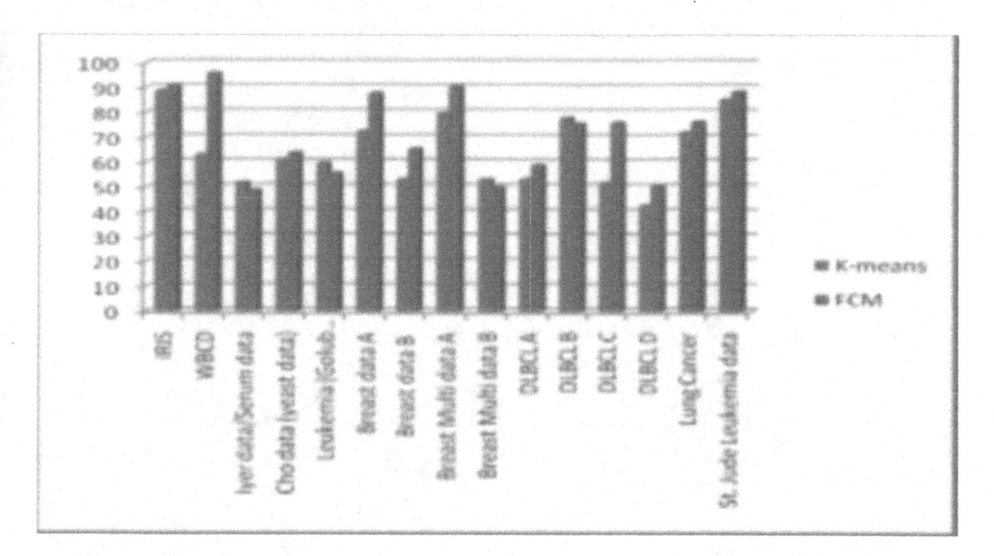

Fig. 1. Comparison of accuracy for k-means and FCM, Horizontal axis represents different Datasets whereas vertical axis represents accuracy in percentage. Blue Bar denotes result for k-means and red indicates FCM.

Table 2 shows the result obtained by K-means and FCM for IRIS data. The total count error in case of K-means algorithm is 17 whereas in case of FCM, it is 14 (This result is equivalent to result obtained in ref. [27]). Percentage increase in accuracy for FCM is by 2 %. 100 % accuracy for cluster 1 is attained in K-means and FCM. Percentage accuracy is increase by 14 % for cluster 2 in case of FCM.

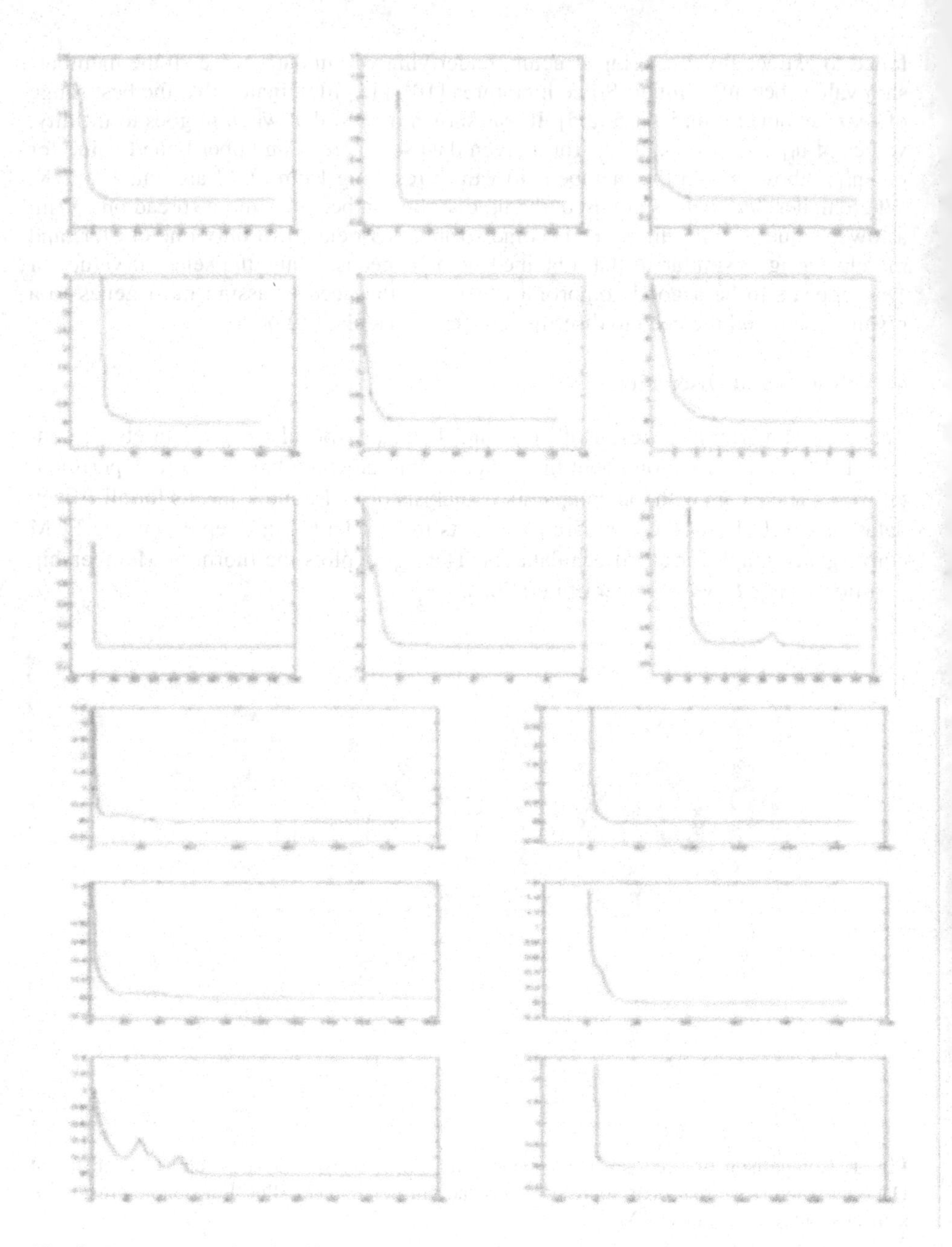

Fig. 2. Convergence Graph of FCM for all Fifteen dataset. Horizontal axis indicates Number of iteration and vertical axis represents fnorm of Membership partition matrix (U_i). The graph are in following order from left to right. First row: Iris, WBCD, Cho; Second row: Leukemia, BreastA, BreastB; Third row: multi BreastA, multi BreastB, DLBCL A; Fourth Row: DLBCL B, DLBCL C; Fifth Row: DLBCL D, Lung Cancer; Sixth row: St. Jude Leukemia, Serum data.

Table 2. Results for IRIS

	# data point	K-means # data point wrongly clustered	# data point correctly clustered	Accuracy	FCM # data point wrongly clustered	# data point correctly clustered	Accuracy
Cluster1	50	0	50	100	0	50	100
Cluster2	50	4	46	92	3	47	94
Cluster3	50	13	37	74	11	39	88
Total	150	17	133	88.67	14	136	90.67

Table 3 shows the result for WBCD data. Percentage increase in accuracy for FCM is by 33% for WBCD. Total count error of K-means is 254, whereas it is reduced to 29 in case of FCM.

Table 3. Results for WBCD

	# data point	K-means # data point wrongly clustered	# data point correctly clustered	Accuracy	FCM # data point wrongly clustered	# data point correctly clustered	Accuracy
Cluster1	444	142	302	68.01	09	435	97.97
Cluster2	239	112	127	53.14	20	219	91.63
Total	683	254	429	62.81	29	654	95.75

Table 4 shows the result obtained by K-means and FCM for SERUM data (IYER data). The data point of one clusters are overlapping with other clusters. This could be the reason to achieve very less accuracy for this datasets. Number of data point correctly clustered for this datasets is 268 for K-means and 252 for FCM. The optimal value of "m" was chosen as 1.92.

Table 4. Results For Serum Data (Iyer Data)

	# data point	K-means # data point wrongly clustered	# data point correctly clustered	Accuracy	FCM # data point wrongly clustered	# data point correctly clustered	Accuracy
Cluster1	33	17	16	48.48	28	5	15.15
Cluster2	100	84	16	16	6	94	94
Cluster3	145	7	138	95.17	69	84	57.93
Cluster4	34	34	0	0	33	1	2.94
Cluster5	43	31	12	27.91	24	19	44.19
Cluster6	7	6	1	14.29	7	0	0
Cluster7	34	17	17	50	17	17	50
Cluster8	14	1	13	92.86	14	0	0
Cluster9	63	28	35	55.56	41	22	34.92
Cluster10	19	12	7	36.84	17	2	10.53
Cluster11	25	12	13	52	17	8	32
Total	517	249	268	51.84	265	252	48.74

Table 5 contains the result for CHO data. Percentage accuracy of K-means is 60.88% and for FCM, it is 63.47 %. Note that percentage accuracy for cluster 4 is increased by 59.26% (72.22 – 12.96), whereas for cluster3 it is decreased by 26.67% (62.67-36).

Table 5. Results for Cho Data

	# data point	K-means # data point wrongly clustered	# data point correctly clustered	Accuracy	FCM # data point wrongly clustered	# data point correctly clustered	Accuracy
Cluster1	67	30	37	55.22	15	52	77.62
Cluster2	135	25	110	81.48	55	80	59.3
Cluster3	75	28	47	62.67	48	27	36
Cluster4	54	47	7	12.96	15	39	72.22
Cluster5	55	21	34	61.82	8	47	85.45
Total	386	151	235	60.88	141	245	63.47

Table 6 holds the results for Leukemia data (Golub's data). Results achieved are accuracy 59.72% in case of K-means and 55.56% in case of FCM. Note that cluster2 is having equal accuracy 44% in case of K-means as well as in case of FCM. The Golub et al.'s microarray datasets is very challenging because the appearance of the two types of acute Leukemia is highly similar in nature. This was the reason we have not achieved much accuracy in this case. One probable solution to deal with this problem is that one can use dimensionality reduction techniques to reduce the number of feature.

Table 6. Results For Leukemia Data (Golub Experiment)

	# data point	K-means # data point wrongly clustered	# data point correctly clustered	Accuracy	FCM # data point wrongly clustered	# data point correctly clustered	Accuracy
Cluster1	47	15	32	68.09	18	29	61.702
Cluster2	25	14	11	44	14	11	44
Total	72	29	43	59.72	32	40	55.56

Table 7 shows the result obtained by K-means and FCM for LUNG Cancer data. Overall accuracy has been raised by 4 % in case of FCM when compared with K-means algorithm. Note that accuracy of cluster is 47.62% and in case of FCM it has been raised up to 61.9 %.

Table 7. Results for lung cancer

	# data point	K-means # data point wrongly clustered	# data point correctly clustered	Accuracy	FCM # data point wrongly clustered	# data point correctly clustered	Accuracy
Cluster1	139	42	97	69.78	36	103	74.1
Cluster2	17	1	16	94.11	2	15	88.24
Cluster3	21	11	10	47.62	8	13	61.9
Cluster4	20	1	19	95	1	19	95
Total	17	55	142	72.08 .	47	150	76.14

Table 8 shows the result obtained by K-means and FCM for St. Jude Leukemia data. The data in this case is of highly separable in nature. As far as K-means is concerned 100 percent accuracy is achieved for cluster 2. But, FCM results 100 % accuracy for cluster 2 as well as cluster 6. Note that the least accuracy was obtained for cluster 5 in K-means as well as in FCM. FCM obtained 16.28% (83.72-67.44) more accuracy compared to K-means algorithm for Cluster 5.

Table 8. Results for St' Jude Leukemia data

	# data point	K-means # data point wrongly clustered	# data point correctly clustered	Accuracy	FCM # data point wrongly clustered	# data point correctly clustered	Accuracy
Cluster1	15	15	0	0	15	0	0
Cluster2	27	0	27	100	0	27	100
Cluster3	64	3	61	95.31	3	61	95.31
Cluster4	20	4	16	80	4	16	80
Cluster5	43	14	29	67.44	7	36	83.72
Cluster6	79	1	78	98.73	0	79	100
Total	248	37	211	85.08	29	219	88.31

Due to constraint of the space we do not present the detailed result obtained in this paper. Summarized result for subtypes of Breast Data as well as for DLBCL data are given in Table 1. As far as Breast data A is concerned, Number of correctly clustered data point has been raised to 86 (FCM) from 71 (K-means) and therefore percentage accuracy is increased by 15.31. In case of Breast data B, percentage accuracy has been increased by 15.32% (87.76-72.44). In case of Breast multi data A, percentage accuracy is increased by 10.68.

As far as DLBCL A is concerned, increase in accuracy in case of FCM is by 5.67%. Note that K-means performs better in case of DLBCL B data. Number of data point correctly clustered in case of K-means is 140 whereas 136 in case of FCM. FCM gives optimal accuracy when" m" is 1.285. As far as DLBCL C and DLBCL D are concerned FCM performs better compared to k-means algorithm. The data of DLBCL B is of highly distinctively separated in nature compare to other DLBCL (A, C, D) and that is the reason to achieve higher accuracy in case of DLBCL B. The reason for inferior performance of FCM compared to K-means for DLBCL B could be solved by finding a better “m” value.

5 Conclusion

The FCM based clustering algorithm is a distinct improvement from the partitioning based clustering algorithm (k-means). Its ability to cluster independent of the data sequence provides a more stable clustering result. When compared to k-means clustering algorithms, the FCM based clustering algorithm is superior in terms of clustering accuracy. As seen from the experiments, the FCM based clustering algorithm was able to provide the highest accuracy and generalization results.

References

1. Anderson, E.: The IRISes of the Gaspe Penisula. Bulletin of the American IRIS society 59, 2–5 (1939)
2. http://archive.ics.uci.edu/ml/datasets
3. http://www.sciencemag.org/feature/data/984559.shl
4. Iyer, V.R., Eisen, M.B., Ross, D.T., Schuler, G., Moore, T., Lee, J.C.F., Trent, J.M., Staudt, L.M., Hudson Jr., J., Bogosk, M.S., et al.: The transcriptional program in the response of human fibroblast to serum. Science 283, 83–87 (1999)
5. Cho, R.J., Campbell, M.J., Winzeler, E.A., Steinmetz, L., Conway, A., Wodicka, L., Wolfsberg, T.G., Gabrielian, A.E., Landsman, D., Lockhart, D.J., Davis, R.W.: A genome-wide transcriptional analysis of the mitotic cell cycle. Mol. Cell. 2, 65–73 (1998)
6. Tavazoie, S., Hughes, J.D., Campbell, M.J., Cho, R.J., Church, G.M.: Systematic determination of genetic network architecture. Nat. Genet. 22, 281–285 (1999)
7. Doulaye, D., Kastner, P.: Fuzzy C-means method for clustering microarray data. Bioinformatics 19(8), 973–980 (2003)
8. http://www.cse.buffalo.edu/faculty/azhang/Teaching/index.html
9. http://www.broad.mit.edu/cgi-bin/cancer/datasets.cgi
10. Alizadeh, A., et al.: Distinct types of diffuse large B-cell lymphoma identified by geneexpression profiling. Nature 43, 503–511 (2000)
11. Golub, T.R., Slonim, D.K., Tamayo, P., Huard, C., Gaasenbeek, M., Mesirov, J.P., Coller, H., Loh, M., Downing, J., Caligiuri, M., Bloomfield, C., Lander, E.: Molecular Classification of Cancer: Class Discovery and Class Prediction by Gene Expression. Science 286(5439), 531–537 (1999)
12. Bhattacharjee, A., Richards, W.G., Staunton, J., Li, C., Monti, S., Vasa, P., Ladd, C., Beheshti, J., Bueno, R., Gillette, M., Loda, M., Weber, G., Mark, E.J., Lander, E.S., Wong, W., Johnson, B.E., Golub, T.R., Sugarbaker, D.J., Meyerson, M.: Classification of Human Lung Carcinomas by mRNA Expression Profiling Reveals Distinct Adenocarcinomas Subclasses. Proceedings of the National Academy of Sciences 98(24), 13790–13795 (2001)
13. Yeoh, E.-J., Ross, M.E., Shurtleff, S.A., Williams, W.K., Patel, D., Mahfouz, R., Behm, F.G., Raimondi, S.C., Relling, M.V., Patel, A., Cheng, C., Campana, D., Wilkins, D., Zhou, X., Li, J., Liu, H., Pui, C.-H., Evans, W.E., Naeve, C., Wong, L., Downing, J.R.: Classification, subtype discovery, and prediction of outcome in pediatric acute lymphoblastic leukemia by gene expression profiling. Cancer Cell 1(2) (2002)
14. http://www-igbmc.u-strasbg.fr/projets/fcm

15. Hoshida, Y., Brunet, J.P., Tamayo, P., Golub, T.R., Mesirov, J.P.: Subclass mapping: identifying common subtypes in independent disease data sets. PLoS ONE 2(11) (2007)
16. Mangasarian, O.L., Wolberg, W.H.: Cancer diagnosis via linear programming. SIAM News 23(5), 1–18 (1990)
17. DeRisi, J., Penland, L., Brown, P.O., Bittner, M.L., Meltzer, P.S., Ray, M., Chen, Y., Su, Y.A., Trent, J.M.: Use of a cDNA microarray to analyze gene expression patterns in human cancer. Nature Genetics 14, 457–460 (1996)
18. Alon, U., Barkai, N., Notterman, D.A., Gish, K., Ybarra, S., Mack, D., Levine, A.J.: Broad patterns of gene expression revealed by clustering analysis of tumor and normal colon tissues probed by oligonucleotide arrays. Proceedings of National Academy of Science 96, 6745–6750 (1999)
19. Ben-Dor, A., Yakhini, Z.: Clustering gene expression patterns. Journal of Computational Biology 6, 281–297 (1999)
20. Eissen, M.B., Spellman, P.T., Brown, P.O., Botstein, D.: Clustering analysis and display of genome wide expression patterns. Proceedings of the National Academy of Sciences 95, 14863–14868 (1998)
21. Ester, M., Kriegel, H.P., Sander, J., Xu, X.: A density-based algorithm for discovering clusters in large spatial databases with noise. In: Proceedings of the 2nd International Conference on Knowledge Discovery and Data Mining (KDD 1996), pp. 226–231 (1996)
22. Guha, S., Rastogi, R., Shim, K.: CURE: an efficient clustering algorithm for large databases. In: Proceedings of the ACM SIGMOD International Conference on Management of Data, pp. 73–84 (1998)
23. Han, J., Kamber, M.: Data Mining: Concepts and Techniques. Morgan Kaufmann, San Francisco (2001)
24. Jain, A.K., Dubes, R.C.: Algorithms for Clustering Data. Prentice Hall, New Jersey (1988)
25. Selim, S.Z., Ismail, M.A.: K-means type algorithms: a generalized convergence theorem and characterization of local optimality. IEEE Trans. Pattern Anal. Mach. Intell. 6, 81–87 (1984)
26. Spath, H.: Cluster Analysis Algorithms. Ellis Horwood, Chichester (1989)
27. Pal, N.R., Bedzek, J.C., Taso, E.C.K.: Generalized Clustering Networks and Kohonen's Self- Organizing Scheme. IEEE Trans. on Neural Networks 3(4), 546–557 (1993)
28. Bezdek, J.C.: Pattern Recognition with Fuzzy Objective Function Algorithms. Plenum Press, New York (1981)
29. Liew, A.W.C., Yan, H., Yang, M., Chen, P.: Microarray Data Analysis. In: Chen, Y.-P.P. (ed.) Bioinformatics Technologies, ch. 12, pp. 353–388. Springer, Heidelberg (2005)
30. Liew, A.W.C., Yan, H., Yang, M.: Data Mining for Bioinformatics. In: Chen, Y.-P.P. (ed.) Bioinformatics Technologies, ch. 4, pp. 63–116. Springer, Heidelberg (2005)
31. Cheng, K.O., Law, N.F., Siu, W.C., Liew, A.W.C.: Identification of coherent patterns in gene expression data using an efficient bi-clustering algorithm and parallel coordinate visualization. BMC Bioinformatics 9(210) (2008), doi.10.1186/1471-2105-9-210
32. Gan, X., Liew, A.W.C., Yan, H.: Discovering biclusters in gene expression data based on high-dimensional linear geometries. BMC Bioinformatics 9(209) (2008), doi:10.1186/1471-2105-9-209
33. Yin, Z.H., Tang Yuangang, Y.G., Sun, F.C., Sun, Z.Q.: Fuzzy Clustering with Novel Separable Criterion. Tsinghua Science and Technology 11, 50–53 (2006)
34. Liu, H.C., Yih, J.M., Liu, S.W.: Fuzzy C-mean Algorithm Based on Mahalanobis Distances and Better initial values. In: 12th International Conference on Fuzzy Theory & Technology, JCIS, Salt Lake City, Utah (2007)

35. Liu, H.C., Yih, J.M., Sheu, T.W., Liu, S.W.: A New Fuzzy Possibility Clustering Algorithms Based On Unsupervised Mahalanobis Distances. In: International Conference on Machine Learning and Cybernetics, Hong Kong, pp. 3939–3944 (2007)
36. Tang, Y., Zhang, Y.-Q., Huang, Z.: FCM-SVM-RFE Gene Feature Selection Algorithm for Leukemia Classification from Microarray Gene Expression Data. In: The 14th IEEE International Conference on Fuzzy Systems (FUZZ 2005), pp. 97–101 (2005)
37. Wang, W., Wang, C., Cui, X., Wang, A.: A Clustering Algorithm Combine the FCM algorithm with Supervised Learning Normal Mixture Model. In: The 19th IEEE International Conference on pattern Recognition (ICPR 2008), December 2008, pp. 1–4 (2008)
38. Bezdek, J.C., Pal, N.R.: Some New Indexes of Cluster Validity. IEEE Transactions Systs., Man Cyberns. 28, 301–315 (1998)
39. Pal, S.K., Bandyopadhyay, S., Ray, S.S.: Evolutionary Computation in Bioinformatics: A Review. IEEE Transactions on Systems, Man, And Cybernetics—Part C: Applications And Reviews 36(5), 601–615 (2006)
40. Jiang, D., Tang, C., Zhang, A.: Cluster Analysis for Gene Expression Data: A Survey. IEEE 16(11) (November 2004)
41. Dhiraj, K., Rath, S.K.: SA-kmeans: A Novel Data Mining Approach to Identifying and Validating Gene Expression Data. In: SPIT-IEEE International conference and colloquium, Mumbai, India, vol. 4, pp. 107–112 (2008)
42. Dhiraj, K., Rath, S.K.: Gene Expression Analysis using Clustering. In: Third IEEE International Conference on Bioinformatics and Biomedical Engineering, to be held on June 11th to 13th in Beijing, China (2009) ISBN: 978-1-4244-2902-8
43. Dhiraj, K., Rath, S.K.: Family of Genetic Algorithm Based Clustering Algorithm for Pattern Recognition. In: 1st IIMA International Conference on Advanced Data Analysis, Business Analytics and Intelligence, to be held on June 6th to 7th in IIM Ahmedabad, INDIA (2009)
44. Dhiraj, K., Rath, S.K.: Comparison of SGA and RGA based clustering algorithm for pattern recognition. International Journal of Recent Trends in Engineering 1(1) (2009)

Enhancing the Performance of LibSVM Classifier by Kernel F-Score Feature Selection

Balakrishnan Sarojini[1,2,*], Narayanasamy Ramaraj[3], and Savarimuthu Nickolas[4]

[1] PhD Research Scholar, Department of Computer Science,
Mother Teresa Women's University, Kodaikanal, India
[2] Working as Professor, K.L.N. College of Information Technology, Madurai, India
balakrishnan.sarojini@gmail.com
[3] Principal, G.K.M. College of Engineering & Technology, Chennai, India
ramaraj_tce@yahoo.co.in
[4] Assistant Professor, Department of Computer Applications,
National Institute of Technology, Tiruchirappalli, India
nickolas@nitt.edu

Abstract. Medical Data mining is the search for relationships and patterns within the medical datasets that could provide useful knowledge for effective clinical decisions. The inclusion of irrelevant, redundant and noisy features in the process model results in poor predictive accuracy. Much research work in data mining has gone into improving the predictive accuracy of the classifiers by applying the techniques of feature selection. Feature selection in medical data mining is appreciable as the diagnosis of the disease could be done in this patient-care activity with minimum number of significant features. The objective of this work is to show that selecting the more significant features would improve the performance of the classifier. We empirically evaluate the classification effectiveness of LibSVM classifier on the reduced feature subset of diabetes dataset. The evaluations suggest that the feature subset selected improves the predictive accuracy of the classifier and reduce false negatives and false positives.

Keywords: Medical data mining, Feature selection, predictive accuracy, false negative, false positive, LibSVM classifier.

1 Introduction

The health care industry maintains databases about demographic and pathological details of the patients to provide quality services. Extracting knowledge from these databases is valuable for decision-making at the point of care for diagnosis, treatment planning, risk analysis, and predictions. Data mining methods review years of medical data to extract valuable information for decision-making at the point of care for diagnosis, treatment planning, risk analysis, and predictions. By interpreting the patient data using the mined knowledge, the healthcare personnel will be assisted to make a guided decision about the clinical case at hand.

* Corresponding author.

S. Ranka et al. (Eds.): IC3 2009, CCIS 40, pp. 533–543, 2009.
© Springer-Verlag Berlin Heidelberg 2009

The Predictive accuracy of any data mining technique is based on the quantity and quality of the data [1]. That too, in the field of medicine a high degree of predictive accuracy is expected. The medical databases are usually large and the problems of inaccurate and/or inconsistent data are inevitable. In a real-world environment, there are many possible reasons why the inaccurate or inconsistent data occur in a database, e.g., equipment mal functioning, the deletion of data instances (or records) due to the inconsistency with other recorded data, not entering data due to misunderstanding, considering the data as an unimportant at the time of entry, etc. the As a result, medical databases may contain redundant, incomplete, imprecise or inconsistent, features which may affect the performance of the data mining techniques [2]. Also, the presence of imprecise, irrelevant and redundant features in the training dataset makes knowledge discovery during training very difficult. So, mining the medical data may require more data reduction and data preparation than data used for other applications [3, 4].

Most data mining methods depend on a set of features that define the behavior of the learning algorithm and directly or indirectly influence the complexity of the resulting models [5]. Feature selection is the process of selecting the most informative features of the dataset that greatly influences the decision-making. Feature selection prior to learning can be beneficial. Reducing the dimensionality of the data reduces the size of the hypothesis space and allows algorithms to operate faster and more effectively [6]. Selecting the right feature set not only improves accuracy but also reduces the running time of the predictive algorithms and lead to simpler, more understandable models. Particularly, in medical data mining, feature selection is appreciable as reduction in the number of features means reduction in the number of clinical measures to be made and diagnosis of the disease with less number of more discriminating features. The quality of the derived feature subset is measured by an estimate of the classification accuracy of a chosen classifier trained on the candidate subset.

We propose a wrapper based feature selection approach to select optimal feature subset that enhances the predictive accuracy of the LibSVM classifier [7]. The most discriminative features of the Pima Indians diabetes dataset [8] are selected using F-score [9], the feature selection tool of LibSVM and Recursive Feature Elimination (RFE) [4] approach is applied to select the optimal feature subset. In the literature, only a few algorithms have been proposed for SVM feature selection [10, 11,12]. However, these papers focus on feature selection but did not deal with parameter optimization for the SVM classifier. Research [13] shows that the training of the SVM classifier with some external parameters (metaparameters or hyperparameters) like SVM kernel parameters (γ in RBF kernel) or regularization (trade-off) parameter (C) may have a strong effect on the behavior of the classifier. Such hyperparameters, if not well chosen, may deteriorate the classification performance of SVM. In addition to feature selection setting appropriate values for the parameters C and γ improves the accuracy of the classifier.

The proposed approach is validated in terms of increase in the accuracy of the classifier and an improved Area Under ROC (Receiver Operating Curve) AUC. The performance of the classifier after feature selection is also validated in terms of sensitivity (positive hit rate) and specificity (negative hit rate).

2 Related Works

As quoted by the senior director of engineering research and development for the Computer-Aided Diagnosis "Due to the extremely poor quality of information in the patient medical record, most of today's healthcare IT systems cannot provide significant support to improve the quality of care"[14]. A smarter and more efficient way to improve quality of care is to use predictive analytics to find patterns in large quantities of data collected in clinical information systems and electronic medical records [15]. The discovery of knowledge from these health care databases can lead to discovery of trends and rules for later diagnostic tools. Consequently, the predictability of disease will become more effective and early detection of disease will aid in increased exposure to require patient care and improved cure rates [16]. Machine learning methods have been applied to a variety of medical domains to improve medical decision-making [17]. Some include predicting breast cancer survivability using data mining techniques [18], application of data mining to discover subtle factors affecting the success and failure of back surgery which led to improvements in care [19]. It is proved that the successful implementation of machine learning methods can help the integration of computer-based systems in the healthcare environment providing opportunities to facilitate and enhance the work of medical experts and ultimately to improve the efficiency and quality of medical care [20].

Data from medical sources are voluminous; they come from many different sources, not all commensurate structure or quality [21]. Researchers and practitioners realize that in order to use data mining tools on these databases effectively data preprocessing is essential for successful data mining [4,22]. Feature selection can also help reduce online computational costs, enhance system interpretability [23] and improve the performance of the learning problems [24]. Several feature selection techniques have been proposed in the literature, including some important survey on feature selection algorithms such Molina et al. [25] then Guyon and Elisseeff [26]. Many researchers involved in studying various important point of feature selection, such as the goodness of a feature subset in determining an optimal one [4]. Koller and Sahami defined discriminant and independent features as relevant features and attempted to obtain relevant features by eliminating features mutually dependent of class labels (irrelevant features) and features related with other features (redundant features) [27]. Supervised learning systems such as classification have been successfully applied in a number of medical domains, for example, in localization of a primary tumor, prognostics of recurrence of breast cancer, diagnosis of thyroid diseases, and rheumatology [28].

Many authors have reported improvement in the performance of the classifier when feature selection algorithms are used [29, 30]. In many pattern recognition applications, identifying the most characterizing features (or attributes) of the observed data, i.e., feature selection [31, 32] is critical to minimize the classification error. The legitimate way of evaluating features is through the error rate/accuracy of the classifier being designed [33]. The classification error rate/accuracy is used as a performance indicator for a mining task, for a selected feature subset; simply conduct the "before-and-after" experiment to compare the error rate/accuracy of the classifier learned on the full set of features and that learned on the selected subset [34].

Support vector machine (SVM) [35] is a classification tool with a great deal of success in a variety of areas from object recognition [36] to classification of cancer morphologies [37]. Support Vector Machines (SVMs), one of the most actively developed classifiers in the machine learning community, have been successfully applied to a number of medical problems [38]-[40]. Selection of the appropriate parameters for a specific kernel is an important research issue in the data mining area [41]. SVM parameter selection may be viewed as an optimization process [42, 43], and optimization techniques may be applied directly to this problem. F-score is a simple technique that measures the discrimination of two sets of real numbers. The larger the F-score is, the more likely this feature is more discriminative [9]. RFE is an iterative procedure to remove non-discriminative features in binary classification problem [4].

For supervised learning, the primary goal of classification is to maximize predictive accuracy; therefore, predictive accuracy is generally accepted and widely used as the primary measure by researchers and practitioners [44]. The performance of a classifier can be visualized by using a Receiver Operating Characteristic (ROC) curve [45]. The area under the ROC curve is a statistically consistent and more discriminating measure than accuracy measure [46].

3 Dataset

The experiments were performed on the Pima Indians diabetes dataset from the UCI (University of California at Irvine) machine learning repository which consists of 768 complete instances described by 8 features labeled as number of times pregnant, glucose tolerance test, diastolic blood pressure, triceps skin fold thickness, 2-hour serum insulin, body mass index, diabetes pedigree function and age and a predictor variable. The class value 1 interpreted as "tested positive for diabetes" is found in 268 numbers of instances and class value 0 in 500 numbers of instances. There is no missing data present in the training dataset.

4 Experimental Analysis

4.1 Experimental Design

To implement our proposed approach, we used the RBF kernel function RBF kernel is advantageous in complex non-separable classification problems due to its ability of nonlinear input mapping. Furthermore, RBF kernel requires only two parameters, penalty parameter C and kernel width γ, to be defined [47]. Research [9] shows that setting proper model parameters can improve the classification accuracy of SVM. The grid search approach is used to find the best C and γ. Optimization by grid search is done by specifying the parameter space, the range of C, gamma and the stopping tolerance. Then, all grid points of (C, γ) are tried and the one which gives the highest cross validation accuracy is chosen as the best parameter to train the training dataset and generate the model.

The F-scores of features of Pima dataset are calculated using the formula [9]

$$F(i) \equiv \frac{\left(\bar{x}_i^{(+)} - \bar{x}_i\right)^2 + \left(\bar{x}_i^{(-)} - \bar{x}_i\right)^2}{\frac{1}{n_+ - 1}\sum_{k=1}^{n_+}\left(x_{k,i}^{(+)} - \bar{x}_i^{(+)}\right)^2 + \frac{1}{n_- - 1}\sum_{k=1}^{n_-}\left(x_{k,i}^{(-)} - \bar{x}_i^{(-)}\right)^2}, \quad (1)$$

Where F(i) is the F-score of the ith feature, n+ and n− number of positive and negative instances, $\bar{x}_i$, $x^{(+)}_i$, $\bar{x}(-)i$ is the average of the ith feature of the whole, positive, and negative data set and x(+)k, i is the ith feature of the kth positive instance, and x(−) k, i is the ith feature of the kth negative instance respectively.

Table -1 show the F-score values for the eight features of the Pima dataset. The features are ranked in the decreasing order of importance.

Table 1. F-scores of the features of Pima Indian Diabetes Dataset

Features	F-scores
a2 :glucose tolerance test	0.278279
a6 :body mass index	0.093697
a8 :age	0.060236
a1: no. of times pregnant	0.051789
a7:diabetes pedigree function	0.031164
a5:2-hour serum insulin	0.017338
a4:triceps skin fold thickness	0.005619
a3:diastolic blood pressure	0.004252

The LibSVM classifier in the machine learning library with Java implementation "WEKA 3.5.2" [48], is used for classification. The classification accuracy is evaluated using 10-fold cross validation test. Cross-validation involves breaking a dataset into 10 pieces, and on each piece, testing the performance of a predictor build from the remaining 90% of the data. The classification accuracy was taken as the average of the 10 predictive accuracy values.

The accuracy of the classifier is used as performance indicator to derive the optimal feature subset. The least rank features are removed one at a time and the accuracy of the classifier before and after feature removal is observed. The feature subset which gives improved accuracy than the whole set of features is the optimal feature subset.

4.2 Measures for Performance Evaluation

4.2.1 Accuracy, Sensitivity, Specificity and AUC

In this study, we used three performance measures: Accuracy (Eq. (1)), Sensitivity (Eq. (2)) and Specificity (Eq. (3)). The accuracy of the classifier is the measure of how effective the classifier identifies the true value of the class label. Sensitivity and specificity are statistical measures of the performance of a binary classification test. They describe how well the classifier discriminates between case with positive and with negative class (with and without disease). The sensitivity measures the proportion of actual positives which are correctly identified as such (i.e. the percentage of sick people who are identified as having the disease); and the specificity measures the proportion of negatives which are correctly identified (i.e. the percentage of well people who are identified as not having the disease).

$$\text{Accuracy} = (TP + TN) / (TP + TN + FP + FN) \quad (2)$$

$$\text{Sensitivity} = TP / (TP + FN) \quad (3)$$

$$\text{Specificity} = TN / (TN + FP) \quad (4)$$

Where TP is the number of true positives (number of 'YES' patients predicted correctly), TN is the number of true negatives (number of 'NO' patients predicted correctly), FP is the number of false positives (number of 'YES' patients predicted as 'NO') and FN is the number of false negatives (number of 'NO' patients predicted as 'YES'). The trade off between specificity and sensitivity as well as the performance of the classifier is visualized using ROC curve.

5 Results and Discussion

The optimal values of C and γ for the Pima dataset obtained using grid search are at 10.0 and 1.0 respectively. These optimal parameters are used to evaluate the LibSVM classifier. The performance of the classifier on the removal each less discriminating feature is shown in table 2. The experimental results show that the accuracy of the classifier improves with the removal of the least ranked features. The peak performance of the classifier is obtained in the fourth iteration for the feature subset {a1, a2, a6, a7, a8}, after the removal of three least rank features a3, a4 and a5. The performance declines after the fourth iteration. The accuracy of the classifier for the whole set of features is 77.0833 and for the feature subset {a1, a2, a6, a7, a8} it is 77.474. The improvement in the classification accuracy is 0.3907.

The experimental results show that the accuracy of the classifier improves with the removal of the least ranked features. The peak performance of the classifier is obtained in the fourth iteration for the feature subset {a1, a2, a6, a7, a8}, after the removal of three least rank features a3, a4 and a5. The performance declines after the fourth iteration. The accuracy of the classifier for the whole set of features is 77.0833 and for the feature subset {a1, a2, a6, a7, a8} it is 77.474. The improvement in the classification accuracy is 0.3907.

Also, the behavior of the parameters True Positive TP, True Negative TN is worth mentioning here. For the optimized feature subset there is a small decrease in the TP value and a notable increase in the TN value compared to TP and TN of whole set of features. The TP and TN values for the whole set of features are 151 and 441 and for the optimized feature subset they are 146 and 449. That is the classifier sacrifices the true positives to increase the True negatives. It shows that the classifier's performance

Table 2. Classification Results on Each Iteration

Iteration	Feature Subset	Classification results				
		Accuracy	True Positive	True Negative	No. of correctly classified instances	No. of incorrectly classified instances
1	All {a1..a8}	77.0833	151	441	592	176
2	A3 removed {a1,a2,a4,a5,a6,a7,a8}	76.0417	142	442	584	184
3	A4 removed {a1,a2,a5,a6,a7,a8}	76.6927	140	449	589	179
4	A5 removed {a1,a2,a6, a7,a8}	77.474	146	449	595	173
5	A7 removed {a1,a2,a6,a8}	76.6927	142	447	589	179
6	A1 removed {a2,a6,a8}	76.6927	146	443	589	179
7	A8 removed {a2,a6}	75.5208	128	452	580	188

in identifying 'NO' instances has improved for the feature subset with highly informative features. In medical domain, it is purely acceptable as it is desirable a healthy patient may be identified as sick but not a sick patient as healthy, which makes the medical diagnosis a failure one. So, it makes a sense to sacrifice the precision of positive classifications in exchange for improving the precision of negative calculations.

Table 3. Sensitivity and Specificity Values

Feature subset	Sensitivity %	Specificity %
All {a1...a8}	71.9	77.03
A3removed {a1,a2,a4,a5,a6,a7,a8}	71	77.82
A4 removed{a1,a2,a5,a6,a7,a8}	73.3	77.82
A5 removed{a1,a2,a6,a7,a8}	74.11	79.00
A7 removed{a1,a2,a6,a8}	72.82	78.01
A1 removed{a2,a6,a8}	71.92	78.41
A8 removed{a2,a6}	72.73	76.35

The sensitivity and specificity values for the whole set of features and on the removal of each feature are shown in Table 3. As it can be seen from the table the sensitivity and specificity values are high for the optimized feature subset. A high value of sensitivity and specificity indicates reduced false positives and false negatives for the derived optimized feature subset.

Another way of evaluating the performance of a classifier is by the analysis of the ROC curve. The two-dimensional ROC curve is defined by the False Positive Rate (FPR) on the x-axis and the True Positive Rate (TPR) (sensitivity) on the y-axis where TPR determines a classifier performance on classifying positive instances correctly among all positive samples and FPR; (1-specificity) on the other hand, defines how many incorrect positive results occur among all negative samples.

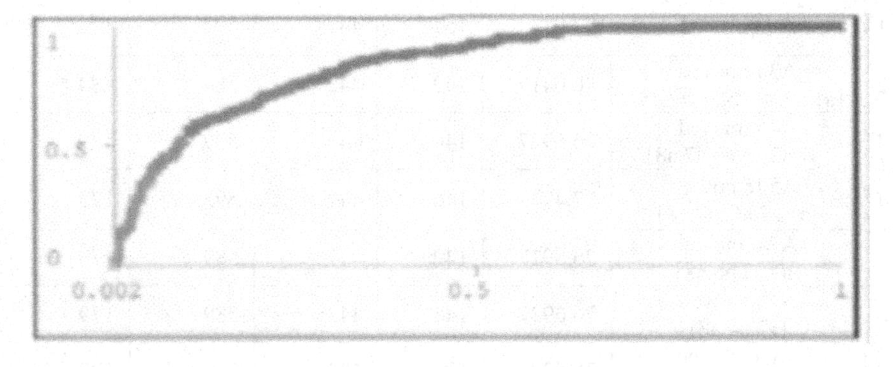

Fig. 1. ROC for whole set of features

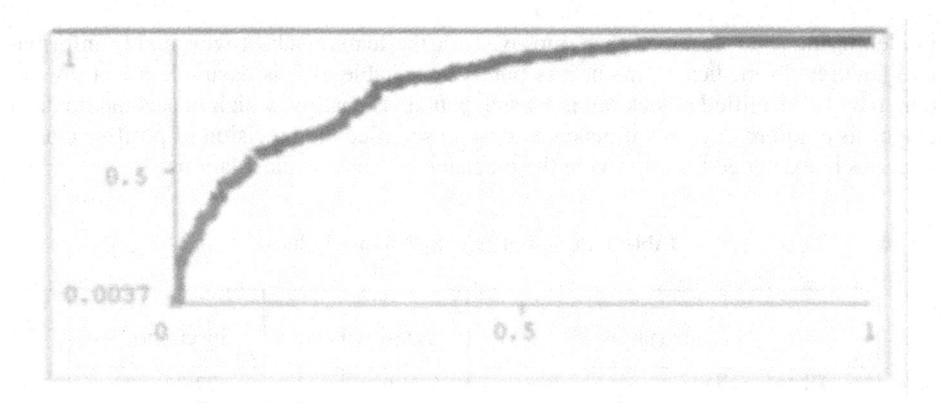

Fig. 2. ROC for the optimized feature set

Figure 1 and Figure 2 depicts the ROC before and after feature selection. The area under ROC for the whole set of features is 0.8285 and for the optimal feature subset it is 0.8377. The improved area of the ROC curve proves that the proposed feature selection approach could enhance the predictive accuracy of the classifier with minimal number of features.

6 Conclusions

The health care organizations face lot of challenges when trying to deal with large, diverse, and often complex health care data source. They implement data mining technologies to improve the efficacy of patient care. At present, many predictive data mining methods have been successfully applied to a variety of practical problems in clinical medicine. The work attempts to emphasize embedding feature selection technique in clinical decision support tool could empower the medical community to improve the quality of diagnosis through the use of technology. The proposed feature selection approach is experimented on the type II diabetes dataset. The performance of the LibSVM classifier is evaluated using the Accuracy of the classifier, Sensitivity and Specificity and AUC. The experimental results prove that the proposed approach produces a feature reduction of 37.5%. The increase in sensitivity and specificity values proves that the discriminating ability of the classifier, to correctly classify positive instances as positive and negative instances as negative, has also improved for the optimal feature set. It means that there is a reduction in the false positives and false negatives. In medical domain it is highly appreciable to achieve reduce in the number of false negatives and false positives as it literally, the difference between life and death. The approach is simple and effective and augments the argument simple methodologies are better for medical data mining.

References

1. Burke, H.B., Goodman, P.H., Rosen, D.B., Henson, D.E., Weinstein, J.N., Harrell Jr., F.E., Marks, J.R., Winchester, D.P., Bostwick, D.G.: Artificial neural networks improve the accuracy of cancer survival prediction. Cancer 79, 857–862 (1997)
2. Lavrac, N.: Selected techniques for data mining in medicine. Artif. Intell. Med. 16, 3–23 (1999)
3. Cios, K.J., Moore, G.: Uniqueness of medical data mining. Artif. Intell. Med. 26, 1–24 (2002)
4. Liu, Motoda, H.: Feature Extraction, Construction and Selection. In: A Data Mining Perspective. Kluwer Academic Publishers, Boston (1998); 2nd Printing (2001)
5. Split, A.M.T., Stegwee, R.A., Teitink, J.A.C.: Business intelligent for healthcare organizations. In: Proceeding of the 35th Annual Hawaii International Conference on System Sciences. IEEE Press, New York (2002)
6. Abraham, R., Simha, J.B., Iyengar, S.: Medical datamining with a new algorithm for feature selection and Naïve Bayesian classifier. In: 10th International Conference on Information Technology
7. Chang, C.-C., Lin, C.-J.: LIBSVM a library for support vector machines (2005), `http://www.csie.ntu.edu.tw/~cjlin/libsvm`
8. Blake, C.L., Merz, C.J.: UCI repository of machine learning databases (1998), `http://www.ics.uci.edu/~mlearn/MLRepository.html`
9. Chen, Y.-W., Lin, C.-J.: Combining SVMs with various feature selection strategies (2005), `http://www.csie.ntu.edu.tw/~cjlin/papers/features.pdf`
10. Bradley, P.S., Mangasarian, O.L.: Feature selection via concave minimization and support vector machines. In: Proceedings of the 13th international conference on machine learning, San Francisco, CA, pp. 82–90 (1998)

11. Weston, J., Mukherjee, S., Chapelle, O., Pontil, M., Poggio, T., Vapnik, V.N.: Feature selection for SVMs. In: Leen, T., Dietterich, T., Tresp, V. (eds.) Advances in Neural Information Processing Systems 13, pp. 668–674 (2001)
12. Guyon, I., Weston, J., Barnhill, S., Bapnik, V.: Gene selection for cancer classification using support vector machines. Machine Learning 46(1–3), 389–422 (2002)
13. Duan, K., Keerthi, S.S., Poo, A.N.: Evaluation of simple performance measures for tuning SVM hyperparameters. Neurocomputing 51, 41–59 (2003)
14. Computer aided diagnosis, data mining combine for improved care, Health care IT (2006)
15. Predicting Health: Jeff Kaplan, managing director at Apollo Data Technologies LLC in Chicago
16. Roshawnna Scales, Mark Embrechts: Computational intelligence techniques for medical diagnostics
17. Kononeko, I., Kukar, M.: Machine learning for medical diagnosis. In: Workshop on Computer-Aided Data Analysis in Medicine, CADAM 1995. IJS Scientific Publishing, Ljubljana (1995)
18. Delen*, D., Walker, G., Kadam, A.: Predicting breast cancer survivability: a comparison of three data mining methods. Artificial Intelligence in Medicine, doi:10.1016/j.artmed.2004.07.002
19. Hedberg, S.R.: The data gold rush. Byte, 83–88 (October 1995)
20. Magoulas, G.D., Prentza, A.: Machine learning in medical applications
21. Lee, S.J., Siau, K.: A review of data mining techniques. Industrial Management and Data Systems 101(1), 41–46 (2001)
22. Pyle, D.: Data Preparation for Data Mining. Morgan Kaufmann Publishers, San Francisco (1999)
23. Boser, B., Guyon, I., Vapnik, V.: A training algorithm for optimal margin classifiers. In: Proceedings of the Fifth Annual Workshop on Computational Learning Theory, pp. 144–152 (1992)
24. Weston, J., Mukherjee, S., Chapelle, O., Pontil, M., Poggio, T., Vapnik, V.N.: Feature selection for SVMs. In: Leen, T., Dietterich, T., Tresp, V. (eds.) Advances in Neural Information Processing Systems 13, pp. 668–674 (2001)
25. Molina, L.C., Belanche, L., Nebot, A.: Attribute Selection Algorithms: A survey and experimental evaluation. In: Proceedings of 2nd IEEE's KDD 2002, pp. 306–313 (2002)
26. Guyon, I., Elisseeff, A.: An Introduction to Variable and Feature Selection. Journal of Machine Learning Research 3, 1157–1182 (2003)
27. Koller, D., Sahami, M.: Towards optimal feature selection. In: 13th International Conference on Machine Learning, Bari, Italy, pp. 284–292 (1996)
28. Richards, G., Rayward-Smith, V.J., Sonksen, P.H., Carey, S., Weng, C.: Data mining for indicators of early mortality in a database of clinical records. Artif. Intell. Med. 22, 215–231 (2001)
29. Siedlecki, W., Sklansky, J.: On automatic feature selection. International Journal of Pattern Recognition and Artificial Intelligence 2(2), 197–220 (1988)
30. Kohavi, R., John, G.: Wrapper for Feature Subset Selection. Artificial Intelligence 97(1-2), 273–324 (1997)
31. Almuallim, H., Dietterich, T.G.: Efficient algorithms for identifying relevant features. In: Proceedings of the Ninth Canadian Conference on Artificial Intelligence. Morgan Kaufmann, Vancouver (1992)
32. Kohavi, R., John, G.: Wrapper for Feature Subset Selection. Artificial Intelligence 97(1-2), 273–324 (1997)

33. Langley, P.: Selection of Relevant Features in Machine Learning. In: Proc. AAAI Fall Symp. Relevance (1994)
34. Liu, H., Yu, L.: Feature Selection for Data Mining
35. Vapnik, V.: Statistical Learning Theory. Wiley, New York (1998)
36. Evgeniou, T., Pontil, M., Papageorgiou, C., Poggio, T.: Image representations for object detection using kernel classifiers. In: Asian Conference on Computer Vision (2000)
37. Mukherjee, S., Tamayo, P., Slonim, D., Verri, A., Golub, T., Mesirov, J., Poggio, T.: Support vector machine classification of microarray data. AI Memo 1677, Massachusetts Institute of Technology (1999)
38. Furey, T.S., Cristianini, N., Duffy, N., Bednarski, D.W., Schummer, M., Haussler, D.: Support vector machine classification and validation of cancer tissue samples using Microarray expression data. Bioinformatics 16, 906–914 (2000)
39. Takeuchi, K., Collier, N.: Bio-medical entity extraction using support vector machines. Artif. Intell. Med. 33, 125–137 (2005)
40. Cohen, G., Hilario, M., Sax, H., Hugonnet, S., Geissbuhler, A.: Learning from imbalanced data in surveillance of nosocomial infection. Artif. Intell. Med. 37, 7–18 (2006)
41. Ali, S., Smith, K.A.: Automatic parameter selection for polynomial kernel. In: Proc. of the IEEE Int. Conf. on Information Reuse and Integration (IRI 2003), Las Vegas, NV, USA, October 27–29, pp. 243–249 (2003)
42. Imbault, F., Lebart, K.: A stochastic optimization approach for parameter tuning of support vector machines. In: Proc. of the 17th Int. Conf. on Pattern Recognition (ICPR 2004), Cambridge, UK, vol. 4, pp. 597–600 (2004)
43. Schittkowski, K.: Optimal parameter selection in support vector machines. Journal of Industrial and Management Optimization 1(4), 465–476 (2005)
44. John, G.H., Kohavi, R., Pfleger, K.: Irrelevant feature and the subset selection problem. In: asnd Hirsh H. Cohen, W.W. (ed.) Machine Learning: Proceedings of the Eleventh International Conference, New Brunswick, N.J., pp. 121–129. Rutgers University (1994)
45. Herron: Machine Learning for Medical Decision Support: Evaluating Diagnostic Performance of Machine Learning classification Algorithms
46. Huang, J., Ling, C.X.: Using AUC and Accuracy in Evaluating Learning Algorithms. IEEE Transactions on Knowledge and Data Engineering 17(3) (2005)
47. Huang, C.-L., Liao, H.-C., Chen, M.-C.: Prediction model building and feature selection with support vector machines in breast cancer diagnosis. Expert Systems with Applications, 578–587 (2008), doi:10.1016/j.eswa.2006.09.041
48. EL-Manzalawy, Y., Honavar, V.: WLSVM: Integrating LibSVM into Weka Environment (2005), http://www.cs.iastate.edu/~yasser/wlsvm

An Electronically Tunable SIMO Biquad Filter Using CCCCTA

Sajai Vir Singh[1,*], Sudhanshu Maheshwari[2], Jitendra Mohan[1], and Durg Singh Chauhan[3]

[1] Department of Electronics and Communications,
Jaypee University of Information Technology, Waknaghat,
Solan-173215, India
sajaivir@rediffmail.com
[2] Department of Electronics Engineering,
Z. H. College of Engineering and Technology,
Aligarh Muslim University, Aligarh-202002, India
[3] Uttarakhand Technical University, Dehradun-248001, India

Abstract. This paper presents an electronically tunable single input multiple output (SIMO) biquad filter using current controlled current conveyor transconductance amplifiers (CCCCTAs). The proposed filter employs only two CCCCTA and three capacitors out of which two are permanently grounded. The proposed filter realizes low pass (LP), band pass (BP) and high pass (HP) responses in voltage form as well as in transadmittance form simultaneously. The circuit can also realize band reject (BR) as well as all pass (AP) responses in transadmittance form, with interconnection of relevant output currents. The circuit enjoys an independent current control of pole frequency and bandwidth as well as quality factor and bandwidth. Both the active and passive sensitivities are no more than unity. The validity of proposed filter is verified through PSPICE simulations.

Keywords: CCCCTA, SIMO, filter.

1 Introduction

The design of continuous time filter using current conveyors has received much attention pertaining to implementation of current and voltage mode filters due to its distinct advantages such as wider signal bandwidth, greater linearity, larger dynamic range and low power consumption over voltage mode counter part [1-3]. The second generation current controlled conveyor (CCCII) introduced by Fabre at [4-5] can be extended to the electronically adjustable domain for different applications.

Recently, a new current mode active building block, which is called as a current controlled current conveyor transconductance amplifier (CCCCTA), has been proposed

* Corresponding author.

S. Ranka et al. (Eds.): IC3 2009, CCIS 40, pp. 544–555, 2009.
© Springer-Verlag Berlin Heidelberg 2009

[6-7]. This device can be operate in both current and voltage modes, providing flexibility and enables a variety of circuit designs. In addition, it can offer several advantages such as high slew rate, high speed, wider bandwidth and simpler implementation [6]. Moreover, the CCCCTA can control the parasitic resistance at X (R_X) port by input bias current so when it is used in some circuit, it can be electronically control and suitable for automatic control system.

In many applications, a need arises to interconnect voltage-mode and current-mode circuits, which causes some difficulties, voltage to current (V–I) converters interface circuits are requested to overcome these difficulties. During V–I interfacing, it is possible to perform the signal processing at the same time, so that total effectiveness of the electronic circuitry can be increased [8]. Thus it is essential to have transadmittance (TA) filters which have wide application areas. The literature survey show that the single input multiple output biquad filter circuits using different active element such as OTAs [9-10], Four Terminal Floating Nullors (FTFNs) [11-12],current conveyors [8,13-19,21] and CDTA [20] have been reported. Unfortunately, these reported circuits suffer from one or more of the following drawbacks

- Lack of electronic tunability [8,11-16, 18-19, 21].
- Excessive use of active and /or passive elements [8- 9, 11-13,15-18,20].
- Can not provide transfer functions in voltage and transadmittance form with the same topology [8-9,11-14,18,20].

The proposed new configuration enjoys the following advantageous features:

(i). Employs only two CCCCTA and three capacitors out of which two are permanently grounded.
(ii). Realizing low pass, high pass, band pass, band reject and all pass in transadmittance form and low pass, high pass, band pass in voltage form..
(iii). Its active and passive sensitivities are low
(iv). The ω_o, Q and ω_o/Q are electronically tunable with bias currents of CCCCTAs
(v). Both ω_o and ω_o/Q, and Q and ω_o/Q are orthogonally tunable.
(vi). Transadmittance gain of low pass, high pass, band pass can be independently tuned by biasing current of CCCCTA.

2 Current Controlled Current Conveyor Transconductance Amplifier (CCCCTA)

The CCCCTA properties can be shown in the following equation

$$\begin{bmatrix} I_Y \\ V_X \\ I_{za} \\ I_{zb} \\ I_{oa} \\ I_{ob} \end{bmatrix} = \begin{bmatrix} 0 & 0 & 0 & 0 \\ R_X & 1 & 0 & 0 \\ 1 & 0 & 0 & 0 \\ 1 & 0 & 0 & 0 \\ 0 & 0 & g_{ma} & 0 \\ 0 & 0 & g_{mb} & 0 \end{bmatrix} \begin{bmatrix} I_X \\ V_Y \\ V_{za} \\ V_{zb} \\ V_{oa} \\ V_{ob} \end{bmatrix} \tag{1}$$

where R_x, g_{ma} and g_{mb} are the parasitic resistance at x terminal and transconductance of CCCCTA. For a bipolar CCCCTA, the R_x, g_{ma} and g_{mb} can be expressed to be

$$R_X = \frac{V_T}{2I_B} \tag{2}$$

$$g_{ma} = \frac{I_{Sa}}{2V_T} \tag{3}$$

and

$$g_{mb} = \frac{I_{Sb}}{2V_T} \tag{4}$$

where I_B, I_{Sa} and I_{Sb} are the bias currents of CCCCTA, , respectively and V_T is the thermal voltage. The schematic symbol is illustrated in Fig.1.

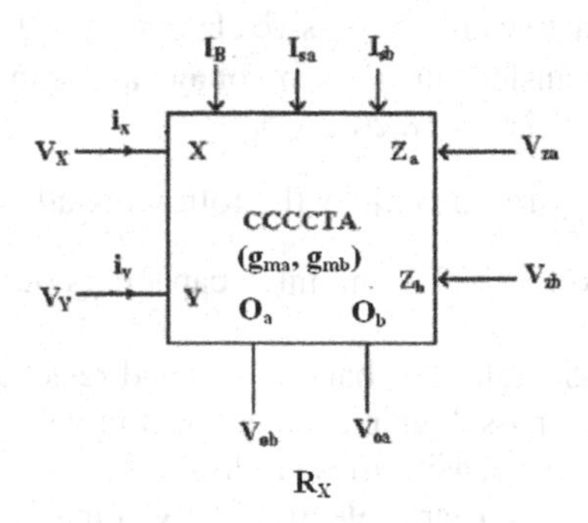

Fig. 1. CCCCTA Symbol

3 Proposed Configuration

The proposed SIMO biquad filter is shown in Fig. 2. It is based on two CCCCTA, three capacitors. Routine analysis yield the circuit transfer function $T_{LP}(s)$, $T_{BP}(s)$ and $T_{HP}(s)$ for the voltage outputs ($V_{LP}(s)$, $V_{BP}(s)$ and $V_{HP}(s)$) and $T'_{LP}(s)$, $T'_{BP}(s)$ and $T'_{HP}(s)$ for the transadmittance outputs($I_{LP}(s)$, $I_{BP}(s)$ and $I_{HP}(s)$), given by

$$T_{LP}(s) = \frac{V_{LP}(s)}{V_{in}(s)} = \frac{-C_1/C_3}{s^2C_1C_2R_{X1}R_{X2} + sC_2R_{X2} + 1} \tag{5}$$

$$T_{BP}(s) = \frac{V_{BP}(s)}{V_{in}(s)} = \frac{-sC_1R_{X2}}{s^2C_1C_2R_{X1}R_{X2} + sC_2R_{X2} + 1} \tag{6}$$

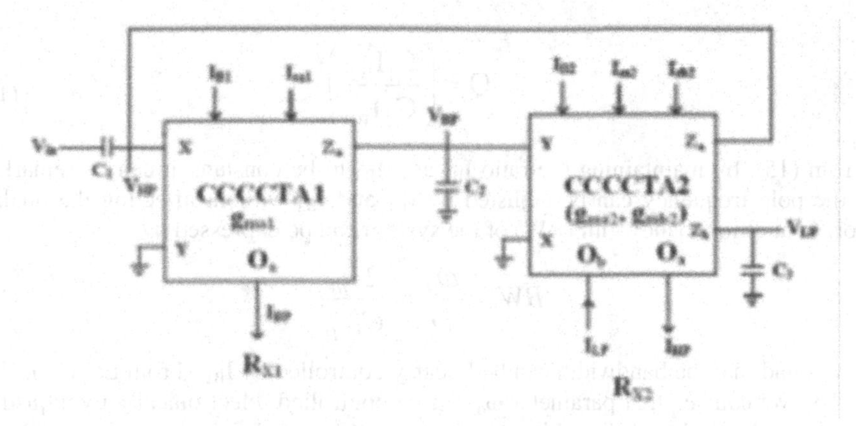

Fig. 2. An electronically tunable SIMO biquad filter using CCCCTA

$$T_{HP}(s)=\frac{V_{HP}(s)}{V_{in}(s)}=\frac{s^2C_1C_2R_{X1}R_{X2}}{s^2C_1C_2R_{X1}R_{X2}+sC_2R_{X2}+1} \tag{7}$$

$$T'_{LP}(s)=\frac{I_{LP}(s)}{V_{in}(s)}=\frac{g_{mb2}C_1/C_3}{s^2C_1C_2R_{X1}R_{X2}+sC_2R_{X2}+1} \tag{8}$$

$$T'_{BP}(s)=\frac{I_{BP}(s)}{V_{in}(s)}=\frac{-sg_{ma1}C_1R_{X2}}{s^2C_1C_2R_{X1}R_{X2}+sC_2R_{X2}+1} \tag{9}$$

$$T'_{HP}(s)=\frac{I_{HP}(s)}{V_{in}(s)}=\frac{s^2g_{ma2}C_1C_2R_{X1}R_{X2}}{s^2C_1C_2R_{X1}R_{X2}+sC_2R_{X2}+1} \tag{10}$$

The pole frequency (ω_o), the quality factor (Q) and Bandwidth (BW) ω_o/Q of each filter response can be expressed as

$$\omega_o=\left(\frac{1}{C_1C_2R_{X1}R_{X2}}\right)^{\frac{1}{2}} \tag{11}$$

$$Q=\left(\frac{C_1R_{X1}}{C_2R_{X2}}\right)^{\frac{1}{2}} \tag{12}$$

$$BW=\frac{\omega_O}{Q}=\frac{1}{C_1R_{X1}} \tag{13}$$

Substituting intrinsic resistances as depicted in (2), it yields

$$\omega_o=\frac{2}{V_T}\left(\frac{I_{B1}I_{B2}}{C_1C_2}\right)^{\frac{1}{2}} \tag{14}$$

$$Q = \left(\frac{C_1 I_{B2}}{C_2 I_{B1}} \right)^{\frac{1}{2}} \tag{15}$$

From (15), by maintaining the ratio I_{B1} and I_{B2} to be constant, it can be remarked that the pole frequency can be adjusted by I_{B1} and I_{B2} without affecting the quality factor. In addition, bandwidth (BW) of the system can be expressed by

$$BW = \frac{\omega_O}{Q} = \frac{2I_{B1}}{C_1 V_T} \tag{16}$$

We found that the bandwidth can be linearly controlled by I_{B1}. From Eq. (14), (15) and (16), we can see that parameter ω_o can be controlled electronically by adjusting bias current I_{B2} with out disturbing parameter ω_o/Q . Furthermore, parameter Q can also be controlled by adjusting the bias current I_{B2} with out disturbing parameter ω_o/Q. The voltage gain of the low pass, high pass and band pass can be expressed as

$$G_{LP} = \frac{C_1}{C_3} \tag{17}$$

$$G_{BP} = \frac{C_1}{C_2} \tag{18}$$

$$G_{HP} = 1 \tag{19}$$

Similarly, transadmittance gain of the low pass, high pass and band pass can be expressed as

$$G'_{LP} = g_{mb2} \frac{C_1}{C_3} \tag{20}$$

$$G'_{BP} = g_{ma1} \frac{C_1}{C_2} \tag{21}$$

$$G'_{HP} = g_{ma2} \tag{22}$$

from Eq. (20), (21) and (22), it can be seen that the transadmittance gain of low pass, high pass and band pass can be independently electronically adjusted without affecting the Q, ω_o and BW. It can be seen that the filter circuit can realize the low pass, band pass and high pass transfer functions in voltage mode as well as transadmittance mode. The band reject transfer function $T'_{BR}(s)$ in transadmittance form can be easily obtained from the currents $I_{BR}(s) = I_{LP}(s) + I_{HP}(s)$.

$$T'_{BR}(s) = \frac{I_{BR}(s)}{V_{in}(s)} = \frac{s^2 g_{ma2} C_1 C_2 R_{X1} R_{X2} + g_{mb2} C_1 / C_3}{s^2 C_1 C_2 R_{X1} R_{X2} + s C_2 R_{X2} + 1} \tag{23}$$

It is clear from Eq.(23) that one can obtain regular band reject filter for $C_3 . g_{ma2} = C_1 . g_{mb2}$ note that since zero and pole frequency can take different values, one can also obtain low pass band reject and high pass band reject filters for $C_3.g_{ma2} < C_1.g_{mb2}$ and $C_3.g_{ma2} > C_1.g_{mb2}$ respectively.

Also, all pass transfer function in transadmittance form can be obtained from the currents $I_{AP}(s) = I_{LP}(s) + I_{BP}(s) + I_{HP}(s)$, by keeping. $g_{ma2} = g_{ma1} = g_{mb2} = 1mS$ and $C_1 = C_2 = C_3$

$$T'_{AP}(s) = \frac{I_{AP}(s)}{V_{in}(s)} = \frac{s^2 g_{ma2} C_1 C_2 R_{X1} R_{X2} - s g_{ma1} C_1 R_{X2} + g_{mb2} C_1 / C_3}{s^2 C_1 C_2 R_{X1} R_{X2} + s C_2 R_{X2} + 1} \tag{24}$$

4 Non Ideal Analysis

For non-ideal case, the CCCCTA can be, respectively, characterized with the following equations

$$V_X = \beta_i V_Y + I_X R_X \tag{25}$$

$$I_{Zai} = \alpha_{ai} I_X \tag{26}$$

$$I_{Zbi} = \alpha_{bi} I_X \tag{27}$$

$$I_{Oai} = \gamma_{ai} g_{mai} V_{Zai} \tag{28}$$

$$I_{Obi} = \gamma_{bi} g_{mbi} V_{Zbi} \tag{29}$$

Where β_i, α_{ai}, α_{bi}, γ_{ai} and γ_{bi} are transferred error values deviated from one, of ith CCCCTA where i=1, 2. In the case of non-ideal and reanalyzing the proposed filter in Fig. 2, it yields the transfer functions as

$$T_{LP}(s) = \frac{V_{LP}(s)}{V_{in}(s)} = \frac{-\alpha_{a1}\alpha_{b2}\beta_2 C_1 / C_3}{s^2 C_1 C_2 R_{X1} R_{X2} + s C_2 R_{X2} + \beta_2 \alpha_{a1} \alpha_{b2}} \tag{30}$$

$$T_{BP}(s) = \frac{V_{BP}(s)}{V_{in}(s)} = \frac{-s\alpha_{a1} C_1 R_{X2}}{s^2 C_1 C_2 R_{X1} R_{X2} + s C_2 R_{X2} + \beta_2 \alpha_{a1} \alpha_{a2}} \tag{31}$$

$$T_{HP}(s) = \frac{V_{HP}(s)}{V_{in}(s)} = \frac{s^2 C_1 C_2 R_{X1} R_{X2}}{s^2 C_1 C_2 R_{X1} R_{X2} + s C_2 R_{X2} + \beta_2 \alpha_{a1} \alpha_{a2}} \tag{32}$$

$$T'_{LP}(s) = \frac{I_{LP}(s)}{V_{in}(s)} = \frac{\gamma_{b2} g_{mb2} \alpha_{a1} \alpha_{b2} \beta_2 C_1 / C_3}{s^2 C_1 C_2 R_{X1} R_{X2} + s C_2 R_{X2} + \beta_2 \alpha_{a1} \alpha_{b2}} \tag{33}$$

$$T'_{BP}(s) = \frac{I_{BP}(s)}{V_{in}(s)} = \frac{-s\gamma_{a1} g_{ma1} \alpha_{a1} C_1 R_{X2}}{s^2 C_1 C_2 R_{X1} R_{X2} + s C_2 R_{X2} + \beta_2 \alpha_{a1} \alpha_{a2}} \tag{34}$$

$$T'_{HP}(s) = \frac{I_{HP}(s)}{V_{in}(s)} = \frac{s^2 \gamma_{a2} g_{ma2} C_1 C_2 R_{X1} R_{X2}}{s^2 C_1 C_2 R_{X1} R_{X2} + s C_2 R_{X2} + \beta_2 \alpha_{a1} \alpha_{a2}} \tag{35}$$

In this case, the ω_o and Q are changed to

$$\omega_o = \left(\frac{\beta_2 \alpha_{a1} \alpha_{a2}}{C_1 C_2 R_{X1} R_{X2}} \right)^{\frac{1}{2}} \tag{36}$$

$$Q = \left(\frac{\beta_2 \alpha_{a1} \alpha_{a2} C_1 R_{X1}}{C_2 R_{X2}} \right)^{\frac{1}{2}} \tag{37}$$

5 Sensitivity Analysis

The ideal sensitivities of the pole frequency, the quality factor and band width with respect to passive components can be found as

$$S_{C_1,C_2}^{\omega_o} = -\frac{1}{2}, S_{I_{B1},I_{B2}}^{\omega_o} = \frac{1}{2}, S_{V_T}^{\omega_o} = -1 \tag{38}$$

$$S_{I_{B1},C_2}^{Q} = -\frac{1}{2}, S_{C_1,I_{B2}}^{Q} = \frac{1}{2}, S_{V_T}^{Q} = 0 \tag{39}$$

$$S_{C_1,V_T}^{BW} = -1, S_{I_{B1}}^{BW} = 1, S_{C_2,I_{B2}}^{BW} = 0 \tag{40}$$

From the above calculations, it can be seen that all sensitivities are constant and equal or smaller than 1 in magnitude.

Using Eq.(36) and (37). The non-ideal sensitivities can be found as

$$S_{C_1,C_2,R_{X1},R_{X2}}^{\omega_o} = -\frac{1}{2}, S_{\alpha_{a1},\alpha_{a2},\beta_2}^{\omega_o} = \frac{1}{2}, S_{\gamma_{a1},\gamma_{a2},\gamma_{b2},\alpha_{b2},\beta_1}^{\omega_o} = 0 \tag{41}$$

$$S_{R_{X2},C_2}^{Q} = -\frac{1}{2}, S_{R_{X1},C_1,\alpha_{a1},\alpha_{a2},\beta_2}^{Q} = \frac{1}{2}, S_{\gamma_{a1},\gamma_{a2},\gamma_{b2},\alpha_{b2},\beta_1}^{Q} = 0 \tag{42}$$

From the above results, it can be observed that all the sensitivities due to non-ideal effects are equal and less than 1 in magnitude.

6 Simulation Results

To verify the theoretical analysis, PSPICE simulation has been used to confirm the validity of the proposed, an electronically tunable SIMO biquad filter using CCCCTA of Fig.2. In simulation, the CCCCTA is realized using BJT implementation [6] with supply voltages V_{CC} =1.64V and V_{EE} = -1.64V shown in Fig. 3 with the transistor model of PR100N (PNP) and NP100N (NPN) of the bipolar arrays ALA400 from AT&T [22].

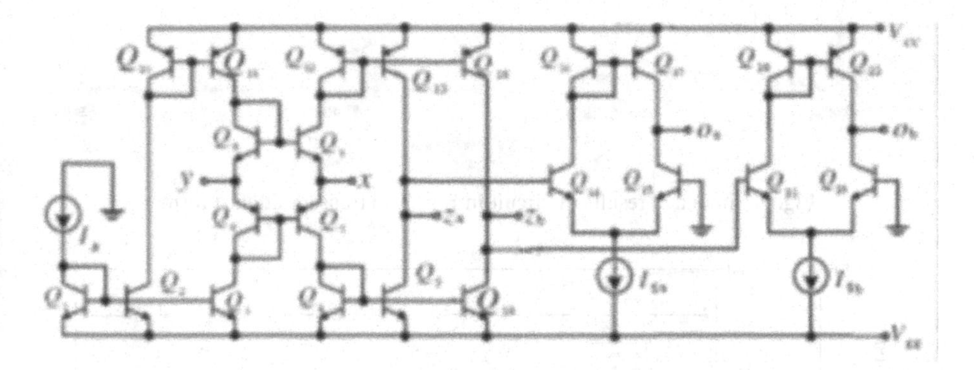

Fig. 3. Internal Topology of CCCCTA

To obtain $f_o=\omega_o/2\pi$=277Khz at Q=1, the active and passive components are chosen as I_{B1}= I_{B2}=45μamp, I_{Sa1}= I_{Sa2}= I_{Sb2}=52μamp and C_1=C_2= C_3=2nf. Fig. 4 Shows the simulated frequency responses of the LP, HP, and BP in voltage form of the proposed circuit in Fig.2. Fig. 5 shows simulated frequency responses of LP, HP, BP, BR and AP in transadmittance form of the proposed circuit in Fig. 2. Fig. 6 Show the gain and phase response of transadmittance all pass filter . From simulation result it can be seen that the simulated pole frequency of 264.51Khz is obtained as compared to the calculated value of 277Khz which show that the proposed electronically tunable SIMO biquad filter using CCCCTA performs the basic filter function well and satisfy the expected filter characteristics.

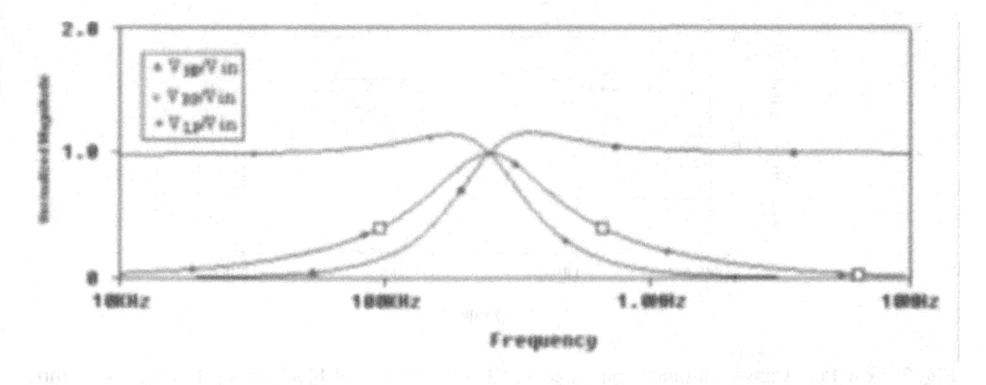

Fig. 4. Simulated results of circuit in Fig. 2. in voltage form

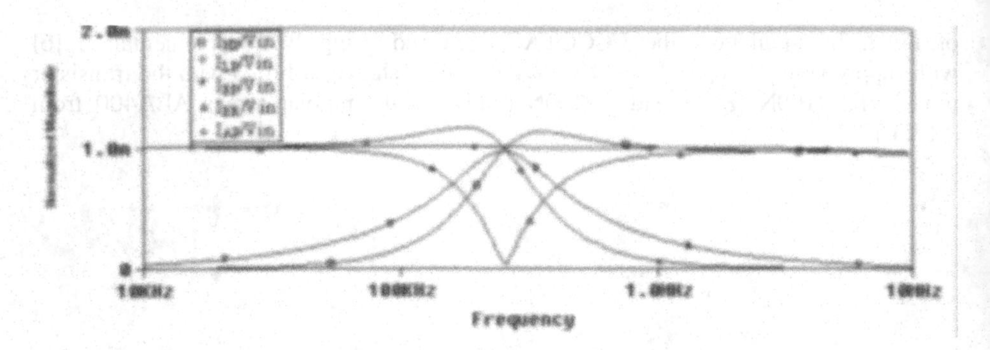

Fig. 5. Simulated results of circuit in Fig. 2. in transadmittance form

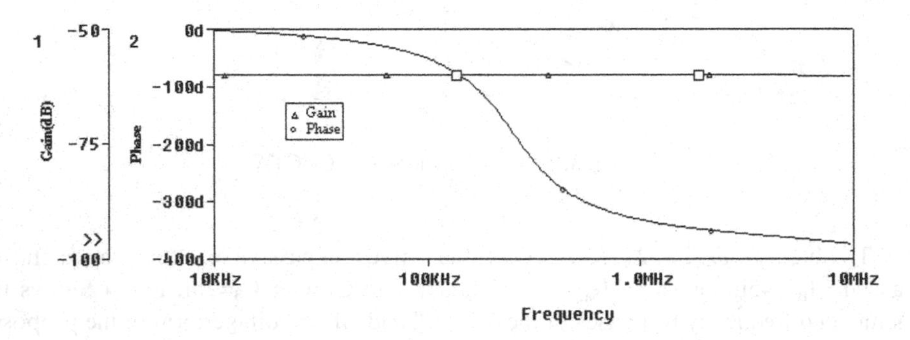

Fig. 6. Simulation result of all pass transadmittance form of circuit in Fig. 2

By adjusting the bias current (I_{Sb2}) of CCCCTA2 in the circuit of Fig. 2, the tunability of the transadmittance gain for the low pass response is tested. Different I_{Sb2} values of 52μamp, 104μamp and 156μamp are selected for CCCCTA2, which result in low pass responses with transadmittance gain of 1, 2 and 3 milliamp/volt, respectively. The low pass frequency responses with different gains are shown in Fig. 7.

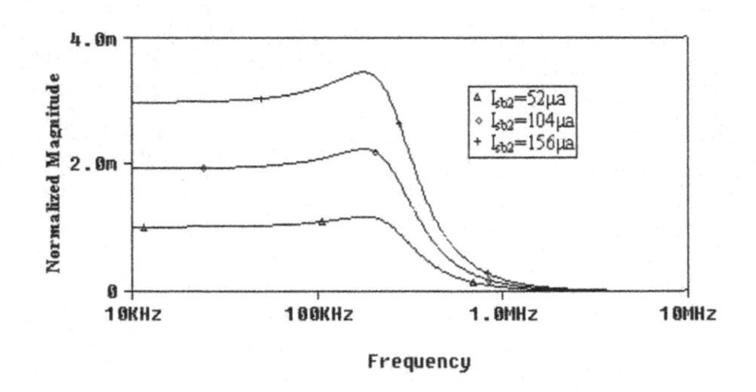

Fig. 7. Low Pass transadmittance responses for different value of I_{Sb2} keeping I_{B1}= I_{B2}=45μamp and I_{Sa1}= I_{Sa2}=52μamp

By adjusting the bias current (I_{Sa1}) of CCCCTA1 in the circuit of Fig. 2 the tunability of the transadmittance gain for the band pass response is tested. Different I_{Sa1} values of 52μamp, 104μamp and 156μamp are selected for CCCCTA1, which result in band pass responses with transadmittance gain of 1, 2 and 3 milliamp/volt, respectively. The band pass frequency responses with different gains are shown in Fig. 8.

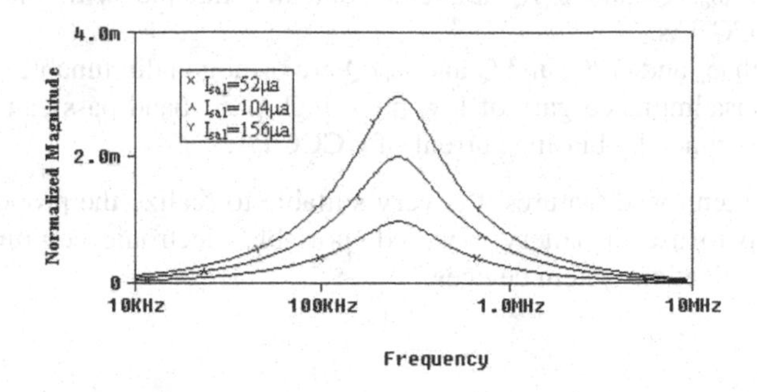

Fig. 8. Band Pass transadmittance responses for different value of I_{Sa1} keeping I_{B1}= I_{B2}=45μamp and I_{Sa2}= I_{Sb2}=52μamp

Similarly, by adjusting the bias current (I_{Sa2}) of CCCCTA2 in the circuit of Fig. 2 the tunability of the transadmittance gain for the high pass response is tested. Different I_{Sa2} values of 52μamp, 104μamp and 156μamp are selected for CCCCTA2, which result in high pass responses with transadmittance gain of 1, 2 and 3 milliamp/volt, respectively. The high pass frequency responses with different gains are shown in Fig. 9.

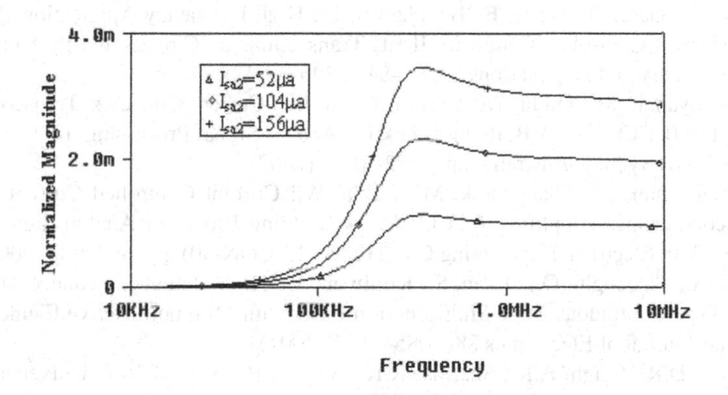

Fig. 9. High Pass transadmittance responses for different value of I_{Sa2} keeping I_{B1}= I_{B2}=45μamp and I_{Sa1}= I_{Sb2}=52μamp

7 Conclusion

An electronically tunable SIMO biquad filter using CCCCTA using only two current controlled current conveyor transconductance amplifiers (CCCCTAs) and three capacitors has been presented. The proposed filter offers the following advantages

(i) Realizing low pass, high pass, band pass, band reject and all pass in transadmittance form and low pass, high pass, band pass in voltage form.
(ii) Low pass and high pass band reject responses can also be obtained.
(iii) Two of the capacitors being permanently grounded
(iv) Low sensitivity figures
(v) The ω_o, Q and ω_o/Q are electronically tunable with bias currents of CCCCTAs
(vi) Both ω_o and ω_o/Q, and Q and ω_o/Q are orthogonally tunable.
(vii) Transadmittance gain of low pass, high pass, band pass can be independently tuned by biasing current of CCCCTA.

With above mentioned features it is very suitable to realize the proposed circuit in monolithic chip to use in battery powered, portable electronic equipments such as wireless communication system devices.

References

[1] Roberts, G.W., Sedra, A.S.: All Current-Mode Frequency Selective Circuits. Electronics Letters 25, 759–776 (1989)
[2] Wilson, B.: Recent Developments in Current Mode Circuits. Proc. IEE, Pt G. 137, 63–67 (1990)
[3] Maheshwari, S.: High Performance Voltage-Mode Multifunction Filter with Minimum Component Count. WSEAS Transactions on Electronics 5, 244–249 (2008)
[4] Fabre, A., Saaid, O., Wiest, F., Boucheron, C.: Current Controlled Band-Pass Filter Based on Translinear Conveyors. Electronics Letters 31, 1727–1728 (1995)
[5] Fabre, A., Saaid, O., Wiest, F., Boucheron, C.: High Frequency Application Based on a New Current Controlled Conveyor. IEEE Transactions on Circuits and System-I: Fundamental Theory and Applications 43, 82–91 (1996)
[6] Siripruchyanun, M., Jaikla, W.: Current Controlled Current Conveyor Transconductance Amplifier (CCCCTA): A Building Block for Analog Signal Processing. In: Proceeding of ISCIT 2007, Sydney, Australia, pp. 1072–1075 (2007)
[7] Siripruchyanun, M., Phattanasak, M., Jaikla, W.: Current Controlled Current Conveyor Transconductance Amplifier (CCCCTA): A Building Block for Analog Signal Processing. In: 30th Electrical Engineering Conference (EECON-30), pp. 897–900 (2007)
[8] Toker, A., Cicekoglu, O., Ozcan, S., Kuntman, H.: High-Output-Impedance Transadmittance Type Continuous-Time Multifunction Filter with Minimum Active Elements. International Journal of Electronics 88, 1085–1091 (2001)
[9] Bhaskar, D.R., Singh, A.K., Sharma, R.K., Senani, R.: New OTA-C Universal Current-Mode/Transadmittance Biquads. IEICE Electron. Express. 2, 8–13 (2005)
[10] Wu, J., El-Masry, E.I.: Universal Voltage and Current-Mode OTAs Based Biquads. International Journal of Electronics 85, 553–560 (1998)
[11] Shah, N.A., Iqbal, S.Z., Parveen, B.: SITO High Output Impedance Transadmittance Filter Using FTFNs. Analog Integrated Circuits and Signal Processing 40, 87–89 (2004)
[12] Liu, S.T., Lee, J.L.: Insensitive Current/Voltage-Mode Filters using FTFNs. Electronics Letters 32, 1079–1080 (1996)
[13] Chang, C.M.: Multifunction Biquadratic Filters Using Current Conveyors. IEEE Transactions on Circuits and Systems-II: Analog and Digital Signal Processing 44, 956–958 (1997)

[14] Chang, C.M., Lee, M.J.: Voltage-Mode Multifunction Filter with Single Input and Three Outputs Using Two Compound Current Conveyors. IEEE Transactions on Circuits and Systems-I: Fundamental Theory and Applications 46, 1364–1365 (1999)
[15] Pandey, N., Paul, S.K., Bhattacharyya, A., Jain, S.B.: A New Mixed Mode Biquad Using Reduced Number of Active and Passive Elements. IEICE Electron. Express 3, 115–121 (2006)
[16] Abuelma'atti, M.T., Bentrcia, A.: A Novel Mixed-Mode CCII-Based Filter. Active and Passive Electronic Components 27, 197–205 (2004)
[17] Abuelma'atti, M.T.: A Novel Mixed-Mode Current-Controlled Current-Conveyor-Based Filter. Active and Passive Electronic Components 26, 185–191 (2003)
[18] Cam, U.: A New Transadmittance Type First-Order All Pass Filter Employing Single Third Generation Current Conveyor. Analog Integrated Circuits and Signal Processing 43, 97–99 (2005)
[19] Hou, C.L., Lin, C.C.: A Filter with Three Voltage-Inputs and One Voltage-Output and One Current Output Using Current Conveyors. Tamkang Journal of Science and Engineering 7, 145–148 (2004)
[20] Shah, N.A., Quadrai, M., Iqbal, S.Z.: CDTA Based Universal Transadmittance Filter. Analog Integrated Circuits and Signal Processing 52, 65–69 (2007)
[21] Maheshwari, S.: Current-Mode Filters with High Output Impedance and Employing Only Grounded Components. WSEAS Transactions on Electronics 5, 238–243 (2008)
[22] Frey, D.R.: Log-Domain Filtering: An Approach to Current-Mode Filtering. IEE Proceedings-G: Circuits, Devices and Systems 140, 406–416 (1993)

An Architecture for Cross-Cloud System Management*

Ravi Teja Dodda[1], Chris Smith[2], and Aad van Moorsel[2]

[1] Department of Computer Science & Engineering,
Indian Institute of Technology Kharagpur, India
ravi.teja@iitkgp.ac.in

[2] School of Computing Science, Newcastle University, United Kingdom
c.j.smith4@ncl.ac.uk, aad.vanmoorsel@ncl.ac.uk

Abstract. The emergence of the cloud computing paradigm promises flexibility and adaptability through on-demand provisioning of compute resources. As the utilization of cloud resources extends beyond a single provider, for business as well as technical reasons, the issue of effectively managing such resources comes to the fore. Different providers expose different interfaces to their compute resources utilizing varied architectures and implementation technologies. This heterogeneity poses a significant system management problem, and can limit the extent to which the benefits of cross-cloud resource utilization can be realized. We address this problem through the definition of an architecture to facilitate the management of compute resources from different cloud providers in an homogenous manner. This preserves the flexibility and adaptability promised by the cloud computing paradigm, whilst enabling the benefits of cross-cloud resource utilization to be realized. The practical efficacy of the architecture is demonstrated through an implementation utilizing compute resources managed through different interfaces on the Amazon Elastic Compute Cloud (EC2) service. Additionally, we provide empirical results highlighting the performance differential of these different interfaces, and discuss the impact of this performance differential on efficiency and profitability.

1 Introduction

Conventionally, compute resources have been purchased as physical entities (servers) based upon some predictions of future usage and the maximization of some objective function, for example, minimized response time. A large fixed cost is associated with the purchase, configuration and deployment of each physical resource. Given this cost, businesses are reliant on the accuracy of their predictions of future usage to ensure that the cost-benefit trade-off of the purchase is resolved in favour of the benefit over some given time period. Accordingly,

* The authors are supported in part by EPSRC Grant EP/F066937/1 ("Economics-inspired Instant Trust Mechanisms for the Service Industry") and UK Department of Trade and Industry Grant P0007E ("Trust Economics").

S. Ranka et al. (Eds.): IC3 2009, CCIS 40, pp. 556–567, 2009.
© Springer-Verlag Berlin Heidelberg 2009

a risk is associated with the purchase of each physical resource. The purchase, configuration and deployment of each physical resource additionally requires significant periods of time, often days or weeks. Businesses must therefore make predictions days or weeks in advance, yielding a heavily deliberative rather than reactive approach to resource provisioning.

The emergence of the cloud computing paradigm promises flexibility and adaptability through the on-demand provisioning of compute resources. In contrast to the conventional approach of provisioning compute resources as physical entities, the cloud computing paradigm facilitates the provisioning of logical compute resources from a given provider accessible over the infrastructure of the World Wide Web. These logical compute resources can be purchased, configured and deployed in a matter of minutes based upon the current state of the system, for example, the current response time. Rather than provisioning resources based upon the maximization of an objective function using some predictions of future usage, businesses can provision resources based upon the maximization of the objective function given the current state of the system. The process of system management is therefore transformed from a deliberative to more reactive and adaptable process. The risk [2,3] associated with the provisioning of additional compute resources is reduced as the resources are regularly billed on a per-hour basis and resources can be de-provisioned once it is determined that their cost has surpassed their benefit, where benefit represents their contribution to the objective function. Accordingly, cloud computing is particularly valuable to small and medium businesses, where an effective and affordable IT infrastructure is critical to aiding productivity, since the large fixed costs and risk for physical, in-house compute resources can be avoided.

The flexibility and adaptability offered by the cloud computing paradigm facilitates the focus of businesses on their core objectives, rather than becoming heavily involved in strategic IT decisions regarding infrastructure. Whilst easing system management in this respect the cloud computing paradigm also presents some issues for system management [11,14]. The conventional approach to the provisioning of compute resources enabled their management in an analogous manner through management interfaces such as Simple Network Management Protocol (SNMP) [4] and Web Based Enterprise Management [5]. The purchase of compute resources from cloud providers does not facilitate such a stipulation. Cloud providers are free to expose their own bespoke interfaces to their logical compute resources utilizing different architectures and implementation technologies, conventionally based around Web services [8,10,13]. Once the utilization of cloud resources by a business extends beyond a single provider, this heterogeneity of management interface presents a significant issue for system management. A business will generally wish to manage the resources in a manner identical to the conventional approach, that is, all resources expose a homogeneous interface. Yet currently, when utilizing resources from different cloud providers this is not possible due to the heterogeneity of the interfaces. In this paper, we address this system management problem.

The remainder of the paper is structured as follow. Section 2 introduces the problem of managing compute resources which are dispersed across different cloud providers and consequently utilize different interfaces. Section 3 presents an architecture for the management of such resources in a cloud agnostic manner standardizing a set of generic operations which can be mapped to any specific interface exposed by a compute resource. This architecture is realized using the Amazon Elastic Compute Cloud (EC2) Service and Local Cloud Cluster and we provide details of this implementation. Section 5 provides empirical results obtained from experimentation utilizing our implementation through Amazon EC2, and illustrates the performance differential that occur when choosing between the different architectures and implementation technologies of cloud providers. Finally, we conclude our work in Section 6.

2 System Management of Compute Resources

In this section we discuss the requirements of system management with regard to compute resources purchased from cloud providers. We make an initial assumption that there is some decision-making component which computes some objective function using the current state of the compute resources in the system. That is, the component decides when to provision and when to de-provision resources based upon the cost-benefit analysis. This component could potentially have human involvement yet more likely would be the utilization of an appropriate decision-making algorithm. The selection of an appropriate decision-making algorithm is beyond the scope of this paper, but such a decision-making component may be based on economic considerations or service levels [6,9,12].

The system management of compute resources requires the fulfilment of a set of functional requirements related to the monitoring and configuration. Firstly, appropriate mechanisms are required for the identification and lookup of resources. Secondly, the manipulation of and introspection on the resource life-cycle should be facilitated. Thirdly, resources should offer implementation-independent, abstract and standardized operations for state management. Finally, the execution of the management control loop and enactment of the management operations implemented by the decision-making component should be supported by all resources.

The compute resources from different cloud providers resolve this set of requirements utilizing different architectures and implementation technologies. As a consequence, whilst these requirements are fulfilled within the domain of a given cloud provider, once the usage of a business extends beyond this single provider this set of requirements are breached. No longer, for example, can resources be assumed to support standardized operations for state management operations. This is particularly important issue when one considers the benefits yielded by extending the pool of compute resources utilized by a business beyond a single provider. This may be beneficial from a business perspective, reducing the danger to be tied into a single provider, and also in terms of fault-tolerance through replication of resources at each different provider, or in terms of performance by balancing load across different providers in accordance with the performance experienced at each provider.

3 An Architecture for Cross-Cloud System Management

In this section we propose an architecture for cross-cloud system management that seeks to resolve the issue with system management requirements discussed in Section 2. We show how the benefits in terms of factors such as fault-tolerance and performance of utilizing compute resources from different cloud providers can be yielded whilst retaining the set of functional requirements for system management.

The primary component of our architecture is the generic compute instance. This component represents a compute resource from any given cloud provider, conforming to any given architecture and utilizing any given implementation technologies. The component therefore abstracts away all the specific implementation details of a compute resource. For this component we define an abstract set of operations which must be exposed, that is, the interface of the component. The definition of these operations encapsulates all the system management requirements defined in Section 2 whilst utilizing generic terminology not specific to any particular cloud provider. We list the supported operations below:

- `getId()`
- `start()`
- `stop()`
- `getState()`
- `getPublicIP()`
- `setPublicIP()`
- `setConfiguration()`
- `getConfiguration()`

These operations facilitate complete state management of any compute resource. The `getId()` operation provides the unique identifer of a resource enabling the resource to be referenced from the purposes of management. The `start()` and `stop()` operations enables the resource to be started and stopped respectively, equivalent to the power-up of a physical resource. Given that compute resource usage is billed on a per-hour basis, the utilization of these operations will be defined by the decision-making component [12]. The `getState()` operation facilitates introspection on the state of a compute resource. This state will include details such as the name of the provider, the time at which the resource was started, pending or terminated and the specification of the resource in terms of CPU and memory. Such details may be utilized by the decision-making component in order to inform to the management control loop. The `getPublicIP()` retrieves the public IP address of the instance set by the cloud provider. The `setPublicIP()` sets a different public IP address for convenience. The `setConfiguration()` and `getConfiguration()` operations enable the configuration of a resource to be set or retrieved respectively. The configuration will include the operating system and any applications required and will usually take the form of some type of machine image.

For each generic compute instance, the operations can be broadly classified into three phases as shown in Figure 1. The launch phase deals with the instantiation of the instances and the association of a generic id, for example, a Universal Unique Identifier (UUID) for each compute resource and starting all the job processes. The monitor phase deals with monitoring the state of the instances, facilitating modification of configurations, and introspection on state. The shutdown phase deals with the shutdown of instances, and the billing by the cloud provider.

Given definition of a generic compute instance, the specific compute resources provided by different cloud providers must be effectively mapped to this generic

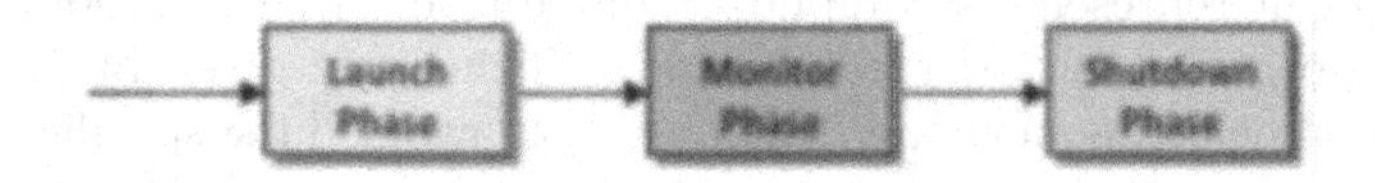

Fig. 1. Phases of a Generic Compute Instance

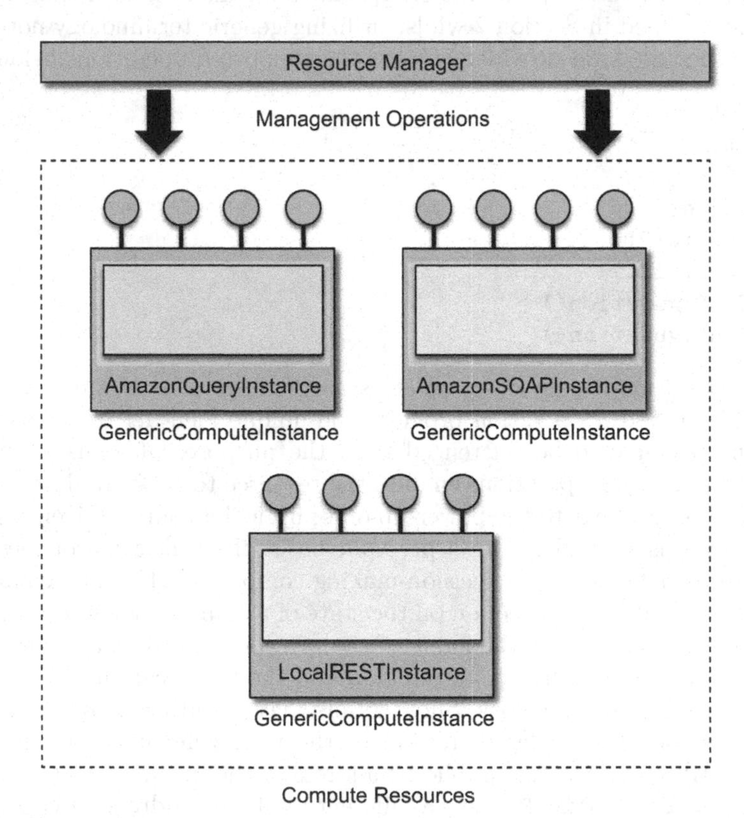

Fig. 2. An Architecture for Cross-Cloud System Management

definition. In order to do so, we define the set of generic operations as an interface (in accordance with Object Oriented Principles) and enable specific implementations of this interface to define the mapping between the generic operations of the generic compute instance and the specific operations of the resources provided by a given cloud provider. An implementation of the interface must be provided for any given interface we wish to utilize from any given cloud provider. For instance, a provider may offer two different interfaces to manage compute resources: i_1 and i_2. An implementation of the interface should be provided for both i_1 and i_2, mapping the management operations exposed by each interface to the set of generic operations. The implementation can then be reused for any compute resources we wish to manage from that provider through that specific interface.

The conformance of each compute resource to the interface of the generic compute instance enables any resource from any provider to be managed in an analogous manner. A resource manager is utilized to enact the generic operations on each compute resource in accordance with directions from the decision-making component. An overview of the architecture is illustrated in Figure 2. The resource manager retains a collection of all the compute resources that currently exist within the system and can reference each specific compute resources using the `getId()` operation. The identification scheme chosen is completely independent from that utilized by the cloud provider for the resource, enabling standardized referencing of resources by the generic id instantiated in the Launch phase of the compute resource. The resource manager is also responsible for the construction and destruction of compute resources as instructed by the decision-making component. The construction of a compute resource would take a form similar to the following:

```
GenericCloudInstance i1 = new CloudInstanceA()
...
GenericCloudInstance iN = new CloudInstanceB()
```

Through the utilization of our generic interface we facilitate system management in a cloud-independent manner. We need only define the mapping from the generic interface to the specific implementation of the cloud provider once, and can then reuse for any compute resource we wish to utilize from that provider. Virtual machine instances could be utilized to further generalize the instances, such that regardless of the cloud provider we utilize, a homogenous machine image can be used across all instances from all providers.

4 An Implementation Using Amazon Elastic Compute Cloud (EC2)

In this section we describe an implementation of the architecture presented in Section 3 using the Amazon Elastic Compute Cloud (EC2) [1] based on Query and SOAP interfaces and Local Cloud Cluster based on REST interface. (REST is the Representational State Transfer architectural style as defined by Fielding

[7], in practice implying that an implementation uses light weight HTTP instead of solutions such as SOAP.)

EC2 supports exposes two different interfaces for the management of compute resources: Query and SOAP. The Query interface utilizes Hypertext Transfer Protocol (HTTP) by enabling operations to be defined in the query string of a URL. A list of supported parameters is defined by the service, and consumers can then simply add these parameters and the appropriate values into the query string of a given URL again defined by the service. A HTTP GET operation is then performed on the constructed URL to enact the desired management operation. The SOAP interface utilizes Simple Object Access Protocol (SOAP) over HTTP to enact operations. The consumer provides within a SOAP envelope the operation desired and any relevant parameters to that operation. This SOAP envelope is then transmitted using a HTTP GET to an endpoint located at a URL defined by the service. The operations supported through the query interface are identical to those supported by the SOAP interface, it is simply the method by which these operations are invoked which differs.

A number of APIs are available which provide a programmatical means by which to invoke operations through each interface. We chose the Java implementation provided by Amazon for the query and SOAP interface. It is important to note that these implementations had significant differences in terms of library dependencies. The implementation of the SOAP interface had a significantly larger number of dependencies than the query interface, referencing many libraries relating to aspects such as XML parsing and serialization. Since the SOAP interface has larger number of dependencies and no open source library being available for it, it is rarely used by the Businesses due to unawareness about it. We implemented the SOAP interface using the ec2javaclient jar file given in the APIs by Amazon EC2 and then including the dependencies for XML parsing and serialisation of the requests. The potential performance impact of these interfaces is investigated in Section 5. We then created a class for each interface which mapped this implementation to the generic interface we defined in Section 3.

In order to demonstrate the applicability of our approach, we utilized in our implementation some local physical resources (local cloud cluster) in addition to those we utilize from the EC2 service. In doing so, we illustrate how compute resources from different cloud providers can be managed in a cloud-independent manner. Each local resource exposed a REST style interface and in the same manner as the resources from Amazon, we defined a mapping from this interface to our generic interface. The implementation therefore comprised of compute resources conforming to three different interfaces yet facilitated the analogous management of all these resources.

We created a resource manager which maintained a list of the compute resources currently utilized and enabled the management operations to be performed on each resource in accordance with the decision-making component. The resource manager instantiated these resources in the manner shown below and performed operations by simply invoking methods on the instantiated java object, for example, `i1.start()`, `i2.getState()`.

```
GenericCloudInstance i1 = new AWSQueryInstance()
GenericCloudInstance i2 = new AWSSOAPInstance()
GenericCloudInstance i3 = new LocalRESTInstance()
```

4.1 User Interface of Architecture for Cross-Cloud System Management

The architecture presented in Section 3 provides a resource manager for companies to create a flexible and efficient infrastructure management to host and manage applications both within and outside its organization. We provide a user interface, which is a web-based resource manager, giving users the ability to perform many virtual server management tasks Our user interface consists of the following features:

- Resource Management
- Creation of instances/ machines
- Dashboard-Control Center
- Cluster Management
- Network Management
- Repository Management
- Security & Protection

These features facilitate complete management of compute resources from user perspective. These features are discussed below in brief.

Resource Management. This feature faciliates management of user's clusters and virtual instances through a single easy to use interface. This provides an interface to start, stop and reboot virtual instances on any host, anywhere in the universe, by selection of single instance or a group of instances. It also facilitates provision of virtual machines on specific hosts through the host management interface. We also provide an interface to edit the virtual instance configurations and also a console to log-in and configure virtual instances.

Creation of instances/machines. This feature faciliates creation of instances/ machines by uploading ISO or XVM2 images and also specific images of cloud providers like Amazon EC2 and installation of unmodified and original operating system images such as Linux or Microsoft Windows and providing parameters required. These fresh installations can be completed using the console provided.

Dashboard-Control Center. Once the user logs in, he will see the Dashboard. It shows all the operations happening in the cloud. It also shows instance command and control operations , repository management operations and also logs the system errors.

Cluster Management. This feature facilitates provision of groups of virtual instances and assigning them to clusters. It also provides an interface to send launch parameters to virtual instances to allow cluster management events which are scripted to be passed to packaged machines.

Network Management. This feature is kept under Resource Management. It facilitates the creation of static networks within the user's cloud to provide predictable addresses to all his critical virtual instances.

Repository Management. This feature facilitates the download and provision from a library of pre-exisiting virtual machine images. It also provides an interface to provide your own URL'S for creating an internal repository of virtaul instances for the enterprise to share. The user can package existing virtual instances by creating snapshots of the images and also download from local or remote repository by casting. It shows all local and remote appliances.

Security & Protection. This feature facilitates the security and protection of virtual instances by providing access at an individual or group level to specific hosts and virtual instances through our security interface. We provide an interface to create and manage groups and users.

5 Empirical Results

In this section we provide empirical results obtained from our implementation of the architecture presented in Section 3. We contrasted the performance of the two interfaces for the management of compute resources exposed by the Amazon Elastic Compute Cloud in terms of response time. Whilst our architecture provides the ability to manage compute resources from different cloud providers in an analogous manner, the choice of which architecture or interface to utilize for compute resources from a given provider has significant performance consequences. Clearly, one wishes to manage a system in the most effective and efficient manner and performance, in terms of response time, of enacting management operations affects both the effectiveness and efficiency of the management process.

Our implementation provided the ability to manage compute resources from EC2 through either the Query or the SOAP interface. We do not include details of the interface specifically, but further information can be found at [1]. We dispatched 1000 identical requests enacting a given monitor operation on a resource managed through the Query interface, and repeated this for the SOAP API. The management operation we chose was the `getState()` operation which provides details regarding the current state of the resource. Each request was time-stamped on dispatch of the request and time-stamped again on receipt of the response enabling the computation of a response time for each request. Figure 3 illustrates the empirical results obtained from the experimentation.

The time taken to retrieve the state depends on a number of factors, including the size of the state representation, the bandwidth of the connecting

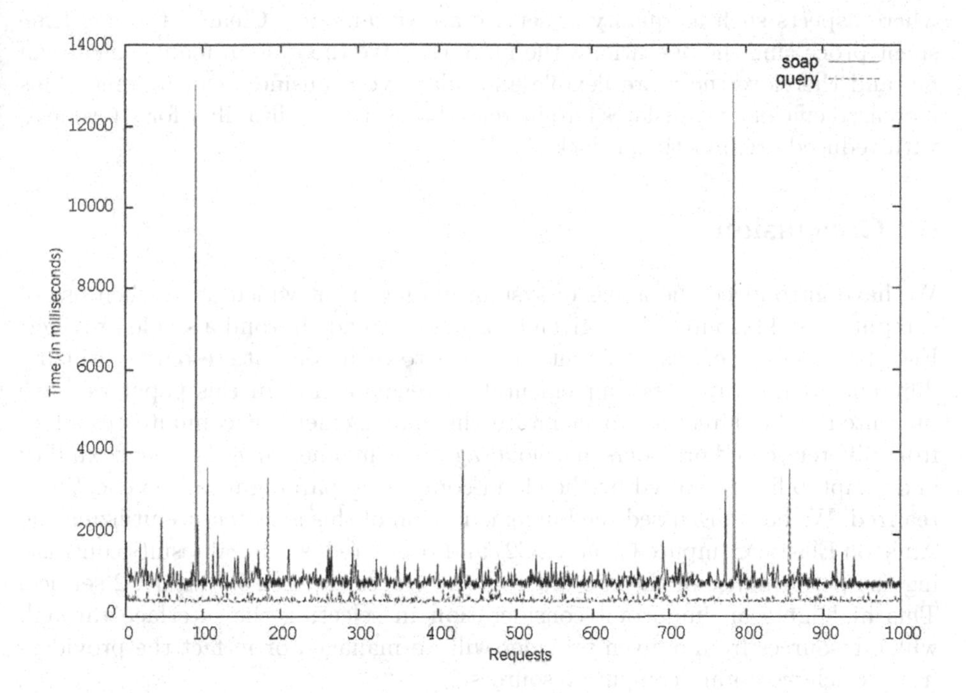

Fig. 3. Comparison between the Response Times of Amazon EC2 Query and Soap Interfaces

infrastructure etc. The reason behind comparing the amazon instances is because we assume these to be constant and therefore the only difference is the processing of requests here. Here we concentrate on the choice of interface by a given cloud provider. From Figure 3, the mean response time for the SOAP interface (905 ms) was nearly double that of the query interface (508 ms), highlighting a significant performance differential in the enactment of management operations, even though Query interface being used by significantly large number of businesses than that of the SOAP interface. Note that the outliers hardly influence the mean response time, as can also be visually understood from Figure 3. Additionally, the variance of the SOAP interface was significantly greater than the query interface introducing uncertainty in any predictions based on the enactment of these management operation, perhaps in relation to downstream Quality of Service (QoS). One can speculate as to the reasons for such a differential, for example parsing increased quantities of XML. Yet, given the black-box nature of each interface provided by a given cloud provider one may also have to consider such factors as cumulative load through that interface and the allocation of resources to the processing of requests through that interface. Clearly, different cloud providers will expose different interfaces and these interfaces will almost certainly exhibit varied performance characteristics. Our work, though, highlights that fact that performance is an important consideration in the selection of an interface and even a cloud provider, particularly in domains where aspects such as quality of service are paramount. Clearly, the less time spent processing the operations, the more reactive the system management can be, and therefore the more flexible and adaptive a business can become. This increased efficiency translates into increased long-run profitability for a business, with reduced organizational slack.

6 Conclusion

We have introduced the issues of system management which arise when use of compute cloud resources by a given consumer extends beyond a single provider. Each provider can expose different interfaces to their compute resources utilizing different architectures and implementation technologies. In this paper we have presented an architecture to facilitate the management of compute resources from different cloud providers in a cloud-agnostic manner such that the flexibility and adaptability promised by the cloud computing paradigm can be effectively realized. We have discussed the implementation of this architecture utilizing the Amazon Elastic Compute Cloud (EC2) and presented empirical results contrasting the performance of the two different interfaces exposed by the EC2 service. This highlights an important consideration in selecting the interface through which resources from a given provider will be managed or in fact the providers from which we obtain compute resources.

References

1. Amazon Elastic Compute Cloud (EC2) (2008), http://aws.amazon.com/ec2
2. Aven, T.: Foundations of risk analysis: A knowledge and decision-oriented perspective. Wiley, Chichester (2003)
3. Baird, B.F.: Managerial Decisions Under Uncertainty. Wiley Series in Engineering and Technology Management. John Wiley and Sons Inc., New York (1989)
4. Case, J., Fedor, M., Schoffstall, M., Davin, J.: Simple Network Management Protocol (SNMP) (May 1990), http://www.ietf.org/rfc/rfc1157.txt
5. DMTF. Web-Based Enterprise Management (WBEM) (October 2005)
6. Eric Bouillet, D.M., Ramakrishnan, K.: The structure and management of service level agreements in networks. IEEE Journal on Selected Areas in Communication 20(4), 691–699 (2002)
7. Fielding, R.T.: Architectural Styles and the Design of Network-based Software Architectures. PhD thesis, University of California, Irvine (2000)
8. Martin-Flatin, J.-P., Doffoel, P.-A., Jeckle, M.: Web services for integrated management: A case study. In (LJ) Zhang, L.-J., Jeckle, M. (eds.) ECOWS 2004. LNCS, vol. 3250, pp. 239–253. Springer, Heidelberg (2004)
9. Molina-Jiménez, C., Pruyne, J., van Moorsel, A.: The role of agreements in it management software. In: de Lemos, R., Gacek, C., Romanovsky, A. (eds.) Architecting Dependable Systems III. LNCS, vol. 3549, pp. 36–58. Springer, Heidelberg (2005)
10. Schönwälder, J., Pras, A., Martin-Flatin, J.-P.: On the Future of Internet Management Technologies. IEEE Communications Magazine (2003)

11. Sloman, M. (ed.): Network and Distributed Systems Management. Addison Wesley Publishers Ltd., Wokingham (1994)
12. Smith, C., van Moorsel, A.: Mitigating provider uncertinty in service provision contracts. In: Proceedings of the Workshop on Economic Models and Algorithms for Grid Systems (EMAGS 2007), Austin, Texas, USA (September 2007)
13. Vambenepe, W.: Web services distributed management: Management using web services (muws 1.0) part 1 (2005)
14. van Moorsel, A.: Grid, management and self-management. The Computer Journal 48(3), 325–332 (2005)

Energy Efficiency of Thermal-Aware Job Scheduling Algorithms under Various Cooling Models

Georgios Varsamopoulos, Ayan Banerjee, and Sandeep K.S. Gupta

The Impact Laboratory,
Arizona State University
Tempe, AZ 85287, USA
http://impact.asu.edu/

Abstract. One proposed technique to reduce energy consumption of data centers is thermal-aware job scheduling, i.e. job scheduling that relies on predictive thermal models to select among possible job schedules to minimize its energy needs. This paper investigates, using a more realistic linear cooling model, the energy savings of previously proposed thermal-aware job scheduling algorithms, which assume a less realistic model of constant cooling. The results show that the energy savings achieved are greater than the savings previously predicted. The contributions of this paper include: i) linear cooling models should be used in analysis and algorithm design, and ii) although the job scheduler must control the cooling equipment to realize most of the thermal-aware job schedule's savings, some savings can be still achieved without that control.

1 Introduction

Large data centers today contain up to tens of thousands of servers, consuming tens of megawatts of electricity annually [1, 2]. There is a growing problem of energy consumption that points toward seeking increasingly sophisticated ways, both in hardware and software, to achieve greater energy efficiency from data center facilities [3, 4, 5, 6, 7, 8]. Recent research has shown, through simulation, considerable savings for high-performance data centers through predictive *thermal-aware job scheduling*, i.e., scheduling that takes its thermal impact into consideration [7]. Previous work used a constant-value cooling model, i.e. cooling at a constant temperature, to estimate the energy savings. Nevertheless, most real-world cooling systems follow a discontinuous step-wise linear cooling model, i.e. cooling that supplies cool air at a temperature linearly dependent on that of the data center. This is pointed by both the technical specifications and *in situ* measurements, presented later in this paper.

This paper re-examines the energy savings of the XInt family [7,9] of thermal-aware job scheduling algorithms under the step-wise linear cooling model. Specifically, it address the question "*what would the energy consumption of job schedules be in a step-wise linear cooling environment when they are derived by the XInt family algorithms under constant cooling by the XInt family algorithms under a step-wise linear cooling environment.*" This is done by constructing an analytical temporal model of cooling and combining it with the heat recirculation as described by the *abstract linear heat interference model* [6] used by the XInt family. Numerical results show that energy savings

S. Ranka et al. (Eds.): IC3 2009, CCIS 40, pp. 568–580, 2009.
© Springer-Verlag Berlin Heidelberg 2009

achieved are greater compared to previous results, while the order of energy savings of the examined algorithms is preserved.

The rest of the paper is organized as follows: Section 2 introduces concepts on data center layout, heat recirculation, cooling system efficiency, thermal maps and thermal-aware job scheduling. Section 3 presents the abstract linear heat recirculation model, and the XInt family of thermal-aware job scheduling algorithms that are evaluated. Section 4 introduces the cooling models, their thermal and power behavior. Section 5 presents simulation results of energy savings under the constant and linear cooling models. Section 6 concludes the paper.

2 Preliminaries

Data center Operation and layout. Figure 1 shows a typical data center's organization. Computing equipment sits on a raised floor plenum, organized into *rows* that separate *aisles*, alternating between cold air intake aisles and hot air exhaust aisles. Cold air from the *computer room air conditioner* (CRAC) is supplied to the room through grated tiles in the raised floor of cold aisles, to keep all servers below a manufacturer-specified *red-line* temperature. A data center is abstracted to consist of n nodes (chassis)[1]. Each node i consists of several processors (cores). Each node i draws air with inlet temperature T_i^{in}, adds heat by consuming power p_i and dissipates hotter air with outlet temperature T_i^{out}. The outlet temperature of a node comes from the combined activity of the servers in that node. The total computing power consumption of a data center is P^{comp}, that being the sum of all node power consumption: $P^{\text{comp}} = \sum_i p_i$. In the formulations in this paper, we use a *vectorized* notation for brevity, i.e. $\boldsymbol{p} = \{P_i\}_n$, $\boldsymbol{T}^{\text{in}} = \{T_i^{\text{in}}\}_n$ etc.

Heat recirculation and cooling efficiency. Contemporary data centers are cooled by chilled-water CRAC units, using conventional air-cooled methods (i.e. fan & heat sink) at the computing equipment. To keep all the equipment at a normal operating temperature, the CRAC has to supply cool air at an adequately low temperature. However, due to the non-linear cooling efficiency, the CRAC's set temperature affects the coefficient of performance (CoP). CoP characterizes the efficiency of heat removal; it is the ratio of the heat removed from a system over the work required to perform the removal. At any given point, the CRAC power can be described as [2]:

$$P^{\text{AC}} = \frac{P^{\text{in}}}{\text{CoP}\,(T^{\text{sup}})}. \tag{1}$$

where P^{in} is the heat rate of the air that enters the CRAC and $\text{CoP}(T^{\text{sup}})$ is the CoP of the cooling system when supplying cold air at T^{sup}.

Using the above equation in an equilibrium state, and assuming that all heat produced by computing equipment enters the CRAC, the total cost is the sum of the power used to run the equipment and the power used to cool it down:

$$P^{\text{total}} = P^{\text{comp}} + P^{\text{AC}} \overset{(1)}{=} \left(1 + \frac{1}{\text{CoP}(T^{\text{sup}})}\right) P^{\text{comp}}. \tag{2}$$

[1] As data centers use blade-based systems and power/ventilation of the blades is shared within each chassis, the abstractions here are chassis-oriented.

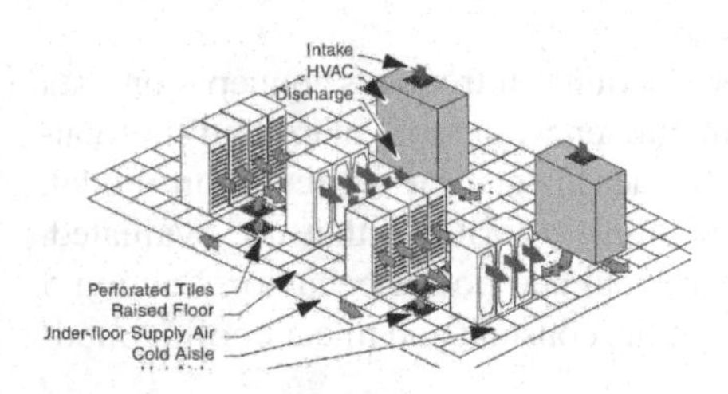

Fig. 1. Typical data center layout (source: ASHRAE [10])

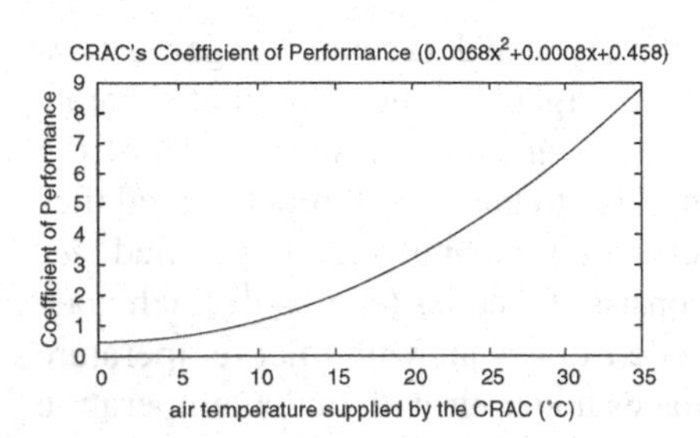

Fig. 2. Coefficient of performance of a typical CRAC unit. (source: [2])

Table 1. Scalar Symbols and Definitions

Symbol	Definition
n	number of server chassis
u_i	utilization of the ith chassis
c_{ij}	temperature rise coefficient at j^{th} chassis caused by heat from i^{th} chassis.
C	the matrix $\{c_{ij}\}$
p_i	power consumed by i^{th} chassis
a_i	power output coefficient of ith chassis
b_i	idle power output of ith chassis
q_j	number of cores requested by job j
$m, \dot{m}$	mass and mass flow rate, respectively
T_i^{in}	temperature of the ith chassis
T^{rise}	the vector of excess temperature above T^{sup} at the computing equipment inlets
T^{sen}	input air temperature of the CRAC
T^{sup}	output air temperature of the CRAC
T^{thres}	T^{sen} threshold between CRAC modes
P^{comp}	sum of the computing power
P^{out}	heat rate removed by CRAC(s)
P^{AC}	the power expended by CRAC(s)
α	slope of the line in linear cooling models
β	offset of the line in linear cooling models
ς	specific heat of air

Thermal Maps. Let $\boldsymbol{u} = \{u_1, u_2, \ldots, u_n\}$ be the utilization vector of n server chassis and $\boldsymbol{T}^{in} = \{T_1^{in}, T_2^{in}, \ldots, T_n^{in}\}$ be the corresponding stable-state temperature vector at the air inlets of these chassis for that utilization. The *thermal map* of the data center is defined as the translation function of $\boldsymbol{T}=F(\boldsymbol{u})$. This definition is a deterministic steady-state model. Such models are preferable in designing job scheduling algorithms mainly because they reduce the complexity of the decision making.

Thermal-aware job scheduling. The goal of thermal-aware scheduling is to allow the CRAC to act with higher efficiency by reducing the temperature it needs to supply to keep all compute equipment below its red-line temperature. The scheduler does so by allocating jobs to servers so that it reaches a *thermal balance* condition where all server inlet temperatures minimal and as equal as possible. As pointed out in previous research, ***the more balanced the inlet temperatures are, the higher the CRAC thermostat can be set at, thus saving more energy*** [7]. Much of the benefit of thermal-aware scheduling comes from dynamically setting the CRAC to maximize the CoP with each thermal map. One of this paper's contributions is that ***energy benefits can be still achieved without re-setting the CRAC***.

3 ALHI Model and XInt Job Scheduling Algorithms

Previous research on thermal-aware modeling and scheduling has produced a heat recirculation model, termed Abstract Linear Heat Interference model (ALHI), and the XInt series spatio-temporal scheduling algorithms based on ALHI. This section will provide description of the ALHI and the XInt algorithms, whose performance will be evaluated in the simulation section.

3.1 The Abstract Linear Heat Interference Model

The ALHI asserts that the thermal map function F is linear [7]. The ALHI's core is a heat recirculation matrix $\boldsymbol{C}=\{c_{ij}\}_{n\times n}$, such that c_{ij} is the *temperature interference coefficient* of chassis i on the inlet temperature of chassis j, i.e., a heat rate of p_i will cause, by recirculation, a *temperature rise* of $c_{ij}p_i$ at the inlet of chassis j. Experimental results in the literature suggest a linear dependence of the power expended to the CPU utilization of the equipment [6, 7, 11]. Translating the power p_i to a server utilization rate u_i using a linear power model, we have $p_i = a_i u_i + b_i$, where b_i is the idle (i.e. zero utilization) power and a_i the linear coefficient of the utilization-to-power relation (a_i and b_i are usually obtained by power measurements). Inserting that to ALHI, we get the temperature rise vector:

$$\boldsymbol{T}^{\text{rise}} = \boldsymbol{C}\boldsymbol{p}^{\text{comp}} = \boldsymbol{C}(\boldsymbol{a} \odot \boldsymbol{u} + \boldsymbol{b}), \tag{3}$$

where $\odot$ is the element-wise product[2].

3.2 The XInt Algorithm Family

The algorithms are divided into *spatial-only*, which decide on the placement (i.e. the server assignment) only, and *spatio-temporal*, which decide on both the start time and placement of the jobs.

Spatial-only. The XInt scheduling algorithm uses Equation 3 to minimize the $\boldsymbol{T}^{\text{rise}}$. Equation 3 deals only with the spatial dimension of server utilization; as such, XInt [7] considers solving for the following job placement optimization problem:

> ***Given an idle data center, the recirculation matrix C, and the power parameters a and b, find a placement for a task of size** q **(i.e. requesting** q **cores) that minimizes*** $\max\{\boldsymbol{T}^{rise}\}$.

The XInt implementations used available software packages to solve this minimax formulation. The two methods used were a genetic algorithm (XInt-GA) and a sequential quadratic programming (XInt-SQP), the latter performed on the continuous version of the problem, and then regressing to a near integer solution [7]. Also, a variant of XInt was provided that performs thermal-aware placement of multiple jobs *simultaneously submitted* to a partially utilized data center [7].

An alternative to optimizing Equation 3 is to minimize $\sum\{\boldsymbol{T}^{\text{rise}}\}$. Minimization of the summed $\boldsymbol{T}^{\text{rise}}$ is effectively minimization of the cumulative heat recirculation, as $\sum \boldsymbol{T}^{\text{rise}} = \sum \boldsymbol{C}\boldsymbol{p}^{\text{comp}} = \sum_i \sum_j c_{ij}p_i$. Using this optimization to perform thermal-aware job scheduling is based on the assertion that if the server with the minimum recirculated heat are selected to execute the jobs, then the $\boldsymbol{T}^{\text{rise}}$ vector will be reduced. The developed

[2] For example, $[a\ \ b\ \ c\ \ d] \odot [w\ \ x\ \ y\ \ z] = [aw\ \ bx\ \ cy\ \ dz]$.

least recirculated heat (LRH) heuristic, fully presented in [9], optimizes this formulation by pre-calculating *server ranks* as follows:

$$r_i = \sum_j v_j c_{ij} p_i^{\max}, \quad \text{where } v_j = \sum_i c_{ij} p_i^{\max},$$

where $p_i^{\max}$ is the maximum power output of node i. The algorithm then assigns tasks to the available (i.e. idle) servers with the minimal r_i values. The ranks v_j act as weight factors to the importance of the heat interference.

Spatio-temporal. The XInt algorithm above considers the spatial placement of the jobs only, with disregard to the duration of the jobs. New algorithms were developed to perform spatio-temporal scheduling and assess the energy savings of thermal-aware scheduling [9]. The spatio-temporal scheduling optimization problem examined was formulated as follows:

> ***Given an idle data center, the recirculation matrix C, and the power parameters a and b, find a spatio-temporal schedule for* a sequence of tasks *(each task i of size q_i, arrival time t_i^{arr} and deadline t_i^{ded}), in order to minimize the energy consumption.***

The algorithms developed are as follows:

FCFS-XInt: The FCFS-XInt is a combination of FCFS temporal placement (with back-filling), with XInt-based spatial placement. Benefits of this algorithm include its on-line nature and its high compatibility with FCFS-based commercial job schedulers.

SCINT: SCINT is a genetic algorithm (GA) implementation of spatio-temporal scheduling. It is an extension of XInt into the time dimension and it is an off-line algorithm. SCINT discretizes the time into ***time slots***, and using a GA approach constructs a schedule that resembles a slot-based server reservation table.

EDF-LRH: The SCINT and XInt algorithms are very slow for on-the-fly scheduling. MATLAB runs take several minutes for the SCINT and XInt-GA, and a few seconds for XInt-SQP. For that matter, a faster heuristic for the problem was developed that used an *earliest deadline first* (EDF) temporal scheduling with a *least recirculated heat* (LRH) spatial placement. The LRH optimization is defined as "minimize $\sum_{i=1}^{n} T_i^{\text{rise}}$," which is a minimization problem, as opposed to a minimax problem such as XInt, and it takes a fraction of a second to compute [9].

4 Cooling Models

This section introduces three cooling models: a) the *constant* model, b) the segmented constant-linear model, and c) the stepwise linear model.

Constant Model. In this model, the CRAC provides a constant output temperature regardless of its input temperature (Figure 3a). Previous research routinely used the constant cooling model because it reaches a converging steady state solution in simulations [2, 7]. It is conceptually the simplest and fastest model to create a thermal map with. This model suppresses heat recirculation through the CRAC; thus it fails to capture any heat that would otherwise pass through the CRAC and not be extracted. This

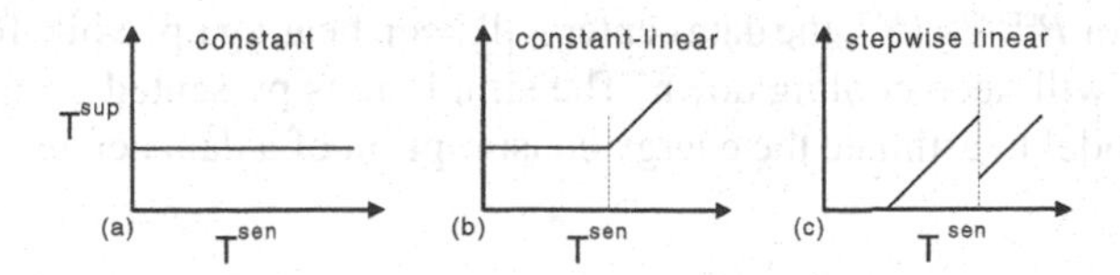

Fig. 3. Cooling models as characterized by their temperature transfer function: a) constant, b) segmented constant-linear, c) step-wise linear

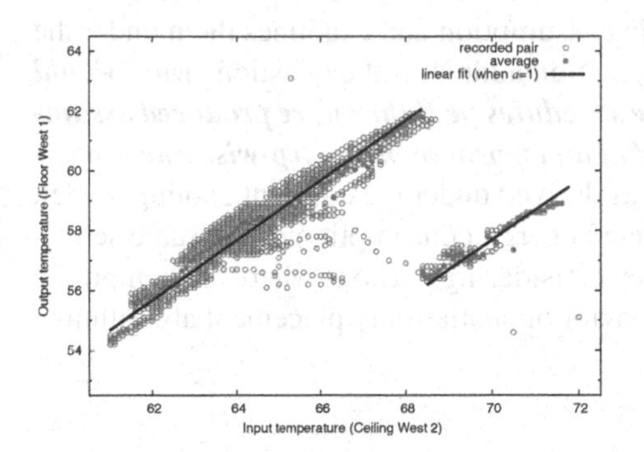

Fig. 4. Recorded pairs of input and output temperatures for an *in situ* CRAC unit, showing distinct operational modes

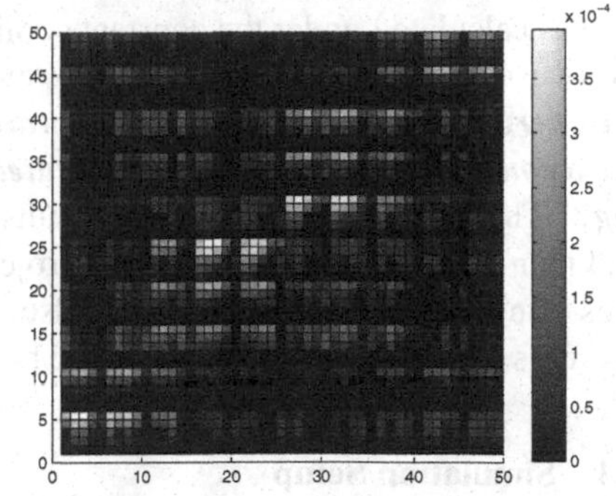

Fig. 5. The matrix C of coefficients as derived for the simulated data center in [7]

significantly alters the calculation of heat interference coefficients in a simulated environment with constant cooling model. It also suggests that the CRAC is capable of unboundedly extracting any heat—there is a maximum load a CRAC can cool. This drawback is addressed by the segmented constant-linear model, discussed in the following subsection.

Segmented constant-linear model. This model functions as the constant cooling model except that a maximum power load can be specified, beyond which all heat received is released back to the room (Figure 3b). It provides a constant temperature until the threshold temperature T^{thres} is reached and then follows a linear rise. This model is used in *computational fluid dynamics* (CFD) simulators to determine whether there exists sufficient cooling for a data center.

Stepwise Linear Model. The linear cooling model suggests that the temperature of the air within the CRAC is reduced in a linear fashion. It consists of linear segments $T^{\text{sup}} = \alpha T^{\text{sen}} + \beta$, where T^{sup} is the CRAC's output air temperature, and T^{sen} is the CRAC input air temperature (Figure 3c). This model has been realized from sensor measurements of *in situ* CRAC equipment, as shown in Figure 4.

Heat-extracting CRACs follow the stepwise linear model, where the CRAC switches between power modes according to the cooling needs, where in each mode the CRAC extracts a constant amount of heat, P^{out}. Each continuous section corresponds to a separate P^{out}. For $P^{comp}>P^{out}$, the data center will keep heating up, while for $P^{comp}<P^{out}$, the data center will keep cooling down. The simulations presented in the next section will use this model to estimate the energy consumption of a data center.

5 Energy Consumption under Various Cooling Models

This section presents an evaluation of the schedulers' effective energy consumption under the constant and the step-wise linear models. The latter evaluation takes the schedules as calculated under the constant cooling assumption and examines them under the step-wise linear model. Effectively, the evaluation addresses the question "***what would the energy consumption of thermal-aware schedules be if they were produced assuming a constant cooling model but executed in an environment of step-wise linear cooling?***" The section first presents the results as derived under the constant cooling model, and then presents new results of the projected energy consumption of the same schedules using a step-wise linear model. Also, by considering portions where the computing power is constant, it also addresses the behavior of spatial-only placement algorithms.

5.1 Simulation Setup

Physical CFD modeling. FloVENT, a CFD simulator by Mentor Graphics, was used to obtain recirculation coefficient matrix used by the XInt, SCINT and LRH algorithms. A model of a data center with physical dimensions 9.6 m × 8.4 m × 3.6 m, was created in FloVENT. It has two rows of industry standard 42U racks arranged in a typical cold aisle and hot aisle layout. The cold air is supplied by one computer room air conditioner, with the air flow rate 8 m^3/s. There are ten racks and each rack is equipped with five 7U (12.25-inch) chassis. The interference coefficient matrix in Figure 5 is derived from this setup. In this section, the *red-line* temperature is assumed at 35°C for all servers. The time slot length selected is 30 minutes.

System and job power profiles. In the simulations, we used 30 Dell PowerEdge 1955 chassis and 20 Dell PowerEdge 1855 chassis.

The algorithms' energy consumption has been evaluated using job traces from the ASU Fulton HPCI data center. The job traces provide: i) the job *arrival* times (i.e. the t_i^{arr}), ii) their corresponding *reservation* times, here treated as *deadlines* (i.e. the t_i^{ded}), iii) the number of servers required (i.e the q_i), and iv) the job start and finish times using the FCFS scheduling with back-filling. The estimates of the job execution times, t_i^{exe}, on the servers are based on the actual execution times in the ASU job traces, calculated as the difference between each job's start and finish times. The job traces are visualized in Figure 6, which shows the arrival time, estimated execution (both on the x-axis) and the number of processors requested (y-axis), and in Figure 7, which shows the arrival time, time reservation (treated as deadline), and the number of processors requested (y-axis).

The job execution time estimates for a type of node other than the one the job was actually run on, we simply multiply the execution time of the original equipment with the average gain in execution time on the other equipment. This gain is calculated as

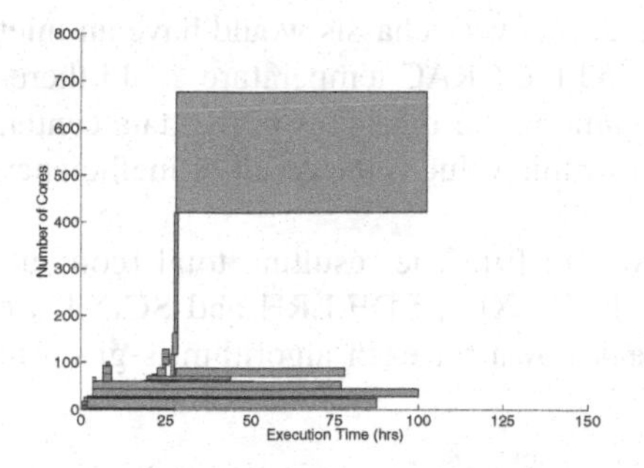

Fig. 6. Arrival, estimated execution time and number of processors of the job trace used in evaluating the scheduling algorithms [9]

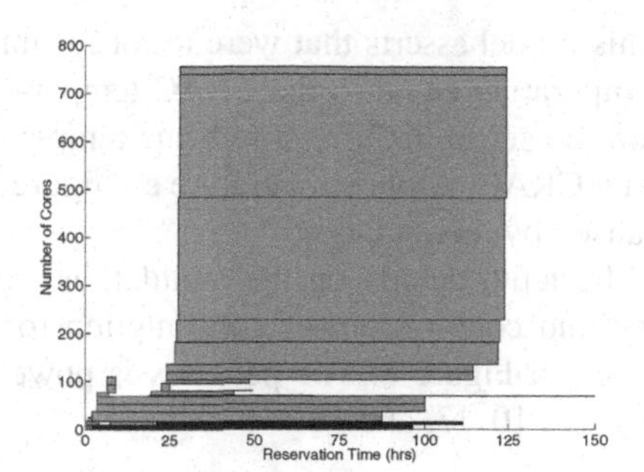

Fig. 7. Arrival, reservation time (in this paper it is treated as deadline) and number of processors of the job trace in Fig. 6

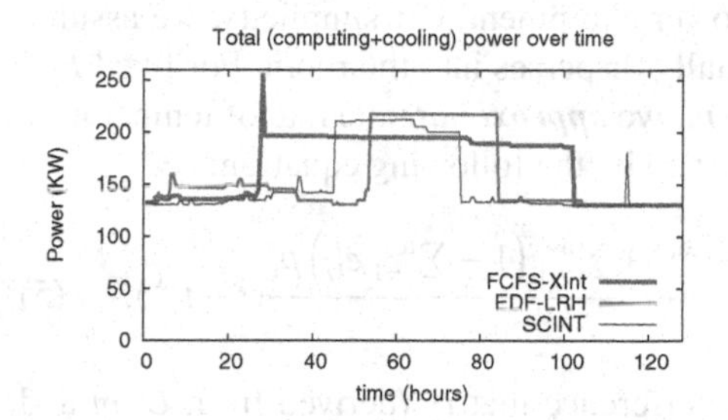

Fig. 8. Computing power consumption over time for the examined algorithms, under constant cooling, for the job trace in Figures 6,7 [9]

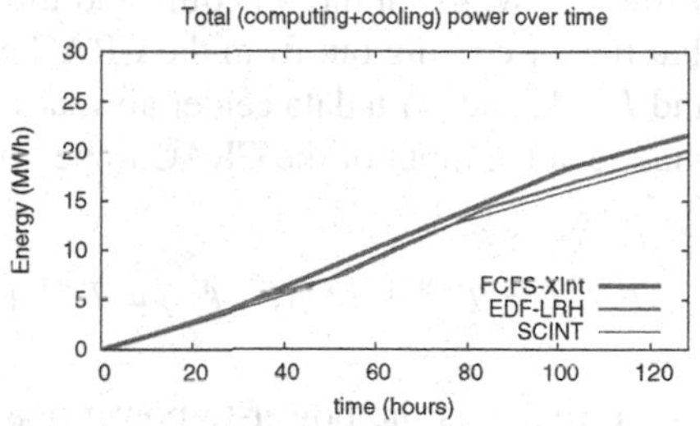

Fig. 9. Cumulative energy consumption (computing and cooling) over time for Figure 8, under the constant cooling

the ratio of the execution times between the two types of equipment measured[3] by Standard Performance Evaluation Corporation (SPEC) [12, 13]. From the estimates we get an average speed up of 2.5 when jobs run on 1955 servers in comparison to 1855 servers. The values of the power model used by FCFS-XInt, SCINT and EDF-LRH are: i) **PowerEdge 1855**: $a = 72$, $b = 1820$, ii) **PowerEdge 1955**: $a = 175$, $b = 2420$.

5.2 Energy Savings of under Constant Cooling Model

The ALHI model, as presented in [7, 9], was used in conjunction with a constant-temperature cooling supply. This means that the CRAC is supplying cool air at some constant temperature, T^{sup}, irrespective of the thermal condition of the data center. Therefore, the thermal map of the data center is analytically expressed as:

$$\boldsymbol{T}^{\mathrm{in}} = \boldsymbol{T}^{\mathrm{rise}} + T^{\mathrm{sup}} = \boldsymbol{C}(\boldsymbol{a} \odot \boldsymbol{u} + \boldsymbol{b}) + T^{\mathrm{sup}} \tag{4}$$

[3] The SPEC tests were run on both the Dell PowerEdge 1855 and 1955 servers using different benchmark applications, e.g. gzip, bzip, gcc, and so on.

This model asserts that were it not for interference, every chassis would have an inlet temperature equal to the CRAC temperature and the CRAC temperature could therefore be set to the lowest red-line temperature among the machines in the data center. Any CRAC output temperature configured below this value is the result of inefficiency caused by recirculation.

In-depth details on the simulation are given in [9]. The resulting total (computing and cooling) power consumption for the FCFS-XInt, EDF-LRH and SCINT are given in Figure 8. The per-chassis power break-down for each algorithm is given in Figures 10, 13, and 16.

5.3 Energy Consumption under the Step-Wise Linear Cooling Model

To calculate the energy consumption under the step-wise linear cooling model of the job schedules produced in the previous subsection, it is needed to predict the behavior of the CRAC given the schedule and the power equipment. For simplicity, we assume that the air coming out from the CRAC equally disperses into the room. For fixed P^{out} and P^{comp}, and for a data center air mass of m, we ***approximate*** the rate of temperature change at the input of the CRAC to be governed by the following equation:

$$T^{\text{sen}}(t) = T^{\text{init}} + \dot{T}^{\text{sen}}(t - t^{\text{sw}}) = T^{\text{init}} + \frac{-P^{\text{out}} + \sum_{i=1}^{n}\left(1 - \sum_{j=1}^{n} e_{ij}\right) p_i}{m\varsigma}(t - t^{\text{sw}}), \quad (5)$$

where $\{e_{ij}\}_{n\times n}$ is the power-to-power heat interference matrix (derived from $\boldsymbol{C}$, m and ς [7]), T^{init} is the starting temperature and t^{sw} is the time the CRAC takes to switch modes (here, it is assumed t^{sw}=10 minutes). The above equation asserts that the heat entering the CRAC is a linear weighted sum, based on the ALHI, of the servers' powers. The equation also accounts for a delay in switching between CRAC modes. If $\dot{m}$ is the CRAC's air mass flow rate, the temperature difference between T^{sen} and T^{sup} is:

$$T^{\text{sup}}(t) = T^{\text{sen}}(t) - \frac{P^{\text{out}}}{\dot{m}\varsigma}, \quad (6)$$

Methodology. Using Equations 5 and 6, we calculate the CRAC input and output temperatures. Then, using the temperature at the end of a time slot (as divided by SCINT) as the starting temperature of the next slot, we can calculate the T^{sup} over time. Using Equation 6 we can calculate the T^{sup}. Then, we can use the CoP to calculate the power consumption of the CRAC as:

$$P^{\text{AC}} = \frac{P^{\text{out}}}{\text{CoP}\,(T^{\text{sup}})} = \frac{P^{\text{out}}}{\text{CoP}\left(T^{\text{sen}} - \frac{P^{\text{out}}}{\dot{m}\varsigma}\right)}. \quad (7)$$

We use a three-mode, two-threshold CRAC with the following specifications:

$$\boldsymbol{P}^{\text{out}} = [5\,\text{W} \quad 75\,\text{W} \quad 250\,\text{W}], \quad \boldsymbol{T}^{\text{thres}} = [16\,^{\circ}\text{C} \quad 20\,^{\circ}\text{C}], \quad P^{\text{comp}} = \sum_{i=0}^{n} a_i u_i + b_i.$$

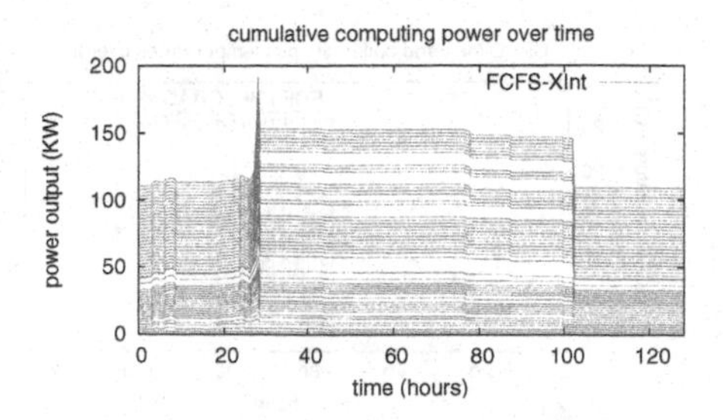

Fig. 10. Power output, divided per chassis, as yielded by FCFS-XInt's schedule

Fig. 11. CRAC input (T^{sen}) and output (T^{sup}) temperatures yielded for Figure 10

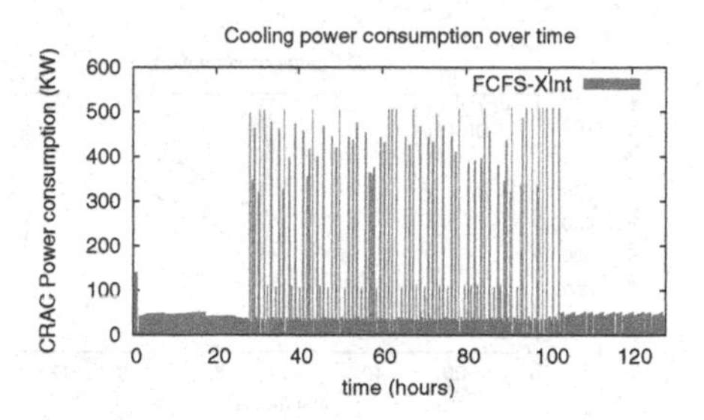

Fig. 12. CRAC consumed power for FCFS-XInt

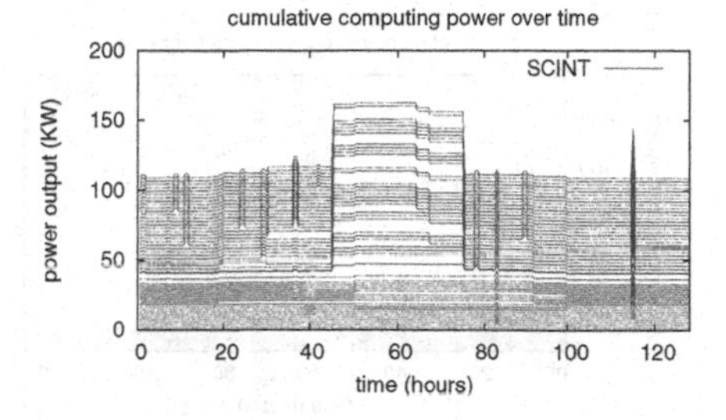

Fig. 13. Power output, divided per chassis, as yielded by SCINT's schedule

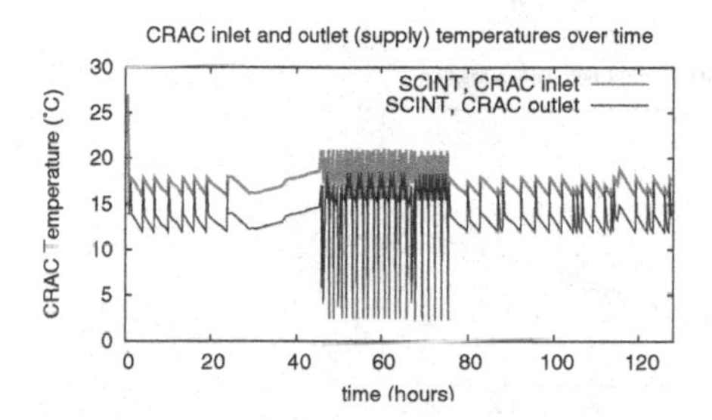

Fig. 14. CRAC input (T^{sen}) and output (T^{sup}) temperatures yielded for Figure 13

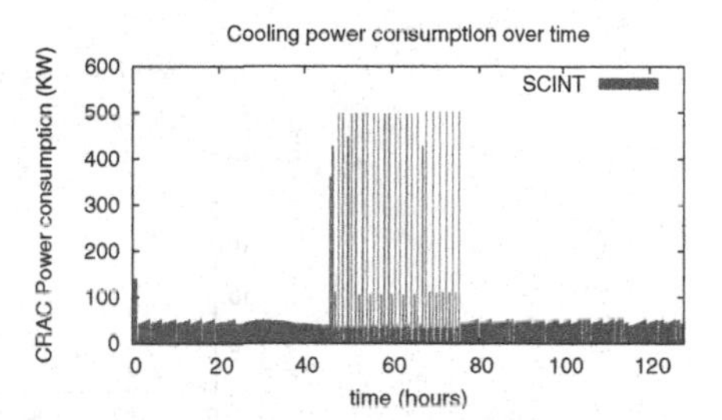

Fig. 15. CRAC consumed power for SCINT

Results. For each of the examined algorithms, we produced a set of figures consisting of 1) the server power graph, 2) the CRAC input and output temperatures, and 3) the CRAC power over the duration of the schedule. Figures 10, 13 and 16 show the power consumption of the produced schedules over time, divided into per-chassis lines (the wider the gap between two lines, the more power is output from that chassis at that moment); the topmost line is identical to the corresponding line in Fig. 8.

Figures 11, 14 and 17 show the resulting T^{sen} and T^{sup} temperatures over time, as calculated by Equations 5 and 6; we can see that the higher the heat rate generated, the faster and wider the oscillations among modes are.

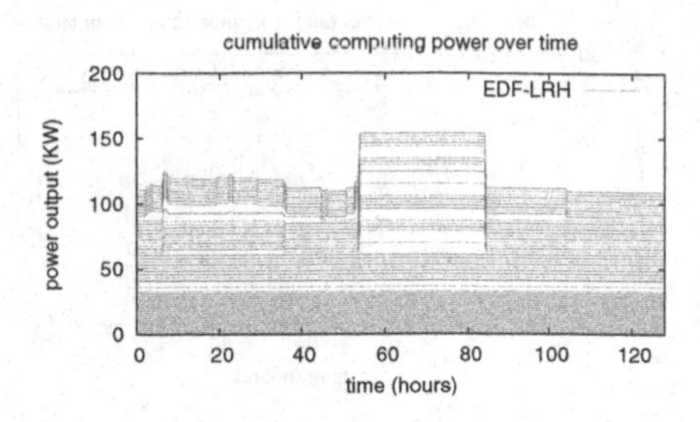

Fig. 16. Power output, divided per chassis, as yielded by EDF-LRH's schedule

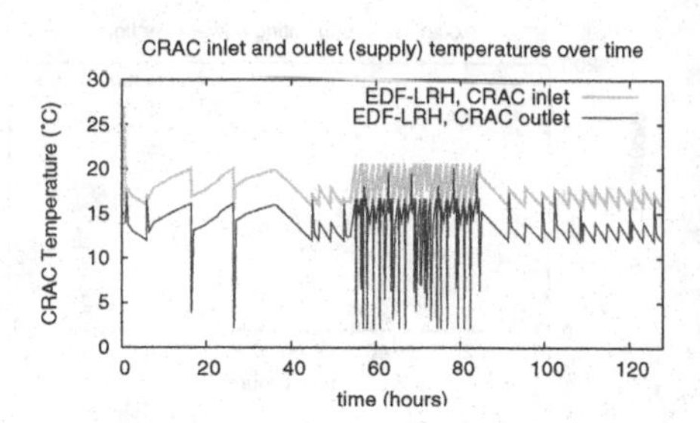

Fig. 17. CRAC input (T^{sen}) and output (T^{sup}) temperatures yielded for Figure 16

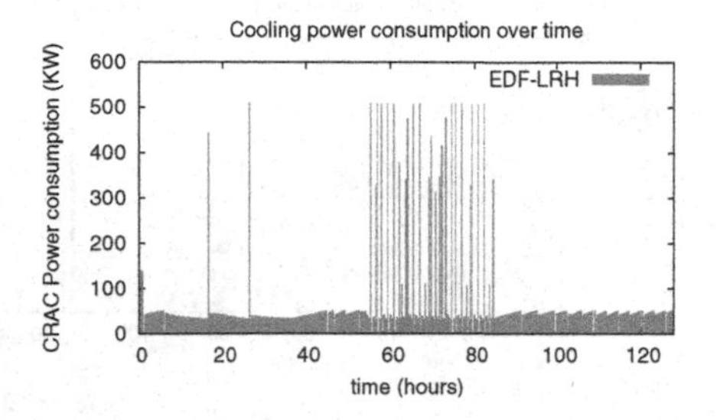

Fig. 18. CRAC consumed power for EDF-LRH

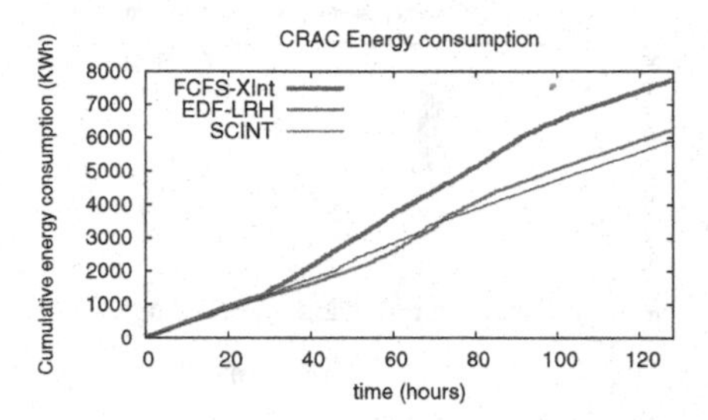

Fig. 19. Cumulative CRAC energy consumption from Figures 12, 15 and 18

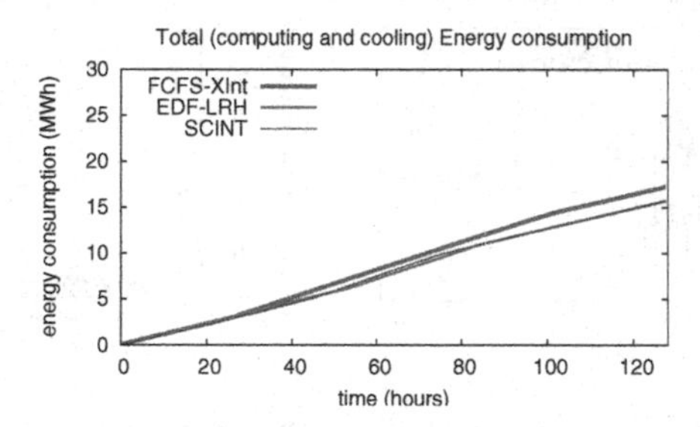

Fig. 20. Cumulative total (computing and cooling) energy consumption from Figures 8 and 19

Figures 12, 15 and 18 show the respective resulting power consumption over time presented as a filled curve (the surface corresponds to the cumulative energy), as calculated by Equation 7; it is clear that the FCFS-XInt algorithm causes a lot of high-power spikes to the CRACs.

Figure 19 provides the CRAC energy consumptions as they accumulate over time, for the three examined algorithms. Figure 20 provides the summed cumulative energy for computing and cooling, yielded from Figures 8 and 19.

6 Discussion and Conclusions

This paper introduced an analytical stepwise linear model to describe the behavior of CRAC systems in a realistic way. It also made analysis-based numerical estimations of the power consumption of schedules under the stepwise linear cooling model, produced by thermal-aware algorithms that assume a constant-value cooling model. The conclusions are summarized as follows:

Order of efficiency is preserved. One observation is that order of efficiency, as obtained in the constant cooling model, is preserved in the stepwise linear cooling model. Figure 19 shows that FCFS-XInt causes the worst cooling performance, followed by EDF-LRH and SCINT, which agrees with the constant-cooling results (Figure 8). Figure 20 shows the total (computing and cooling) energy as it accumulates over time.

Cooling oscillation seems to save energy. Comparing Figures 9 and 20, we see that the linear cooling model nominally reduces the total energy consumption; a preliminary explanation is that the oscillatory behavior of step-wise linear cooling in conjunction with the different power levels makes the CRAC conserve energy. This observation merits further investigation to be confirmed.

Savings can be achieved without re-setting the CRAC thermostat. In [7, 9], the optimizations rely on re-setting the CRAC thermostat to the highest allowed point, by solving Equation 4 for T^{sup} and replacing T^{in} with the red-line temperature. However, Figure 20 clearly shows that SCINT and EDF-LRH have lower energy consumption that FCFS-XInt, which means that energy savings are achieved without resetting the thermostat. We project that by re-setting the CRAC thermostat to appropriate levels over time, greater energy savings can be yielded. Therefore, realistic knowledge of the cooling system behavior helps create more accurate predictions and, in consequence, more efficient job schedules.

Acknowledgments

The authors thank the director of ASU's HPC Lab, Dan Stanzione, and his crew for granting access to ASU Fulton HPCI center, Mary Murphy-Hoye at Intel Corp. for donating the sensors to perform the measurements, and Michael Jonas for assisting into classifying and describing the cooling models. This work was partly funded by NSF grants #0649868 and #0834797, Intel Corp. and Science Foundation of Arizona.

References

1. Moore, J., Sharma, R., Shih, R., Chase, J., Patel, C., Ranganathan, P.: Going beyond CPUs: The potential of temperature-aware data center architectures. In: First Workshop on Temperature-Aware Computer Systems (June 2004)
2. Moore, J., Chase, J., Ranganathan, P., Sharma, R.: Making scheduling "cool": Temperature-aware resource assignment in data centers. In: 2005 Usenix Annual Technical Conference (April 2005)

3. Boucher, T.D., Auslander, D.M., Bash, C.E., Federspiel, C.C., Patel, C.D.: Viability of dynamic cooling control in a data center environment. ASME Journal of Electronic Packaging 128, 137–144 (2006)
4. Fontecchio, M.: Companies reuse data center waste heat to improve energy efficiency (May 2008)
5. Donald, J., Martonosi, M.: Techniques for multicore thermal management: Classification and new exploration. SIGARCH Comput. Archit. News 34(2), 78–88 (2006)
6. Tang, Q., Gupta, S.K.S., Stanzione, D., Cayton, P.: Thermal-aware task scheduling to minimize energy usage for blade servers. In: 2nd IEEE Int'l Dependable, Autonomic, and Secure Computing (DASC 2006) (September 2006)
7. Tang, Q., Varsamopoulos, G., Gupta, S.K.S.: Thermal-aware task scheduling for data centers through minimizing peak inlet temperature. IEEE Transactions on Parallel and Distributed Systems, Special Issue on Power-Aware Parallel and Distributed Systems (TPDS/PAPADS) 19(11), 1458–1472 (2008)
8. Heath, T., Diniz, B., Carrera, E.V., Meira, W.J., Bianchini, R.: Energy conservation in heterogeneous server clusters. In: Proceedings of the Symposium on Principles and Practice of Parallel Programming (PPoPP) (2005)
9. Mukherjee, T., Banerjee, A., Gupta, S.K.S.: Spatio-temporal thermal-aware job scheduling to minimize energy consumption in virtualized heterogeneous data centers (Elsevier) Computer Networks, Special Issue on Resource Management in Heterogeneous Data Centers (accepted for publication) (2009)
10. ASHRAE: Thermal guidelines for data processing environments
11. Heath, T., Centeno, A.P., George, P., Ramos, L., Jaluria, Y.: Mercury and Freon: temperature emulation and management for server systems. In: ASPLOS-XII: Proceedings of the 12th international conference on Architectural support for programming languages and operating systems, pp. 106–116. ACM Press, New York (2006)
12. SPEC: Standard Performance Evaluation Corporation – CINT2000 Result: Dell PowerEdge (1955),
http://www.spec.org/cpu2000/results/res2006q3/cpu2000-20060626-06297.html
13. SPEC: Standard Performance Evaluation Corporation – CINT2000 Result: Dell PowerEdge (1855),
http://www.spec.org/cpu2000/results/res2005q3/cpu2000-20050902-04544.html

Predicting Maintainability of Component-Based Systems by Using Fuzzy Logic

Arun Sharma[1], P.S. Grover[2], and Rajesh Kumar[3]

[1] Amity Institute of Information Technology,
Amity University, Noida, India
arunsharma@aiit.amity.edu
[2] Guru Tegh Bahadur Institute of Technology,
Guru Gobind Singh Indra Prastha University, Delhi, India
groverps@rediffmail.com
[3] School of Mathematics and Computer Applications,
Thapar University, Patiala, India
rakumar@thapar.edu

Abstract. Software maintenance is a very broad activity in software development that includes error corrections, enhancement of capabilities, optimization, and deletion of obsolete capabilities and so on. Maintenance includes all changes to the product, once the client has agreed that it satisfied the specified document. Maintenance in case of component-based systems requires several different activities than in other legacy applications. Also, to measure maintainability for component-based systems as a single variable is still unexplored. Present paper discusses several maintainability related issues and proposes a fuzzy logic based approach to estimate the maintainability for component-based systems. It also validates the proposed approach by using Analytical Hierarchy Process by considering two class room based case studies.

Keywords: Component-based Systems, Maintainability, Reusability, Interaction Complexity, Fuzzy Logic.

1 Introduction

The maintenance of existing software can account for 70% of the total efforts put-in application development [1]. It is also estimated that this may go up to 80%, if the approaches used in the development are not improved. A survey by Lientz and Swanson [2] discovered that about 65% of maintenance was concerned with implementing new requirements, 18% with changing the system to adapt it to a new operating environment and 17% to correct system faults. Because change is inevitable, mechanism must be developed for evaluating, controlling and making modifications. Therefore a different and effective approach is needed to reduce the overall maintenance efforts. Maintenance of component-based systems may require several different activities than normal applications, such as, upgrading the functionality of black-box components (for which code may not be available), replacement of older version components with the new ones for better and improved functionality, tracing the problem of compatibility between the new components with system, and so on.

S. Ranka et al. (Eds.): IC3 2009, CCIS 40, pp. 581–591, 2009.
© Springer-Verlag Berlin Heidelberg 2009

Present paper discusses various issues and challenges, related with the maintainability of component-based systems. The present work proposes a Fuzzy Logic based approach to estimate the maintainability of these systems. We identified the factors influencing maintainability and then categorized them into different fuzzy sets. Classroom projects are considered to estimate and validate the proposed maintainability model.

2 Maintainability

Maintainability is a prediction of the ease with which a system can evolve from its current state to its future desired state. It is termed as the most difficult and costliest activity due to it's inherently involvement in making predictions about the future.

Maintainability is defined as [3]:

"The ease with which a software system or component can be modified to correct faults, improve performance or other attributes, or adapt to a changed environment"

2.1 Maintainability Challenges for CBS

In traditional software systems, we relate maintainability with the term MTTR (mean time to repair), which should be as low as possible for good maintainability. However, this traditional measure, which relies on source code visibility, is insufficient to contend the maintenance demands of component-based development (CBD). Component-based systems consist of black- boxes; the maintainers of these systems do not have access to the source code of the components. Here the main difficulty is to identify whether the problem is in the component itself or in the system or may be due to the interaction between the two. Moreover, if system needs to be upgraded for some advanced additional features, the compatibility of the new system with the existing components may vanish. In this case, components will also need the upgradation or may require installation of new components with the desired compatibility. Therefore, we may have to consider both components and component-based systems separately while dealing with the maintenance issues. Voas [4] discusses several aspects of maintainability of CBS, which includes difficulties in maintenance activities due to frozen functionality, incompatible upgrades, defective and complex components or defective middleware such as wrappers or the glue code. The work also suggests some guidelines for maintenance activities, which are:

- Use CBD approach only for large and complex systems where reusability may be a great concern. For small systems, CBS may not be the right choice.
- For better understanding of the component, keep detailed documentation for each component.
- In CBS, suitable component selection with better performance and functionality is the main challenging job. Component repositories (within and outside the organization) should have the adequate number of suitable components so that developer can pick the best according to its requirements.
- If multiple applications share a component, but cannot tolerate changes any one of these applications need, keep separate similar components in the repository.

Voas's study on maintenance of CBS, though theoretical, but is a step ahead towards our understanding and measuring maintainability for software components.

2.2 Estimating Maintainability

Quantitative measurement of an operational system's maintainability is desirable both as an instantaneous measure and as a predictor of maintainability over the time. Efforts to estimate and track maintainability are intended to help reduce or reverse a system's tendency towards degraded integrity and to indicate when it becomes cheaper and/or less risky to rewrite the code than to change it.

Khairuddin and Elizabeth [5] proposed a maintainability model by considering two separate qualities; reparability and evolvability. Reparability involves corrective maintenance, while evolvability involves preventive and adaptive maintenance. The model describes software related factors affecting maintainability of software components. These factors include Modularity, Readability, Programming Language, Standardization, Level of Validation and Testing, Complexity and Traceability.

Arsanjani *et al.* [6] addresses the issue of software maintenance of component-based systems by identifying encapsulating and externalizing the variations around design decisions. The approach is based on the notation of Enterprise Component (EC), which is defined as an architecture patterns that provide a uniform mechanism for management of component boundaries between systems. The process includes the identification of requirements for the system and components, formalizing and abstracting them into a domain-specific language's grammar. This approach enables a highly re-configurable architectural style to help build and maintain reusable components that are responsive and resilient to changing requirements.

UML is a language for specifying, constructing, visualizing and documenting artifacts of software-intensive systems. It can be used to represent key parts of the internal structures of the components without relying on the source code and is now the industry standard for software modeling. Wu and Offutt [7] present a new UML-notation based approach for maintaining CBS. Authors used a static analysis to identify the interface events and dependence relationship that would be affected by the modification in the maintenance activity. It also provides a UML-based framework to evaluate the similarities among old and new components. The given approach seems to be quite reasonable but needs further study in real environment to check the validity and assumptions made.

Ardimento *et al.* [8] reports the results of empirical study aimed at understanding how characterization of components affects the maintenance effort of the component-based systems. They have made the assessment that (i) functionality of each component should be as concentrated as possible over a single aspect of the application domain; (ii) the training time offered by the component's producer usually indicates the complexity of understanding it and if a component is difficult to understand, then it is also difficult to maintain; and (iii) a deep knowledge of the component is necessary for the organization before its adoption, therefore, a trial usage of components is advised before the final decision about their adoption. Kajko-Mattsson *et al.* [9] discussed the problems faced by software community towards maintenance and elaborate some concepts for proposing a maintainability model. This model considered both, the product aspects as well as process aspects related with the maintenance.

Artificial Neural Network (ANN) based approach is adopted by Singh *et al.* [10] to predict the maintainability of the systems. They considered Readability of Source Code (*RSC*), Documentation Quality, Understanding of Software (*UOS*) and Average Cyclomatic Complexity (*ACC*) as independent variables to measure maintainability, which is considered as dependent variable. These variables were used to train the ANN by using MatLab. The results obtained were quite appreciable with the prediction quality of 91.42%.

Similar approach is adopted by Aggarwal *et al.* [11] to predict the maintainability of the object-oriented (OO) systems. They considered principal components of eight OO metrics as independent variables. These include Lack of Cohesion (*LCOM*), Number of Children (*NOC*), Depth of Inheritance (*DIT*) and others. Maintainability was considered as dependant variable. Results obtained by training the network using back propagation algorithm show that independent variables chosen for the study were able to predict the maintenance efforts with a mean absolute relative error (*MARE*) of 0.265. Shukla and Mishra [12] also used Neural Network based approach to estimate software maintenance efforts. They chose 14 factors as cost drivers for their study and conducted the experiment by taking various options of number of hidden layers and number of hidden nodes. Input data selected for training was 60% of the total, while 20% each were used for validation and testing. *MRE* obtained from the experiment was around 5%. Results concluded that neural network was able to successfully model the maintenance effort.

Aggarwal *et al.* [13] used Fuzzy model to measure the maintainability of the software system. The authors considered four factors affecting maintainability, namely, average number of live variable, average life span of variables, average cyclomatic complexity and the comments ratio. All inputs are classified into fuzzy sets viz. low, medium and high, while maintainability is classified as very good, good, average, poor and very poor. All the inputs and outputs are fuzzified and in total 81 rules are proposed for the model. Model is validated against the software projects developed by undergraduate engineering students. However, the model is not validated on real life complex projects.

3 Proposed Fuzzy-Based Approach for Estimating Maintainability of CBS

In this paper, we propose a Fuzzy Logic based approach for estimating maintainability for component-based systems. It is often impossible to estimate software quality attributes directly. For example, attributes (say, maintainability) are affected by many different factors, and there is no straightforward method to measure them. To estimate maintainability of CBS, one needs to establish a relationship of the factors with maintainability to achieve the desired goal.

As discussed earlier, maintenance of CBS may require a change either in the component or in the system or in the integration code. Emphasis in the present work is to estimate maintainability, if the changes require customization/replacement of the component or integration code has to be modified. Following factors have been identified, which will influence maintainability of CBS:

i) Interaction Complexity among components
ii) Reusability of the component
iii) Testability
iv) Understandability
v) Trackability

The brief description of all these factors is as follows:

Component-based systems are developed by integrating a number of components in the system. Interaction among components in a CBS can be characterized by the use of component's interface or through other components interactions. Due to the integration of these components, more and more interconnections exist in the system. Interactions among components result in dependency, which leads to the complex system and results in poor understanding and high cost of maintenance. Interaction happens when a component provides an interface and other components use it, and also when a component submits an event and other components receive it. Interactions promote dependencies [14].

Software reuse refers to the utilization of a software component with in a product, where the original motivation for constructing this component was not known. Here, reuse is seen as black-box reuse, where the application developer sees the interface, not the implementation of the component. The interface contains public methods, user documentation, requirements and restrictions of the component [15]. Testability is the capability of the component to be tested with simplified test operations and reduced test cost. For better testability, it is important to test how easy it is to observe the component in terms of its behavior, input parameters and outputs. To achieve all these, proper test suites and test cases have to be provided to test the component thoroughly before being used in the system. Here the maintainers have to keep in mind also the environment or the platforms in which the component has to be tested for the compatibility purpose.

Documentation of the component is considered for understanding the behavior of the component, as source code of these components are not available to the maintainers. Documentation provides the ease with which a user can learn to operate, prepare inputs for, and interpret outputs of a system. A good quality document must include functional description, installation details, system administrator's guide, system reference manuals etc. Finally, trackability may be estimated on the basis of initial information stored for performance, portability, conformance, security issues and other non-functional aspects. It helps in understanding and implementing the maintenance activities with less effort [16].

3.1 Fuzzy Logic

Fuzzy Logic is a mathematical tool for dealing with uncertainty and also it provides a technique to deal with imprecision and information granularity [17]. Fuzzy logic offers a particularly convenient way to generate a mapping between input and output spaces by using natural expressions [18]. In direct contrast to neural networks, which take training data and generate opaque models, fuzzy logic is based on if-then rules, which are designed by considering the opinion of experts from that domain. It has been found that the most accurate prediction models are based on analogy and experts opinion. Expert-based estimation was also found to be better than all regression-based models [19]. Henceforth the use of fuzzy logic in maintenance prediction is desirable since expert knowledge can be incorporated into the fuzzy maintenance prediction models.

Major advantage of this approach is that it is less dependent on historical data. Fuzzy logic models can be constructed without any data or with little data [20,21]. This makes fuzzy logic superior over data-driven model building approaches such as neural network, regression and case-based reasoning. In addition, fuzzy logic models can adapt to new environment when data become available [22].

3.2 Implementation of Fuzzy Logic

Implementing a fuzzy system requires that the different categories of the different inputs be represented by fuzzy sets which, in turn, is represented by membership functions. The domain of membership function is fixed, usually the set of real numbers, and whose range is the span of positive numbers in the closed interval [0,1]. There are total 11 membership functions available in MatLab. We considered Triangular Membership Functions (TMF) for our problem, because of its simplicity and heavy use by researchers for prediction models. It is a three-point function, defined by minimum α, maximum β and modal value m i.e. TMF (α, m, β), where $(\alpha \leq m \leq \beta)$. This process is known as fuzzification. These membership functions are then processed in fuzzy domain by inference engine based on knowledge base (rule base and data base) supplied by domain experts and finally the process of converting back fuzzy numbers into single numerical values is called defuzzification [23].

3.3 Experimental Design

We propose that maintainability of component-based system is a measure of five factors mentioned above. These combined factors can be used to estimate the maintainability, as it cannot be measured directly. The proposed fuzzy logic based model considers all five factors as inputs and provides a crisp value of maintainability using the Rule Base. All inputs can be classified into fuzzy sets viz. Low, Medium and High. The output maintainability is classified as Very High, High, Medium, Low and Very Low. All possible combinations (3^5 i.e. 243) of inputs are considered to design the rule base. Each rule corresponds to one of the five outputs based on the expert opinions. Some of the proposed rules are shown as:

- If Interaction Complexity among components is High, Reusability of component is Low, Testability is Low Understandability is Low and Trackability is Low then it is very difficult to maintain the system i.e. Maintainability will be Very Low.
- If Interaction Complexity among components is High, Reusability of component is Low, Testability is Low, Understandability is Medium and Trackability is Medium then Maintainability will be Low.
- If Interaction Complexity among components is Medium, Reusability of component is Medium, Testability is Medium, Understandability is Medium and Trackability is Medium then Maintainability will be Medium.
- If Interaction Complexity among components is Low, Reusability of component is High, Testability is High, Understandability is Medium and Trackability is Medium then Maintainability will be High.

- If Interaction Complexity among components is Low, Reusability of component is High, Testability is High, Understandability is High and Trackability is High then Maintainability will be Very High.

All 243 rules are inserted into the proposed model and a rule base is created. Depending on a particular set of inputs, a rule is fired. Using the rule viewer, output i.e. maintainability is observed for a particular set of inputs using the MATLAB Fuzzy tool box. Table 1 shows the values of various parameters set for inputs, outputs for fuzzification process. Fuzzification of inputs into output, membership functions for reusability, interaction complexity and maintainability and rules for defuzzification process are shown in Fig. 1, Fig. 2 and Fig. 3 respectively. By using rule viewer, the value of maintainability can be estimated by taking all five values for input factors.

Table 1. Parameter Values for Inputs and Output

System	Name='maintainability', Type='mamdani', Version=2.0, NumInputs=5, NumOutputs=1, NumRules=243, AndMethod='min', OrMethod='max' ImpMethod='min', AggMethod='max', DefuzzMethod='centroid'
Input1	Name='Reusability', Range=[0 1], NumMFs=3, MF1='Low':'trimf',[0 0.2 0.35], MF2='medium':'trimf',[0.3 0.5 0.68], MF3='high':'trimf',[0.65 0.8 1.0]
Input2	Name='Intercation_Complexity', Range=[0 1], NumMFs=3, MF1='low':'trimf',[0 0.2 0.35], MF2='medium':'trimf',[0.32 0.5 0.68], MF3='high':'trimf',[0.62 0.8 1.0]
Input3	Name='Testability', Range=[0 1], NumMFs=3 MF1='low':'trimf',[0 0.16 0.35], MF2='medium':'trimf',[0.3 0.5 0.68], MF3='high':'trimf',[0.62 0.85 1.0]
Input4	Name='understandability', Range=[0 1], NumMFs=3 MF1='low':'trimf',[0 0.16 0.35], MF2='medium':'trimf',[0.3 0.53 0.68], MF3='high':'trimf',[0.63 0.8 1.0]
Input5	Name='trackability', Range=[0 1], NumMFs=3 MF1='low':'trimf',[0 0.15 0.35], MF2='medium':'trimf',[0.32 0.55 0.66], MF3='high':'trimf',[0.63 0.76 1.0]
Output1	Name='output1', Range=[0 1], NumMFs=5 MF1='Very_Low':'trimf',[0 0.1 0.21], MF2='Low':'trimf',[0.19 0.3 0.42], MF3='Medium':'trimf', [0.38 0.5 0.62], MF4= 'High' :'trimf', [0.59 0.7 0.81], MF5= 'Very_High' :'trimf',[0.78 0.9 1.0]

3.4 Empirical Evaluation

We considered a class room based project, which is a Billing System for Hotel and is developed in Java. This application consists of three components, namely, Mathematical Calculator, Tax Calculator and Calendar. All these components are already built-in components in JavaBeans and are available online. A good documentation of these components is also available on websites to understand the functionalities of these components. These components are used as black-box components. The students were asked to replace the old version Tax Calculator component with the new upgraded version. Values of all five input factors are measured by using appropriate metrics, for the modifications required in the system. We measured reusability by Neural Network based approach discussed in [15]. Interaction complexity is measured based on the number of dependent components, discussed in [14]. Understandability, Testability and Trackability are measured based on the information available on the website and provided by the developers of these components. After normalizing the values of these factors, defuzzification is applied for these values using the proposed model. The maintainability for these values comes out to be 0.72, which is under High category. It means that the system is highly maintainable and can adopt the changes/modifications in the component.

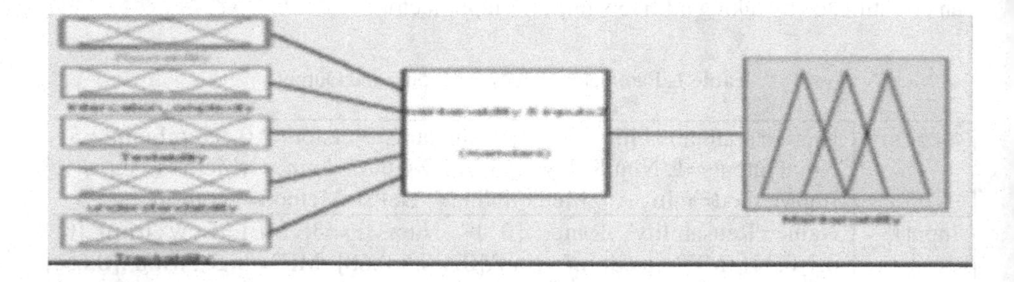

Fig. 1. Fuzzification of Inputs into Output (Maintainability)

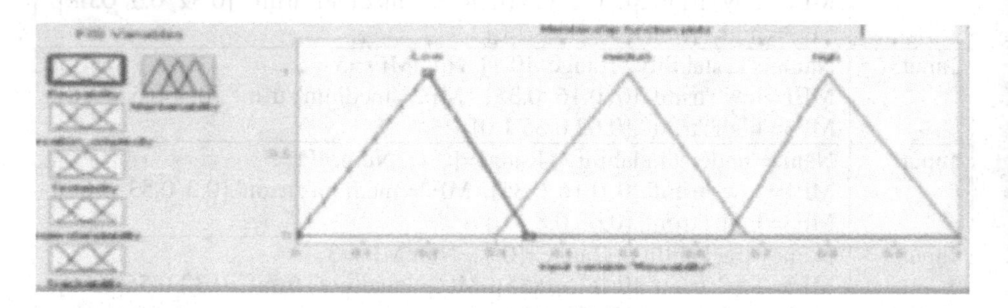

Fig. 2. Membership Functions for Reusability

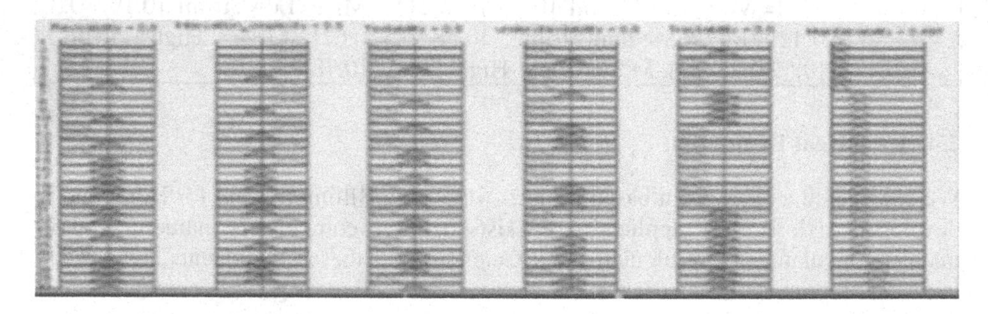

Fig. 3. Snapshot of Rule Viewer Showing Some of the Rules

3.5 Validation of the Proposed Model

The proposed methodology is validated by using Analytical Hierarchical Process (AHP). It is a quantitative technique that facilitates structuring a complex multi-attribute problem. It provides an objective methodology for deciding among a set of solution strategies for solving that problem. Users of the AHP decompose their decision problem into a hierarchy of more easily comprehended sub-problems, each of which can be analyzed independently [24].

Two class room projects are considered for the validation purpose. Proposed fuzzy logic based approach is applied to estimate the maintainability for these two projects. For validation, a survey is conducted on software professionals working on maintenance projects in different software organizations. They were asked to give their preferences of maintainability factors discussed earlier on a scale of 1 (never required) to 4 (always required). This data is then analyzed by using AHP. The weight values for all the factors were obtained on a scale of 0 to 1, given in Table 2 as:

Table 2. Weight Values of Maintainability Factors

Interaction Complexity	Reusability	Testability	Understandability	Trackability
0.26	0.18	0.13	0.32	0.11

The metrics for all the four factors of maintainability were measured for both the projects. These metric values were then multiplied by their corresponding weight values to get the values of overall maintainability for these projects. The values of maintainability, obtained from these two approaches are shown in Table 3.

Table 3. Maintainability Values by Fuzzy Logic and AHP

Project	Fuzzy Logic	AHP
P1	0.72	0.66
P2	0.43	0.32

From the Table 3, it is clear that values obtained from Fuzzy Logic based approach for Maintainability is almost near to that of AHP approach. It indicates that proposed approach may be able to predict the maintainability for CBS and can be used by the application developers.

4 Conclusions

Maintaining a component-based system may require several different activities than other legacy systems. Due to the use of black-box components in the system, it is difficult to identify whether the maintenance activity is to be performed in the system or in the components (which is very limited due to the non-availability of the source code) or in the integration of components in the system. This paper proposes a Fuzzy Logic based model to estimate maintainability for CBS. Fuzzy Logic based approach has several advantages over other methods including Neural Network and others. One major advantage is that it may also work without the data. We empirically evaluated the proposed model on a classroom project. The proposed model is able to evaluate the maintainability of the system. The model is validated against two classroom projects by using AHP and results obtained are very close in both the approaches. However, it still requires the validation of the proposed model for real-life projects.

References

1. Pressman, R.S.: Software Engineering: A Practitioner's Approach, 6th edn. McGraw Hill Book Co., New York (2005)
2. Lientz, B.P., Swanson, E.B.: Software Maintenance Management. Addison-Wesley, Reading (2000)
3. IEEE Standard for Software Maintenance, IEEE Std 1219-1998. The Institute of Electrical and Electronics Engineers, Inc. (1998)
4. Voas, J.: Maintaining Component-Based Systems. IEEE Software 15(4), 22–27 (1998)
5. Khairuddin, H., Elizabeth, K.: A Software Maintainability Attributes Model. Malaysian Journal of Computer Science 9(2), 92–97 (1996)
6. Arsanjani, A., Zedan, H., Alpigini, J.: Externalizing Component Manners to Achieve Greater Maintainability through a Highly Re-configurable Architectural Style. In: Proceedings of International Conference on Software Maintenance, pp. 628–637 (2002)
7. Wu, Y., Offutt, J.: Maintaining Evolving Component-Based Software with UML. In: Seventh European Conference on Software Maintenance and Reengineering (CSMR 2003), Benevento, Italy, pp. 133–142 (2003)
8. Ardimento, P., Bianchi, A., Visaggio, G.: Maintenance-oriented Selection of Software Components. In: Proceedings of 8th European Conference on Software Maintenance and Reengineering, pp. 115–124 (2004)
9. Kajko-Mattsson, M., Canfora, G., Chorean, D., van Deursen, A., Ihme, T., Lehmna, M., Reiger, R., Engel, T., Wernke, J.: A Model of Maintainability – Suggestion for Future Research. In: Proceedings of International Multi-Conference in Computer Science & Computer Engineering (SERP 2006), pp. 436–441 (2006)
10. Singh, Y., Kaur, A., Sangwan, O.P.: Neural Model for Software Maintainability. In: Proceedings of International Conference on ICT in Education and Development, pp. 1–11 (2004)
11. Aggarwal, K.K., Singh, Y., Kaur, A., Malhotra, R.: Application of Artificial Neural Network for Predicting Maintainability using Object-Oriented Metrics. Transactions on Engineering, Computing and Technology 15, 285–289 (2006)
12. Shukla, R., Mishra, A.K.: Estimating Software Maintenance Effort- A Neural Network Approach. In: Proceedings of the 1st conference on India Software Engineering Conference, Hyderabad, India, pp. 107–112 (2008)
13. Aggarwal, K.K., Singh, Y., Chandra, P., Puri, M.: Sensitivity Analysis of Fuzzy and Neural Network Models. ACM SIGSOFT Software Engineering Notes 30(4), 1–4 (2005)
14. Sharma, A., Kumar, R., Grover, P.S.: Dependency Analysis for Component-Based Software Systems. accepted for publication in ACM SIGSOFT Software Engineering Notes 34(3) (in press) (May 2009)
15. Sharma, A., Kumar, R., Grover, P.S.: Reusability Assessment for Software Components. accepted for publication in ACM SIGSOFT Software Engineering Notes 34(2), 1–6 (2009)

16. Sharma, A., Kumar, R., Grover, P.S.: Few Useful Considerations for Maintaining Software Components and Component-Based Systems. ACM SIGSOFT Software Engineering Notes 32(5), 1–5 (2007)
17. Sivanandam, S.N., Sumathi, S., Deepa, S.N.: Introduction to fuzzy logic using MATLAB. Springer, Heidelberg (2007)
18. Zadeh, L.A.: From Computing with numbers to computing with words-from manipulation of measurements to manipulation of perceptions. International Journal of Applied Mathematics and Computer Science 12(3), 307–324 (2002)
19. Musilek, P., Pedrycz, W., Succi, G., Reformat, M.: Software Cost Estimation with Fuzzy Models. ACM SIGAPP Applied Computing Review 8, 24–29 (2000)
20. MacDonell, S.G., Gray, A.R., Calvert, J.M.: FLSOME: Fuzzy Logic for Software Metric Practitioners and Researchers. In: The Proceedings of the 6th International Conference on Neural Information Processing ICONIP 1999, Perth, pp. 308–313 (1999)
21. Ryder, J.: Fuzzy Modeling of Software Effort Prediction. In: Proceedings of IEEE Information Technology Conference, Syracuse, New York, pp. 53–56 (1998)
22. Sailu, M.O., Ahmed, M., AlGhamdi, J.: Towards Adaptive Soft computing Based Software Effort Prediction. In: Fuzzy Information, Processing NAFIPS 2004, pp. 16–21 (2004)
23. Aggarwal, K.K., Singh, Y., Kaur, A., Malhotra, R.: Software Reuse Metrics for Object-Oriented Systems. In: Proceedings of the Third ACIS Int'l Conferencc on Software Engineering Research, Management and Applications, pp. 48–55 (2005)
24. Sharma, A., Kumar, R., Grover, P.S.: Estimation of Quality for Software Components - an Empirical Approach. ACM SIGSOFT Software Engineering Notes 33(5), 1–10 (2008)

Energy-Constrained Scheduling of DAGs on Multi-core Processors

Ishfaq Ahmad[1], Roman Arora[1], Derek White[1], Vangelis Metsis[1], and Rebecca Ingram[2]

[1] University of Texas at Arlington, Computer Science and Engineering
iahmad@cse.uta.edu, roman.arora@mavs.uta.edu, derekwwhite@mavs.uta.edu, meci@uta.edu
[2] Trinity University, Computer Science Department
ringram@trinity.edu

Abstract. This paper proposes a technique to minimize the makespan of DAGs under energy constraints on multi-core processors that often need to operate under strict energy constraints. Most of the existing work aims to reduce energy subject to performance constraints. Thus, our work is in contrast to these techniques, and it is useful because one can encounter numerous energy-constraint scenarios in real life. The algorithm, named Incremental Static Voltage Adaptation (ISVA), uses the Dynamic Voltage Scaling technique and assigns differential voltages to each sub-task to minimize energy requirements of an application. Essentially, ISVA is a framework, rather than yet another DAG scheduling algorithm, in that it can work with any efficient algorithm to generate the initial schedule under no energy constraints. Given the initial schedule, ISVA efficiently identifies tasks' relative importance and their liabilities on energy. It then achieves the best possible new schedule by observing its energy budget. The algorithm marginally degrades the schedule length with extensive reduction in energy budgets.

Keywords: Parallel processing, multi-cores, scheduling, energy.

1 Introduction

Multi-core processors are becoming increasingly popular. By integrating several cores on a single chip and by running multiple threads in parallel on the same chip, considerable performance gains are expected. However, the lack of generally applicable software and tools for allocating tasks to cores remains a key challenge. With an increasing number of cores, multi-core processors are becoming progressively complex and heterogeneous in nature. Moreover, aggressive scalability of these architectures can lead to significant power and heat dissipation, making the adjustment of the voltage and frequency of cores essential considerations in scheduling. For these reasons, research is now being done to design energy-aware scheduling algorithms that rely on voltage scaling to reduce processors' power usage.

In this paper, we address the problem of power-aware scheduling/mapping of tasks onto homogeneous processor architectures. The objective is to minimize the energy

S. Ranka et al. (Eds.): IC3 2009, CCIS 40, pp. 592–603, 2009.
© Springer-Verlag Berlin Heidelberg 2009

consumption and the makespan of complex computationally intensive scientific problems, subject to energy constraints. Most of the existing work aims to reduce energy subject to performance constraints. Thus, our work is in contrast to these techniques, and it is useful because one can encounter numerous energy-constraint scenarios in real life. Energy-constraints are often the result of limited battery supply in mobile environments where multi-core processors are being launched. Moreover, there are thermal constraints that can also place a limit on the maximum energy consumption.

Most energy minimization techniques are based on Dynamic Voltage Scaling (DVS) [3]. Specifically, we propose an algorithm by utilizing DVS and evaluate its effectiveness. The algorithm, named Incremental Static Voltage Adaptation (ISVA), uses the DVS technique and assigns differential voltages to each sub-task to minimize energy requirements of an application. Given the initial schedule, ISVA efficiently identifies tasks' relative importance and their liabilities on energy. It then achieves the best possible new schedule by observing its energy budget.

The rest of the paper is organized as follows. In Section 2, we present related work on scheduling and energy optimization techniques. In Section 3, we present task and energy models as well as existing scheduling algorithms. Section 4 describes the proposed energy constrained scheduling methodology. Section 5 discusses the testing methodology. In Section 6 we present our experimental results and give some concluding remarks in Section 7.

2 Related Work

Static scheduling of a parallel application on a multiprocessor system for minimizing the total completion time (or meeting deadlines) is a well-known problem that is known to be NP-complete. Researchers have devised a plethora of heuristics using a wide spectrum of techniques, including branch-and-bound, integer-programming, searching, graph-theory, randomization, genetic algorithms, and evolutionary methods [10]. Furthermore, there has been recent research on DVS scheduling algorithms for assigning tasks on parallel processors [1, 5, 16], but mostly for real-time systems where performance constraints are given. Recently, several DVS-based algorithms for slack allocation have been proposed for independent tasks and DAGs in a multiprocessor system [6, 10, 12, 14]. We have proposed static DVS scheduling algorithms for slack allocation on parallel machines and multi-core processors [7], outperforming other techniques, and in terms of time and memory requirements over Linear Programming based formulations for minimizing the energy presented in [15]. A few runtime approaches for slack allocation have

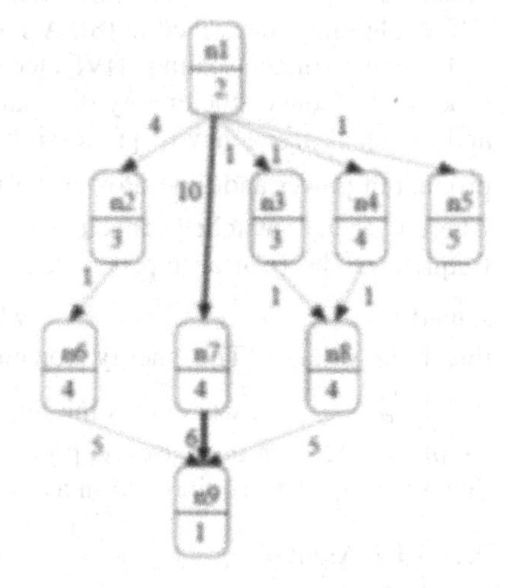

Fig. 1. Sample task graph with the critical path in bold

been studied in the literature for precedence-free tasks and for tasks with precedence relationships in a real-time system [1, 5, 10, 12, 16].

However, this work differs from existing research in that we approach the issue of energy efficiency from a different perspective. We consider an initial valid schedule and an energy constraint, and then optimize to reduce the increase in schedule length given that we have a limit on energy. Such a scenario is useful when one is working under energy budgets.

3 Task and Energy Model

The task model is a directed, acyclic graph (DAG) in which the nodes represent tasks, and the edges represent dependencies [9]. Tasks which have no predecessors are called entry nodes, and those which have no successors are called exit nodes. Additionally, each task has an associated computation cost, which is an estimate of the amount of time required to complete the task. There is also a communication cost for tasks with dependencies, which represents an estimate of the amount of time required for a task to send the results of its computation to a successor if the two tasks are scheduled to separate processors. In our model, we make the common assumption that when a task and its successor are scheduled to the same processor the communication cost becomes zero. Refer to Figure 1 for a sample task graph. Each task has a specific level in the graph, which is the maximum number of nodes from it to an entry node. Entry nodes have level 0. In the sample task graph in Figure 1, n4 is at level 1, and n9 is at level 3. The critical path of a task graph is the longest path from an entry node to an exit node, including communication costs. In Figure 1 the critical path (in bold) has length 23 and consists of nodes n1, n7, and n9.

The general problem of finding an optimal schedule is NP-complete, except for a few special cases [9]. Although several algorithms that create initial schedules based on task graphs have been proposed, this paper will focus on two of them: the traditional Level by Level algorithm (LBL) described later and the Dynamic Critical Path (DCP) algorithm described in [8]. A summary of these algorithms is presented below.

Dynamic voltage scaling (DVS) technique, available on most current and emerging processors, reduces the energy dissipation by dynamically scaling the supply voltage and the clock frequency of processing cores [3]. Mathematically, energy is simply a product of power and time. Power dissipation, P_d, is represented by $P_d = C_{ef} \cdot V_{dd}^2 \cdot f$, where C_{ef} is the switched capacitance, V_{dd} is the supply voltage, and f is the operating frequency. The relationship between the supply voltage and the frequency is represented by $f = k \cdot (V_{dd} - V_t)^2 / V_{dd}$, where k is a constant of the circuit and V_t is the threshold voltage. The energy consumed to execute task T_i, E_i, is expressed by $E_i = C_{ef} \cdot V_{dd}^2 \cdot c_i$, where c_i is the number of cycles to execute the task. The reduced supply voltage can decrease the processor speed in a linear manner, consequently reducing the energy requirement in a quadratic fashion [1].

3.1 LBL Algorithm

The LBL algorithm assigns tasks to processors level by level, so that all predecessors of a task have already been assigned to a processor when the task is being scheduled. If a task has no predecessors, it will be scheduled to the processor that finishes first. If instead a task has one or more predecessors, then the *critical predecessor* is identified as the one with the maximum value of end time + communication cost, referred to later as the latest start time. If the critical predecessor is scheduled to the processor that finishes first, then the task is scheduled to that processor. Otherwise, two processors are considered for scheduling the task: the one that finishes first and the one containing the critical predecessor. The potential start time of the task is calculated for each of the two (taking into account end times and communication costs for all predecessors), and the task is scheduled to the processor which allows for an earlier start time.

The advantage of LBL is its simplicity and speed. It requires $O(v(v + p))$ time to complete, where v is the number of tasks, and p is the number of processors. This is because finding the critical predecessor has complexity $O(v)$ in a graph with v nodes, and locating the first ending processor has complexity $O(p)$, where p is the number of processors. In comparison, some similar algorithms described in [9] have complexities $O(v(v + e))$ or $O(e(v + e))$, and DCP has complexity $O(v^3)$. The disadvantage is that the schedules generated by LBL use a simple heuristic that yields suboptimal schedules. The pseudo-code for the LBL algorithm is shown in Figure 2.

```
for all levels in the graph do
  for all tasks t in the level do
    fep ← processor that finishes first
    if t has no predecessors then
      schedule t to fep
    else
      cpp ← processor containing the critical prede-
      cessor of t, the task with the largest value of
      end time + communication cost
      if cpp = fep then
        schedule t to fep
      else
        consider start time of t if it were scheduled
        to cpp or fep
        schedule t to the processor that would give
        it an earlier start time
      end if
    end if
  end for
end for
```

Fig. 2. Pseudo code for the LBL scheduling algorithm

3.2 DCP Algorithm

Rather than scheduling tasks level by level, DCP schedules tasks in their order of importance, which is determined by calculating the dynamic critical path. Furthermore, tasks' start times are not fixed until all tasks are scheduled, which allows for more flexibility in determining start times.

First, the absolute earliest start time (AEST) and the absolute latest start time (ALST) are calculated for each task. These values take into account the start and end times of predeccessors and successors and also the communication costs if tasks are scheduled to different processors. Tasks that are on the dynamic critical path have AEST = ALST. Such tasks are scheduled first, with priority going to those that have the smallest AESTs. AEST and ALST values are recomputed after each task is scheduled.

If a task is on the dynamic critical path, then the only processors considered as candidates for scheduling the task are those containing its predecessors and successors and one additional processor. Otherwise, if the task is not on the dynamic critical path, the only processors considered are those that already have tasks scheduled on them. Each processor in the list of candidates is considered to determine whether or not there is room for the task. If there is space, the task is tentatively scheduled there, and then the task's critical child is determined. This is the successor with the smallest difference between ALST and AEST (i.e., closest to the dynamic critical path). Then, a composite AEST is computed, which is the sum of the task's AEST on the candidate processor and the critical child's AEST on the same processor. The task will be scheduled to the processor which offers the best (i.e., the lowest) composite AEST. According to [8], the complexity of the DCP algorithm is $O(v^3)$. The advantage is that it produces much shorter schedules than many other scheduling algorithms. The corresponding pseudo-code is illustrated in Figure 3.

```
while not all nodes scheduled do
  n_i ← the highest node with the smallest difference
  between its ALST and AEST; break ties by choos-
  ing the one with the smaller AEST
  if n_i's ALST is equal to its AEST then
    call Select_Processor(n_i, On_DCP)
  else
    call Select_Processor(n_i, Not_On_DCP)
  end if
  Update AEST and ALST for all nodes
end while
Make all nodes' start time to be their respective
AESTs
```

Fig. 3. Pseudo code for DCP algorithm

4 ISVA Techniques

The Incremental Static Voltage Adaption algorithm (ISVA) is proposed as an energy-aware scheduling technique. The goal is to minimize as much as possible the schedule length degradation caused by having to execute tasks with a reduced energy budget. Dynamic voltage scaling is used to adjust the voltage levels at which individual tasks are executed.

The algorithm initially takes a DAG, the number of processors, the available voltage levels, and an energy budget. An initial schedule is created at the lowest voltage using a scheduling algorithm such as LBL or DCP. The energy consumed is computed and, if the energy budget has not been exceeded, the algorithm proceeds. The processor that finishes last in the generated schedule is identified, we call this processor the *critical processor*. A list L is constructed with the tasks in the critical processor. For each task on list L, a list of parent tasks is constructed, which includes the task's predecessor. If a task has multiple predecessors, only the one with the latest end time is added to the list, we call this the *immediate predecessor* task. Once we have finished processing all tasks in list L, we move the contents of L to the beginning of list E and store the predecessor tasks onto list L. We repeat this process of processing tasks on list L until no tasks are available in L. This will happen when we reach the entry nodes of the DAG. From the list E of parent tasks, a task is selected which has the lowest voltage level and is as high as possible in the graph. If the selected task is not running at the maximum voltage level, then its voltage level is incremented. The task incremented is called the *candidate task*. If there is no candidate task for voltage adjustment, the algorithm will select the earliest starting task that is at the lowest voltage level from within all processors as the candidate task and adjust this one. The algorithm eventually stops when any task voltage adjustment would cause us to exceed the energy budget. The corresponding pseudo-code is illustrated in Figure 4.

```
for all tasks t_i ∈ T do
  Voltage(t_i) ←lowest voltage level
end for
Schedule tasks in T to processors in P
Recompute energy_consumed
if energy_consumed > available_energy then
  break // Not enough energy to run tasks
else
  while energy_consumed < available_energy do
    p_j←Last_processor_to_finish(P) // Critical processor
    List L← All tasks in p_j
    List E ← φ
    while L ≠ φ do
      E ← L + E
      List M ← φ
      for all tasks t_k ∈ L do
        Task t_p ←Immediate_predecessor(t_k)
        Add t_p to M
      end for
      L ← M
    end while
    Task t_c ← Get_earliest_start_lowestVL_task(E)
    if t_c = null then
      t_c ← Get_earliest_start_lowestVL_task(T)
    end if
    if t_c ≠ null then
      Increment Voltage(t_c) by 1 level
      Schedule tasks in T to processors in P
    else
      break // No valid tasks for voltage adjustment remain
    end if
  end while
end if
```

Fig. 4. Pseudo code for ISVA algorithm

5 Testing Methodology

In order to test the ISVA algorithm, we propose a comparison of schedule length increase to energy budget used for a series of DAGs. The energy budgets used for the test ranged from 40% - 80% and the schedule of tasks to processors was generated by using the LBL and DCP algorithms. The ISVA algorithm was used in order to identify the order and tasks whose voltage levels would be adjusted. The method chosen to test this algorithm was to generate several task graphs based on a series of input parameters and to compare the results generated by using different scheduling algorithms, different voltage levels, and different energy budgets.

5.1 Workload Generation

In generating task graphs, there are two primary aspects to be considered, namely: the characteristics relevant to the graphs and the method for ensuring that graphs are acyclic. The first parameter is the number of nodes in the graph. It is important to test the algorithm on a wide range of graph sizes to ensure that it is scalable. Another important variable is the communication to computation ratio (CCR). Graphs which are communication-intensive will tend to have longer schedules because either processors will have longer idle time slots waiting for communication to complete or more tasks will be scheduled to a single processor in an attempt to avoid creating idle time slots.

The branching factor is the average number of successors per node. This value reflects the number of edges in the graph relative to the number of nodes. Graphs with large numbers of edges can be more limited in terms of the order in which nodes are scheduled, and this reduction in flexibility could result in an increase in the schedule length. The overall "shape" of the graph is described by α. For graphs with $\alpha = 1.0$, the shape will be approximately "square." That is, the graph will have $\sqrt{v}$ levels and $\sqrt{v}$ nodes per level, where v is the number of nodes. This concept of α is described in [8]. In general, the number of levels in the graph is equal to $\alpha\sqrt{v}$. For smaller values of α, the graph will have fewer levels, which will mean that more tasks may be executed in parallel, allowing for shorter schedules.

In addition to variables related to the graphs generated, there are certain parameters describing the hardware that could be used, and these parameters should also be tested. One of these is the number of processors (cores) used. Ideally, as the number of processors increases, the schedule length will decrease because more tasks will be executed simultaneously. The performance of a scheduling algorithm should scale with the number of processors. The final variable considered in these tests is the number of voltage levels available to the processors. The question is whether or not schedule length varies when processors have several voltage levels to choose from.

When one generates task graphs randomly, it is vital that graphs remain acyclic; otherwise, it will be impossible to honor the dependencies, and the algorithms will probably not complete as expected. In these tests, the graph generator was given four parameters: number of nodes, CCR, branching factor, and α. The number of levels was calculated first (using α), and a random number of nodes was chosen for each level, so that the total number of nodes was equal to that specified. Next, nodes were each assigned an identification number in ascending order by level. Thus, all nodes on the same level have consecutive IDs, and nodes with higher level numbers (closer to the exit nodes) have larger IDs than nodes near the top of the graph (with lower level numbers). With this structure, maintaining an acyclic graph is simple: a node may not have a successor with ID less than its own ID or that is on the same level that it is.

5.2 Variable Values Used in Testing

In choosing the ranges of values to test, the goal is to try as wide a range as possible; however, time limitations constrained the addition of more values. A compromise has to be made between choosing several values and being able to complete the tests in a reasonable amount of time.

The ranges of values that are used for the graph variables are as follows: Number of nodes = {50, 100, 150}; Branching factor = {4}; CCR = {0.1, 1.0, 10.0}; α= {2.0}. The ranges of values that are used for the system variables are as follows: Number of voltage levels = {2, 3, 4, 5} and Number of processors = {Unbounded}.

The actual voltage levels ranged from 2 to 8V, and the value for each corresponding level was calculated based on how many voltage levels are being used. For example, with 2 voltage levels, we use {2V, 8V}, with 3 voltage levels, we use {2V, 5V, 8V}, and so on. The experiments were run on a 2.4 GHz PC.

6 Experimental Results

The results collected indicate that on average, running ISVA on DCP results in less schedule length degradation when compared to ISVA on LBL. Overall, using DCP as the scheduling algorithm resulted in 3-6% smaller schedules than using LBL. As expected, greater energy budgets results in shorter schedule length degradation when compared to lower energy budgets. Results show that operating at lower energy budgets has a small impact on schedule length degradation. For example, while an 80% energy budget produces a 20% schedule degradation, operating at half (40%) energy budget results in roughly an additional 10% in degradation increase. This can be explained from the linear relationship between energy and time, and the quadratic relationship between voltage level and energy.

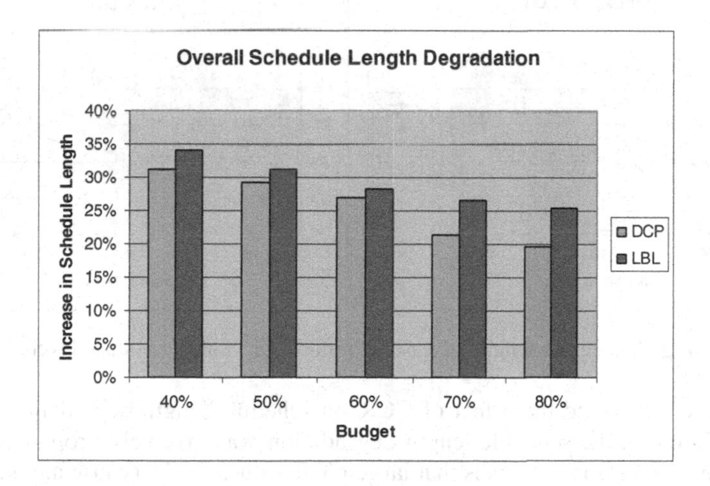

Fig. 5. Overall Schedule Length Degradation for ISVA on LBL and ISVA on DCP

Running at higher voltage levels increases power consumption quadratically but produces only a linear improvement in execution time. Results for various energy budgets are illustrated in Figure 5 with details of different energy budgets applied to task sizes of 50, 100, and 150 shown in Figure 6.

Furthermore, we observed no significant correlation between the number of voltage levels and the schedule length increase. However, operating with 4 voltage levels created significantly different results from the other configurations. Resulting schedule length increases using DCP and LBL variations of ISVA are shown with 2, 3, 4, and 5 available voltage levels in Figure 7.

As for the number of processors used by both algorithms, we noticed that ISVA on DCP used significantly less processors than ISVA on LBL and achieved better results. This is illustrated in Figure 8.

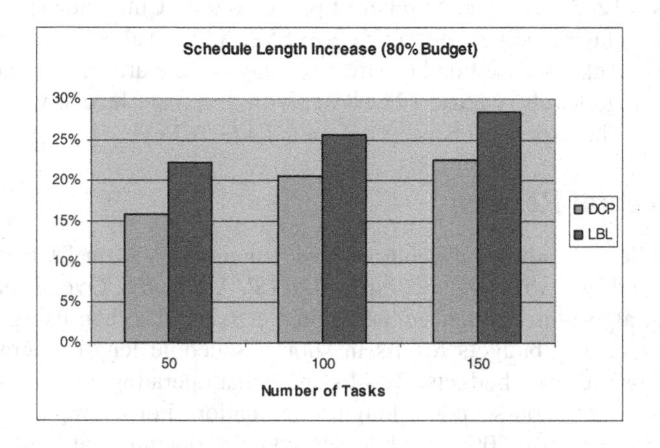

Fig. 6. Schedule length degradation of 50, 100, and 150 tasks with 80% energy

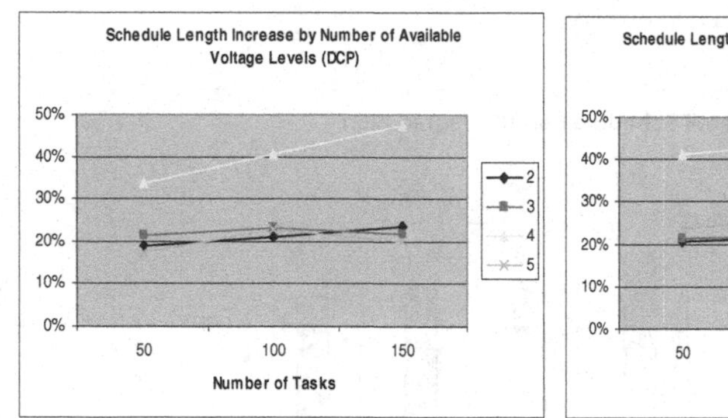

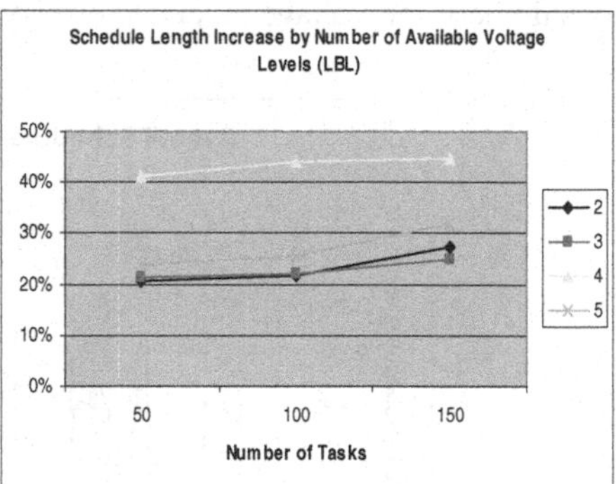

Fig. 7. Schedule length increase by number of available voltage levels

Finally, we analyzed the effect of CCR on schedule length degradation. As illustrated in Figure 9, the schedule length degradation was inversely proportional to the CCR value. The reason for this is that large CCR values tend to cause any schedule to

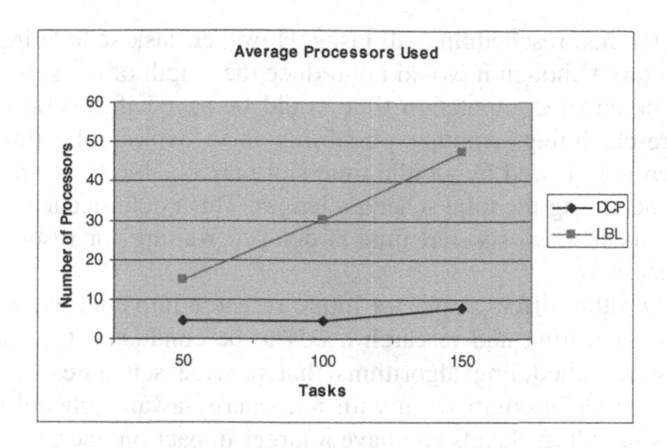

Fig. 8. Average number of processors used

be longer because there are large idle time slots for communication costs and larger numbers of tasks clustered on a single processor. This leaves more room for voltages to decrease, causing individual tasks to execute more slowly, without causing an overall increase in schedule length. In contrast, schedules that have lower communication costs are more easily parallelizable in the sense that tasks will require very little additional time to complete when they are on different processors than their predecessors. Such graphs will be more prone to increases in schedule length when task execution time increases, as tasks are assigned to lower voltage levels.

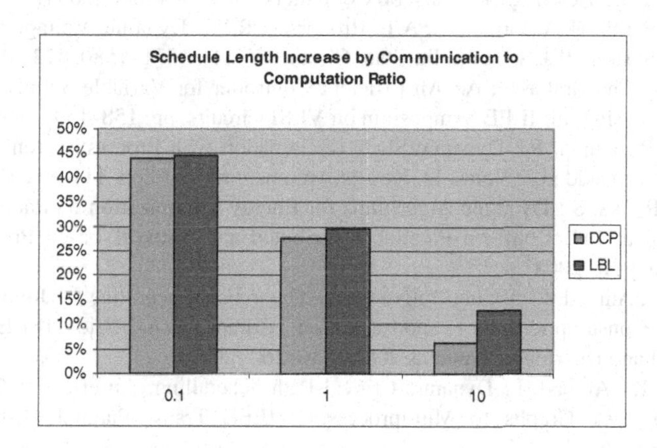

Fig. 9. Schedule length increase by communication to computation ratio

7 Future Research and Conclusions

Clearly, the ideal energy-aware algorithm would allow for reduced energy consumption while maintaining or only slightly increasing schedule length. Although this may seem like a goal unlikely to be achieved, there are some ways in which this algorithm might be improved. First, the algorithm works by choosing a task, incrementing its voltage level, and then rescheduling all tasks. However, task scheduling tends to be a complex operation. Although it would not reduce the length of the schedule produced, a significant amount of computation time could be saved if several tasks were adjusted before rescheduling. Another possibility is to exploit idle time slots. Tasks whose execution is followed by an idle time slot may be able to run at lower voltage levels without increasing the total schedule length. This could occur if the task has no successor, or if its successor's start time is delayed waiting for results from another one of its predecessors.

The ISVA algorithm offers a spur for future research into energy-aware scheduling algorithms. Further testing and research needs to be conducted to design more efficient energy-aware scheduling algorithms that provide schedules of equal or only slightly greater length in comparison with non-energy-aware scheduling algorithms. Small decreases in voltage levels can have a larger impact on energy usage, and even small but consistent energy savings can accumulate over time. For this reason, it is important to continue searching for better energy-aware algorithms.

References

1. Ahmad, I., Khan, S.U., Ranka, S.: Using Game Theory for Scheduling Tasks on Multi-Core Processors for Simultaneous Optimization of Performance and Energy. In: Workshop on NSF Next Generation Software Program in conjunction with the International Parallel and Distributed Processing Symposium (IPDPS 2008), Miami, FL (2008)
2. Aydin, H., Melhem, R., Moss, D., Meja-Alvarez, P.: Power-Aware Scheduling for Periodic Real-Time Tasks. IEEE Trans. on Computers 53(5), 584–600 (2004)
3. Burd, D., Pering, T.A., Stratakos, A.J., Brodersen, R.W.: Dynamic Voltage Scaled Microprocessor System. IEEE J. of Solid-State Circuits 35(11), 1571–1580, 473–484 (2000)
4. Gutnik, V., Chandrakasan, A.: An Efficient Controller for Variable Supply-Voltage Low Power Processing. In: IEEE Symposium on VLSI Circuits, pp. 158–159 (1996)
5. Jejurikar, R., Gupta, R.: Dynamic Slack Reclamation with Procrastination Scheduling in Real-Time Embedded Systems. In: Design Automation Conf., pp. 111–116 (2005)
6. Kang, J., Ranka, S.: Dynamic Algorithms for Energy Minimization on Parallel Machines. In: Euromicro Intl. Conf. on Parallel. Distributed and Network-based Processing (PDP 2008) (to appear, 2008)
7. Khan, S.U., Ahmad, I.: A Cooperative Game Theoretical Technique for Joint Optimization of Energy Consumption and Response Time in Computational Grids. IEEE Transactions on Parallel and Distributed Systems 20(3), 346–360 (2009)
8. Kwok, Y.-K., Ahmad, I.: Dynamic Critical-Path Scheduling: An effective Technique for Allocating Task Graphs to Multiprocessors. IEEE Trans. Parallel Distributed. Systems 7(5) (1996)
9. Kwok, Y.-K., Ahmad, I.: Static Scheduling Algorithms for Allocating Directed Task Graphs to Multiprocessors. ACM Computing Surveys 31(4), 406–471 (1999)
10. Luo, J., Jha, N.K.: Power-Conscious Joint Scheduling of Periodic Task Graphs and Aperiodic Tasks in Distributed Real-time Embedded Systems. In: Int. Conf. on Computer-Aided Design, pp. 357–364 (2000)
11. Luo, J., Jha, N.K.: Static and Dynamic Variable Voltage Scheduling Algorithms for Real-time Heterogeneous Distributed Embedded Systems. In: ASP-DAC 2002: Proceedings of the 2002 conference on Asia South Pacific Design Automation/VLSI Design, p. 719. IEEE Computer Society, Washington (2000)

12. Schmitz, M.T., Al-Hashimi, B.M.: Considering Power Variations of DVS Processing Elements for Energy Minimisation in Distributed Systems. In: Int'l Sym. on System Synthesis, pp. 250–255 (2001)
13. Schmitz, M.T., Al-Hashimi, B.M.: Considering Power Variations of DVS Processing Elements for Energy Minimisation in Distributed Systems. In: Int'l Sym. on System Synthesis, pp. 250–255 (2001)
14. Yu, Y., Prasanna, V.K.: Power-Aware Resource Allocation for Independent Tasks in Heterogeneous Real-Time Systems. In: 9th IEEE International Conference on Parallel and Distributed Systems (2002)
15. Zhang, Y., Hu, X. (Sharon) ., Chen, D.Z.: Task Scheduling and Voltage Selection for Energy Minimization. In: Design Automation Conf., pp. 183–188 (2002)
16. Zhu, D., Melhem, R., Childers, B.R.: Scheduling with Dynamic Voltage/Speed Adjustment Using Slack Reclamation in Multiprocessor Real-Time Systems. IEEE Trans. on Parallel and Distributed Systems 14(7), 686–700 (2003)

Evaluation of Code and Data Spatial Complexity Measures

Jitender Kumar Chhabra and Varun Gupta

Department of Computer Engineering, National Institute of Technology,
Kurukshetra, Kurukshetra-136119 India
jitenderchhabra@rediffmail.com, varun3dec@yahoo.com

Abstract. The comprehension of a program is largely dependent on the spatial abilities of the programmers. Some spatial complexity measures have been proposed in the literature to estimate the effort required in the process of software comprehension. Two of the widely discussed and important spatial measures for procedure-oriented software are code spatial complexity and data spatial complexity. This paper evaluates these two measures using Weyuker's properties. The paper further validates these two measures on the basis of one of the most acceptable framework proposed by Briand et al. The results of this study show that these two spatial metrics satisfy all properties and parameters required by these two formal evaluation frameworks and thus these spatial measures are practical and useful.

Keywords: Spatial complexity, software complexity, software psychology, software metrics.

1 Introduction

Majority of the software development effort is spent on maintaining existing software systems rather than developing new ones. Most of the time of the software maintenance process gets spent in understanding the system being maintained. This implies that in order to improve software development process, maintenance activities should be made more efficient, which in turn requires an improvement in the process of program comprehension. Program comprehension is a complex cognitive skill, which involves grasping of a mental representation of the program structure and functionality [1]. In order to understand the process of comprehension, the role of human factors and cognitive process of human mind needs to be studied, to understand the software psychology [2]. For the maintenance of software, understanding of human skills is necessary in order to facilitate the understanding of the source code of a program. Program comprehension task consists of studying a program and then remodeling it based on the retention in memory [3]. After completion of remodeling task, the reader analyzes the rebuilt code in terms of information chunks, which gives insight to the kinds of internal information structures contained in the code. It is in terms of these internal structures, in which reader understands the code. Thus, spatial representations of various program segments greatly affect comprehensibility of a program and this type of software complexity is known as the spatial complexity of the software [4], [5], [6].

S. Ranka et al. (Eds.): IC3 2009, CCIS 40, pp. 604–614, 2009.
© Springer-Verlag Berlin Heidelberg 2009

The major contributions of this paper are evaluation and validation of spatial complexity measures proposed in [4] for procedure-oriented programs. These measures are evaluated using well-defined and widely accepted properties and evaluation criteria given by Weyuker [7] and Briand et al. [8] that a software complexity measure should satisfy. Since, a program is made up of two types of entities: code and data. The code represents the processing logic of the software and data recognize the inputs and the outputs of the software. Thus, both code and data contribute to the complexity of software and overall spatial complexity of software is integration of code spatial complexity and data spatial complexity. Two spatial complexity measures proposed in [4] are based on code as well as data. These two types of spatial measures are described in next two sections.

2 Code Spatial Complexity Measures

The module is considered as the basic unit of processing for the purpose of computation of code spatial complexity. The functionality of the module is visible in the definition part, while the use of that module is understood through various calls of that module. A module's code spatial complexity is defined in terms of distance between the definition and usage of the module. The cognitive effort required to understand a module depends on the distance (in terms of Lines of Code (LOC)) between the definition and usage of the module [6]. If a module is being used immediately after its definition, the comprehension process for the software containing the particular module will be easier as no searching for that module is to be done as details of that module may still be fresh in the working memory of the human being. Alternatively, if a module is being used after a large number of lines from its definition, lot of thinking has to be done, as details of that module may not be present in the working memory of the human being [4]. The code spatial complexity of a module (MCSC) was defined as average of distances (in LOC) between the use and definition of the module: -

$$MCSC = \frac{\sum_{i=1}^{n} \text{Distance}_i}{n}$$

where n represents count of uses of that module and Distance is number of lines of code between the call of module and the corresponding definition. In case of multiple files,

Distance = (Distance of call from bottom of file containing call) + (Distance of definition of the module from top of the file containing definition)

Code spatial complexity of software containing various modules was defined as the average of code spatial complexities of all modules.

$$CSC = \frac{\sum_{i=1}^{m} \text{MCSC}_i}{m}$$

where m is the number of modules in the software.

3 Data Spatial Complexity Measures

Every data member (variable or constant) is defined once, but is used at many places throughout the software. All inputs, intermediate results, and outputs are represented through data members. Different data members are processed at different places to generate the final output. The functionality of any software is dependent on the inputs, the use of various processing steps to generate the intermediate results, and then ultimately computing the final outputs. Thus, these data members are used to compute the data spatial complexity of software. The data spatial complexity is used to measure the cognitive effort required for the comprehension of the software. The sequence of usages of a data member in a program also plays a role in determining the cognitive effort required to understand the purpose of that particular data member in the program. When a data member is being used for the first time, definition of the variable is of use, but when a data member is re-used in the program, the previous usage of the data member becomes more important than definition of the variable. Thus, cognitive effort required to understand the purpose of a data member depends on the number of lines of code between the successive usages of the data member [4]. Thus, the data spatial complexity of a data member was defined as

$$DMSC = \frac{\sum_{i=1}^{p} \text{Distance}_i}{p}$$

where p is the total number of usages of that data member and Distance is the number of lines of code between the current use of the data-member from its just previous use or definition of data member in case of first usage of data member. In case of data members being used in multiple files,

Distance = (Distance of first use of the data member from top of current file) + (Distance of definition of the data member from top of file containing definition.)

As reported by the authors, only global data members are required to be considered for computing data spatial complexity. Thus, data spatial complexity of the software was defined as average of data-spatial complexity of all global data members i.e. all global variables and constants.

$$DSC = \frac{\sum_{i=1}^{q} DMSC_i}{q}$$

where q is the total number of global variables and constants in the software.

4 Evaluation of Spatial Measures Using Weyuker Properties

Weyuker [7] developed a formal list of nine properties for evaluating software complexity metrics. These properties are used to determine the usefulness of various software complexity measures and a good complexity measure should satisfy most of the Weyuker's properties. These properties have been used by many eminent researchers for evaluating their measures [9, 10, 11, 12, 13]. The above described code spatial complexity and data spatial complexity measures are evaluated here using these nine properties proposed by Weyuker. While describing these properties, a program body is denoted by P and its complexity is designated as |P|, which will always be a non-negative number.

***Property 1*:** This property states that (∃P), (∃Q) such that (|P| ≠ |Q|)
This property states that a complexity measure must not be "too coarse".

Proof: This property states that a measure should not rate all programs as equally complex. This property is satisfied by the code and data spatial metrics, as two programs P and Q may generally have different amounts of code complexity and data complexity.

Let CSC_P and CSC_Q be the code spatial complexity for program P and Q respectively. Now, $CSC_P \neq CSC_Q$ will generally hold true, because code spatial complexity of a program depends on the LOC between the usages and definitions of all the modules in the program and most probably, program P and Q may have different number of modules contained in them. Even if they have same number of modules, their code spatial complexity values will differ, because of different number of LOC between the different uses and definitions of modules in them.

Similarly, let DSC_P and DSC_Q be data spatial complexity for a program P and Q respectively. Also, $DSC_P \neq DSC_Q$ will generally be true. Data spatial complexity of a program is the average of data spatial complexities of all its global data members and data spatial complexity of a data member is the average of distances of various usages of that data member from their previous usages. Thus, data spatial complexity of a program depends on number of global data members and distances in LOC between their usages. Therefore, program P and Q will have different data spatial complexities because of presence of different global data members and their different use.

Hence, code and data spatial complexity measures satisfy Property 1.

***Property 2*:** Let c be a non-negative number, and then there is only finite number of programs of complexity c.

This property asserts that a complexity measure must not be "too coarse".

Proof: This property is satisfied by the metrics under consideration as there can only be a finite number of programs having the same values of code and data spatial complexity, because values of these measures depend on a number of factors. For instance, code spatial complexity for a program depends on count of modules in that program and distances in LOC between the different uses and definitions of modules in them. Similarly, data spatial complexity for a program depends on count of global data members defined in that program and distances in LOC between their successive usages in the program. There can be always a finite number of programs having same value of these factors/metrics and thus, property 2 is well satisfied by the code and data spatial complexity measures.

***Property 3*:** There are distinct programs P and Q such that (|P| = |Q|)
This property emphasizes that a complexity measure must not be "too fine".

Proof: This property ensures that a complexity metric should not be so fine that any particular value of the metric is only given by a single program. This property requires that $CSC_P = CSC_Q$ where CSC_P and CSC_Q are the code spatial complexities for two different and unrelated programs P and Q respectively and $DSC_P = DSC_Q$ where DSC_P and DSC_Q are the data spatial complexities for programs P and Q respectively. This is quite possible that two different and totally unrelated programs P and Q may come out with the same values of spatial complexity measures after performing the calculations involved in the measurement of spatial complexity. It is always possible to have same average distance (in terms of LOC) between function's definition and use in two different unrelated programs. Thus property 3 is satisfied by the above measures.

***Property 4*:** $(\exists P)\ (\exists Q)$ such that $(P \equiv Q\ \&\ |P| \neq |Q|)$
There is no one-to-one correspondence between functionality and complexity.

Proof: This property emphasizes that two programs having same functionality can have different values for complexity measures i.e.
$CSC_P \neq CSC_Q$ and $DSC_P \neq DSC_Q$ hold true for any two programs P and Q such that $P \equiv Q$.

The above-described metrics satisfy this property, because values given by these measures do not depend on the functionality of the programs but upon the implementation details of these programs. This is very much possible that two programs P and Q having same functionality may have different implementations and different implementations of the programs P and Q will result in different spatial complexities of the two programs.

***Property 5*:** $(\forall P)\ (\forall Q)\ (|P| \leq |P;Q|$ and $|Q|| \leq |P;Q|)$
Property 5 states the concept of monotonicity with respect to composition.

Proof: This property demands that complexity for the concatenation of two programs can never be less than the complexity for either of the programs.
Let CSC_P and CSC_Q be the code spatial complexities of programs P and Q respectively and CSC_{PQ} be the code spatial complexity of the concatenated program of P and Q. Then, according to the definition of code spatial complexity, the resultant code spatial complexity of the combined program would be the summation of code spatial complexities of individual programs, if they were independent and if the code of program Q was just appended after the code of program P, without disturbing the individual distances of definition and usage of modules. However, the combining the main function of Q into the main function of P might result into a slight increase in the CSC of P as well as Q i.e.

$$CSC_{PQ} \approx CSC_P + CSC_Q$$

Similarly, if DSC_P and DSC_Q are the data spatial complexities of independent programs P and Q respectively and DSC_{PQ} is the data spatial complexity of the concatenated program of P and Q, then, concatenating the program Q in program P may force a combined declaration of global variables of Q along with program P's global declarations in some languages (e.g. C), which might increase the DSC of program Q

$$DSC_{PQ} \approx DSC_P + DSC_Q$$

If the programs P and Q were not independent, then the common code and data may appear once in the concatenated program. In that case the code/data spatial complexity of common portion will contribute once in the metrics and independent portions of both P and Q will continue to have their original contribution towards CSC as well as DSC. In that situation

$CSC_{PQ} = CSC_P + CSC_Q - CSC_{common\text{-}code}$
and $DSC_{PQ} = DSC_P + DSC_Q - DSC_{common\text{-}data}$

It is obvious that $CSC_{common\text{-}code}$ can never be greater than either CSC_P or CSC_Q. So in both cases whether P and Q are independent or not, it is evident that

$CSC_P \leq CSC_{PQ}$ and $CSC_Q \leq CSC_{PQ}$
Also, $DSC_P \leq DSC_{PQ}$ and $DSC_Q \leq DSC_{PQ}$

Thus, the code and data spatial complexity metrics satisfy the Property 5.

***Property 6*:** 6a: $(\exists P)\ (\exists Q)\ (\exists R)\ (|P| = |Q|\ \&\ |P; R| \neq |Q; R|)$
6b: $(\exists P)\ (\exists Q)\ (\exists R)\ (|P| = |Q|\ \&\ |R; P| \neq |R; Q|)$

Proof: As already stated in Property 3, programs having different implementations may have the same values for spatial complexity measures. When these two different programs are combined with the same program, this may result into different spatial complexities for the two different combinations. This means,

$$(\exists P)\ (\exists Q)\ (\exists R)\ (CSC_P = CSC_Q\ \&\ CSC_{PR} \neq CSC_{QR} \text{ and } CSC_{RP} \neq CSC_{RQ})$$

where CSC_P and CSC_Q are the code spatial complexities for two different and unrelated programs P and Q respectively and CSC_{PR} and CSC_{QR} represent code spatial complexities of programs obtained after concatenating program P with R and, then Q with R respectively. Similarly, CSC_{RP} and CSC_{RQ} represent code spatial complexities of programs obtained after concatenating program R with P and, then R with Q respectively.

Also, $(\exists P)\ (\exists Q)\ (\exists R)\ (DSC_P = DSC_Q\ \&\ DSC_{PR} \neq DSC_{QR} \text{ and } DSC_{RP} \neq DSC_{RQ})$ where DSC_P, DSC_Q, DSC_{PR}, DSC_{QR}, DSC_{RP} and DSC_{RQ} represent the data spatial complexities for programs P, Q, P concatenated with R, Q concatenated with R, R concatenated with P, and R concatenated with Q respectively.

Thus, property 6a and 6b are well satisfied the code and data spatial complexity measures.

***Property 7*:** This property says that there are two programs P and Q such that Q is formed by permuting the order of the statements of P, and $|P| \neq |Q|$ that means a complexity measure should be sensitive to the permutation of statements.

Proof: The spatial complexity measures are defined in terms of distances in lines of code between different program elements such as modules and data members. Thus, spatial complexity of a program depends on the order of statements of the program. When program Q is formed by permuting the order of the statements of the program P, then values of code and data spatial complexity measures of the program Q will also change from the values obtained from the program P due to change in LOC between different program elements. Thus, for programs P and Q,

$CSC_P \neq CSC_Q$ and $DSC_P \neq DSC_Q$ where program Q is formed by permuting the order of the statements of P

Hence, the code and data spatial measures also satisfy property 7.

***Property 8*:** If P is a renaming of Q, then $|P| = |Q|$

Proof: Since, names of a program and its elements do not have any role in the definitions of the code and data spatial complexity, thus renaming of a program doe not have any effect on the spatial complexity measures i.e.

$CSC_P = CSC_Q$ and $DSC_P = DSC_Q$ where P is a renaming of Q.

Thus property 8 is well satisfied by the measures.

***Property 9*:** $(\exists P)$ $(\exists Q)$ such that $(|P| + |Q| < |P; Q|)$
This property states the concept of module monotonicity.

Proof: According to this property, the spatial complexity of a new program obtained from the combinations of two programs, can be greater than the sum of spatial complexities of two individual programs. As a program grows from its component modules' bodies, increased value of spatial complexity is always possible due to the increased distance between the definition and use of these components. This is true for both code as well as data spatial complexity measures as modules and data members are brought into a single program from two different programs, this may increase distances (in LOC) between different program elements. Moreover, as global data members can only be declared in the beginning of the program and then used in the program body, this will automatically increase the distance between their use and definition in the combined program. Thus, same data member when used in combined program will contribute more to data spatial complexity than its use in individual program. Hence, the data spatial complexity of a combined program would be more than the sum of the data spatial complexities of individual programs. Similarly, the code spatial complexity of a combined program would be more than the sum of the code spatial complexities of individual programs, because the definitions of all modules of both P and Q may be written together and put ahead of the all possible usage of these modules, in the source code of combined program. This will result in to the increased distance in terms of LOC among the definition and us of modules and thus leads to increased value of CSC i.e.

$(\exists P)$ $(\exists Q)$ such that $(CSC_P + CSC_Q < CSC_{PQ})$ where CSC_P, CSC_Q and CSC_{PQ} denote the code spatial complexities of programs P, Q and P; Q respectively.

Similarly, $(\exists P)$ $(\exists Q)$ such that $(DSC_P+ DSC_P < DSC_{PQ})$ where CSC_P, CSC_Q and CSC_{PQ} represent data spatial complexities of programs P, Q and P; Q respectively.

Thus, property 9 is also satisfied by the code and data spatial complexity measures.

5 Validation of Spatial Measures Using Briand's Criteria

In this section, we validate the spatial measures by using the formal evaluation framework given by Briand et al [8]. In this framework, Briand et al have given five properties, which a complexity measure must satisfy to be useful. Before applying the spatial measures against these properties, it seems appropriate to provide the basic definitions for complexity measures given in the framework [8].

System: A system S is represented as a <E, R>, where E represents the set of elements of S, and R is a binary relation on E ($R \subseteq E \times E$) representing the relationships between elements of S. For the purpose of evaluation of code and data spatial complexity measures, E is defined as the set of modules and global data members and R as the set of usages of modules and global data members.

Module: For a given a system S = <E, R>, a system m = <Em, Rm> is a module of S if and only if $Em \subseteq E$, $R \subseteq Em \times Em$ and $Rm \subseteq R$. A module m may be a code segment or a subprogram.

Complexity: The complexity of a system S is a function Complexity(S) that is described by Property 1 to Property 5.

Property 1 (Nonnegative): *The complexity of a system S = <E, R> is nonnegative if Complexity (S)* ≥ 0.

Proof: Since code and data spatial complexity measures are defined in terms of distances (in LOC), which are always non-negative numbers, thus, this property is well satisfied by both the measures.

Property 2 (Null Value): *The complexity of a system S = <E, R> is null if R is empty i.e.* $R = \emptyset \Rightarrow Complexity\ (S) = 0$.

Proof: As defined above, R is a set of usages of modules and global data members in the context of spatial measures. If there were no usage of any module in the program, value of the code spatial complexity (CSC) measure would be zero. Similarly, if there were no usage of any global data member in the program, value of the data spatial complexity (DSC) measure would be zero. Thus, Property 2 is also satisfied by the spatial measures.

Property 3 (Symmetry): *The complexity of a system S = <E, R> does not depend on the convention chosen to represent the relationships between its elements i.e.* ($S = \langle E, R\rangle$ *and* $S^{-1} = \langle E, R^{-1}\rangle) \Rightarrow Complexity(S) = Complexity(S^{-1})$.

Proof: According to this property, a complexity measure should not be sensitive to representation conventions used for relationships. A relationship can be represented by two equivalent representation conventions "active" (R) or "passive" (R^{-1}) form [8]. As defined in the beginning of this section, relationships for the code and data spatial complexity measures represent the set of usages of modules and data members. The convention used to represent these relationships will not have definitely any effect on the values of the code and data spatial complexity measures for a program, because only the count of these relationships is considered while computing the values for the spatial measures. Thus, the spatial measures satisfy this property.

Property 4 (Module Monotonicity): *The complexity of a system S = <E, R> is no less than the sum of the complexities of any two of its modules with no relationships in common i.e.*

$S = <E, R>$ *and* $m_1 = <Em_1, Rm_1>$ *and* $m_2 = <Em_2, Rm_2>$ *and* $m_1 \cup m_2 \subseteq S$ *and* $Rm_1 \cap Rm_2 = \emptyset) \Rightarrow Complexity(S) \geq Complexity(m_1) + Complexity(m_2)$.

Proof: This property is analogous to Weyuker's property 9 and as already shown in section 4, the code and data spatial complexity measures satisfy this property. Thus, property 4 is also satisfied by the spatial measures.

Property 5 (Disjoint Module Additivity): *The complexity of a system S = <E, R> composed of two disjoint modules* m_1, m_2, *is equal to the sum of the complexities of the two modules i.e.*

$(S = <E, R>$ *and* $S = m_1 \cup m_2$, *and* $m_1 \cap m_2 = \emptyset) \Rightarrow Complexity(S) = Complexity(m_1) + Complexity(m_2)$.

Proof: If two independent subprograms are combined together in a single program, then there will be no calling of one subprogram or data from the other subprogram as the two subprograms are totally independent. Hence their spatial complexity will not get affected due to the combination of the two. Then spatial complexity value of the combined program will be the sum of spatial complexity values of the subprograms. Thus, the spatial complexity measures also satisfy this property.

6 Related Work

In the literature, a number of software complexity measures have been proposed such as McCabe's cyclomatic complexity measure [14], Halstead programming effort measures [15], Oviedo's data flow complexity measures [16], Basili's measure [17], [18] cognitive complexity measures [19], [20], [21], [22] and spatial complexity measures [4], [5], [6], [23]. Some of them are useful to measure the structural complexity and others help in computing the cognitive complexities. Cognitive complexity metrics are always more difficult to be trusted and validated due to less-understood effect of human factors and human mind's working on the computation of such metrics. In order to improve the faith in some of such metrics, there are certain necessary (may not be sufficient) properties and evaluation criterion suggested by several researchers that software complexity metrics should satisfy [24], [25], [7], [8]. Amongst available evaluation criteria, the frameworks proposed by Weyuker [7] and Briand et al. [8] are widely used. Weyuker has developed a formal list of 9 properties for software complexity metrics and has evaluated a number of existing complexity metrics using these properties [7]. Briand et al. have proposed five properties to be satisfied by a complexity measure to be practical and more useful [8]. Several efforts have been made to evaluate and compare complexity metrics using this formal set of properties such as [12], [21], [26], [27], [28]. Similarly, in this work, we have chosen Weyuker' properties and Briand et al. criteria for the evaluation of code spatial complexity metrics and data spatial complexity metrics proposed in [4].

7 Conclusion

The paper has concentrated on the evaluation of two important spatial complexity metrics named as code spatial complexity metrics (CSC) and data spatial complexity metrics (DSC) proposed in [4]. These spatial complexity metrics have been studied against formal evaluation frameworks such as Weyuker's nine properties [7] and Briand et al evaluation criteria [8]. These metrics are found to satisfy all nine properties given by Weyuker and all the five complexity properties required by Briand et al evaluation criteria. The theoretical and formal evaluation carried out in this paper indicates that these spatial complexity measures are robust and useful.

References

1. Mayrhauser, A.V., Vans, A.M.: Program Comprehension during Software Maintenance and Evolution. IEEE Computer 3, 44–55 (1995)
2. Shneiderman, B.: Software Psychology: Human Factors in Computer and Information Systems. Winthrop Publishers Inc. (1980)
3. Baddeley, A.: Human Memory: Theory and Practice- Revised edn. Psychology Press, Hove (1997)
4. Chhabra, J.K., Aggarwal, K.K., Singh, Y.: Code and Data Spatial Complexity: Two Important Software Understandability Measures. Information and Software Technology 45(8), 539–546 (2003)
5. Chhabra, J.K., Aggarwal, K.K., Singh, Y.: Measurement of Object Oriented Software Spatial Complexity. Information and Software Technology 46(10), 689–699 (2004)
6. Douce, C.R., Layzell, P.J., Buckley, J.: Spatial Measures of Software Complexity. Technical Report, Information Technology Research Institute, University of Brighton, UK (1999)
7. Weyuker, E.J.: Evaluating Software Complexity Measure. IEEE Transactions on Software Complexity Measure 14(9), 1357–1365 (1988)
8. Briand, L., Morasca, S., Basili, V.: Property-based Software Engineering Measurement. IEEE Transactions of Software Engineering 22(1), 68–86 (1996)
9. Chidamber, S., Kemerer, C.: A Metrics Suite for Object-Oriented Design. IEEE Transactions on Software Engineering 20(6), 476–493 (1994)
10. Tegarden, D.P., Sheetz, S.D., Monarchi, D.E.: The Effectiveness of Traditional Metrics for Object-Oriented Systems. In: Twenty-Fifth Hawaii International Conference on System Sciences, vol. IV. IEEE Computer Society Press, Los Alamitos (1992)
11. Tegarden, D.P., Sheetz, S.D., Monarchi, D.E.: A Software Complexity Model of Object-Oriented Systems. Decision Support Systems: the International Journal 13, 241–262 (1995)
12. Misra, S., Misra, A.K.: Evaluation and Comparison of Cognitive Complexity Measure. ACM SIGSOFT Software Engineering Notes 32(2), 1–5 (2007)
13. Chhabra, J.K., Aggarwal, K.K., Singh, Y.: A Unified Measure of Complexity of Object-Oriented Software. Journal of the Computer Society of India 34(3), 2–13 (2004)
14. McCabe, T.J.: A Complexity Measure. IEEE Transactions on Software Engineering 2(4), 308–319 (1976)

15. Halstead, M.H.: Elements of Software Science. North Holland, New York (1977)
16. Oviedo, E.I.: Control Flow, Data and Program Complexity. In: IEEE COMPSAC, Chicago, IL, pp. 146–152 (1980)
17. Basili, V.R.: Qualitative Software Complexity Models: A Summary in Tutorial on Models and Methods for Software Management and Engineering. IEEE Computer Society Press, Los Alamitos (1980)
18. Basili, V.R., Selby, R.W., Phillips, T.Y.: Metric Analysis and Data Validation across FORTRAN Projection. IEEE Transactions on Software Engineering 9(6), 652–663 (1983)
19. Wang, Y., Shao, J.: Measurement of the Cognitive Functional Complexity of Software, In: IEEE International Conference on Cognitive Informatics, ICCI 2003, pp. 67–71 (2003)
20. Wang, Y., Shao, J.: New Measure of Software Complexity based on Cognitive Weights. Canadian Journal of Electrical & Computer Engineering 28(2), 69–74 (2003)
21. Misra, S.: Modified Cognitive Complexity Measure. In: Levi, A., Savaş, E., Yenigün, H., Balcısoy, S., Saygın, Y. (eds.) ISCIS 2006. LNCS, vol. 4263, pp. 1050–1059. Springer, Heidelberg (2006)
22. Misra, S.: A Complexity Measure based on Cognitive Weights. International Journal of Theoretical and Applied Computer Science 1(1), 1–10 (2006)
23. Chhabra, J.K., Gupta, V.: Towards Spatial Complexity Measures for Comprehension of Java Programs. In: International Conference on Advanced Computing and Communications, pp. 430–433 (2006)
24. Lakshmanian, K.B., Jayaprakash, S., Sinha, P.K.: Properties of Control-Flow Complexity Measures. IEEE Trans. Software Eng. 17(2), 1289–1295 (1991)
25. Tian, J., Zelkowitz, M.V.: A Formal Program Complexity Model and its Application. J. Systems Software 17, 253–266 (1992)
26. Baker, A.L., Zweben, S.H.: A Comparison of Measures of Control Flow Complexity. IEEE Transaction on Software Engineering 6(6), 506–511 (1980)
27. Misra, S., Misra, A.K.: Evaluating Cognitive Complexity Measure with Weyuker Properties. In: Third IEEE International Conference on Cognitive Informatics (ICCI 2004), pp. 103–108 (2004)
28. Misra, S.: Validating Modified Cognitive Complexity Measure. ACM SIGSOFT Software Engineering Notes 32(3), 1–5 (2007)

Pitcherpot: Avoiding Honeypot Detection

Vinod K. Panchal[1], Pramod K. Bhatnagar[2], and Mitul Bhatnagar[3]

[1] Defence Terrain Research Lab, Metcalfe House, Delhi 10054, India
vkpans@ieee.org
[2] Institute for system studies and analyses, Metcalfe House, Delhi 10054, India
pramod_wesee@yahoo.co.in
[3] Swami Keshwanand Institute of technology, University of Rajasthan, India
solar_winds_planet@yahoo.com

Abstract. This paper explores the various ways honeypots could be detected by the malicious attacker. This includes the different prevalent criteria and characteristics for honeypots generation & their weaknesses. Further this paper proposes a new way of implementation of a honeypot (Pitcher pots Systems) that effectively facilitate its identity avoidance and yet offers better ways to study the attacker.

Keywords: Pitcherpots, Honeypots, Honeypot firewalls, TCP/IP protocol, Black Hat community.

1 Introduction

The computers are always perfect but people designing them are not. Every system which is developed has some vulnerabilities in it. Sometimes these vulnerabilities are detected and corrected. If not, the software user keeps using the same software without any idea of vulnerability in it. If the same vulnerability is used by outside person (or computer expert) to compromise data or computer it is termed as security breached. As the world is becoming more and more network centric, security breaches are becoming more common but much severe. It is very difficult to study the vulnerabilities systems have and methods used by attackers. These attackers never reveal there methods of attacking and it is very difficult to analyze them when they are not expected.

To study attackers and the kind to techniques they use, a technology called 'honeypots' was developed. A honeypot is basically a system that emulates a weak or venerable system in an effort to try and attract a potential attacker to try and crack or break the machine while the honeypot is logging all the activities that occur [1]. Honeypots are of two main types: research honeypots and production honeypots. Research honeypots are the systems which give full resources to the attacker and are used to study them. Production honeypots are just used to find out about the attacker and the type of breach. Research honeypots are generally there in universities and military organizations and require a great deal of maintenance. They give the attacker full services thus they are easy to compromise and used against other secure systems[6].

S. Ranka et al. (Eds.): IC3 2009, CCIS 40, pp. 615–624, 2009.
© Springer-Verlag Berlin Heidelberg 2009

The black hat community at present is working on finding out the honeypots in the internet and put them in there blacklist which is updated to internet. So as soon as honeypots are setup attackers detect them and black list them [7]. This paper will discuss various detection techniques of the current honeypot systems by an attacker and propose a system (pitcherpots) which counters these techniques to a great extent. These systems have the ability to change themselves according to the attacker commands and give them full resources on any system. At the same time the system prevent the target system from being harmed in any way.

This paper is organized into five sections. The section following the introduction is talking about types of Honeypot and there identification. The third section describes the proposed architecture of a pitcherpot. The forth section summarizes the important findings. The last section concludes and talks about future works and improvements.

2 Honeypot Identification

Honeypot technology is still very under developed compared to other security solutions currently available. People do not like to use them as they are still treated as research application. The current honeypots are very naive and can be easily detected and tricked by an advanced attacker. These detections are the results of characteristic architecture of the honeypot and few mistakes made by the developers developing them. As honeypots are kept at the most appropriate place in the network if they are compromised it becomes very easy to hack in the network. The major operations applied for honeypot detection are:

2.1 Traffic Scans in the Network

Honeypots are mainly present in the main networks and are kept on the most appropriate place in the same. This is there to mainly attract the attacker. But as these systems are useless to the normal network there is no genuine traffic to them. It gives attackers a way to approximate honeypot in the network. If the attacker starts scanning the traffic in the network he can detect honeypots in it (as shown in Fig. 1). The honeypots are passive and there is no traffic directed to them. This gives an illusion that either this system is not useful or a honeypot.

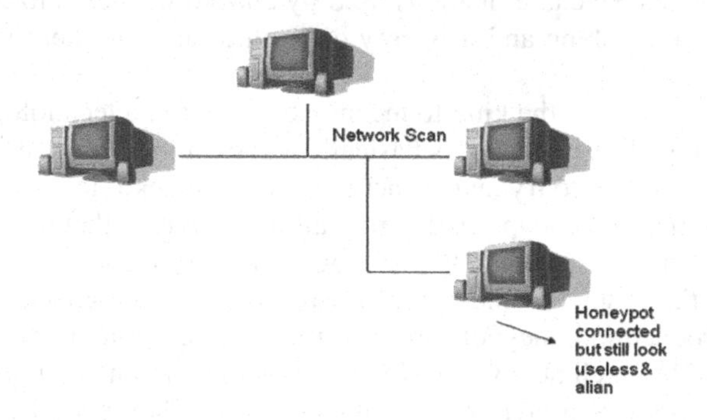

Fig. 1. Isolated honeypot in network

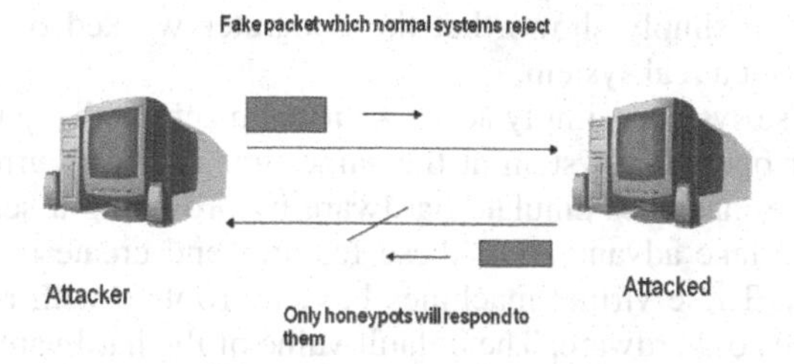

Fig. 2. Honeypot reply to a fake packet

2.2 Send Fake Packets to Honeypots

Most of the honeypots designed are not given all the resources because the type of threat is not defined. Honeypot do not check packets taking them to be malicious attacks accept them. The packet might have been send by the hacker to just test the system and acceptance gives him an idea of system being a honeypot (Fig. 2).

2.3 Connection to System Results in No Acknowledgement

The firewall of the honeypot systems allows any data inflow in the system from the network but restricts the outflow of data. This is because the system doesn't want to be used as a mediator in attacks on other systems. When a connection is established on the honeypot it simply accepts the connection but will block the acknowledgement depending on the rule designed. Also if the attacker tries connecting more then one system through it the firewall blocks them.

2.4 Detection of Virtual Computers

Most Linux based systems used as honeypots are making virtual systems in themselves [2].

The virtual systems are simply a fake representation of a system. The major tools used to make virtual systems are:

i. *UML (User Mode Linux)*: UML is a Linux kernel that can be run from within another running Linux system. This gives user the ability to forward any packets or network activity to the UML to give the person attempting a connection the feel that he or she is actually inside a live Linux system. But there are few problems with these systems that can easily be picked up by an attacker that knows what he/she is looking for. Firstly it uses a virtual drive to store data. Thus data stored in drives mounted on the UML system as /dev/ubd/*. The simplest way is to see mounting of drive by the attacker. Secondly the CPU information in the computer is given in as follows:

    cat/proc/cpuinfo

    processor: 0

    vendor_id: User

    model name: UML

    mode: tt

This information simply shows that the computer worked on is a UML virtual computer and not a real system.

ii. *VMWare*: These systems simply act as a virtual machine that gave user the ability to run another operating system at the same time as their current working environment. These machines emulate hardware by providing a set of virtual device drivers. People take advantage of these features and create a honeypot machine using this tool. These virtual machines have a problem with respect to masking virtual identities of hardware. The default value of the hardware is simply kept as

    Hardware: Reported type

    Video Card: VMWare Inc [VMware SVGA II] PCI Display Adopter

    Network Interface Card: Advanced MicroDevices [AMD] 79c970

    Hard Disk: VMWare Virtual IDE Hard Disk

    CD Drive: NECVMWar VMware IDE CDR10

    SCSI Controller: VMware SCSI Controller

    It has to be changed manually which is most of the time not done by the person setting this system. Secondly the vendor part of the MAC address is always one of the following values:

    00-05-69-xx-xx-xx   00-0C-29-xx-xx-xx   00-50-56-xx-xx-xx

    There are also some backdoors in the VMware tool to handle the settings in running system.

iii. *Use of chroot*: This is a similar approach to the virtual machine and run on the same host kernel as the main OS. A chroot basically limits a user to a specific area of the system but at the same time it gives them the impression that they are at the top level of the filesystem. This allows for the creation of a small, simulated environment in an area of a system that won't damage anything else. We can create a honeypot environment say in /tmp/honeypot. It has all the files and folders like /dev and /proc. It do not have /usr and /etc folders which can be generated as per requirement. The major problem these systems have is they do not imitate the inode value of the important files. The attacker simply runs command ls –lia on the root directory. This command gives the value of inode of the '.' And '..' directories. On a standard system you will notice that the inode of these 2 directories are

    2 drwxr-xr-x   21   root   root   2096 Oct 16 18:17 .

    2 drwxr-xr-x   21   root   root   2096 Oct 16 18:17 ..

    If you are in chroot and you run this command the inode will be quite different:

    1553552 drwxr-xr-x   6 1000   100   4096 Dec 14 12:58 .

    1553552 drwxr-xr-x   6 1000   100   4096 Dec 14 12:58 ..

    The inode in this example are the same inodes as the directory that has been chrooted to.

iv. Honeyd honeypots: Honeyd has a different functionality then honeypots but still they have their own weaknesses. The major problem they have is they tend to respond to those packets they have no business responding to.

2.5 Proxy Checking through a Third Source

The attackers generally check a proxy through a third source (Fig.3). They simply make a fake mail server with themselves. After this he connects honeypot server to port 25 of fake mail server. If the proxy gives a connection but the server does not respond at all the proxy is most probably a honeypot [3].

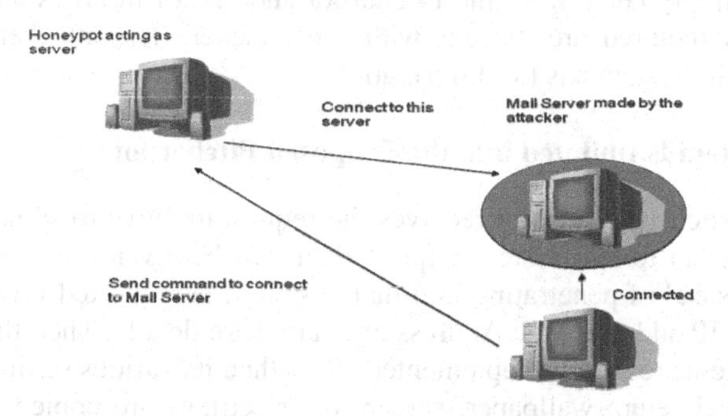

Fig. 3. Attacker checking through a fake mail server

2.6 Cannot Handle Multiple Attacks at the Same Time

The honeypots as are logging the whole system they can barely sustain more then one attack at a time. This is the best way to apply a DOS attack on the honeypot system it will give you a very slow response [4].

3 Architecture Proposed for Pitcherpot Systems

The Pitcherpot system should be as a useful component in the network. It should be scanning the whole network traffic and should send fake packets to balance the packet contents. The system sends fake packet in such a way that the opposite system returns a reply packet. And the packets should be changing their structure and size randomly. This makes the system look like a simple network administrator node. The network administrator system is the most valuable system as it got the rights to enter anywhere in the network. And if it is compromised the network can be stopped as well as controlled. This makes this system more tempting as well as attractive to the attacker.

The system has the simple approach. It can copy another system's structure and pose as the same system. The OS file system is a type of a tree with a root folder. The windows system has a root folder named as '/Desktop' whereas Linux system has the root folder named '/root'. This file system tree of other system is used from our system in case of various operations on the OS. The remote logging is one of the applications in which the computer can look at the file system and directory structure of other system and can change it. This file system if copied and implemented on the Pitcherpot, it will start acting as the same system. The system also takes up a fake MAC address and IP address of the original system. And also it took up all the settings of the host system. This will simple remove the problem of Pitcherpot detection by various tools as we are not posing as a fake system but we are a real system on the network.

The attacker when tries to compromises the pitcher pot it starts logging the attacker operations. If the attacker tries to connect to some other system the pitcher pots firewall will simply transfer the file system and settings of that system (in honeypot cases generally it is blocked) and convert it into same system. As we transfer the file system we simply compare the transferred file system and directory structure with pitcher pots own file system. All the files which are of the operating system or are there in Pitcherpot are simply taken from the Pitcherpot and rest of the files are just images. All the services required are present with the attacker. The attacker is no where blocked and all the system has full information.

3.1 Way a System Is Imitated into the Proposed Pitcherpot

As soon as the Pitcherpot firewall receives the request to move to some other system it will send a packet to that system requested. It will be given remote access to that system. Then instead of penetrating into that system it is requested for its details like computer name, IP address, Mac Address and hardware details. Then the registry settings of that system are simply implemented. Also then its various settings like resolution and graphical setup, wallpaper, screen saver settings are copied. Then the file system tree is simply copied from that system and it is compared to the current Pitcherpot and integrated to this system. The time this operation is being done is shown to the attacker as if his attack is taking that time. As most of the attacks take a lot of time to complete or to give results the Pitcherpot get a lot of time to cover up whole process as back work.

The Pitcherpot system will operate on these systems according to the type of there Operating System.

For windows Operating system (Fig 4) first the system takes up the BIOS settings from the computer. Then the different software drivers (Taken as Pitcherpot drivers in Fig. 4) are taken. In this operation network related operations are also taken. Then resident system programs are not changed as OS is same. Application programs can be changed according to rules set by user..

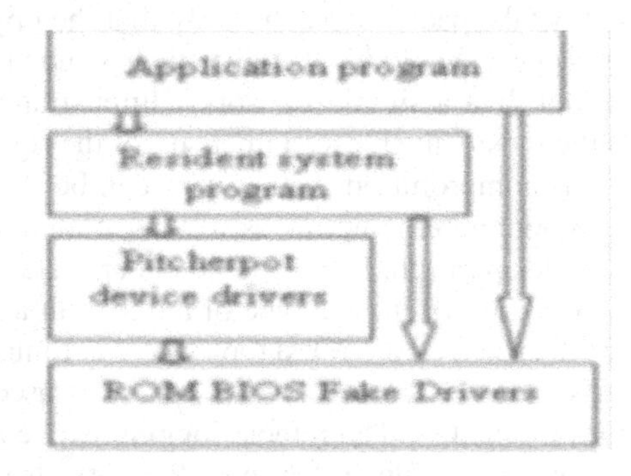

Fig. 4. DOS System based Pitcherpot ([5])

The UNIX system treats every thing as text files thus it is very easy to do changes in it. The kernel of a UNIX system is used for handling interrupts and manages resources so it is no need to change this. At the kernel level only changing the way system calls interface is to be modified to log the attacker. The major modification is to be done with Shell. The system files can be easily copied and compared with the existing files. And user files are simply copied from the original system.

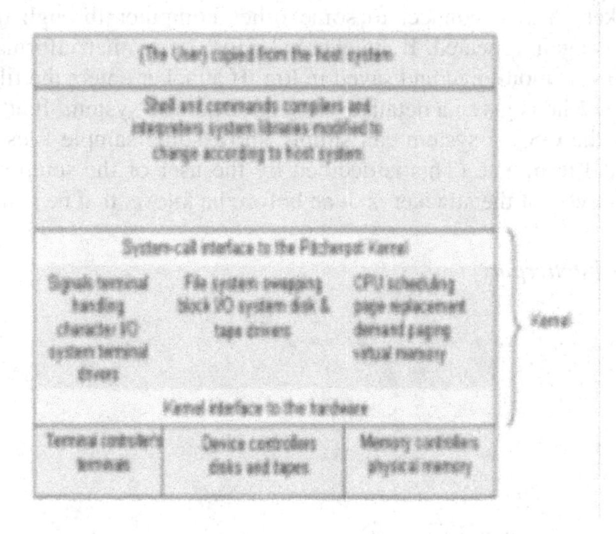

Fig. 5. UNIX System based Pitcherpot ([5])

4 Detail of Flow Graph

The proposed system is designed to scan network after every fixed interval of time. This is going to provide two major functionalities. Firstly, it will act as an administrative node handling the traffic in the network. Secondly, it will help introduce fake packets in the network making it a part and not an isolated node. If there is an attack on pitcherpot it moves forward otherwise continue with this process.

In case of attack, system starts the logging process. This process is integrated into the system process thus will keep interrupting and save the detail into a log file. The pitcherpot then connects to other computer (in the start that computer is an independent imitation itself). The pitcherpots do not have any self identity and they change their identity every time someone enters. This is to protect these systems from manufacturing errors or signatures (in vmware type honeypots there is a method to detect by the default values it has). It also helps the system save itself from being blacklisted.

The pitcherpot sends a request to targeted computer and then establish a special connection with it. Then the pitcherpot ask for detailed information of that computer. This operation is dependent on the OS of the target computer (the detailed explanation for Windows and UNIX based computers is given in section 3). In general this is a step by step process. First, the details of the target firmwarc are implemented. Then the detailed structure of the targeted Kernel is imitated. The Kernel has the file system details, page scheduling algorithms and character I/O system terminals, then the details of the system soft wares are defined to the pitcherpots, process details and libraries the system uses. Then the details of the application programs are taken. It implements these details on itself and starts pretending as that same computer.

If the attacker tries to connect to some other computer through this computer, whole process is again repeated. If attacker asks for some other information from the target computer it is monitored and saved in log. If attacker enters the file hierarchy in the pitcherpot, he/she is given a detailed view of target file system. If attacker asks for the files not in the OS the system can simply show some sample files predefined or give an error and terminate (This is decided by the user or the setup administrator). The detailed analysis of the attacker is done before he knows that he is in a pitcherpot.

***Flow Graph of Pitcherpot*:**

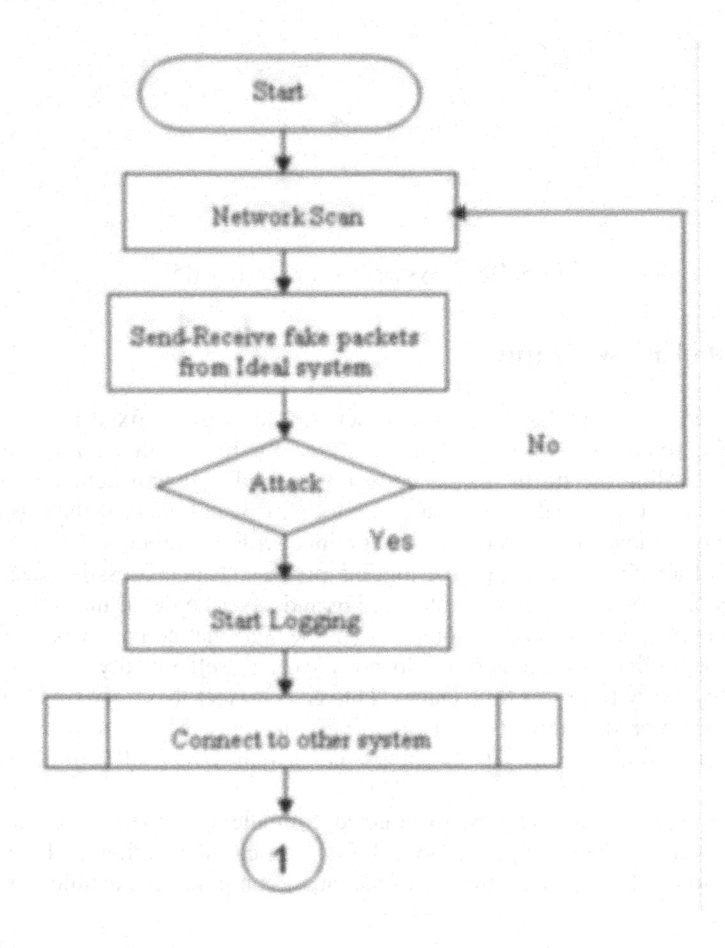

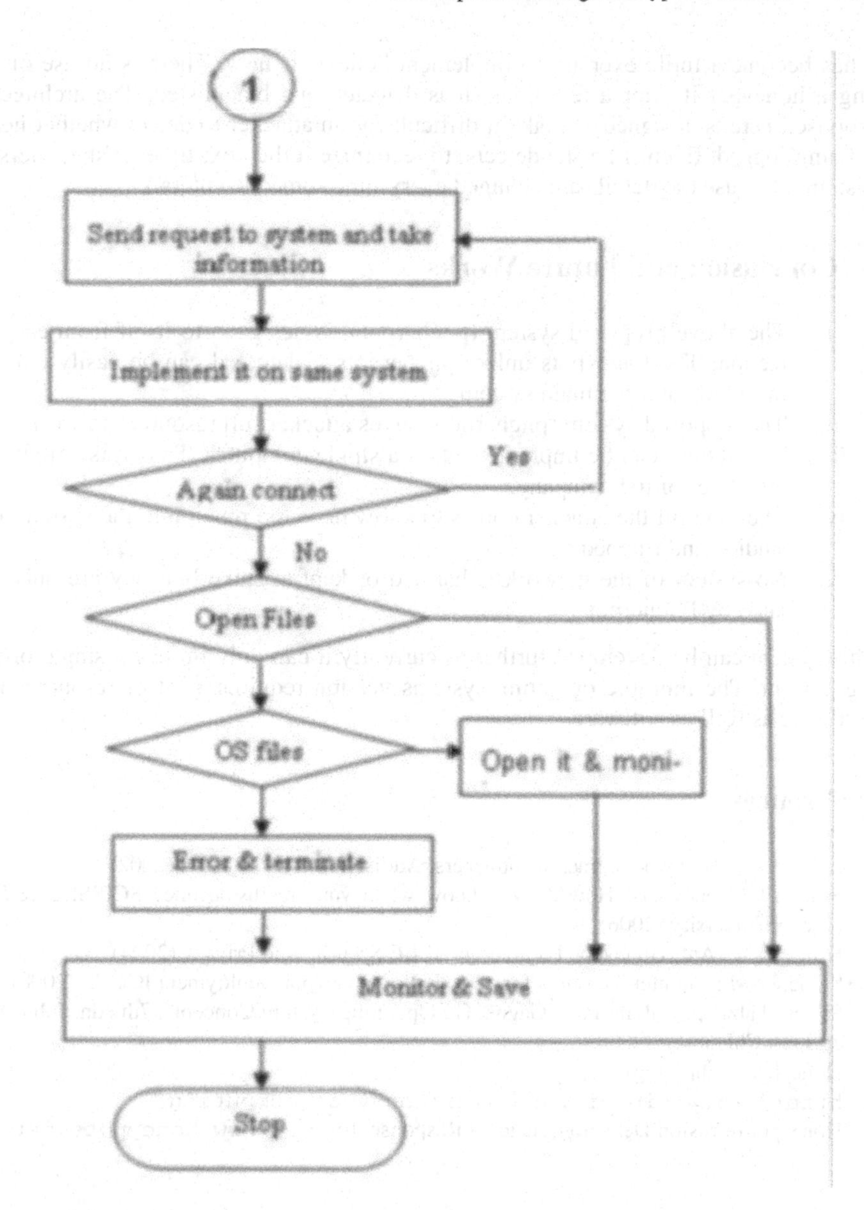

5 Important Findings

Currently used Honeypot systems are simply the unused machines in the company. They are applied just for testing the efficiency of company's network systems and furnish the details of the places they are being attacked from. Usually the implementation is not done properly. The trend is to hide maximum details from the attacker simultaneously not making the attacker suspicious. The hacker community has strongly come against this practice by detecting honeypot systems and blacklisting them. Thus it has become a futile exercise to implement honeypots now. There is no use of putting a honeypot if after a few hours it is detected and blacklisted. The architecture proposed here is designed to make it difficult for an attacker to detect whether he/she is being logged. Even if he/she detects, to recognize it the next time he/she enters the system (Because the details are changed every time someone enters.)

6 Conclusion and Future Works

i. The above proposed system (pitcherpots) is designed to itself from easy detection. The honeypots unlike pitcherpots if detected can be easily compromised to enter the main system.
ii. The proposed system (pitcherpots) gives attacker full resources to work on.
iii. The system can be implemented on a single computer. Thus it is simply cost effective for the company.
iv. The time till the attacker comes to know he is too much into the system to be studied and trapped.
v. No system of the network is harmed or kept at stake but they are only imitated by Pitcherpot.

This system can be developed further as currently it can only imitate a single operating system. The multiple operating systems version requires a lot of resources both hardware as well as software.

References

1. Spitzner, L.: Honeypots: tracking attackers. Addison-Wesley, Boston (2002)
2. Simon, I.: Honeypots: How do you know when you are inside one? SCISSEC & Edith Cowan University (2006)
3. Krawetz, N.: Anti Honeypot Technology. IEEE Security and Privacy (2004)
4. Shiue, L.-M.: Counter measures for detection of Honeypot Deployment ICCCE (2008)
5. Silberschatz, A., Galvin, P.B., Gagne, G.: Operating System Concepts, 7th edn. John Wiley & Sons, Chichester (2005)
6. Honeypot (computing), http://en.wikipedia.org/wiki/Honeypot_computing
7. Honeypot Intrusion Detection, Incident Response, http://www.honeypots.net/

Verification of Liveness Properties in Distributed Systems

Divakar Yadav[1] and Michael Butler[2,*]

[1] Institute of Engineering and Technology, U P Technical University, Lucknow, India
[2] School of Electronics and Computer Science, University of Southampton, UK
divakar.yadav@ietlucknow.edu, mjb@ecs.soton.ac.uk

Abstract. This paper presents liveness properties that need to be preserved by Event-B models of distributed systems. Event-B is a formal technique for development of models of distributed systems related via refinement. In this paper we outline how enabledness preservation and non-divergence are related to the liveness properties of the B models of the distributed systems. We address the liveness issues related to our model of distributed transactions and outline the construction of proof obligations that need to be discharged to ensure liveness.

Keywords: Formal Methods, Distributed Systems, Event-B, Liveness Properties.

1 Introduction

Distributed systems are hard to understand and verify due to complex execution paths and unanticipated behavior. A rigorous reasoning of these systems is required to precisely understand behavior of such systems. Safety and liveness are two important issues in the development of the distributed systems [9]. The distinction between safety and liveness properties is motivated by different tools and techniques for proving them and various interpretations of these properties are discussed in [8]. Informally, as described in [9], the safety property expresses that something *bad* will not happen during a system execution. A liveness property expresses that something *good* will eventually happen.

Event-B [10], a variant of classical B [1], is a formal technique for developing models of distributed systems. Event-B supports refinement and provides a complete framework for development of models of distributed systems. Existing B tools provides an environment for the generation of proof obligations for consistency and refinement checking and to discharge them. This technique consists of first describing the abstract problem, introducing solutions or details in refinement steps to obtain more concrete specifications and verifying that proposed solutions are valid. A specific development in this approach is made of a

* Michael Butler's contribution is part of EU projects IST project IST 511599 RODIN (Rigorous Open Development Environment for Complex Systems) and ICT 214158 DEPLOY (Industrial deployment of system engineering methods providing high dependability and productivity www.deploy-project.eu/).

S. Ranka et al. (Eds.): IC3 2009, CCIS 40, pp. 625–636, 2009.
© Springer-Verlag Berlin Heidelberg 2009

series of more and more refined models. Each model is made of static properties (invariants) and dynamic properties (events). A list of state variables is modified by activation of finite list of events. The events are guarded by predicates and these guards may be strengthened at each refinement step. An approach to incremental design of distributed systems and guidelines for construction of Event-B models are outlined in [3,4,5].

Existing B tools provide a strong proof support to aid reasoning about the safety properties by generating the proof obligations and providing an environment to discharge the proof obligations. In addition to the safety properties, it is also useful to verify that the models of distributed systems are live. In this paper we outline issues related to liveness in Event-B models and present the guidelines to address these issues in the Event-B development of the distributed systems. A case study of distributed transactions [11] for replicated database is used to outline construction of proof obligations.

2 Safety and Liveness in the Event-B Models

With regard to the safety, the most important property which we want to prove about the models of distributed systems is *invariant preservation*. The invariant is a condition which must hold permanently on the state variables. By *invariant preservation* we mean proving that the actions of the events do not violate the invariants. Existing B tools generate proof obligations for following.

1. *Consistency Checking*: Consistency of a machine is established by proving that the actions associated with each event modifies the variables in such a way that the invariants are preserved under the hypothesis that the invariants hold initially and the guards are true. Discharging proof obligations generated due to consistency checking proves that the machine is consistent with respect to the invariants.
2. *Refinement Checking*: The refinement of a machine consists of refining its state and events. The gluing invariants relate the state of the refinement, represented by the concrete variables, to the abstract state represented by the abstract variables. An event in the abstraction may be refined by one or more events, and discharging proof obligations generated due to refinement checking ensure that gluing invariants are preserved by actions of the events in the refinement.

In order to ensure that the models of the distributed are *live* and eventually makes *progress* we need to prove that Event-B models are *non-divergent* and *enabledness* preserving [10].

3. *Non-Divergence*: In an incremental development approach using Event-B, new events and the variables can be introduced in the refinement steps. Each new event of the refinement refines a *skip* event in the abstraction and defines actions on the new variables. Proving the non-divergence requires us to prove that the new events do not take control forever.

4. *Enabledness Preservation*: By enabledness preservation, we mean that whenever some events (or group of events) is enabled at the abstract level then the corresponding events (or group of events) are eventually enabled at the concrete level.

Existing B tools [2,10] generate proof obligations due to consistency and refinement checking and they provide an environment to discharge them using automatic prover or by interaction. To ensure liveness in Event-B models we need to prove that models of distributed system are *Non-Diergent* and *Enabledness* preserving.

3 Event-B Models of Distributed Transactions

Our system model consist of sets of sites and data objects.We considered two types of transactions namely, Read-only and Update Transaction. In our model we considered *read anywhere, write everywhere* approach. A read-only transaction read the data locally while the update transactions are processed within the framework of a two phase commit protocol [7] to ensure global atomicity. Two update transactions are said to be in *conflict* if atleast one data object appears in *write set* of both transactions. To meet the strong consistency requirements, *conflicting* transactions need to be executed in isolation. We ensure this property by not *starting* a transaction at a site if any conflicting update transaction is *active* at that site. The commit or abort decision of a global transaction T_i is taken at the coordinator site within the framework of a two phase commit protocol. A global transaction T_i *commits* if *all* T_{ij} *commit* at S_j. The global transaction T_i *aborts* if *some* T_{ij} *aborts* at S_j.

A formal refinement based approach using Event-B to model and analyze distributed transaction is given in [11]. In our abstract model, an update transaction modifies the abstract one copy database through a single atomic event. In the refinement, an update transaction consists of a collection of interleaved events updating each replica separately. The transaction mechanism on the replicated database is designed to provide the illusion of atomic update of a one copy database. Through the refinement proofs, we verify that the design of the replicated database conforms to the one copy database abstraction despite transaction failures at a site. We assume that the sites communicate by a reliable broadcast [6] which eventually deliver messages without any ordering guarantees.

3.1 Abstract Transaction States

In our abstract model of transactions, the global state of an update transactions is modelled by a variable *transstatus*. The variable *transtatus* is defined as $transtatus \in trans \rightarrow TRANSSTATUS$, where $TRANSSTATUS = \{COMMIT, ABORT, PENDING\}$. The *transstatus* maps each transaction to its global state. With respect to an update transaction, activation of following events change the transaction states.

- *StartTran(tt)*: The activation of this event *starts* a fresh transaction and the state of the transaction is set to *pending*.
- *CommitWriteTran(tt)*: A *pending* update transaction commits by atomically updating the abstract database and it status is set to *commit*.
- *AbortWriteTran(tt)*: A *pending* update transaction aborts by making no change in the abstract database and it status is set to *abort*.

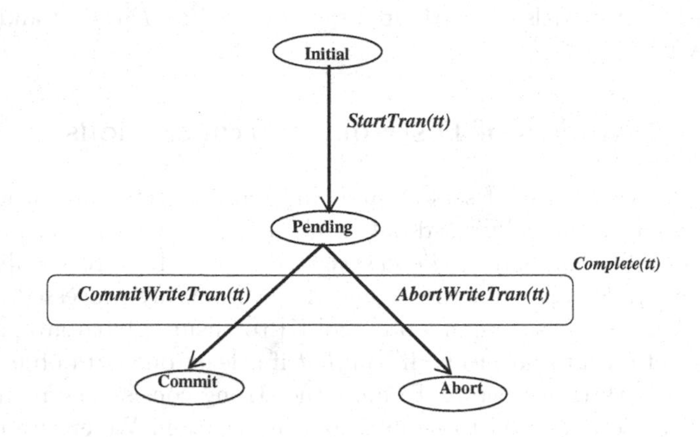

Fig. 1. Transaction States in the Abstract Model

The transitions in the transaction states due to the activation of events in the abstract model of the transactions are outlined in the the Fig. 1. The *CommitWriteTran(tt)* and *AbortWriteTran(tt)* together are represented as *Complete(tt)*, as both of these event models the completion of a update transaction.

3.2 Refined Transaction States

In the refined model, a global update transaction can be submitted to any one site, called the coordinator site for that transaction. Upon submission of an update transaction, the coordinating site of the transaction broadcasts all operations of the transaction to the participating sites by an *update* message. Upon receiving the update message at a participating site, the transaction manager at that site starts a sub-transaction. The activity of a global update transaction at a given site is referred as a sub-transaction.

The *BeginSubTran(tt,ss)* event models starting a sub-transaction of *tt* at participating site *ss*. In this refinement, the state of a transaction at a site is represented by a variable *sitetransstatus*. The variable *sitetransstatus* maps each transaction, at a site, to transaction states given by a set *SITETRANSTATUS*, where *SITETRANSTATUS*={*pending*, *commit*, *abort*, *precommit*}. A transaction t is said to be active at a site s if it has acquired the locks on the object set at that site. Our model prevents starting sub-transaction at a site if any conflicting transaction is already active at that site. Following guard of *BeginSubTran(tt)* event ensures that a sub-transaction of *tt* starts at a site *ss* when no active transaction *tz* running at *ss* is in *conflict* with *tt*:

$$(ss \mapsto tz) \in activetrans \Rightarrow objectset(tt) \cap objectset(tz) = \varnothing$$

As a consequence of the occurrence of this event, transaction *tt* becomes *active* at site *ss* and the *sitetransstatus* of *tt* at *ss* is set to pending. Instead of giving the specifications of all events of the refinement in the similar detail, brief descriptions of the new events in this refinement are outlined below.

- *BeginSubTran(tt)*: This event models *starting* a sub-transaction at a site. The status of the transaction *tt* at site *ss* is set to *pending.*
- *SiteAbortTx(ss,tt)*: This event models *local abort* of a transaction at a site. The transaction is said to complete execution at the site. The status of the transaction *tt* at site *ss* is set to *abort.*
- *SiteCommitTx(ss,tt)*: This event models *precommit* of a transaction at a site. The status of the transaction *tt* at site *ss* is set to *precommit.*
- *ExeAbortDecision(ss,tt)*: This event models *abort* of a *precommitted* transaction at a site. This event is activated once the transaction has globally aborted. The status of the transaction *tt* at site *ss* is set to *abort.* The transaction is said to complete execution at the site.
- *ExeCommitDecision(ss,tt)*: This event models *commit* of a *precommitted* transaction at a site. This event is activated once the transaction has globally committed. The status of the transaction *tt* at site *ss* is set to *precommit.* The replica at the site is updated with the transaction effects and the transaction is said to complete execution at this site.

4 Non-divergence

New events and the variables can be introduced in the refinement. Each new event of the refinement refines a *skip* event and defines a computation on new variables. In order to show that the model of system is non-divergent, we need to show that the new events introduced in the refinement do not together diverge, i.e., run forever. For example, if the new events in the refinement, such as, *BeginSubTran*, *SiteAbortTx* or *SiteCommitTx* take the control forever then the events of global commit/abort are never activated and a global commit decision may never be achieved. In order to prove that the new events do not diverge, we use a construct called *variant.* A variant V is a variable such that $V \in \mathbb{N}$, where $\mathbb{N}$ is a set of natural numbers. For each new event in the refinement we should be able to demonstrate that the execution of each new event decrease the variant and variant never goes below zero. This allow to us prove that a new event cannot take control forever, since a variant can not be decreased indefinitely.

The state diagram for a concrete transaction state transitions at a site is shown in the Fig. 2. A transaction state at each participating site is first set to *pending* by the activation of *BeginSubTran.* The activation of the event *SiteCommitTx* changes the status from *pending* to *precommit* while the activation of

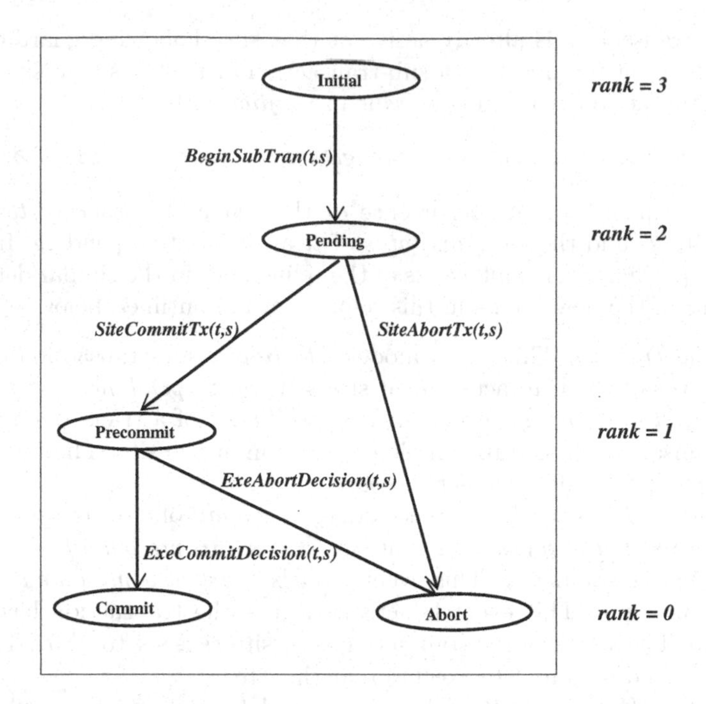

Fig. 2. Concrete Transaction States in the Refinement

SiteAbortTx sets the status from *pending* to *abort* at that site. A transaction in the *precommit* state at a site changes the state to either *commit* or *abort* by the activation of event *ExeCommitDecision* or *ExeAbortDecision* respectively. As shown in the Fig. 2, each state is represented by a rank. The *initial* state represents a state of a transaction *tt* at a participating site *ss* when it is not active, i.e., the *sitetranssstatus* of *tt* at *ss* is not defined ($ss \notin dom(sitetranstatus(tt))$). After submission of a transaction, a transaction first become *active* at the coordinator site. Subsequently, due to the activation of the event *BeginSubTran(tt,ss)*, subtransactions are started separately at different sites, i.e., at each activation of this event, *tt* becomes active at participating sites *ss*. As shown in the figure, new events in the refinement change the state of a transaction at a site such that each time the rank is decreased.

A variant in the refinement is defined as a variable *variant*:

$$variant \in trans \rightarrow Natural$$

and initialized as $variant := \varnothing$.

When a fresh transaction *tt* is submitted by the activation of the event *StartTran(tt)*, the initial value of variant is set as $varinat(tt) := 3 * N$, where N is total number of the sites in the system. All new events in the refinement decrease the variant by one. For example, events *SiteCommitTx(tt,ss)* and *SiteAbortTx(tt,ss)* decrease the variant by one and change the status of a transaction from a *pending* state to *precommit* or *abort* state. Since activation of the new events in the refinement decrease the variant, the rank of state is changed from three to zero such that *variant(tt)* will always be greater than or equal to zero.

In order to prove that the activation of the new events given in the Fig. 2 do not diverge, we need to prove that the changes in the state of a transaction at a site corresponds to the decrement in the rank from three to zero. Also, the variable *variant* is decreased each time a new event in the refinement is activated. Thus, we construct invariant property involving variable *variant* that need to be satisfied by the action of the events in the refinement. This property is given the Fig 3.

$$
\begin{array}{l}
\forall t \cdot (t \in trans \Rightarrow \\
\quad variant(t) \geq (\; 3 * card(SITE - activetrans^{-1}[\{t\}] \\
\qquad\qquad +2 * card(sitetransstatus(t)^{-1}[\{pending\}] \\
\qquad\qquad +1 * card(sitetransstatus(t)^{-1}[\{precommit\}] \\
\qquad\qquad +0 * card(sitetransstatus(t)^{-1}[\{commit, abort\}] \\
\qquad\qquad) \\
\quad)
\end{array}
$$

Fig. 3. Invariant used in variant Proofs

In this expression, $activetrans^{-1}[\{t\}]$ returns a set of sites where transaction t is in active state. Similarly, $sitetransstatus(t)^{-1}[\{pending\}]$ returns a set of site where a transaction t is in *pending* state. In order to prove that the new events in the refinement do not diverge, we have to show that the above invariant property on a variable *variant* holds on the activation of the events in the refinement. In order to prove this invariant property we need to add invariants (1) and (2) to the model. The invariant (1) states that if a transaction t is not *active* at a site s then the variable *variant* is greater than or equal to zero. The invariant (2) state that the variable *variant* is greater than or equal to zero if the status of a transaction t at site s either *precommit, pending, abort* or *commit*.

$$\forall (s,t) \cdot (t \in trans \wedge s \in SITE \wedge (s \mapsto t) \notin activetrans \Rightarrow variant(t) \geq 0) \quad (1)$$

$$\forall (s,t) \cdot (t \in trans \wedge s \in SITE \wedge sitetransstatus(t)(s) \in \{pending, precommit, abort, commit\} \Rightarrow variant(t) \geq 0) \quad (2)$$

Once the above invariants are added to the model, the B Tool generate the proof obligations associated with the events in the refinement. These proof obligations may be discharged using automatic/interactive prover of the tool. By discharging the proof obligations we ensure that the model of distributed system is non-divergent. The same strategy can be followed for each level of the refinement chain of Event-B development to show non-divergence in development of distributed systems.

5 Enabledness Preservation

With respect to the liveness, the freedom from deadlock is an important property in a distributed database system. In our model of transactions, it require us to prove that each transaction eventually *completes* the execution, i.e., either it commits or aborts. It require us to prove that if a transaction *completes* the execution in the abstract model of a system, then it must also *complete* the execution in the concrete model. We ensure this property by enabledness preservation. The enabledness preservation in Event-B requires us to prove that the guards of the one or more events in the refinement are enabled under the hypothesis that the guard of one or more events in the abstraction are also enabled [10]. Precisely, let there exist events E_1^a, E_2^a ..., E_n^a in the abstraction and a corresponding event E_i^r in the refinement refines the abstraction event E_i^a. The events H_1^r, ..., H_k^r are the new events in the refinement. A weakest notion of enabledness preservation can be defined as follows:

$$\begin{aligned}&Grd(E_1^a) \lor Grd(E_2^a) ... \lor Grd(E_n^a)\\&\Rightarrow\\&Grd(E_1^r) \lor Grd(E_2^r) ... \lor Grd(E_n^r) \lor Grd(H_1^r) \lor Grd(H_2^r) ... \lor Grd(H_k^r) \qquad (3)\end{aligned}$$

Weakest notion of enabledness preservation given at (3) state that if one of the event in the abstraction is enabled then one or more events in refinement are also enabled. The strongest notion of the enabledness is defined at (4). It state that if the event E_i^a in the abstraction is enabled then either the refining event E_i^r is enabled or one of the new events are enabled.

$$Grd(E_i^a) \Rightarrow Grd(E_i^r) \lor Grd(H_1^r) \lor Grd(H_2^r) ... \lor Grd(H_k^r) \qquad (4)$$

A concrete model outlined in Fig. 2 may be deadlocked due to the race conditions, i.e., if updates are delivered to the sites without any order. To ensure that all updates are delivered to all sites in the same order, we need to *order* the update transactions such that all sites deliver updates in the same order. It may achieved if a site broadcasts an update using a total order broadcast [12]. In next section, we outline how the enabledness preservation properties relates to our model of transactions in the presence of abstract ordering on the update transactions.

Abstract Ordering on Transactions: To ensure that our concrete model of transactions does not block and makes progress, we introduce a new event *Order* in the refinement. The very purpose of introducing new event *Order(tt)* is to ensure that the transactions are executed at all sites in a predefined abstract order on the transactions. The abstract ordering on the transaction can be realized by introducing explicit *total ordering* [6] on the messages in the further refinements. To model an abstract order on the transactions we introduce new variables *tranorder* and *ordered* typed as follows:

$$\begin{aligned}&tranorder \subseteq trans \leftrightarrow trans\\&ordered \subseteq trans\end{aligned}$$

A mapping of the form $t1 \mapsto t2 \in tranorder$ indicate that a transaction *t1* is ordered before *t2*, i.e., at all sites *t1* will be processed before *t2*. The set variable *ordered* contains the transactions that has been ordered.

Proof Obligations for Enabledness Preservation: In this section, we outline the proof obligations to verify that the refinement is enabledness preserving. Our objective is a prove that if a transaction *completes* in the abstraction then it also *completes* in the refinement. The weakest notion of enabledness preservation[1] given at (3) requires us to prove following:

$$\begin{aligned}
&Grd(StartTran(t) \vee CommitWriteTran(t) \vee\ Grd(AbortWriteTran(t))\\
&\Rightarrow Grd(StartTran^{*}(t))\\
&\vee\ Grd(Order(t))\\
&\vee\ Grd(BeginSubTran(t,s))\\
&\vee\ Grd(SiteCommitTx(t,s))\\
&\vee\ Grd(SiteAbortTx(t,s))\\
&\vee\ Grd(CommitWriteTran^{*}(t))\\
&\vee\ Grd(AbortWriteTran^{*}(t))
\end{aligned} \tag{5}$$

The property given at (5) is not sufficient as it state that if one of the events in *StartTran*, *AbortWriteTran* or *CommitWriteTran* are enabled in the abstraction then one of the refined event or the new events are enabled in the refinement. It does not guarantee that if a transaction t completes in abstraction then it also completes in the refinement. What we need to prove is that if either *AbortWriteTran* or *CommitWriteTran* in the abstraction is enabled then one of the refined event or new event in the refinement are enabled. According to the strongest notion of enabledness preservation given at (4), it requires us to prove (6), (9) and (10).

$$\begin{aligned}
&Grd(StartTran(t))\\
&\Rightarrow Grd(StartTran^{*}(t))\\
&\vee\ Grd(Order(t))\\
&\vee\ Grd(BeginSubTran(t,s))\\
&\vee\ Grd(SiteCommitTx(t,s))\\
&\vee\ Grd(SiteAbortTx(t,s))\\
&\vee\ Grd(CommitWriteTran^{*}(t))\\
&\vee\ Grd(AbortWriteTran^{*}(t))
\end{aligned} \tag{6}$$

The property at (6) state that if the guard of *StartTran* event is enabled then the guard of refined *StartTran* or the guards of new events are enabled. This property is provable due to following observations.

$$Grd(StartTran(t) \Rightarrow Grd(StartTran^{*}(t)) \tag{7}$$

[1] An event E in the abstract model is defined as E^{*} in the refinement.

In order to prove this property, the following proof obligation needs to be discharged. This proof obligation is trivial and can be discharged by the automatic prover of the tool.

$$\forall t(t \in TRANSACTION \wedge t \notin trans \Rightarrow t \notin trans) \tag{8}$$

$$\begin{aligned} & Grd(CommitWriteTran(t)) \\ & \Rightarrow Grd(StartTran^*(t)) \\ & \vee\ Grd(Order(t) \\ & \vee\ Grd(BeginSubTran(t,s)) \\ & \vee\ Grd(SiteCommitTx(t,s)) \\ & \vee\ Grd(SiteAbortTx(t,s)) \\ & \vee\ Grd(CommitWriteTran^*(t)) \\ & \vee\ Grd(AbortWriteTran^*(t)) \end{aligned} \tag{9}$$

The property at (9) state that if the guard of *CommitWriteTran* event is enabled than the guard of refined *CommitWriteTran* or the guards of new events are enabled. This property is too storing to prove due to following reasons. A transaction may not *commit* in the refinement until some other transaction either do not *commit* or *abort*. It may happen due to the interleaved action of the events in refinement such that some other transaction has started at some site and the commit of the transaction t depends on commit or abort of other transaction. Also, due same reasons the property at (10) is also not provable.

$$\begin{aligned} & Grd(AbortWriteTran(t)) \\ & \Rightarrow\ Grd(StartTran^*(t)) \\ & \vee\ Grd(Order(t) \\ & \vee\ Grd(BeginSubTran(t,s)) \\ & \vee\ Grd(SiteCommitTx(t,s)) \\ & \vee\ Grd(SiteAbortTx(t,s)) \\ & \vee\ Grd(CommitWriteTran^*(t)) \\ & \vee\ Grd(AbortWriteTran^*(t)) \end{aligned} \tag{10}$$

We observe that the proof obligations constructed due to the weakest notion of the enabledness preservation are not sufficient to prove that if a transaction completes in abstraction then it also completes in the refinement. Also, we observe that strongest notion of enabledness preservation is too strong to prove this property, therefore unprovable. What we really need a notion of enabledness preservation that is stronger than the weakest notion(see property 3) and weaker than the strongest notion(see property 4). This can be defined as below.

1. If the event *StartTran* is enabled in the abstraction then it is also enabled in the refinement.

2. If the completion event, i.e.,either *CommitWriteTran* or *AbortWriteTran* events are enabled in the abstract model then these completion events are also enabled in the refinement.

We have already outlined that the first property is preserved by the our model of transaction given as property (7). For second property, we further construct the property given at (11).

$$\begin{aligned}
&Grd(CommitWriteTran(t)) \vee Grd(AbortWriteTran(t)) \\
&\Rightarrow Grd(Order(t)) \\
&\vee Grd(BeginSubTran(t,s)) \\
&\vee Grd(SiteCommitTx(t,s)) \\
&\vee Grd(SiteAbortTx(t,s)) \\
&\vee Grd(CommitWriteTran^{*}(t) \\
&\vee Grd(AbortWriteTran^{*}(t)
\end{aligned} \quad (11)$$

We observe that property (11) is also not provable because a transaction t cannot complete its execution until some other conflicting transaction, already active, do not complete. Using the same strategy outlined above we construct a new property to illustrate that if the events corresponding to a completion of a transaction t_x in the abstraction are enabled then the new events *Order*, *BeginSubtran*, *SiteCommitTx*, *SiteAbortTx* are enabled for other transactions t_y or the refined *Complete* events are also enabled for t_x. Once these properties are added to the Event-B model of distributed transactions, the B Tool generate proof obligations due to these additions. In order to discharge new proof obligations we are required to add fresh invariants until the B prover is able to discharge all proof obligation.

The proof obligations outlined above are specific to the model of the transactions, however, using the same strategy the proof obligations for the other models of the distributed systems may be generated. Discharging these proof obligations ensures that the model of distributed system is enabledness preserving. It can be noticed that same strategy should be used to formulate the proof obligations for each level of refinement.

6 Conclusions

In this paper, we addressed the issue of liveness in the B models of distributed system. The safety and liveness are two important issues in the design and development of distributed systems. The safety properties express that something *bad* will not happen during system execution. A liveness property expresses that something *good* will eventually happen during the execution. With regards to the safety properties, the existing B tools generate proof obligations for consistency and refinement checking. These proof obligations may be discharged by using automatic/interative provers of B Tools. To ensure livness in the models of distributed systems, it is useful to state that the model of the system under development is *non-divergent* and *enabledness* preserving. In order to show that the new event in the refinement do not take control forever, i.e., models are non-divergent, we outlined a method for construction of an invariant property on variant. To ensure that a concrete models also make progress and do not block more often than its abstraction, it is necessary to prove that if an abstract model makes a *progress* due to the activation of events then the concrete model also make a progress due to the activation of the events in the refinement. We ensure this property by enabledness preservation. A method for construction of proof obligations for ensuring enabledness preservation is outlined in this paper. The proof obligations corresponding to both *non-divergent* and *enabledness* preserving can be discharged using automaic/interactive prover of B Tools to ensure that models of distributed system are live.

References

1. Abrial, J.R.: The B Book. Assigning programs to meanings. Cambridge University Press, Cambridge (1996)
2. Abrial, J.-R., Butler, M., Hallerstede, S., Voisin, L.: An open extensible tool environment for Event-B. In: Liu, Z., He, J. (eds.) ICFEM 2006. LNCS, vol. 4260, pp. 588–605. Springer, Heidelberg (2006)
3. Butler, M.: Incremental Design of Distributed Systems with Event-B, Marktoberdorf Summer School 2008 Lecture Notes (2008), `http://eprints.ecs.soton.ac.uk/16910`
4. Butler, M.: An approach to the design of distributed systems with B AMN. In: Bowen, J., Hinchey, M., Till, D. (eds.) ZUM 2008. LNCS, vol. 1212. Springer, Heidelberg (2008)
5. Butler, M., Yadav, D.: An incremental development of mondex system in Event-B. Formal Aspects of Computting 20(1), 61–77 (2008)
6. Defago, X., Schiper, A., Urban, P.: Total order broadcast and multicast algorithms: Taxonomy and Survey. ACM Computing Survey 36(4), 372–421 (2004)
7. Gray, J., Reuter, A.: Transaction Processing: Concepts and Techniques. Morgan Kaufmann, San Francisco (1993)
8. Kindler, E.: Safety and Liveness Properties: A Survey. Bulletin of the European Association for Theoitical Computer Science 53, 268–272 (1994)
9. Lamport, L.: Proving the Correctness of Multiprocess Programs. IEEE Transactions on Software Eng. 3(2), 125–143 (1977)
10. Metayer, C., Abrial, J.R., Voison, L.: Event-B Language. RODIN deliverables 3.2 (2005), `http://rodin.cs.ncl.ac.uk/deliverables/D7.pdf`
11. Yadav, D., Butler, M.: Rigorous Design of Fault-Tolerant Transactions for Replicated Database Systems Using Event B. In: Butler, M., Jones, C.B., Romanovsky, A., Troubitsyna, E. (eds.) Rigorous Development of Complex Fault-Tolerant Systems. LNCS, vol. 4157, pp. 343–363. Springer, Heidelberg (2006)
12. Yadav, D., Butler, M.: Formal Development of a Total Order Broadcast for Distributed Transactions Using Event B. LNCS, vol. 5454, pp. 152–176. Springer, Heidelberg (2009)

InfoSec-MobCop – Framework for Theft Detection and Data Security on Mobile Computing Devices

Anand Gupta, Deepank Gupta, and Nidhi Gupta

Information Technology Department, Netaji Subhas Institute of Technology, New Delhi, India
omaranand@nsitonline.in, deepank.gupta@nsitonline.in, nidhi.gupta@nsitonline.in

Abstract. People steal mobile devices with the intention of making money either by selling the mobile or by taking the sensitive information stored inside it. Mobile thefts are rising even with existing deterrents in place. This is because; they are ineffective, as they generate unnecessary alerts and might require expensive hardware equipments. In this paper a novel framework termed as InfoSec-MobCop is proposed which secures a mobile user's data and discovers theft by detecting any anomaly in the user behavior. The anomaly of the user is computed by extracting and monitoring user specific details (typing pattern and usage history). The result of any intrusion attempt by a masquerader is intimated to the service provider through an SMS. Effectiveness of the used approach is discussed using FAR and FRR graphs. The experimental system uses both real users and simulated studies to quantify the effectiveness of the InfoSec-MobCop (Information Security Mobile Cop).

Keywords: Mobile Device Security, Masquerade Detection, Opaque Authentication, Typing Pattern.

1 Introduction

Every second person on this planet owns a mobile phone for basic communication [1]. Apart from providing basic communication, mobile phones with their ever-increasing capabilities, store a lot of sensitive information about the user. Usually expensive high-end mobile phones are often loaded with sensitive information of the users and thus become a lucrative target for thieves. Mobile theft, like an online crime, provides instant gains and is very easy to get away with. Harrington et al.[2] show how mobiles act as criminal currency. It has been observed that a thief usually extracts sensitive information, removes the original SIM and sells the device in the black-market. Thus, it is important not only to prevent theft, but also to secure the user's data from being accessed by unauthorized malicious user.

In this paper, we present a solution to detect and catch mobile thefts while preventing the sensitive data from going into the wrong hands. In order to achieve this aim, a novel framework, termed as InfoSec-MobCop has been proposed. The framework proposes a layered security model in which the software resides on the ROM of the mobile device. The framework supports both GSM and CDMA based Mobile Phones.

S. Ranka et al. (Eds.): IC3 2009, CCIS 40, pp. 637–648, 2009.
© Springer-Verlag Berlin Heidelberg 2009

It provides a layer of user-authentication based on their typing patterns before allowing the access to encrypted data. It also stores the current location of the phone along with the IMEI number. When a theft is discovered (based on user authentication), SMS is sent to the service provider containing the current location of the mobile device. There are some implementations by BPL [4], MTNL [5] and Samsung [6] to track mobile phones in cases of possible thefts, but they do not provide user authentication mechanisms.

Moreover, the proposed method of user authentication does not require any extra hardware like contemporary biometric authentication mechanisms such as fingerprinting, iris recognition techniques suggested by Snelick et.al. [3] and Saleem et.al. [14]. The framework has been designed keeping in mind the requirements of minimal memory and computation footprint, thus making it suitable for mobile devices. It also authenticates the user using multiple mechanisms thereby providing higher accuracy rates than the individual methods. The validity of the authentication model is proven by experimental results.

To explain the InfoSec-MobCop system and its validation with experiments, the rest of the paper is organized as follows. Section 2 gives an overview of the previous work done in this field. Section 3 explains the motivation behind this approach and our contribution in it. Section 4 explains the terms and notations used throughout the paper. Section 5 describes the system design for the proposed InfoSec-MobCop framework. In Section 6, the results are presented in the form of graphs. Section 7 concludes the paper and paves the roadmap for the future work.

2 Previous Works

Some of the major telecom and mobile manufacturing companies in India like BPL, MTNL and Samsung have released software which sends an SMS to some predefined number when a SIM change occurs on the phone. BPL was the first to provide such a service and it requires the user to identify two numbers where the SMS will be sent in case of a theft as explained in [4]. Whenever there is a change in the SIM, two SMS are sent to the two numbers without any notification on the sending mobile phone. The SMS is sent after every 60 minutes, notifying the user that the phone is active. This scheme does not have location identification and is not very useful in catching the thief.

MTNL released another software [5] which stores IMEI number, SIM card number and registers the phone number of the other mobile to which the SMS needs to be sent to in case the mobile gets stolen. Whenever the SIM is changed, an SMS is sent to the registered number along with the IMEI number and the current location of the handset. Although, this software solves the previous problem, but still it does not protect the user's data from the thief. Innova Technologies also released similar software [6] with the added advantage of having the phonebook and other details being password protected. This software also suffers from a large number of false positives as the previous two.

Our scheme introduces user authentication using mobile key-pad based typing patterns and call history. No existing authentication based solution using keyboard typing patterns for mobile phones has been proposed till date, but the problem has been studied for computer Keyboards extensively in [7], [9], [10]. Proposed solution uses the user typing pattern for authentication in a pre-event scenario. This approach has been applied for keyboards and gives a high accuracy according to Zhuang et.al. [7]. Here we try to apply this machine learning capability for mobile keypads, thus differentiating the users based on their typing behavior on mobile keypads by Silfverberg et al. [8].

Other solutions include biometric authentication using fingerprinting [14] and protecting the user data on mobile phones using Transient Authentication [11].The latter scheme uses a small wireless radio range device which authenticates the device on the behalf of the user. But, the cached data on the device or the physical possession of the device or fooling the device by multiple repeaters can help the attacker gain the knowledge related to the passwords. Any communication loophole between the device and the handheld can further worsen the situation.

3 Motivation and Contribution

As discussed above, most of the existing solutions are preventive measures which do not try to do user authentication and theft discovery. Other solutions like [11] providing user authentication requires expensive hardware equipment which makes them less usable. There have been some studies [12], [13] doing masquerade detection by monitoring the behavior and environment of the user.

We apply an approach of learning from the user behavior (viz typing pattern and call history) to form patterns to detect a masquerader. To the best of our knowledge, no existing work caters to the needs of data security and theft discovery on mobile devices effectively. Hence it motivated us to present InfoSec-MobCop which is an endeavor to study the effectiveness of a user pattern based learning system used for prevention of mobile thefts and data security. The detailed work has been presented in the subsequent sections.

4 Terms and Notations

The following terms will be frequently used in this paper:

- *False Acceptance Rates (FAR)*: This number represents the cases when the user has been classified as a legitimate user when he is actually an attacker.
- *False Rejection Rates (FRR)*: This number represents the cases when a legitimate user has been classified as an attacker.
- *Masquerade Detection Rates (MDR)*: This number represents the fraction of detected attacks to the total number of attacks.
- *Time Taken To Detect (TTD)*: This number represents the time taken by the system to detect an intruder into the system.
- *Sample Depth*: The number of samples required for the algorithm to learn the individual pattern of the user and to start giving effective results is known as sample depth.
- *Binary Authentication Indicator (BAI)*: This n bit vector contains information regarding the authenticity of the user based on his typing behavior.

- *Hamming Distance*: The difference in the bits of 2 BAIs.
- *Time Window*: Time Window describes a moving window in time which has been taken into consideration for generating a pattern.
- *Incognito mode* occurs when some the user wishes to lend the device to a known person.

We propose following parameters which will be monitored by InfoSec-MobCop continuously for user authentication

- *Typing speed*: Refers to the number of characters (not the number of words) typed per unit of time.
- *Accuracy*: Refers to the number of correct characters typed per unit of time. This is calculated by subtracting the number of backspaces pressed by the user from the total number of characters typed, (giving the number of correct characters typed) and then dividing it by the total time taken.
- *Inter-symbol duration (ISD)*: It is the time elapsed between the pressing of one key and the pressing of another key by the user.

5 InfoSec-MobCop System

5.1 System Design

The framework consists of the following components as shown pictorially in Figure 1.

Interface: When the mobile is switched on, it stores the registration details of the user in a file which is sent to the server. It keeps track of the user preferences. If, for instance, the incognito mode is on, the samples are not tracked or monitored and access is granted unconditionally. This information is stored in User Preferences database table given in Table 1.

Sample collector: All the activities of the user, viz. call history and typing patterns, are tracked and the samples are collected. It consists of two components, viz. *Typing Pattern Collector and Usage Pattern Collector*. The collected parameters include the typing speed, accuracy and various ISDs. These computed values are stored in the

Database every time the user types in the message. Similarly user calls are logged into the database. The database is stored on the mobile device itself.

Table 1. User Preference Database

Request ID	Incognito Mode	Date/Time
60683ff5-14d0-4ac1-b5b7-0024e9ad8d36	Y	11-01-2009 10:10:34

Databases: The software maintains the samples of the user and starts working only when sampled depth is met. This storage happens in the mobile device itself and it cannot be deleted even after factory reset. The databases for the typing pattern and Call History Pattern are given in Table 3 and Table 4 respectively.

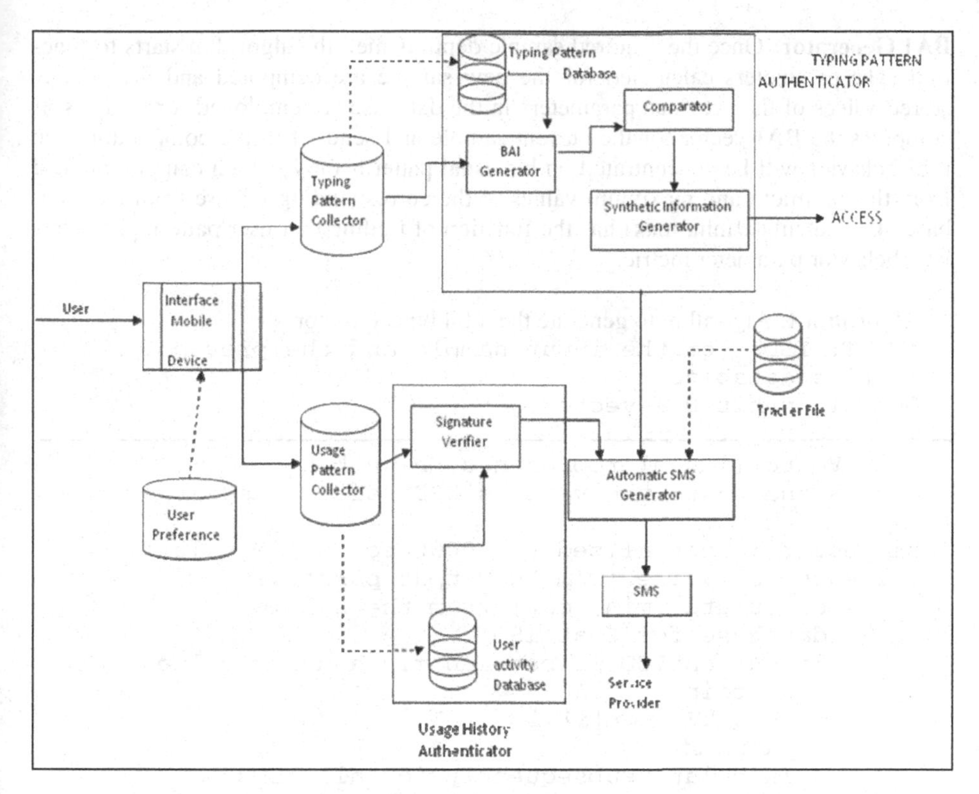

Fig. 1. System Design

Table 2. Tracker File

Flag	Location ID	Date/Time	Phone Number	IMEI Number
N	*1*[1]	11-01-2009 10:10:34	9052342227	*12345*
N	2	11-01-2009 10:10:34	9052342227	*12345*

Table 3. Typing Pattern Database

Username	Password	Speed	Accuracy	Backspaces	ISD Array
User1	Pass1	13	100	0	{55, 43, 10, 109}
User1	Pass1	14	100	0	{50, 21, 5, 104}

Table 4. User Activity Database

ID	Contact No	Type of Call	Date/Time	OnRoaming
60683ff5-14d0-4ac1-b5b7-0024e9ad8d36	9052342227	D	11-01-2009 10:10:34	N
ea526a3e-9c97-4a59-85da-01b9e8774fc5	9052345621	R	11-01-2009 13:45:21	N

[1] The information mentioned in italics is not real due to confidentiality purposes.

BAI Generator: Once the required sample depth is met, the algorithm starts to function. The parameters calculated for the new sample are computed and the already stored values of the previous parameters in the database are employed for analysis. It computes the BAI vector for the current sample and sends it to the comparator. The user behavior will be concentrated on his usual patterns only, which can be obtained from the minimum and maximum values of the corresponding feature from the database. This calculate[min, max] has the function of finding that user pattern characteristic behavior parameter metric.

Algorithm 1. Algorithm to generate the n bit binary vector

```
INPUT: ISDs for the input sample and the previous
samples database
OUTPUT: n bit BAI vector
```

```
//bitVector[1.....n] represents the n bit
corresponding to the various ISDs of the user

bitvector[] initialized to 00000 for 5 ISD values
for each ISD value from the input passcode
      calculate [min, max] from the sample
      database for that ISD
      if currentISD is between min and max value
      so obtained
           bitVector[i]=1
          end if
        calculate subsequently for all ISDs
end for
return bitVector
```

Comparator: The comparator takes into consideration the current BAI vector and the one predefined. It then computes the hamming distance between the two according to the algorithm mentioned as below.

Algorithm 2. Algorithm for comparison of the BAI for authentication

```
INPUT: n bit BAI vector
OUTPUT: Authorization status of the user/ attack of a
masquerader
```

```
//current hamming distance = h'
//predefined hamming distance = h
//authVector defines the authenticated vector
grant = false
h' = calculateHammingDist(BAI, authVector)
if h' < = h
          grant = true
          current sample is added to the database
else
          grant = false
end if
```

```
if grant = true then
        grant access to the user
end if

function calculateHammingDist(BAI, authVector)
h' = 0
for each bit i in the BAI
        if BAI[i] = authVector[i]
         h' ++
        end if
end for
return h';
end function
```

If the hamming distance falls within the accepted limits the sample is accepted and added to the database and the user is granted access. If the hamming distance is greater than the defined limits, then the current user is deemed a masquerader and the sample is not added. Once an attacker is detected, the flag in the tracker file is set and a sms containing the file details is sent to the service provider.

The service provider keeps tracking the mobile device and if the user complains of a mobile theft, he is notified. The system fails when a new mobile device is stolen since the software doesn't have enough sample depth for validation and hence assumes the thief as the authenticated user.

Synthetic Information Generator: In case the framework fails to verify the user based on the typing patterns, it calls Synthetic Information Generator to generate false information. Thus, the masquerader does not know that he has been detected since he is able to navigate around. Due to synthetic false default records placed instead of real data, no information is compromised to the masquerader. This concept has been termed as *"Opaque Authentication"*.

Signature Verifier: Once the required sample depth is met, the algorithm starts to function. The parameters calculated for the new sample are computed and the already stored values of the previous parameters in the database are employed for analysis. It computes the BAI vector for the current sample and compares it with the threshold to verify the user.

Algorithm 3. Signature Verifier Algorithm

```
INPUT: User Pattern Database, Current time window t,
Previous Time Window t'.
A time window has a start time denoted by t.start and
end time denoted by t.end.
OUTPUT: Boolean value
```

```
 //This function calculates the new contacts acquired
during the period of current time window as compared to
the total user history.
 newNos = NewNos(t.start, t.end)
```

```
  //This function compares the frequency of calls made
during the time window t with the previous time window t'
  freqFraction = CompareFreq(t, t')
  signature = w1 * newNos + w2 * freqFraction
  if(signature < threshold)
          return true
  else
          return false
```

Automatic SMS Generator: This component keeps updating the tracker file every time some activity happens in the form of mobility of the phone or some access denied responses. The SMS is generated in case of theft detection or unauthorized access to sensitive information.

The tracker file information is given in Table 2.

5.2 Validation Components

The flow diagram explains the control flow of the validation system which consists of the following components as given in Fig. 2.

Synthetic Data Generator (SDG): This component consists of a synthetic data generator which constructs a set of user profiles based on the user's characteristic parameters which determine the user pattern. The parameters are given as follows:

- *Correlation Coefficient(x = f(i, j))*: This number represents the fraction of the common contacts between two users(i and j). The value of x ranges from 0 to 1. A value of 0 indicates unrelated people while 1 represents that i and j are strongly correlated.
- *Frequency of Calling (freq = f(i, c, v, d)):* Represents number of calls freq made by a user i on the day d in the range [c – v, c + v].
- *Received/Dialed/Missed (RDM)* Numbers Fraction: This parameter is a characteristic of the user which represents the fraction of received, dialed and missed calls on an average by a user.
- *IsNewPhase* : This parameter is a Boolean value which represents if the user is currently going through a new phase of his life. This means that a person has changed his normal routine and he might be on roaming.
- *Frequency of NewContacts (freqnc : f(i, c', v', n)):* Represents number of new contacts acquired in n days in the range [c'-v', c'+v'].
- *Frequency Factor (freqfac=f(i,n)):* This fraction freqfac represents the ratio of the maximum calls made to a number to minimum calls made to any other number by a user i in d n days.

The task of SDG is to generate synthetic data for the specified number of users based on the specified parameters in the form of various data files which can further be used for validation purposes.

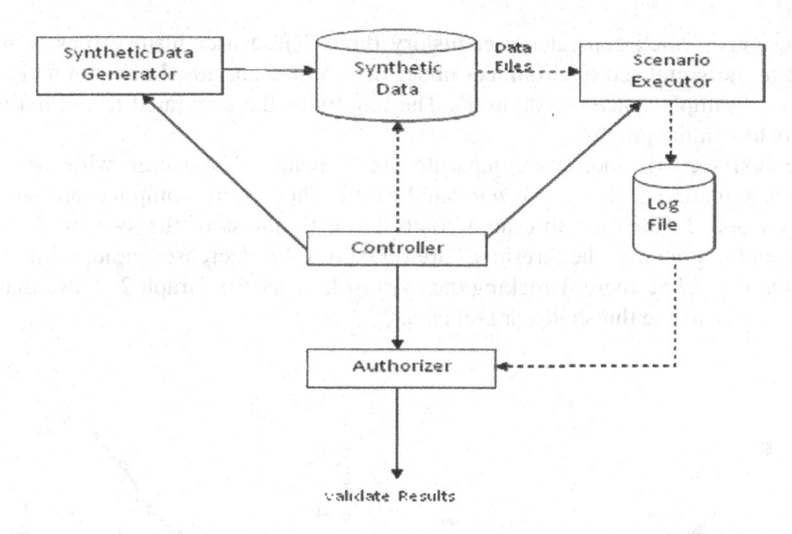

Fig. 2. Validation System Design

Scenario Executer: The scenario executer takes the data file generated by the SDG and executes the corresponding set of operations for a particular test scenario. The test scenarios are based on the fact that the thief can both be related and unrelated to the user. Based on this criterion, the test scenarios are divided into test against related masqueraders and against unrelated masqueraders. The following describes the scenarios in a detailed manner:

- *Unrelated users*: Similarity between the samples of a user and an unrelated masquerader is very less. This characteristic is exploited and various graphs varying the threshold, time window and sample depth are plotted and the corresponding FRR, FAR and MDR are analysed.
- *Related Users*: The theft can be undertaken by a related person of the user and hence the similarity in their behavior is more as compared to the unrelated masquerader. Threshold, Time Window and Sample Depth are varied and the corresponding FRR, FAR and MDR analyzed graphically.

Controller: The controller is the main program which requests the SDG to generate the data files for a large number of users. The control then passes on to the Scenario Executer to execute the specified test scenarios. The logs generated by the scenario executer are then analyzed for correctness of the results.

Authorizer: This component considers the logs generated by the controller and establishes the validity of the results of the software.

6 Experiment and Results

User behavior data was monitored and sample data was collected by observing the usage habits of a set of 100 real users. The parameters were abstracted out and SDG was designed which generates user history data. These user history files were then placed in the simulated environment under different scenarios. Each simulation consisted of a sample space of 100 users. The results of the simulated test scenarios are given below in the graphs.

The first scenario includes comparing user's behavior signature with user's own previous signature and with an unrelated thief. The results compiled are shown in Graphs 1 and 2. The base threshold controls the tightness of the system. As can be seen from the graph, as the thresholds are relaxed, FAR decreases rapidly, but there is an increase in FRR, thereby making the system less secure. Graph 2 shows that TTD also increases as the thresholds are relaxed.

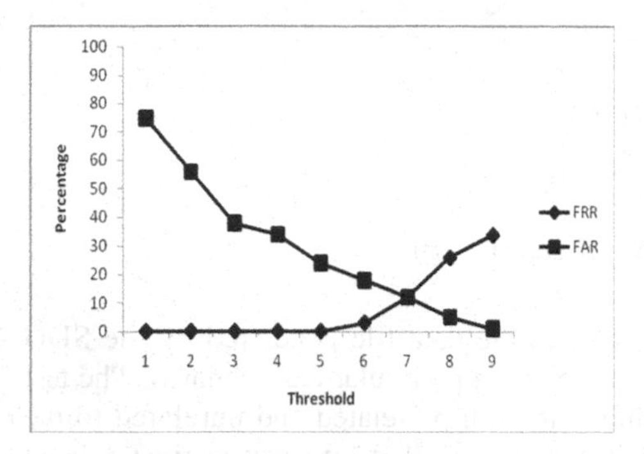

Graph 1. Effect of Threshold on FAR and FRR in Unrelated User Scenario

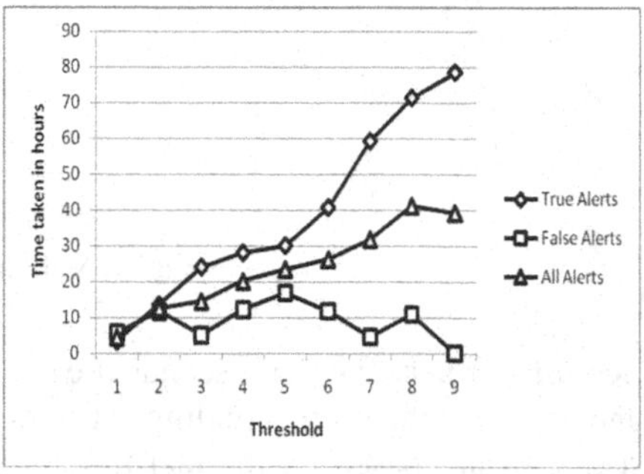

Graph 2. Effect of threshold on TTD for masquerader

The next scenario shown in Graph 3 & 4 examines the variation in FAR ratios for correlated users. The more closely the thief is correlated with the user in its signature, the higher are the FAR ratios. It particularly degenerates into an open-system with relaxed thresholds, lower time windows and higher correlated masqueraders.

A Windows Mobile based application was created for test on Windows Mobile 6.0 based devices by real users. The application, when deployed, takes around 4 MB while the database takes around 1.5 MB of storage space. The execution time of the application increases with the increase in the time-window, but the over-head is not very large.

The results of the tests done are compiled in the graph 5. As shown by the graph, the results get more and more accurate with the increase in the sample depth of the user. At sufficiently high sample depths, the user authentication is quite reliable using this method even on the mobile phones.

Finally, we collate the results of the two approaches in the form of Graph 6. The points in the I Quadrant represent instances when the user has been authenticated by both Typing Pattern and Call History. Quadrant II represents instances which have been authenticated by typing pattern but fail by call history. Quadrant III represents instances which fail by both the methods. Quadrant IV represents instances which have been authenticated by call history but fail by typing pattern.Thus, it can be seen that only Quadrant III points are the cases where the system fails to secure the user completely. With optimum values of system parameters, the Quadrant III values are sufficiently low.

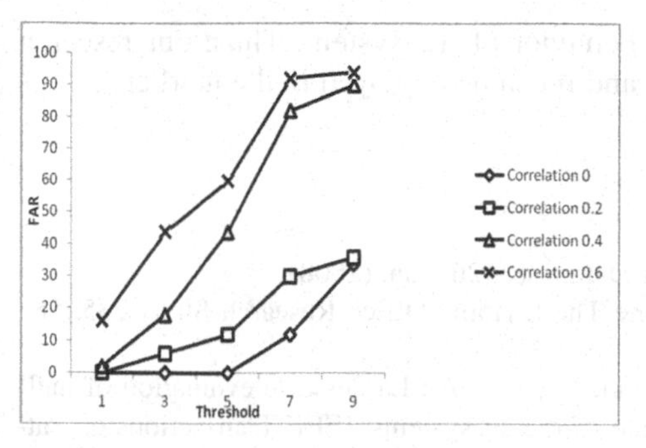

Graph 3. FAR with correlated users with variation in Threshold

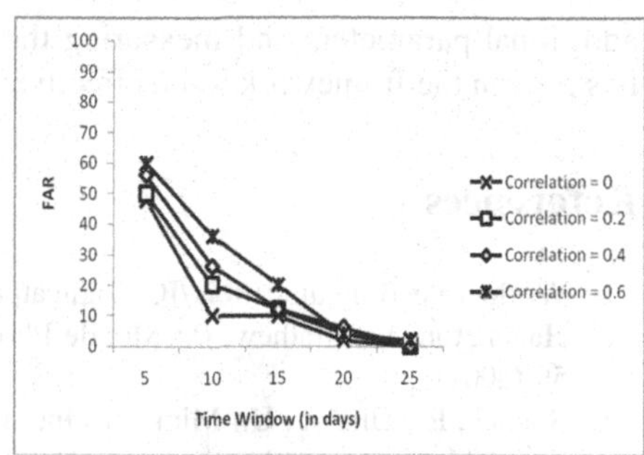

Graph 4. FAR with correlated users variation in time window

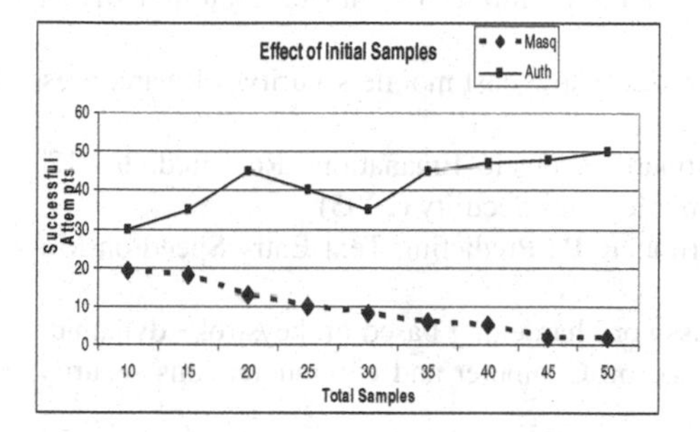

Graph 5. Effect of Sample Depth on Masquerade Detection

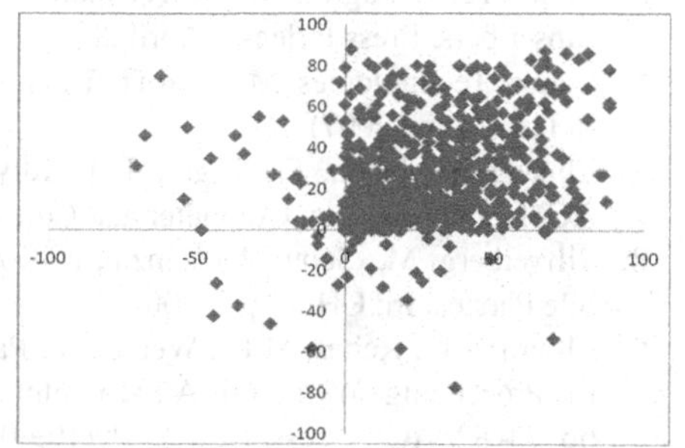

Graph 6. Scatter Graph combining User Pattern and Typing Pattern

7 Conclusions

As discussed above, the framework does not require expensive hardware equipment. Also, the computational and memory footprint is small enough to be deployed effectively on mobile computing devices. The algorithm has a very high MDR and is able to detect the masquerade attacks in reasonable times.

This analysis brings three key points to our focus. Firstly, the system collects an adequate sample depth to learn the patterns of any user to distinguish him/her effectively from a masquerader.

Secondly, as shown by the graphs, we need to strike a balance between the MDR and TTD of the system by optimizing the following characteristics viz Time Window, Sample Depth, Threshold and BAI. Varying these characteristics can make the system behave according to the user preferences. For instance, a corporate user might keep smaller time windows, higher sample depths, higher BAI and tighter thresholds so as to provide a higher security. Thus, the system is able to adjust itself to the user preferences and privacy.

Lastly, as shown by the scatter graph, the two systems combined together provide MDRs comparable to conventional biometric security measures. Some of the future work can include the addition of application history and speech recognition as additional parameters and measuring the behavior of the system. Thus, our research has proven the framework to be effective; and it can be deployed in the market.

References

1. World Telecommunication /ICT Indicator Database, 12th edn. (2008)
2. Harrington, V., Mathew, P.: Mobile Phone Theft. Home Office Research Study 235, 58–59 (2001)
3. Snelick, R., Uludag, U., Mink, A., Indovina, M., Jain, A.: Large-scale evaluation of multimodal biometric authentication using state-of-the-art systems. IEEE Transactions on Pattern Analysis and Machine Intelligence (2005)
4. Spiel Studios. Spiel Studios & BPL Mobile release MobileTracker. Press Release (August 2007)
5. Micro Technologies. Micro Technologies to offer Mobile Security to 2 million MTNL subscribers. Press Release (April 2008)
6. Innova Technologies. Mumbai IT firm introduces low-cost mobile security software. Press Release (June 2007)
7. Zhuang, L., Zhou, F., Tygar, J.D.: Keyboard Acoustic Emanations Revisited. In: 12th ACM Conference on Computer and Communications Security (2005)
8. Silfverberg, M., Scott MacKenzie, I., Korhonen, P.: Predicting Text Entry Speed on Mobile Phones. In: CHI (April 2000)
9. Monrose, F., Reiter, M.K., Wetzel, S.: Password hardening based on keystroke dynamics. In: Proceedings of the 6th ACM conference on Computer and communications security, pp. 73–82. ACM Press, New York (1999)
10. Lau, E., et al.: Enhanced user authentication through keystroke biometrics. 6.857 Computer and Network Security final project report (2004)
11. Nicholson, A.J., Corner, M.D., Noble, B.D.: Mobile Device Security Using Transient Authentication. IEEE Transactions on Mobile Computing (2006)
12. Mazhelis, O.: Masquerader Detection in Mobile Context based on Behaviour and Environment Modelling, Jyvaskyla (2007)
13. Bhukya, W.N., Suresh Kumar, G., Negi, A.: A study of effective-ness in masquerade detection. IEEE, Los Alamitos (2006)
14. Saleem, A., Al-Akaidi, M.: Security system based on fingerprint ID for mobile phone. In: 6th Middle East Simulation Multiconference, pp. 97–101 (2004)
15. Biddlecombe, E.: Crimes of fashion [mobile phone theft]. Communications International, pp. 38–39, 41, 43 (May 2002)

Multi-scale Modeling and Analysis of Nano-RFID Systems on HPC Setup

Rohit Pathak[1] and Satyadhar Joshi[2]

[1] Acropolis Institute of Technology & Research
[2] Shri Vaishnav Institute of Technology & Science
Indore, Madhya Pradesh, India
{rohitpathak,satyadhar_joshi}@ieee.org

Abstract. In this paper we have worked out on some the complex modeling aspects such as Multi Scale modeling, MATLAB Sugar based modeling and have shown the complexities involved in the analysis of Nano RFID (Radio Frequency Identification) systems. We have shown the modeling and simulation and demonstrated some novel ideas and library development for Nano RFID. Multi scale modeling plays a very important role in nanotech enabled devices properties of which cannot be explained sometimes by abstraction level theories. Reliability and packaging still remains one the major hindrances in practical implementation of Nano RFID based devices. And to work on them modeling and simulation will play a very important role. CNTs is the future low power material that will replace CMOS and its integration with CMOS, MEMS circuitry will play an important role in realizing the true power in Nano RFID systems. RFID based on innovations in nanotechnology has been shown. MEMS modeling of Antenna, sensors and its integration in the circuitry has been shown. Thus incorporating this we can design a Nano-RFID which can be used in areas like human implantation and complex banking applications. We have proposed modeling of RFID using the concept of multi scale modeling to accurately predict its properties. Also we give the modeling of MEMS devices that are proposed recently that can see possible application in RFID. We have also covered the applications and the advantages of Nano RFID in various areas. RF MEMS has been matured and its devices are being successfully commercialized but taking it to limits of nano domains and integration with singly chip RFID needs a novel approach which is being proposed. We have modeled MEMS based transponder and shown the distribution for multi scale modeling for Nano RFID.

Keywords: Nanotechnology, RFID (Radio Frequency Identification), Multi-Scale Modeling, HPC (High Performance Computing).

1 Introduction

The central point though of any technology remains its market acceptance and social trust. Many nations have taken special care to develop and synchronize research in RFID. Emerging Issues, challenges and policy options from a European prospective

S. Ranka et al. (Eds.): IC3 2009, CCIS 40, pp. 649–659, 2009.
© Springer-Verlag Berlin Heidelberg 2009

for RFID technologies has been discussed in details in the reports published in 2007 [1] thus implying the need for cheaper and reliable RFID systems. The report described in details about the strength, weakness, opportunities and threat. The main issues that still remains to be addressed are implementation of nanotechnology based RFID systems. Nanotechnology business report describes the realization of nanotechnology in business technologies for Australian and Global areas [2]. The architecture described in [3] supported by, some remarkable achievements and patents in nano electronics, wireless communication and power transmission techniques, nanotubes, lithography, biomedical instrumentation, genetics, and photonics, thereby making the arena of molecular machinery realistic [3]. RFID integration with Biotechnology will benefit medical science a great deal where we can use the developments in this area to benefit all humans. Biomedical sensors will monitor important body functions and status (i.e. blood sugar level, heartbeat rate, presence of toxic agents), and advanced algorithms adapted to each individual may trigger alarms when non-normal values are encountered are described in detail in report by Forbes [4].The design, simulation, implementation, testing and characterization of several antennas to be used in on-the-body sensors is given in [5]. On-the-body sensors are a promising component in Wireless Body Sensor Networks (WBSN). The antennas will provide the wireless communication links between nodes in the WBSN. The wireless nature of the network increases the flexibility and mobility of a subject wearing them [5].

ICT implants may be used to repair deficient bodily capabilities they can also be misused, particularly if these devices are accessible via digital networks which have been expanded in [6] again reflects the need for nanotechnology enabled RFIDs. The idea of letting ICT devices get under our skin in order not just to repair but even to enhance human capabilities gives rise to science fiction visions with threat and/or benefit characteristics still remains an issue to be resolved. May Electronics giants are also working on RFID, HP's vision for RFID, its leadership role in global standards development, RFID-related research conducted at HP Labs [7]. In [8] the U.S. Government's adoption process for the electronic passport as a case study for identifying the privacy and security risks that arise by embedding RFID technology in everyday things, the main reasons why the Department of State did not adequately identify and address these privacy and security risks, even after the government's process mandated a privacy impact assessment are given. Privacy and security concerns for individuals and analyze how these concerns were handled in the procurement and developments of the e-Passport are discussed [8]. Fully effective development of RFID is not possible without the consideration of issues related to data protection and consumer privacy. Beside this the ethical and sociological impacts of any innovation are important to consider alongside with economic and technological issues which has been described in details in [9].

RFID applications in fields such as manufacturing, highway toll management, building access badges, mass transit, and library check-out, making RFID more accessible for small and medium size enterprise and speeding up technology diffusion amongst users has been worked out. Avoiding distortion of competition, low cost systems are discussed in European Policy Outlook on RFID published in 2007[10]. This RFID is an emerging area where we can realize and apply developments of Nanotechnology and HPC to get solutions of the problems that exists.

2 Modeling of RFID MEMS Devices

Modeling and simulation plays a very important role today in Nano enabled devices. Modeling today is not just about getting computation done from computer but it is far more complex where each justified and non justified phenomenon needs to covered when we talk about simulating nano scale devices. In this we have we have tried to implement simulations and computations on an HPC setup. We have taken in account the recent developments in this regard that have taken place in the last 2 years to develop our library for simulation of Nano-RFID. A Remotely powered addressable UHF RFID integrated system 2.45 GHz with achieved operating rate of 4 W working at has been discussed in [11]. A UHF Near-Field RFID System with Fully Integrated Transponder has been discussed in [12]. Thus we can see that the main proposals are being made in UHF RFIDs. Techniques to reduce energy consumption remain the most challenging and important. Some elementary work in this regard has been discussed in [13]. A proposal for Batterless RFID systems has been given in [14], which can be solved by developing low power nano RFID system as shown in our work. WSN is areas that will benefit the most from Nano FRID based innovations. MEMS-Based Inductively Coupled RFID Transponder for Implantable Wireless Sensor Applications has been discussed in [15]. Unobtrusive long-range detection of passive RFID tag motion has been expanded in [16]. Movement tracking algorithms and monitoring system has been substantiated.

In this part we have shown the modeling of MEMS in Sugar on for the device proposed in recent years for modeling RFID. Three models have been put forth recently in this regard that will help in understanding and realizing the proposals made earlier [17]. We have included them in building the library for multi scale simulation of Nano RFID.

Battery less-Wireless MEMS Sensor System with a 3D Loop Antenna has been proposed earlier by Sho Sasaki and Tomonori Seki in 2007 which was been proposed

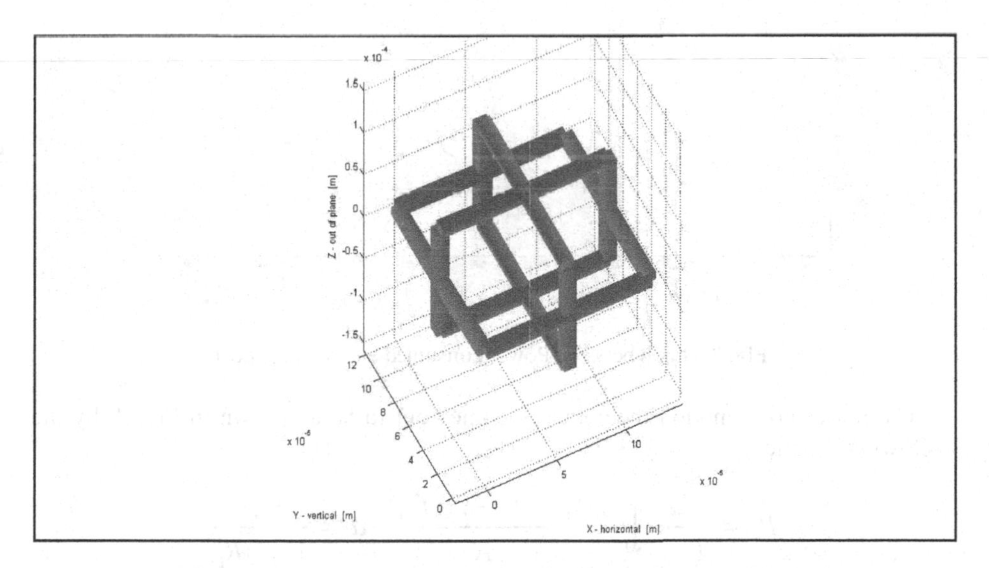

Fig. 1. 3D antenna modeled in MEMS SUGAR using MATLAB

in [17] has been shown in Fig. 1. In this part SUGAR will help in the development of the mathematical proposal made earlier since it can predict the parameters of MEMS devices accurately. The structure can be used in comprehensive analysis of the device and calculate other parameters. 3D Loop Antenna has been discussed with the importance of MEMS based circuitry eminent which requires modeling to be worked upon especially when we are using other Nano enabled circuitry [17]. Future will see many on-chip RFID systems with the integration of CNT, MEMS devices to make RFID better than the current generation. In our work we have shown some of the aspects that need to model to accelerate research in this area.

3 Modeling of Abstraction Level RFID

In the last part we have shown the library development to show how the latest works in this area have been included in our library. We know that modeling is generally done under a domain of abstraction layer. In this part we have shown the modeling of abstraction based basic parameters of RFID. The basic modeling for generic properties of RFID which needs to be taken in accounts my working on any RFID systems. Integration of this part is essential in calculating various other parameters for RFID. Though this part is based on some abstraction layer it is also useful in multi scale modeling which is shown in the next part.

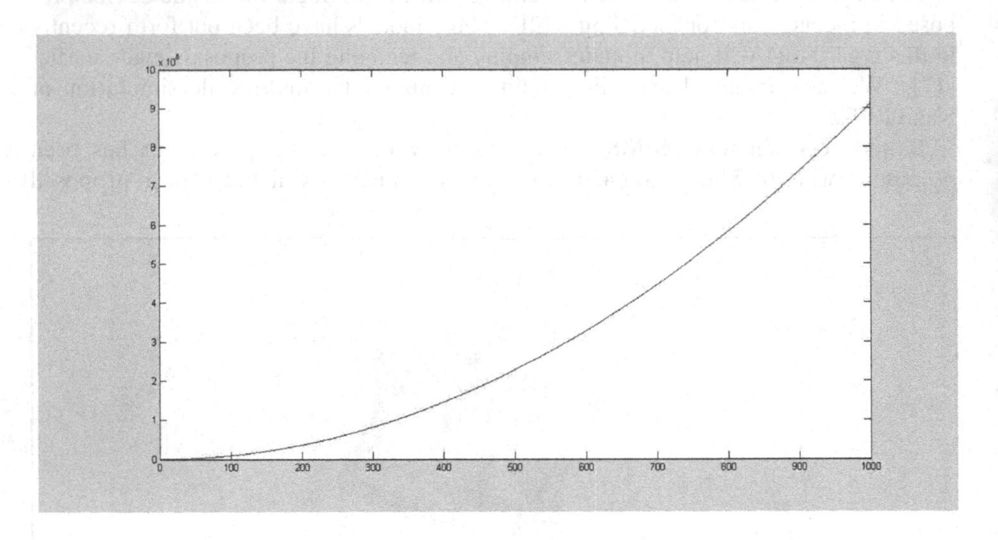

Fig. 2. Graph between Power consumed and Voltage [18]

The power consumption theoretically comes out to be as shown in Fig. 2. by the following equation

$$P_{R1} = \frac{2}{T_{CLK}} \int_0^{T_{CLK}/2} \frac{[v_A(t) - v_B(t)]^2}{R_1} \cdot dt = \frac{V_{DD}^2}{\ln(3) R_1}.$$

4 Implementation of Multi-scale Modeling on MCCS, Designing the Framework for NANO RFID

Nano composites, CNT, CMOS, Sensors are in nano domain are being developed in recent years that will see commercialization in the upcoming years. Transistors based on array of SWNT which current remains as low as 16μA per channel has been described in [19]. Thus high performance low power Carbon Nanotube is the best candidate for Nano-RFID systems where we have few Watts of power to use for passive Nano-RFIDs. MEMS are in Micro Domain devices which have been modeled in SUGAR in our library. Packaging is in macro domain phenomenon.

Multi scale modeling needs formulae from different level from Micro to Macro level of the same phenomenon so we have computed different formulas that truly reflect the multi level formulas which are given in the table below. Micro level MEMS based devices to Macro level calculations of abstraction layer are being proposed but a synchronized study is never done using the power of HPC. To study the effect of Nano scaled devices Modeling needs to be optimized so that different Levels of abstraction can be taken in account.

Table 1. Description of various abstraction scales synchronized for multi-scale modelling

Abstraction Scale	Properties
Nano CMOS 30-100nm	Capacitance in CMOS
RF MEMS 100-500nm	Properties of antenna
Nano composites 10-30 nm	Particles
Packaging, Interconnect above 1000nm	Models for Packaging

Several parameter were deduced from measurement of C(V) characteristics. Comparison between different CNTFET SBCNTFET, MOSCNTFET, and SI MOSFET has been discussed and comparisons of SBCNT with parasitic, SI MOSFET and SB CNTFET without parasitic. Thus parasitic low with low capacitance CNT is an area of research which is under high scanner [20]. Nano composite are poised to be the most important application, simulation capacitance plays a very important role and nano composites when used makes multi scale modeling more relevant and needful.

The interpretation of this conductance can be obtained in terms of the Poole–Frenkel equation [21].

$$G = G_0 \exp\left(\frac{\beta \sqrt{V / d_i}}{kT} \right)$$

The Poole–Frenkel field-lowering coefficient is given by

$$\beta = \left(\frac{q^3}{\pi \varepsilon_r \varepsilon_0} \right)^{1/2}$$

Combination CMOS interface chip and CNT (sensing medium) where CNT act as highly sensitive material at room temperature has been discussed in [22]. Poly-silicon Gate Depletion Effects plays a very important role in Nano CMOS design. It's because of the capacitive effect the most important in nano CMOS simulation. The gate depletion width as given in [21]:

$$X_{gd} = \frac{\varepsilon_{Si}\varepsilon_0}{C_{ox}}\left[\sqrt{1+\frac{2C_{ox}^2}{\varepsilon_{Si}\varepsilon_0 qN_{gate}}\left(V_g - V_{th} + \gamma s\sqrt{2\varphi_b} - 1\right)}\right]$$

$$\approx \frac{C_{ox}\left(V_g - V_{th} + \gamma s\sqrt{2\varphi s}\right)}{qN_{gate}}$$

here are various tunneling mechanism given, where direct-tunneling current is formulated as

$$J = \frac{q^3}{8\pi h\phi_b\varepsilon_{ox}} C\left(V_G, V_{ox}, t_{ox}, \phi_b\right)\exp\left\{-\frac{8\pi\sqrt{2m_{ox}\phi_b^3}}{3hq\,|\,E_{ox}\,|}\left[1-\left(1-\frac{V_{ox}q}{\phi_b}\right)^{3/2}\right]\right\}$$

C is basically an empirical shape factor

$$C\left(V_g, V_{ox}, t_{ox}, \phi_b\right) = \exp\left[\frac{20}{\phi_b}\left(\frac{|V_{ox}|-\phi_b}{\phi_{b0}}+1\right)^{\alpha}\left(1-\frac{|V_{ox}|}{\phi_b}\right)\right]\frac{V_g}{t_{ox}}N$$

The general form for the inversion and accumulations is given by the mathematical equation

$$N = \frac{\varepsilon_{OX}}{t_{OX}}\left\{S\ln\left[1+\exp\left(\frac{V_{ge}-V_{th}}{S}\right)\right] + v_t\ln\left[1+\exp\left(-\frac{V_g - V_{fb}}{v_t}\right)\right]\right\}$$

We know that the basic essence of MEMS devices is their ability of Mechanical motions at small scales. At this abstraction scale these formulas are used. These are the part of mechanics that needs to be incorporated. We know that MEMS comes into the domain of 100-500 nm – RF MEMS design [23].

M_A (Nm) is the reaction moment at the left end, and R_A (N) is the vertical reaction at the left end and P is load.

$$M_A = -\frac{Pa}{l^2}(l-a)^2$$

$$R_A = \frac{P}{l^3}(l-a)^2\left(l+2a\right)$$

$$y=\frac{2}{EI}\int_{l/2}^{l}\frac{\zeta}{48}\left(l^3-6l^2a+9la^2-4a^3\right)da$$

In the above equations ζ is load per unit length, E is young's modulus I being moment of inertia.

$$k_a'=-\frac{P}{y}=-\frac{\xi l}{y}=32Ew\left(\frac{t}{l}\right)^3$$

E is young's modules, w is width, t is thickness and l is length.

Micro machined antenna simulation is one of the most important parts of our library. The important calculations needed are resonant frequency which is given by the equation in which L and w are the length and width of the entity. This formula is used for electrical based calculation at this scale. It is important at multi scale simulation is eminent. $f_{0,n,p}^{res}$ is given by [23].

$$f_{0,n,p}^{res}=\frac{1}{2\pi(\mu\varepsilon)^{1/2}}\left[\left(\frac{n\pi}{L}\right)^2+\left(\frac{p\pi}{w}\right)^2\right]^{1/2}$$

The radiant field for a specific arrangement as shown in [23] for two slots arrangement separated by distance L is

$$E_\phi=j\frac{2hE_0}{\pi}\frac{e^{-jkr}}{r}\tan\theta\sin\left(\frac{kw}{2}\cos\theta\right)\cos\left(\frac{kL_{eff}}{2}\sin\theta\sin\phi\right)$$

We know that micro strip antennas are the essence of RF MEMS devices the total directivity for cavity model the radiation mechanism is given by:

$$D_W=\frac{2D}{1+g_{12}}$$

Where mutual inductance is:

$$g_{12}=\frac{1}{120\pi^2}\int_0^{\pi}\frac{\sin^2\left(\frac{\pi W\cos\theta}{\lambda}\right)\tan^2\theta\sin\theta J_0\left(\frac{2\pi L}{\lambda}\sin\theta\right)}{G}d\theta$$

Thus these can be used to calculate the quality factor of the antenna which is the most important parameter of antenna that's needs to taken in account.

Code 1. Sudo PVM code to spawn four processes to perform computations of Nano-Scale, Micro-Scale, Macro-Scale and Miscellaneous computations.

```
#include<pvm/pvm3.h>
#include<stdio.h>
int main(){
          int ID, ps1ID, ps2ID, ps3ID, ps4ID;
          printf("Master Process ID: %x - Status:
running\n", pvm_mytid());
```

```
            if((ps1ID = pvm_spawn("nanoScale", (char**)0,
0, "", 1, &ID)) == 1)
             printf("Process one started. ID:t%x\n", ID);

            else
             printf("cant start process one");
            if((ps2ID = pvm_spawn("microScale", (char**)0,
0, "", 1, &ID)) == 1)
             printf("Process two started. ID:t%x\n", ID);

            else
             printf("cant start process two");
            if((ps3ID = pvm_spawn("macroScale", (char**)0,
0, "", 1, &ID)) == 1)
             printf("Process three started. ID:t%x\n",
ID);
            else
             printf("cant start process three");
            if((ps4ID = pvm_spawn("misc", (char**)0, 0,
"", 1, &ID)) == 1)
             printf("Process four started. ID:t%x\n",
ID);
            else
             printf("cant start process four");
             pvm_exit();
            return 0;}
```

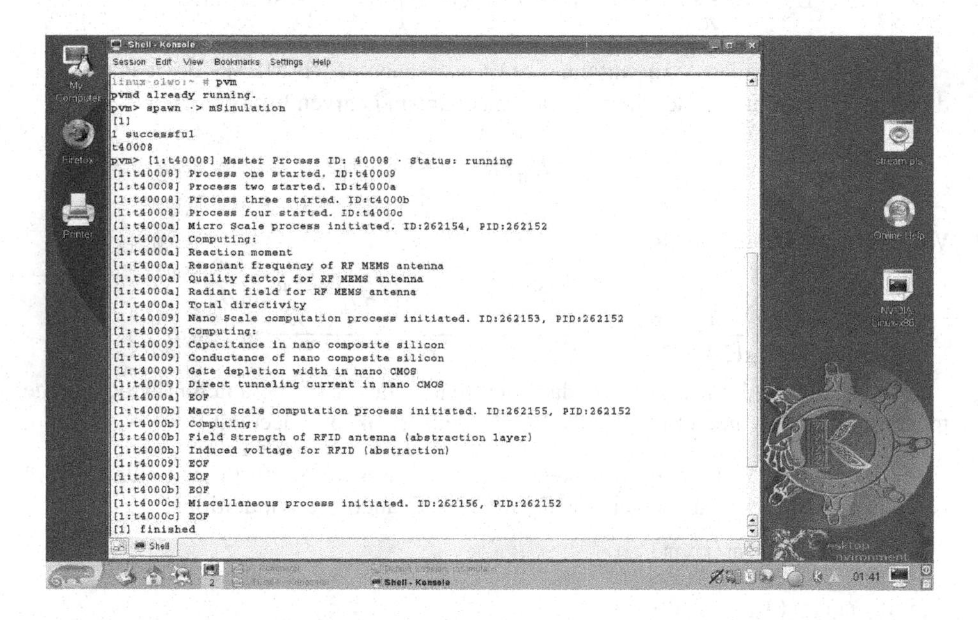

Fig. 3. Output of the Multi scale modeling computations done on SUSE 11.1, PVM version 3

Code 2. Sudo PVM code for the process which performs computation of Nano-Scalesuch as Capacitance of Nano-Composite Silicon, Conductance of Nano-Composite Silicon, Depletion Width in Nano-CMOS and Tunneling Current in Nano-CMOS.

```
#include "pvm/pvm3.h"
#include<stdio.h>
#include"comp.h"
int main()
{
        int ptid = pvm_parent();        printf("Nano
Scale computation process initiated. ID:%d, PID:%d\n\
Computing:\n\
Capacitance in nano composite silicon\n\
Conductance of nano composite silicon\n\
Gate depletion width in nano CMOS\n\
Direct tunneling current in nano CMOS\n", pvm_mytid(),
ptid);
        /*Calc Capacitance*/
        comp("NANO_CAPA");
        /*Calc Conductance*/
        comp("NANO_COND");
        /*Calc Depletion width*/
        comp("NANO_GATE_DPLE");
        /*Calc Tunneling Current*/
        comp("NANO_TUNN_CURR");
        pvm_exit();
        return 0;
}
```

The implementation has been shown in Fig. 3. which has been carried out on SUSE v11.1 using PVM (Parallel Virtual Machine) v3.0. Code 1. & Code 2. are pseudo codes of the implementation. Reliability of RFID based Nanotech systems will depend on many factors where multi scale modeling will play the most important role where we needs to correlate things from different abstraction layers to accurately predict aspects of reliability.

5 Conclusion

Thus we have tried to address the main issues of implementation of simulation of Nano RFID that remains the major challenges for their practical implementation in real devices. We have shown the modeling of transponders, antenna and components of RFID that are recently proposed. Aspects of our library for Nano RFID have been shown where abstraction and multi scale modeling are covered comprehensively. RFID devices need to be low power and their implantation in body needs makes it important to use legacy circuitry so that all issues regarding compatibility and ethics needs to be taken in account. We have modeled the current Nano-RFID under the current tools available as to unify the research going in RFID in the last decade. HPC will benefit multi scale modeling and will help in realization of true limits of this technology. PVM was used for multi scale modeling. Successfully modeling RFID in the current technological and modeling developments and accelerate their realization in business. Multi scale calculation integration and optimization has been shown. Multi scales modeling with various scales are shown and computations are being done.

References

1. Cabrera, M., Burgelman, J.-C., Boden, M., da Costa, O., Rodríguez, C.: eHealth in 2010: Realising a Knowledge-based Approach to Healthcare in the EU Challenges for the Ambient Care System. European Commission Joint Research Centre (April 2004), http://www.stop-project.eu/portals/1/publications/eur22770en.pdf
2. Nano Technology Business Research Report: A study of business' understanding of and attitudes towards nanotechnology. Dandolopartners (July 11, 2005), http://www.innovation.gov.au/Industry/Nanotechnology/Documents/FinalBusReportJuly20051021103038.pdf
3. Cavalcanti, A., Shirinzadeh, B., Freitas Jr., R.A., Kretly, L.C.: Medical Nanorobot Architecture Based on Nanobioelectronics. Recent Patents on Nanotechnology 1(1) (Feburary 2007), http://www.bentham.org/nanotec/samples/nanotec1-1/Cavalcanti.pdf
4. Ausen, D., Westvik, R., Svagård, I., Österlund, L., Gustafson, I., Vikholm-Lundin, I., Winquist, F., Lading, L., Gran, J.: Foresight Biomedical Sensors. Nordic Innovation Centre (NICe) project number: 04247 (September 2007), http://www.sintef.no/project/FOBIS/FOBIS_final%20report_web.pdf
5. Cui, Y.: Antenna Design for On-the-body Sensors. M.Sc. Thesis. MIC - Department of Micro and Nanotechnology Technical University of Denmark (Feburary 2006), http://www.nanotech.dtu.dk/upload/institutter/mic/forskning/mems-appliedsensors/publications/master_thesis/2006%20ying%20cui.pdf
6. Opinion on the ethical aspects of ICT implants in the human body: No. 20. The European Group on Ethics in Science and New Technologies to the European Commission (March 16, 2005), http://ec.europa.eu/european_group_ethics/docs/avis20compl_en.pdf
7. Radio frequency identification (RFID) at HP. White paper, Hewlett-Packard Development Company (2004), http://h71028.www7.hp.com/ERC/downloads/5982-4290EN.pdf
8. Meingast, M., King, J., Mulligan, D.K.: Security and Privacy Risks of Embedded RFID in Everyday Things: the e-Passport and Beyond. Journal of Communications 2(7) (December 2007), http://www.academypublisher.com/jcm/vol02/no07/jcm02073648.pdf
9. Srivastava, L.: Ubiquitous Network Societies: The Case of Radio Frequency Identification. International Telecommunication Union (2005), http://www.itu.int/osg/spu/ni/ubiquitous/Papers/RFID%20background%20paper.pdf

10. European Policy Outlook RFID. European Commission, Brussels, Federal Ministry of Economics and Technology (BMWi) (July 2007), http://www.iot-visitthefuture.eu/fileadmin/documents/roleofeuropeancommision/European_Policy_Outlock_RFID.pdf
11. Curty, J.-P., Joehl, N., Dehollain, C., Declercq, M.J.: Remotely powered addressable UHF RFID integrated system. IEEE Journal of Solid-State Circuits 40(11), 2193–2202 (2005)
12. Shameli, A., Safarian, A., Rofougaran, A., Rofougaran, M., Castaneda, J., De Flaviis, F.: A UHF Near-Field RFID System With Fully Integrated Transponder. IEEE Transactions on Microwave Theory and Techniques. Part 2,56(5), 1267–1277 (2008)
13. Jiang, B., Smith, J.R., Philipose, M., Roy, S., Sundara-Rajan, K., Mamishev, A.V.: Energy Scavenging for Inductively Coupled Passive RFID Systems. IEEE Transactions on Instrumentation and Measurement 56(1), 118–125 (2007)
14. Kaya, T., Koser, H.: A New Batteryless Active RFID System: Smart RFID. In: 1st Annual RFID Eurasia, 2007, September 2007, pp. 1–4 (2007), doi:10.1109/RFIDEURASIA.2007.4368151
15. Lu, H.M., Goldsmith, C., Cauller, L., Lee, J.-B.: MEMS-Based Inductively Coupled RFID Transponder for Implantable Wireless Sensor Applications. IEEE Transactions on Magnetics 43(6), 2412–2414 (2007)
16. Jiang, B., Fishkin, K.P., Roy, S., Philipose, M.: Unobtrusive long-range detection of passive RFID tag motion. IEEE Transactions on Instrumentation and Measurement 55(1), 187–196 (2006)
17. Sasaki, S., Seki, T., Imanaka, K., Kimata, M., Toriyama, T., Miyano, T., Sugiyama, S.: Batteryless-Wireless MEMS Sensor System with a 3D Loop Antenna. In: IEEE Sensors, October 2007, pp. 252–255 (2007), doi:10.1109/ICSENS.2007.4388384
18. Finkenzeller, K.: RFID Handbook. Wiley, Hoboken (1999)
19. Chen, C., Xu, D., Kong, E.S.-W., Zhang, Y.: Multichannel Carbon-Nanotube FETs and Complementary Logic Gates With Nanowelded Contacts. IEEE Electron Device Letters 27(10), 852–855 (2006)
20. Keshavarzi, A., Raychowdhury, A., Kurtin, J., Roy, K., De, V.: Carbon Nanotube Field-Effect Transistors for High-Performance Digital Circuits— Transient Analysis, Parasitics, and Scalability. IEEE Transactions on Electron Devices 53(11), 2718–2726 (2006)
21. Wong, B., Mittal, A., Cao, Y., Starr, G.W.: NANO-CMOS Circuit and Physical Design. IEEE Wiley-IEEE Press, ISBN: 978-0-471-46610-9
22. Cho, T.S., Lee, K.-j., Kong, J., Chandrakasan, A.P.: The design of a low power carbon nanotube chemical sensor system. In: Proc. 45th ACM/IEEE Design Automation Conference, 2008. DAC 2008, June 8-13, pp. 84–89 (2008)
23. RF MEMS Theory, Design, and Technology Gabriel M. Rebeiz

Author Index